Inland Fishes of California

Inland Fishes of California

Revised and Expanded

PETER B. MOYLE

Illustrations by Chris Mari van Dyck and Joe Tomelleri

UNIVERSITY OF CALIFORNIA PRESS
Berkeley   Los Angeles   London

University of California Press
Berkeley and Los Angeles, California

University of California Press, Ltd.
London, England

© 2002 by the Regents of the University of California

Library of Congress Cataloging-in-Publication Data
Moyle, Peter B.
    Inland fishes of California / Peter B. Moyle ; illustrations by Chris
Mari van Dyck and Joe Tomelleri.— Rev. and expanded.
        p.   cm.
    Includes bibliographical references (p.).
    ISBN 0-520-22754-9 (cloth : alk. paper)
    1. Freshwater fishes—California.   I. Title.
QL628.C2 M68   2002
597.176′09794—dc21   2001027680

Manufactured in Canada
11   10   09   08   07   06   05   04   03   02
10   9   8   7   6   5   4   3   2   1

*With gratitude, I dedicate this book to Marilyn A. Moyle,
my wife, partner, and friend of more than 35 years*

Special Thanks

The illustrations for this book were made possible by grants from the following:

California-Nevada Chapter, American Fisheries Society

Western Division, American Fisheries Society

California Department of Fish and Game

Giles W. and Elise G. Mead Foundation

We appreciate the generous funding support toward the publication of this book by the United States Environmental Protection Agency, Region IX, San Francisco

Contents

Preface

This book is the first revision of the one I boldly began writing in 1972, when I had lived in California for just two years. Writing it was my way of getting to know a fish fauna that was a mixture of familiar and unfamiliar elements. The familiar parts were introduced fishes, most of them native to the eastern part of this country, where I had received training as a fish biologist. The unfamiliar parts were native fishes, most of them occurring only in California. The first edition was published in 1976, and its principal message was that we knew astonishingly little about many of the fishes, especially native fishes. Since that time, I have been collecting information to fill in knowledge gaps and to correct errors in the first edition. The job is far from finished, but, given the precarious state of the native fishes, I thought it important to summarize once again what we know about them. I sometimes wonder if complete accounts of the systematics and natural history of many native fishes can be completed before they go extinct. Species accounts for several fishes are already obituaries, and others may become so in the near future. I can only hope that the information provided in this book will help to reduce the loss of our native fishes. At the same time, managing the altered aquatic ecosystems of California requires knowledge of the alien fishes that now dominate many of them, including favorite sport fishes. The adaptations of alien fishes to the California environment and their impact on native fishes is therefore also a major theme.

The species accounts are the most important part of this book. They are preceded by chapters providing overviews of the distribution, ecology, and conservation of the fishes, followed by a key to make identification easier. Each species account is organized as follows:

Common name, *Scientific name*
Identification
Taxonomy
Names
Distribution
Life history
    Habitat
    Nonbreeding behavior
    Feeding habits
    Age and growth
    Reproduction
    Early life history
Status
    Rating
    Abundance
    Management
References

Identification This is not a complete species description but a compilation of features useful for separating the species from other California fishes. Terminology is defined in the introduction to the key.

Taxonomy This section is especially important for species for which there is controversy or uncertainty about systematics or that have a confusing taxonomic history. It is used to discuss advances in our understanding of the systematics of the species. Minor questions of name changes or long-settled taxonomic questions are usually mentioned in the Names section of each species account.

Names The common and scientific names used here, with a few exceptions, are from the American Fisheries Society's

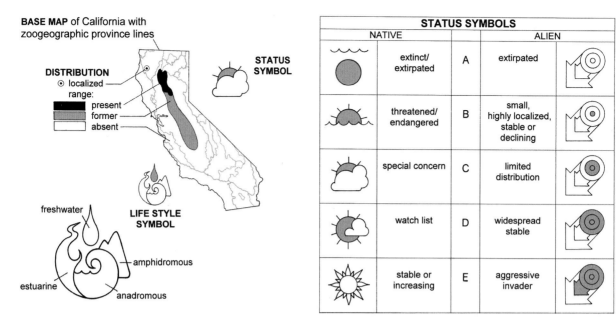

Figure 1. Symbols used on distribution maps to indicate distribution, status, and life style of each fish species.

1990 *List of Common and Scientific Names of Fishes from the United States and Canada.* The origins of the scientific names come from many sources, but most frequently from Jordan and Evermann (1896).

Distribution This section describes the distribution of each species, focusing on California. The distribution maps provided are designed only to give a general idea of the distribution of each species, not site-specific information (Fig. 1). Larger distribution maps for each species are available through the Information Center for the Environment at the University of California, Davis (http://ice.ucdavis.edu). Even these maps should be regarded as snapshots of the present distribution of each species, because distributions are changing constantly, as the landscape changes under human influence, native species decline and introduced species expand.

Life history Much of the information on the habits of California fishes is hidden in the "gray literature" of unpublished theses and reports. I have tried to be as comprehensive as possible, but no doubt I have overlooked some of these important sources of information. (If you are aware of a document I have missed containing useful tidbits, please send me a copy. Maybe I can use the information in the *next* edition!) Life history information that is not referenced is based on unpublished data or observations of my own.

Status In this section, I rate the status of each species in the state and then discuss abundance trends and management needs. My rating system is as follows:

I. Native species

A. Extinct/extirpated. The species is gone from California (extirpated) or gone from the planet (globally extinct).

B. Threatened or endangered. The species is likely to become extinct or extirpated in the near future (<25 years) unless steps are taken to save it. An endangered species is on a more rapid path to extinction than a threatened species. Most of these species are formally listed by either the state or the federal government; some are not (but probably should be). The formal status of each threatened species is given in the account.

C. Special concern. The species is in decline or has a very limited distribution, so special management is needed to keep it from becoming threatened or endangered.

D. Watch list. The species appears to be declining but is not yet in serious trouble. Its populations must be monitored to see if special protective action is necessary.

E. Stable or increasing. The species is abundant or increasing in population.

II. Alien species

A. Extirpated in California. The species was once established but the introduction failed. These species are mentioned only in family accounts.

B. Small, highly localized populations. The species is established in just a handful of localities and is stable or declining in numbers.

C. Localized likely to become more widespread or already widespread but not abundant in most areas. Alternately, it may be fairly common but is declining. The species is usually a recent introduction and is just starting to expand its range, or it is a long-established species that is only region-ally abundant.

D. Widespread and stable. The species is widely distributed but seems to have reached the limits of its range. Presumably such species are integrated into local ecosystems.

E. Widespread and expanding. These fish are aggressive invaders that are still expanding their range to all suitable habitats in the state.

Incorporated into each Status section are opinions, usually my own, on the management needs of each species. You will note that I have a strong bias in favor of native fishes over alien fishes.

References In the species accounts, the references are numbered and listed for the most part in the order in which they are cited in the text, by author and date, in an effort to save space and make the text more readable. Thus a listing like "3. Rutter 1908" is a citation near the beginning of an account, with a more complete citation to be found in the References section at the end of the book.

Terminology The classification system used follows the fourth edition of Moyle and Cech (2000), which in turn follows mostly Nelson (1994). The result is a fairly major re-ordering from the first edition. The terminology used to describe all aspects of fish biology is also based on Moyle and Cech (2000), reflecting new understanding of various aspects of ichthyology. For example, I use the term *shoaling* where most American biologists would use the word *schooling*. I reserve *schooling* as the word referring to aggregations of fishes (shoals) that are polarized and swimming in synchrony (schools).

To improve readability, scientific names of resident California species are in most cases used just twice: once in the key and once in the account of the species. The common names are in any case increasingly more stable through time than the scientific names.

The word *lake* in this book is reserved for true lakes and is not used to refer to reservoirs, no matter what the agencies who build reservoirs call them. This usage is consistent and reflects the fact that reservoirs are very different ecologically from natural lakes.

I use the term *amphidromous* to describe the basic life history of coastal sculpins that live and spawn in streams but have larvae that rear in an estuary (Moyle and Cech 2000).

Abbreviations Some common abbreviations found in the book, referring to agencies, are as follows:

BLM, Bureau of Land Management

CDFG, California Department of Fish and Game

DWR, California Department of Water Resources

NMFS, National Marine Fisheries Service

TNC, The Nature Conservancy

USFS, U.S. Forest Service

USFWS, U.S. Fish and Wildlife Service

USGS, U.S. Geological Survey

For length designations, the following abbreviations are used: SL, standard length; TL, total length; FL, fork length. All are defined in the introduction to the key in the Identification chapter.

Illustrations Most of the pen and ink drawings in this book are copyrighted by the artist, Chris M. van Dyck. These drawings are available to be used for nonprofit purposes at no cost by members of the American Fisheries Society and others, provided a request is made in writing to the author and the artist (1123 Kerria Avenue, McAllen, TX 78501). Other uses should be arranged with the artist.

Acknowledgments

This book would not have been possible without the collaboration of dozens of biologists who shared information and insights with me over the years. The large group who helped me develop an understanding of California's fascinating fishes during my first five years in the state are acknowledged in the first edition, published in 1976. I have many memories of their patient kindness and willingness to dig through files to enlighten me. For this edition I acknowledge most individuals who helped me by listing them as a "pers. comm." embedded in the text. Those oft-repeated two words represent hundreds of e-mails, letters, and conversations—and they convey the important point that this book is in some ways a project of the entire community of fisheries and fish biologists in California. For those whose knowledge or insights I absorbed without acknowledgment, my apologies. Some special acknowledgments are nevertheless required.

Jerry J. Smith and Larry R. Brown patiently reviewed many species accounts and improved them with critiques based on their impressive knowledge of the fishes. The entire manuscript was reviewed by Wendell L. Minckley and Robert A. Fisher. Minckley's humbling comments on my writing style were very helpful and resulted in the deletion of the word *the* at least 7,000 times. Selected sections benefited from comments by William A. Bennett, Eric Gerstung, David W. Kohlhorst, Robert L. Leidy, Scott A. Matern, Phil Pister, Donald Sada, Gary Scoppetone, and Christina Swanson.

The timely book on introduced fishes in California by William A. Dill and Almo J. Cordone (1997) saved me immense amounts of time uncovering information and checking facts. Their wry comments on inaccuracies in my first edition regarding times and places of introductions are reflected in improvements in this edition, even though I still rely on a secondary source (their book).

Patrick K. Crain, more than anyone else, kept my research projects going while I was in the throes of writing the final draft, doing much more than I had any right to expect him to do. Ronald A. Yoshiyama worked with me on a number of reports and review papers important to this book, including the second edition of *Fish Species of Special Concern in California,* which is the foundation of many of the native species accounts. His exhaustive research has given many accounts increased depth and accuracy. The support of the Giles W. and Elise G. Mead Foundation for this work is greatly appreciated.

In the ten years or so in which I worked on the manuscript many graduate students and postdoctoral scholars who were part of my fish ecology group contributed by providing information and ideas, reviewing sections, or just serving as sounding boards for thoughts that struck me during my reading and writing. They include Donald Baltz, William Bennett, Anne Brasher, Elizabeth Campbell, Gail Dethloff, Joaquin Feliciano, Leslie Ferguson, Mark Gard, Nathan Goedde, Bruce Herbold, Jeff Kozlowski, Robert Leidy, Theo Light, Michael Marchetti, Scott Matern, Lesa Meng, Paul Randall, Robert Schroeter, Ted Sommer, Elizabeth Strange, Bruce Vondracek, and Rolland White.

Others who responded repeatedly and enthusiastically to my requests for information included Randall Baxter, Carl Bond, Joseph J. Cech, Jr., Barry Costa-Pierce, Walter R. Courtenay, Jr., Michael H. Fawcett, Tim Ford, Dan Gale, Michael Giusti, Sharon Keeney, Tom T. Kisanuki, Dennis Lee, Stafford Lehr, Douglas Markle, J. D. McPhail, W. L. Minckley, Linda Pardy, Stewart Reid, Terry Roelofs, Gary Scoppetone, Ramona Swenson, Camm C. Swift, Thomas L. Taylor, and David Vanicek. I am grateful to members of the Executive Committee of the California-Nevada Chapter of the American Fisheries Society, who arranged to provide funding for the black-and-white drawings, including Michael Meinz, Kathy Heib, Jean Baldrige, Ramona Swenson, Camm C. Swift, Pat Coulston, Alice Low, and Dennis McEwan. For the color insert Kenneth Hashagen arranged funding from the Western Division of the American Fisheries Society; Chuck Knutson, from the California Department of Fish and Game; and Robert A. Leidy, from the U.S. Environmental Protection Agency.

It has been a continuing pleasure to work with Chris M. van Dyck, who did most of the artwork, including the wonderful drawings of salmonids created originally for the 1976 edition. It is hard to imagine this project succeeding without her skill, energy, and suggestions for improving the presentations. Those interested in using her copyrighted drawings should see the section on illustrations in the Preface.

It has also been a pleasure working with Joseph Tomelleri to arrange use of his superb color portraits of native fishes. Alan Marciochi created 27 of the fish portraits, holdovers from the first edition. Archival drawings by H. L. Todd were made available by the National Marine Fisheries Service (www.photolib.noaa.gov). Other drawings were provided by David S. Lee, Paul Vecsei, Walter R. Courtenay, Jr., Camm C. Swift, Carl Bond, Michael Bell, and Reeve Bailey, while Christopher M. Dewees permitted me to use three of his fish prints. My thanks to each of them.

I have been fortunate to work with a great group of faculty and staff in the Department of Wildlife, Fish, and Conservation Biology at the University of California, Davis, who define the word *collegiality*. I particularly appreciate the leadership of the three departmental chairs during the long gestation of this book: Daniel Anderson, Joseph Cech, and Deborah Elliott-Fisk. I am also grateful for the assistance of those members of the departmental staff who helped iron out the many problems, major and minor, I faced in putting the book together: Marjorie Kirkman-Iverson, Della Nunes, and Peggy Davis.

Finally, I appreciate the patience and help of Doris Kretschmer, executive editor at the University of California Press, in putting this book together, and the wonderfully detailed editing of the final manuscript by Peter Strupp of Princeton Editorial Associates.

Conversion Factors

Degrees °C	°F
0	32
5	41
10	50
15	59
20	68
25	77
30	86
35	95
40	104

mm	in.
10	0.4
25	1.0
50	2.0
75	3.0
100	3.9
125	4.9
150	5.9
175	6.9
200	7.9
225	8.9
250	9.9
275	10.8
300	11.8
325	12.8
350	13.8
375	14.8
400	15.8
500	19.7
600	23.6
700	27.6
800	31.5
900	35.5
1000	39.4

1 cm = 0.39 in.
1 m = 3.28 ft
1 km = 0.62 mi
454 g = 1 lb
1 kg = 2.21 lb

Distribution Patterns

The highly endemic fish fauna of California is scattered through a diverse landscape with an incredibly complicated geologic history. Present zoogeographic patterns must be regarded as snapshots in time of a fauna that has shifted about through the millennia in response to geologic and climatic events. Major events such as volcanic eruptions, earthquakes, and movements of the earth's crust have altered entire drainage systems, creating or destroying streams, lakes, and estuaries. Fluctuations in climate have caused streams to flow or not flow; lakes to fill in, dry up, or overflow; and sea level to rise and fall, alternately separating and connecting nearby coastal drainages.

Complicating our understanding of distribution patterns is the fact that California is a tough place for a freshwater fish species to persist through time. Local and regional extinctions have probably been common, especially in the past 10,000 years as the postglacial climate became drier. As a result, the state contains only about 66 native freshwater, estuarine, or anadromous species within its huge area (Table 1). On the other hand, the frequency with which populations of fish become isolated through natural events promotes creation of new species. The faunal count is nearly doubled when incipient species are counted: subspecies, marine fishes that enter fresh water on an irregular basis, and distinctive runs of anadromous species. In particular, migratory species such as threespine stickleback, river lamprey, and rainbow trout generate numerous isolated populations of nonmigratory forms in upstream areas, which often behave as distinct species. In recent years, natural speciation processes have been overwhelmed by a combination of water diversions, habitat alterations, introduced species, and climate change. Massive, human-caused changes to the waterscape occurred before the fish fauna was well documented, adding another level of confusion to the zoogeographic patterns. Nevertheless, figuring out why each native species lives where it does remains a fascinating exercise.

California contains all or part of six ichthyological provinces: Klamath, North Coast, Great Basin, Sacramento–San Joaquin, South Coast, and Colorado River (Fig. 2). Each province contains a group of endemic species, demonstrating long isolation. All can be further divided into subprovinces that contain one or more endemic species or subspecies. Each fauna is a mixture of species that arrived in the province by different means (Moyle and Cech 2000).

Euryhaline marine species are fishes that enter the lower reaches of streams from the ocean. A freshwater sojourn is not essential for these species to complete their life cycles. Usually the individuals that move into fresh water are juveniles. Examples include starry flounder, staghorn sculpin, and shiner perch.

Saltwater dispersants are species that spend much of their life history in fresh water but either can move through salt water themselves or have immediate ancestors that did so. Thus their distribution patterns are explained in part by movements through the ocean. All species of this type in California are anadromous or had ancestors that were anadromous. Examples include rainbow trout, threespine stickleback, chinook salmon, and all lampreys.

Freshwater dispersants are species that arrived at their present locations by freshwater routes or evolved in place from a distant marine ancestor. They are incapable of moving long distances through salt water. Thus they have to colonize new areas by moving through streams, and this may not be possible until a mountain range erodes to connect two drainages or until sea level falls, allowing streams to become connected on a coastal plain. Most of California's endemic fishes are freshwater dispersants, including all the minnows (Cyprinidae) and suckers (Catostomidae). Some freshwater dispersant species, such as tule perch and riffle sculpin, are members of families that contain mostly saltwater dispersants, but their own distribution patterns reflect dispersal entirely through fresh water.

Table 1

Native Fishes of the Inland Waters of California

Species	Life style[a]	Regions[b]	Status[c]
Pacific lamprey	AN, F	KL, NC, SC, SJ	IB, IC
Pit-Klamath brook lamprey	F	SJ	ID
River lamprey	AN	KL, NC, SJ	ID
Kern brook lamprey	F	SJ	IC
Western brook lamprey	F	KL, NC, SC, SJ	ID
Klamath River lamprey	F	KL	ID
White sturgeon	AN	KL, NC, SJ	IE
Green sturgeon	AN	KL, NC, SJ	IC
Tui chub	F	GB, KL, SJ	IA-1E
Thicktail chub	F	SJ	IA
Blue chub	F	KL	IC
Arroyo chub	F	GB*, SC	IC
Bonytail	F	CL	IA
Lahontan redside	F	GB, SJ*	IE
Hitch	F	SC*, SJ	IC-ID
California roach	F	SC*, SJ, NC	IB-IE
Sacramento blackfish	F	GB*, SC*, SJ	IE
Sacramento splittail	E, F	SJ	IB
Clear Lake splittail	F	SJ	IA
Hardhead	F	SJ	ID
Sacramento pikeminnow	F	NC*, SJ, SC*	IE
Colorado pikeminnow	F	CL	IA
Speckled dace	F	GB, KL, NC, SC, SJ	IB-IE
Mountain sucker	F	GB, SJ*	ID
Santa Ana sucker	F	SC	IB
Sacramento sucker	F	SC*, SJ, NC	IE
Modoc sucker	F	SJ	IB
Tahoe sucker	F	GB	IE
Owens sucker	F	GB, SC*	ID
Klamath largescale sucker	F	KL	IC
Klamath smallscale sucker	F	KL	IE
Lost River sucker	F	KL	IB
Shortnose sucker	F	KL	IB
Razorback sucker	F	KL	IB
Flannelmouth sucker	F	CL	IA
Delta smelt	E	SJ	IB
Longfin smelt	E	NC, SJ	IC
Eulachon	AN	KL, NC	IC
Coho salmon	AN	KL, NC	IA-IB
Chinook salmon	AN	KL, NC, SJ	IB-ID
Pink salmon	AN	KL, NC, SJ	IA
Chum salmon	AN	KL, NC, SJ	IA-IB
Rainbow trout	AN, F	GB, KL, NC, SC, SJ	IB-IE
Cutthroat trout	AN, F	GB, KL, NC	IB-IC
Bull trout	F	SJ	IA
Striped mullet	E	SC	IE
Topsmelt	E	NC, SC, SJ	IE
California killifish	E	SC	IE
Desert pupfish	F	GB	IB
Owens pupfish	F	GB	IB.
Amargosa pupfish	F	GB	IB
Salt Creek pupfish	F	GB	IC
Threespine stickleback	AN, E, F	GB*, KL, NC, SC, SJ	IB-IE
Prickly sculpin	AM, E, F	KL, NC, SC, SJ	IE

Table 1 (Continued)

Species	Life style[a]	Regions[b]	Status[c]
Coastrange sculpin	AM	KL, NC, SC	IE
Riffle sculpin	F	NC, SJ	IE
Pit sculpin	F	SJ	IE
Reticulate sculpin	F	KL	IC
Marbled sculpin	F	KL, SJ	ID-IE
Paiute sculpin	F	GB	IE
Rough sculpin	F	SJ	IC
Sacramento perch	F	GB*, KL*, SJ	IC
Tule perch	E, F	NC, SJ	IC-IE
Shiner perch	E	NC, SC, SJ	IE
Tidewater goby	E	SC, SJ, NC	IB
Longjaw mudsucker	E	CL*, SC, SJ	IE
Starry flounder	E	NC, SC, SJ	IE

Note: Only species that occur in fresh or brackish water on a regular basis are included.

[a]Abbreviations: AM, amphidromous; AN, anadromous; E, estuarine resident; F, freshwater resident.

[b]Abbreviations: CL, Colorado; GB, Great Basin; KL, Klamath; NC, North Coast; SC, South Coast; SJ, Sacramento–San Joaquin. An asterisk after the basin indicates that the species is introduced rather than native.

[c]For codes, see the Preface.

In the sections that follow, explanations of distribution patterns are based in large part on the detailed study of Minckley et al. (1986), which in turn owes a debt to the work of Robert R. Miller and Carl L. Hubbs, who spent years wandering about the West collecting fishes and inspecting streams, lakes, and land forms (Hubbs and Miller 1948; Miller 1948, 1961b, 1965, 1981; Hubbs et al. 1974; Miller et al. 1991).

Klamath Province

The Klamath Province has three distinct subprovinces in California: (1) the upper Klamath River basin above Klamath Falls, including the Lost River; (2) the Klamath River below the falls, including the Trinity River; and (3) the Rogue River, represented by only a few tributary headwaters in the state. In addition, for convenience, I include a large area (1d in Fig. 2) in this province that is largely covered with old lava flows and was historically fishless. Including Rogue River fishes, there are only 30 native species in the province, 8 of them endemic (10, if those shared with the Pit River are counted) (Table 2). Fish faunas of the upper and lower Klamath Subprovinces are surprisingly distinct from one another, presumably because the connection between the two regions is geologically recent and because their major habitats are quite different. The upper Klamath Subprovince is dominated by large, shallow lakes and sluggish rivers, whereas the lower subprovince is dominated by large, swift rivers mostly confined between steep canyon walls. The importance of habitat is indicated by the fact that, when Iron Gate Dam was built across the lower Klamath River, the reservoir created was colonized by lake-dwelling fishes from the upper basin. Historically, the two provinces were connected by movement of anadromous salmon and steelhead into the tributaries to the large lakes.

Upper Klamath Subprovince The native fish fauna (15 species) of the Upper Klamath Subprovince consists primarily of freshwater dispersants (12 species), most having their closest relatives in the Great Basin. This reflects the complex geologic history of the region, in which a large river (the ancestor of the Snake River, now a tributary to the Columbia River) originating in Idaho flowed into the ocean in the Klamath region during the Eocene period and again during the Pliocene period (Aalto et al. 1998). Some of the species in the subprovince have related species in the Pit River of the neighboring Sacramento watershed, indicating ancient past connections as well. In addition, three of the species are saltwater dispersants that could have invaded at almost any time. The fishes belong to just five families—Catostomidae, Cyprinidae, Cottidae, Salmonidae, and Petromyzontidae—and each species has its own affinities to fishes of other provinces.

The suckers (Catostomidae) consist of three endemic species (shortnose sucker, Lost River sucker, and Klamath largescale sucker) usually placed in three different genera

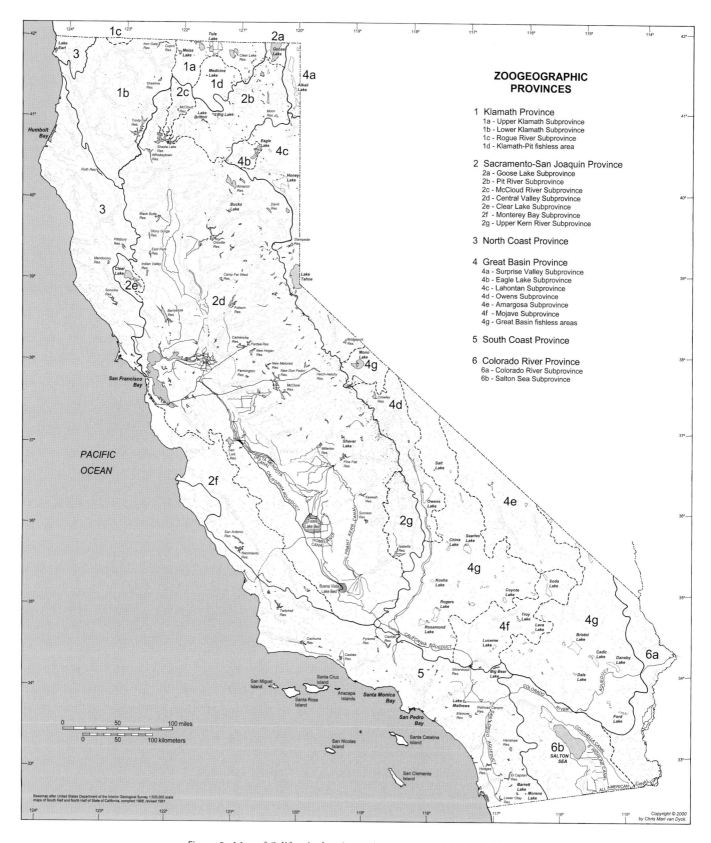

Figure 2. Map of California showing major zoogeographic subdivisions.

(*Chasmistes, Deltistes,* and *Catostomus,* respectively). These species have maintained their morphological distinctiveness despite extensive hybridization among them. To further complicate matters, the Klamath smallscale sucker has also contributed genes to this hybrid complex (Tranah 2001), although it is not included here as part of the upper Klamath fauna because of its extreme rarity in the basin. The shortnose sucker is similar to other species, living and fossil, of the genus *Chasmistes,* all adapted for life in large lakes of the Great Basin and having a long fossil history (Miller and Smith 1981). The Lost River sucker is another lake-adapted fish that seems related to the *Chasmistes* group. Similar suckers are found as fossils in the Great Basin, in a region (ancient Bonneville Lake, Utah) that also had connections to the ancestral Snake River. The Klamath largescale sucker is a typical riverine sucker, similar to riverine species in the Columbia and Sacramento drainages. Together, these species represent a remarkable experiment in evolution as they struggle to maintain their identities in a highly altered environment. It is possible that the increased genetic diversity resulting from hybridization increases the ability of each form to persist under adverse conditions.

The three cyprinids, like the suckers, seem to have Great Basin–Bonneville connections. The blue chub and the Klamath tui chub are upper Klamath endemics. The blue chub is quite distinctive, and its relationships to other members of the genus *Gila* are uncertain. The Klamath tui chub, on the other hand, is part of a species complex widespread throughout the Great Basin. The speckled dace occurs in both the upper and lower rivers and is regarded as one subspecies. However, a careful analysis of dace from different parts of the province will probably reveal two or more subspecies, as have been found for marbled sculpin.

The three sculpins (Cottidae) of the upper Klamath are all endemic. All three are freshwater dispersants, with benthic larvae rather than the pelagic larvae of sculpins capable of dispersing through salt water. The slender sculpin (*Cottus tenuis*) and the Klamath Lake sculpin (*C. princeps*) are both found only in Oregon, although the slender sculpin is closely related to the rough sculpin of the Pit River drainage (see discussion under Pit River). Likewise, the marbled sculpin occurs in both the Klamath and Pit drainages, with subspecies in the Pit, upper Klamath, and lower Klamath Rivers (Daniels and Moyle 1984).

Trout (Salmonidae) native to the upper Klamath represent two or three separate invasions by these vagile species. Bull trout, found in a few Oregon tributaries, are otherwise native to the Columbia River drainage and the McCloud River of California. They presumably are holdovers from times when the ancient Snake River flowed through the region (Minckley et al. 1986). There are two forms of rainbow trout in the upper Klamath, redband trout and coastal rainbow trout. The redbands are presumably relics of one or more early invasions, whereas the coastal rainbows initially invaded as steelhead (anadromous rainbow trout) after the upper and lower rivers became connected in fairly recent times (Pleistocene).

The lampreys (Petromyzontidae) are another fascinating part of the upper Klamath fauna, with a complex evolutionary history. Four species of lamprey are now recognized from the region, but they may represent a complex of forms that have some gene flow among them. The Miller Lake lamprey (*Lampetra minima*) is a tiny species from the Williamson and Sycan Rivers, Oregon; the Pit-Klamath brook lamprey is found in the Pit River as well; the Klamath River lamprey is confined to the Upper Klamath Subprovince; the dwarf Pacific lamprey is a landlocked form of a widespread anadromous species. The anadromous Pacific lamprey ultimately gave rise to all these forms, but it is not at all clear how this occurred. Presumably there were multiple invasions during the various episodes of marine connections of the ancestral rivers. The Pit-Klamath brook lamprey and the distinctive "Pacific" lampreys in Goose Lake (now connected to the Pit River) indicate ancient invasions. Further complicating the picture is the fact that the Miller Lake lamprey and the Pit-Klamath brook lamprey are closely related, suggesting that one is derived from the other (D. Markle, pers. comm.).

Overall, the fish fauna of the Upper Klamath Subprovince is remnant of a more widespread fauna that occupied the Great Basin region in wetter times, combined with descendants of anadromous fishes that invaded during times of ocean connection. Not surprisingly, the fishes have long, independent evolutionary histories as well. The suckers and lampreys in particular show evidence of unusual arrangements of shared genes, presumably improving the ability of each form to adapt to changing, often severe, local conditions. Superimposed on these fishes are descendants of anadromous fishes that invaded at various times.

Lower Klamath Subprovince This region contains 21 native species, of which 17 are saltwater dispersants, mainly anadromous lamprey (two species), sturgeon (two species), salmonids (six species), smelt (two species), and stickleback (one species) plus two amphidromous sculpins (Table 2). The only freshwater dispersants are Klamath speckled dace, lower Klamath marbled sculpin, Klamath smallscale sucker, and Pacific brook lamprey. The dace and marbled sculpin presumably invaded from upstream during the Pleistocene, when water spilling from Upper Klamath Lake eroded a permanent connection to the lower river. The smallscale sucker has uncertain taxonomic affinities, but it is tied somehow to the suckers of the Upper Kalmath Subprovince.

Table 2

Presence of Fish Species in Major Watersheds of the Klamath and North Coast Aquatic Zoogeographic Regions of California

Watershed name	Klamath		North Coast													
	Lower Basin	Upper Basin	Tomales Bay	Russian River	Gualala River	Garcia River	Navarro River	Big River	Noyo River	Matolle River	Bear River	Eel River	Mad River	Little River	Redwood[a] Creek	Smith River
Subprovince number	1a	1b	3	3	3	3	3	3	3	3	3	3	3	3	3	3
Pacific lamprey	N	N	N	N	N	N	N	N	N	N	N	N	N	N	N	N
River lamprey	N*	—	—	N*	—	—	—	—	—	—	—	N*	N*	—	—	—
Pacific brook lamprey	N	—	—	N*	—	—	—	—	—	—	—	N*	—	—	—	—
Pit Klamath brook lamprey	—	N	—	—	—	—	—	—	—	—	—	—	—	—	—	—
Klamath River lamprey	N*	N*	—	—	—	—	—	—	—	—	—	—	—	—	—	—
White sturgeon	N*	—	—	—	—	—	—	—	—	—	—	—	—	—	—	—
Green sturgeon	N*	—	—	—	—	—	—	—	—	—	—	E	—	—	—	—
American shad	I	—	—	I	—	—	—	—	—	—	—	I	—	—	—	—
Threadfin shad	—	—	—	I	—	—	—	—	—	—	—	I	—	—	—	—
Common carp	—	I	I	I	—	—	—	—	—	—	—	—	—	—	I	—
Goldfish	—	I	—	I	—	—	—	—	—	—	—	—	—	—	I	—
Golden shiner	I	I	—	I	—	—	—	—	—	—	—	I	I	—	I	—
Sacramento blackfish	—	—	—	I	—	—	—	—	—	—	—	—	—	—	—	—
Hardhead	—	—	—	N*	—	—	—	—	—	—	—	—	—	—	—	—
Hitch	—	—	—	N*	—	—	—	—	—	—	—	I	—	—	—	—
Sacramento pikeminnow	—	—	—	N	—	—	—	—	—	—	—	—	—	—	—	—
Blue chub	—	N	—	—	—	—	—	—	—	—	—	—	—	—	—	—
Tui chub	—	N	I	—	—	—	—	—	—	—	—	—	—	—	—	—
California roach	N	—	N	N	N	—	N	—	—	—	—	I	—	—	—	—
Speckled dace	—	N	N	N	—	—	—	—	—	—	—	I	—	—	—	—
Fathead minnow	—	I	—	I	—	—	—	—	—	—	—	I	—	—	—	—
Lost River sucker	—	N*	—	—	—	—	—	—	—	—	—	—	—	—	—	—
Shortnose sucker	—	N*	—	—	—	—	—	—	—	—	—	—	—	—	—	—
Klamath smallscale sucker	N	N	—	—	—	—	—	—	—	—	—	—	—	—	—	—
Klamath largescale sucker	—	N	—	—	—	—	—	—	—	—	—	—	—	—	—	—
Sacramento sucker	—	—	N	N	—	—	N	N	—	—	N	N	N	—	N	—
Channel catfish	—	—	—	I	—	—	—	—	—	—	—	—	—	—	—	—
White catfish	—	—	—	I	—	—	—	—	—	—	—	—	—	—	—	—
Brown bullhead	I	I	—	I	—	—	—	—	—	—	—	I	I	—	I	—
Black bullhead	—	I	—	I	—	—	—	—	—	—	—	—	—	—	—	—
Eulachon	N*	—	—	—	—	—	—	—	—	—	—	—	N*	—	N*	N*
Wakasagi	I	—	—	—	—	—	—	—	—	—	—	—	—	—	I	—
Longfin smelt	N*	—	—	N*	—	—	—	N*	N*	—	—	E	—	—	—	—
Pink salmon	E	—	—	E	—	E	—	—	—	—	—	E	E	—	—	—
Chum salmon	N*	—	—	—	—	—	—	—	—	—	—	—	—	—	—	N*
Coho salmon	N*	—	N*	N*	N*	N*	N*	N*	N*	N*	N*	N*	N*	N*	N*	N*
Chinook salmon	N	E	—	N*	—	—	—	—	N*	N*	N*	N*	N*	N*	N*	N*

Species	1	2	3	4	5	6	7	8	9	10	11	12	13	14	15	16
Kokanee	—	—	—	—	—	—	—	—	—	—	—	I	—	—	I	—
Rainbow trout	N	N	N	N	N	N	N	N	N	N	N	N	N	N	N	N
Cutthroat trout	N*	N	N	N*	N*	N	N	N	N*	N*	N*	N*	N*	N*	N*	N*
Brown trout	—	—	I	—	—	—	—	—	—	—	—	I	—	—	—	—
Brook trout	—	—	—	—	—	—	—	—	—	—	I	I	—	—	—	—
Bull trout	N^{b*}	—	I	—	—	—	—	—	—	—	—	—	—	—	—	—
Mosquitofish	—	I	I	—	—	—	—	—	—	—	—	I	I	—	—	—
Topsmelt	0	0	0	0	0	0	0	0	0	0	0	0	0	0	0	0
Inland silverside	—	0	I	—	—	—	—	—	—	—	—	—	—	—	—	—
Threespine stickleback	N	N	N	N	N	N	N	N	N	N	N	N	N	N	N	N
Brook stickleback	—	—	—	—	—	—	—	—	—	—	—	—	—	—	—	—
Striped bass	0	0	0	—	—	—	—	—	—	—	—	—	—	—	—	—
Sacramento perch	—	—	?	—	—	—	—	—	—	—	—	—	—	—	—	—
Black crappie	—	I	I	—	—	—	—	—	—	—	—	I	—	—	—	—
White crappie	—	I	I	—	—	—	—	—	—	—	—	I	—	—	—	—
Green sunfish	—	I	I	—	—	—	—	—	—	—	—	I	—	—	—	—
Bluegill	—	I	I	—	—	—	—	—	—	—	—	I	—	—	—	—
Pumpkinseed	—	—	I	—	—	—	—	—	—	—	—	—	—	—	—	—
Redear sunfish	—	I	I	—	—	—	—	—	—	—	—	I	—	—	—	—
Largemouth bass	—	I	I	—	—	—	—	—	—	—	—	I	—	—	I	—
Smallmouth bass	—	I	I	—	—	—	—	—	—	—	—	I	—	—	—	—
Spotted bass	—	I	I	—	—	—	—	—	—	—	—	—	—	—	—	—
Yellow perch	—	I	I	—	—	—	—	—	—	—	—	—	—	—	—	—
Shiner perch	0	0	0	0	0	?	0	0	0	0	?	0	0	0	0	0
Tule perch	0	—	N*	—	—	—	—	—	—	—	N*	N*	N*	N*	N*	N*
Tidewater goby	—	N*	?	—	—	—	—	—	—	—	—	—	—	—	—	—
Yellowfin goby	—	I	I	—	—	—	—	—	—	—	—	—	—	—	—	—
Staghorn sculpin	N	N	N	N	N	N	N	N	N	N	N	N	N	N	N	N
Slender sculpin^b	N^{b*}	—	—	—	—	—	—	—	—	—	—	—	—	—	—	—
Klamath lake sculpin^b	N	N	N	N	N	N	N	N	N	N	N	N	N	N	N	N
Coastrange sculpin	N	N	N	N	N	N	N	N	N	N	N	N	N	N	N	N
Prickly sculpin	N	N	N	—	—	—	—	—	—	—	—	—	—	—	—	—
Marbled sculpin	N	N	N	N	—	—	—	—	—	—	—	—	—	—	—	—
Riffle sculpin	—	—	—	—	—	—	—	—	—	—	—	—	—	—	—	—
Starry flounder	N	N	N	N	N	N	N	N	N	N	N	N	N	N	N	N
No. native species	21	15	20	8	8	9	8	7	9	9	8	15	14	9	12	12
No. introduced species	14	14	21	0	0	0	0	0	0	0	0	10	8	0	6	0
Total species	35	29	41	8	8	9	8	7	9	9	8	25	22	9	18	12
Species at risk	7	7	7	1	1	1	1	1	2	2	1	4	5	2	5	6
Extinct species	1	0	1	1	0	0	0	0	0	0	0	1	3	0	0	0

Notes: Upper Klamath is also in Oregon. North Coast watersheds listed are the largest watersheds; they do not differ sufficiently from one another to be recognized as subprovinces. Records are only for species known to have reproducing populations. Abbreviations: E, extinct native; I, introduced; N, native; ?, status uncertain (not counted in totals); 0, occasional marine visitor (not counted in totals); *, population at risk of extinction.

^a Redwood Creek watershed includes Freshwater Lagoon, now isolated from it, which contains all the exotic species.

^b Oregon only.

Rogue River Subprovince The Klamath Province contains this subprovince because the only native freshwater dispersant is the Klamath smallscale sucker, which may be distinct from the smallscale sucker in the Klamath River. The Rogue River is also the southernmost drainage containing reticulate sculpin, abundant in most coastal streams in Oregon and Washington. Otherwise, the Rogue contains the same saltwater dispersant species found in the lower Klamath River.

Klamath-Pit fishless area This is a large region that is covered with lava and scrubby forests. It contains no real watersheds and was presumably without fish historically. Much of the water from the region's limited rain percolates through the lava and emerges as the big springs that form the Fall River, a tributary to the Pit River. The area contains Medicine Lake, an old caldera into which trout have been planted for recreational fishing.

Sacramento–San Joaquin Province

The Sacramento–San Joaquin drainage system dominates central California (Fig. 2). Historically, about half of all California's water flowed out through its estuary. Its large size, diverse habitats, and isolation have made it a center of fish speciation. This speciation was facilitated by a complex geologic history that isolated various sub-basins or caused neighboring basins to connect to it. Within this complex province are 17 endemic species (including those that have colonized a few neighboring watersheds). The number of endemic forms increases to 40–50 when subspecies and distinct runs of chinook salmon are counted as well. In addition, there are 18 species shared with neighboring drainages, plus 5 euryhaline marine species that occur in lower reaches of streams on a regular basis. In all, 40 native species inhabit the province (Table 3). The Sacramento–San Joaquin Province can be divided into seven subprovinces, each supporting one or more distinct fish taxa: *(1)* Central Valley, *(2)* Goose Lake, *(3)* Pit River, *(4)* McCloud River, *(5)* Clear Lake, *(6)* Monterey Bay, and *(7)* Upper Kern River (Table 3).

The **Central Valley Subprovince** is drained by the Sacramento and San Joaquin Rivers. The Kern, Tule, Kaweah, and Kings Rivers of the southern end of the San Joaquin Valley originally connected to the San Joaquin River only during exceptionally wet years, when former lakes Buena Vista and Tulare flooded into one another and overflowed into the river. The Central Valley has been the center of speciation for the province because of its large size, varied habitats, and ancient age. Its freshwater dispersant fauna presumably became isolated from the rest of the fish fauna of western North America 10–17 million years ago (Minckley et al.

1986), resulting in a fauna that is very different from that of other isolated Western basins. The relationship of this fauna to others is complicated and obscure, as shown when native species are discussed individually.

The Sacramento perch is the only member of the family Centrarchidae native west of the Rocky Mountains. It is distinct enough to be placed in a separate genus (*Archoplites*). The fossil record indicates the genus was once widespread in the West. Some of the earliest fossils are known from Pliocene lake deposits in the Snake River Plain (in modern Idaho), which also contain catfish (Ictaluridae) fossils (G. Smith 1981). Curiously, no catfish are native to any of the modern faunas of California, although introduced species have done well.

The tule perch is the only freshwater species in the family Embiotocidae, marine fishes found along the North American and Asian coasts of the North Pacific. The distribution of tule perch within the province shows that they are freshwater dispersants. Other freshwater embiotocids, now extinct, are known from Pleistocene deposits in central California (Casteel 1976).

The Sacramento blackfish and hardhead have modern and fossil distributions similar to that of Sacramento perch, presumably because they both are found in warm lakes and slow-moving streams (Casteel and Hutchison 1973). Both are the only species in their genera (*Orthodon, Mylopharodon*), but the hardhead shares a common ancestry with pikeminnows (*Ptychocheilus*) (Carney and Page 1990).

Hitch and California roach also belong to an endemic genus (*Lavinia*). Neither has a fossil record outside the Sacramento–San Joaquin Province. Within the province, hitch are largely confined to lowland and lacustrine habitats, whereas roach are the most widely distributed species in small streams. Genetic studies indicate that some roach populations in different subprovinces may deserve designation as species, resurrecting species names given by J. O. Snyder in the early 20th century (J. Jones, pers. comm. 2001).

The Sacramento splittail also has no known fossil record, but it is one of the most distinctive of the native minnows, with possible affinities to Asiatic cyprinids (Howes 1984). It is a benthic feeder with an unusual capacity (for a cyprinid) to live in brackish water.

The Sacramento pikeminnow has relatives in the same genus (*Ptychocheilus*) in the Columbia and Umpqua Rivers to the north and in the Colorado River to the east and south. It is most closely related to the Colorado pikeminnow, which in turn is similar to fossil pikeminnows from the Miocene of Arizona (G. R. Smith 1981; Carney and Page 1990). A southern source for Sacramento pikeminnow fits with their absence from the Klamath and Rogue Rivers, which lie between the Sacramento and Umpqua drainages. The recent successful introduction of northern pikeminnow into the Rogue River indicates that lack of suitable

Table 3

Presence of Fish Species in Major Watersheds of the Sacramento–San Joaquin Aquatic Zoogeographic Region of California

Watershed name Subprovince number	Goose Lake 2a	Pit River 2b	McCloud River 2c	Central Valley 2d	Clear Lake 2e	Monterey Bay 2f	Kern River 2g
Pacific lamprey	N*	—	E	N	E	N	—
River lamprey	—	—	—	N*	—	—	—
Pacific brook lamprey	—	—	—	N	N*	N*	—
Pit Klamath brook lamprey	N	N	—	—	—	—	—
Kern brook lamprey	—	—	—	N*	—	—	—
White sturgeon	—	—	—	N	—	—	—
Green sturgeon	—	—	—	N*	—	—	—
American shad	—	—	—	I	—	—	—
Threadfin shad	—	—	—	I	I	I	I
Common carp	—	I	—	I	I	I	—
Goldfish	—	—	—	I	I	I	—
Golden shiner	—	I	—	I	I	I	I
Sacramento blackfish	—	—	—	N	N	N	—
Hardhead	—	N	—	N	—	—	N
Hitch	—	—	—	N*	N*	N*	—
Sacramento pikeminnow	—	N	—	N	N	N	N
Tui chub	N*	N	—	I	—	—	—
Thicktail chub	—	—	—	E	E	E	—
Sacramento splittail	—	—	—	N*	—	—	—
Clear Lake splittail	—	—	—	—	E	—	—
California roach	N*	N*	—	N	N	N	—
Speckled dace	N	N	—	N	—	N	—
Lahontan redside	—	—	—	I	—	—	—
Red shiner	—	—	—	I	—	—	—
Fathead minnow	I	I	I	I	I	I	—
Mountain sucker	—	—	—	I	—	—	—
Sacramento sucker	N	N	N	N	N	N	N
Modoc sucker	N*	N*	—	—	—	—	—
Blue catfish	—	—	—	I	—	—	—
Channel catfish	I[a]	I	—	I	I	I	I
White catfish	—	—	—	I	I	I	I
Brown bullhead	I	I	—	I	I	I	—
Black bullhead	—	—	—	I	I	I	—
Delta smelt	—	—	—	N*	—	—	—
Wakasagi	—	—	—	I	—	—	—
Longfin smelt	—	—	—	N*	—	—	—
Coho salmon	—	—	E	E	—	N*	—
Chinook salmon	—	E	E	N	—	—	—
Kokanee	—	—	I	I	—	—	—
Rainbow trout	N	N	N	N	N	N	N*
Cutthroat trout	—	—	—	I	—	—	—
Brown trout	I	I	I	I	I	I	I
Brook trout	I	I	I	I	—	—	I
Lake trout	—	—	—	I	—	—	—
Bull trout	—	—	E	E	—	—	—
Rainwater killifish	—	—	—	I	—	—	—
Mosquitofish	—	I	—	I	I	I	—
Topsmelt	—	—	—	0	—	0	—
Inland silverside	—	—	—	I	I	I	—
Threespine stickleback	—	—	—	N	N	N	—
Striped bass	—	—	—	I	—	0	—
White bass	—	—	—	I	—	I	—

Table 3 (Continued)

Watershed name Subprovince number	Goose Lake 2a	Pit River 2b	McCloud River 2c	Central Valley 2d	Clear Lake 2e	Monterey Bay 2f	Kern River 2g
Sacramento perch	—	—	—	N*	N*	E	—
Black crappie	I[a]	I	—	I	I	I	I
White crappie	I[a]	—	—	I	I	I	I
Warmouth	—	—	—	I	—	—	—
Green sunfish	—	I	—	I	I	I	I
Bluegill	I	I	—	I	I	I	I
Pumpkinseed	I[a]	—	—	I	—	—	—
Redear sunfish	—	I	—	I	I	I	—
Largemouth bass	I	I	—	I	I	I	I
Spotted bass	—	I	—	I	—	—	—
Smallmouth bass	—	I	—	I	—	I	—
Redeye bass	—	—	—	I	—	—	—
Yellow perch	I[a]	—	—	I	—	—	—
Bigscale logperch	—	—	—	I	—	—	—
Shiner perch	—	—	—	0	—	0	—
Tule perch	—	N	—	N	N	E	—
Tidewater goby	—	—	—	E	—	N*	—
Yellowfin goby	—	—	—	I	—	—	—
Longjaw mudsucker	—	—	—	0	—	0	—
Shimofuri goby	—	—	—	I	—	—	—
Chameleon goby	—	—	—	0	—	—	—
Staghorn sculpin	—	—	—	N	—	N	—
Rough sculpin	—	N	—	—	—	—	—
Coastrange sculpin	—	—	—	—	—	N	—
Prickly sculpin	—	—	—	N	N	N	—
Pit sculpin	N	N	—	?	—	—	—
Marbled sculpin	—	N	—	—	—	—	—
Riffle sculpin	—	—	N	N	—	N	—
Starry flounder	—	—	—	N	—	0	—
No. native species	9	14	7	28	14	19	4
No. introduced species	11	15	5	40	18	20	12
Total species	20	29	12	68	32	39	16
Species at risk	4	2	0	8	3	4	1
Extinct species	0	1	3	3	3	3	0

Notes: Records are only for species known to have reproducing populations. Abbreviations: E, extinct native; I, introduced; N, native; ?, status uncertain (not counted in totals); 0, occasional marine visitor (not counted in totals); *, population at risk of extinction.
[a]Oregon only.

habitat is not a good explanation for their absence from intervening rivers.

Other freshwater dispersants also have close relatives in nearby drainages. The thicktail chub was apparently closest to the arroyo chub of the Los Angeles basin, and other Southwestern species in the genus *Gila* (Barbour and Miller 1978). The speckled dace occurs in all drainages surrounding the Sacramento–San Joaquin Province and probably has at least subspecies in each zoogeographic province. This fish occurs in headwater streams, and so can more easily move (or be moved) between drainages than most other species.

The Sacramento sucker belongs to a genus (*Catostomus*) widespread throughout North America, with species that are very similar to one another. Its closest relative is probably the Tahoe sucker of the Lahontan Province (G. R. Smith 1992).

The closest relatives of the Central Valley riffle sculpin are probably sculpin (*Cottus*) species with low dispersal abilities in the Great Basin and Klamath Provinces, rather than the sculpins considered to be riffle sculpins in Oregon. The Pit sculpin of the Pit River is a recent derivative of the riffle sculpin.

The freshwater fish fauna of the Central Valley Subprovince has been enriched by species with fairly recent saltwater dispersant ancestors but that now show evidence of speciation within the drainage. The delta smelt is confined to the subprovince but belongs to a genus (*Hypomesus*) widespread in estuaries, lagoons, and lakes along the Pacific coast of both North America and Asia. The more euryhaline longfin smelt also belongs to a widespread genus (*Spirinchus*), and the species itself seems to be present in a number of Pacific coast estuaries. Nonpredatory lampreys, most notably the Kern brook lamprey, have evolved from anadromous Pacific and river lampreys.

The runs of chinook salmon and rainbow trout/steelhead show adaptations to the unusual conditions of Central Valley streams and are genetically distinguishable from runs in other systems. Particularly distinctive is the winter-run chinook salmon, which spawns in cold spring-fed streams in the upper Sacramento River drainage.

Overall, the present Central Valley fish fauna shows evidence of long isolation and limited ancestry (Avise and Ayala 1976), with complex origins. The distinctive morphology, physiology, and life history patterns of the species reflect an evolutionary history of adaptation to a region where extended droughts are common, as are massive floods.

The **Goose Lake Subprovince** is a large, arid drainage basin that straddles the California-Oregon border and centers on Goose Lake, an enormous shallow lake. Historically, the lake has overflowed into the Pit River and also nearly dried up. The fishes of the lake are morphologically and genetically distinct, reflecting adaptations for life in its rich, alkaline, and muddy waters and survival in remnant habitats during periods of severe drought. The tui chub and Sacramento sucker have been described as subspecies. The most distinctive fishes are the undescribed Goose Lake lamprey and the Goose Lake redband trout. The lamprey is a bronze-colored predatory form (or forms) related to the lampreys of the Upper Klamath Subprovince (see that account). The redband trout is a rainbow trout that has two distinct life history strategies: one strategy is to live in the lake, grow to large size, and spawn in the streams, and the other is to be a small resident of headwater streams. When the lake dries up and then fills again, it can be quickly recolonized by fish from headwater populations. In addition, streams in the basin support Pit-Klamath brook lamprey, speckled dace, Pit sculpin, Modoc sucker, and California roach. The systematics of all eight native species have yet to be worked out, especially in relation to those of similar forms in the upper Pit River region.

The **Pit River Subprovince** contains 14 native species, an interesting mixture of fish of Sacramento and Klamath origin, including three endemic sculpins (Pit, rough, and "bigeye" marbled sculpin) and the endemic Modoc sucker. The province consists of the Pit River drainage of the northeast-ern corner of the state, a region subject to intense mountain building and vulcanism during the Pliocene and Pleistocene Periods. Lava flows repeatedly changed the face of the landscape, creating the desolate Devil's Garden area of today. In the late Pliocene (two million years ago), the upper Pit River drained north and west, into the upper Klamath River, which in turn connected to the ancient Snake River, which drained from the Great Basin. In the early Pleistocene (about one million years ago), the Klamath connection was dammed by lava, creating a deep lake (Lake Alturas) where a shallow lake had previously existed. Lake Alturas eventually spilled over a gap in the Adin Mountains, eroding a connection to the Sacramento drainage, and its bed was later largely obliterated by more lava flows (Pease 1965).

As a result of these dramatic changes in drainage connections, the Pit River Subprovince contains fishes derived from both the Sacramento–San Joaquin and Klamath Provinces. The Sacramento–San Joaquin fishes are all Pleistocene invaders that were able to pass the falls and rapids in the deep canyon of the lower Pit River: Sacramento pikeminnow, hardhead, California roach, and Pit sculpin (derived from riffle sculpin). Tule perch are present in the lower river but have not traversed Pit Falls. Fishes with ancestors in common with the modern Klamath fauna are Pit-Klamath brook lamprey, marbled sculpin, rough sculpin, tui chub, and redband trout. The rough sculpin is very similar to the slender sculpin of the Klamath lakes of Oregon. The redband trout is found in isolated headwaters and shares a common ancestry with redband trout of the McCloud River and upper Klamath Subprovince in Oregon (Behnke 1992; Nielsen et al. 1999).

Another species found only in scattered headwaters is the endemic Modoc sucker. It appears to be most closely related to the Sacramento sucker. It is also found in a few Oregon headwaters of Goose Lake.

Overall, the ichthyological history of the Pit River Subprovince can be described as follows. The ancestral, pre-Pleistocene drainage was part of the ancestral upper Klamath drainage, which connected to a large river flowing from the Great Basin. The ancestral fish fauna was part of a widespread western fauna that became fragmented through the complex geologic activity described by Minckley et al. (1986). Just prior to its divorce from the Klamath drainage, the Pit drainage included one or more lakes containing fishes similar to those that now live in the Klamath Lakes of Oregon (and large lakes of the Great Basin). It also contained a stream fauna of speckled dace, marbled sculpin, Pit-Klamath brook lamprey, Modoc sucker, and redband trout. When the Pit and Klamath drainages became isolated from one another, the fishes in each drainage began their independent evolutionary journeys. In the Pit drainage this evolution was perhaps hastened by two events: elimination of the large lakes and invasion of riverine fishes from the

Sacramento River. The lacustrine fishes either became extinct (e.g., lake suckers, *Chasmistes*) or adapted to the lake-like environments of large, clear, spring-fed streams, Fall River and Hat Creek (rough sculpin, marbled sculpin, tui chub). Invading fishes seem to have eliminated the native stream fauna, except brook lamprey and speckled dace. Pit sculpin largely replaced marbled sculpin, except in Hat Creek and Fall River, where the Pit sculpin seems less able to avoid predators than the other sculpins. Sacramento sucker replaced Modoc sucker except in streams isolated by natural barriers. Elimination of the remaining barriers by humans has been a major cause of endangerment of Modoc sucker (Moyle and Marciochi 1975).

The **McCloud River Subprovince** contains only the McCloud River and its tributaries, sandwiched between the Pit River and the upper Sacramento River drainages. Although the river has two large falls that have helped to isolate its upper watershed, the main factor responsible for its distinctive fish fauna (seven native species) is the unusual nature of the river itself. It has fairly constant year-round flows of cold water from Mt. Shasta, much of which emanates from giant springs. Other water from the mountain enters through creeks of glacial meltwater that contains glacial silt, giving the lower river a green or milky color. The river flows through a deep forested canyon, with trees and amphibians reminiscent more of the North Coast than of the hot California interior. Historically, its numerous deep pools provided refuges for coldwater fishes, even at low elevations: spring-run and winter-run chinook salmon, steelhead trout, bull trout, and riffle sculpin, as well as McCloud River redband trout in the main river and tributaries above the falls.

From a zoogeographic perspective, the most distinctive element of the McCloud River fish fauna is (or rather, was) the bull trout, for which the closest other populations are in tributaries to the upper Klamath River in Oregon. It is common in the Columbia River drainage farther north. Presumably the bull trout was found throughout the original upper Klamath-Pit River drainage during the cooler and wetter Pleistocene and managed to colonize the McCloud River after the Pit River became connected to the Sacramento River. It then disappeared from the rest of region after the climate became warmer and drier, although it may have just gone unnoticed in the spring-fed waters of the upper Sacramento and Pit Rivers within recent times. The unique coldwater conditions of the McCloud River also made it the principal home of two distinctive runs of chinook salmon (both now gone from the river as the result of Shasta Dam). Most distinctive genetically is the winter-run chinook, which entered the river in winter and spawned in spring; this strategy was possible only because cold water allowed the embryos to incubate in the gravel during summer. They could then hatch in late summer and move into the Sacramento River and out to sea when river temperatures were low. Spring-run chinook entered the river in spring but did not spawn until early fall. The cold waters and deep pools enabled large numbers of adults to summer in the river and juveniles to rear for a year or more.

The riffle sculpin in the McCloud River is distinctive enough to have been described as a separate species (*Cottus shasta*), but its taxonomic status has never been properly evaluated. Curiously, the closest drainage to the McCloud River, Squaw Creek just to the east, contains Pit sculpin. The McCloud River redband trout lives in the upper parts of the drainage, above the reach of spawning steelhead trout, which will hybridize with it. There are at least two distinct forms, one of them confined to tiny Sheepheaven Creek (Nielsen et al. 1999).

The **Clear Lake Subprovince** is centered on Clear Lake, which occupies only a small drainage basin in the Coast Range, although it is one of the largest natural lakes in California. It is regarded as the oldest lake in North America; organic sediment has been deposited continually in one basin for about 480,000+ years (Casteel et al. 1977; Casteel and Rymer 1981; Hearn et al. 1988). There are also remnants of a more ancient ancestral lake in the area, dating back 1.8–3.0 million years. Subsidence of the faulted block on which the lake rests has kept up with the sediment deposition, resulting in over 320 m of sediment deposits. Coring samples of the sediment have allowed scientists to recreate the history of the lake and the local climate by examining remains of algae, zooplankton, and fish deposited through time (Casteel 1976).

The native fish fauna of the lake is dominated by species otherwise found mainly in quiet waters of the Central Valley floor. These fishes are incapable of moving up the lake's outlet stream, Cache Creek, as it exists today, a fast-moving stream flowing through a steep, narrow canyon. They could only have entered the lake when the gradient between it and the valley floor was not as steep. The fishes have thus been isolated from the main system for a long time, and their remains are present in sediment deposits going back hundreds of thousands of years (Casteel et al. 1977). A number of the fishes have diverged morphologically from the ancestral valley forms and are recognized as separate species or subspecies: Clear Lake splittail, Clear Lake hitch, Clear Lake tule perch, and, possibly, Clear Lake prickly sculpin (Hopkirk 1973). Hopkirk also described another cyprinid species (*Endemichthys grandipinnis*) from the lake, but its status is uncertain.

The geologic events that lead to the formation of Clear Lake and to the establishment of its fish fauna are complex (Anderson 1936; Hinds 1952; Brice 1953; Hodges 1966; Swe and Dickinson 1970; Hopkirk 1973). In the early or middle Pleistocene, when the Coast Range was much lower, the Clear

Lake basin was a valley connected by a low-gradient stream (Cache Creek, or possibly Putah Creek) to the Sacramento system. The basin may also have drained via Cold Creek into the Russian River. The basin at this time contained one or more lakes that provided suitable habitat for invading Sacramento fishes. As the Coast Ranges rose, the gradient of Cache Creek increased, isolating the fishes in the basin. Tectonic activity, or perhaps deposition of alluvial deposits from Scotts Creek, may also have blocked outflow through Cold Creek. Meanwhile, faulting caused the northwest portion of the basin to subside, resulting in a depression containing the main portion of Clear Lake. Volcanic activity in middle and late Pleistocene, including that creating Mt. Konocti, further modified the lake basin. Most dramatic was a lava flow that blocked Cache Creek near its exit from the lake, raising the lake level and making Cold Creek the main outlet. This change may have permitted the Russian River to be colonized by some Clear Lake fishes. Finally, in the Pleistocene a landslide (or alluvial debris from Scotts Creek) blocked Cold Creek, allowing the lake to spill over the Cache Creek lava flow, reestablishing Cache Creek as the outlet.

The streams of this province contain Sacramento pikeminnow, Sacramento sucker, California roach, and rainbow trout, which appear indistinguishable from those of the Central Valley Subprovince. In addition, presumed Pacific brook lamprey are present in at least one stream, Kelsey Creek. Prior to construction of a dam on the outlet of Clear Lake, both steelhead rainbow trout and Pacific lamprey apparently ascended Cache Creek to spawn in tributaries to the lake.

The **Monterey Bay Subprovince** consists mainly of three major streams flowing into Monterey Bay: the San Lorenzo, Pajaro, and Salinas Rivers. For convenience, it also includes the small coastal drainages from Santa Cruz to San Francisco. One of these (Pescadero Creek) contains California roach. The drainages are also the southernmost habitats for coho salmon. The Pajaro and Salinas Rivers had (until historical times) almost a full complement of freshwater dispersant fishes characteristic of the Central Valley Subprovince: Sacramento sucker, California roach, hitch, Sacramento blackfish, Sacramento pikeminnow, speckled dace, thicktail chub, Sacramento perch, tule perch, and riffle sculpin. The only species missing were hardhead and splittail. Snyder (1913) failed to collect Sacramento perch, thicktail chub, and pikeminnow from the Salinas River, but remains of all three are present in prehistoric archaeological sites (Gobalet 1990), and pikeminnow are common in the river today. This is not surprising, because the Pajaro was a tributary of the Salinas River in the late Pleistocene. The San Lorenzo River contains only suckers, roach, and dace. Of fishes present in the Monterey Bay Subprovince, only sucker, roach, and hitch may be well enough differentiated

to justify calling them subspecies. The hitch was originally described as a separate species by Snyder (1913), but his description was based in part on hybrids between hitch and roach (Miller 1945b). However, Monterey hitch do have fewer dorsal and anal fin rays than those from the Sacramento drainage, even at sites where roach are absent, so subspecific designation is probably warranted.

The nature of the freshwater dispersant fish fauna indicates that this subprovince probably had two separate connections to the Central Valley during the middle or late Pleistocene: (1) a headwater connection between the San Benito River (a tributary of the Pajaro River) and the San Joaquin River, and (2) a lowland connection between Coyote Creek and Llagas Creek (also a Pajaro tributary). The San Benito connection came earlier and permitted California roach, Sacramento sucker, and speckled dace to enter the system (Murphy 1948c). The main pieces of evidence for this early connection are (1) the degree of differentiation of roach and sucker, compared with other fishes, (2) the similarity of the two species to their counterparts in the San Joaquin system, and (3) the presence of populations of roach above impassable falls in the San Benito River (Murphy 1948c). Other fishes native to the Pajaro-Salinas system are mainly lowland forms. They presumably entered by way of Coyote Creek, which now flows into San Francisco Bay. There is strong geologic evidence that the upper portion of Coyote Creek changed course several times in the past to flow into Llagas Creek, a Pajaro tributary (Branner 1907). Coyote Creek also makes a plausible source for the lowland species because it contains (or did until recently) a nearly full complement of Central Valley fishes, despite having long since been cut off by salt water from the main system. The absence of hardhead from Coyote Creek helps to explain their absence in Monterey Bay drainages.

From the Pajaro River, freshwater fishes presumably spread to the Salinas and San Lorenzo Rivers through lowland connections that existed when sea level was lower, or through recent estuarine connections between the Pajaro and Salinas Rivers when flooding makes the surface waters nearly fresh. The freshwater dispersant fauna of these rivers is supplemented with saltwater dispersant fishes, mainly Pacific lamprey, threespine stickleback, prickly sculpin, steelhead, and coho salmon.

The **Upper Kern River Subprovince** is the upper Kern River basin that contains the river and its tributaries above the present site of Isabella Reservoir. Only two species of fish are native to the basin, Sacramento sucker and endemic golden trout, now regarded as three subspecies of rainbow trout. The sucker is apparently a recent invader from the lower Kern River, but the golden trout evolved from rainbow trout isolated in the Upper Kern basin. Three distinct types of trout are currently recognized, which apparently evolved in isolation from one another: Volcano Creek golden trout, Little Kern River golden trout, and Kern River rainbow trout. The latter may have resulted from hybridization between an ancestral "redband" trout and later-arriving coastal rainbow trout.

North Coast Province

The North Coast Province includes coastal drainages from the Golden Gate on San Francisco Bay to the Smith River on the Oregon border, but excludes the mouth of the lower Klamath River. It is a collection of coastal streams and rivers with largely independent zoogeographic histories but with more faunal similarities than differences (Table 2). The exception is the Russian River, a coastal stream that has "captured" much of the Sacramento–San Joaquin fauna; 9 of 20 native species in the river are otherwise endemic to the Sacramento–San Joaquin basin. Some other drainages contain California roach, Sacramento sucker, or both, indicating past headwater connections to streams of the Central Valley, but overall anadromous and other saltwater dispersant fishes dominate the faunas (15 of 25 species; 16 of 21 if the Russian River is excluded). There are no endemic species to define this province, so it is basically a province of convenience.

The Mad, Eel, and Bear Rivers share one native freshwater dispersant, the Sacramento sucker. This sucker has been recognized as a separate species, but there seems little reason to consider it a distinct taxon (Ward and Fritzsche 1987). It presumably moved from the Eel River to the Mad River (or vice versa) through their once-common estuary (Humboldt Bay) and into the Bear River from the Eel River by way of headwater connections. It is curious that only the sucker managed to invade these drainages, because in recent years California roach, speckled dace, and Sacramento pikeminnow have all been successfully introduced into the Eel River.

The next major drainage southward, the Navarro River, contains both Sacramento sucker and California roach. South of the Navarro, the Gualala River contains only roach. The taxonomic identity of the two roach populations is uncertain; they have been variously listed as separate species, as subspecies, and as not being distinct from roach of the Central Valley. The same is true for roach from the Russian River and tributaries to Tomales Bay (Walker, Lagunitas, and Olema Creeks). It is likely, however, that all these populations have been isolated from one another long enough to merit recognition as distinct taxa at one level or another.

By far the largest collection of freshwater dispersant fishes in coastal drainages occurs in the Russian River, which is inhabited by California roach, hitch, Sacramento pikeminnow, hardhead, and tule perch. The tule perch is distinctive enough to be described as a subspecies (Hopkirk 1973). Just how these fishes got into the Russian River has been debated ever since Holway (1907) suggested the river was the ancestral home of the entire Sacramento–San Joaquin fauna, an idea quickly rejected by Snyder (1908d). There are two geologically possible routes by which the Sacramento–San Joaquin fauna could have entered the Russian River, either through Clear Lake (Lake County) or through drainage connections with San Francisco Bay.

Transfer of fish from Clear Lake was possible as a result of a complex but well-documented series of geologic events. Clear Lake first drained into the Sacramento River through Cache Creek. Cache Creek was blocked by a lava flow, raising the level of the lake so that it spilled into Cold Creek, a tributary to the Russian River. Cold Creek was then blocked by a landslide, and the drainage down Cache Creek was reopened, as is discussed in more detail under the Clear Lake Subprovince.

Transfer of fishes to the Russian River from San Francisco Bay is possible because the bay was a river valley until the late Pleistocene and only low divides today separate two of its tributaries (Copeland Creek and Petaluma River) from two Russian River tributaries (Santa Rosa and Sonoma Creeks). This region is extremely active geologically (it is on the San Andreas fault), so dramatic shifts in drainages are possible (Wahrshaftig and Birman 1965).

A close examination of the fish fauna supports the hypothesis that both routes were involved. Hardhead and riffle sculpin are present in the Russian River drainage, but absent from the Clear Lake basin. Sacramento perch and Sacramento blackfish, once two of the most abundant species in Clear Lake, were absent from the Russian River until introduced, an indication that lack of suitable habitat would not have kept them from becoming established in more ancient times. However, the Sacramento perch is no longer present in the river. The California roach of the Russian River seems to be most similar to the form in the Clear Lake basin. Although Russian River tule perch bear greater morphological similarity to Clear Lake perch than to Sacramento–San Joaquin perch (Hopkirk 1973), genetically it is divergent from both forms (Baltz and Loudenslager 1984). All other freshwater dispersants in the Russian River are adapted for stream living and could have entered through either route.

Great Basin Province

The Great Basin is the vast, arid region of western North America between the Sierra Nevada and the Rocky Mountains, divided into numerous smaller basins. During the Pleistocene and before, many of these basins contained large lakes that often had aquatic connections to one another. Today these lakes are either dry, reduced to remnants, or too alkaline to support fish. The basins are now largely isolated, and their remnant fishes have evolved into forms adapted to local conditions. These conditions range from cold mountain creeks, to warm highly fluctuating streams at low elevations, to alkaline lakes, to tiny desert springs. Each basin therefore tends to have one or more endemic species or subspecies, as is evident in basins (subprovinces) all or partly in California: Surprise Valley, Eagle Lake, Lahontan, Owens, Amargosa, and Mojave. Altogether these basins contain only 13 native species, 6 endemic to the Great Basin, including 4 endemic to the California portions of the Great Basin (Table 4). In addition, there are a number of large areas, including the Mono Lake basin, that were historically fishless.

The **Surprise Valley Subprovince** contains two basins, Surprise Valley and Cowhead Lake, in the extreme northeastern corner of the state. The floor of Surprise Valley contains three large, highly alkaline lakes that periodically dry up. As far as is known, streams draining the Warner Mountains on the California side of this valley had no native fishes, although it is possible that redband rainbow trout were present before nonnative rainbows were introduced. There are also tui chubs in at least one farm pond in the basin, but their origin is uncertain. On the Nevada side, Wall Canyon Creek contains an undescribed sucker (*Catostomus* sp.) and speckled dace. Surprise Valley and the Cowhead Lake basin have not been connected in recent times (if ever), and the Cowhead Lake drainage should probably be treated as a separate subprovince, or as part of the Warner Valley drainage of Oregon. It contains an endemic tui chub subspecies in a lowland slough and speckled dace in the streams. It is also possible that redband trout were (or are) present.

The **Eagle Lake Subprovince** is centered around Eagle Lake, a large terminal lake that once drained into Lake Lahontan (see the next section). It contains an endemic subspecies of rainbow trout (rather than cutthroat trout), the only rainbow trout native to the Great Basin. Its ancestors presumably crossed one of the low divides separating the Eagle Lake drainage from the Pit River. The only other species present are Lahontan redside, tui chub, speckled dace, and Tahoe sucker. The tui chub may be an endemic subspecies. Conspicuous by their absence are Lahontan cutthroat trout, Paiute sculpin, mountain sucker, and mountain whitefish.

The **Lahontan Subprovince** consists of four watersheds in California on the east side of the Sierra Nevada, north to south: Susan River, Truckee River, Carson River, and Walker River (Table 4). Collectively, they have by far the most diverse fish fauna of any Great Basin subprovince (eight species in California, four of which are shared by all watersheds). During the Pleistocene, these basins all drained into Lake Lahontan, which occupied much of the northwestern third of Nevada and the Honey Lake region of California. The main remnants of that lake today are Pyramid and Walker Lakes, Nevada. In Nevada, the principal watershed in this subprovince is the Humboldt River, although there are numerous smaller ones as well, such as the isolated Soldiers Meadow drainage, which contains desert dace (*Eremichthys acros*). The major drainages share endemic Lahontan cutthroat trout, Tahoe sucker, Lahontan redside, Lahontan speckled dace, and tui chub (various subspecies). Other shared species—Paiute sculpin, mountain sucker, and mountain whitefish—are also found in zoogeographic regions outside California. These three species are either recent invaders of the system (which seems unlikely given their isolation from their nearest relatives on the opposite side of the Great Basin) or cryptic species in need of taxonomic reevaluation. Another species endemic to the subprovince not found in California is cui-ui sucker (*Chasmistes cujus*), which is endemic to Pyramid Lake (sink for the Truckee River).

The Lahontan fauna has been in place for a long time; fossils of most modern species are present in deposits that date at least to the Miocene. Related species are found in other parts of the Great Basin, the Columbia River drainage (which now includes the ancient Snake River), and the Klamath drainage (Minckley et al. 1986). In short, much of the Lahontan fauna descends from a fauna that was widespread in western North America when climate and landscape were less rugged—although some species (e.g., mountain whitefish) could have invaded later from the Columbia drainage. Because various basins within the subprovince also have been isolated from one another, some localized differentiation of fishes has also taken place. For example, Silver King Creek in Alpine County contains the Paiute cutthroat trout, essentially a Lahontan cutthroat trout with few spots. Likewise, Pyramid Lake and Lake Tahoe (Truckee watershed) contain lake-adapted forms of Lahontan cutthroat and tui chub.

The **Owens Subprovince** consists of the Owens River and its tributaries, which ultimately flow into now-dry Owens Lake. The native fish fauna consists of five endemic forms: Owens sucker, Owens tui chub, Owens speckled dace (two undescribed subspecies), and Owens pupfish. The sucker and tui chub are very closely related to species in the Lahontan Subprovince, and most likely were part of the Lahontan fauna when the Owens drainage and the Mono Lake basin (which is between the Owens and Lahontan Subprovinces) were all connected to the Lahontan drainage. On the other hand, the pupfish is most closely related to the desert pupfish of the Colorado Province, suggesting ancient connections. This region is still active geologically, and much of its past history has been obscured by lava flows and other geologic events, making the ancient history of the fauna difficult to work out (Minckley et al. 1986).

Table 4

Presence of Fish Species in Major Watersheds of the Great Basin Aquatic Zoogeographic Region of California

Watershed name Subprovince number	Great Basin									
	Surprise Valley 4a	Eagle Lake 4b	Susan River 4c	Truckee River 4c	Carson River 4c	Walker River 4c	Owens 4d	Amargosa 4e	Mojave 4f	Mono Lake 4g
Threadfin shad	—	—	—	—	—	—	—	—	—	—
Common carp	—	—	—	I[a]	I[a]	I	—	—	—	—
Goldfish	—	—	—	—	I	I	—	—	—	—
Golden shiner	—	—	—	I	I[a]	—	—	—	—	—
Sacramento blackfish	—	—	—	I[a]	I[a]	—	—	—	—	—
Hitch	—	—	—	—	—	—	—	—	—	—
Tui chub	N*	N	N	N	N	N	N*	—	N*	I
Arroyo chub	N*	—	—	—	—	—	N*	—	I	—
Speckled dace	N*	N	N	N	N	N	N*	N	—	—
Lahontan redside	—	N	N	N	N	N	—	—	—	I
Fathead minnow	—	—	—	I[a]	I[a]	I[a]	—	—	—	—
Mountain sucker	—	N	N*	N*	N*	N*	—	—	—	—
Tahoe sucker	—	N	N	N	N	N	—	—	—	—
Owens sucker	—	—	—	—	—	—	N*	—	—	—
Channel catfish	—	—	—	—	—	—	I	—	I	—
White catfish	—	—	—	—	—	—	—	—	I	—
Brown bullhead	—	I	—	I[a]	I[a]	I[a]	I	—	I	—
Black bullhead	—	—	N	N	N	N	I	—	I	—
Mountain whitefish	—	—	—	I	I	N	—	—	I	—
Kokanee	—	—	I	I	I	I	—	—	I	—
Rainbow trout	N*	N*	I	I	I	I	I	—	I	I
Cutthroat trout	I	—	E	N*	N*	N*	I	—	—	—

Species	C1	C2	C3	C4	C5	C6	C7	C8	C9	C10
Brown trout	—	—	I	I	I	I	I	—	I	—
Brook trout	I	I	I	I	I	I	I	—	I	—
Owens pupfish	—	—	—	—	—	—	N*	—	—	—
Amargosa pupfish	—	—	—	—	—	—	—	N*	I	—
Salt Creek pupfish	—	—	—	—	—	—	—	N*	—	—
Mosquitofish	—	—	—	I	I	—	—	I	I	—
Threespine stickleback	—	I	—	—	—	—	—	—	I	—
White bass	—	—	I[a]	I[a]	I[a]	—	—	—	—	—
Sacramento perch	—	—	I[a]	I[a]	I	I	—	—	I	—
Black crappie	—	—	—	—	I	I	—	—	—	I
White crappie	I	—	—	—	—	—	—	—	—	—
Green Sunfish	—	—	I[a]	I[a]	I[a]	I	?	—	I	I
Bluegill	I	I	I[a]	I[a]	I[a]	I	I	—	—	I
Pumpkinseed	I	I	—	—	—	—	—	—	—	—
Largemouth bass	I	I	I[a]	I[a]	I[a]	I	I	I	I	—
Smallmouth bass	—	—	I[a]	I[a]	I[a]	I	—	—	—	—
Bigscale logperch	—	—	—	—	—	—	—	—	I	—
Tule perch	—	—	—	—	—	—	—	—	I	—
Prickly sculpin	—	—	—	—	—	—	—	—	I	—
Paiute sculpin	N	—	N	N	N	N	—	—	—	—
No. native species	3	5	8	8	8	8	4	3	1	0
No. introduced species	2	1	11	15	14	13	14	2	23	6
Total species	5	6	19	23	22	21	18	5	24	6
Species at risk	3	1	1	1	1	1	4	3	1	0
Extinct species	0	0	1	0	0	0	0	0	0	0

Notes: The Susan, Truckee, Carson, and Walker Rivers are part of the Lahontan subprovince, so it is placed with the other fishless regions of the Great Basin (4g). Records are only for species known to have reproducing populations. Abbreviations: E, extinct native; N, native; ?, status uncertain (not counted in totals); *, population at risk of extinction.

[a]Nevada only.

Table 5

Presence of fish species in major watersheds of the Southern California and Colorado River aquatic zoogeographic regions of California

Watershed name	Southern California											Colorado River	
	San Diego	San Luis Rey	Santa Margarita	Los Angeles	Santa Clara	Santa Inez	Santa Maria	San Luis Obispo	Morro	Big Sur	Carmel River	Colorado River	Salton Sea
Subprovince number	5	5	5	5	5	5	5	5	5	5	5	6a	6b
Pacific lamprey	E	E	E	E	N*	N*	N*	N*	N*	N	N	—	—
Pacific brook lamprey	—	—	—	E	—	—	—	—	—	—	—	—	—
Threadfin shad	—	—	—	—	—	—	?	—	—	—	—	—	—
Common carp	—	—	—	—	—	—	?	—	—	—	—	—	—
Goldfish	—	—	—	—	—	—	—	—	—	—	—	—	—
Golden shiner	—	—	—	—	—	—	?	—	—	—	—	—	—
Sacramento blackfish	—	—	—	—	—	—	—	—	—	—	—	—	—
Hitch	—	—	—	—	—	—	—	—	—	—	N	—	—
Sacramento pikeminnow	—	—	—	—	—	—	—	—	—	—	—	E	E
Colorado pikeminnow	—	—	—	—	—	—	—	—	—	—	—	E	E
Bonytail	—	—	—	—	—	—	—	—	—	—	—	E	—
Arroyo chub	—	—	N*	N*	—	—	—	—	—	—	—	—	—
California roach	—	—	—	—	—	—	N	—	—	—	—	—	—
Speckled dace	—	—	—	N*	—	—	—	N*	—	—	—	—	—
Red shiner	—	—	—	—	—	—	—	—	—	—	—	N*	—
Fathead minnow	—	—	—	N	—	—	—	—	—	—	—	—	—
Santa Ana sucker	—	—	—	—	—	—	—	—	—	—	—	N*	—
Razorback sucker	—	—	—	—	—	—	—	—	—	—	—	N*	E
Owens sucker	—	—	—	—	—	—	—	—	—	—	—	—	—
Sacramento sucker	—	—	—	—	—	—	—	—	N	—	—	—	—
Blue catfish	—	?	—	—	—	—	—	—	—	—	—	—	—
Channel catfish	—	—	—	—	—	—	—	—	—	—	—	—	—
White catfish	—	—	—	—	—	—	—	—	—	—	—	—	—
Yellow bullhead	—	—	—	—	—	—	—	—	—	—	—	—	—
Brown bullhead	—	—	—	—	—	—	—	—	—	—	—	—	—
Black bullhead	—	—	—	—	—	—	—	N	—	—	—	—	—
Flathead catfish	—	—	—	—	—	—	—	—	—	—	—	—	—
Rainbow trout	N*	N*	N*	N*	N*	N*	N*	N*	N*	N*	N*	—	—
Brown trout	—	—	—	—	—	—	—	—	—	—	N	—	—
Rainwater killifish	N	N	N	N	—	?	—	—	—	—	—	—	—
California killifish	—	N	N	N	N	—	—	—	N	—	—	—	—
Desert pupfish	—	—	—	—	—	—	—	—	—	—	—	E	N*
Mosquitofish	—	—	—	—	—	—	—	—	—	—	—	—	—
Sailfin molly	—	—	—	—	—	—	—	—	—	—	—	—	—

Shortfin molly	—	—	—	—	—	—	—	—	—	—	I
Porthole livebearer	—	—	—	—	—	—	—	—	—	—	I
Topsmelt	0	0	0	0	0	0	0	0	0	—	—
Inland silverside	N*	N*	N	N	N	N	N	—	N	N	—
Threespine stickleback	N*	E	N*	N	N	N	N	N	N	—	—
Striped bass	—	—	I	—	—	—	—	—	—	—	I
White crappie	I	I	I	I	I	I	I	—	I	I	—
Black crappie	I	I	I	I	I	I	I	—	I	I	—
Warmouth	I	—	I	I	—	—	—	—	—	—	—
Green Sunfish	I	I	I	I	I	I	I	—	I	I	—
Bluegill	I	I	I	I	I	I	I	—	I	I	I
Pumpkinseed	—	—	I	—	—	—	—	—	—	—	—
Redear sunfish	I	I	I	I	I	I	I	—	I	I	I
Largemouth bass	I	I	I	I	I	I	I	—	I	I	I
Spotted bass	—	—	I	—	I	—	—	—	—	—	—
Smallmouth bass	I	—	I	I	—	—	—	—	—	—	—
Redeye bass	—	—	I	I	I	I	I	—	—	—	—
Bigscale logperch	—	—	—	—	—	—	—	—	—	—	—
Mozambique mouthbrooder	—	—	I	I	—	—	—	—	—	I	I
Redbelly tilapia	—	—	I	I	—	—	—	—	—	I	I
Blue tilapia	—	—	—	—	—	—	—	—	—	—	?
Nile tilapia	—	—	—	—	—	—	—	—	—	—	?
Shiner perch	0	0	0	0	0	0	0	0	0	—	—
Tule perch	—	—	I	I	—	—	—	—	—	—	—
Striped mullet	N*	N	N	N	N	N	N	0	0	N	N
Tidewater goby	N*	E	N*	N*	N*	N*	N*	N*	N*	—	—
Yellowfin goby	I	I	I	I	I	I	I	—	—	I	I
Longjaw mudsucker	0	0	0	0	0	0	0	0	0	—	—
Shimofuri goby	I	—	I	I	I	I	I	—	—	—	I
Staghorn sculpin	N	N	N	N	N	N	N	0	0	N	N
Prickly sculpin	N	N	N	N	N	N	N	N	N	N	N
Coastrange sculpin	—	—	—	—	—	—	—	N	N	—	—
Starry flounder	—	—	0	0	0	0	0	0	0	—	—
No. native species	6	5	9	6	8	8	8	5	5	4	4
No. introduced species	25	15	34	24	7	8	9	0	13	23	27[a]
Total species[a]	31	20	43	30	15	16	17	5	18	27	31
Species at risk	3	2	5	2	3	2	3	1	1	2	1
Extinct species	1	2	3	0	0	0	0	0	0	2	3

Notes: Because of overlapping distributions of native fishes, Southern California watersheds are not grouped into subprovinces. Records are only for species known to have reproducing populations. Abbreviations: E, extinct native; I, introduced; N, native; ?, status uncertain (not counted in totals); *, population at risk of extinction; 0, occasional marine visitor (not counted in totals); *, population at risk of extinction.

[a] Includes three species of marine fish introduced into the Salton Sea.

The **Amargosa Subprovince** covers Death Valley and the Amargosa River, which flows into it. Death Valley is the lowest and most arid place in North America, yet it still contains fishes, remnants of the fauna that inhabited Pleistocene lakes and streams. These fishes are in Amargosa River, Salt Creek, and numerous springs flowing from fault lines along the mountains (four species total). The water is typically warm, often saline, and ancient in origin, perhaps 8,000 to 12,000 years old before it emerges. All the fishes are small in size and capable of withstanding environmental extremes. In California, the fauna consists of Salt Creek pupfish, with two subspecies (one in the creek, one on the hypersaline marshy floor of Death Valley); Amargosa pupfish, with three subspecies (one in the Amargosa River, two in tributary springs); and speckled dace. There are additional subspecies of the Amargosa pupfish in Nevada, in the Ash Meadows spring system, including the Devils Hole pupfish (*Cyprinodon diabolis*). The geology of this area is complex, but its ancient connections are presumably to the ancestral Colorado River. Fossil pupfish, dating back to the Miocene, are known from the region, however, so other possibilities exist. Exhaustive and often speculative analyses of the origin of the fauna in relation to geology are reviewed in Soltz and Naiman (1978), R. R. Miller (1981), and Minckley et al. (1986).

The **Mojave Subprovince,** which is basically the Mojave River drainage, contains just one species of native fish, Mojave tui chub. The nearest relative of the chub is presumably Owens tui chub. It is likely that they are both derived from tui chubs that lived in the large, interconnected Pleistocene lakes that occupied the desert regions of southern California.

Great Basin fishless areas are the large regions of desert and mountain that historically contained no fish and for the most part still lack fish. The best known such area is the Mono Lake basin. The fishes that once inhabited the streams flowing into highly alkaline Mono Lake presumably were wiped out by vulcanism during the past million years, up to and including historic times (Hart 1996). Because the basin has a number of permanent streams (which maintain the lake), it has been subjected to numerous introductions of fish, and at least six species now inhabit the basin. High-elevation streams elsewhere (e.g., Rock Creek, San Bernardino County) also contain introduced trout.

South Coast Province

This province is both arid and active geologically, so it has a somewhat limited fish fauna with a rather long and complex history (14 species occur in fresh water on a regular basis). It contains about ten large watersheds and many more smaller coastal drainages from Baja California north to Monterey Bay (Table 5). Uncertainties over the historical distributions of some native species, distributional overlaps in others, and the presence of a few widespread species have made the designation of subprovinces problematic, although arguments can be made for placing watersheds from the Santa Margarita River south in one subprovince (San Diego), the Los Angeles basin in another, and all the remaining watersheds north to the Carmel River in a third. Only streams of the Los Angeles basin (Santa Ana, San Gabriel, and Los Angeles Rivers) have an endemic group of freshwater dispersant fishes (arroyo chub, Santa Ana sucker, speckled dace), although arroyo chub are apparently also native to the neighboring Santa Margarita watershed, and there is a mysterious speckled dace in San Luis Obispo Creek. Most of the watershed is (or was) dominated by salt water dispersants. The larger streams are (were) used for spawning by anadromous rainbow trout, Pacific lamprey, and possibly threespine stickleback. The trout, lamprey, and stickleback left isolated populations in the headwaters of a few streams, creating landlocked forms, most notably the rainbow trout (*O. mykiss nelsoni*) of Baja California, Mexico; an undescribed (and now extinct) nonpredatory lamprey of the Los Angeles basin; and unarmored threespine sticklebacks of the Los Angeles basin. Sticklebacks are present in ancient fossil deposits in the region, so it is possible that some unusual populations in this province had inland origins.

Numerous euryhaline marine species are found in the lagoons and lower reaches of the streams, but two species are found only in such habitats: tidewater goby and California killifish. The goby is endemic to lagoons of the California coast, north to the Oregon border, but southern California populations are genetically distinct from the rest. They seem to disperse mainly when neighboring streams are connected by low sea level or high-outflow events that create coastal waters with low surface salinities. The killifish has presumably become distributed along the coast by moving through salt water, but its ultimate origins are inland because the genus is widespread in North American fresh and brackish waters and is common as fossils in what is now the Great Basin (Minckley et al. 1986).

The origins of arroyo chub, Santa Ana sucker, speckled dace, and California killifish in the region have long puzzled zoogeographers. Their closest relatives are in the Colorado River drainage (sucker, dace) and in Mexico (chub, killifish). Minckley et al. (1986) argue persuasively that the arroyo chub and California killifish rode into the region on a shifting continental plate that split from the continent farther south and that supports the fishes of the Mexican plateau. Both are lowland species that were unlikely to have ancestors capable of moving into the region through connections of upland tributaries. Both speckled dace and the group of suckers containing the Santa Ana sucker (subgenus

Pantosteus) are fishes capable of living in small, swift streams, and both have distribution patterns throughout the West that suggest dispersal through streams. Therefore, it is likely that these two species entered the region by way of stream connections to the ancient Colorado River drainage.

Colorado River Province

The Colorado River drains much of the arid interior of western North America, about 650,000 km². The river itself is huge and muddy, fed by numerous tributaries with past histories as independent drainages. Despite the size of the drainage and river, this ichthyological province contains only 32 native fishes, 16 of them widespread. Most species are endemic; the few that are not considered endemic are probably in need of taxonomic reevaluation. In many respects the most remarkable part of this fauna is the big river fishes with curious morphological adaptations that allow them to thrive in a warm, muddy, fluctuating river. The large minnows of the genus *Gila* in particular show a morphological diversity that reflects a wonderfully complex evolutionary history. Sporadic hybridization among the *Gila* species has enhanced genetic diversity within species while increasing genetic similarity among them. They nevertheless maintain morphological distinctness, allowing them to occupy diverse niches (Minckley and De-Marais 2000).

The endemic cyprinid and catostomid fishes of the Colorado River are related to those of both the modern Sacramento–San Joaquin and modern Columbia River faunas (shared genera). The big rivers in the three basins have been isolated from one another for millions of years, so they mainly have in common derivatives of a once widely distributed ancestral fauna (Minckley et al. 1986). The desert pupfish is also a relict of a more ancient fauna, indicated by the broad distribution of the genus *Cyprinodon* across the southwestern United States, Mexico, the Caribbean, and the Gulf and eastern coasts of North America. Fossil *Cyprinodon* from Miocene deposits of Death Valley indicate that pupfish were widely distributed even then, and they developed isolated populations by being carried off on shifting continental plates or by having coastal populations isolated or uplifted by tectonic activity (Minckley et al. 1986). Pupfish, with their small sizes and astonishing physiological tolerances, survive where few other species of aquatic organisms can, making it easy to envision them persisting in regions of intense geologic activity and little permanent water.

I divide the California portion of the Colorado River Province into two subprovinces: Lower Colorado River and Salton Sea. The **Lower Colorado River Subprovince** is basically the river as it flows along the California border for about 400 km. It was originally a fairly uniform section of river with no permanent tributaries in California. Therefore it supported only a few of the endemic riverine fishes: Colorado pikeminnow, bonytail, razorback sucker, and flannelmouth sucker. In addition, desert pupfish lived in riparian marshes and springs, and a few euryhaline marine species, such as striped mullet and machete, invaded the river from the Gulf of California.

The **Salton Sea Subprovince** originally contained only desert pupfish when Euro-Americans arrived on the scene. The Salton Sea was created by the inadvertent diversion of the Colorado River into the dry basin in the early 1900s. However, archaeological evidence indicates that the sea naturally filled on occasion with water from the Colorado River, bringing fishes with the inflowing water. Over 500 years ago the sea was a freshwater lake (Lake Cahuilla); its abundant bonytail, razorback sucker, striped mullet, and machete were an important source of food for the people living around the lake (Gobalet and Wake 2000). Today it still contains pupfish in a few places, but otherwise its fishes are a diverse collection of alien species, many found nowhere else in California.

Ecology

The native freshwater fishes of California have been evolving in isolation for millions of years. The general environment to which they have adapted is a harsh one. The climate has fluctuated tremendously over both long and short periods of time, from very wet to very dry. The landscape is geologically unstable, with rapidly rising and eroding mountain ranges, active volcanoes, and shifting continental plates (a major cause of earthquakes). Even on a seasonal basis, streams fluctuate from raging, cold torrents in spring to warm trickles in autumn. Not surprisingly, the few fishes that have managed to persist in this environment show adaptations in their morphology, physiology, behavior, and life history patterns to deal with environmental extremes (Moyle and Li 1979; Moyle et al. 1982; Moyle and Herbold 1987). The distinctive nature of the fish fauna, as well as the assemblages (communities or zones) of which they are a part, is shown by the following generalities that characterize it. Examples of these generalities—as well as exceptions to them—will be found in later sections of this chapter that describe the ecology of fish assemblages in and around the state.

1. **A majority of native fishes have a life history strategy characterized by large body size and high fecundity.** About 52 percent of all inland fishes have an adult body size greater than 20 cm SL with associated high fecundity (egg production) in females. This pattern is particularly prevalent among the freshwater dispersant minnows and suckers (20 of 26 species) and anadromous salmonids and sturgeons (all species). All of these fish have potential life spans in excess of 5 years (some in excess of 30 years). In terms of numbers and biomass, in most environments these large fishes are the dominant species, even during their early life history stages. In contrast, a majority of fishes in streams of eastern North America are small and short-lived (Moyle and Herbold 1987).

2. **Local fish faunas are morphologically diverse.** Except for early life history stages, native fishes in each drainage are relatively easy to distinguish from one another. This characteristic reflects distinctive morphologies, related to feeding habits and habitat preferences, and a high degree of ecological segregation among the species. In contrast, fish assemblages in Eastern streams tend to have large numbers of similar species, particularly among minnows and darters (Percidae), although the overall morphological diversity is just as great or greater because of the large number of species.

3. **Local species richness is low.** A typical assemblage of native fishes contains one to seven species, although richness may be higher in large lakes and some rivers. In contrast, Eastern streams and lakes often have assemblages in excess of 25 species.

4. **In streams with access to the sea, anadromous fishes are important members of fish assemblages.** Fourteen anadromous fishes, with numerous independent runs, spawn in coastal streams and rivers. They, and their juveniles, are often among the most abundant stream fishes, and they can be important sources of energy for stream ecosystems. Other saltwater dispersant fishes—such as mullet, sculpins, and gobies—are also frequently important.

5. **Almost all species spawn in the spring (March, April, May).** Most precipitation in California falls in winter and spring. Much of it falls as snow in the Sierra Nevada and becomes runoff in spring. Most California fishes have reproductive cycles keyed to this seasonal abundance of water and spawn within a three-month period. Most apparent exceptions to the spring-spawning rule are fishes that have extended spawning seasons and can spawn a month earlier or later if conditions are right. Some runs of anadromous fish (e.g., fall run chinook salmon) also spawn at different times to take advantage of special conditions.

6. **Most species exhibit little parental care.** Most California fishes (75%) do not guard their embryos or young; 56 percent are broadcast spawners over open substrates and 19 percent bury their embryos and then abandon them. The

nonguarders include all but eight freshwater dispersant species and all anadromous species except threespine stickleback. The broadcast spawners include all species of sturgeon, minnows, suckers, and smelt. Only two species (3%) are livebearers. All sculpins, gobies, and pupfish show parental care, as do threespine stickleback and Sacramento perch. With the exception of Sacramento perch, all species with parental care have small (<100 mm SL) body size, and most live in fairly permanent habitats (coldwater streams, lagoons, springs). This characteristic suggests that, from an evolutionary perspective, it pays to invest energy in producing lots of young when times are good, spawning in environments that are likely to have relatively low densities of potential predators on the young.

7. **Different life history stages of each species tend to be ecologically segregated.** This generality is true of most Eastern fishes as well, but the segregation seems to be better developed in Western fishes, among which juvenile fishes often behave ecologically like species different from the adults. This characteristic allows juveniles to avoid predation by adults and use resources not available to adults.

8. **Most species have physiological or behavioral mechanisms that allow them to survive or avoid extreme environmental conditions.** In the species accounts in this book, there are repeated references to the amazing ability of various species to survive high temperatures, high alkalinities, and low oxygen levels—conditions common in the summer waters of California. Other species, especially anadromous ones, avoid the extreme conditions by migrating either out to sea or up into consistently cold water in the mountains.

9. **Most species have well-developed dispersal abilities.** In a region where streams dry up or change course frequently, the most successful species are those that can quickly colonize new habitats. Most native fishes have tremendous dispersal abilities as both juveniles and adults. Smith (1982) found that reaches of the Pajaro River that went dry during a prolonged drought were recolonized by native minnows and suckers within a few months once water returned. In the Eel River, Sacramento pikeminnow colonized most of the suitable habitat in over 400 km of stream in less than 15 years, from a single introduction into a headwater region (Brown and Moyle 1993). All the saltwater dispersant species have considerable capability to colonize coastal streams.

10. **The more a stream or lake has been altered by human activity, the more likely it is to be dominated by introduced fishes.** Over a third of the fish species found in California's inland waters were introduced into them, mostly from the eastern United States. Introduced fishes dominate many bodies of water in the state because they are better adapted than native fishes to warm, impounded, and often nutrient-rich waters that are the by-product of civi-

lization. Native fishes that can survive in such waters are often eliminated by predation, exotic diseases, and, perhaps, competition. Fish assemblages in relatively undisturbed streams, in contrast, often show a remarkable ability to resist invasions by introduced species (Baltz and Moyle 1992). It is interesting to note that the number of native species inhabiting a watershed has little impact on the ability of alien species to invade it. Some of the most species-rich watersheds (e.g., the Central Valley) and some of the least species-rich watersheds (e.g., the Colorado River) are among the most invaded watersheds.

Assemblages and Faunal Filters

A local fish assemblage is very dynamic, changing from year to year or season to season. In relatively undisturbed bodies of water, these changes can be fairly predictable, provided there is adequate understanding of the life histories and ecological tolerances of the fishes. Unfortunately, we rarely have such an understanding and so are continually surprised by "sudden" changes in fish assemblages, especially when such changes mean that a stream no longer supports good fishing for a favored species. The first step in developing an understanding of how stream fish assemblages are structured is to realize that the fauna present in a given area has passed through a series of selective zoogeographic and ecological "filters" that progressively reduce the number of species at a locality from the total present in a zoogeographic province (Smith and Powell 1971).

The broadest filters are zoogeographic (Fig. 3). In the case of California, the faunas of the different zoogeographic provinces had their ancestral origins mostly in a widespread fauna of the early Pliocene, an era when much of the western landscape was less fragmented by high mountain ranges than it is today and perhaps was drained by one or two large river systems. As regions became subdivided, the faunas in each region were filtered to a smaller subset of the original fauna, a subset created by a combination of adaptation to local environmental conditions and zoogeographic accidents. Thus regions with lakes throughout their history retained specialized lacustrine fishes (e.g., suckers of the genus *Chasmistes*) as part of the fauna; those without lakes, even for a short period of geologic time, lost many of those elements. Within a region or zoogeographic province, local barriers serve as selective filters for faunal expansions or invasions. For example, falls on the Pit River prevented invasion by the lowland Sacramento River fauna (e.g., hitch, blackfish) into the Big Valley region, where plenty of suitable habitat for these fishes exists.

On a more local and shorter temporal scale, there are physiological filters—environmental conditions that prevent a species from moving into a reach of stream or into a

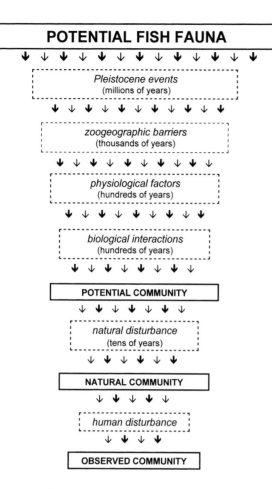

Figure 3. Each local fish fauna is the result of the screening of a regional fish fauna through a series of filters (dotted boxes) that act on different time scales. Most modern fish assemblages have been altered to a greater or lesser degree by human activity, even assemblages in apparently pristine environments.

lake because the species lacks the physiological capability to survive under those conditions. The result of these filters is division of local faunas into warmwater and coldwater fish assemblages. Rainbow trout are physiologically incapable of living in warm waters preferred by Sacramento blackfish, just as blackfish are physiologically incapable of persisting in the cold, swift streams favored by trout. A yet finer filter is behavioral—the interactions among species that affect local distribution patterns. Avian predators, for example, may exclude small fish from some shallow streams; competition from Sacramento suckers may prevent Modoc suckers from using lowland streams to which they would seem well adapted. A final filter, often unrecognized, is the human (anthropogenic) filter—human activities that change the nature of local environments or bring in new species.

In short, every local fauna is a product of both ancient and recent events and continues to change. Fishes that have been living together for eons are usually morphologically distinct, and this circumstance is presumed to be the result of evolution to minimize energetically expensive competitive interactions. How fishes divide available food and space among coexisting species is therefore often predictable through the study of morphology. The bladelike pharyngeal teeth of Sacramento pikeminnow, for example, reflect their piscivorous nature, while the molarlike pharyngeal teeth of hardhead, a species usually found with pikeminnow, reflect reliance on hard-shelled invertebrates and algae. Despite the usefulness of morphology in predicting ecological segregation, dissimilar species may still compete for limited resources. Thus riffle sculpin exclude the morphologically dissimilar speckled dace from riffles as the result of competition for hiding places under rocks (Baltz et al. 1982). Where sculpins are absent, dace are abundant in riffles. Clearly, understanding how a fish assemblage is structured requires taking very little for granted—even obvious morphological differences among species.

The previous chapter on distribution dealt largely with the zoogeographic filters through which each regional assemblage has passed. This chapter therefore deals largely with the physiological, ecological, and behavioral filters that structure assemblages. The sections that follow are brief descriptions of selected fish assemblages, the interactions among species making up the assemblages, and relationships of the species to their local environments. They represent only a small fraction of the assemblages present in the state and were chosen because they contain many, if not most, of the species in each zoogeographic province, and also because I was familiar enough with them to describe them with some confidence. The descriptions are generalities, like the statements in the introduction to this section. Anyone who has spent time studying any one of the assemblages will realize that each assemblage rarely conforms precisely to the picture presented here. Assemblages vary in species composition from year to year and from place to place. Furthermore, the behavior of each species is flexible, in relation both to other species and to the environment.

Sacramento–San Joaquin Province

Central Valley Streams

Streams of the Central Valley have headwaters, historically without fish, in mountainous areas. They plunge downward through steep canyons and deep pools in the foothills before flowing into sluggish rivers or lakes on the valley floor. The distinct habitats found in mountains, foothills, and valley floor contain distinct assemblages of fish that can have wide or narrow zones of overlap, depending on the gradient of the stream and other environmental conditions. In streams of the San Joaquin Valley, distributional overlap among assemblages is narrow enough to be mapped with some confidence (Fig. 4), but in tributaries to the Sacramento River, the overlap among regions with distinct assemblages (often called zones) is fairly broad (Fig. 5). Usually four fish assemblages can be recognized in Central Valley streams: *(1)* rainbow trout assemblage, *(2)* California roach assemblage, *(3)* pikeminnow-hardhead-sucker assemblage, and *(4)* deep-bodied fishes assemblage.

Rainbow trout assemblage. This assemblage is found in clear streams at high elevations, where stream gradients are high (usually a total drop of 3.0 m or more for every kilometer of stream). The water is swift and permanent, with more riffles than pools. The water is also cold, seldom exceeding 21°C, and is saturated with oxygen. The bottom materials are predominantly cobbles, boulders, and bedrock. The banks are well shaded and frequently undercut; logs and root wads often extend into the water, creating pools and other cover. Aquatic plants, submerged or emergent, are few, except where the streams flow through boggy alpine meadows. The dominant native fish are rainbow trout, but sculpin (usually riffle sculpin), Sacramento sucker, and speckled dace are often part of this assemblage, together or separately. In some streams they may be joined by California roach.

When trout, sucker, dace, and sculpin are found together, the resulting assemblage shows a high degree of structure (species segregation in use of food and space). Sculpin and speckled dace feed by picking invertebrates from the bottom, whereas rainbow trout feed primarily on drifting insects, both terrestrial and aquatic (Li and Moyle 1976). The trout also capture larger or more active benthic prey than the other two species, and they will prey on other fishes if given the opportunity. The aggressive and predatory behavior of large trout presumably regulates the distribution and abundance of sculpin and dace (unless the trout are regularly removed by anglers). Sculpin segregate from dace by ambushing larger invertebrates among the rocks, whereas dace browse on smaller forms. Sculpin also typically live and feed in swifter water than dace, although this is partly because dace are excluded from productive riffle areas by sculpins (Baltz et al. 1982). Suckers live by grazing on attached algae, detritus, and associated aquatic insects. They have few direct interactions with other fish species, but small trout will follow large suckers around, picking up small insects disturbed by the suckers' feeding. The fact that small suckers are largely confined to shallow water suggests they are avoiding predatory trout (Baltz and Moyle 1984).

The rainbow trout assemblage has been extended by humans in streams of the Sierra Nevada. Prior to extensive trout planting programs in the late 19th and early 20th centuries, most streams and lakes in elevations above 1,800 m were without fish. The only major exceptions to this were the upper reaches of the Kern River, where golden trout evolved, and those tributaries to the Pit and McCloud Rivers that contained redband rainbow trout. The rainbow trout assemblage has now been extended, through planting, to include most streams and lakes of the Sierras; only rarely are species other than salmonids present in these waters. At lower elevations the presence of this assemblage has occasionally been extended downstream into sections normally inhabited by the pikeminnow-hardhead-sucker assemblage as the result of poisoning operations followed by planting of hatchery trout. These extensions normally last only a few years, after which the treatment has to be repeated if artificially large trout populations are to be maintained (Moyle et al. 1983). Rainbow trout habitat has also been created at low elevations in cold waters flowing from dams. Often these waters, because of their low temperatures and swift currents, naturally exclude native minnows and suckers without further human intervention.

A further result of human manipulation of the rainbow trout assemblage has been to increase its complexity through the introduction of brook trout and brown trout. Brook, brown, and rainbow trout compete for food and space but may coexist by living in slightly different places and by adopting different feeding strategies. When all three species occur together, brook trout tend to be found in cold, spring-fed tributaries of the main stream, feeding equally on surface and bottom foods. Brown trout tend to be found in pools of main streams, feeding mostly on bottom invertebrates and other fish, while rainbow trout are more likely to be in the riffles, feeding on surface insects and drift. Different breeding times and places may also allow the species to coexist.

California roach assemblage. Streams containing this assemblage are small, warm tributaries to larger streams that flow through open foothill woodlands of oak and foothill pine. In the San Joaquin Valley, these streams are located in a narrow elevational band in the foothills in much of the same region that contains the pikeminnow-hardhead-sucker assemblage (Fig. 4). The streams are usually intermittent during summer, so fish are often confined to stagnant pools that may exceed 30°C during the day. During winter and spring the streams are swift and subject to flooding. The main permanent native resident is California roach. Because of their small size and tolerance of low oxygen levels and high temperatures, roach survive where most other fishes cannot. However, predatory green sunfish have replaced California roach in some areas, such as tributaries to the upper San Joaquin and Fresno Rivers. During winter and spring, Sacramento suckers, pikeminnows, and other native minnows may use these streams for spawning. If the pools are sufficiently large and deep, their young-of-year will survive the summer in them.

Pikeminnow-hardhead-sucker assemblage. Most of the streams inhabited by the fishes of this zone have average summer flows of >300 liters/sec; deep, rocky pools; and

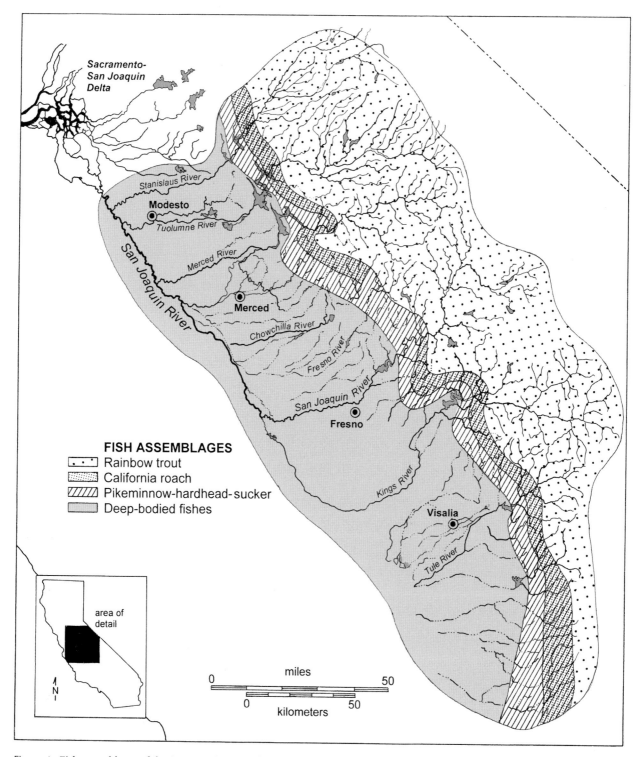

Figure 4. Fish assemblages of the San Joaquin River drainage. The lower reaches of the Merced, Tuolumne, and Stanislaus Rivers are included in the deep-bodied fishes assemblage because in-channel gravel pits and other artificial habitats contain alien fishes. However, these reaches are regulated by releases from reservoirs to enhance salmon spawning and rearing, so they contain many native stream fishes as well, in a zonelike progression downstream from the dam. After Moyle and Nichols (1974).

wide, shallow riffles (Moyle and Nichols 1973; Brown and Moyle 1993). Water quality is usually high (high clarity, low conductivity, high dissolved oxygen, summer temperatures 19–22°C), with complex habitat created by stream meanders and riparian vegetation (Brown 2000; Marchetti and Moyle 2000a,b). Some streams, however, may become intermittent in summer, or at least have such reduced flows that fish are confined to pools. Summer water temperatures in such streams may exceed 25°C and may track air temperatures closely. In Sierra Nevada foothill streams of the San Joaquin drainage, the pikeminnow-hardhead-sucker assemblage occupies a narrow altitude range, from 27 to 450 m above sea level (Fig. 4). The range is much wider in streams of the Sacramento Valley foothills (Fig. 5).

Sacramento pikeminnows and Sacramento suckers are usually the most abundant fishes of this assemblage. Hardhead are largely confined to cooler waters in reaches with deep, rock-bottomed pools. Where they are found, however, they are abundant. Other native fishes that may live here are tule perch, speckled dace, California roach, riffle sculpin, and rainbow trout. Introduced species (especially smallmouth bass and green sunfish) may colonize this zone, but they generally become abundant only if dams stabilize the flow regime, because native fishes are better adapted for living through periods of extreme high flow and extended cool flows. Presumably native fishes find instream refuges from high-velocity water or move to stream edges to avoid being flushed downstream.

In the San Joaquin drainage this assemblage can be sharply separated from assemblages above and below it, largely because most streams occupied by the assemblage become warm or intermittent (or both) in summer. In more permanent streams of the Sacramento Valley, however, species replacement is not as common as species addition.

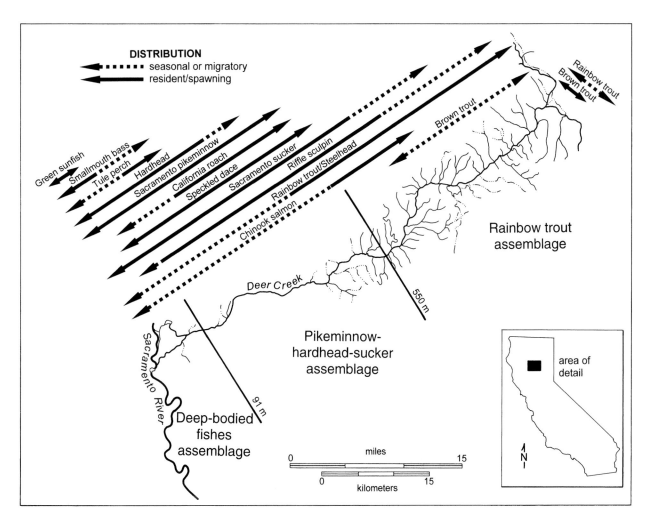

Figure 5. Distribution of fishes in Deer Creek, Tehama County, the largest tributary to the Sacramento River without a major dam in its upper reaches. The different fish assemblages are regions of overlap of the distributions of different sets of native species. Note that introduced species are present in abundance in only two highly disturbed areas: Deer Creek Meadows in the upper reaches, and the lowermost reaches, where water has been diverted for irrigation.

Thus rainbow trout live in much of the zone in the larger and colder streams. Many anadromous fishes (mainly chinook salmon, steelhead rainbow trout, and Pacific lamprey) have (or had) major spawning grounds in the zone, and their young are often part of the assemblage. Juvenile fall-run chinook salmon, however, usually move downstream within a few months after hatching to avoid high summer temperatures, but young spring-run chinook and steelhead may spend a year or more in the cooler upper reaches of this zone. Pacific lamprey spend the entire five to seven years of the ammocoete (larval) stage of their life cycle in muddy backwaters, migrating downstream only when they metamorphose into the predaceous adult stage.

Species in the assemblage show a high degree of segregation in their use of space and food (Figs. 6 and 7). Large Sacramento suckers stay on the bottom in deep pools feeding on algae, detritus, and associated small invertebrates. They may move into shallower or swifter water to feed at night. Juvenile suckers and cyprinids remain throughout the day and night in shallow water of stream edges, the smallest fish in the shallowest water. The distribution of small fishes is a careful balancing act between avoidance of predatory pikeminnow in deep water and avoidance of predatory herons and kingfishers in shallow water. Fish less than 3 cm long are too small for most vertebrate predators to eat, but fish between 3 and 15 cm are perfect prey for both large fish and predatory birds. They thus tend to congregate in water of intermediate depth (50–90 cm) close to deep cover.

Small pikeminnow feed mainly on aquatic insects from both the bottom (benthos) and the surface and water column (drift). Small schools of juvenile pikeminnow are commonly seen swimming close to the edges of pools and runs, foraging on anything small that falls into the water. Large pikeminnows are hunters of large invertebrates, especially crayfish and small fish, including sculpins, juvenile cyprinids, and suckers. They feed most intensively around dawn and dusk, when prey have a hard time seeing them coming, and cruise about large pools during the day, capturing occasional prey with a sudden rush. They will also feed on moonlit nights.

Hardhead poke about the bottom for aquatic insect larvae, occasionally rising to the surface to take drifting insects. The feeding habits of large (≥20 cm TL) adult hardhead are similar to those of smaller fish, but they are more omnivorous, often browsing on filamentous algae and large hard-shelled invertebrates, especially crayfish. Like pikeminnows, they spend a great deal of time cruising about deep pools, but they are usually closer to the bottom.

Rainbow trout, when present, are most abundant in the riffles, where they take advantage of large rocks that break the flow. Usually a favorable spot behind a rock will be defended as a feeding territory by one trout against others of its kind (and probably against other species as well). The trout feed primarily on drifting insects, but they also pick up a few bottom invertebrates and small fish. In pools trout are found mostly in turbulent inflowing waters where they have first chance at insects that float in. Like trout, sculpins and speckled dace are found mostly in riffles and behave as they do in the rainbow trout assemblage, although sculpins tend to be absent from lower elevations and may be replaced in warmwater riffles by dace. Another bottom-oriented fish found in this assemblage at times is tule perch; individuals hang out under deep cover in pools but often forage in faster water.

This description of resource subdivision by the fishes is obviously an idealized picture of interactions in un-

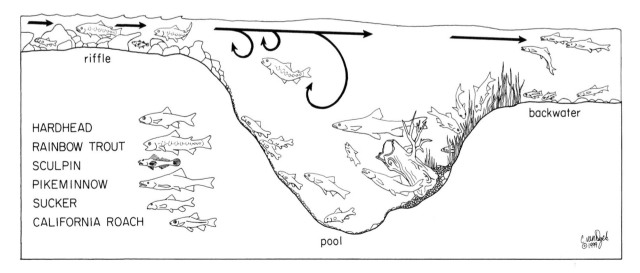

Figure 6. Cross section of a pool containing the pikeminnow-hardhead-sucker assemblage in the Sacramento–San Joaquin drainage.

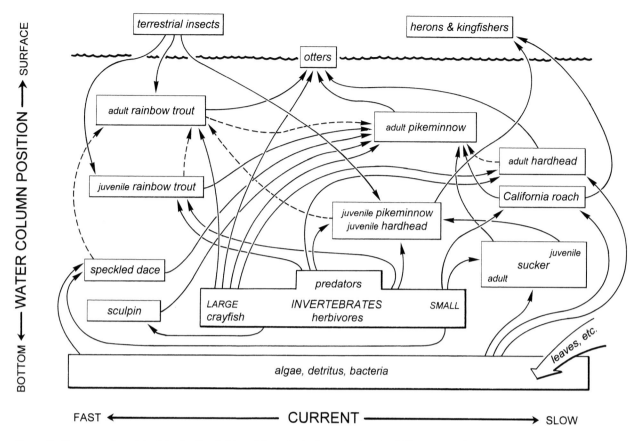

Figure 7. Conceptual model of the feeding habits of the principal species in a foothill stream in the Sacramento–San Joaquin watershed.

disturbed sections of stream that are without introduced fishes or heavy fishing pressure. The actual relationships among the species vary from place to place with the relative abundance of each species.

Deep-bodied fishes assemblage. Before the Sacramento and San Joaquin Rivers were reduced in flow and confined between levees, a unique assemblage of fishes occupied the warm waterways of the valley floor, including sluggish river channels, oxbow and floodplain lakes, swamps, and sloughs. The fishes of this assemblage were found in a variety of habitat types ranging from stagnant backwaters and shallow tule beds to deep pools and long stretches of slow-moving river. Deep-bodied fishes (Sacramento perch, thicktail chub, tule perch) and juvenile fishes predominated in the weedy backwaters while specialized adult cyprinids (hitch, blackfish, splittail) occupied the large stretches of open water. Large pikeminnows and suckers also lived here in abundance, migrating upstream to spawn in tributaries in spring. Anadromous salmon, steelhead, and sturgeon passed through the zone on their way upstream to spawn.

A key habitat contributing to the abundance of the native fishes was the floodplains along the Sacramento and San Joaquin Rivers and their larger tributaries. These areas,

supporting dense riparian forests and a wide variety of wetlands, filled with water in response to winter rains and spring snowmelt. In most years, inundation occurred between February and April, sometimes extending well into summer in wet years. The flooded areas were presumably immensely productive of small invertebrates with rapid life cycles, such as chironomid midges and water fleas (Cladocera) (as are now found on the limited areas still available for flooding). Not surprisingly, the native fishes were adapted for using the flooded areas. Small salmon moving downstream would tarry until the waters started to recede, growing rapidly and protected from predation by the dense vegetation. Juveniles of stream-spawning cyprinids and suckers also moved in and out of the floodplain to feed and grow. Adult splittail, Sacramento blackfish, and perhaps thicktail chub moved onto flooded areas to spawn, their embryos sticking to the vegetation, hatching in time to take advantage of the abundance of small prey.

Perhaps the most productive year-round habitats for adult deep-bodied fishes historically occurred in Kern, Buena Vista, and Tulare Lakes of the San Joaquin Valley floor. These were huge, shallow, interconnected lakes that filled each year with snowmelt waters from the Kern, Tule,

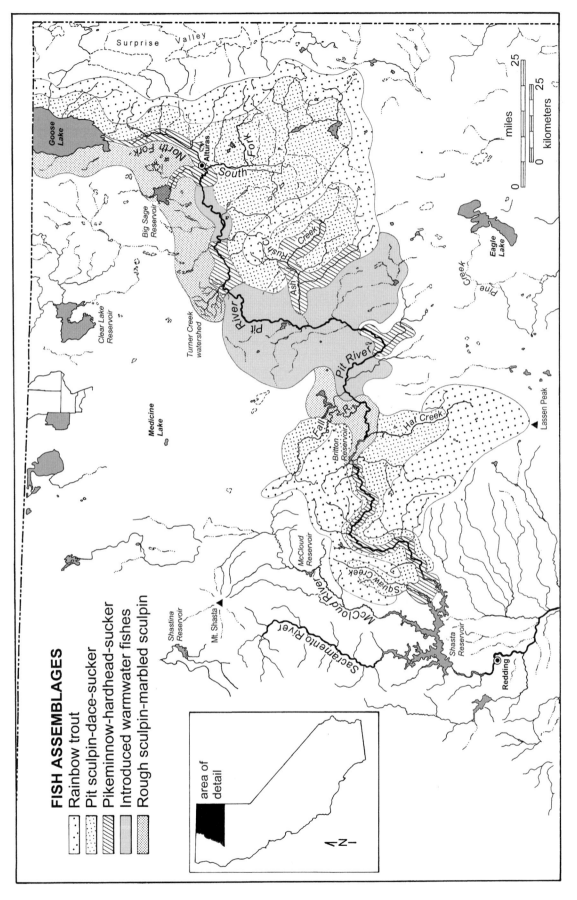

FISH ASSEMBLAGES

- Rainbow trout
- Pit sculpin-dace-sucker
- Pikeminnow-hardhead-sucker
- Introduced warmwater fishes
- Rough sculpin-marbled sculpin

area of detail

Figure 8. Distribution of fish assemblages of the Pit River drainage. After Moyle and Daniels (1982).

Kaweah, and Kings Rivers. During the wettest years, they would be united as one giant lake, but even during moderately wet years, Tulare Lake would cover roughly 80,000 ha (120 × 40 km) (Haslam 1989). In years of extreme drought, the lakes may have dried up completely or nearly so. Most of the time, however, they supported immense populations of fish, providing a steady source of food for the native peoples and huge flocks of piscivorous birds. Archaeological and anecdotal evidence indicates that Sacramento perch, thicktail chub, Sacramento blackfish, Sacramento pike-minnow, and Sacramento sucker were the most abundant fishes in the lake (Ellis 1922; Gobalet and Fenenga 1993). The pikeminnows, suckers, and blackfish apparently migrated up the inflowing streams to spawn in spring and were harvested there by the Yokut people and by early Euro-American settlers. Despite the presence of commercial fisheries for turtles, frogs, and fish, the lakes were diked and drained for agriculture in the late 19th century (Haslam 1989). The fish were confined to ditches and sloughs and then largely replaced by alien species, such as white catfish and common carp. The lakes reappear in exceptionally wet years when floods rush down the old river channels again, and they are quickly colonized by fish, mostly alien species.

The other habitats once occupied by this assemblage have also changed drastically. Most of the water flows through human-modified channels, and the once vast tule beds have been reduced to remnants. The native fishes have consequently either been extirpated or else reduced to a minor part of the fauna, living mostly in the least disturbed sloughs. The dominant fishes today are all alien species: largemouth bass, white and black crappie, bluegill, threadfin shad, striped bass, bigscale logperch, red shiner, inland silverside, white catfish, black and brown bullhead, and common carp. Other alien fishes are present in lesser numbers. The alien fishes feed on alien invertebrates, such as *Corbicula* clams and crayfish, and live among alien plants as well. The fishes still form distinct assemblages associated with different sets of habitat conditions (Brown 2000) but the assemblages cannot be regarded as stable entities because the waters they occupy are continually changing in quality and quantity and the assemblages shift as other alien species become established.

Streams of Pit River Subprovince

The Pit River has a fish fauna similar to that of the Central Valley, but some species are lacking while other species are endemic to the watershed. There are five definable assemblages, mostly variations on the Central Valley theme, but with one that is an original composition: *(1)* rainbow trout assemblage, *(2)* Pit sculpin–dace–sucker assemblage,

(3) pikeminnow-hardhead-sucker assemblage, *(4)* introduced warmwater fishes assemblage, and *(5)* rough sculpin–marbled sculpin assemblage (Fig. 8).

The **rainbow trout assemblage** is basically the same as the various combinations of one to three species that make up this assemblage in the Central Valley. It occupies cold high-elevation tributaries, and rainbow (or redband) trout are the most abundant species. The trout are often joined by Pit sculpin and Sacramento sucker.

The **Pit sculpin–dace–sucker assemblage** occupies the small, numerous second- and third-order streams in the drainage; it is similar to the one described as part of the rainbow trout assemblage in Central Valley streams, where four species are present. However, in Pit streams, trout are usually a minor part of the assemblage (perhaps a recent development caused by removal of riparian vegetation by grazing livestock). The streams have summer temperatures of 20–25°C, moderate gradients, and numerous pools. They may become intermittent in dry years. The most abundant fishes are speckled dace in pools and Pit sculpin in riffles, but they are usually joined by Sacramento suckers and rainbow or brown trout. Sometimes the local assemblage also contains California roach and juvenile Sacramento pikeminnow. In a few small, isolated streams, Modoc suckers are present rather than Sacramento suckers. These streams are also characterized by unusually high densities of speckled dace.

The **pikeminnow-hardhead-sucker assemblage** is virtually the same as that found in the Central Valley. This assemblage is characteristic of the canyon sections of the main Pit River and the lower reaches of its larger tributaries. In most areas, it is characterized by rainbow trout, Pit sculpin, and speckled dace as well as the distinguishing species. This assemblage once occupied the Big Valley reaches of the Pit River as well, but there it has been replaced by an **introduced warmwater fishes assemblage,** which consists of largemouth bass, golden shiner, bluegill, green sunfish, brown bullhead, channel catfish, and Sacramento sucker. It is similar in composition to the present-day deep-bodied fish assemblage of the Central Valley.

The **rough sculpin–marbled sculpin assemblage** is the most distinctive fish assemblage of this region. It occurs in spring-fed streams that are cold, deep, and clear and that are extraordinarily constant in their characteristics, reminiscent of lacustrine habitats. The largest examples of these streams are Fall River and lower Hat Creek. Other species characteristic of this assemblage are tui chub, rainbow trout, and Sacramento sucker. Species such as pikeminnow, hardhead, and Pit sculpin are remarkably rare. Pit sculpin are apparently excluded from the rivers because of their inability to avoid predators on pale, sandy stream bottoms, while rough and marbled sculpins not only avoid predation but segregate from each other in microhabitat use and diet (Daniels 1987; Brown 1991). Streams with this assemblage also contain a number of endemic invertebrates, including the endangered Shasta crayfish (*Pascifasticus fortis*).

San Francisco Estuary

The San Francisco Estuary (Sacramento–San Joaquin Estuary) is the largest estuary in California; it has a unique and complicated physical structure, which influences how it is used by fish. It consists of three distinct segments: the Delta, Suisun Bay, and San Francisco Bay (Fig. 9). The Delta is the uppermost part of the estuary, the footprint of what was once a vast, varied wetland, dissected by meandering channels of the united waters of the Sacramento and San Joaquin Rivers. The Delta narrows between two headlands before connecting with Suisun Bay, a large, shallow, and highly productive expanse of brackish water, strongly influenced by tides. The bay and its associated marshes (mainly Suisun Marsh on its north side) have been major nursery areas for fishes living in the estuary. Suisun Bay is connected to San Pablo Bay, as the upper portion of San Francisco Bay is called, through a long, narrow channel, Carquiniz Straits. San Francisco Bay is basically a marine environment, although salinities can be appreciably diluted by fresh water during high-outflow years, allowing freshwater fishes to move into tributary streams.

When river flows were high in spring, the historical Delta was a morass of flooded islands and marshes. In late summer, when river flows were low, the islands and marshes, protected by natural levees deposited by floods, were often surrounded by saline water pushed upstream by tides. The Delta merged imperceptibly with freshwater marshes that once covered the valley floor; its fishes were a mixture of fresh- and saltwater species. Besides native freshwater fishes such as thicktail chub, hitch, blackfish, and pikeminnows, it contained fishes that live nowhere else in the system (delta smelt), anadromous fishes that spent part of their life cycle there (white sturgeon, chinook salmon, longfin smelt, Pacific lamprey), marine fishes that spent juvenile stages there (staghorn sculpin, starry flounder), and freshwater fishes that could tolerate salinities of 15–20 ppt or higher (Sacramento perch, tule perch, splittail, prickly sculpin). Most fishes fed on abundant crustaceans, especially opossum shrimp (*Neomysis mercedis*), amphipods (*Corophium* spp.), and cyclopoid copepods. Because some native fishes are extinct and all others are reduced in numbers, and because the Delta of today bears only a superficial resemblance to the Delta of yesteryear, we have only limited understanding of how native fishes interacted with each other and their environment. We know only that they were enormously abundant, and so were important as food to native peoples and supported the commercial fisheries of the 19th century.

Today's Delta still consists of islands surrounded by leveed channels. The islands are intensively farmed and the channels are dredged. The levees surrounding each island are artificially maintained to keep out floodwaters, a task made increasingly difficult because most Delta islands are now below sea level. In places, it is possible to stand on the deck of a high boat and peer over a levee to see farmland several meters below water level. The islands are "sinking" because agricultural practices over the past century have allowed the peaty soil to oxidize, turning organic matter into carbon dioxide and contributing to the "greenhouse effect" that is leading to global warming. Every year a few centimeters of soil vaporize or blow away as dust, and every year island surfaces become lower. The probability of island flooding has been reduced somewhat by numerous upstream dams that store much of the runoff (except during really wet years). The dams release their captured water during summer, so flows through the Delta are higher than they would have been historically. Much of this water does not flow in a normal downstream pattern through the Delta but instead flows *across* the Delta thanks to the insatiable thirst of the huge pumps of the State Water Project and the federal Central Valley Project in the south Delta. This peculiar flow pattern makes the Delta a freshwater environment all year round in most years. At times it also results in the lower San Joaquin River actually having a net flow backwards, toward the pumps, for many days. As if change in flow patterns were not enough, there are also hundreds of unscreened irrigation diversions within the Delta, constant addition of pollutants (especially agricultural chemicals), and continual invasions of alien species. Overall, the Delta and the rest of the estuary have become a suboptimal environment for most native fishes, as well as an environment that is likely to keep changing dramatically if diversions, pollution, and invasions are not better regulated (Herbold et al. 1992; Bennett and Moyle 1996).

Delta fishes are virtually the same as those in Suisun Bay, although the bay is more likely to contain euryhaline marine species and the early life history stages of estuarine-dependent species such as striped bass, delta smelt, and longfin smelt. The importance of Suisun Bay as a rearing area for the fishes is related to its salinity, which in turn is tied to freshwater outflow. The annual success of a number of species is tied to the amount of low-salinity water in Suisun Bay, as measured by the position of the 2-ppt bottom salinity isohaline (Jassby et al. 1996); the further "downstream" the isohaline, the more likely the young of Delta fishes will be to have high survival rates. Unfortunately, the value of Suisun Bay as a nursery area has been compromised by invasions of alien copepods, amphipods, shrimp, crabs, and clams, which now dominate both the benthos and the plankton. In particular, the overbite clam, *Potamocorbula amurensis*, has become so abundant in Suisun Bay in recent

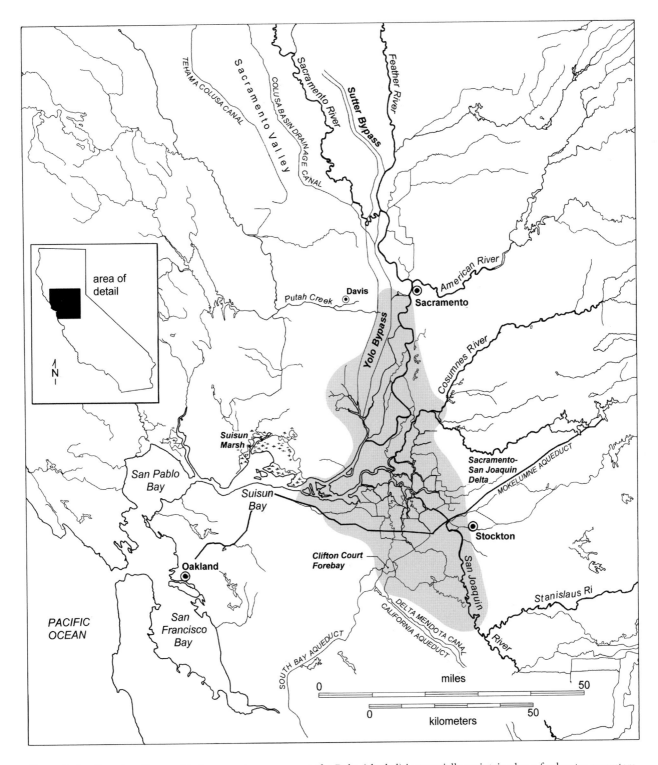

Figure 9. San Francisco Estuary. Under present management, the Delta (shaded) is essentially maintained as a freshwater ecosystem, Suisun Bay and Marsh as a brackish water ecosystem, and San Francisco Bay as a marine system. Yolo Bypass becomes part of the Delta when it floods during wet years, and it then becomes a major spawning and rearing area for fish. A major factor affecting the way fresh water moves through the estuary is pumping in the south Delta to send water down the California Aqueduct and the Delta Mendota Canal.

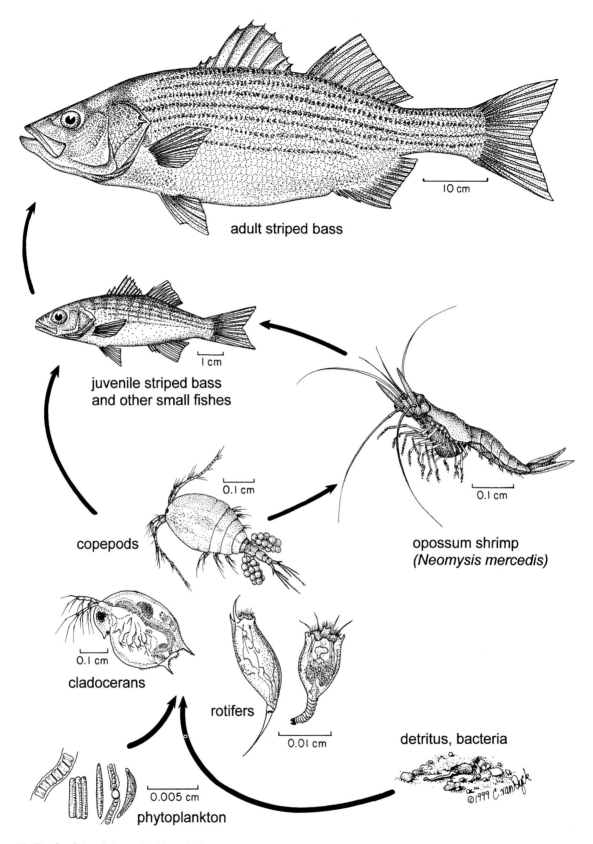

adult striped bass

10 cm

juvenile striped bass
and other small fishes

1 cm

copepods

0.1 cm

opossum shrimp
(*Neomysis mercedis*)

0.1 cm

cladocerans

0.1 cm

rotifers

0.01 cm

detritus, bacteria

phytoplankton

0.005 cm

©1999 C. van Dyck

Figure 10. Food web involving striped bass in the Sacramento–San Joaquin estuary. Although adult bass will eat virtually any fish in the estuary, their principal prey is juvenile striped bass, which in turn depend heavily on opossum shrimp and other planktonic crustaceans. The opossum shrimp is a predator on small zooplankton, which in turn feed largely on algae, bacteria, and detritus. From Kegley et al. (1999).

years that it has converted the system from one in which most energy flows through plankton to one in which it flows through the benthos (and is tied up in the large biomass of clams). As a result there is less zooplankton available for fish or mysid shrimp, probably resulting in decreased growth and survival of a number of species.

Not surprisingly, the fish fauna of the Delta and Suisun Bay is today in a general state of decline. Even if the Delta environment was more conducive to fish life, it is unlikely that fish assemblage structure and composition would remain predictable through time. The present fauna is a conglomeration of 40 or so freshwater, estuarine, and euryhaline marine species, about half of them introduced. The introduced species tend to be the most abundant fishes. Native fishes are an increasingly minor part of the fauna, although the estuary is the principal or only habitat for delta smelt, longfin smelt, and Sacramento splittail. Species segregation is not well developed, given the changing nature of both environment and fauna, and groups of co-occurring species are at best temporary alliances. However, well-established species do differ somewhat in salinity preferences, feeding habits, distribution patterns, and seasonal movements. Because ecological differences among species in the upper estuary are poorly defined, it is easiest to describe the fishes in terms of loose feeding guilds: planktivores, small benthic predators, bottom-feeding omnivores, and piscivores.

There are currently six principal **planktivores** in the Delta, besides larval fishes. Delta smelt and threadfin shad feed in open water on copepods in freshwater regions whereas longfin smelt feed in open water on copepods and opossum shrimp in brackish areas. Delta smelt tend to live in main channels (or Suisun Bay); threadfin shad tend to concentrate in the warmer backwaters in the upper Delta. The ecology and feeding habits of juvenile striped bass are similar to those of longfin smelt, but they eventually switch to feeding on other fish (Fig. 10). American shad are also plankton feeders, but they only enter the upper estuary on a seasonal basis, whereas hitch and inland silversides consume plankton in shallow sloughs or along the edges of channels. Silversides may move offshore to feed at times and compete directly with smelt and other pelagic species. They also prey on eggs and larvae of other fish.

Small benthic predators include native prickly sculpin, tule perch, starry flounder, juvenile white sturgeon, juvenile splittail, and staghorn sculpin, as well as introduced yellowfin goby, shimofuri goby, bigscale logperch, and juvenile catfishes. Important prey for this group are amphipods (especially *Corophium* species) and opossum shrimp. The native and introduced fishes have the potential to be in direct competition if any of their benthic prey becomes limiting, because their habitat requirements and feeding habits overlap widely. The invading shimofuri goby may owe its astonishing success in part to its exploitation of food sources not used by either natives or other exotics, mainly nonnative barnacles and hydroids (Matern 1999).

Three **bottom-feeding omnivores** in the system are common carp, adult splittail, and Sacramento sucker. Their diets contain a large amount of detritus of uncertain food value, as well as a variety of small benthic invertebrates.

Among the more abundant **piscivores** in the Delta are striped bass, white catfish, channel catfish, and largemouth bass. This group preys on smaller resident and migratory fishes, such as juvenile salmon and steelhead. They presumably replaced a suite of native piscivores including Sacramento perch, thicktail chub, Sacramento pikeminnow, and steelhead.

In some respects, the limited feeding and habitat segregation among the fishes, native and nonnative, reflects their ability to adapt to the presence of other fishes. Indeed, most native and alien fish populations show some concordance in their fluctuations in response to long-term environmental variation (Meng and Moyle 1995). Persistent, predictable assemblages of fishes are lacking, however, and there is little evidence of strong interactions among most species. Even striped bass, the top predator in the system (with the exception of humans), feeds largely on its own young under most circumstances. Historically, juvenile striped bass and many other fishes fed mainly on abundant opossum shrimp. When opossum shrimp declined, juvenile bass and other alien species switched to a more generalized diet (Feyrer 1999). In contrast, native fishes switched to alternative prey species, suggesting greater specialization. Nevertheless, the natives still seemed to suffer greater declines in abundance than the aliens.

Clear Lake

Clear Lake is now the largest natural freshwater lake completely within California's borders. It is perched in the coast range at an elevation of 402 m, with a surface area of about 17,670 ha, an average depth of 6.5 m, and a maximum depth of 18 m. Sediment deposits show the lake to have been highly productive for thousands of years, the result of its warm (summer temperatures of 20–25°C), shallow waters, well mixed by summer winds. Heavy summer blooms of algae were no doubt present even before the arrival of civilization, making the lake belie its modern name. Although the name Clear Lake may have reflected wishful thinking on the part of early real estate salesmen, it is likely that historically the algae blooms were not as severe or as persistent as they are today. Native peoples who lived by the lake knew better and called it Konocti (woman mountain) after the dormant volcano that sits along one shore. They appreciated the lake's green productivity and harvested the abundant fishes and birds.

The original native fish fauna consisted of ten resident

species, distributed among three broad habitat types: (1) shallow-water habitat, from the shore down to the limits of rooted aquatic plant growth, probably seldom deeper than 4 m; (2) offshore benthic habitat, consisting of the bottom below the limits of aquatic plant growth; and (3) open water habitat, the water column away from shore, from surface to bottom. Native fishes living in the three habitats were basically lake-adapted variants of species that originally made up the deep-bodied fishes assemblage in the Central Valley. They probably formed distinct assemblages, unlike the modern, more amorphous conglomeration of species.

The **shallow water assemblage** was dominated numerically by large numbers of young-of-year cyprinids: hitch, Sacramento blackfish, thicktail chub, and Clear Lake splittail. These "greenback minnows" and "silversides" greatly impressed early visitors with their large, flashing shoals. Presumably these fish fed on small planktonic organisms or invertebrates associated with the large beds of tules and other aquatic plants. Not surprisingly, three other fish species living here were piscivores: Sacramento perch, thicktail chub, and Sacramento pikeminnow. Young-of-year tule perch were also common, picking small invertebrates from aquatic plants and the bottom. Threespine sticklebacks may have been abundant among the plants and in the tule beds, as were the larvae and small juveniles of species like hitch and splittail.

The **offshore benthic assemblage,** consisted mainly of prickly sculpin (an invertebrate predator), Sacramento sucker (a grazer on algae, detritus, and invertebrates), and tule perch (a benthos picker). These fishes presumably subsisted on huge populations of midge larvae that once occupied the bottom. They were preyed upon by Sacramento perch.

The **open water assemblage** was made up of schools of juvenile and adult hitch, splittail, blackfish, and Sacramento perch. The hitch, splittail, and perch fed on zooplankton and emerging midges, whereas blackfish fed almost exclusively on phytoplankton. All were pursued by large pikeminnows and thicktail chub.

Besides these year-round residents, early records indicate that anadromous steelhead rainbow trout and Pacific lamprey entered through the lake's outlet, Cache Creek, and then spawned in tributaries. Such migrations were halted by the construction of Rumsey Dam in 1914.

Today native assemblages of fish in each habitat have been largely replaced by poorly defined assemblages of introduced species. At least 16 introduced fishes are now established in the lake, and only 4 of the native species still maintain large populations: hitch, blackfish, tule perch, and prickly sculpin. Although each introduced species has definite habitat and food preferences, both the lake habitat and composition of the fish fauna are still changing. For example, the inland silverside, introduced in 1967, quickly became the most abundant fish in the lake. In shallow water it largely replaced bluegill as the dominant fish, just as bluegill apparently replaced the small minnows once so abundant there. The most recent introduction (1985) has been a pelagic planktivore, threadfin shad (Anderson et al. 1986), which has become enormously abundant, causing major changes to the ecosystem and possibly threatening the persistence of Clear Lake hitch (Colwell et al. 1997). The shad died off in a cold winter but reestablished and at times is even more abundant than the silverside.

Central Valley Reservoirs

Ever since Europeans settled in California, the rivers of its great Central Valley have been a source of both admiration and frustration. They were admired for their abundant flows and potential for making the rich soils of the valley floor yield crops, but their fluctuations from raging spring floods to quiet summer trickles made the success of farming endeavors frustratingly unpredictable. The settlers' response was to build dams and store the water in reservoirs. Construction of dams, always a major activity in the Central Valley, gained momentum with the advent of major dam building by the federal Central Valley Project starting in the 1940s and the State Water Project in the 1960s. Reservoirs are now one of the major fish habitats in California, although one of the least studied from a community or ecosystem perspective. The nature of each reservoir and its fish fauna is determined by its elevation, size, location, and water quality. In general, reservoirs are less productive per unit surface area than are lakes, because their deep, steep-sloped basins and fluctuating water levels greatly limit habitat diversity and productivity. Although the agencies that build reservoirs may call them lakes (e.g., Lake Shasta), such names are deceptive and raise expectations that the reservoirs will be as productive of fish as are natural lakes. Because reservoirs are decidedly *not* lakes, in this volume they are labeled truthfully (e.g., Shasta Reservoir).

California reservoirs vary from clear, oligotrophic, coldwater impoundments at high elevations to turbid, eutrophic, warmwater impoundments at low elevations. Most lie at middle elevations in the foothills. These reservoirs often support warmwater fishes in surface and edge waters and salmonids in deeper, cooler water. Salmonid populations can be lost, however, during periods of drought, when reservoir levels are low. In some warm reservoirs they are maintained mainly by planting trout or salmon to create a winter fishery. The midelevation reservoirs are of two main types, with different fish communities: water supply reservoirs and power supply reservoirs.

Water supply reservoirs have many purposes but mainly supply water for irrigation and urban uses. They are filled during winter and spring and drained during summer. The size of the minimum pool left at the end of each year is determined by the balance between water supply and demand. These reservoirs support mainly introduced fishes, although Sacramento sucker usually manage to remain abundant in them. In many cases, native hardhead and pikeminnow were extremely abundant in these reservoirs for the first ten years or so after filling. These fish colonized from the dammed streams and developed large populations because of the initial scarcity of introduced predators and competitors. As populations of introduced fishes, especially centrarchid basses, grew, hardhead and pikeminnow populations showed little recruitment and eventually died out, even though they remained abundant in streams feeding the reservoirs. In a few reservoirs, hitch or tui chubs, often introduced as forage for game fish, have remained abundant. The exact species composition of each reservoir varies with the history of the introductions, but some nonnative species are now almost universal in occurrence: bluegill, green sunfish, largemouth bass, spotted bass, smallmouth bass, common carp, golden shiner, threadfin shad, black crappie, brown bullhead, white catfish, channel catfish, western mosquitofish, and rainbow trout (hatchery strains). It is possible to divide typical midelevation reservoirs into four broad habitats, each with a more or less distinct summer fish assemblage: (1) littoral, (2) epilimnetic, (3) hypolimnetic, and (4) deepwater benthic. These assemblages are not stable entities but change in response to reservoir drawdowns, which can affect reproductive success or force species from their normal habitats.

Littoral habitat occurs along the edges, down to the depth of light penetration or to the upper limits of the thermocline, whichever comes first. It is the habitat most severely affected by fluctuating water level, because it may be alternately flooded or exposed within relatively short periods of time. Despite the fluctuations, large numbers of fish are found here. Bluegill, largemouth bass, and golden shiners (or occasionally tui chubs, hitch, or inland silversides) live close to the surface near shore. Mosquitofish stay in the flooded grass in very shallow areas. Brown bullheads, white catfish, channel catfish, and carp stay near the bottom. Black crappie cluster around submerged boulders and logs during the day, moving out into open water to feed on plankton and fish in the evening. Reproduction is a problem for most fishes, because a sudden drop in water level may expose a nest of embryos, and a sudden rise can submerge it to unfavorable depths. The types of fishes occupying this habitat may change in an upstream direction, because most reservoirs become more riverine near their main inflowing river. This is particularly noticeable among centrarchid

basses; smallmouth bass tend to be dominant at the upper end, largemouth bass in more lacustrine areas, and spotted bass in intermediate habitats.

Epilimnetic habitat occupies the well-lighted, well-oxygenated surface waters away from shore and above the thermocline. The fish fauna here is perhaps the most variable from reservoir to reservoir. Because its primary means of supporting fishes is the zooplankton to which it is home, it contains three main types of fish: (1) plankton-feeding larvae of littoral fishes, especially bluegill and other centrarchids; (2) plankton-feeding adult fishes; and (3) fishes that prey on the plankton feeders. The population biology of planktonic larval fishes in reservoirs is poorly understood, but it is likely that plankton-feeding fishes, notably threadfin shad, reduce their numbers through predation or through the reduction of zooplankton populations. Threadfin shad are the typical plankton-feeding residents of this habitat despite the fact that they were not introduced into the Central Valley until 1959. Other zooplankton grazers that may occupy this zone, mostly in reservoirs that lack threadfin shad, are hitch, tui chub, wakasagi, and American shad. Striped bass are the chief epilimnetic predator in some reservoirs, although their inability to spawn in most means that they must be introduced on a regular basis. Fish from other zones also prey on epilimnetic fish, especially those that venture close to shore.

Hypolimnetic habitat occupies the cold (>20°C) water below the thermocline in reservoirs deep enough to stratify during summer months. The main inhabitants are rainbow trout, which often enter the epilimnion in the evening or at night to feed on whatever forage fish are most abundant. Kokanee salmon are also commonly present, but they stay in the cold depths in the summer months, feeding on zooplankton.

Deepwater benthic habitat is on the bottom, below the thermocline and usually below the limits of light penetration. It is the one zone in which native fishes, especially prickly sculpin and Sacramento sucker, may predominate. White and channel catfish also may live in this zone, but they usually move into littoral areas to feed at night.

Power supply reservoirs are uncommon compared with water supply reservoirs because they are dedicated solely to providing a constant flow of water for running electric generators. Examples include the chain of five reservoirs on the lower Pit River (Britton is the largest) and Kerckoff and Redinger Reservoirs on the San Joaquin River. These reservoirs typically are not drawn down during summer but are maintained at a fairly constant level, although this level may fluctuate by 1–3 m on a daily or weekly basis. Short-term fluctuations in water level inhibit the development of an assemblage of introduced littoral fishes because there is limited habitat for nesting or cover for juveniles. Because of the rapid turnover of the water, these reservoirs may also have lower summer temperatures than water supply reservoirs at the same elevations. In many respects, they are like giant stream pools, and, as a consequence, they may favor native stream fishes (Vondracek et al. 1988b). The most abundant fishes are hardhead, Sacramento pikeminnow, and Sacramento suckers, all of which spawn in inflowing streams. Their young are abundant in littoral areas of the reservoirs, often cruising about in large schools and preyed upon by adult pikeminnows. In Britton Reservoir, tule perch are abundant as well, with adults feeding mainly on benthos and young-of-year on zooplankton; rough sculpin and marbled sculpin live on the reservoir bottom, also feeding on benthic insects.

North Coast Streams

North of San Francisco Bay there are dozens of streams that flow directly into the ocean without entering a major river system. These streams are highly variable in physical characteristics, ranging from warm, intermittent streams to permanent, cold-flowing streams. Because they drain low mountain ranges that do not develop snow packs, North Coast streams have flow patterns that reflect rainfall. They may be raging torrents in winter and spring (in response to rainstorms) but quiet trickles in the summer. Most also have high gradients and flow rapidly to the sea, although a few larger streams meander across floodplains in their lower reaches. All North Coast streams were drastically altered by the mammoth rainstorms of 1955 and 1964, which caused massive erosion of heavily logged, steep slopes all along the coast, burying streambeds and estuaries with gravel and debris. Many deep, narrow, meandering channels were converted overnight to wide, shallow, braided channels, with little habitat for pool-dwelling fishes such as juvenile coho salmon.

Despite variation in temperature regime, flow, and locality, North Coast streams are similar in the composition of their fish faunas, which consist largely of anadromous species and euryhaline freshwater and marine species. The major exception is the Russian River, which contains most of the freshwater dispersant species found in Central Valley streams. However, other streams also contain freshwater dispersants (California roach, Sacramento sucker, or both) that have entered coastal drainages through former connections with interior systems. Usually three intergrading fish assemblages may be recognized: *(1)* resident trout, *(2)* anadromous fishes, and *(3)* estuarine fishes.

The **resident trout assemblage** occupies the uppermost reaches of larger watersheds. Typically, it occurs above natural barriers that halt upstream migration of anadromous fishes or in streams accessible only to steelhead. The water is cold, swift, and well oxygenated; rocky riffles are the pre-

dominant habitat type. Rainbow trout are the most common fish, although cutthroat trout occur in a few streams from the Eel River northward. Usually no other species are present. Smaller streams that contain only juvenile steelhead or coastal cutthroat trout (or both species) are similar.

The **anadromous fishes assemblage** exists as far upstream as fishes can migrate and downstream to reaches influenced by tidal action. Although the water in stream reaches occupied by this assemblage is also cold and fast flowing, pools become increasingly large and frequent as the streams approach the sea. Between pools there are long stretches of shallow riffles over rock, gravel, or sand, used for spawning by coho salmon (and chinook salmon in larger streams, such as the Eel and Mattole Rivers), rainbow trout (steelhead), and Pacific lampreys. Young coho salmon and trout usually spend a year or two in streams before migrating to sea, but ammocoetes of lampreys live in silty backwaters and stream edges for at least four to five years. This assemblage may also contain nonmigratory threespine stickleback, as well as prickly and coastrange sculpin. The sculpins are most abundant close to the stream mouths because both have larval stages that live in estuaries or large, quiet pools. Large prickly sculpins, however, are often found many kilometers upstream, although in low numbers.

The only native freshwater dispersant species likely to be part of this assemblage are California roach and Sacramento sucker. They are found in creeks tributary to Tomales Bay, Gualala River (roach only), Navarro River, Eel River (sucker only), Bear River (sucker only), and Mad River (sucker only). California roach, however, have been introduced into the Eel River, as have pikeminnow and speckled dace (Brown and Moyle 1996).

At present, ecological interactions among species in the anadromous fishes assemblage appear minimal, presumably because environmental fluctuations (especially the cycle of floods and droughts) may keep the populations of most fishes from reaching numbers at which food and space are limiting. There is some broad segregation by habitat. Juvenile steelhead and coho are found mainly in the smaller, colder streams, whereas coho are usually most abundant in pools and steelhead in the riffles. These species segregate in part as a result of aggressive interactions and in part by size. In larger reaches of the Eel River, California roach, threespine stickleback, Sacramento sucker, and juvenile steelhead showed wide overlaps in diet and use of space until pikeminnows invaded. These predators now keep smaller fishes out of much of the pool habitat they previously used, limiting them to pool edges and riffles. As a result, a greater degree of spatial segregation (less overlap in microhabitat use) has developed among the four species (Brown and Moyle 1991). The pikeminnows also appear to be depressing chinook salmon populations through predation on outmigrating young.

Prickly and coastrange sculpin are two similar species with similar life history strategies (amphidromy). In most streams they seem to show little ecological segregation and occupy the same riffles in about equal numbers. In the Smith River, however, they segregate by both depth and velocity, with prickly sculpins concentrating in deep, slow pools and coastrange sculpin concentrating in shallow, swift riffles (White and Harvey 1999).

The **estuarine fishes assemblage** occupies reaches of streams influenced daily by tides. The fishes consequently experience reversing currents, fluctuating temperatures, and salinity gradients on a daily basis. In some streams, such as the Navarro River, the zone with the assemblage may be 4–5 km long, but more often than not it is less than 1 km in length, usually ending at the first rocky riffle. The middle sections are generally slow moving and shallow, but they occasionally have depths of 2–3 m. At the lower ends there are almost invariably lagoons behind wind-and-wave-piled sand bars. Often wave action will seal the lagoons in summer, separating them from the sea. The bottoms are mostly sand or mixed sand and silt.

Species most common here (although not necessarily all in one stream) are threespine stickleback, prickly sculpin, coastrange sculpin, staghorn sculpin, topsmelt, starry flounder, and tidewater goby. Marine species are frequently present as well in the lowermost reaches. The sticklebacks are usually migratory forms that spend much of their life in the estuary or ocean migrating into fresh water to spawn.

In each stream, species tend to segregate according to salinity tolerances, as illustrated by fishes found in this zone of the Navarro River in August 1973. Starry flounder, the sculpins, and threespine stickleback were common throughout, from completely fresh water to the mouth. Sacramento suckers disappeared before the salinity reached 1 ppt, although the largest concentration of adults observed was just above the reach of salt water. California roach dropped out at about 3 ppt, where shiner perch and topsmelt started to become common. At 9–10 ppt, bay pipefish suddenly appeared, living in beds of filamentous algae. Staghorn sculpins were also first found here. Closer to the ocean, at salinities of 23–28 ppt, staghorn sculpin, shiner perch, and bay pipefish were abundant, and two marine species, penpoint gunnel and saddleback gunnel, made their appearances. Although no attempt was made to sample the lagoon just above the mouth, later sampling indicated that it contained more marine and euryhaline fishes, together with young salmonids. In the spring, the brackish parts are used for spawning by marine fishes such as Pacific herring. Although the fish species found in the Navarro River may not be typical of those in every coastal stream, a downstream change in species is typical of every stream with lower reaches long enough to possess a salinity gradient. Lagoons are also frequently important rearing areas for juvenile steelhead, cutthroat trout, and coho salmon.

Klamath Province

Lower Klamath River

The lower Klamath drainage consists of the Klamath River below Klamath Falls, the Trinity River, and more than 200 smaller tributary streams. The system is, on the basis of its physical characteristics and fish fauna, essentially a large coastal stream. Although second in size in California only to the Sacramento River, it lacks the warm, lowland habitat that fostered evolution of the more complex fauna of the Sacramento–San Joaquin system. Instead, it contains cold, fast-flowing, rocky-bottomed streams throughout most of the watershed. In addition, the river's geologic history has made colonization by freshwater dispersant fishes difficult. Thus the fish fauna is dominated by anadromous and amphidromous fishes: Pacific lamprey, threespine stickleback, green sturgeon, American shad (introduced), eulachon, chinook salmon, coho salmon, steelhead rainbow trout, coastal cutthroat trout, coastrange sculpin, and prickly sculpin. As indicated for North Coast streams, salmonids in the lower Klamath system presumably segregate by various means, feeding on aquatic and terrestrial invertebrates in different microhabitats. They spend anywhere from a few months to two years in the streams before moving out to sea. The roles of other anadromous fish are less well understood. Eulachon larvae are rather quickly washed into the estuary, whereas young green sturgeon and American shad may spend a year or more in the deep pools of the main river before going to sea.

In addition to anadromous fishes, there are abundant species that spend all or most of their life cycle in fresh water: nonanadromous rainbow and cutthroat trout, marbled sculpin, brown trout (introduced), speckled dace, and Klamath smallscale sucker. The assemblages are as described for coastal streams: (1) a resident trout assemblage in the upper reaches of tributaries, (2) a mixed anadromous fish–resident fish assemblage in the main river and most tributaries, and (3) an estuarine fishes assemblage in the lower 5–6 km of river. A fairly typical combination of species making up the assemblage in tributary streams is juvenile steelhead, suckers, dace, and both species of sculpin, although marbled sculpin replace the coastal species upstream. The four to five species segregate much as species do in the rainbow trout assemblage of Central Valley streams. Juveniles of other anadromous species may join on a seasonal basis, with actual numbers varying considerably from year to year, depending on the number of adult spawners in the previous year. The carcasses of spawned-out adult salmon and lampreys are an important source of energy for the food webs of these tributary streams, so the number of spawning fish may also indirectly affect the abundance of resident species, as well as the food available to their own young. In short, for biological reasons alone, the fish assemblages of Klamath tributaries are highly dynamic.

Construction of reservoirs on the main river and gravel pits along its side have permitted invasion of warmwater fish assemblages in recent years—a combination of introduced species (e.g., yellow perch, fathead minnow, pumpkinseed sunfish, largemouth bass, and brown bullhead), native species washed downstream from the upper Klamath River, and the original resident fishes.

Upper Klamath River

The upper Klamath drainage has fish assemblages that are very different from those of the lower Klamath drainage. The fauna is dominated by freshwater dispersant fishes rather than anadromous fishes. This makeup is due in part to the geologically recent connection between the two systems and in part to large, shallow lakes of the upper Klamath basin (Upper and Lower Klamath Lakes and Tule Lake), which have no counterparts in the lower Klamath River. Historically, chinook salmon and steelhead entered this region, spawning in tributaries to the large lakes in Oregon. They can now reach only the base of Iron Gate Dam in California. The dams that created Copco and Iron Gate Reservoirs have, however, extended downstream the habitat suitable for upper Klamath fishes.

Four species of upper Klamath fishes are primarily lake dwellers: Klamath Lake sculpin (*Cottus princeps),* slender sculpin (*C. tenuis*), shortnose sucker, and Lost River sucker. The two sculpins are not yet recorded in California but can be expected from Klamath River reservoirs. The two suckers spawned in large numbers in the Lost and Klamath Rivers, but the young were quickly washed into the lakes, presumably to assume the planktonic and benthic feeding habits of the adults. Native fishes that are found in streams as well as lakes include a complex of nonmigratory lampreys related to the Pacific lamprey, rainbow trout, Klamath largescale sucker, blue chub, Klamath tui chub, speckled dace, and marbled sculpin. The lampreys include both nonpredatory brook lampreys and predatory forms adapted for living in large lakes and rivers and preying on large suckers and minnows. The Klamath largescale sucker is the typical bottom-feeding sucker of the system. Blue and tui chubs are (or were) the most abundant fishes in Klamath and Tule Lakes. Just how the two rather similar species segregate ecologically is not clear, because both are opportunistic omnivores. Blue chubs, however, will ascend farther up small tributary streams than tui chubs. Speckled dace and marbled sculpin are primarily stream dwellers but will also live in rocky-bottomed shallows of lakes, where conditions are

similar to riffle habitat. In recent years, introduced species have become more important than natives in the lakes and reservoirs: wakasagi, yellow perch, and pumpkinseed in Copco and Iron Gate Reservoirs, and Sacramento perch in Clear Lake Reservoir and the Lost River. Fathead minnows especially have experienced a population explosion in the lowland lakes in recent years, so the ecosystem may be undergoing further dramatic changes. The key to restoring the health of the lakes and streams of the upper Klamath basin is restoration of conditions that favor native fishes, especially improving stream flows, reducing nutrient input from the watershed, and restoring marshlands and riparian areas.

Great Basin Province

Lahontan Streams

Streams of the drainage of ancient Lake Lahontan rush down the steep eastern slopes of the Sierra Nevada, slowing occasionally to meander through alpine meadows. Eventually they empty into large lakes or desert sinks. The turbulent flows ensure that stream water temperatures remain low enough to support trout even at low elevations, and the low temperatures have limited the success of introduced warmwater fishes. In streams, the native fish assemblages are largely intact, although native cutthroat trout have been largely replaced by rainbow, brown, and brook trout. The ecology of the native fishes is fairly well understood, primarily because of intensive studies of two small streams: Sagehen Creek (Seegrist and Gard 1972; Erman 1973, 1986; Gard and Flittner 1974; Decker 1989) and Martis Creek (Moyle and Vondracek 1985; Strange et al. 1992).

Fish assemblages are hard to define because, as streams increase in size and habitat diversity, native fish species are added but seldom removed. In addition, the single native trout, Lahontan cutthroat trout, has been replaced by three nonnative species. Headwaters usually contain only trout, most commonly brook trout that are replaced by rainbow and brown trout at lower elevations. Usually the first species other than trout to appear in a downstream direction is Paiute sculpin. As gradients decrease and pools and runs become more common, Tahoe sucker and speckled dace join in, followed by Lahontan redside in deeper pools. In larger streams, the assemblage is filled out by mountain sucker, mountain whitefish, and tui chub.

The native fishes of Lahontan streams are morphologically diverse, and this characteristic presumably reduces competition for food and space among species and results in a well-defined assemblage structure (Fig. 11). In Martis Creek, sculpin are primarily found in swift riffles, where fast, shallow water seems to exclude other fish except trout. They consume aquatic insects, especially mayflies (Ephe-

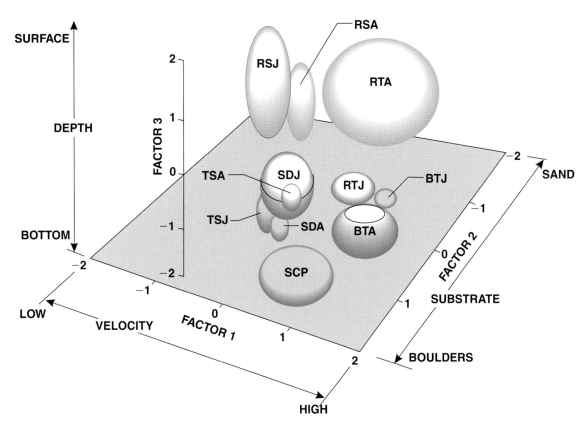

Figure 11. Three-dimensional diagram, generated by principal components analysis, of spatial niches of fishes of Martis Creek, Placer and Nevada Counties. Each globe represents the spatial niche of each species or life history stage as defined by three habitat axes (factors). Factor 1 represents position in the water column, with the top (2) being the surface and the bottom (–2) being the stream bottom. Factor 2 represents water column velocity, where 2 is high-velocity habitat and –2 is low-velocity habitat. Factor 3 represents substrate, where 2 is rocks and boulders (coarse substrate) and –2 is silt and sand (fine substrate). The dark plane in the middle represents median conditions for factor 1, so fish above the plane will be at least halfway up the water column while fish below it will be close to or on the bottom. Thus Paiute sculpin (SCP) may be seen to live on the bottom in high-velocity areas among rocks and boulders. Abbreviations: A, adult; J, juvenile; BT, brown trout; RS, Lahontan redside; RT, rainbow trout; SCP, Paiute sculpin; SD, speckled dace; TS, Tahoe sucker. From Moyle and Vondracek (1985).

meroptera) and stoneflies (Plecoptera). Speckled dace are found, often in large numbers, in the slower water of shallow riffles and runs, where they feed on the bottom on larval dipterans and early instars of mayflies and caddisflies (Tricoptera). Joining dace in these habitats are juvenile suckers, which hug the bottom in small schools, feeding on crustaceans and small insects. Larger suckers live in deeper water, especially on the bottoms of pools, feeding on algae, detritus, and small insect larvae. Lahontan redsides also favor pools and concentrate in swift water at the upstream ends of pools, where they eat drifting insect larvae and winged adult insects. Juvenile redsides are found in slower, shallower water at pool edges or in runs. Brown trout and rainbow trout juveniles live in all habitats except deeper pools occupied by predatory adult brown trout. Juveniles of the two trout species use essentially the same microhabitats and food (drifting insects) and so probably compete for space and food. In contrast, adult rainbow trout tend to live

more in open water than adult brown trout, and they feed on drifting terrestrial and aquatic insects.

The structure of this assemblage may have made it persistent through time and resilient in the face of natural disasters. However, addition of brown trout to the system seems to have made more than one "steady state" possible. In the original assemblage all species spawned in spring, as water levels rose from melting snow. As a consequence, their numbers probably increased and decreased in synchrony; if a year or series of years had poor conditions for spawning or survival of early life history stages, all would suffer. Replacement of spring-spawning cutthroat trout by rainbow trout probably did not alter the assemblage much because rainbows also spawn in spring. However, brown trout (which were introduced after rainbow trout were established) spawn in late fall. If their embryos survive the scouring of winter floods (Erman et al. 1988), juveniles will emerge from the gravel sooner than those of spring-spawning trout. As a result, they have a competitive advantage over other juvenile trout because they are larger and have established territories. More important, they will be relatively immune from the factors causing poor reproductive success in spring spawners. Thus when other species have depressed populations, brown trout may flourish (Strange et al. 1992). Furthermore, brown trout predation on other fishes may keep populations of native fishes from rebounding even when favorable conditions for spawning return. The native fish assemblage can resume its dominance only if brown trout reproduction fails for several winters in succession or if heavy fishing significantly reduces the numbers of adults. In Martis Creek, the ascension of brown trout resulted in the near elimination of speckled dace and Lahontan redside from the stream and a great reduction in the populations of other species (Strange et al. 1992; Strange 1995).

Lake Tahoe

Lake Tahoe is one of the largest high-mountain lakes in the world (surface area, 304 km²), remarkably deep (maximum depth, 501 m; mean depth, 313 m) and clear (the bottom formerly could be seen at a depth of 20–30 m). It is 36.4 km long and 20.9 km wide, and it lies at an altitude of 1,899 m above sea level. The total area of its watershed, including the surface of the lake, is only 830 km². It drains through the Truckee River into Pyramid Lake, Nevada.

The native fishes are the same as those that occur in Lahontan streams, except that a plankton-feeding form (*pectinifer*) of tui chub is present, as well as a benthic-feeding form (*obesa*), and the stream-adapted mountain sucker is absent. Major changes in the fish community wrought by humans so far have been complete replacement of Lahontan cutthroat trout with alien lake trout, rainbow trout, and brown trout and addition of kokanee salmon. Introduction of opossum shrimp (*Mysis relicta*) also caused profound changes in the ecosystem, which affected fish populations (Fig. 12). Despite similarities between the fish fauna of Lake Tahoe and Lahontan streams, the ecological relationships among species in the lake are somewhat different from those in the streams. This fact was first revealed by R. G. Miller (1951), who recognized three distinct fish assemblages: *(1)* shallow water, *(2)* deepwater benthic, and *(3)* midwater (Fig. 12).

The **shallow water assemblage** lives mostly in water less than 10 m deep in rocky-bottomed areas. It is composed of six species: speckled dace, Lahontan redside, Paiute sculpin, Tahoe sucker, rainbow trout, and brown trout. Dace live among rocks, swimming about in loose aggregations. They feed on invertebrates, such as small snails and blackfly larvae, that live on the surface of the rocks. They tend to hide during the day, becoming active at night. In contrast to dace, redsides are diurnal and surface oriented, and they swim about in large schools. They feed equally on bottom, surface, and midwater invertebrates and are perhaps the most numerous fish in the lake. Paiute sculpin live under rocks during the day but come out to forage at night on larger benthic invertebrates, especially midge and caddisfly larvae. Tahoe suckers are present mostly as juveniles (<10 cm TL). They are also most active at night, browsing on detritus, algae, and small invertebrates. They are the one species that seems to feed on a regular basis in more exposed sandy-bottomed areas, as well as in rocky areas. Rainbow trout and brown trout are the main piscivores, moving in to forage in the evening. They capture mostly suckers and redsides, the two species most likely to be out in the open. Dace and sculpin form only a very small part of their diet.

Besides these permanent inhabitants of shallow water, young-of-year of most other fishes can be found here at one time or another. Large aggregations of young-of-year fishes are especially likely to be found along marshy shores, where the emergent plants provide a measure of protection.

The **deepwater benthic assemblage** has two distinct types of habitat: thin beds of aquatic plants and plant-free areas. The aquatic plants—mostly *Chara*, filamentous algae, and aquatic mosses—grow on lower-gradient slopes down to depths of about 150 m. Most plants are present at depths of between 67 and 116 m, with the largest concentrations at 100–116 m (Frantz and Cordone 1967). The plant-free habitat is in water deeper than 150 m, on steep-sloped areas at intermediate depths, and on sandy bottoms at depths of less than 33 m.

Fishes that make up this association are lake trout, Paiute sculpin, the *obesa* form of the tui chub, large Tahoe sucker, and mountain whitefish. Lake trout mostly cruise about near the bottom, foraging among aquatic plants as well as in plant-free areas. Their usual prey are other deepwater fishes, in the following order of importance: Tahoe sucker, Paiute sculpin, tui chub, and mountain whitefish (although opossum shrimp have become a major component of their diet since the introduction). Suckers are probably the most common fish taken, because they are large and almost continuously active, grazing the bottom in schools on algae, detritus, and invertebrates. Sculpins are abundant wherever they can capture detritus-feeding invertebrates (snails, amphipods, chironomid larvae) and each other. Some *obesa* tui chubs move into this association during the day, returning to shallower water (<15 m) at night. Their food is predominately snails, which live in large numbers on the aquatic plants, although various bottom-dwelling invertebrates are also common in their diet. Mountain whitefish are also probably found in association with beds of aquatic plants, and they seldom venture into deep, plant-free areas. Feeding is mostly during the day, on snails, dragonfly larvae, and other plant-dwelling or bottom-living invertebrates.

The **midwater assemblage** consists of two plankton

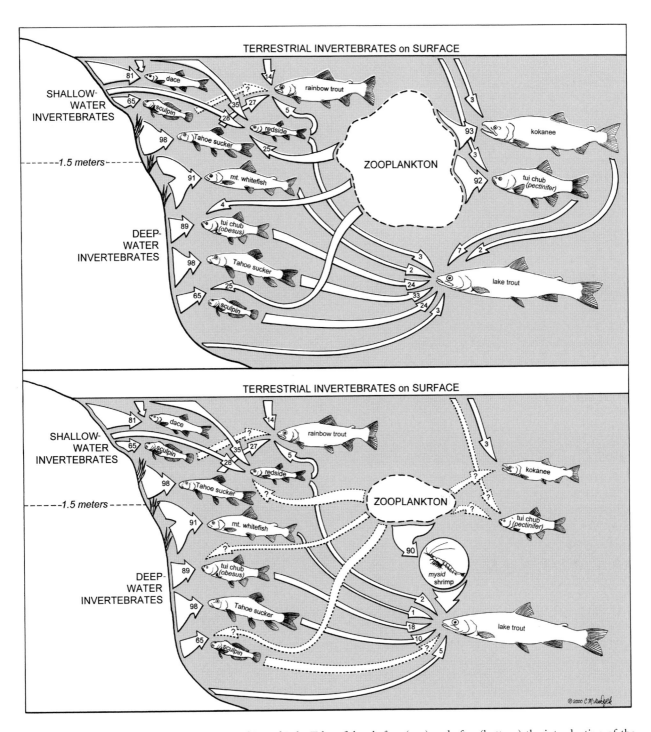

Figure 12. Habitat zones and feeding relationships of Lake Tahoe fishes before (top) and after (bottom) the introduction of the plankton-feeding opossum shrimp, *Mysis relicta*. The major food categories are benthic organisms, flying insects, zooplankton, and fish. There are two forms of tui chub in the lake, bottom-feeding *obesus* and zooplankton-feeding *pectinifer*. The food data are modified from Miller (1951) and other sources.

feeders (kokanee salmon and *pectinifer* tui chub) and one predator (rainbow trout) that live in open waters. The relationship between introduced kokanee and native chub needs to be explored in detail because they are both pelagic planktivores, especially on cladocerans (mostly *Daphnia*

pulex) and copepods (*Epischura* and *Cyclops*). From the evidence available, however, it appears that the two species occupy slightly different habitats. Tui chubs seldom venture far from shore and appear to make regular, diurnal, vertical migrations, possibly following diurnal migrations of zooplankton. They are in deep waters (but off the bottom) during the day, moving into surface waters at night. This movement in part follows the contours of the bottom, since they are also closer to shore at night than they are during the day. Kokanee, on the other hand, seem to be widely distributed in open waters, remaining close to the surface continually except when surface waters become too warm in August and September. During these months large schools of kokanee are found at depths of 15–40 m (Cordone et al. 1971). Rainbow trout are also widely distributed in open waters, where they feed partly on plankton and partly on fish, especially tui chubs. The trout commonly move into shallow water to feed on the abundant minnows during evening.

The long-term stability of these assemblages is not known because the dominant species are aliens and because additional species keep being introduced. Thus the dominant predator is the alien lake trout, the dominant planktivore is the alien kokanee, the dominant zooplankter is the opossum shrimp, and the dominant benthic grazer is the signal crayfish (*Pascifastacus lenuisculus*). Largemouth bass are now found in the shallow, warm marginal habitats, where they may be an important predator on juvenile native minnows.

Eagle Lake

Eagle Lake is the only large natural lake in California, besides Lake Tahoe, that contains Lahontan fishes, and it may be the only large lake that contains solely native species. The second largest freshwater lake completely within California (8,900 ha), Eagle Lake is alkaline (pH 8.4–9.6) and mostly less than 5 m deep, although it has a maximum depth of 23 m. It is fairly productive, supporting large beds of aquatic plants in shallow water. The surface waters usually reach 21°C in the summer, and the lake surface often freezes in winter. Strong winds prevent development of a permanent, well-defined thermocline in summer, but the deep water nevertheless normally remains less than 21°C.

Only five species of fish live in the lake: Eagle Lake rainbow trout, tui chub, Tahoe sucker, Lahontan redside, and speckled dace. The redside and dace inhabit the waters close to the shore, especially where there is cover (rocks, tule beds). Dace feed mainly on small benthic invertebrates, mainly amphipods (*Hyalella azteca*) and chironomid larvae, whereas redsides concentrate on zooplankton (Table 6). Large shoals of young-of-year tui chubs are also found here beginning in mid-July, and they also feed on zooplankton. Large tui chubs live in open waters, feeding mainly on benthic invertebrates and organic debris. The chubs in turn are the main food of trout, especially in late summer when high surface temperatures confine trout to deeper areas. Trout also consume large numbers of leeches and larger zooplankton species. The only species that shares

Table 6
Diets (Percent Volume) of Adult Eagle Lake Fishes, July 1986

	Rainbow trout	Lahontan redside	Tui chub	Speckled dace	Tahoe sucker
Number of fish	121	104	104	32	48
Plankton					
Daphnia	9	60	3	17	3
Leptodora	23	0	0	0	1
Hyalella	34	12	12	17	47
Benthos					
Ephemeroptera	7	0	0	0	2
Helobdella	6	0	0	0	0
Trichoptera	1	22	1	60	22
Other	15	1	0	6	12
Fishes	5	0	0	0	0
Algae	0	5	2	0	6
Detritus	0	0	82	0	7

Source: P. B. Moyle (unpublished data).

deep water with the trout is the Tahoe sucker, which is seldom preyed upon by the trout, apparently as a consequence of its bottom-dwelling habits. It feeds largely on benthic invertebrates.

At the present time, Eagle Lake trout populations are entirely maintained by hatchery plantings. Spawning fish are trapped as they run up Pine Creek, the lake's only permanent tributary. This operation is necessary because flows of the creek have been greatly reduced by a long history of poor land management, making it difficult for adult trout to ascend to good spawning areas and for juveniles to make it back down again. Major restoration work is now under way. In any case, key spawning and rearing areas now contain a large population of introduced brook trout. Tahoe sucker and Lahontan redside also spawn in Pine Creek, but they do not have to ascend so far; they may also be capable of spawning in the lake itself, like tui chubs.

Colorado Province

Colorado River

The short section of the Colorado River that borders California bears little resemblance to the great river of a hundred years ago. Flows have been reduced and confined behind dams, forming large impoundments, such as Havasu Reservoir. The formerly heavy silt load is reduced, the reservoirs acting as settling basins, but in its place are salts, fertilizers, and other products of irrigated agriculture. Not surprisingly, the fish fauna has changed drastically, more so than in any other river system in California.

The original fauna was simple because the California

portion of the river was an ecologically uniform, deep, sluggish channel with fluctuating flows and no large tributaries. The bottom was presumably shifting sand, supporting few benthic organisms. In the main channel were bottom-feeding razorback sucker and pelagic bonytail, both species with bizarre body shapes adapted for moving about in strong currents. The unusual morphology of these fishes may have allowed them to feed in places where food was most abundant, such as on logs and rocks swept clean of fine material by swift currents or in the water column (Stanford and Ward 1986). Preying on these two species, as well as on their own young, were giant Colorado pikeminnow. Desert pupfish may have been found in the shallow backwaters and marshes on the river's edge, along with juveniles of the native riverine species. The only other fishes present were rare stragglers from upstream—such as woundfin (*Plagopterus argentissimus*), speckled dace, and flannelmouth sucker—and euryhaline wanderers from the Gulf of California—such as striped mullet and machete.

Today these native fishes are extinct or rare in the California portion of the river. The river and reservoirs contain instead a conglomeration of at least 44 introduced species. About 20 of these species are common, including common carp, red shiner, threadfin shad, several catfishes, largemouth bass, smallmouth bass, striped bass, bluegill, green sunfish, warmouth, black crappie, mosquitofish, and tilapia of mixed origins. Obviously this is an unstable, artificial assemblage of fishes that will keep changing as long as humans keep changing the nature of the river and introducing new species into it. However, Minckley (1982) found that the complex of species used most of the food resources available and showed some segregation by diet. Ohmart et al. (1988) indicated that there was also considerable segregation by habitat, with a distinct group of species found in the main channel and another in backwaters. Within these habitats there is further segregation by depth, water velocity, and substrate. Nevertheless, overlaps among species in both diet and habitat are more the rule than the exception.

Salton Sea

The Salton Sea is the largest inland body of water within California, with a surface area of about 980 km². It fills the bottom of the Salton Sink in the Imperial Valley at an elevation of 71 m below sea level. The sea is shallow (maximum depth, 15 m; mean depth, 10 m), warm (summer temperatures, 26–33°C), and saline (1999 salinity, 44 ppt). Although overflows from the Colorado River have filled the sink many times in the past, the bodies of water so created have eventually dried up, given an evaporation rate of about 1.8 m/year. The most recent natural predecessor, Lake Cahuilla, supported Native American fisheries before it dried up about 500 years ago (Gobalet 1992). The present

sea was created in the summer of 1905 when, during a flood, the entire Colorado River started flowing through and enlarging the Alamo Channel, a canal dug to bring irrigation water to the Imperial Valley. The river continued to empty into the sink until February 1907, when its flow was finally diverted back into its former channel through a massive earth-moving effort. The level of the sea is maintained through inflow of agricultural wastewater from the Imperial and Coachella Valleys. Accumulation of nutrients from 100 years of agricultural drainage has made the sea extremely eutrophic, with high levels of nitrogen and phosphorus (Gonzalez et al. 1998).

In addition to nutrients, the water being drained into the sea has a high salt content. Rapid evaporation rates result in steadily increasing salinity, although wet years or increased irrigation runoff may temporarily cause it to decrease or stabilize. The change in water chemistry through time is reflected in changes in the sea's fish fauna. In 1915, the fishes were the same freshwater species found in the Colorado River. At present, they are mainly saltwater species introduced from the Gulf of California, plus tilapia species that can tolerate high salinities (Table 7). Given that salinity is currently increasing at a rate of 0.5 ppt/year, the marine species are likely to die out in the near future, initially as the result of salinities too high for survival of eggs and larvae (45–50 ppt). Ultimately, tilapia and perhaps sailfin mollies will become the principal species and will remain abundant in the sea until about the mid-2000s, assuming they are not first wiped out by pollution-related events. Once tilapia and mollies disappear, the sea will become a high-salinity system without fish. Numerous nonnative fishes—including subtropical species such as porthole livebearers, mollies, and tilapia—will continue to exist, however, in low-salinity drains and streams that flow into the sea and show shifting segregation from one another by habitat and temperature preferences (Schoenherr 1979). Native pupfishes are likely to continue to exist only in special, intensely managed refuges.

The three main sport fishes in the Salton Sea—bairdiella or Gulf croaker (*Bairdiella icistia*), orangemouth corvina (*Cynoscion xanthulus*), and sargo (*Anisotremus davidsoni*)—were introduced between 1949 and 1956 from the Gulf of California. They will not be treated in this book beyond the brief discussion here because they are marine fish with no tolerance of low salinities and because their long-term persistence in the Salton Sea is unlikely. Brocksen and Cole (1972) demonstrated that embryos and larvae of these fishes do not survive well at salinities greater than 40 ppt. Stephens (1990) has shown that they cannot spawn at salinities greater than 45 ppt. At present, these fishes still support a fishery, but its maintenance until the sea becomes too salty even for adults will require a hatchery program.

At the same time the three saltwater sport fishes were introduced, two other marine introductions were also suc-

Table 7

Changes in the Fish Fauna of the Salton Sea

Year	Ca.1400	1916	1929	1942	1957	1976	1999
Salinity (ppt)	<20?	<20	34	35	35	40	44
Number of species	6	6	6	6	8	10	10
Colorado pikeminnow	C	?	—	—	—	—	—
Bonytail	A	C	—	—	—	—	—
Razorback sucker	A	C	C	—	—	—	—
Rainbow trout	—	R	R	—	—	—	—
Common carp	—	A	C	C	—	—	—
Striped mullet	C	A	C	A	R	R	R
Desert pupfish	C	R	A	A	C	R	R
Western mosquitofish	—	—	A	A	R	—	—
Longjaw mudsucker	—	—	—	?	C	C	C
Machete	R	—	—	C	—	—	—
Threadfin shad	—	—	—	—	A	R	R
Sargo	—	—	—	—	C	A	C
Bairdiella	—	—	—	—	A	A	A
Orangemouth corvina	—	—	—	—	A	A	C
Sailfin molly	—	—	—	—	—	A	C
Mozambique tilapia	—	—	—	—	—	A	A
Redbelly tilapia	—	—	—	—	—	C	C?

Sources: Evermann (1916); Coleman (1929); Dill (1944); Walker (1961); S. Keeney, CDFG (pers. comm. 1999). The information for 1400 is based on fish from archaeological sites (Gobalet 1994).

Notes: Abbreviations: A, abundant; C, common; R, rare. Species found only in freshwater drains or streams feeding the sea are not included. Mozambique tilapia may represent a hybrid complex of forms.

cessful: longjaw mudsucker (*Gillichthys mirabilis*), a small bottom fish, and pile worm (*Neanthes succinea*), a major food organism for fish. In the early 1970s, Mozambique tilapia (now a presumptive hybrid with other tilapia species), redbelly tilapia, and sailfin mollies invaded. The two tilapias became very abundant and apparently eliminated desert pupfish—the one native fish still present—from the sea itself. Large die-offs of tilapia in 1988–1990 gave pupfish another temporary foothold in the sea (K. Nicol, CDFG, pers. comm. 1991), but they are now gone again, barely persisting in inflowing streams and drains (S. Keeney, CDFG, pers. comm. 1999).

The food web established deliberately through introduction of marine fishes and other organisms is relatively simple (Walker 1961). Primary production is by abundant planktonic algae, mainly diatoms, dinoflagellates, and green algae. These are fed upon by zooplankton, mostly rotifers, copepods, and larval stages of bottom invertebrates. Young tilapia presumably feed directly on abundant zooplankton, although adults are more omnivorous and feed on algae and benthos as well. Tilapia in turn are important prey of corvina, providing a plankton-based food web. However, the base of the food web leading to corvina, sargo, and bairdiella usually appears to be organic matter, which decays and forms fine detrital ooze, the main food of pileworms. Pile-worms are the main item in the diet of bairdiella and sargo, which are in turn fed on by orangemouth corvina. The latter species, achieving weights of 14.5 kg in the sea, is an important object of the sport fishery, although tilapia harvest may now be more important in terms of numbers and biomass (S. Keeney, CDFG, pers. comm. 1999).

At present, tilapia (mainly Mozambique tilapia) are the most abundant fish in the sea. Their populations undergo enormous fluctuations as the fish die in huge numbers from various causes (S. Keeney, pers. comm. 1999). In winter, die-offs may occur because of stress induced by low temperatures (11–14°C). When temperatures of the sea are high, die-offs of tilapia and bairdiella are related to oxygen depletion, although the immediate cause of death is often stress-induced diseases and parasitic infections. Toxins released from algal blooms may also cause death, as may agricultural and industrial wastes entering via the drains. The fish kills are of concern not only for aesthetic reasons (tilapia populations at least have amazing powers of recovery) but also because the fish, dead and alive, are eaten by large numbers of migratory waterfowl. Living tilapia carrying type C botulism organisms in their guts have been implicated in the deaths of thousands of birds, including brown and white pelicans, grebes, and cormorants. Massive die-offs of birds and fish are indicative of a very unstable ecosystem that is on a trajectory toward simplification, one that ultimately will be without fish. Major studies are under way to find ways to save the "dying" sea, although until it actually dries up completely it will continue to be rich in life, if not in fish.

Because demand for fresh water by humans outside the basin is increasing, conservation measures are likely to reduce the amount of water flowing in, accelerating the increase in salinity. Proposed solutions to the problems, however, involve making all or part of the sea less saline through such schemes as exchange pumping of water from the Salton Sea with water from the Gulf of California or diking off large sections of the sea to contain fresher inflowing water. Such solutions are enormously costly in money and energy and are unlikely to be sustainable. They also do not really address the ever-increasing nutrient levels. In the short run, the sea is likely to shift to a system dominated by herbivorous or omnivorous fishes with high salinity tolerances, mainly tilapia and mollies, which will be preyed upon mainly by birds. Gonzalez et al. (1998) suggest that eutrophication of the sea could be alleviated at least temporarily by harvesting tilapia in large amounts, because the fish have the capacity to take up large amounts of nitrogen and phosphorus. In the long run, the sea is likely to turn into an ecosystem based on brine shrimp and brine flies, like the Great Salt Lake or Mono Lake (University of California–Mexus Border Water Project 1999).

Change

California has undergone, and continues to undergo, massive changes in its aquatic ecosystems. Resilient natural systems, reasonably predictable in structure, are rapidly being replaced by highly altered systems, unpredictable in structure and dominated by alien species. Native fishes, the best-known components of the natural aquatic systems, are rapidly being lost (Moyle and Williams 1990). Of 67 species, 7 (10%) are extinct in the state or globally, 13 (19%) are officially listed as threatened or endangered (as of 2001), and 19 (29%) are listed as Species of Special Concern, which will need to be listed soon if present trends continue (Table 1). This means that 58% of all inland fish species of California are extinct or in serious decline. In addition, a number of subspecies of more widely distributed species are in trouble, including two that are extinct and nine formally listed as threatened or endangered. These numbers can change rapidly because species can decline in abundance and go extinct within very short periods of time. The last thicktail chub was seen in 1957, the last Tecopa pupfish in 1965, the last Colorado pikeminnow in California in 1967, the last Clear Lake splittail in 1972, the last bull trout in California in 1975, the last High Rock Spring tui chub in 1989. In the same period at least 16 species of fish were successfully introduced into the state. In short, California is losing about one native species or subspecies of fish every five or six years, on the average, and gaining an alien species about once every two years! Introduced species are abundant in their native ranges as well, so the result of this "trade" is a net loss of species worldwide. The changes in the California fish fauna are reflected in the changes in the fishes in Clear Lake and the San Joaquin River at Friant, localities for which long-term records exist (Tables 8 and 9).

The rapid decline of the native fish fauna is caused by interactions among natural and human factors. The main natural factor that makes species prone to extinction in California is their limited range; most are confined to one drainage or one body of water. Species that are in the most trouble today come from a wide variety of habitats, but a majority occur either in small, isolated springs and creeks or in big rivers, especially in drier regions of the state. The reason for this is that very small and very large aquatic systems are most vulnerable to damage by humans (Moyle and Williams 1990). The human factors that have negative effects on the abundance of native fishes are, in order of overall importance, *(1)* water diversions, *(2)* habitat modification, *(3)* pollution, *(4)* alien species, *(5)* hatcheries, and *(6)* exploitation. The main purpose of this section is to describe how these factors have affected the fishes.

Water Diversions

From our society's perspective, water in California is poorly distributed. Most precipitation falls in the northern half of the state in mountainous or coastal areas, whereas most people live in the southern half of the state in deserts and dry valleys. Furthermore, most precipitation occurs during winter and spring, whereas the greatest demand for water, for irrigation and power production, occurs during the long, hot summer. The solution to this distribution problem has been to build dams, diversions, and aqueducts, to store the water and carry it to distant places for use as needed. From the Gold Rush era onward, dam building has been a major activity in California, with a major peak in the early 20th century, although the biggest dams were built in the interval from the 1940s through the 1960s (Fig. 13). The Los Angeles metropolitan area, for example, imports its water from the Mono Lake basin (about 430 km distant), the Owens Valley (about 380 km), the Colorado River (about 390 km), and the Feather River (about 600 km). The most massive alterations took place in the Central Valley, where the federal Central Valley Project and the State Water

Table 8

Changes in the Fish Fauna of Clear Lake, Lake County

	1872	1894	1929	1941	1950	1963	1973	1998
Native species								
Pacific lamprey	N	P	—	—	—	—	—	—
Threespine stickleback*	N	P	—	—	—	—	—	—
Rainbow trout	C	C	C	—	—	—	—	—
Thicktail chub	A	C	C	P	—	—	—	—
Clear Lake splittail	A	A	A	A	R	R	—	—
Sacramento pikeminnow*	A	A	A	P	R	R	—	—
Sacramento sucker*	A	A	A	P	C	R	—	—
Sacramento perch	A	C	A	C	C	R	R	R
Sacramento blackfish	A	A	A	A	A	A	A	A
Hitch	A	A	A	A	A	A	A	C
Tule perch	A	A	N	P	N	C	C	C
Prickly sculpin	C	C	N	P	N	A	A	A
Introduced species								
Common carp	—	A	A	A	A	A	A	A
Brown bullhead	—	A	A	A	A	A	A	A
White catfish	—	A	A	A	A	A	A	A
Channel catfish	—	—	C	N	N	C	C	C
Largemouth bass	—	—	C	C	C	C	C	A
Bluegill	—	—	A	A	A	A	A	A
Black crappie	—	—	A	P	A	A	A	C
Mosquitofish	—	—	—	C	P	C	A	C
Green sunfish	—	—	—	P	C	C	C	C
Goldfish	—	—	—	—	P?	P	C	C
White crappie	—	—	—	—	—	A	A	C
Golden shiner	—	—	—	—	—	C	C	R
Redear sunfish	—	—	—	—	—	—	R	R
Inland silverside	—	—	—	—	—	—	A	A
Threadfin shad	—	—	—	—	—	—	—	A
Total number of species	12	15	17	18	18	20	19	20
Percent native species	100	80	59	44	44	40	26	25

Sources: Based on information from Stone (1876); Jordan and Gilbert (1894); Coleman (1930); Lindquist et al. (1943); Murphy (1951); Cook et al. (1966); Colwell et al. (1997); and P. B. Moyle and CDFG (unpublished records).

Notes: Abbreviations: A, abundant; C, common; N, not recorded but probably present; P, present; R, rare; *, native species found in inflowing streams that are likely to be in the lake on occasion.

Project, together with power companies and urban water agencies, have dammed virtually every large stream. The only drainage with more extensive alterations in the West is the Colorado River, but most of its dams and diversions are upstream from California. Even so, a good chunk of Colorado River water goes to California farms and cities.

The biggest single consumer of water in California is irrigated agriculture, which takes 70–80 percent of stored water in the state and also pumps great volumes of groundwater. Large amounts are wasted because of an antiquated system of water laws, especially those governing riparian water rights. A landowner with riparian rights along a stream can divert as much water as desired to water crops on his or her own land, but the water cannot be sold. This doctrine results in large amounts of water being diverted to flood-irrigate pasture and alfalfa in summer, an extravagant use of water in a desert climate (Reisner 1986). Likewise, large quantities of water are needed to flush salts from irrigated land on the west side of the San Joaquin Valley and in the Coachella and Imperial Valleys, sending salts and toxic materials such as selenium into the rivers. The result is much less water available to fish and reduced quality of the water that remains.

Dams and diversions affect fish in many ways, usually simultaneously, so faunal changes are inevitable when a water project is built. Among the ways in which they affect fish are the following: *(1)* blocking migrations, *(2)* dewatering streams and lakes, *(3)* changing temperature and flow

TABLE 9

Changes in the Fish Fauna in the San Joaquin River at Friant, Fresno County

	1898	1934	1941	1971	1985
Native species					
Splittail	X	—	—	—	—
Hitch	X	X	X	—	—
California roach	X	X	X	—	—
Hardhead	X	X	X	—	—
Sacramento pikeminnow	X	X	X	—	—
Sacramento blackfish	X	X	X	—	—
Chinook salmon	X	X	X	—	—
Tule perch	X	X	X	—	—
Sacramento sucker	X	X	X	X	X
Rainbow trout	X	X	X	X	X
Prickly sculpin	X	X	X	X	X
Threespine stickleback	X	X	X	X	X
Kern brook lamprey	N	N	N	X	X
Pacific lamprey	N	N	N	X	X
Introduced species					
Brown trout	—	X	X	X	X
Common carp	—	X	X	X	X
Bluegill	—	X	X	X	X
Smallmouth bass	—	X	X	N	X
Brown bullhead	—	—	—	X	X
Mosquitofish	—	—	—	X	X
Green sunfish	—	—	—	X	X
Largemouth bass	—	—	—	X	X
Total number of species	14	17	21	14	14
Percent native species	100	77	62	43	43

Sources: Based on information from Rutter (1903); Needham and Hanson (1935); Dill (1946); Moyle and Nichols (1974); and Brown and Moyle (1993).

Notes: This was originally a transitional reach between valley floor and foothills, so it had a high diversity of native fishes. After 1941 flow in the reach was regulated by releases from Friant Dam, converting it into a coolwater trout stream containing trout that are mostly of hatchery origin. Abbreviations: N, probably present but not recorded; X, present.

regimes, *(4)* entrainment, *(5)* creation of reservoirs, *(6)* altering upstream areas, and *(7)* altering estuaries.

Blocking Migrations

One of the most immediate effects of dams is in blocking up- and downstream movements of fish. In the Sacramento–San Joaquin watershed, dams deny chinook salmon access to >1,800 km of stream they once used—more than 80 percent of their former habitat (Fig. 14; Yoshiyama et al. 1996).

Amounts of stream lost to steelhead are even greater because they spawn in smaller tributaries to main rivers, but their former distribution is too poorly known to estimate the actual number of kilometers lost. The culmination of these blockages were Friant and Shasta Dams. Friant Dam, finished in 1948, completely prevented a large run of spring-run chinook salmon from reaching their holding and spawning grounds in the upper San Joaquin River. This dam completed a process of blocking upstream access by salmon in the San Joaquin drainage that began with the construction of LaGrange Dam on the Tuolumne River in 1894. No attempt was made to find ways to get the salmon over or around these dams, so a run that was probably in excess of 500,000 fish per year was completely lost.

In the Sacramento River, closing of Shasta Dam in 1942 cut off access by both winter- and spring-run chinook salmon to major spawning areas; however, the two runs were saved from extinction by coldwater releases from the dam, creating some new habitat. This fortuitous circumstance was largely negated by completion of Red Bluff Diversion Dam in 1964, which diverted Sacramento River water into canals of the Tehama-Colusa Irrigation District. This dam had salmon ladders to allow fish to pass. Unfortunately they were poorly designed, making it difficult for upstream migrants to find them. Peculiarities of construction also made the dam a major cause of death of young salmon that had to pass over it on their way to sea. The result was a steady decline in wild Sacramento River salmon. Attempts to reverse the decline have involved leaving the dam gates open during periods of salmon migration, allowing free passage of fish. Similarly, Copco Dam cut off access by chinook salmon and steelhead to the upper Klamath basin, resulting in extirpation of the runs that went into Oregon.

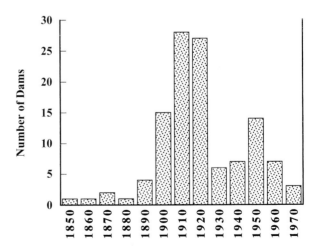

Figure 13. Number of large dams constructed in California, 1850–1980, by decade. From Yoshiyama et al. (1998).

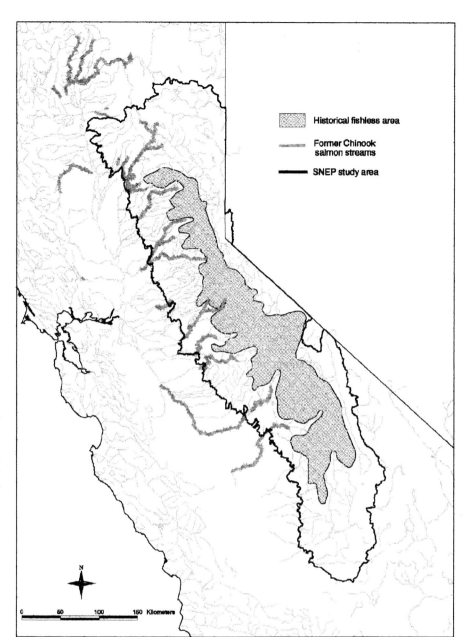

Figure 14. Two major changes in fish distribution in Central California. The dark lines show areas formerly accessible to chinook salmon and steelhead rainbow trout that are now blocked by dams, while the shaded area indicates the formerly fishless region of the Sierra Nevada now occupied by alien fish, mainly trout. The Sierra Nevada Ecosystem Project (SNEP) study area roughly delimits the Sierra Nevada range. From Moyle and Randall (1998); reprinted by permission of Blackwell Science, Inc.

Even blockage of within-river migrations may create problems. Blockage of the migrations by numerous dams on the Colorado River may have been responsible for the extirpation in California waters of Colorado pikeminnow, and blockage of spawning migrations of bull trout by McCloud Dam on the McCloud River may have led to the extirpation of bull trout in the state.

Dewatering Streams and Lakes

One of the main reasons for the construction of dams, reservoirs, and irrigation diversions is to catch runoff and send it, via aqueducts, to locations where it can be used for

irrigation or industrial and municipal consumption. This naturally leaves less water available for fish downstream from dams. Friant Dam cut off virtually all flow to the lower San Joaquin River, effectively turning it into an agricultural drain, largely unsuitable for native fishes or for passage of migratory fishes. Closure of the dam was the final and major blow to San Joaquin spring-run chinook salmon. In the words of George Warner, a biologist involved in the desperate efforts to save this run, "the trickle of water [in the San Joaquin River] soon disappeared in the sand, stranding salmon migrants more than one hundred miles from the sea. The tragic conclusion to the history of the 1948 spring run was that the only beneficiaries of our efforts to salvage a valuable resource were the raccoons, herons, and egrets" (Warner 1991, p. 65).

Less dramatic but perhaps just as devastating to native fishes have been the cumulative effects of the dewatering of small streams by many smaller dams scattered around California. For example, construction of Hidden Valley Dam on the Fresno River in the 1970s converted the stream below the dam from a rather attractive sandy-bottomed stream dominated (95%) by native fishes to a series of stagnant pools dominated by common carp and other introduced species (81%) in 1985 (D. L. Miller et al. 1988).

The effects of dewatering often take a long time to be felt, especially if flows are reduced but not cut off completely. One of the most dramatic examples of such a delayed outcome was the fall in the level of Pyramid Lake, Nevada, following diversion of most of the flow of the Truckee River (in California) for irrigation. The sandy delta exposed at the mouth of the river by the declining lake level prevented both Lahontan cutthroat trout and cui-ui sucker (*Chasmistes cujus*) from spawning in the river. The trout are present in the lake only because of the planting of a nonnative strain; the suckers are listed as endangered. The suckers survived only because they are extraordinarily long lived, with life spans of 40–50 years (Scoppetone 1988). Only massive conservation efforts, including restoration of flows during the spawning period, have permitted them to reproduce in recent years. Similar reduction in flows of inlet streams during periods when they are used for spawning was at least partially responsible for the extinction of Clear Lake splittail in Clear Lake, Lake County. Splittail were either stranded as adults during spawning runs or stranded as newly hatched juveniles, unable to return to the lake (Cook et al. 1966).

Changing Temperature and Flow Regimes

Rivers below dams inevitably have altered temperature and flow regimes. Dams on the Sacramento and Colorado Rivers made river flows below the dams more constant, eliminating flood flows in winter or spring and converting the turbid, warm rivers of summer into cold, clear streams suitable for trout and salmon. In the Colorado River, the result was creation of an endangered fish fauna, with extinction of most native species in the California portion of the river. In the Sacramento River, cooler summer waters have made both juvenile and adult salmon year-round residents; distinctions between fall-, late fall-, winter-, and spring-run races have become increasingly blurred as a result. These runs evolved to take advantage of special conditions in tributaries and the predictable, highly seasonal patterns of flow in the river—conditions and patterns that are now significantly altered. The continuing decline of all four runs in the Sacramento River indicates that, overall, the altered flow and temperature regimes have not improved conditions for chinook salmon (Yoshiyama 1999; Yoshiyama et al. 2000). These runs increasingly depend on fish of hatchery origin.

In some regulated streams, a small change in temperature regime can result in a major change in the fish fauna. Development of the North Fork Feather River for hydroelectricity resulted in a series of dams that raised summer temperatures in parts of the river. Reaches that were probably once dominated by rainbow trout and anadromous fishes now favor native coolwater fishes (hardhead, pikeminnow, sucker), and attempts to alter this situation by periodically poisoning native fishes and planting large numbers of trout have largely failed (Moyle et al. 1983).

In a few streams, the altered flow and temperature regimes can benefit fisheries. For example, 12 km of Putah Creek (Solano and Yolo Counties) are used to convey water from Berryessa Reservoir to Putah Diversion Dam, where most is diverted into Putah South Canal. The 12-km stretch has low but constant flows in winter and high flows in summer, when agricultural and urban water demand is highest. The result is a coldwater stream that supports a substantial population of large, wild rainbow trout, as well as abundant riffle sculpin, threespine stickleback, and Sacramento sucker. Increased flows in summer allow CDFG to plant the stream heavily with hatchery trout, making the stream one of the most popular fishing spots in the region. The summer bait fishery for hatchery trout does not seem to affect the populations of wild trout in the creek.

Entrainment

Fish are entrained by a diversion when they are carried away in the diverted water, usually to some place where chances of survival are low, such as the cooling system of a power plant or an irrigation ditch. Entrainment of outmigrating salmon and steelhead smolts has long been recognized as a factor contributing to the decline of fisheries. A great deal of effort has therefore been devoted to designing, installing, and maintaining fish screens on water diversions—with limited success. Young salmonids are actually more easily screened from diversions than most other juvenile fishes because they are fairly large (usually >50 mm) and are strong swimmers. Species with a helpless larval stage can suffer large losses of the larvae to entrainment. This seems to be one of the main reasons why populations of most fishes have declined in the Sacramento–San Joaquin Delta since 1970 or so. Large numbers of young are entrained in *(1)* pumping plants of the State Water Project and the Central Valley Project, *(2)* hundreds of small unscreened diversions taking water to irrigate Delta islands, and *(3)* the cooling intakes of power plants. The John F. Skinner Fish Protection Facility at the pumps of the State Water Project screens hundreds of thousands of larger fish from the California Aqueduct, but it cannot retain larval fish. Even its success at screening larger fish is limited. Mortality rates of "rescued" fish are probably high, if not during transport then to predators after the fish have been trucked back to the estuary. Managers of salmonid hatcheries on the Sacramento and San Joaquin Rivers have long recognized the problems juvenile salmon and steelhead have in migrating through the Delta; they achieve higher survival rates of their fish by trucking them around the Delta and releasing them in such places as Berkeley Marina on San Francisco Bay. A problem with diversions, including the pumps in the Delta, is that their direct effects are hard to distinguish from the indirect effects of water removal, such as a change in hydraulics (Bennett and Moyle 1996).

Creation of Reservoirs

Reservoirs are hard on the native fish fauna because they favor lake-adapted alien species over native stream-adapted forms. Thus pikeminnows and hardhead became rare in most water supply reservoirs in the Sacramento–San Joaquin drainage after an initial 5–10 years of abundance. Young that were trapped in reservoirs when they filled managed to grow up, but their young were unable to survive, presumably because they were devoured by introduced predators, especially largemouth and smallmouth bass.

Reservoirs have benefited some native fishes. Prickly sculpin and Sacramento sucker are permanently established in a number of Central Valley reservoirs, as are hitch and tui chub. Sacramento perch, virtually extinct in their native habitat, are extremely abundant in a number of alkaline reservoirs outside their native range, such as Crowley Reservoir on the Owens River. Reservoirs operated solely for power production may actually favor native fishes because they usually remain full and create conditions that might be found in a giant riverine pool. Thus Britton Reservoir on the Pit River is dominated by Sacramento sucker, hardhead, Sacramento pikeminnow, tule perch, and other native fishes, despite the presence of introduced species such as largemouth bass and white crappie (Vondracek et al. 1988b).

Altering Upstream Areas

A subtle effect of dams is their isolation of stream reaches upstream of the reservoir. If a stream located above a dam should lose its native fish fauna through natural or human-made disasters, there is no way it can be naturally recolonized from other nearby systems. For example, California roach are now largely absent from the small streams of the upper San Joaquin River above Friant Dam, with no hope of natural recolonization (Moyle and Nichols 1974). When salmon runs are blocked, a stream loses a major source of nutrients (from salmon carcasses) as well as other major components of the ecosystem, such as juvenile salmon. Sometimes these juveniles are replaced in part by progeny of trout that live in the reservoir and use the stream for spawning. In the McCloud River, brown trout, rainbow trout, and kokanee from Shasta Reservoir use the river for spawning (Sturgess and Moyle 1978). Other upstream migrants are less welcome. A barrier was constructed on Hat Creek (Shasta County) to prevent Sacramento suckers from moving up from Britton Reservoir. There was indirect evidence that the grazing activities of large suckers dislodged aquatic plant beds, which are prime habitat for the invertebrates eaten by the creek's famous trout.

Altering Estuaries

One common justification for building dams is that "water flowing into the ocean is wasted." This attitude reflects a profound ignorance of the value of estuaries, which require large amounts of fresh water to function. They are major nursery areas for juvenile salmonids and other fishes; invertebrate food organisms are abundant, so the fish can grow rapidly before going to sea. Species such as longfin smelt, white sturgeon, and striped bass spend all or most of their lives in estuaries. Their early life history stages often grow and survive best in the zone where fresh water and salt water mix, where food production is high. In the San Francisco Estuary, reduced inflows of fresh water move this mixing zone upstream, away from the productive shallows of Suisun Bay and into the deeper and less productive river channels. The result is reduced survival of young, coupled with their increased vulnerability to entrainment when they are in the river channels (Jassby et al. 1995).

The decline of fishes in the San Francisco Estuary can be observed on a smaller scale in numerous small coastal estuaries. The tidewater goby, which lives only in small coastal lagoons, is disappearing as populations blink out one at a time, usually following diversion or alteration of inflowing streams needed to maintain estuarine conditions. The same lagoons are increasingly unsuitable for rearing of juvenile salmonids (such as steelhead and coastal cutthroat trout), accentuating decline of these fishes caused by other factors operating upstream.

Habitat Modification

Most of California's major inland waterways today bear little resemblance to the streams and lakes encountered by the first European explorers and settlers. The once turbulent and muddy lower Colorado River is now a giant irrigation ditch and drain, carrying salts and other agricultural wastes to Mexico and occasionally to the Gulf of California. The former giant lakes of the San Joaquin Valley are today vast cotton farms. The Sacramento–San Joaquin Delta, once an enormous tule marsh dissected by meandering river channels, has been transformed into islands of farmland protected by high levees from water that flows by in dredged channels. Much of the Los Angeles River is a cement-lined drainage canal. Streams in mountain meadows have been stripped of riparian vegetation by livestock, and their banks collapsed by sharp hooves. Other small streams have been turned into straight ditches through channelization. Thus it is not surprising that habitat modification is a major cause of changes in California's fish fauna. Different species are affected by different types of habitat change, however, so it is worthwhile to consider separately the effects of *(1)* stream channel alterations, *(2)* draining of streams and lakes, *(3)* grazing livestock, *(4)* logging, *(5)* mining, and *(6)* watershed changes.

Stream Channel Alterations

Humans have been altering the channels of California's streams ever since the first Spaniard stepped off a boat, shovel in hand. Today straightening and dredging of stream channels is being carried out in the name of flood control. The idea is to move water as fast as possible, so it will not flood lands surrounding the channel (the floodplain)—ignoring the fact that this increases the probability of flooding downstream. Channelized sections of Rush Creek, Modoc County, when compared with nonchannelized sections, contain fewer fish overall, much smaller trout, and fewer individuals of the rare Modoc sucker; only Pit sculpin and speckled dace manage to maintain large populations in channelized sections (Moyle 1976). The decrease in size and numbers of fish was caused by reduction of habitat diversity, especially the elimination of pools.

A classic example of a stream much abused in the name of flood control is lower Putah Creek (Yolo and Solano Counties). Flooding of surrounding lands was a natural annual event for this creek, resulting in the rich alluvial soils prized by farmers. The flooding, of course, was otherwise unacceptable to farmers and to inhabitants of farming towns, such as Davis. Over the course of a century, the creek was increasingly straightened and confined between levees, although in the first half of the 20th century it maintained a reputation as a fine fishing stream, especially for introduced smallmouth bass. Some farmers actually fed their workers sturgeon, salmon, and other fish caught from the creek. In 1957, Monticello Dam was finished, capturing most of the flow in Berryessa Reservoir. About 12 km of creek below the dam were maintained as a water delivery channel to Putah Diversion Dam and Putah South Canal. Valley reaches below the diversion dam, however, were largely written off as fish habitat. Bulldozers were regularly used in the channel to keep vegetation cleared between the levees; gravel was mined from the bed; car bodies, waste concrete, and other trash were dumped on the levees.

Despite all this activity, fish populations managed to maintain themselves in the little water remaining (from sewage effluent and other similar sources), and they staged a spectacular comeback when the University of California began maintaining its portion of the channel as a natural area. Regrowth of willows and other vegetation provided cover for fish and food for beaver, which built numerous dams that created additional pools favored by fish. The fish populations that built up included not only alien game fishes such as largemouth bass, bluegill, and white catfish, but also native fishes such as Sacramento blackfish, pikeminnow, sucker, hitch, and tule perch., The long-term survival of these fish depend on releases from upstream dams to provide enough water to keep the stream alive. In the drought years of 1990 and 1991, flows were turned off and most fish perished. Only action by a local environmental group, the Putah Creek Council (working with the university and the city of Davis), kept the creek from drying up completely (Moyle et al. 1998). In the late 1990s, a series of wet years led to recovery of native resident fish and to return of chinook salmon, Pacific lamprey, and steelhead to spawn successfully. These fish are now protected by an agreement that will keep the stream flowing even during drought years.

Dredged channels of the Sacramento–San Joaquin Delta are examples of stream channel alterations on a mammoth scale. The channels are inhabited by a variety of fishes, but it is mainly introduced species that survive in such altered environments. When levees are breached and floodplains restored, flooded areas are heavily used by juvenile salmon, splittail, and other native fishes. Similar negative effects were observed when sloughs along the lower Colorado River were drained as part of a large channelization project (Beland 1953a).

Draining of Streams and Lakes

The ultimate reduction in fish habitat in California through dewatering was the drainage of Tulare, Buena Vista, and Kern Lakes on the floor of the San Joaquin Valley. These huge, shallow lakes supported a small commercial fishery for turtles and native minnows in the 19th century. Unfortunately, they were drained for farmland before anyone was able to take a close look at the fish fauna.

On a smaller scale, continuous drainage and diking of wetlands that border lakes and streams have negative effects on fish populations. Some, such as splittail, require flooded vegetation for spawning, whereas others, such as hitch, use flooded marshes as cover for their young. Marshlands, with their large biomass of plants, are also a source of nutrients for aquatic systems, such as the San Francisco Estuary, supporting food chains that lead to fish.

Grazing Livestock

Grazing by livestock in riparian areas has severely damaged thousands of miles of California streams. It has been going on for 300–400 years, so in many areas undamaged streams hardly exist, and the public perception of a "natural" stream is often of one that is denuded of much of its riparian cover. Willow Creek is a common name for California streams—yet creeks with this name often have few willows along their banks. Although livestock densities on rangeland are usually expressed in terms of acres per animal, in fact the animals concentrate along streams, where there is water and succulent vegetation (Minshall et al. 1989). The effects of livestock are many and far reaching:

- They remove the riparian plants that provide cover for fish, are a major source of insect food, stabilize streambanks, and keep water temperatures cooler through shading.
- They eat aquatic plants, removing cover for fish and invertebrates in the process, and stir sediments from the stream bottom, lowering the ability of algae to capture sunlight by decreasing water clarity and covering rocks with sediment.
- They trample banks, causing undercuts (important as cover for fish) to cave in. Bank collapse also increases erosion, filling pools and riffles with silt. This results in shallower, more uniform stream channels and less habitat for fish.
- They compact soils in meadows around streams, reducing their ability to hold water and increasing the rapidity of runoff. This results in downcutting of the streambed, in some cases by as much as 2–4 m below its original level, replacing a meandering stream with a gully. In some areas, the compaction changes wet meadows into dry sagebrush flats and permanent streams into intermittent ones.
- They pollute the water with their feces and urine.

Not surprisingly, streams with heavy grazing pressure have reduced fish populations, especially of the larger fish favored by anglers. A classic example of this is Pine Creek (Lassen County), the principal tributary of Eagle Lake and spawning stream of Eagle Lake rainbow trout. More than a century of heavy grazing of meadows around the stream converted most of them to sagebrush flats and caused much of the stream to cut a channel 1–3 m deep, with rounded, sloping banks. The lower reaches became warm and inter-mittent, unsuitable for downstream passage of juvenile trout from more permanent spawning and rearing areas upstream. As a result, Eagle Lake trout survive only because CDFG captures most fish attempting to move upstream to spawn, spawns them artificially, and rears their young in hatcheries for 1–2 years for reintroduction into the lake.

For many streams, such as Pine Creek, the damage done by livestock is reversible, provided animals are excluded from using the creek area on a continuous basis and other well-known stream restoration techniques are applied (Minshall et al. 1989). This type of restoration is increasing in California, despite the reluctance of some managers of public land to reduce grazing allotments or engage in the expensive fencing of stream corridors.

Logging

Like grazing, continuous logging activity in some areas has altered streams to such an extent that we hardly know what a natural stream looks like. Logging, and the road building on steep slopes associated with it, can alter flow regimes (usually exaggerating both high and low flows); increase erosion, sedimentation, and turbidity; compact streambeds; increase water temperatures; create barriers to fish migration (e.g., by causing landslides); and reduce the amount of logs and other debris in streams that are important for creating habitat structure. Removal of trees and compaction of soil by logging equipment tend to increase winter and spring runoff, resulting in more damaging floods. At higher elevations, snow melts more quickly in the absence of shade; this reduces the length of the runoff season and increases peak flows.

In some situations, vegetation removal may actually create year-round flows in normally intermittent streams, improving the streams for some fish species. Large spring floods, however, may offset any gains by increasing streambank erosion, silting in pools and riffles (or, alternately, by scouring and compacting them), decreasing water clarity, and creating barriers of fallen trees and logs. Poor logging practices—such as using streambeds for roadways or clear-cutting steep hillsides—exaggerate these effects, just as careful logging practices—such as leaving a wide buffer of uncut forest along streams (including fishless seasonal tributaries) and selective cutting of timber stands—can minimize them. Thus Burns (1972) found that careless logging along the Noyo River (Mendocino County) caused a 42 percent decrease in young steelhead biomass and a 65 percent decrease in young coho salmon biomass, yet careful logging along other similar streams temporarily increased production of juveniles of these two species. However, the continued decline of coho salmon in the Noyo and other rivers, even in areas that have not been clear-cut, reflects the need to leave large trees in the riparian zone. These trees eventually fall into creeks, creating cover that is especially important during periods of high flow in winter. Indeed, there is growing realization that overwintering habitat is one of the key limiting factors for coho salmon and presumably other fishes (see the coho salmon account, p. 247).

An example of a stream devastated by logging is Bull Creek, now in Humboldt Redwoods State Park. It originally flowed through a large watershed heavily forested by coast redwoods and other old-growth trees and had a fairly narrow channel full of deep pools. It supported large runs of coho salmon and steelhead, as well as other native fishes. Virtually all the large redwoods on the floodplain, except for some groves near the Eel River, were removed first, creating a sunny, exposed area with a shallow stream flowing through a braided channel. Then in the 1950s most trees were removed from the steep slopes of the upper drainage, and large-scale erosion of hillsides took place, sending huge quantities of rock and gravel downstream and making reforestation of the hillsides extremely difficult. The eroded material was deposited in the downstream reaches, creating an even more extensive exposed, gravelly floodplain and eliminating most large pools.

The massive nature of the erosion can be easily seen in Cuneo Creek, a tributary to Bull Creek, where it is possible to stand on the buried remains of an old bridge and look at a newer bridge several meters overhead; there is reportedly another bridge buried several meters below the bridge in the stream channel! As a result of habitat burial, coho salmon disappeared from the drainage, steelhead numbers were reduced, and introduced California roach and Sacramento pikeminnow invaded. Following the devastation, private owners of the watershed generously sold it to the California state park system, which is now undertaking to restore Bull Creek. Restoration will have been accomplished when a large run of coho salmon again spawns in the creek.

The need for such restoration attempts on other coastal streams is indicated by the fact that at least half have lost their coho populations in the past 50 years, and there are now fewer than 5,000 wild coho spawning in the state in most years (Brown et al. 1994). Virtually all former coho streams have a history of heavy logging in their drainages.

Mining

The first really drastic alterations of California streams were those of gold miners, who, in their frantic search for tiny bits of metal, despoiled hundreds of miles of streambed by placer and hydraulic mining. In the process of digging up the streambeds and banks, they destroyed large salmon runs in Sierra Nevada streams and turned shady, pool-and-riffle trout streams into long, shallow, exposed runs. Some streams are still nearly barren of fish. The South Fork Yuba

River at Malakoff Diggins, for example, contains only sparse populations of pikeminnow, hardhead, and suckers, and few rainbow trout; other species that should be found there are not (Gard 1994). Hydraulic mining also sent millions of cubic meters of gravel and debris into the Sacramento River, raising its bottom by as much as 9 m. Not surprisingly, this practice increased flooding of surrounding lands and resulted in a ban on hydraulic mining in 1884. Curiously, the influx of all this material was probably responsible for the astonishingly rapid establishment of striped bass and American shad in the river, because both species produce semibuoyant embryos that seem adapted for silt-laden environments, unlike the embryos of the native fishes, which stick to the bottom or are buried in gravel.

Today many streams are once again attracting gold miners, using suction dredges to extract tiny bits of gold from worked-over river gravels. In most areas, these activities are highly localized and brief in duration, and they seem to have little effect on resident fishes, except where dredgers burrow (illegally) into streambanks (Harvey 1986). Where adult spring-run chinook salmon and summer steelhead hold over summer, dredging can disturb the fish, causing them to swim about and use energy reserves needed for spawning. When they do spawn, redds built on the gravel spoils from dredging are more likely to be scoured during high flows than redds built on undredged gravel areas (Harvey and Lisle 1999). Where dredging activity is common, these fishes tend to disappear, although poaching by dredgers (who usually camp by the streams) may be a major factor as well.

Another well-established mining activity in streams is gravel removal. In low-gradient reaches of large streams, gravel is an abundant, valuable, and even renewable resource, washed in with each flood. Dams, however, reduce or eliminate recruitment of gravel, and modern extraction techniques can remove enormous amounts fairly quickly. Although most gravel mining takes place in summer, when flows are low, it nevertheless can alter streambeds and channels, eliminate fish from the extraction areas, and send silt downstream. In some areas, such as lower Tuolumne and Merced Rivers, gravel extraction has created big pits in the channel, which remain because dams upstream eliminated most floods and gravel recruitment. These pits are inhabited by largemouth bass, smallmouth bass, channel catfish, and other alien fishes, which support a local fishery. Unfortunately, they are also major predators on juvenile salmon, which must pass through the pits on their way downstream. One study estimated that 67 percent of juvenile salmon passing through the lower Tuolumne were consumed by such predators (EA Engineering, Science, & Technology 1990, unpubl. study).

A major long-term consequence of hard rock mining is the leaching of heavy metals and acidic water from abandoned mines; these substances become a permanent source of pollution in streams. The Coast Range, for example, is riddled with mercury mines from the 19th century, which continue to leach toxic metals into creeks, contaminating the fish and food webs of which they are part. In Clear Lake, Lake County, spoils from the Sulphur Bank Mine rest on the shore and are a major source of mercury in the lake. Concern over its potential effects on human health and on the Clear Lake ecosystem were significant enough for the mine to became a USEPA Superfund site in 1991 (Webber and Suchanek 1998).

Watershed Changes

The reduction or alteration of stream fish faunas rarely has a single cause. Often it is hard to identify exactly why a stream once rich in life has become relatively barren. The causes are usually rooted in long-term, multiple abuses of the entire watershed: too much grazing by livestock, removal of trees by logging, road building on unstable slopes, poorly regulated mining, heavy use by off-road vehicles, urban development, dams and diversions, and so on. Coastal drainages of southern California contain many streams degraded by debris torrents. These are semiliquid landslides that rush down mountain watercourses following heavy rains on lands that have been destabilized by multiple factors and from which much of the vegetation has been removed by intense fires (also of human origin). To a certain extent such torrents are natural, but their frequency has undoubtedly increased with increased human abuse of the land.

In the San Francisco Bay area, the multiple effects of urbanization have drastically changed both stream habitats and the fish fauna (Leidy and Fiedler 1985). At upper elevations of the streams, where watersheds are protected for water supply purposes, native fishes predominate in well-shaded streams with high water quality. At low elevations, streams are often confined to concrete channels or are unshaded, silt-bottomed ditches containing polluted water. Such habitats are dominated by alien species.

In northern California, coastal streams, such as the Eel and Trinity Rivers, are still recovering from the disastrous floods of 1955 and 1964. These floods resulted from extraordinarily heavy winter rains that ran quickly off landscapes that naturally do not retain much water. The natural tendencies to shed water quickly and erode were accentuated by years of overgrazing, poor logging practices, and road building on unstable slopes. The result was massive landslides, which filled streambeds and pools with loose gravels throughout the drainages. Enormous flows greatly widened stream channels and eliminated most riparian vegetation. Habitat for anadromous fish was greatly reduced when sections of stream subsequently became too warm

and shallow for juveniles during the summer. Most holding habitat for adult spring-run chinook and summer steelhead was eliminated. In South Fork Trinity River, the spring run of chinook salmon abruptly decreased from around 11,000 fish to 0–350 fish (Campbell and Moyle 1991). Deep pools in these drainages are gradually being scoured out again, but because land management practices have not changed much, devastating floods can be expected again.

The fact that fish declines are tied to multiple and cumulative abuses of the land and water has encouraged a growing watershed protection movement. Increasingly, agencies such as USEPA and CDFG are working with watershed-based citizen groups to solve problems, as those living within watersheds come to recognize that protection and restoration of watershed processes are in their own best interest. The symbol of a healthy watershed is often the return of native fishes—especially spectacular forms like coho and chinook salmon (Moyle et al. 1998; House 1999).

Pollution

One of the sad realities of California is that water not used directly for one purpose or another is likely to be polluted to some degree. Pollution is especially hard on the native fishes. In foothill streams of the San Joaquin Valley, most native fishes are able to live only in clear, unpolluted sections. The exception is California roach, which can live in large numbers in streams polluted with effluent from small-town sewage disposal systems. Fish kills from various types of pollution are common:

- In 1971, fishes inhabiting the lower Pajaro River, including a run of steelhead, were virtually wiped out by failure of the sewage treatment plant at Watsonville, which released large amounts of raw sewage.

- Three years earlier, a similar kill took place in the Pajaro when a farmer washed his crop-spraying gear in the river, releasing highly toxic pesticides (Lollock 1968). This disaster apparently was responsible for eliminating the last tule perch living in Monterey Bay drainage streams.

- Bury (1972) recorded a kill of more than 2,500 Pacific lampreys, rainbow trout, Klamath smallscale suckers, and speckled dace in a small stream in Trinity County, due to a spill of 2,000 gallons of diesel oil.

- A kill of several hundred rainbow trout in Mill Creek, Mendocino County, occurred in August 1973, when an airplane carrying a load of fire-retardant chemicals and clay accidentally dumped the load into the stream rather than on a small wildfire burning nearby (H. W. Li, pers. comm.).

- The biggest fish kill in recent years was the 1991 Cantera spill on the upper Sacramento River, where a railroad tank car of soil fumigant plunged into the Sacramento River, spilling its highly toxic cargo. About 65 km of high-quality trout water was denuded of its animal life.

Fish kills such as these, with a variety of causes, can, if repeated in one stream system, permanently alter the nature of its fish fauna. Streams do have remarkable powers of recovery from spills of toxic materials—provided the material is not persistent and the spills are not chronic (Payne and Associates 1998). However, "rapid" recovery of a fishery may take several years, and such a long interval can be devastating to a local economy dependent upon the fishery.

Although direct fish kills by pollution are common, more significant to fish populations are chronic, nonlethal forms of pollution that decrease growth, inhibit reproduction, or prevent migration. Laboratory studies of persistent pesticides, such as DDT, have shown that low levels can have such effects on salmon and trout, but the subtle nature of the effects usually makes it difficult to link the decline of a fish population to pesticide levels. Thus an increase in pesticide levels from rice paddies draining into the Sacramento River during the 1980s was, according to laboratory toxicity studies, enough to account for the continuing decline of striped bass populations (Bailey et al. 1994). Larval striped bass are sensitive to the rice pesticides, which were present in the water, and many of them showed deformed livers, indicative of toxicity (Bennett et al. 1995). However, when pesticide levels dropped owing to a change in agricultural practices, the bass did not recover, suggesting that multiple factors were suppressing the bass population.

Unfortunately, some of the biggest pollution-related disasters may be yet to come, thanks to pollutants from toxic waste sites. Particularly worrisome is Iron Mountain Mine on Spring Creek, a tributary to the Sacramento River. Water leaching from this mine is highly acid and laced with heavy metals, including copper, zinc, and cadmium. Large amounts are retained behind an earthen dam, from which the water is allowed to trickle into the river. If the dam should fail or be overwhelmed by flood, an enormous kill of Sacramento River fishes, including salmon and steelhead, would almost certainly result.

Alien Species

The introduction of alien species into California was inevitable, both because Europeans have seldom been satisfied with the flora and fauna native to newly settled areas (Crosby 1986) and because a fundamental Western value seems to be that nature can always be improved upon. To a certain extent, introductions of fishes may also have been necessary, because so many aquatic habitats altered or created in the past 150 years are poorly suited for native fishes. The 51 alien freshwater fishes of California have a worldwide origin, although most of them (36) are from other parts of North America (Table 10). There are four species from other parts of western North America, four species from Africa, three species from Europe, and seven species from eastern Asia.

The first official introduction into California was made in 1871, when American shad were carried across country on the newly completed transcontinental railroad. The next decade brought a spate of introductions from the East Coast, carried in special railroad cars, the largest of their day, specially built to transport fish. On return trips from California, the cars usually carried rainbow trout and chinook salmon from the McCloud River and other localities. Most introductions were sponsored by the U.S. Fish Commission and its state counterpart, the California Fish Commission, with help from groups such as the California Acclimatization Society and entrepreneurs such as Julius Poppe, who brought in common carp (Dill and Cordone 1997). Members of these organizations were convinced that California fisheries would be greatly improved with the introduction of "superior" nonnative fishes. In the 1870s, 11 species were successfully introduced, and many other introductions failed. In following decades, there was a steady stream of official and unofficial introductions into the state, with a peak (13 species) in the 1960s. However, introductions have increasingly been deliberate, unauthorized actions or by-products of other human activity, mainly trade. CDFG has not authorized any since 1972, except for the use of sterile, triploid grass carp for weed control in canals of the Coachella Valley. CDFG did give a permit in 1982 to a Lassen County rancher, allowing him to raise Mozambique tilapia in High Rock Spring. Technically, this was not an introduction because tilapia were already present in southern California waters. Yet the result was extinction of tui chub and speckled dace endemic to the spring.

Illegal introduction of fishes—not only bringing in new species but also transferring already established species to new localities—is a growing problem in the state. Thus white bass were moved by anglers to Kaweah Reservoir in the San Joaquin drainage from Nacimiento Reservoir on the coast. Because of the potential of this predator to devastate populations of salmon and other fishes, several million dollars were spent on its eradication (N. Villa, CDFG, pers. comm.). A similar operation was necessary to eradicate northern pike from Frenchman Reservoir on the Feather River. Soon after this eradication effort, pike appeared in Davis Reservoir (1994), on another Feather River tributary. In 1997 the reservoir was poisoned with rotenone, in an enormously costly and contentious procedure, but the pike reappeared in 1999. Present plans are to contain the pike within Davis Reservoir

Table 10

Alien Species Established in California

Species	Year of introduction	Origin	Principal reason for introduction	Present status
Goldfish	1860s(?)	Japan	Ornamental	IID
American shad	1871	E USA	Food	IIC
Brook trout	1871 or 1872	E USA	Sport	IID
Common carp	1872	Europe	Food	IID
Brown bullhead	1874	M USA	Food	IID
White catfish	1874	E USA	Food	IID
Smallmouth bass	1874	M USA	Sport/food	IID
Striped bass	1879	E USA	Food/sport	IIC
Lake trout	1889(?)	M USA	Sport/food	IIC
Yellow perch	1891	M USA	Sport/food	IIC
Channel catfish	1891(?)	M USA	Food/sport	IID
Golden shiner	1891(?)	E USA	Forage	IIE
Warmouth	1891(?)	M USA	Sport/food	IIC
Largemouth bass	1891 or 1895	M USA	Sport/food	IID
Black crappie	1891 or 1908	M USA	Sport/food	IID
White crappie	1891 or 1908	M USA	Sport/food	IID
Green sunfish	1891 or 1908	M USA	Mistake	IID
Brown trout	1893	Europe	Sport	IID
Arctic grayling	1906 and 1970	M USA	Sport	IIA
Bluegill	1908	M USA	Sport	IID
Tench	1922	Europe	Food	IIB
Western mosquitofish	1922	E USA	Insect control	IIE
Spotted bass	1936	SE USA	Sport	IIE
Black bullhead	1930s	E USA	Sport/food	IID
Kokanee	1941	W Canada	Sport	IID
Yellow bullhead	Ca. 1940	E USA	Sport/food	IIC
Redear sunfish	Ca. 1950 and 1954	SE USA	Sport	IID
Red shiner	Ca. 1950	M USA	Bait	IIE
Bigscale logperch	1953	SW USA	Hitchhiker	IID
Fathead minnow	1953(?)	M USA	Forage/bait	IIE
Threadfin shad	1954	SE USA	Forage	IID
Rainwater killifish	1950s	E USA	Hitchhiker	IIC
Wakasagi	1959	Japan	Forage	IIE
Blue tilapia	Early 1960s	Africa	Aquaculture	IIC
Nile tilapia	Early 1960s	Africa	Aquaculture	IIC
Mozambique tilapia	Early 1960s	Africa	Aquaculture	IIE
Redeye bass	1962	SE USA	Sport	IIC
Flathead catfish	1962(?)	SE USA	Sport	IID
Yellowfin goby	Early 1960s	E Asia	Ballast water	IIE
Sailfin molly	Early 1960s	SE USA	Ornamental	IIC
Shortfin molly	Early 1960s	Mexico	Ornamental	IIB
White bass	1965	E USA	Sport	IIE
Redbelly tilapia	Late 1960s	Africa	Weed control	IIC
Inland silverside	1967	SE USA	Insect control	IIE
Oriental weatherfish	Late 1960s	E Asia	Ornamental	IIB
Blue catfish	1969	M USA	Sport	IIC
Porthole livebearer	Early 1970s	Mexico	Ornamental	IIB
Shimofuri goby	Ca. 1980	Japan	Ballast water	IIE
Grass carp	1985	E Asia	Weed control	IIB
Northern pike	1980s	M USA	Sport	IIB
Shokihaze goby	Ca. 1995	Japan	Ballast water	IIB

Source: Based on Dill and Cordone (1997).

Notes: The list is in chronological order. Source codes: E, eastern; M, Midwestern; SE, southeastern; W, western. Status codes are defined in the Preface; A, recently extirpated; E, abundant and invading new areas.

rather than to try to eliminate it (CDFG 2000). It is too late for an eradication program for Sacramento pikeminnow and California roach, which were introduced, probably by anglers, into the Eel River drainage, where they have major effects on native fishes (Brown and Moyle 1996).

Increasingly, fishes are being introduced into new areas by aqueducts that bridge drainages. The aqueduct connecting the Owens Valley to the Los Angeles basin has transferred Owens suckers to the Santa Clara River, where they have hybridized with Santa Ana suckers. The California aqueduct, which takes water from the Sacramento–San Joaquin Delta, has successfully transported a wide variety of fishes to southern California, including native species such as tule perch, hitch, blackfish, and prickly sculpin. The aqueduct has also contributed to the rapid spread of alien species. For example, the inland silverside was introduced into Clear Lake in 1967 and was present in southern California by 1984 (Fig. 15). The spread of silversides was enhanced by anglers who moved them to numerous reservoirs on the unproven assumption that they are good forage fish for bass.

Despite the importance of water projects in distributing fish across California, most species have been introduced

deliberately because of American perceptions that the native fish fauna is inadequate to satisfy the needs of a growing state. This perception was dominant during the late 19th and early 20th centuries despite the abundance of salmon, trout, and large cyprinids, all of which were harvested in large numbers. It is still a common attitude among anglers, although increasingly uncommon among fisheries biologists. Reasons given for introducing fish fall into the following categories: *(1)* improving fishing by introducing new and better species; *(2)* improving fishing by improving the forage base for harvested species; *(3)* providing bait for anglers; *(4)* providing biological control of aquatic pests; *(5)* providing better species for aquaculture; and *(6)* providing homes for pet fish. In addition, a number of small species have been transported into the state as a by-product of other human activities, such as dumping of ballast water. By-product introductions, however, must now be regarded as deliberate introductions because the industries and individuals involved have no excuse for not knowing their activities may be bringing in new species. These fishes are best regarded as a form of pollution, discharged into the environment.

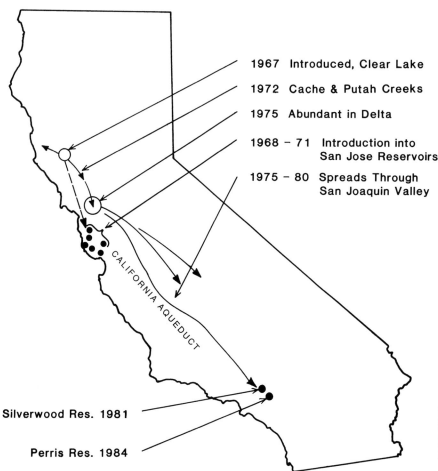

1967 Introduced, Clear Lake

1972 Cache & Putah Creeks

1975 Abundant in Delta

1968 – 71 Introduction into San Jose Reservoirs

1975 – 80 Spreads Through San Joaquin Valley

CALIFORNIA AQUEDUCT

Silverwood Res. 1981

Perris Res. 1984

Figure 15. Spread of inland silverside from its site of first introduction (Clear Lake) to southern California, 1967–1984.

Fishing

Most of the deliberate introductions into California were meant to improve sport and commercial fishing and to provide cheap food for the people of the state. One of the most successful introductions of this type was common carp, which was considered in the late 1800s to be superior in both sporting and culinary qualities to most other fish (Moyle 1984). It is curious that this fish was introduced into waters already supporting large numbers of native, carplike fishes, just as it is curious that brook trout, lake trout, brown trout, kokanee, and grayling were introduced into a state with perhaps the most diverse salmonid fauna in North America. More understandable were the introductions of catfishes, basses, and sunfishes, which now form the backbone of California's warmwater fisheries, because native cyprinids were simply not acceptable to Euro-American anglers. The only widely accepted warmwater game fish native to California is Sacramento perch, which declined quickly during the 20th century. Unfortunately, some anglers still consider bringing in new fish to be a good way to improve fishing. This misconception has resulted in the disastrous introduction of northern pike and the spread of other nonnative fishes to new waters, such as white bass to Kaweah and Pine Flat Reservoirs and yellow perch to Lafayette Reservoir.

An area in which fish introductions have had major—but until recently largely unnoticed—impacts has been trout introductions into high-elevation lakes and streams. With a few exceptions (e.g., Lake Tahoe, the upper Kern River), waters at elevations over 2,000 m were originally fishless, including over 4,000 lakes in the Sierra Nevada (Fig. 14). Thanks to continuous introduction programs from the 19th century (by coffeepot and horseback) to the present (by airplane), trout are now abundant in all alpine areas, radically changing the ecology of lakes and streams (Moyle and Randall 1998). The most conspicuous result has been the decline of amphibians such as mountain yellow legged frog (*Rana muscosa)* and Yosemite toad (*Bufo canorus*) that depend on deep lakes for overwintering; they presumably are eaten when they venture too far from shore.

Forage

The results of introducing game fishes have often been disappointing, especially in reservoirs. In many instances, the disappointed fisheries managers concluded that growth and survival of the game fishes would be improved if more food was provided. Additional fishes were therefore introduced as forage. These fishes have generally been small zooplankton feeders (such as threadfin shad, wakasagi, and inland silverside), although native fishes (such as tui chub, hitch, and threespine stickleback) have also been tried. Their success

in improving fisheries has been mixed, and in some cases forage fishes may actually decrease the growth and survival of young game fishes by competing with them for food.

Bait

Golden shiner, red shiner, and fathead minnow are the only legal bait fishes in California, and they have become widespread as the result of repeated introductions by irresponsible anglers who dump their leftover minnows into whatever water they are fishing. Golden shiners are especially successful, and their establishment in small lakes often leads to the decline of trout and other species, because of the shiners' tendency to reduce the amount of zooplankton and other available food. Various native minnows (such as California roach, hitch, and tui chub) have also become established in waters outside their native ranges, presumably as a result of illegal use as bait.

Biological Control

Western mosquitofish and, more recently, inland silverside were introduced to help control mosquitoes and gnats by feeding on the larvae. The success of both species in insect control is a subject for debate, although careful use of mosquitofish in rice paddies as well as in urban ponds and ditches has proven to be an acceptable alternative to insecticides. Mosquito control is likewise one reason given for introducing the Mozambique mouthbrooder, although it has also been justified as a sport fish, weed control agent, aquarium fish, and aquaculture species. The weakest of these reasons is probably weed control, and as a result other fishes (mainly other tilapia species and grass carp) have been introduced to check aquatic weeds in ponds and canals. Because aquatic plants that are weeds in one body of water can be essential habitat for fish in another, there is considerable concern over the introduction of fishes, especially grass carp, for weed control. In California so far, the only grass carp permitted are sterile triploids in the Coachella Valley.

Aquaculture

Aquaculture is a growing industry in California, and there are frequent proposals to bring in new species for culture purposes or to move species already present to new areas. The main fishes raised in artificial systems in California are channel catfish, striped bass (or striped bass–white bass hybrids), and rainbow trout, although golden shiners, fathead minnows, and red shiners are raised for bait, and goldfish, koi, and various tropical fishes are reared for the aquarium industry. The problem with fish farms is that they leak fish; invariably whatever species is being grown escapes into local waters. This is the most likely method by which Mozambique tilapia, sailfin molly, and porthole livebearer became established in southern California and blue catfish became established in the Central Valley. As mentioned, the extinction of High Rock Spring tui chub and speckled dace in 1989 seems to have resulted from the establishment of tilapia in the spring—escapees from an aquaculture operation.

Pets

Owners of aquarium fishes who have tired of their charges and released them in (or flushed them into) the nearest lake or stream are probably responsible for most wild goldfish populations and for guppies that frequent sewage treatment plants. The single records of a number of tropical fishes from waters around the state are also the result of such introductions. These fishes rarely survive for long, either killed by unfavorable environmental conditions or eaten by predators. However, "escaped" pets may occasionally survive for long periods of time; an example is the 1.3-m-long alligator gar caught in the Delta in 1991. Releases of pet fish, such as sailfin mollies, into warm desert springs where they can survive have posed a major problem for desert fish conservation, because the alien fishes compete with or prey on native fishes and eat endemic invertebrates as well.

By-product Introductions

At least five species of fish and numerous invertebrates have been introduced into the state as by-products of human activity, and more can be expected. Bigscale logperch came in with a shipment of largemouth bass. Rainwater killifish first probably arrived as eggs on oyster shells. Yellowfin and shimofuri gobies apparently were flushed into estuaries with ballast water from cargo ships, as were numerous invertebrates. Rapid transport of organisms around the world in clean ballast water is a growing problem, resulting in major changes in estuarine and bay ecosystems. The problem is particularly acute in San Francisco Bay, where a new species becomes established, on average, once every 12 weeks; the Bay has been called the "most invaded estuary in the world" (Cohen and Carlton 1998, p. 555). Within the state, the transport of fishes by canals is also a type of by-product introduction, with far-reaching consequences.

Impacts of Alien Fishes

Alien fishes have radically changed the nature of California's fish fauna because they are the most abundant fishes in many waterways. Nevertheless, the invaders have been only partially responsible for reduction of the native fish fauna. By and large, alien species are most abundant in aquatic habitats modified by human activity, whereas native fishes persist in undisturbed areas. In the San Joaquin River system, for example, the aggressive, predatory green sunfish is widely distributed in foothill streams. In undisturbed regions they occur only as scattered large adults, while native minnows remain abundant. If a stream section is dammed, bulldozed, or otherwise changed, the sunfishes quickly take over and native fishes become uncommon (Moyle and Nichols 1974). In Deer Creek, Tehama County, introduced warmwater fishes dominate on the Sacramento Valley floor, where the channel has been altered and water diverted, and in a section of heavily grazed and fished meadow where introduced brown trout dominate. In other reaches of the stream, native fishes predominate (Fig. 5) and apparently actively "resist" the invasions of nonnative species (Baltz and Moyle 1993). The most abundant fishes in most reservoirs are aliens, even though the streams feeding them may be dominated by natives. All these disturbed habitats would contain native species if alien species were absent, indicating that biotic interactions between the two groups in altered habitats favor introduced species. These interactions include (1) competition, (2) predation, (3) habitat interference, (4) disease, and (5) hybridization.

Competition

Competition between two species for a resource (usually food or space) in limited supply, which results in one species being eliminated, is frequently invoked as a cause for faunal changes. Yet it is in fact very difficult to demonstrate. If an alien species can survive in an undisturbed environment, it is likely to reach some sort of population equilibrium with species already present, reducing populations of the native fishes but not eliminating them. Thus the introduction of golden shiners into a California trout lake usually results in decreased growth and reproduction of the trout population, but the trout seldom disappear altogether. However, native California fish species that seem to have been eliminated from their natural ranges because of competition from introduced species include Sacramento perch and Lahontan cutthroat trout. The disappearance of the perch from the Sacramento–San Joaquin system was gradual (not obviously correlated with environmental changes), yet the species is very successful in a wide variety of ponds, reservoirs, and lakes into which it has been introduced. The common denominator among these waters is the absence of ecologically similar but more aggressive species, particularly black crappie and bluegill. It is likely that competition takes place for nest sites, food, or both. Predation on young-of-year perch may also be involved. Elimination of Lahontan cutthroat trout from its native streams is apparently also due to aggressive competition for space and food from introduced brown, brook, and rainbow trouts, although disease, predation, and hybridization may also have played a role.

Predation

Predation by alien species on native fishes is another mechanism commonly invoked to explain the disappearance of species. In reservoirs, this is the most likely mechanism by which smallmouth and largemouth bass eliminate pikeminnows and hardhead. Before bass are introduced, these cyprinids can be abundant, but once bass are established they gradually disappear, because no young are recruited. In the South Yuba River, young-of-year hardhead are found mainly above a barrier to smallmouth bass invasion, although larger hardhead are common below the barrier. The young cyprinids school in shallow water and are thus extremely vulnerable to bass predation. Predation by green sunfish is probably responsible for local extinctions of California roach, although habitat change may also play a role. The sunfish invade intermittent roach streams, which are ecologically similar to their native Midwestern streams, and become trapped with the roach in summer pools. Under these circumstances they can easily eliminate the roach. In the Eel River, predation by introduced pikeminnow is responsible for major changes in community structure and seems to be a significant factor in depressing chinook salmon populations.

Particularly vulnerable to predation by alien species are larval and early juvenile stages, during the first few days to weeks after hatching. In the Colorado River, natural reproduction of native cyprinids and suckers seems to be largely prevented by the abundance of alien fishes, such as red shiner, in habitats required as nursery areas by larval fish (Minckley 1991a). Likewise, in Putah Creek, recruitment of juvenile fish from larvae seems to occur mainly in reaches where alien fishes are scarce (Marchetti and Moyle 2000).

Habitat Interference

Habitat interference occurs when an alien species changes habitat characteristics by its activities and the change forces native forms to leave or suffer reduced populations. Common carp are the main villains in this category because they root bottoms, digging up aquatic plants and greatly increasing the amount of suspended matter in the water. Fishes (including many game fishes) that require clear water for feeding or breeding may have their populations reduced or eliminated. In California the effect of carp is difficult to assess because they live mostly in disturbed habitats. Habitat alteration continues to be the main objection to the introduction of herbivorous fishes (e.g., grass carp, redbelly

tilapia) into natural waters, because they may eliminate or change the composition of aquatic plant communities important in the life cycles of other fishes.

Disease

Disease is a poorly understood mechanism by which one species can replace another. Alien species, unless they have gone through several generations of quarantine, are likely to bring their diseases and parasites with them. These in turn may kill or weaken native fishes not immune to them. This outcome has been especially noted in salmonids; even moving strains of one species from one place to another can have severe effects on native populations. Disease or parasites are often suspected as causes of fish declines, but are rarely documented, especially in California.

Hybridization

Hybridization between two closely related species or subspecies has been a problem primarily when fish are transferred from one drainage system in California to another. The Mojave tui chub is now an endangered species because it has hybridized in most of its natural range with introduced arroyo chub, and the hybrids are almost identical with pure arroyo chubs (Hubbs and Miller 1943). Results are similar when Lahontan cutthroat trout or golden trout hybridize with introduced coastal rainbow trout.

Hatcheries

Fish hatcheries have long been a solution for maintenance of fisheries in the face of massive water development and heavy exploitation. The basic assumption behind hatcheries is that they can produce fish to replace those lost through human machinations and thereby permit activities to continue that deplete wild populations. Historian Michael Black (1995) has found that salmon and steelhead hatcheries are part of the failed *serialistic policy* of fisheries management agencies, which have tacitly agreed to keep trying to find new technological solutions to the problem of declining fish populations (including better hatcheries, fish ladders, trucking fish around problem areas, and genetic engineering), rather than addressing the root causes. When one policy fails another is tried, until the fish are gone (which, of course, is one solution to the problem). Because salmon and steelhead populations in the state have collapsed despite the presence of many hatcheries, large and small, the value of hatcheries has been questioned. In fact, there is growing recognition that the decline of wild stocks of salmon and steelhead, or their failure to recover from decline, may be partially due to the negative effects of hatchery-reared fish on wild fish and fisheries. This section deals mainly with the problems created by hatcheries for anadromous fish.

The ways in which hatchery fish and wild fish interact are complex, and negative effects of hatchery fish on wild fish are not always intuitively obvious; this may explain why it has taken so long to figure them out. The effects of hatchery fish on wild fish can be divided into ten categories: *(1)* genetic effects, *(2)* spawning interference, *(3)* spread of disease or parasites, *(4)* juvenile predation, *(5)* juvenile competition, *(6)* life history effects, *(7)* oceanic effects, *(8)* harvest effects, *(9)* other management effects, and *(10)* changes in public attitudes. These factors rarely operate independently of one another or in the absence of other outside effects.

Genetic Effects

Genetic effects are generally divided in turn into *(1)* direct effects of hatchery fish on wild fish, *(2)* indirect effects of hatchery fish on wild fish, and *(3)* genetic effects of hatcheries on hatchery fish (Waples 1991b). The *direct genetic effects* of hatchery fish are mainly the result of interbreeding and introgression with wild fish. These effects are still not as well understood as they need to be, but there is good reason to think that the genetic distinctiveness of local wild stocks or runs may be lost when there is massive intrusion of hatchery fish. Indeed NMFS refused to list coho salmon from the lower Columbia River as a threatened species because of evidence of extensive introgression of domestic and wild stocks. For wild fish genetic distinctiveness is presumed to reflect local adaptation (Taylor 1991), which is important for long-term survival of populations. Hatchery populations may be either less diverse genetically than local wild populations (because of hatchery practices) or more diverse (because of the use of fish from outside sources). In either case, an artificially changed genetic makeup of local stocks may make it harder for them to adapt to a changing environment, an important characteristic in an era of climate change. For example, alteration of genetic material that "programs" juvenile coho salmon to emerge a few days or weeks later than is optimal for a system could potentially greatly decrease survival rates. Such problems are likely to be especially severe when natural populations are already low. It is important to recognize, however, that local adaptation may not be as precise as it is sometimes made out to be and that regional adaptations with considerable variation are probably the norm. Indeed these are partly the basis for the Evolutionarily Significant Unit (ESU), the preferred currency of salmonid conservation. An ESU is a geographic group of populations that share common genetic, life history, ecological, and other traits and that seem to be on a common evolutionary trajectory (Waples 1991a,b).

Member populations (often runs in different streams) are assumed to be more likely to interbreed or interact with other populations within an ESU than with neighboring populations outside the ESU. If climatic and geologic conditions were stable for a long enough period, each ESU would presumably become a classic biological species.

One method adopted to maintain the genetic distinctiveness of local stocks is to use streamside hatcheries that spawn only local fish. Unfortunately, if survival rates to adulthood in the hatchery-reared fish are lower than those for wild fish and the wild fish population is small, the hatchery may wind up becoming a sink for wild fish, resulting in decreased spawning in the wild. This has happened in a number of instances in Idaho and Oregon and no doubt also in California, especially with coho salmon. Once the populations become low, of course, there is the added temptation to bring in outside fish to meet production quotas.

The above scenario might be best regarded as an *indirect genetic effect* because any factor that reduces population size in wild fish creates the danger of reducing genetic diversity within the population. Other problems discussed here—such as competition, predation, and disease—have the added complication of reducing genetic diversity when they reduce effective population size (the number of wild spawners) to extremely low levels.

Genetic changes in hatchery stocks are important to understand because they affect the nature of the interactions with wild fish. Hatchery workers and the hatchery environment select for fish that are adapted for survival in hatcheries; five to seven generations of hatchery rearing are usually enough to cause major changes in the ability of a fish species to survive in the wild. Despite the lower ability of hatchery fish to survive (and, if they do survive, their poor ability to compete with wild fish while spawning), their sheer numbers can overwhelm even strong differences in fitness between hatchery and wild stocks. There is certainly a greater awareness than ever before of the genetic changes that hatcheries wreak on salmon and steelhead, and more and more effort is being made to use breeding techniques that maximize genetic diversity. Nevertheless, the selective pressures in a hatchery are always going to be different from those in the wild, and the results of these differences will manifest themselves in the behavior and survival of fish that are released.

Spawning Interference

Fish of hatchery origin that come to natural streams to spawn compete with wild fish for mates or for spawning sites. Fleming and Gross (1994) indicate that coho salmon of hatchery origin may have much lower spawning success and embryo survival than wild fish in the same stream. Hatchery males are generally less aggressive and less successful at gaining mates than wild males. Nevertheless, hatchery fish, especially if they make up a high percentage of the spawners, can disrupt the breeding systems of wild fish through their activities, depressing production of wild fish. The net result is an overall decrease in production.

Spread of Disease or Parasites

The crowded conditions in which hatchery fish live make them exceptionally vulnerable to epidemics of diseases and parasites, which may spread to wild populations. Use in hatcheries of fish from outside a region may introduce new diseases, as has happened with the spread of whirling disease among trout populations in the western United States. Hatchery fish selected for disease resistance may carry disease into the wild to infect wild fish that are not resistant. The spread of disease through hatchery effluent or from aquaculture operations (especially salmon net-pen operations) is always a possibility, no matter how "clean" a hatchery operation seems to be.

Juvenile Predation

Hatchery juveniles released into streams may cause predation mortality of wild fish to increase directly or indirectly. Juvenile salmon and steelhead released from hatcheries are typically larger that their wild counterparts and may therefore prey directly on wild fish in streams. For example, Sholes and Hallock (1979) monitored the release of 532,000 yearling chinook salmon in the Feather River and estimated that they consumed perhaps 7.5 million smaller wild fish. More indirectly, the presence of large numbers of hatchery juveniles in a stream or estuary may also help to sustain large populations of other predators (such as striped bass, rainbow trout, or pikeminow), resulting in increased predation on wild juveniles. This effect may be particularly important for salmonids that spend a year or more in fresh water before going to sea. It is worth noting that juveniles of hatchery origin are generally more vulnerable to predation in the wild than their wild counterparts, so successful hatchery operations depend on releases either of huge numbers of small juveniles or of juveniles of large size in order to sustain fisheries. Recent studies in British Columbia, for example, have indicated that mortality rates of wild juvenile salmonids greatly increased once large numbers of hatchery smolts were released; the principal cause of the increased mortality was the large numbers of small sharks attracted to the estuary by concentrations of naive hatchery fish.

Juvenile Competition

Juvenile hatchery salmon and steelhead released into a stream may compete with wild fish for food and space and disrupt social hierarchies in wild fish. The closer a stream or estuarine rearing area is to carrying capacity, the more likely hatchery fish are to have a negative effect. They may displace wild fish to areas where they are more vulnerable to predation or force them to emigrate at smaller sizes than they would normally.

Life History Effects

Hatcheries often select for particular phenotypes (e.g., early spawning) or have practices (e.g., timing of release of juveniles) that change the life history traits of local wild populations as the result of interactions between wild and hatchery fish. In New Zealand, there is evidence that repeated releases of large numbers of hatchery chinook salmon (of California origin!) into a stream caused wild populations to shift from a stream-type life history strategy to an ocean-type life history strategy, with potentially lower survival. Essentially, the flood of hatchery fish into the stream, and the resultant low survival of fish that stayed in the stream and had to compete with them, selected for juveniles of wild fish that went out to sea at a younger age.

Populations with a strong hatchery influence may also produce more small jack males than those without such influence, although the reason for this may be related more to heavy size-selective fishing on hatchery stocks than to any other factor. Given that being a jack male is an evolutionarily viable alternative life history strategy in salmon, and that jack males are usually not spawned in hatcheries, selection should be in the opposite direction. A related problem is that wild populations of salmon often contain runs or subpopulations with different life history strategies. Hatcheries typically focus on the run with the life history strategy that is easiest to rear in a hatchery. In California, hatcheries have long focused on fall-run chinook salmon because of their comparative ease of culture—perhaps at the expense of other runs. For example, the Feather River Hatchery has supposedly been rearing both spring- and fall-run chinook, but hatchery practices have pretty much allowed the two runs to merge, to the point that they are no longer truly distinguishable in the Feather River (Yoshiyama et al. 1998).

Oceanic Effects

Ocean conditions seem to affect the growth and survival of hatchery and wild fish in similar ways, although survival at any given size or age is usually lower in hatchery fish. However, it is possible that, during times of low ocean productivity, competition for limited resources by large numbers of hatchery fish may further reduce growth and survival of wild populations, especially those whose levels are already depressed (e.g., endangered stocks).

Harvest Effects

Salmon of hatchery origin can sustain much higher harvest rates than those of wild origin, so high harvest rates in mixed-stock fisheries can result in further depression of depleted natural stocks. The presence of large numbers of hatchery fish can create a demand for a fishery in order to avoid the "waste" of fish of hatchery origin, making it difficult to manage mixed-stock fisheries to sustain wild populations of salmon and steelhead. This may be what is happening in rivers of the Central Valley today, now that harvest restrictions, intended to protect endangered stocks, are returning large numbers of fall-run chinook of presumed hatchery origin to rivers and streams. Small streams that have not seen salmon for decades have suddenly produced spawners, and this seems to be a positive development. However, the potential exists for hatchery fish to overwhelm remaining wild stocks in the rivers. This is regarded as a major problem in Norway, where Atlantic salmon escaping from aquaculture operations are entering spawning streams in large numbers to compete with native strains for spawning sites.

Other Management Effects

Because of their availability in large numbers, fry and smolts from hatcheries are often used as the principal experimental animals to assess emigration and survival rates in response to regulated flows or other manipulations of regulated streams. Management recommendations based on these studies may not be suitable for wild fish and may thereby cause further declines. This outcome is currently a major subject of discussion in the management of outflows in San Joaquin River tributaries and for Delta outflows on the San Joaquin side, where all studies of smolt survival have been carried out with hatchery fish. An additional complication is that many of the fish used have come from a hatchery (Nimbus) on the Sacramento side, and some of these fish have later returned to spawn in the San Joaquin tributaries. This situation further complicates efforts to save native San Joaquin strains of chinook salmon (if any still exist).

Changes in Public Attitudes

The presence of hatcheries can be a deterrent to restoration of self-sustaining populations of salmon and steelhead because voters often view hatcheries as permanent solutions for saving them (Black 1995). Their presence has reduced the likelihood that expensive alternative solutions, such as habitat restoration and the removal of dams, will be instituted. This is still a problem (although less so than was formerly the case). Thus there is a major ongoing program to artificially rear striped bass to plant in the Delta, fueled by the frustration of anglers over the slowness of ecosystem recovery efforts—even though there is little evidence that the program will actually do any good. In contrast, with salmon and steelhead restoration there is a growing appreciation of the need for watershed conservation as a long-term solution.

Benefits of Hatcheries

Although this discussion has focused on negative aspects of hatcheries, they do have their benefits, if used wisely. Hatcheries that rear domesticated trout to plant in roadside streams, reservoirs, and urban ponds provide angling opportunities that might otherwise be lacking, and they do little damage to wild populations of trout. Such fish, in fact, are designed to be caught by virtue of their genetic background and methods of rearing. Small-scale streamside hatcheries can be a useful tool for rehabilitating runs of anadromous fish depleted by habitat destruction, provided habitat restoration is taking place at the same time. Such hatcheries can become local institutions, increasing awareness of problems and involving local people in conservation efforts (House 1999). The assumption, of course, is that streamside hatcheries will be abandoned once runs are again healthy. Even large salmon and steelhead hatcheries may still have their place for maintaining fisheries, provided all fish released are marked and means are developed to harvest selectively those of hatchery origin. An undeniable benefit of such hatcheries is public education. The large runs of fall-run chinook salmon generated by the Nimbus hatchery in the lower American River, for example, create a public spectacle in an urban area, both in the river and at the hatchery. Such events can be used to create public interest in salmon conservation in general.

Exploitation

Overexploitation of a species always has the potential to drive its populations to very low levels, perhaps even to extinction, especially if other factors are also causing them to decline. One of the most dramatic examples of this tendency in California was the fishery for white sturgeon in the late 19th century, which caused a severe depletion of the population. The fishery was shut down in 1916 and not reopened until 1954. The sturgeon was exceptionally vulnerable to overfishing because of its large size, longevity, and late age of maturity. In recent years, fisheries have probably contributed to the continuing decline of both striped bass and chinook salmon. In the case of striped bass, removal of large females from the population by fishing has reduced the number of eggs produced, during a time when survival of eggs and larvae is low because of diversions and the presence of pollutants. A similar situation has existed for chinook and coho salmon taken by commercial and sport fisheries off the California coast. The fishery maintained a high rate of exploitation of wild salmon populations already stressed by water diversions and degradation of their spawning streams. A major problem has been that larger and older fish are captured in fisheries, so runs consisted mainly of three-year-old fish. If spawning should fail, owing to natural or unnatural conditions, there would be few fish left to return in following years as four- or five-year-olds, which are needed to keep the run viable. Reductions in the salmon fishery in recent years have resulted in a positive response in some populations, especially in chinook salmon, but the lack of recovery of coho salmon demonstrates the importance of other factors in their decline.

Sport fishing and (to a lesser extent) commercial fishing can also be major factors shaping freshwater fish communities. Fishing is highly selective for both species and size of fish. Sport fishing is aimed primarily at large carnivores, whereas freshwater commercial fishing is aimed at large fishes not reserved for sport fishing, such as common carp and Sacramento blackfish. If sport fishing removes a large percentage of fish at the top of a food chain, the population structure of the species making up the lower links is bound to change. In simple systems, such as farm ponds containing only largemouth bass and bluegill, excessive harvesting of top carnivores (bass) may irreversibly change the system, unless fishing imbalances are continuously corrected. Thus the harvesting of large-size largemouth bass from a pond may cause a population explosion among their prey (bluegill). The bluegill in turn may greatly reduce the insect and zooplankton populations needed to support young bass, resulting in fewer bass than before and large numbers of stunted bluegill.

Conclusions

The fish fauna of California is changing rapidly. Streams, lakes, and estuaries that once supported a unique and valuable collection of native fishes are being replaced by canals, ditches, reservoirs, and polluted lagoons that support mainly hardy exotic fishes—often with flesh so laced with toxic residues they are unfit to eat. Rich and self-sustaining

fisheries have been sacrificed in favor of wasteful irrigation practices, urban sprawl, and logging, grazing, and mining practices that degrade the environment rather than sustain it. Fish and fisheries are even sacrificed to recreation, because streams are diverted to water golf courses in the desert, casinos and hotels are built alongside delicate alpine lakes and streams, and hillsides wash into streams after being scarred by road building and off-road vehicles.

For years the extravagant use of California's limited water at the expense of its natural fish populations was justified using a number of rationalizations:

- The native fishes were mostly trash fish, either of no use to humans or, worse, competitors or predators of useful fish.

- Fishing in human-made habitats such as reservoirs was more productive than fishing in natural streams and lakes.

- Fish hatcheries could sustain fisheries for salmon, trout, striped bass, or any other species deemed important enough to rear.

- Modern technology and human ingenuity could fix any problem and even improve upon nature: fish could be screened from diversions, brought over dams with fish ladders, encouraged with artificial reefs of old car bodies and tires, or even genetically engineered to survive water of poor quality.

Unfortunately these rationalizations have not held up well. The promise that fish and fisheries would be maintained in the face of continued water development has not been kept. The problems are exacerbated during long periods of drought, when fish populations are naturally stressed and human competition for limited supplies of water is most intense. Even though water supplies to cities and farms may be drastically cut back, streams and rivers still suffer the most. Fish populations decline and often do not recover well, even when wet years return. The results of such a short-sighted water policy can been seen in the plummeting sales of sport-fishing licenses, the closure of sport and commercial fisheries for salmon and steelhead, the increased number of endangered species, and the rapid rate at which the native fish fauna is being depleted by extinctions.

The following chapter describes how the native fish fauna, and the fisheries it supports, can be restored.

A Conservation Strategy

As the human population of California grows, native fish populations decline, reflecting a general deterioration of aquatic habitats. But this downward trend does not have to continue. In fact, it is reasonable to assume that a California supporting healthy populations of native fishes will be a much healthier state for humans as well—with water safe for drinking and swimming, and fish that are safe to observe and eat.

Because the state's native fishes are most abundant and diverse in relatively healthy environments (Moyle et al. 1998), they can serve as surrogates for most (but by no means all) native aquatic biota in conservation actions. The use of fish as a focus for aquatic conservation is necessary because of the sheer size of California and the enormous diversity of its aquatic environments. Fish also tend to rouse greater public sympathy for conservation actions than do plants, insects, or even amphibians. At present, however, they are not doing very well: more than 70 percent of the native fishes have less than 10 percent of their habitat in waters under some kind of formal protection (Moyle and Williams 1990). For most fishes, "less than 10 percent" means "none." The native fish fauna is in decline because hundreds of local actions, large and small, have degraded unprotected habitat. These actions are so pervasive that change is taking place very rapidly. Consequently, protection of aquatic diversity statewide requires hundreds of localized conservation actions, which will be most effective if they are carried out within the context of a statewide strategy. Otherwise there are likely to be, for example, hundreds of kilometers of trout streams protected but very few kilometers of streams for California roach.

This chapter presents a conservation strategy by discussing *(1)* why it is important to protect native fishes, *(2)* how to prevent future problems, and *(3)* how to protect native fishes statewide.

Why Protect Native Fishes?

Of all California's native fishes, only 11 species, mostly salmonids, contribute to important fisheries today. Another 12 once harvested are now in such low numbers that they no longer have much economic value. Most of the rest are known mainly to ichthyologists and sometimes to fisheries managers (usually as pests, forage, or endangered species). If most of California's native fishes—but especially the rarer species—became extinct tomorrow, no fisheries or ecosystems would collapse due to their absence. So why bother to protect them? Many arguments have been developed at length (e.g., Norton 1987; Moyle and Moyle 1995), but some of the more salient reasons fall into five overlapping categories: *(1)* economics, *(2)* ecosystem protection, *(3)* genetic diversity, *(4)* aesthetics, and *(5)* morality.

Economics

The perception that most native fishes are valueless is narrowly European-American, the product of a culture that seems to regard only boneless fillets of large fish as fit to eat. Native Americans ate most local fishes and especially favored the large cyprinids and suckers (Schulz and Simons 1973; Lindstrom 1996). Asian immigrants found these same fishes similar to species they were accustomed to eating in Asia and thus have a long tradition of harvesting native fishes. Commercial fisheries for Sacramento blackfish harvest thousands of pounds each year for Asian-American markets. In short, the value of many fishes is simply not appreciated, although this view is likely to change in the future given the increased popularity of fish as food for all segments of society. Indeed this is a good reflection of the concept of *safe minimum standard*, which translates in this situation to the idea that we should not let any species go extinct because we cannot predict their economic value in the future.

One reason to expect "worthless" fishes to increase in value is that most conventional sport and commercial fishes are in decline. This is particularly true of anadromous fishes—even fall-run chinook salmon, steelhead, and white sturgeon, which are mainstays of fisheries. Runs of these three fishes, increasingly supported by hatchery production, are remnants of what was once an astonishingly diverse fishery for anadromous fishes: four species of salmon, two species of sea-run trout, three species of smelt, and two species of sturgeon. There were separate fisheries for distinct runs of these species as well, such as the four runs of chinook salmon in the Sacramento and San Joaquin Rivers, or summer and winter steelhead in the Eel River. Each species and each run used riverine resources in a different way, greatly increasing total production of fish. In the 1800s and early 1900s, before anyone was aware of the complexity of California's anadromous fish populations, almost continuous fisheries existed for "salmon," "sturgeon," and "smelt." If one run or species had low returns as a result of natural disaster, other runs or species would not, and fisheries were thus able to remain economically viable.

Today most of these options are gone. Not only is total yield a fraction of what it was, but dependence on a few runs of fish means that fisheries are much more likely to suffer irregular fluctuations between "boom" and "bust" years. In short, restoring and maintaining a diversity of species and runs results in more fish and more stable fisheries.

Another important economic argument is the long-term value of fisheries. In California, fisheries have consistently been sacrificed for mining, logging, grazing, and farming. In the short term, trading off fisheries for these other industries might seem worthwhile, because their annual returns in dollars are enormous compared with the annual values of fisheries. Yet mines are depleted, often becoming toxic waste sites; many logged areas regenerate slowly or not at all; and overgrazed hillsides become gullied. Even irrigated agriculture eventually declines as soils become saline; salinization in many areas is inevitable, whether it takes 5, 50, or 500 years. In contrast, fisheries can go on indefinitely, climate permitting. Anadromous fish keep coming back, year after year, bringing the productivity of the ocean to streams and to human society. In fact, fisheries and other industries that depend on wildlands and water are not necessarily mutually exclusive. But other industries must give more consideration to how their operations affect fisheries *now* and in the future.

Finally, it is worth noting the value of many small species, such as the three smelt species (Delta smelt, longfin smelt, and eulachon). All three were harvested by commercial fisheries in the 19th century, and similar species are still highly valued as food around the world. The eulachon long supported dipnet fisheries and was (and still is, to a limited extent) an important traditional food for Native Americans. A more immediate value of delta and longfin smelts, given their comparatively low populations, is that their requirements are similar to those of other fishes of the San Francisco Estuary, such as striped bass, that have high economic value. Thus protecting smelt may also protect the fisheries for striped bass, shad, and other species, because all require a functioning estuary. In short, protecting obscure fishes can help keep ecosystems functioning—even disturbed ones. This and other more general economic arguments are discussed in Moyle and Moyle (1995) and Moyle and Cech (1999).

Ecosystem Protection

Fishes are the most noticeable components of aquatic ecosystems, and their declines reflect ecosystem deterioration. Protection and restoration of ecosystems are desirable because of the myriad benefits provided by intact aquatic ecosystems, such as clean water, flood control, recreation, fisheries, and spiritual renewal. Thus protecting smelt and splittail can help protect and restore estuarine ecosystems. Protecting southern races of steelhead and Santa Ana suckers provides incentive to restore some of the most degraded streams in California. Protecting coho salmon provides additional protection for old-growth coastal forests. Protecting summer steelhead and spring chinook salmon necessitates protection for the remote canyons in which they spend the summer, as well as for long stretches of stream between the canyons and the ocean. Protecting tidewater gobies protects coastal lagoons. Protection of coastal cutthroat trout provides additional protection for the unique Smith River, as well as other North Coast streams and coastal lagoons.

In short, the health of these species is closely tied to the health of some of the most important aquatic ecosystems in California. Protecting species can therefore provide motivation and symbolism for broad environmental conservation, desirable for the sake of many other species, including humans.

Genetic Diversity

Conservation biologists are increasingly recognizing that protecting genetic diversity within species is important for conserving them. Genetic diversity is needed to enable species to adapt to environmental change, and the adaptiveness represented by genetic diversity can be of immense value to humans. This relationship is especially easy to see in anadromous fishes, which all have their southernmost populations in California. Their populations have adapted to the often harsh conditions that naturally exist here: warm water, fluctuating flows, extended droughts, extreme seasonality of suitable habitats. Such hardy fish were created through thousands of years of evolution, and their genetic heritage cannot be recreated or even maintained in hatcheries. They are valuable not only because they can survive in the increasingly stressed habitats of California but also because they may be needed to help maintain fisheries in more northern areas.

Global warming is occurring so rapidly that most species will not be able to adapt through local genetic changes; they will need genes from populations already adapted to warmer conditions. California fishes are clearly a reservoir of such valuable genetic information; losing populations of these species is thus like throwing out a valuable insurance policy for fisheries in Oregon and Washington as well as California. Wild stocks are also valuable for the growing aquaculture industry, because they contain the genetic information needed to develop strains of fish with disease resistance and other characteristics.

Aesthetics

Among the best reasons for saving species are aesthetic ones. We want them to be around so that we and our descendants can glimpse them in natural settings. Our culture has a particularly strong appreciation for salmon, dating back at least 10,000 years to the time when the first images of salmon appeared on the walls of European caves. Chinese culture has a similar appreciation of carp. The strength and beauty of these fish and their struggle upstream to spawn, in the face of waterfalls, predators, and fishermen, have long been a source of inspiration. A stream packed with spawning salmon is awe inspiring; an encounter with wild salmon or steelhead in a forest stream or remote canyon pool can be an unforgettable experience. Even quiet encounters with species like hardhead and tule perch in a clear, warm, rockbound pool can be fascinating. Hardhead and tule perch have an additional aesthetic consideration: they are species that occur *only* in California—part of a unique fauna that helps define why California is such a special place for humans to live. To understand and appreciate endemic fishes is to understand the dynamic and severe nature of California's environment and to appreciate the evolutionary forces that created its present-day fauna. Such understanding can help us to live *with* the environment rather than constantly trying to control it.

Morality

For centuries the dominant ethic of our society toward wild creatures was, for the most part, if it does not have value to humankind, it can be ignored or destroyed. There is a growing movement to change that basic ethic, a movement rooted in religions of both the East and the West. Books have been written on the subject (e.g., Ehrenfeld 1981; Norton 1987; G. Snyder 1990), but the often beautiful and complex arguments boil down to deep-seated feelings that it is simply wrong to eliminate species and ecosystems from this earth when we have the knowledge and power to prevent their loss.

Prevention

One of the first steps in any conservation strategy is to prevent the development of new problems that are likely to confound other efforts. In general, prevention is best accomplished by applying the *precautionary principle* to new initiatives: do not undertake new actions or policies unless it has been proven they will do no permanent, irreversible harm to aquatic environments. This approach also applies to "new" actions under old policies, such as constructing homes in floodplains. Obviously—given the state's massive urbanization, high demand for stored water, and intense use of agricultural, forest, and range lands—the precautionary approach is difficult to adopt. Yet some actions lend themselves to *immediate* application of this principle better than others. Such immediate actions could include halting invasions, reducing the use of pesticides and other pollutants, adopting sensible land use practices, and improving water distribution and allocation practices.

Halting Invasions

Aquatic ecosystems in California are continually disrupted by invasions of alien species. Expensive habitat restoration efforts can be negated by an invasion, and the costs of recovering endangered species are greatly increased when alien species suppress their populations. Some steps that should be taken to halt new invasions include the following:

- Prevent the discharge from ships of ballast water that contains estuarine or freshwater organisms. At the same time make the shipping industry and port authorities responsible for damage caused by new ballast water invaders. A major step in this direction was a state law passed in 1999: AB 703, the Ballast Water Management for Control of Nonindigenous Species Act.

- Ban the use of live fish as bait in the inland waters of California, especially commercially raised minnows.

- Limit the planting of trout in alpine lakes to reservoirs and lakes within easy walking distance of roads; eradicate fish from selected high-elevation watersheds to permit recovery of amphibians and invertebrates.

- Educate anglers about the dangers and costs of moving fish around; strongly enforce existing laws against unauthorized movement of fishes.

- Set up an interagency Alien Species Response Team with funding and authority to quickly take appropriate action to halt new invasions while they are still controllable.
- Require the aquaculture, aquarium, and horticultural industries to take responsibility for the potentially invasive species they sell by, as appropriate, banning some species, labeling others, making contributions to invasive species control and prevention programs, and providing facilities where people can return unwanted fishes and invertebrates.

Reducing the Use of Pesticides and Other Pollutants

In some respects, the waters of California are cleaner than they were 30 years ago, thanks to the federal Clean Water Acts of 1960, 1965, and 1972 and related state acts. These acts resulted in dramatic reduction of point-source pollution, especially industrial waste and sewage. Unfortunately, heavy metals, pesticides, and other toxic contaminants continue to pour into our waters, mainly from such nonpoint sources as farms, mines, construction sites, logging areas, and urban and suburban drains. The myriad ways to prevent further toxic effects and to reduce the amount and variety of contaminants are covered in many other documents (e.g., Kegley et al. 1999), and it is clearly in our best interests to do so. Healthy fish indicate healthy waters.

Adopting Sensible Land Use Practices

Any human activity on land has the potential to affect water in the streams and lakes into which the land drains. Many of our practices—such as channelization, construction of levees, development on floodplains, destabilization of hillsides through vegetation removal, and ditching and draining of marshlands—cause direct and dramatic changes in the way streams and rivers work, usually to our long-term detriment (Mount 1995). There is a growing realization that "business as usual" in use of the land cannot continue, especially if we value fish, riparian areas, and wetlands. The best signs of this awareness are the citizen-based watershed groups that have sprung up around the state, even for such seemingly lost causes as the Los Angeles River. Such groups need to be nourished, especially with funding, so they can work to improve land use practices. On a bigger scale, the multiagency CALFED organization recognizes that restoration of the San Francisco estuarine ecosystem will require changing land use practices throughout the Central Valley, the Sierras, and the San Francisco Bay region, in part by preventing uses that have been permitted in the past.

Improving Water Distribution and Allocation Practices

Prevention of the wasteful use of water, particularly on agricultural lands, must be an important part of any strategy to protect aquatic ecosystems. Unfortunately California water law, combined with heavy state and federal subsidies of developed water, encourages its extravagant use, for example to flood-irrigate alfalfa during times of drought. Landowners with riparian water rights are allowed to use as much water as they need on their land but are not allowed to sell it, so they have little incentive to conserve. Water from federal and state water projects is typically sold to farms and cities at prices far below the actual costs of storage and delivery (including the costs of dams and other infrastructure). Prevention of water waste will be most effective if there are financial incentives not to waste it, requiring major changes in the way water is valued and allocated. Various proposals exist for reform of water law and water allocation (e.g., water marketing), but none is likely to be instituted until California faces another drought-induced crisis. An additional motivation for reform has been recent mandates requiring federal projects to provide large amounts of water for environmental purposes, beyond minimum downstream flow schedules, such as the 800,000 acre-ft/year required from the federal Central Valley Project for fish and wildlife. How this water should be used is still a matter of controversy, but at least its allocation sets a precedent in acknowledging that prevention of further declines and extinctions of native fishes depends on having sufficient water in the system.

Protection

Although stopping or reducing environmentally destructive practices is important, such action must be combined with active protection of species, faunas, habitats, watersheds, and regions. The proposal put forth in this section (Moyle and Yoshiyama 1994; Moyle 1995) covers five tiers of protection, each offering progressively more protection, but also being more difficult to implement, than the preceding one. The tiers are not mutually exclusive; they are interactive and complementary.

Tier 1: Endangered species. Protect under state and/or federal endangered species acts (ESAs) or other legislation all aquatic taxa likely to be extirpated from California within the next 20–30 years. This includes those native fishes classified as status IB in this book (Table 1).

Tier 2: Species clusters or assemblages. Provide special management for clusters of declining species (including reptiles, amphibians, and invertebrates) that inhabit the same aquatic habitats or watersheds. The cluster could also be a natural assemblage of organisms in which the assemblage is disappearing even if the component species are still fairly common (Moyle et al. 1998).

Tier 3: Habitats. Development and implementation of a system of protected aquatic habitats, called Significant Natural Areas (SNAs), that provides systematic, statewide protection of aquatic biodiversity. Examples of all habitats listed in Moyle and Ellison (1991) should be included.

Tier 4: Watersheds. Develop a statewide system of protected watersheds, called Aquatic Diversity Management Areas (ADMA), to enhance biodiversity through protection of natural processes in complete ecological units. Eventually *all* watersheds in the state should be managed in ways that include some element of protection for aquatic life.

Tier 5. Bioregions. Develop and implement management schemes for multiple watersheds in a region with unifying biological features (bioregion). This approach would involve managing entire landscapes or ecosystems for natural values, recognizing humans as part of the landscape.

Although this strategy has been developed specifically for California, it is applicable to other regions as well, especially in the western United States. The number of tiers could be increased to encompass state, country, continent, and planet, but higher tiers are beyond the scope of this book.

Tier 1: Endangered Species Protection

The federal ESA is one of the strongest environmental laws ever written (Orians 1993). The California state law is weaker but still provides considerable protection for listed species. The power of the two ESAs is being tested continuously in conflicts over water diversion and land use, but resolution of the conflicts has usually resulted in improved habitats for fishes. Examples include the following:

- The listing of delta smelt as threatened in both the state and federal ESAs has been a major factor in motivating disparate interest groups to join together to find ways to restore habitats and natural hydraulics to the San Francisco Estuary. The smelt is endemic to the estuary, from which large quantities of water are diverted southward for agricultural and urban users (Moyle et al. 1992).

- The listing of tidewater goby provides significant protection for coastal lagoons up and down the state and may ultimately provide some protection for watersheds that drain into the lagoons.

- The listing of coho salmon has focused attention on the poor condition of hundreds of coastal watersheds and has been a key factor in the settling (more or less in favor of fish) of a number of disputes over logging and land use practices.

- The listing of various species and subspecies of pupfish has been a key factor in protecting desert spring and stream ecosystems in California and Nevada.

The federal ESA is a powerful tool for conservation of aquatic species. Yet relying on it has several disadvantages:

- The act comes into play only when a species is on the verge of extinction and recovery is likely to be expensive and controversial.

- The uncompromising nature of many of the act's provisions almost automatically leads to confrontation over methods of implementation.

- Measures taken to protect listed species have precedence over measures to protect unlisted species, even though the unlisted species may be in trouble as well.

- Measures to save listed species are likely to focus on "quick fixes" and technological solutions, such as transplants and captive rearing, rather than on ecosystem protection measures.

- Recovery of a species under the ESA means only that it has achieved a population size such that it can reasonably be expected not to go extinct; it does not mean that it has self-sustaining populations that are ecologically significant.

- The number of species qualifying for listing in places such as California generally exceeds the capacity of state and federal agencies to handle the complex listing process for all species, especially if the number is constantly increasing (as it is).

Systematic protection of biodiversity beyond what the ESAs can provide is clearly needed.

Tier 2: Management of Species Clusters or Assemblages

One response to criticisms of the ESAs is to intensively manage groups of declining species that seem to have broadly similar ecological requirements and that co-occur in limited geographic areas. If a number of species are protected simultaneously, the ecosystem of which they are part will also be protected, along with poorly known or less charismatic organisms that also live there. Three basic strategies use the cluster approach to protect threatened ecosystems: *(1)* have multiple species in the cluster listed under the ESA, *(2)* develop a management plan to prevent listing, and *(3)* protect the cluster as a threatened community or assemblage of organisms.

Clusters with Listed Species Moyle et al. (1995) recommended 15 clusters of California fishes for joint management. Each of these clusters contains species that usually co-occur on a regular basis; they include not only species recommended for listing, but also species already listed as threatened or endangered, declining species recommended for "special concern" status, and species not yet in serious trouble but indicative of special habitat conditions. Although their recommendations dealt only with fishes, they also recommended that the clusters be expanded to include other aquatic vertebrates (especially amphibians) and invertebrates. These are situations in which Habitat Conservation Plans, a special tool for dealing with the management of endangered species on private land under the ESA, might be especially appropriate.

Unfortunately, even with the best of intentions, mandated protection of listed species can result in measures that may harm unlisted species. For example, managing the flows of the Sacramento River for endangered winter-run chinook salmon may reduce the amount of water needed to support the other three runs of chinook salmon in the river (all of which are in decline), as well as other native fishes (Moyle et al. 1995). USFWS recognized this dilemma, and, following the 1993 listing of delta smelt as a threatened species, it appointed a Delta Native Fishes Recovery Team (rather than a delta smelt recovery team). The charge to the team was to "address the Delta ecosystem as a whole, considering the declines of other native fishes in addition to delta smelt, . . . [which] may require active management to restore sustainable populations" (M. L. Plenert, USFWS, letter to P. B. Moyle, 31 March 1993). A cluster plan was developed that included delta smelt, longfin smelt, Sacramento splittail, green sturgeon, Sacramento perch, spring-run chinook salmon, San Joaquin fall-run chinook salmon, and late fall-run chinook salmon (USFWS 1996). Even in this situation, however, actions to protect listed species legally have precedence over actions to protect nonlisted species. Thus one justification for listing additional species, such as Sacramento splittail (listed in 1999), is that such listings provide a stronger legal foundation for a multispecies or ecosystem approach to management of the estuary (NHI 1992; Fiedler et al. 1993).

Occasionally clusters of species may be treated together when fishes in the cluster are all threatened by one major factor, even if the species are not all parts of the same ecosystem. For example, the formal listing of all species of fish and endemic invertebrates in the springs and streams of the Amargosa River region (California and Nevada) is recommended by Moyle and Yoshiyama (1992) because most depend on the outflows of springs fed by deep and ancient aquifers. The water in these aquifers is now being mined by local agriculture and is proposed to be mined on a massive scale by the city of Las Vegas (McPhee 1993). Such mining may dry up many, or all, of the spring sources (Moyle et al. 1995).

Clusters of Declining Species Ideally clusters of declining species should be managed together, before any become listed under the ESA. However, the threat of listing clusters of species may be needed to provide motivation to undertake necessary ecosystem recovery efforts. An example of a cooperative arrangement to protect a species cluster is the ongoing restoration of the fishes of Goose Lake, a large alkaline lake that straddles the California-Oregon border. The lake and its tributaries contain four endemic fishes (see the Distribution Patterns chapter). In 1992, after a prolonged drought, Goose Lake dried up. As the lake desiccated, USFWS staff began a status review of the four species, preparatory to recommending their emergency listing as endangered, based on species accounts in Moyle and Yoshiyama (1992) and observations of local biologists (N. Kanim, USFWS, pers. comm. 1993). However, the listing was held in abeyance while the Goose Lake Fishes Working Group (an informal association of regional agency biologists) worked with local landowners, interest groups, university biologists, and representatives of land management agencies to see if alternatives to listing could be found. The cooperation of landowners was essential for protection of the fishes, because most possible refuges were on private land or on public land leased for grazing. The efforts of the working group were successful in demonstrating that *(1)* there was general willingness to cooperate with recovery efforts, *(2)* there were more refuges for the fishes than had been previously supposed, and *(3)* funding was available for stream restoration and other recovery programs (G. M. Sato, BLM, pers. comm. 1993). When the drought ended and the lake refilled, the four Goose Lake fishes quickly recovered, demonstrating that formal listing of them may not be necessary.

Threatened Assemblages Assemblages of species, such as those discussed in the Ecology chapter, represent natural biotic units on which protection efforts can focus. As populations of species and habitats become increasingly fragmented, not only do species become threatened with extinction but the natural assemblages of fishes (and other organisms), with all their interactions, become threatened as well. Thus in Putah Creek, Yolo and Solano Counties, there exists a compressed transitional (foothill-mountain) assemblage of fishes below Putah Diversion Dam, containing eight resident species and three anadromous species. This rich assemblage has become increasingly rare in the Central Valley. When flows below the dam were reduced during a period of drought, extirpation of the assemblage became likely. In deciding a court case brought in an effort to increase flow to the stream, the judge ruled that the integrity of this assemblage, even though it contained no endangered species, was protected under both Section 5937 of the Fish and Game Code (fish must be maintained in "good condition" below a dam) and the Public Trust Doctrine (Moyle et al. 1998). In May 2000, all the parties involved in the lawsuit signed an accord that provided for increased flows down the creek to protect the native fish assemblage.

Tier 3: Habitats

Small areas of unusual or exceptionally pristine habitat have long been protected by public agencies (e.g., the Research Natural Areas of USFS) and private groups (e.g., the preserves of The Nature Conservancy), although the focus of such areas is usually a terrestrial (plant) feature. These are traditional nature preserves. Many of these areas, protected and unprotected, have been catalogued by CDFG as SNAs. I have adopted the term to apply to small aquatic habitats or habitat segments that merit special protection because of their native fauna and flora. Aquatic SNAs are of two basic types: *(1)* small, isolated, and fairly pristine waters and *(2)* segments of streams, often below dams, that are dominated by native fishes or that contain important native elements not protected elsewhere. Examples of the first type include spring systems, small intermittent tributary streams, vernal pools, and small isolated lakes. Because of their size, hence vulnerability, these SNAs need special and nearly complete protection, often including fencing. Protection is likely to include fairly intensive management to keep out invasive species and livestock or to restore populations extirpated through natural processes. The latter approach may be necessary if the SNA is isolated from similar areas that normally would have been a source for natural recolonization. The size of these SNAs, however, also makes them relatively easy to protect. Some examples of potential SNAs of the first type include the following:

- Indian Creek, is a small tributary to a northern California stream. Because of its location it has been relatively inaccessible to livestock that roam the area, and as a consequence it has maintained a lush riparian community. The stream itself contains abundant native fishes (mainly California roach and rainbow trout) as well as large numbers of the increasingly rare foothill yellow-legged frog (*Rana boylei*) and the Pacific pond turtle (*Clemmys marmorata*).

- Crystal Spring (Shasta County) is a large spring area that overflows a lava dike into Hat Creek. The spring area contains a diverse aquatic flora and fauna, including endemic rough and marbled sculpins and Shasta crayfish (*Pascifastacus fortis*). It is privately owned, by a power company.

- Stump Spring (El Dorado County) is a seasonal spring that flows into the Cosumnes River. It contains no fish, but it is one of the few localities known for an endemic genus of stonefly (*Cosumnoperla*), which has larvae that are subterranean for most of the year. It is located in Stanislaus National Forest.

- Six Bit Gulch (Tuolumne County) is the principal reach of the Horton Creek drainage, which contains Red Hills roach, a peculiar but undescribed subspecies (or species) of roach. Although on BLM land, the area is unprotected and is used for off-road vehicle recreation. Horton Creek flows into New Don Pedro Reservoir.

The second type of SNA typically has highly disturbed aquatic habitats above and below it, but circumstances (usually fortuitous and artificially maintained) make the stream segment important for native aquatic organisms. Some examples are the following:

- The McCloud River between McCloud Dam and Shasta Reservoir (Shasta County) is a large, cold river that supports mostly native organisms, including rainbow trout, riffle sculpin, and many invertebrates and amphibians. Even though some key components of the system (chinook salmon, bull trout) are missing, the natural elements remaining merit special protection; much of the river is in a TNC preserve.

- Putah Creek below Putah Diversion Dam (Yolo and Solano Counties) depends on flow releases to maintain its habitats, yet it manages to support a remarkably diverse native fish fauna, including tule perch and small runs of chinook salmon, Pacific lamprey, and steelhead.

- Big Tujunga Creek below Big Tujunga Dam (Los Angeles County) is the only place left in the Los Angeles River watershed that supports native fishes, mainly Santa Ana suckers and arroyo chubs.

- Lagunitas Creek (Marin County), despite having several dams in the watershed, supports remnant runs of coho salmon and steelhead, plus a rare native shrimp (*Syncarus pacificus*) and a largely native fish fauna (Moyle and Smith 1995).

SNAs are included as a tier in order to promote recognition of the fact that some aquatic systems or areas do need the intense protection and management normally associated with traditional nature preserves. Most SNAs are small but can protect unusual fish or invertebrates and associated communities of organisms that might otherwise be overlooked or that are part of watersheds that are otherwise in poor condition.

Tier 4: Watersheds

Protection of biodiversity has traditionally centered around setting up preserves and refuges. *Preserves* are areas, usually small (like SNAs), set aside to protect communities of native organisms in order to ensure the survival of species by minimizing negative human impacts. Historically preserves have been envisioned as museums that freeze present conditions and exclude all human use except scientific study. Conceptually, they are based on equilibrium models of ecology that have been largely replaced by more dynamic (stochastic) models (Fiedler et al. 1993). *Refuges,* in contrast, are areas intensively managed for select groups of species, such as waterfowl, or areas set aside to protect economically important or endangered species without too much concern for maintaining native biotic communities (Williams 1991). In practice, areas labeled "preserves" and "refuges" run the gamut from highly artificial environments to highly protected natural areas. The two terms are used rather loosely, often meaning different things to different agencies and people. Therefore I prefer the term *Aquatic Diversity Management Area* (ADMA) (Moyle and Yoshiyama 1992, 1994; Moyle 1995).

An ADMA is a watershed that has as its top management priority the maintenance of aquatic biodiversity.[1] Other uses are permitted, but they are secondary to, and must be compatible with, the primary goal. The key to maintenance of ADMAs is flexibility, recognizing that active management is needed to maintain or enhance biodiversity and that an ADMA is likely to change through time. ADMAs are not necessarily pristine environments, but they are usually reasonable approximations of them.

The characteristics of ADMAs given here are derived from the ongoing debate on how nature preserves should be designed (Moyle and Sato 1991). Unfortunately, most debate over preserve design has centered on terrestrial systems and has paid little attention to the special problems of protecting aquatic environments. Therefore, the six criteria listed here are those used for design of preserves in general, although they are discussed in the context of aquatic systems (Moyle and Sato 1991). These ideas owe much to the concept of key watersheds developed by Thomas et al. (1993) for streams of the Pacific Northwest that produce anadromous fish.

1. An ADMA must contain resources and habitats necessary for persistence of the species and communities it is designed to protect. This criterion assumes that all life history stages of all organisms (not just fish) are known—a degree of knowledge that is simply not attainable. Design of an ADMA therefore should be based on the largest and most mobile species, on the assumption that their habitat needs will also encompass those of less well-known species. This means that ADMAs will largely be based on the needs of fish, amphibians, and macroinvertebrates (including migratory species that are present for only part of their life cycle), and on the needs of conspicuous riparian organisms (trees, birds, mammals).

2. An ADMA must be large enough to contain the range and variability of environmental conditions necessary to maintain natural species diversity. An ADMA that is too small will ultimately fail even if correct environmental conditions are present. The actual size of an ADMA will depend on the biota being protected, but 50 km² seems like a reasonable minimum size for most watersheds. There is no maximum size. A riverine biota may require several thousand square kilometers, encompassing much of a drainage. ADMAs must have their water sources protected, including aquifers and extreme headwaters.

3. ADMA integrity must be protected from edge and external threats. Reducing edge and external threats is a continual challenge to designers of natural areas, and it is largely in order to reduce these threats that ADMAs should encompass entire watersheds. *Edge threats* result from the gradient of habitat quality between the ADMA and adjacent areas. The less distinct the boundary, the more likely the ADMA will suffer from habitat degradation (due to, e.g., access roads or the aerial spread of pesticides) and invasions of unwanted species. Edge threats are likely to be particularly severe in low-elevation ADMAs, where watershed boundaries are not sharp or defined by steep, rocky ridges.

External threats do not recognize boundary lines and include such factors as pollutants, diseases, and introduced species. They pose a particularly severe problem for ADMAs because agents that affect the biota in any part of a drainage may eventually be carried by the water throughout its entirety (Moyle and Sato 1991). A particularly insidious external threat to aquatic systems is pumping of groundwater from aquifers distant from the springs and streams that the aquifers feed. Thus pumping of groundwater in Nevada may eventually dry up springs essential for survival of pupfish and spring snails (Hydrobiidae) in California. Species that are typically good invaders—such as green sunfish, common carp, red shiners, and bullfrogs (*Rana catesbeiana*) —have tremendous dispersal abilities, can work their way over low barriers, and can survive in a wide variety of habitats, even those in fairly good condition.

Edge and external threats will always be problems for ADMA management, but they can be reduced by creating large ADMAs, improving management of adjacent watersheds, and constructing barriers to prevent invasions of unwanted species. Ideally barriers should block entry of nonnative species but not of native migrants. For California streams, the best barrier to invasion is often a natural flow regime, because native species are generally well adapted to living under fluctuating conditions (Baltz and Moyle 1993; Moyle et al. 1998).

---

1. In previous publications I have included SNAs as ADMAs, with SNA-ADMAs having areas of less than 50 km² and watershed ADMAs having areas of 50 km² or more. I have subsequently decided that is it less confusing to have ADMAs apply only to watersheds.

4. An ADMA should have interior redundancy of habitat to reduce effects of localized species extinctions due to natural processes. This criterion to a degree reiterates criterion 2, but the need for local redundancy cannot be overemphasized. Aquatic species frequently occur as small populations in narrow habitat types, where populations come and go in relation to natural events and demographic processes. Adequate local redundancy therefore will allow recolonization to occur quickly and naturally. Thus the best ADMAs are those large enough to include multiple examples of all habitat types covered in Moyle and Ellison (1991).

5. Each ADMA should be paired with at least one other ADMA that contains most of the same species but is far enough distant that both are unlikely to be affected by a regional disaster. Large disasters—volcanic eruptions, earthquakes, pesticide spills, forest fires—can fundamentally alter the integrity of an ADMA. Therefore, sources of species must exist for biotic reconstruction, if necessary. For streams, this requirement means creating ADMAs in separate drainages with similar characteristics and biotas. For species inhabiting temporary ponds, this may mean protecting ponds at widely separated localities. Thus the Conservancy fairy shrimp (*Branchinecta conservatio*), endemic to central California, is well protected because TNC has several widely separated vernal pool preserves in Tehama, Merced, and Solano Counties (Eng et al. 1990). Greater replication of ADMA types increases chances for long-term survival of native organisms. However, some ADMAs will not be replicable if they contain highly localized endemics (e.g., desert springs with pupfish subspecies, Goose Lake).

6. An ADMA should support populations of organisms large enough to have a low probability of extinction because of random demographic and genetic events. Small populations of organisms can become extinct as a result of natural fluctuations. Small populations can also experience "bottlenecks" that greatly reduce genetic variability and, consequently, their ability to adapt to changing environmental conditions. This is particularly a problem in setting up small watershed ADMAs, where fish and invertebrate populations may frequently be driven to low levels by extreme high flow events or droughts. Under natural conditions, populations from different watersheds eventually mix again—something that is not possible in an isolated ADMA unless enough of a drainage is included to permit natural recolonization events (Zwick 1992). For some California fishes, localized extinctions caused by artificial isolation are already occurring (L. R. Brown et al. 1992).

The foregoing rules imply that California watersheds vary widely in their suitability for becoming ADMAs. Very few watersheds contain all their native organisms living under relatively natural conditions, especially natural flow regimes; many are highly degraded and contain only fragments of their native biota. The ideal ADMA is a pristine en-

vironment, but realistically all watersheds have been altered by humans in some manner—some severely so. If highly altered watersheds are all that are available to protect certain species or habitats, they should be included in a system of ADMAs, and efforts should be made to restore them to more natural conditions, even if such efforts might involve removal of dams. Such ADMAs, however, will probably contain a remnant native biota coexisting with introduced species. A rating system developed by Moyle and Sato (1991) recognizes the need for managing habitats that range from pristine to degraded, with highest priority given to the most pristine areas as ADMAs to prevent their further degradation.

A more systematic way of rating the suitability of watersheds for ADMAs was developed by Moyle and Randall (1998) for the Sierra Nevada, using a watershed-based index of biotic integrity (W-IBI). The W-IBI is essentially a composite score of ratings for six variables that indicate the resemblance of present conditions in a watershed to presumed pristine conditions. The variables are *(1)* abundance of native ranid frogs, *(2)* abundance of native fishes, *(3)* presence of native fish assemblages, *(4)* distribution of anadromous fishes, *(5)* distribution of trout, and *(6)* abundance of stream fishes, native and introduced. The W-IBI permits scoring of large watersheds on a 100-point scale—with 100 representing pristine conditions (not achieved by any watershed)—and therefore their ranking in terms of suitability for large-scale conservation efforts (Fig. 16). Similar IBIs can be developed for other California regions (Moyle and Marchetti 1998).

I regard creation of a system of ADMAs (or some similar system of protected watersheds) as essential to provide *minimum* protection for California's aquatic biodiversity for the next 50–100 years. ADMAs are needed to ensure that we have the Leopoldian pieces available for ecosystem restoration, when and if our society changes its dominant value system and decides to live with nature rather than constantly contending with it (G. Snyder 1990). ADMAs should also serve as standards against which degradation of other areas can be measured. For these functions to be realized on a statewide basis, a system of ADMAs must be established that includes representatives of the 160 habitat types described in Moyle and Ellison (1991).

The first step in the process of systematically creating an ADMA system is to identify potential ADMAs in each region of the state. This can best be done using expert opinion combined with a systematic method of identifying the "best" watersheds, such as the use of a specially developed IBI. The list of potential ADMAs would be a source of information for management agencies and for concerned citizens who want to form watershed conservation groups or find support for existing ones. Highest priorities should be given to assigning ADMA status to watersheds that *(1)* are

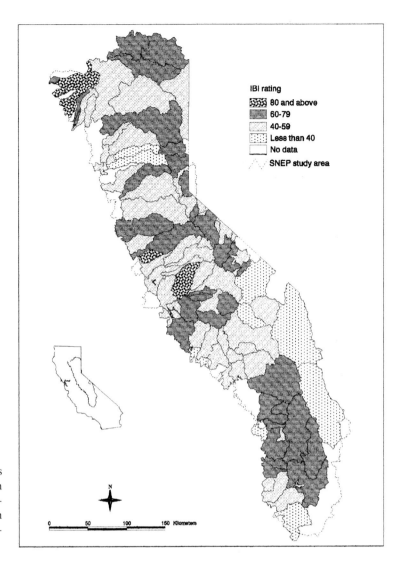

IBI rating
▨ 80 and above
▓ 60-79
░ 40-59
▒ Less than 40
☐ No data
〰 SNEP study area

N

0   50   100   150 Kilometers

Figure 16. A rating of Sierra Nevada watersheds using an index of biotic integrity (W-IBI). High scores indicate greater suitability for special management to benefit native aquatic organisms. From Moyle and Randall (1998); reprinted by permission of Blackwell Science, Inc.

unique ecosystems with endemic organisms, such as Eagle Lake (Lassen County), (2) are critical habitat for threatened or endangered species, (3) have high IBI scores, and (4) have the right combination of size, low degree of human disturbance, location, and intact fish assemblages to be the best representative of a particular aquatic ecosystem. Each ADMA description should include a statement of how much of it is already protected under de facto ADMAs (e.g., watersheds in parks and natural areas) and what parts are most threatened with degradation, so that limited personnel, time, and money can be used most efficiently for acquisition and management. As many ADMAs (or parts of them) as possible should be incorporated into established systems of protection, such as wilderness areas or national parks. Ideally the formation of a system of ADMAs should be a statewide effort, coordinated by the state Resources Agency, but ADMA designation does not have to wait for such official blessing. I suggest that regional environmental

groups make their own ADMA recommendations in order to encourage systematic, official efforts at watershed protection.

No matter how complete, a system of ADMAs by itself will not protect California's aquatic biodiversity in the long run. This is because the ADMA system as proposed is a fragmented one, with pieces scattered across the landscape, mostly unconnected to one another. Such fragmentation of aquatic habitats ultimately leads to loss of biodiversity through local extinctions without recolonization (Zwick 1992). There is also the danger that conferring special protection on selected watersheds will justify the granting of less protection to other watersheds. What an ADMA system can do (as can protecting endangered species singly or in clusters) is provide a minimum level of biodiversity insurance until biodiversity can be managed on a broader scale. As Noss (1992, p. 241) points out, "Biodiversity can be conceived of as a nested hierarchy of elements at several levels of biological organization. Familiar levels of organization are genetic, population-species, community-ecosystem, and landscape. Generally speaking, as level of organization ascends from gene to landscape (and beyond, to biosphere), so does the spatial scale at which these elements occur." The first four tiers of biodiversity protection provide for protection only at the lower three levels of this nested hierarchy, over a short (50–100 years) time frame. Real and lasting protection, however, can only occur at higher levels of organization (Franklin 1993), represented by the fifth and sixth tiers.

Watersheds are the next logical unit on which to focus conservation efforts (Reeves and Sedell 1992; Naiman et al. 1993). In California, DWR has divided the state into hydrologic basins that can be used as a basis for watershed-oriented landscape management. Each watershed should be evaluated at some scale for its natural attributes and have a management plan that can be used by citizens and various levels of government to assist in making land use decisions. Watersheds could also be managed as clusters, preferably associated with an ADMA watershed in order to maximize protection of aquatic biodiversity.

Tier 5: Bioregions

To be truly successful, biodiversity protection must be integrated within landscape-scale environmental protection based on the understanding that human health and well-being are tied to environmental health (Noss 1992; Barnes 1993). One way to approach biodiversity protection at this scale is through the use of the bioregion as the unit of management. Bioregions are human constructs. We look at a broad area of land and decide that internal similarities in biological and human-created features combined with differences from surrounding areas merit its recognition as a distinct entity. Examples include the Sierra Nevada, Central Valley, North Coast, or Klamath bioregions. Obviously bioregions can overlap, with boundaries that are deliberately vague, although if drawn on maps they usually follow major watershed boundaries. One key aspect of a bioregion is that the people living there identify with it and its attributes, for example, with the Klamath bioregion as having

coastal rain forests, fog, and big runs of salmon. Ideally an artistic or literary tradition has developed as part of this bioregional identity, such as the rich literature, from Mark Twain to Gary Snyder, that focuses on the Sierra Nevada. Art and literature help local people identify with a region and with its natural attributes, and such identification in turn leads to an increased desire for ecosystem protection on a broad scale.

One of the best examples of an ongoing attempt at bioregional planning and restoration is CALFED, the massive joint federal-state-stakeholder effort to solve the ecological problems of the San Francisco Estuary. CALFED planning encompasses the entire watersheds of the Sacramento and San Joaquin Rivers, although it focuses mainly on the areas below major dams, between Shasta Dam on the Sacramento River and Friant Dam on the San Joaquin River. CALFED recognizes the region as an ecosystem with integrated parts; thus restoration of estuarine biota and function requires "fixing" problems upstream as well. CALFED anticipates spending several billion dollars reversing past damage (e.g., moving back levees to recreate floodplains) and appreciates that local watershed groups must be involved in the process. Although the driving force behind CALFED is providing a reliable water supply for the San Joaquin Valley and southern California, it may well have the beneficial effect not only of improving environmental conditions in major waterways of northern California but also of increasing regional awareness of the great value, aesthetic and economic, of natural habitats and naturally functioning ecosystems. CALFED is a grand experiment that, if successful, may well provide an example of bioregional restoration for other regions to follow.

Although the development of CALFED, the numerous watershed groups, favorable environmental laws, and other recent actions give reason for optimism, I feel obligated to end this chapter on a darker note. In the long run no conservation scheme will work if the astonishing growth rate of California's human population is not curtailed and if we do not implement more sustainable methods of managing our wild, agricultural, and urban lands. We, as individuals, must be willing to get by on much less, so fish (and other creatures) can have more. In the long run, this policy will benefit us and our descendants as well, by keeping the planet livable.

Identification

Identification of fish taken from California's waters is often tricky. Some groups of species, such as sculpins or juvenile cyprinids, are naturally hard to tell apart. Individuals occasionally lack supposedly definitive characteristics because of injury, colors that fade in turbid waters, or simply natural variation. Hybridization among species is common, especially in disturbed waters or among introduced species. In Oroville Reservoir, for example, a "black" bass caught by an angler may be a smallmouth, largemouth, spotted, or redeye bass—or potentially any cross between members of the four species! Location is often a good clue for identification, but it is not as reliable as it might be because so many fishes have been moved around or because similar species have overlapping ranges. A sculpin caught on the east side of the Sierra Nevada, for example, is a Paiute sculpin, but one caught in a North Coast stream has about an equal probability of being either a prickly sculpin or a coastrange sculpin.

The following are some suggestions for identifying fish in California:

- If it is crucial for identification to be accurate, specimens (preferably more than one) of the species should be kept for careful identification in the laboratory, using a microscope or hand lens to make important counts of fin rays or scales. It is a good idea to keep voucher specimens, preserved in formalin or alcohol. Photographs of freshly caught fish can also be helpful. The use of digital cameras to take photographs of fish in the field is an increasingly useful practice because of the ease with which photos can be compared on a computer screen. For small fish, counts of fin rays and other structures are sometimes more easily performed from digital photos than from the actual specimen!

- Identify the fish using more than one source. A good backup for the keys and descriptions in this book is Page and Burr (1991).

- Recognize that keys in this book (or elsewhere) will not work every time; there is simply too much variability in fish morphology and in human perception. Accuracy can be improved by comparing the specimen to pictures and to the more detailed descriptions in each species account.

- Accuracy of identification improves with familiarity with fish and with keys, so practice using keys on common fish, including making scale and fin ray counts.

- For larval and juvenile fish, consult Wang's (1986) monumental work or its promised successor.

Using the Key

The characters needed to identify California fishes are presented in Fig. 17, appear within the key itself, or are described below. For more precise definitions of the characters used in taxonomy, Hubbs and Lagler (1958) should be consulted.

Standard length (SL) is the distance from the tip of the snout or lower jaw (whichever sticks out farther) to the end of the vertebral column. The end of the vertebral column can be found by flexing the tail and noting the slight projecting ridge that is present just in front of the caudal fin.

Total length (TL) is the greatest length that can be measured, from the tip of the snout or lower jaw to the end of the longest ray of the caudal fin when the upper and lower lobes are squeezed together. Total length must be used carefully, because the tips of the caudal fin can be frayed or broken, especially in preserved fish.

Fork length (FL) is the distance from the tip of the snout or lower jaw to the middle of the fork of the caudal fin. This measurement is commonly used by fisheries workers because it is easier to measure than standard length and less variable than total length. However, many fish lack forked tails.

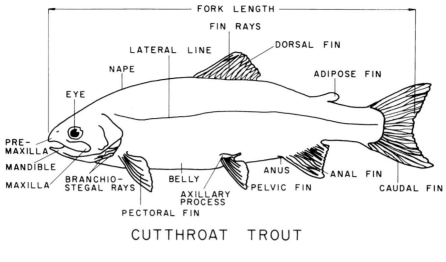

CUTTHROAT TROUT

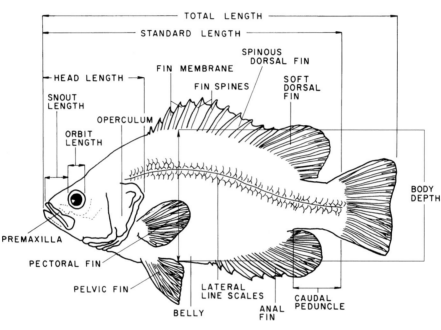

SACRAMENTO PERCH

Figure 17. Important features and measurements of a soft-rayed fish (top) and a spiny-rayed fish (bottom).

Body depth is the greatest depth that can be measured, excluding the dorsal and anal fins.

Head length is the distance from the tip of the snout to the most distant point at the edge of the operculum.

Fin spines are unbranched, unsegmented supports for fins that, if present, are on the leading edge of the fin. The smallest spines and most anterior spines may be hard to see. In sculpins the main spine in the pectoral fin is fused with the first ray, so counts are given as fin "elements" to avoid double-counting the first structure.

Fin rays are counted at the base of each ray to avoid counting branches (rays tend to fan out toward the fin edge). In soft-rayed fins that have an angular shape and a straight anterior edge, as in minnows and suckers, only principal rays

are counted; the one or two rudimentary rays that may be present in front of the first principal ray are ignored.

Lateral line scales are the scales bisected by the lateral line, extending from the edge of the opercular opening to the base of the tail. The count represents the number of body scale rows, so it may be taken even if the lateral line is not visible. The count in such cases is called *scales in the lateral series.* In fishes that lack scales but that possess a visible lateral line (sculpins), *lateral line pores* may be counted; they are small openings visible with a hand lens.

Scales above the lateral line are counted from the origin of the dorsal fin (first dorsal fin if there is more than one) down to the lateral line, not including the lateral line scale.

Scales below the lateral line are counted from the origin of the anal fin up to the lateral line, preferably by following one scale row, not including the lateral line scale.

Scales before the dorsal fin are the total number of scale rows that cross the back of the fish before the dorsal fin and behind the posterior dorsal end of the head. The end of the head is often marked with a line that separates the scaled from the unscaled portion.

Pharyngeal tooth counts can be important for the definitive identification of cyprinids but are difficult to perform, because the teeth have to be dissected out of fresh specimens. They are present on pharyngeal arches in the lower half of the pharyngeal region, just behind the gill rakers. There are two sets of pharyngeal teeth in each fish, one on each side. The teeth are in one or two rows, and their numbers are presented as a formula, for example 1,4-4,1, where the 1 is the number of teeth in the upper (minor) row and 4 is the number in the lower row. Jenkins and Burkhead (1994) present a good discussion on using pharyngeal teeth for identification of cyprinids.

Key to the Inland Fishes of California

Family Key

1a. Mouth with true jaws; gill cover (operculum) present 2
1b. Mouth jawless, a round sucking disc; no operculum present .... Petromyzontidae (lampreys), p. 84

2a. Sides with 5 rows of bony plates; upper lobe of tail much longer than lower (heterocercal) ... Acipenseridae (sturgeons), p. 85
2b. Sides without 5 rows of bony plates; tail lobes about equal (homocercal) 3

3a. More than 30 branchiostegal rays (fanlike bones) on underside of lower jaw; machete, *Elops affinis*[2] .... Elopidae (tarpons)
3b. Fewer than 30 branchiostegal rays (fanlike bones) on underside of lower jaw 4

4a. Scales on belly form a sharp, sawtoothed ridge; vertical, adipose eyelids present .... Clupeidae (herrings), p. 85
4b. Belly smooth and usually rounded; no vertical, adipose eyelids 5

5a. Adipose fin present 6
5b. Adipose fin absent 8

6a. Scales absent, chin barbels present .... Ictaluridae (catfishes), p. 85
6b. Scales present, chin barbels absent 7

7a. Small fleshy or scaly appendage (axillary process) present at base of each
    pelvic fin .... Salmonidae (trout, salmon, whitefish), p. 88
7b. Axillary processes absent .... Osmeridae (smelts), p. 88

8a. One side of body unpigmented; both eyes on one side of head; starry flounder, *Platichthys
    stellatus* .... Pleuronectidae (flounders)
8b. Both sides of body pigmented; eyes on opposite side of head 9

9a. Body encased in bony plates; snout long and tubular; bay pipefish, *Syngnathus leptorhynchus* .... Syngnathidae (pipefishes)
9b. Body not encased in bony plates; snout blunt 10

10a. Body smooth, long and slender (eel-like) 11
10b. Body not eel-like 13

11a. 5–6 barbels on each side of jaw; oriental weatherfish, *Misgurnus anguillicaudatus*[3] (Fig. 18) .... Cobitidae (loaches)
11b. Barbels absent 12

12a. Dorsal fin extends from tail region to head .... Pholididae (gunnels), p. 91
12b. Dorsal fin extends from tail region to middle of body .... Anguillidae (eels)[4]

13a. Pelvic fins united to form a sucking disc .... Gobiidae (gobies), p. 93
13b. Pelvic fins separated, not forming a sucking disc 14

14a. Dorsal fin consists of 3–5 unconnected spines followed by a soft-rayed fin; caudal peduncle
    narrow .... Gasterosteidae (sticklebacks), p. 91
14b. Dorsal fin spines and rays connected to others by membrane; caudal peduncle various 15

15a. Scales absent; pectoral fins large and rounded .... Cottidae (sculpins), p. 93
15b. Scales present; pectoral fins not as above 16

16a. Two distinct, widely separated dorsal fins present 17
16b. Dorsal fin single or divided into two sections that touch or are barely separated 19

17a. Caudal fin rounded; pelvic fins in front of pectoral fins; spotted sleeper, *Eleotris picta* (Fig. 19)[5] .... Eleotridae (sleepers)
17b. Caudal fin forked; pelvic fins well behind pectoral fins 18

18a. In head-on view, mouth shaped like wide, inverted V with distinct peak in center of lower jaw; stripes on side, if
    visible, multiple and narrow; striped mullet, *Mugil cephalus* .... Mugilidae (mullets)
18b. Mouth not a distinct V in head-on view; single wide band on sides .... Atherinopsidae (silversides), p. 91

19a. No spines present in dorsal fin 20
19b. Spines present in dorsal fin 25

---

2. Machete are marine fish that occasionally enter the lower Colorado River. They were once common in the Salton Sea (Walker et al. 1961).
3. Oriental weatherfish have been reported as established in the Westminster Flood Control Channel, Orange County (St. Amant and Hoover 1969). I have no recent confirmation that they are still present.
4. Eels of various species of *Anguilla* have been captured in California waters, but they are undoubtedly nonbreeding animals that have escaped from ponds where they were being raised for food (McCosker 1989).
5. A single spotted sleeper, normally found in streams and estuaries of Mexico and Central America, was taken from a canal in the Imperial Canal (Hubbs 1953).

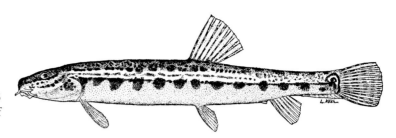

Figure 18. Oriental weatherfish (Cobitidae), 95 mm SL, China. From Nichols (1943); courtesy of The American Museum of Natural History.

Figure 19. Spotted sleeper (Eleotridae), 57 mm SL, Mexico. CAS 51006.

Figure 20. Northern pike (Esocidae), 540 mm SL, Lake County, Minnesota. Fish print by Christopher M. Dewees.

20a. Snout strongly flattened (like duck's bill); mouth lined with sharp teeth; body elongate; northern pike, *Esox lucius* (Fig. 20)[6] .... **Esocidae (pikes)**

20b. Snout not flattened; teeth in mouth small or absent; body various 21

21a. Scales present on head; caudal fin rounded or square 22

21b. Scales absent from head; caudal fin forked, often only slightly 24

22a. Third ray of anal fin unbranched; anal fin modified to intromittent organ in males (Fig. 21) ... **Poeciliidae (livebearers), p. 90**

22b. Third ray of anal fin branched; anal fin not modified in males 23

23a. Body deep, depth divisible into body length less than 3.5×; caudal peduncle deep and compressed; back behind head arched .... **Cyprinodontidae (pupfishes), p. 90**

23b. Body depth divisible into body length more than 3.5×; caudal peduncle not deep and compressed; back behind head straight or slightly rounded .... **Fundulidae (killifishes), p. 90**

24a. Mouth usually subterminal, with fleshy papillose lips;[7] dorsal fin with 10 or more principal rays .... **Catostomidae (suckers), p. 87**

24b. Mouth usually terminal, with smooth lips; dorsal fin usually with fewer than 10 principal rays .... **Cyprinidae (minnows), p. 86**

25a. Distinct scaled ridge present along base of dorsal fin .... **Embiotocidae (surfperches), p. 93**

25b. No such ridge present 26

26a. Anal fin spines 1–2 27

26b. Anal fin spines 3 or more 28

27a. One stout dorsal fin spine; dorsal fin single .... **Cyprinidae (carp and goldfish), p. 86**

27b. Four or more spines on dorsal fin; dorsal fin divided into two distinct sections .... **Percidae (perches and darters), p. 93**

6. Northern pike became established in Davis Reservoir, Plumas County, from an illegal introduction. A major effort by CDFG is under way to control and eventually eliminate them because of their potential hazard to salmon populations.

7. Both the bigmouth buffalo and the shortnose sucker have terminal mouths, but they can be distinguished from cyprinids by their comb-like pharyngeal teeth, lack of spines in the dorsal and anal fins (possessed by carp and goldfish), lack of barbels (carp), large size, and distinctive appearances.

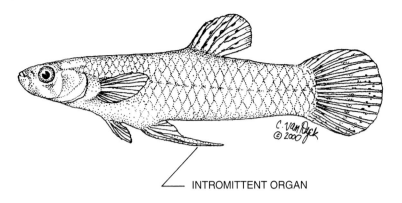

INTROMITTENT ORGAN

Figure 21. Western mosquitofish, showing intromittent organ (gonopodium).

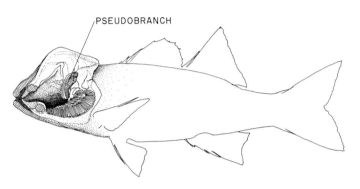

PSEUDOBRANCH

Figure 22. Pseudobranch on the inner surface of a striped bass operculum.

28a. Well-developed pseudobranch (gill-like structure) present on inner surface of operculum (Fig. 22) . . . . **Moronidae[8] (temperate basses), p. 92**
28b. Pseudobranch absent or inconspicuous 29
29a. Dorsal and anal fins long and pointed at rear; one nostril present on each side of head; lateral line interrupted . . . . **Cichlidae (cichlids), p. 93**
29b. Dorsal and anal fins rounded; two nostrils present on each side of head; lateral line continuous . . . . **Centrarchidae (sunfishes and basses), p. 91**

Petromyzontidae, Lamprey Family

1a. Eyes and sucking absent or poorly developed (ammocoetes) 2
1b. Eyes and sucking disk well developed (adults) 5
2a. Trunk myomeres (segments) more than 66; body and head darkly pigmented; light spot in center of tail . . . . Pacific lamprey, *Lampetra tridentata*
2b. Trunk myomeres fewer than 67; body and head not darkly pigmented; no light spot in center of tail 3
3a. Trunk myomeres 58–67 4
3b. Trunk myomeres 51–57 . . . . nonpredatory brook lampreys, *Lampetra* spp.
4a. Trunk myomeres 63–67 . . . . river lamprey, *Lampetra ayersi*
4b. Trunk myomeres 58–65 (usually 60–63), upper Klamath River . . . . Klamath River lamprey, *Lampetra similis*
5a. Tooth plates on oral disc conspicuous and well developed, with distinct points 6
5b. Tooth plates poorly developed and blunt 9
6a. TL greater than 28 cm . . . . Pacific lamprey, *Lampetra tridentata*
6b. TL less than 28 cm 7
7a. Supraoral tooth plate (in center of disc) with 3 cusps, 4 inner lateral tooth plates on each side (Fig. 23) 8
7b. Supraoral tooth plate with 2 cusps, 3 inner lateral tooth plates on each side . . . . river lamprey, *Lampetra ayersi*
8a. Trunk myomeres 66 or more . . . . Pacific lamprey, *Lampetra tridentata*
8b. Trunk myomeres fewer than 66 . . . . Klamath River lamprey, *Lampetra similis*[9]

8. Also listed in some references as "Percichthyidae."
9. The Klamath brook lamprey, *L. folletti*, is now included within the Klamath River lamprey.

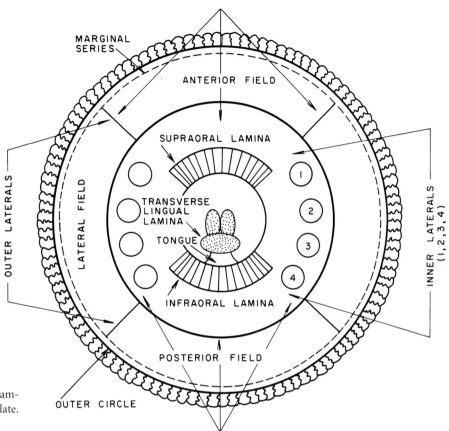

Figure 23. Diagrammatic disc of a lamprey, showing position of tooth plate. After Vladykov and Follett (1962).

9a. Caudal fin pale, pigmentation along edge only, San Joaquin drainage .... Kern brook lamprey, *Lampetra hubbsi*
9b. Caudal fin dark, evenly pigmented except for margin 10
10a. Supraoral tooth plate usually with 3 cusps, Pit and upper Klamath River
    drainage .... Pit-Klamath brook lamprey, *Lampetra lethophaga*
10b. Supraoral tooth plate with 2 cusps, coastal drainages .... western brook lamprey, *Lampetra richardsoni*

Acipenseridae, Sturgeon Family

1a. 1–2 middorsal scutes (bony plates) behind dorsal fin; 23–30 scutes in row on each side
    of body .... green sturgeon, *Acipenser medirostris*
1b. No middorsal scutes behind dorsal fin; 38–48 scutes in row on each side .... white sturgeon, *Acipenser transmontanus*

Clupeidae, Herring Family

1a. Last ray of dorsal fin long and threadlike; single black spot near operculum .... threadfin shad, *Dorosoma petenense*
1b. Last ray of dorsal fin not elongated; either more than one or no black spots present near operculum 2
2a. Row of black spots on side; scales in lateral series more than 55 .... American shad, *Alosa sapidissima*
2b. No black spots on side; scales in lateral series fewer than 55; marine .... Pacific herring, *Clupea harengeus pallasii*

Ictaluridae, Catfish Family

1a. Tail forked 2
1b. Tail square or rounded 4
2a. Anal fin rays 30–36; margin of anal fin nearly straight .... blue catfish, *Ictalurus furcatus*
2b. Anal fin rays fewer than 30; margin of anal fin rounded 3
3a. Anal fin rays 24–29; small dark spots usually present on sides .... channel catfish, *Ictalurus punctatus*
3b. Anal fin rays 19–23; no dark spots on sides .... white catfish, *Ameiurus catus*

4a. Anal fin rays 23–27; chin barbels whitish . . . . yellow bullhead, *Ameiurus natalis*

4b. Anal fin rays fewer than 25; chin barbels dark 5

5a. Anal fin rays 12–15; lower jaw projects beyond upper jaw . . . . flathead catfish, *Pylodictis olivaris*

5b. Anal fin rays 17–24; jaws even 6

6a. Membranes between anal fin rays black; body not mottled; whitish bar present at base
of tail . . . . black bullhead, *Ameiurus melas*.

6b. Membranes between anal fin rays same color as or lighter than rays; body mottled; no whitish bar present at base
of tail . . . . brown bullhead, *Ameiurus nebulosus*

Cyprinidae, Minnow Family

1a. Dorsal fin long, with stout, serrated "spine" at front 2

1b. Dorsal fin short, without "spine" 3

2a. Conspicuous barbels present on each side of mouth; 32 or more lateral line scales present . . . . . common carp, *Cyprinus carpio*

2b. Barbels absent or tiny; lateral line scales fewer than 32 . . . . goldfish, *Carassius auratus*

3a. Fleshy keel (ridge) present between pelvic and anal fins . . . . golden shiner, *Notemigonus crysoleucas*

3b. No such keel 4

4a. Barbels present (may be tiny, at end of maxilla) 5

4b. Barbels absent 7

5a. More than 90 scales along lateral line; deep bodied . . . . tench, *Tinca tinca*

5b. Fewer than 90 scales along lateral line; slender bodied 6

6a. Upper lobe of caudal fin longer than lower; anal fin rays 7–9; body silvery, no speckles . . . splittail, *Pogonichthys macrolepidotus*

6b. Caudal fin symmetrical; anal fin rays 6–7; body not silvery, usually speckled . . speckled dace, *Rhinichthys osculus* (see also 17a)

7a. Small ridge of skin (frenum) connects upper lip to snout (Fig. 24) . . . . hardhead, *Mylopharodon conocephalus*[10]

7b. Frenum absent 8

8a. Caudal peduncle extremely long and narrow . . . . bonytail, *Gila elegans*

8b. Caudal peduncle normal 9

9a. Upper lobe of caudal fin longer than lower lobe . . . . Sacramento splittail, *Pogonichthys macrolepidotus*[11]

9b. Caudal fin symmetrical 10

10a. Mouth large, straight, and terminal; maxillary reaches middle of eye or beyond; snout pointed 11

10b. Mouth small, usually subterminal or angled upward; maxillary does not reach middle of eye; snout blunt 12

11a. Dorsal and anal fin rays 9; Colorado River drainage . . . . Colorado pikeminnow, *Ptychocheilus lucius*

11b. Dorsal and anal fin rays 7–8; Sacramento–San Joaquin drainage . . . . Sacramento pikeminnow, *Ptychocheilus grandis*

12a. Lateral line scales tiny (>90); mouth terminal; head of adult flattened . . . . Sacramento blackfish, *Orthodon microlepidotus*

12b. Lateral line scales small to large (<60), mouth and head various 13

13a. Head wide between eyes, flattened on top; mouth terminal; lateral line scales with dark edges and spot
at base . . . . grass carp, *Ctenopharyngodon idella*

13b. Head narrow between eyes, not flattened on top; mouth various; lateral line scales plain 14

14a. Scales along lateral line fewer than 40 . . . . red shiner, *Cyprinella lutrensis*

14b. Scales along lateral line more than 40 15

15a. Anal fin rays 10–14; posterior edge of extended anal fin forms oblique angle to lateral line . . . . hitch, *Lavinia exilicauda*

15b. Anal fin rays 7–9; posterior edge of extended anal fin forms perpendicular angle to lateral line 16

16a. Maximum body depth less than 2× caudal peduncle width; scales on back distinctly outlined; scales behind head
crowded; adults with horizontal dark bar on dorsal fin . . . . fathead minnow, *Pimephales promelas*

16b. Maximum body depth more than 2× caudal peduncle width; scales on back various; scales behind head uniform;
no dark bar or spot on dorsal fin 17

17a. Eyes small; distance between eye and tip of snout more than 1.5× width of eye; body usually speckled; dark band usually
connects eye to snout . . . . speckled dace, *Rhinichthys osculus*

17b. Eyes moderate to large, distance between eye and tip of snout less than 1.5× width of eye; body not speckled;
no dark band on snout 18

18a. Caudal peduncle moderate, depth of peduncle divisible more than 3× into distance from insertion of anal fin to base
of tail at midline; in fish over 50 mm SL, greatest body depth divisible more than 4× into SL 19

10. A few speckled dace without barbels may key out here, but note the thick caudal peduncle and overhanging snout.

11. Clear Lake splittail, *P. ciscoides*, now extinct, will also key out here.

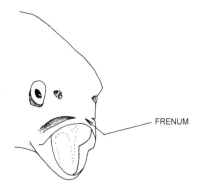

Figure 24. Frontal view of a hardhead, showing frenum.

18b. Caudal peduncle short and thick, depth of peduncle divisible less than 3× into distance from insertion of anal fin
to base of tail at midline; in fish over 50 mm SL, greatest body depth divisible less than 4× into SL 20

19a. Sides with wide dark band between 2 pale bands; anal fin rays 8–10; intestine, when viewed from side, has one
S-shaped bend . . . . Lahontan redside, *Richardsonius egregius*

19b. Sides usually without 3 bands; anal fin rays 6–9, usually 6–7; intestine, when viewed from side, has 2–3 S-shaped
bends . . . . California roach, *Lavinia symmetricus*

20a. Lateral line scales more than 65; body depth usually one-firth or less of TL; Klamath river system . . . . blue chub, *Gila coerulea*

20b. Lateral line scales fewer than 65; body depth usually one-fifth or more of TL 21

21a. Snout does not overhang mouth; anal fin rays usually 8 (7–9); gill rakers usually more than 8 . . . . tui chub, *Siphateles bicolor*[12]

21b. Snout overhangs mouth; anal fin rays usually 7; gill rakers 9 or fewer; southern California . . . . arroyo chub, *Gila orcutti*

Catostomidae, Sucker Family

1a. Mouth terminal; lips thin, with few or no papillae 2

1b. Mouth subterminal; lips usually thick, with distinct papillae 3

2a. Dorsal fin long, 23–30 rays; lateral line scales 36–39; southern California . . . . bigmouth buffalo, *Ictiobus cyprinellus*[13]

2b. Dorsal fin short, 11–12 rays; lateral line scales 73–88; Klamath system . . . . shortnose sucker, *Chasmistes brevirostris*

3a. Upper and lower lips separated by deep indentations at corners of mouth; median notch of lower lip shallow (Fig. 25B) . . 4

3b. Upper and lower lips not separated by deep indentations; margin of lip continuous; median notch of lower lip moderate
to deep (Fig. 25A) 5

4a. Pigmentation present on membranes between rays of caudal fin; axillary process at base of pelvic fins a simple fold;
southern California . . . . Santa Ana sucker, *Catostomus santaanae*

4b. Pigmentation absent or very sparse on membranes between rays of caudal fin; axillary process at base of pelvic fins well
developed; Great Basin . . . . mountain sucker, *Catostomus platyrhynchus*

5a. Well-developed, sharp-edged ridge on back before dorsal fin; Colorado River . . . . razorback sucker, *Xyrauchen texanus*

5b. No such ridge present 6

6a. Distinct hump on snout; lips thin, papillae only moderately developed; Klamath Basin . . Lost River sucker, *Catostomus luxatus*

6b. Snout without hump; lips thick and papillose 7

7a. Lateral line scales more than 80 8

7b. Lateral line scales fewer than 80 11

8a. Median indentation on lower lip moderate, 2 or more rows of papillae crossing its midline; 5–6 rows of papillae on
upper lip; lower Klamath River . . . . Klamath smallscale sucker, *Catostomus rimiculus*

8b. Median indentation on lower lip deep, usually only 1 row of papillae crossing its midline; 2–6 rows of papillae on
upper lip 9

9a. Dorsal fin rays 12–13, fin falcate; adults with large fleshy lobes on lower lips; caudal peduncle narrow;
Colorado River . . . . Flannelmouth sucker, *Catostomus latipinnis*[14]

12. Lateral line scales in Klamath tui chubs are fewer than 54. The extinct thicktail chub, *Siphateles crassicauda*, of the Sacramento–San Joaquin
River drainage keys out here.

13. Bigmouth buffalo were introduced into southern California reservoirs and the lower Colorado River, but are probably no longer present.

14. Flannelmouth suckers are present below Davis Dam, Nevada-Arizona, and may occasionally be found in the California reach of the Colorado River above Havasu Reservoir.

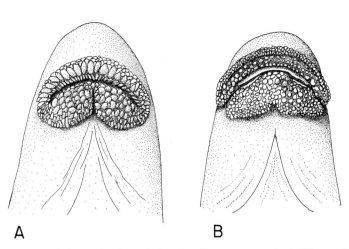

A   B

Figure 25. (A) Mouth of a typical sucker (Sacramento sucker); (B) mouth of a *Pantosteus*-type sucker (mountain sucker).

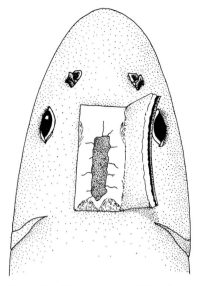

Figure 26. Cutaway view of dorsal surface of Tahoe sucker head, showing well-developed frontoparietal fontanel.

9b. Dorsal fin rays 9–11, fin not falcate; adults without large fleshy lobes on lower lips; caudal peduncle moderate in width 10

10a. Skin-covered opening on top of head (frontoparietal fontanel) small or absent; adults usually less than 20 cm TL; middle Pit River . . . . Modoc sucker, *Catostomus microps*[15]

10b. Frontoparietal fontanel well developed (Fig. 26); adults usually greater than 18 cm TL; Great Basin . . . . Tahoe sucker, *Catostomus tahoensis*

11a. Dorsal fin rays usually 10 or fewer; belly dusky; Owens River . . . . Owens sucker, *Catostomus fumeiventris*

11b. Dorsal fin rays usually 11 or more; belly white to yellow 12

12a. Dorsal fin rays usually 11, occasionally 12; Klamath River . . . . Klamath largescale sucker, *Catostomus snyderi*[16]

12b. Dorsal fin rays usually 12 or more, rarely 11; Sacramento–San Joaquin basin . . . . . Sacramento sucker, *Catostomus occidentalis*

Osmeridae, Smelt Family

1a. Mouth small, maxilla does not reach past middle of eye 2

1b. Mouth large, maxilla usually reaches beyond posterior margin of eye 4

2a. Head length more than 4× eye diameter and more than 2.5× longest anal fin ray; scales in lateral series 66–73; marine . . . . surf smelt, *Hypomesus pretiosus*

2b. Head length less than 4× eye diameter and less than 2.5× longest anal fin ray; scales in lateral series 53–60 3

3a. One or no chromatophores (pigment spots) between mandibles: dorsal fin rays 9–10, anal fin rays 15–17 . . . . delta smelt, *Hypomesus transpacificus*

3b. Ten or more chromatophores between mandibles; dorsal fin rays 7–9; anal fin rays 13–15 . . . . wakasagi, *H. nipponensis*

4a. Pectoral fin, when depressed, reaches, or nearly reaches, pelvic fin base; operculum without concentric striations . . . . longfin smelt, *Spirinchus thaleichthys*

4b. Pectoral fin, when depressed, reaches about halfway to pelvic fin base; operculum with concentric striations . . . . eulachon, *Thaleichthys pacificus*

15. Modoc suckers are sympatric only with Sacramento suckers, from which they can be readily differentiated by their short dorsal fin (10–11 rays) and generally small size at maturity.

16. Klamath largescale suckers, Lost River suckers, and shortnose suckers are highly variable in morphology and are often hard to tell apart at small sizes.

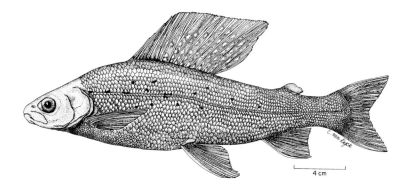

Figure 27. Arctic grayling, 25 cm SL, Alaska.

Salmonidae, Salmon and Trout Family

1a. Dorsal fin long (17+ rays); dorsal fin base longer than head length . . . . Arctic grayling, *Thymallus arcticus*[17] (Fig. 27)

1b. Dorsal fin short (<17 rays); dorsal fin base shorter than head length 2

2a. Mouth small, subterminal, maxilla does not reach middle of eye; scales large
(<100 in lateral line) . . . . mountain whitefish, *Prosopium williamsoni*

2b. Mouth large, terminal, maxilla reaches past middle of eye; scales small (>100 in lateral line) 3

3a. SL less than 12 cm (juveniles) 4

3b. SL greater than 12 cm (adults) 14

4a. Parr marks absent, maximum size 5 cm SL . . . . pink salmon, *Oncorhynchus gorbuscha*

4b. Parr marks present, maximum size 10–12 cm SL 5

5a. Anal fin rays 8–12; anal fin higher than length of base 6

5b. Anal fin rays 13–19; anal fin longer than high 11

6a. Dorsal fin with conspicuous dark spots or with darkened anterior ray 7

6b. Dorsal fin without dark spots or darkened anterior ray 10

7a. Parr marks wide (combined width greater than or equal to combined width of spaces between parr marks); red or
yellow spots present on live wild fish 8

7b. Parr marks narrow (combined width less than combined width of spaces between parr marks); no red or yellow
spots present on live wild fish 9

8a. Parr marks 8–9; adipose fin of live fish plain; tip of chin with dark pigment . . . . brook trout, *Salvelinus fontinalis*

8b. Parr marks 10–12; adipose fin of live fish orange; tip of chin plain . . . . brown trout, *Salmo trutta*

9a. Mouth large, maxillary extending beyond posterior margin of eye; teeth present on rear of tongue; dorsal fin rays
8–11 (usually 10); red slash present along inner edge of lower jaw . . . . cutthroat trout, *Oncorhynchus clarki*

9b. Mouth moderate, maxillary does not extend beyond posterior margin of eye; no teeth on tongue; dorsal fin rays 10–13
(usually 11–12); slash marks usually absent from lower jaw . . . . rainbow trout, *Oncorhynchus mykiss*

10a. Distance from tip of snout to base of dorsal fin about one-half SL; parr marks narrow vertical bars; central Sierra
Nevada . . . . lake trout, *Salvelinus namaycush*

10b. Distance from tip of snout to base of dorsal fin less than one-half SL; parr marks irregular blotches;
McCloud River (extinct) . . . . bull trout, *Salvelinus confluentus*

11a. Parr marks short, only a few reaching below lateral line, if at all 12

11b. Parr marks large, most reaching below lateral line 13

12a. Parr marks small and faint, usually entirely above lateral line; sides of living fish below lateral line iridescent green;
uncommon . . . . chum salmon, *Oncorhynchus keta*

12b. Parr marks sharply defined, usually a few extending slightly below lateral line; sides of living fish below lateral
line silvery . . . . sockeye salmon and kokanee, *Oncorhynchus nerka*

13a. Parr marks wider than interspaces; adipose fin with clear area at base . . . . chinook salmon, *Oncorhynchus tshawytscha*

13b. Parr marks narrower than interspaces; adipose fin completely speckled . . . . coho salmon, *Oncorhynchus kisutch*

14a. Anal fin rays 13–19; anal fin base longer than length of longest ray 15

14b. Anal fin rays 8–12; anal fin base shorter than length of longest ray 19

15a. Large black spots on back and tail 16

17. Grayling, an introduced species, is most likely now extirpated from California. The last known population was in Lobdell Reservoir, Mono
County.

15b. No such spots on back and tail (but fine speckling may be present on back) 18

16a. Spots on back large and oval; more than 160 scales in lateral line; exaggerated hump on back of adult
males . . . . pink salmon, *Oncorhynchus gorbuscha*

16b. Spots on back small and round; fewer than 150 scales in lateral line, hump of spawning males low 17

17a. Gums of lower jaw black; spots present on both lobes of tail; anal fin rays 15–17 . . . . chinook salmon, *Oncorhynchus tshawytscha*

17b. Gums of lower jaw white to gray; spots present on upper lobe of tail only, or absent; anal fin rays
12–15 . . . . coho salmon, *Oncorhynchus kisutch*

18a. Gill rakers short and stout; 19–26 on first gill arch; uncommon . . . . chum salmon, *Oncorhynchus keta*

18b. Gill rakers long and slender; 30–40 on first gill arch . . . . sockeye salmon and kokanee, *Oncorhynchus nerka*

19a. Body with dark spots on light background; teeth present on shaft of vomer (detectable as line of teeth running down
middle of roof of mouth) 20

19b. Body with light spots (e.g., red, orange, green) on dark background; teeth absent from shaft of vomer 23

20a. Dark spots on sides, each surrounded by pale halo; spots usually absent from caudal fin (a few may be present on
dorsal edge) . . . . brown trout, *Salmo trutta*

20b. Dark spots on sides without halos; caudal fins usually heavily spotted[18] 21

21a. Basibranchial teeth present;[19] red slash marks present along inner edges of lower jaw; scale rows between lateral line
and base of dorsal fin 32–48; maxillary extends well beyond posterior edge of eye . . . . cutthroat trout, *Oncorhynchus clarki*

21b. Basibranchial teeth absent; red slash marks absent from lower jaw; scale rows between lateral line and base of dorsal
fin 25–32; maxillary does not extend beyond posterior edge of eye except in some large
(50+ cm) fish . . . . rainbow trout, *Oncorhynchus mykiss*[20]

22a. Tail deeply forked; leading edges of pelvic and anal fins not distinctively pigmented; central Sierra
Nevada . . . . lake trout, *Salvelinus namaycush*

22b. Tail not deeply forked; leading edges of pelvic and anal fins white or cream colored 23

23a. Back mottled with wormlike markings; dorsal and caudal fins marbled . . . . brook trout, *Salvelinus fontinalis*

23b. Back with pale spots, not mottled; dorsal and caudal fins not marbled . . . . bull trout, *Salvelinus confluentus*

Fundulidae, Killifish Family

1a. Number of scales in lateral series more than 30; SL up to 115 mm . . . . California killifish, *Fundulus parvipinnis*

1b. Number of scales in lateral series fewer than 30; SL less than 41 mm . . . . Rainwater killifish, *Lucania parva.*[21]

Cyprinodontidae, Pupfish Family

1a. Dorsal fin equidistant between base of caudal fin and snout; pelvic fins small, usually with 7 rays 2

1b. Dorsal fin closer to base of caudal fin than to snout; pelvic fins reduced or absent, usually with 6 or fewer rays 3

2a. Scales with spinelike projections on circuli; interspaces between circuli not reticulated (Fig. 28A); southern
California . . . . desert pupfish, *Cyprinodon macularius*

2b. Scales without spinelike projections on circuli; interspaces between circuli reticulated (Fig. 28B);
Owens Valley . . . . Owens pupfish, *Cyprinodon radiosus*

3a. Scales in lateral series 27–34; scales before dorsal fin 22–33, usually 25–30 . . . . Salt Creek pupfish, *Cyprinodon salinus*

3b. Scales in lateral series 25–26; scales before dorsal fin 15–24, usually 17–19 . . . . Amargosa pupfish, *Cyprinodon nevadensis*

Poeciliidae, Livebearer Family

1a. Four to eight large black spots on each side . . . . porthole livebearer, *Poeciliopsis gracilis*

1b. Sides without black spots 2

2a. Scales in lateral series 29–32; anal fin rays 6–7; intestine short, without coils; origin of dorsal fin behind origin
of anal fin . . . . western mosquitofish, *Gambusia affinis*[22]

18. Paiute cutthroat trout, *O. clarki seleniris,* have few spots anywhere on the body but possess parr marks as adults.
19. If basibranchial teeth are present, they can be detected by gently feeling the base of the trout's tongue between the gills, with a finger.
20. Golden trout and redband trout are now considered to be subspecies of rainbow trout and will key out here.
21. Rainwater killifish resemble female mosquitofish (Poeciliidae), from which they can be distinguished by number of dorsal rays (9–14 versus 6–7 on mosquitofish).
22. Eastern mosquitofish, *Gambusia holbrooki,* may also be present in the state. It has 7 dorsal fin rays, as opposed to 6 dorsal fin rays on the western mosquitofish.

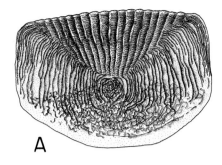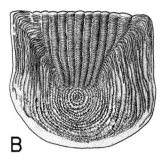

Figure 28. Scales of pupfish: (A) desert pupfish; (B) Owens pupfish. After Miller (1948).

2b. Scales in lateral series 28 or fewer; anal fin rays 8–10; intestine long and coiled; origin of dorsal fin in front of origin of anal fin 3
3a. Dorsal fin with 12 or more rays . . . . sailfin molly, *Poecilia latipinna*
3b. Dorsal fin with fewer than 12 rays 4
4a. Dorsal fin rays usually 10–12; scales in lateral series usually fewer than 26 . . . . variable platyfish, *Xiphophorus variatus*
4b. Dorsal fin rays usually 7–9; scales in lateral series 26–28 5
5a. Mature fish usually greater than 40 mm TL; males nearly equal in size to females; no red or green on body or fins . . . . shortfin molly, *Poecilia mexicana*
5b. Mature fish usually less than 40 mm TL; males much smaller than females; males usually with red or green on caudal fin . . . . guppy, *Poecilia reticulata*[23]

Atherinopsidae, Silverside Family

1a. SL more than 10 cm; marine . . . . topsmelt (adults), *Atherinops affinis*
1b. SL less than 10 cm 2
2a. Pigment spots on bottom of caudal peduncle between anal fin base and caudal fin base in 2 rows; fewer than 3 dorsal scale rows outlined by pigment . . . . inland silverside, *Menidia beryllina*
2b. Pigment spots on bottom of caudal peduncle not in distinct rows; more than 3 dorsal scale rows outlined by pigment; coastal estuaries . . . . topsmelt (juveniles), *Atherinops affinis*

Gasterosteidae, Stickleback Family

1a. Three dorsal spines . . . . threespine stickleback, *Gasterosteus aculeatus*
1b. Five dorsal spines . . . . brook stickleback, *Culea inconstans*

Pholididae, Gunnel Family[24]

1a. Pelvic fins present; V-shaped markings on back, marine . . . . saddleback gunnel, *Pholis ornata* (Fig. 29A)
1b. Pelvic fins absent; back plain, marine . . . . penpoint gunnel, *Apodichthys flavidus* (Fig. 29B)

Centrarchidae, Sunfish Family

1a. Anal fin spines 5 or more 2
1b. Anal fin spines 3 4
2a. Dorsal fin spines 11–13; dorsal fin base much longer than anal fin . . . . Sacramento perch, *Archoplites interruptus*
2b. Dorsal fin spines 5–10; dorsal and anal fin bases about equal in length 3
3a. Dorsal fin spines 7–8; length of dorsal fin base equal to or greater than distance from origin of dorsal fin to eye . . . . black crappie, *Pomoxis nigromaculatus*
3b. Dorsal fin spines 5–6; length of dorsal fin base less than distance from origin of dorsal fin to eye . . . . white crappie, *Pomoxis annularis*

23. Guppies can be expected almost anywhere in the state where there is warm water. The presence of breeding populations in natural or seminatural waters has not been confirmed, but substantial populations exist in some sewage treatment ponds, such as that on the campus of the University of California, Davis.
24. Gunnels are maine fish that occasionally occur in the upper reaches of coastal estuaries.

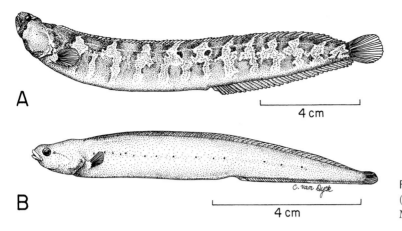

Figure 29. (A) Saddleback gunnel, 14 cm SL; (B) penpoint gunnel, 9 cm SL, both Navarro River, Mendocino County.

4a. Scales large, 53 or fewer in lateral series; body depth usually more than one-third SL; sunfishes[25]   5
4b. Scales small, 58 or more in lateral series; body depth usually less than one-third SL; basses   9
5a. Teeth present on tongue; upper jaw (maxilla) extends beyond middle of eye . . . . warmouth, *Lepomis gulosus*
5b. No teeth on tongue; upper jaw does not extend beyond middle of eye   6
6a. Pectoral fins short and rounded, contained about 4× in SL; mouth large, upper jaw extends to middle
    of eye . . . . green sunfish, *Lepomis cyanellus*
6b. Pectoral fins long and pointed, contained less than 3× in SL; mouth small, upper jaw does not reach middle of eye   7
7a. Dorsal fin with black blotch on base of last few rays; gill rakers long and slender (>2× longer than
    wide) . . . . bluegill, *Lepomis macrochirus*
7b. Dorsal fin without black blotch; gill rakers short and stubby (about 2× longer than wide)   8
8a. Rear portion of dorsal fin speckled; living adults with scarlet spot on opercular flap, and blue and orange stripes
    on cheek . . . . pumpkinseed, *Lepomis gibbosus*
8b. Rear portion of dorsal fin without speckles; living adults with orange or red margin on opercular flap, and without
    conspicuous stripes on cheek . . . . redear sunfish, *Lepomis microlophus*
9a. SL more than 10 cm   10
9b. SL less than 10 cm (young-of-year)   13
10a. Upper jaw extends behind eye; soft and spiny portions of dorsal fin with narrow connection, so spiny portion appears
    strongly convex; lateral stripe well developed . . . . largemouth bass, *Micropterus salmoides*
10b. Upper jaw does not extend behind eye; soft and spiny portions of dorsal fin with broad enough connection that spiny
    portion appears gently rounded; lateral stripes various   11
11a. Lateral band present; lateral line scales fewer than 67 . . . . spotted bass, *Micropterus punctulatus*[26]
11b. Lateral band absent; lateral line scales more than 67   12
12a. Rays in rear portion of dorsal fin usually 13–15; 12–13 scale rows above lateral line . . . smallmouth bass, *Micropterus dolomieu*
12b. Rays in rear portion of dorsal fin usually 11–12; 7–10 scale rows above lateral line . . . . redeye bass, *Micropterus coosae*
13a. Distinct lateral band of botches present   14
13b. No distinct lateral band present   15
14a. Caudal fin without strong banding or bicolored, with dark band running along outer edge; no orange coloration
    present on fins . . . . largemouth bass, *Micropterus salmoides*
14b. Caudal fin tricolored, with black band in the middle and tips of fin pale; orange on tail usually present near caudal
    peduncle . . . . spotted bass, *Micropterus punctulatus*
15a. Narrow vertical bars present on sides, extending below lateral line; dorsal fin rays 11–12 . . . . redeye bass, *Micropterus coosae*
15b. Narrow vertical bars on sides absent or indistinct; dorsal fin rays 13–15 . . . . smallmouth bass, *Micropterus dolomieu*

Moronidae, Temperate Bass Family

1a. Body depth less than one-third SL; head 5× longer than second anal fin spine . . . . striped bass, *Morone saxatilis*
1b. Body depth more than one-third SL; head 3× longer than second anal fin spine . . . . white bass, *Morone chrysops*

25. Sunfishes that seem to be intermediate in their characteristics between two species may be hybrids. Hybrids most likely to be encountered
    are warmouth-bluegill, green sunfish–bluegill, green sunfish–redear sunfish, bluegill–redear sunfish, and green sunfish–pumpkinseed.
    The hybrids are usually dark but highly colored sterile males.
26. Spotted bass are easily confused with largemouth bass when the jaw reaches the margin of the eye; spotted bass have regular rows of spots
    below the lateral stripe, a small patch of teeth on the tongue, and small irregular scales along the bases of the dorsal and anal fins. These
    characters are lacking in largemouth bass.

Embiotocidae, Surfperch Family

1a. Dorsal spines 10 or fewer; estuaries .... shiner perch, *Cymatogaster aggregata*
1b. Dorsal spines 15 or more; fresh water .... tule perch, *Hysterocarpus traski*

Percidae, Perch and Darter Family

1a. Mouth small, upper jaw (maxilla) does not reach to below eye; snout overhangs
    upper lip .... bigscale logperch, *Percina macrolepida*
1b. Mouth large, upper jaw extends to or past eye; snout does not overhang upper lip .... yellow perch, *Perca flavescens*

Cichlidae, Cichlid Family[27]

1a. 8–12 gill rakers on lower half of first arch; lateral line scales 28–30; in adults, head wider than body;
    egg layer .... redbelly tilapia, *Tilapia zilli*
1b. More than 13 gill rakers on lower half of first arch; lateral line scales 30–35; head not wider than body in adults;
    mouthbrooders 2
2a. Mouth in breeding males enlarged, reaching eye, so top of head becomes concave; caudal fin plain; dorsal fin without
    pale upper edge; dark blotches or no markings on sides .... Mozambique tilapia, *Oreochromis mossambicus*
2b. Mouth in breeding males not enlarged, top of head not concave; caudal fin with irregular pigment pattern; dorsal fin
    with pale upper edge; sides of adults plain but juveniles often have 7–10 vertical bars .... Blue tilapia, *Oreochromis aureus*

Gobiidae, Goby Family

1a. Maxillary bone usually does not extend past posterior margin of eye 2
1b. Maxillary bone extends past posterior margin of eye, nearly reaching opercular opening 6
2a. Numerous barbels around mouth .... Shokihaze goby, *Tridentiger barbatus*[28]
2b. No barbels present around mouth 3
3a. Dark bands present on leading edge of dorsal fin 4
3b. No dark bands on leading edge of dorsal fin 5
4a. First ray of pectoral fin separated from rest of fin for about half length of ray (Fig. 30); edges of ray with tiny
    serrations .... chameleon goby, *Tridentiger trigonocephalus*
4b. First ray of pectoral fin separated from rest of fin only at tip; edges of ray smooth ...... shimofuri goby, *Tridentiger bifasciatus*
5a. First dorsal fin with pigmented tip and 8 spines; scales large, fewer than 50 in lateral
    line .... yellowfin goby, *Acanthogobius flavimanus*
5b. First dorsal fin with clear tip and 6–7 spines; scales tiny, more than 60 in
    lateral line .... tidewater goby, *Eucyclogobius newberryi*
6a. Dorsal fins widely separated; anal fin elements 9–14; second dorsal fin elements
    9–14 .... longjaw mudsucker, *Gillichthys mirabilis*
6b. Dorsal fin edges nearly touching; anal fin elements 15–18; second dorsal fin elements 14–18 .... arrow goby, *Clevelandia ios*[29]

Cottidae, Sculpin Family[30]

1a. Spine on operculum large, branched, and sharp .... staghorn sculpin, *Leptocottus armatus*
1b. Spine(s) on operculum small and simple (Fig. 31) 2
2a. Pelvic rays 3 3
2b. Pelvic rays 4 4
3a. Cirri (small soft tufts) present on head; marine .... sharpnose sculpin, *Clinocottus acuticeps*[31]
3b. No cirri on head; middle Pit River drainage .... rough sculpin, *Cottus asperrimus*

27. The three species in this key are apparently the ones most widely distributed in southern California. The Nile tilapia is apparently also present but difficult to tell from the blue and Mozambique tilapia. Hybrids among the species are common, so identification is difficult.
28. Shokihaze gobies appeared in the San Francisco Bay estuary in the 1990s in brackish water. They are uncommon (so far). Chameleon gobies are marine (introduced) and are included because of their past confusion with shimofuri gobies.
29. The arrow goby is an occasional marine visitor to the lower reaches of coastal streams along the entire coast. It rarely reaches 5 cm SL, whereas the longjaw mudsucker reaches 20 cm SL.
30. Sculpins are highly variable. Keying results should be carefully checked with species descriptions and distributions.
31. Sharpnose sculpins have been collected once in fresh water in Del Norte County.

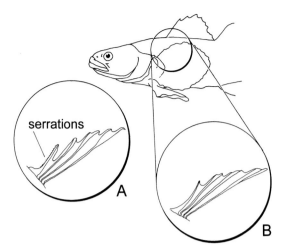

Figure 30. Upper part of pectoral fin in two species of *Tridentiger*.

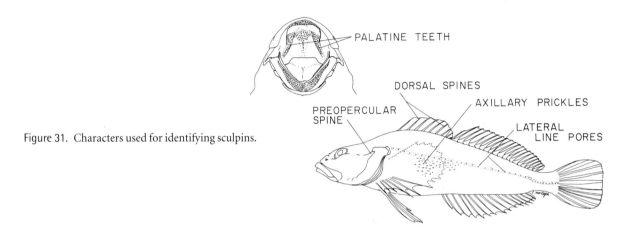

Figure 31. Characters used for identifying sculpins.

4a. Long anal fin (15–18 rays) and second dorsal fin (19–23 rays) . . . . prickly sculpin, *Cottus asper*

4b. Short anal fin (usually <15 rays) and second dorsal fin (<20 rays)   5

5a. No axillary prickles present   6

5b. Patch of tiny prickles present underneath pectoral fin (axillary prickles; Fig. 31)   7

6a. Dorsal fins separated; dark patch present on front of first dorsal fin; two median pores on chin; east side of
     Sierra Nevada . . . . Paiute sculpin, *Cottus beldingi*

6b. Dorsal fins joined; no dark patch on front of first dorsal fin; one median pore on chin; Klamath and Pit
     drainages . . . . marbled sculpin, *Cottus klamathensis* (some)

7a. Two conspicuous saddle marks on back; no dark spot on first dorsal fin; coastal
     drainages . . . . coastrange sculpin, *Cottus aleuticus*

7b. Saddle marks absent or diffuse; dark spot on first dorsal fin; interior drainages   8

8a. Dorsal fins obviously connected; lateral line does not reach end of caudal peduncle (incomplete); dorsal spines
     usually 7 . . . . marbled sculpin, *Cottus klamathensis* (some)

8b. Dorsal fins not connected or connected only at base; lateral line usually complete; dorsal spines usually 8–9   9

9a. Mouth large, usually wider than body behind pectoral fins 10

9b. Mouth small, narrower than body behind pectoral fins; Rogue River drainage . . . . reticulate sculpin, *Cottus perplexus*

10a. First dorsal fin with 8–9 spines; dorsal fins separate (may touch at base); lateral line pores usually more than 32;
     palatine teeth absent; Pit River drainage . . . . Pit sculpin, *Cottus pitensis*

10b. First dorsal fin with 7–8 spines; dorsal fins connected; lateral line pores fewer than 32; palatine teeth usually
     present (Fig. 31); Central Valley drainages . . . . riffle sculpin, *Cottus gulosus*

Lampreys, Petromyzontidae

Lampreys are specialized aquatic vertebrates, eel-like in form but lacking the jaws and paired fins of true fishes. They are distantly allied to the long-extinct ostracoderms, among the earliest known vertebrates, which were heavily armored creatures that sucked organic ooze from ocean, lake, and river bottoms (Moyle and Cech 2000). Like these ancient jawless fishes, lampreys have a persistent notochord, a cartilaginous skeleton, a single nostril, a small brain, and two semicircular canals in each side of the head, rather than the usual three.

Survival of lampreys into modern times has depended on their ability to prey on the jawed fishes that replaced their ancestors. An adult lamprey will latch onto the side of a large fish with its suckerlike mouth and rasp a hole with its powerful tongue, which is covered with sharp, horny plates ("teeth"). The feeding lamprey extracts blood and body fluids from fish and drops off when satiated. Although the gaping wound left by the lamprey may be fatal, many fish do survive lamprey attacks. It is not unusual to find fish with two or more lamprey scars. Under normal conditions lamprey and their prey coexist successfully; lampreys maintain their populations without destroying those of their prey. However, when sea lampreys (*Petromyzon marinus*) invaded the Great Lakes, they nearly succeeded in wiping out the large fishes, presumably because the fish were not adapted to their style of predation.

The predatory portion of the lamprey life cycle is usually short (6–19 months) compared with the portion spent as larvae (ammocoetes) in streams (3–7 years). Generally, adults migrate upstream from a large body of water into a tributary stream to spawn. They build a nest in a gravel-bottomed area, spawn, and usually die. The embryos hatch and the ammocoetes are carried downstream to mud- or sand-bottomed backwaters and stream edges. They burrow into the bottom and spend the next few years growing on a diet of detritus and algae. The role of ammocoetes in the ecology of streams remains largely unstudied, although they are often found in the stomachs of predatory fishes.

One of the most fascinating aspects of lamprey biology is the frequent evolution of nonpredatory species from predatory ones. The nonpredatory species are generally small as adults, and their rasping plates are reduced in size and number. The larval portion of their life cycle is like that of the predatory forms except that it tends to last longer, and the ammocoetes thus tend to grow larger (Hardisty and Potter 1971). The adults, however, do not migrate after metamorphosis but remain in their home streams, where they spawn and die without feeding. The nonpredatory adult stage allows lampreys to live in small streams, where few large fishes are present for food or where distances to large bodies of water are great. Both predatory and nonpredatory lampreys are common in California, but their taxonomy is complex. Most nonpredatory lampreys on the Pacific coast are derived from river lamprey, which are small in size and capable of living in freshwater as adults. However, in the upper Klamath drainage there is a taxonomically difficult group of predatory and nonpredatory lampreys that are all derived from Pacific lamprey. Pacific lamprey normally requires a period in salt water to complete its life history, but freshwater populations are known (Beamish 1980).

Classification and identification of lampreys depend largely on the number, structure, and position of the horny plates (usually labeled teeth or laminae) on the sucking disc. The plates are named according to their position (anterior, posterior, or lateral) in the three concentric circles that can be visualized on the disc (Fig. 32). They are described in detail by Vladykov and Follett (1962) and Hubbs and Potter (1971). Lamprey identification, particularly of small adults, should be performed with care. Ammocoetes can be identified with the aid of Richards et al. (1982) and Wang (1986).

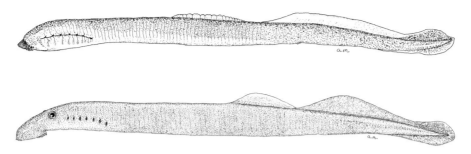

Figure 32. Pacific lamprey. Top: Ammocoete, 12 cm TL, San Joaquin River, Fresno County. Middle: Dwarf form, 24 cm TL, Clear Lake Reservoir, Modoc County. Bottom: Sucking disc, adult, after Vladykov and Follett (1962).

Pacific Lamprey, *Lampetra tridentata* (Gairdner)

Identification Any large (>40 cm TL) adult lamprey in California belongs to this species. However, dwarf (15–30 cm TL) landlocked populations also exist, and these should not be mistaken for recently transformed, silvery individuals of anadromous populations. Sharp, horny plates (teeth) are present in all areas of the sucking disc, more than in any other California lamprey (4). The most distinctive plate is the crescent-shaped supraoral lamina with three sharp cusps, the middle cusp smaller than the two lateral cusps. There are also four large, inner lateral plates on each side. The middle two are tricuspid, the outer two bicuspid (formula 2-3-3-2). The tongue ends in 14–21 small points (transverse lingual lamina), the middle one slightly larger than the others. The two dorsal fins are slightly separated.

The second dorsal is continuous with the caudal fin. Adults have 62–71 body segments (myomeres). The horizontal diameter of the eye is 2–4 percent of the total length, and the length of the oral disc is usually 6–8 percent of the total length. The dorsal fins are higher in males than in females, and males lack an anal fin, which is conspicuous in females. Males also possess small genital papillae. Ammocoetes have 68–70 segments between the anus and the last gill opening (15). The body and lower half of the oral hood are usually dark and well pigmented, although there is typically a pale area associated with a ridge in the caudal region.

Spawning adults are typically dark (usually a greenish-black color) on top but paler on the belly, frequently a golden color. Newly metamorphosed individuals are silvery in color. Adults in Goose Lake are a shiny bronze color.

Taxonomy Pacific lampreys have given rise to landlocked populations throughout their range, despite the difficulties adults have in living in fresh water (1). A number of these populations have been described as separate species, some predatory (e.g., *L. similis, L. minima*), some nonpredatory (e.g., *L. lethophaga*). There is often considerable overlap in characters among the Pacific lamprey and its derivatives, as well as between predatory and nonpredatory forms, so the interrelationships among the species require close examination, especially in the upper Klamath basin (2, 3, 29). Studies using mitochondrial DNA show promise in resolving the issues (27). A particular problem is the taxonomic status of the dwarf predatory lamprey inhabiting the isolated waters of Goose Lake, first noted by Carl Hubbs in 1925 (17). Studies by C. Bond (unpublished) indicate that the Goose Lake lamprey represents a distinct taxon. Given its long isolation from other Pacific lamprey populations, as well as its distinctive appearance and ecology, it is quite likely that the Goose Lake lamprey deserves recognition as a full species. Molecular studies also suggest that this lamprey is distinct from lampreys in the Klamath River (25).

It is possible that Pacific lampreys within one stream system have more than one run (22) or that some upstream populations have individuals that remain resident, rather than going to sea, much like rainbow trout. In the Trinity River, for example, there may be two distinct forms of Pacific lamprey, one smaller and paler than the other, that represent either separate runs or resident versus migratory individuals (26).

The Pacific lamprey was formerly placed in the genus *Entosphenus,* now recognized as a subgenus that includes the Pacific lamprey and its nonpredatory derivatives. However, studies of mitochondrial DNA indicate that the genus should probably be resurrected for the group that includes Pacific lamprey, Pit-Klamath brook lamprey, and Klamath River lamprey (27). If this designation is adopted, river lamprey, western brook lamprey, and Kern brook lamprey would remain in the genus *Lampetra.*

Names *Lampetra* is apparently derived from the Latin words *lambere,* to suck, and *petra,* stone, although the Oxford English Dictionary indicates that it may just be an "etymologizing perversion" of the word *lamprey,* of uncertain origin. The words refer to the lamprey habit of clinging to stones in streams with their suckerlike mouths. *Tridentatus* (three-toothed) is a reference to the structure of the supraoral laminae. Lampreys are frequently called eels by fishermen, and large runs of lampreys are responsible for the name Eel River.

Distribution Pacific lampreys are found in Pacific coast streams from Hokkaido Island, Japan (16), through Alaska, and down to Rio Santo Domingo in Baja California (18). Malibu Creek, Los Angeles County, seems to be the southernmost point of regular occurrence in California, despite some records from the Santa Ana River (5) and a single ammocoete taken from the San Luis Rey River (San Diego County) in 1997 (28). However, there are also recent records from Rio Santo Domingo, Baja California (18). In general, lampreys today have a scattered or disjunct distribution south of San Luis Obispo County (5), although there are regular runs in the Santa Clara River (19). In the ocean they have been captured from waters near Japan to Baja California (6, 7). Dwarf, landlocked forms have been identified from the upper Klamath River (4, 8) and from Goose Lake, Modoc County; these forms may be separate species. A recently (1963) landlocked population exists in Clair Engle Reservoir on the Trinity River, Trinity County.

Life History Pacific lampreys, with the exception of landlocked populations, spend the predatory phase of their life in the ocean. They attack a wide variety of fishes, including various salmon and flatfishes (9). In British Columbia 14–45 percent of salmon in different runs had scars from lamprey attacks (9), but similar data are not available from California. Lampreys themselves are often observed with parts of their tails missing, indicating that they are prey for

other fishes, especially sharks. In the mouth of the Rogue River, Oregon, sea lions consume migrating lampreys in large numbers (10). Despite far-flung oceanic records, it is unlikely that Pacific lampreys normally wander far from the mouths of their home spawning streams, because their prey is most abundant in estuaries and other coastal areas. The oceanic phase apparently lasts 3–4 years in British Columbia (9), but it may be shorter in more southern waters. Landlocked forms spend the predatory phase (of unknown duration) in lakes or reservoirs, feeding on suckers and other large fishes (11). In Goose Lake the major prey seems to be tui chubs, although redband trout were presumably once important prey as well (17).

Adults, 30–76 cm TL, usually move up into spawning streams between early March and late June. However, upstream movements in January and February have also been observed (19, 21), and movements into July have been observed in northern streams. In the Trinity River some migration has also taken place in August and September (12). It is quite possible that Pacific lamprey in large river systems, such as the Klamath and Eel, have a number of distinct runs, like salmon. One indication is that many lampreys migrate upstream several months to a year before they spawn (9, 19), hiding under stones and logs until fully mature. In the Klamath River there may be at least two distinct runs: a spring run that spawns immediately after the upstream migration and a fall run, which holds over and spawns in the following spring (22).

Most upstream movement takes place at night and tends to occur in surges, although small numbers may move upstream more or less continuously over a two- to four-month period. In the Santa Clara River (Ventura County) first movement occurs after winter rains breach the sand bar blocking the lagoon at the mouth in January, February, or March; within 6–14 days, the first lampreys reach a fish ladder 16.8 km upstream (19). Although lampreys typically move upstream during periods of high flow, they will migrate under a wide range of flows—25 to 1,700 m³/min— in the Santa Clara River (19). Lampreys can move considerable distances, stopped only by major barriers, such as Friant Dam on the San Joaquin River and Scott Dam on the Eel River. How far upstream lampreys originally migrated in California is not known, but I have observed them spawning in Deer Creek (Tehama County), about 440 km from salt water. Presumably migrations of 500–600 km were once not unusual.

The remarkable ability of Pacific lampreys to surmount less formidable barriers is described by Kimsey and Fisk (20, p. 6):

Great wriggling masses of lampreys are often seen ascending barriers and fish ladders on coastal streams in the early spring. . . . In many cases the flow is too great for the fish to move across the barrier in one attempt. They solve the problem by swimming until tired, then attaching themselves to the bottom and sides and resting for a while. When recovered, they make another attempt and move upstream several more feet. In this manner, by successive spurts and resting periods, they move over various obstructions until they reach their spawning grounds.

Both sexes help construct a crude nest, 35–60 cm in diameter, by removing the larger stones from a gravelly area where current is fairly swift and depths are 30–150 cm. Water temperatures are typically 12–18°C. On 10 April 1991, I observed lampreys spawning in a rocky riffle of the lower American River; the mean depth of 34 nests was 59 cm (range, 30–82 cm), and the mean water column velocity over nests was 64 cm/sec (range, 24–84 cm/sec). Another lamprey nest was observed among silt-covered cobbles in a backwater, where the mean water column velocity was only 11 cm/sec (depth 44 cm). In Putah Creek, on 5 May 1999, the mean depth of 26 nests on a gravelly road crossing was 50 cm (range, 36–73 cm), and mean water column velocity was 29 cm/sec (range, 17–45 cm/sec). In Deer Creek I observed nest construction at depths up to 1.5 m. To remove a stone during nest construction, the lamprey latches on to the downstream side and swings vigorously in reverse. Sometimes, two will pull simultaneously on the same stone. Usually the combination of lamprey pulling and current pushing is enough to move the rock downstream. The final result is a shallow depression with a pile of stones at the downstream end.

For the spawning act, the female attaches to a rock on the upstream edge of the nest, while the male attaches himself to the head of the female, wrapping his body around hers. Occasionally, they may both attach to rocks, but remain side by side (15). Both lampreys then vibrate rapidly, and a small white cloud of eggs and milt is released. The fertilized eggs are washed into the gravel, especially at the downstream end of the nest, where they adhere to the rocks. After spawning the lampreys loosen rocks from above the nest, causing silt, sand, and gravel to cover the eggs. Spawning is repeated on the same nest a number of times until both sexes are spent. Because several pairs often spawn in the same area, males may mate with more than one female (15). Usually, both sexes die shortly thereafter. However, some adults were found to survive and spawn again a year later (at a larger size) in Washington streams (24). The presence of live adult lampreys in downstream migrant traps on the Santa Clara River (19) suggests that repeat spawning also occurs in California. If the fecundity of Pacific lampreys is similar to that of eastern sea lampreys, each female, depending on her size, lays 20,000–200,000 eggs.

The embryos hatch in about 19 days at 15°C. After hatching ammocoetes spend a short time in the nest gravel. Eventually they swim up into the current and are washed down-stream to a suitable area of soft sand or mud. Ammocoetes burrow tail first into the sand or mud and begin lives as filter feeders, sucking organic matter and algae off the substrate surface. They do not stay in one area for their entire growth period. Active ammocoetes can be trapped at almost any time of the year (12, 13). In the Trinity River ammocoetes of sizes down to 16 mm colonized areas from which they had been eradicated during the winter high-water period (12). Most movement takes place at night. The length of the ammocoete stage is uncertain, but it probably lasts 5–7 years. Ammocoetes reach 14–16 cm TL when they start the dramatic metamorphosis from reclusive, detritus-feeding larvae to active, predatory adults. They develop large eyes, a sucking disc, silver sides, and dark blue backs; they also demonstrate radical changes in internal anatomy (7). There are dramatic changes in physiology, such as development of the ability to tolerate abrupt transfer into sea water, which is lethal to ammocoetes (23). Downstream migration begins when transformation is completed, seemingly during high-outflow events in winter and spring, perhaps coincident with the upstream migration of the adults.

Status ID (anadromous form). IC. (Goose Lake form). Anadromous Pacific lampreys are still present in most of their native areas, but large runs that once characterized streams such as the Eel River seem to have largely disappeared. Certainly the once-common "great wriggling masses" are rarely seen. Unfortunately, little attention has been paid to lampreys, and there is only anecdotal evidence (mainly from Native American fishermen) that runs in North Coast streams are much smaller than they used to be. They have been eliminated from many streams in the urbanized southern end of their range, but they are remarkably persistent, as indicated by the continuing runs up the Santa Clara River (19), which has relatively undisturbed upper reaches. In Putah Creek (Yolo and Solano Counties) they managed to maintain small runs following construction of the Solano Project, which dried up much of the lower creek. Pacific lamprey are usually absent from highly altered or polluted streams. In October 1979 Wang (15) collected lampreys from the Napa River that were "intoxicated" with wine spilled into the river! Presumably other pollutants have had worse effects.

Despite their predaceous habits, they seem to have little effect on fish populations and are at times themselves important prey of sea lions. Lampreys were highly esteemed as food by a number of Native American tribes in California (14) and are still considered a delicacy in some European countries. There is a major need to examine the status of the species throughout its range, as well as to study its biology to see, for example, if multiple runs exist in some rivers, like those of chinook salmon and steelhead, or if landlocked strains are present in larger river systems.

Populations of the Goose Lake lamprey should be monitored because Goose Lake is susceptible to drying during periods of drought and because its tributary streams are all altered and diverted. Fortunately, small populations persist in at least one reservoir in the drainage (25). It is of major importance to develop an understanding of the taxonomy and life history requirements of this form for conservation purposes. Likewise, the landlocked "Pacific" lampreys of the upper Klamath drainage must be both studied and monitored.

References 1. Beamish and Northcote 1989. 2. Bailey 1980, 1982. 3. Bond and Kan 1973. 4. Vladykov and Kott 1979. 5. Swift et al. 1993. 6. Hubbs 1967. 7. McPhail and Lindsey 1970. 8. Hubbs 1971. 9. Beamish 1980. 10. Jameson and Kenyon 1977. 11. Coots 1955. 12. Moffett and Smith 1950. 13. Long 1968. 14. Kroeber and Barrett 1960. 15. Wang 1986. 16. Morrow 1980. 17. Hubbs 1925. 18. Ruiz-Campos and Gonzalez-Guzman 1996. 19. ENTRIX 1996. 20. Kimsey and Fisk 1964. 21. Trihey and Associates 1996a. 22. Anglin 1994. 23. Richards and Beamish 1981. 24. Michael 1984. 25. M. Docker, University of Northern British Columbia, pers. comm. 1999. 26. T. Healey, CDFG, pers. comm. 1995. 27. Docker et al. 1999. 28. C. Swift, pers. comm. 1999. 29. Lorion et al. 2000.

Pit-Klamath Brook Lamprey, *Lampetra lethophaga* Hubbs

Identification This is a small (<21 cm TL), nonpredatory lamprey (1). Their disc resembles that of Pacific lamprey, but the plates (teeth) are smaller and fewer. The lateral circumoral plates typically number 1-2-2-1 or 2-3-3-2, but cusps are frequently missing. The posterior circumoral plates number 9–15, many with just one cusp. The supraoral plate has 3 cusps, although the middle one may be degenerate or missing. Infraoral teeth are usually 5. The cusps on the transverse lingual lamina are filelike and difficult to see. The mouth is small and puckered, with disc length less than 5 percent of total length. When the disc is expanded, it is narrower than the head (3). Trunk myomeres number 60–70. The gut is atrophied in mature specimens. Adults tend to be dark gray on top but brass to bronze ventrally.

Taxonomy This nonpredatory species was described by Hubbs (1) from specimens collected in scattered localities in two drainages, the Pit and the Klamath. It is closely related to the Pacific lamprey (4). Populations in the two drainages, however, may have been independently derived from a predatory member of the Pacific lamprey complex and thus may represent separate taxa. A form from the Klamath River was described as a species, *L. folletti* (5), but the species has not been widely recognized (6). Technically, *L. folletti* should continue to be recognized as a species until its designation has been formally refuted in a thorough analysis. C. Bond (8) indicated that brook lampreys in the Goose Lake drainage differ from those in the Pit River drainage and may deserve separate taxonomic recognition. The Pit-Klamath brook lamprey may have given rise to the predatory dwarf Miller Lake lamprey (*L. minima*) of the Klamath basin (7).

Names "The name *lethophaga,* figuratively referring to the elimination of feeding as adults, is formed by combining the Latinized expression *leth* . . . a forgetting or forgetfulness . . . [and] *phag-*, to eat" (1, p. 151). Other names are as for the Pacific lamprey.

Distribution The Pit-Klamath brook lamprey is limited to the Pit River system in northeastern California and the upper Klamath River of south-central Oregon, above the Klamath lakes (1, 2). In Oregon the only recorded populations seem to be in Crooked Creek, a tributary to Agency Lake, and the Sprague River system, a tributary to the Williamson River. However, distributional records should be treated with a certain amount of skepticism until the taxonomy of Klamath-region lampreys has been worked out.

Life History The principal habitats of this species are in low-gradient reaches of clear, cool (summer temperatures rarely reach 25°C) rivers and streams with sand-mud bottoms or edges. Trout are frequently in the same waters, as are marbled and rough sculpins and speckled dace. The ammocoetes burrow into soft bottoms, often among aquatic plants (2), where they presumably feed on algae and detritus. The time spent

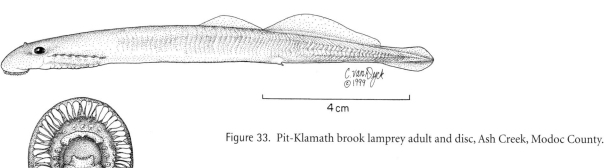

Figure 33. Pit-Klamath brook lamprey adult and disc, Ash Creek, Modoc County.

as ammocoetes seems to be at least 4 years, based on an analysis of size classes. Maximum size is about 21 cm.

Metamorphosis probably takes place in autumn. Spawning does not begin until early spring but may occur anytime during summer. Some populations, although transforming into the adult form, do not develop nuptial features characteristic of "normal" spawners: dark, contrasting coloration of back and belly; united, thick, and frilled dorsal fins; and enlarged anal fin.

Status IE. The Pit-Klamath brook lamprey is widely distributed in at least the Pit River watershed and seems to be in no danger. Some human changes in streams may actually benefit them; in Rush Creek, Modoc County, large numbers of ammocoetes were found in a silty-bottomed pool immediately below a channelized section. They were also common in muddy-bottomed irrigation diversions from the creek.

References 1. Hubbs 1971. 2. Moyle and Daniels 1982. 3. Page and Burr 1991. 4. Docker et al. 1999. 5. Vladykov and Kott 1976b. 6. Robins et al. 1991. 7. Lorion et al. 2000. 8. C. Bond, pers. comm.

Klamath River Lamprey, *Lampetra similis* (Vladykov and Kott)

Description The Klamath River lamprey is a small (14–27 cm TL, mean 21 cm) predatory lamprey with strong, sharply hooked cusps on the oral plates (2). Like the Pacific lamprey, it has 3 strong cusps on the supraoral plate. It has 13 teeth in the anterior field above the mouth, 4 inner lateral plates on each side with the typical cusp formula of 2-3-3-2, 20–29 cusps on the transverse lingual lamina (tongue plate), and 18 teeth in the posterior field below the mouth. There are 8 velar tentacles. Trunk myomeres number 58–65 (usually 60–63). The disc length is about 9 percent of the total length, and the disc is as wide or wider than the head. The horizontal diameter of the eye is about 2 percent of total length. Coloration is similar to that of the Pacific lamprey, although this lamprey is often more heavily pigmented. Ammocoetes have not been described.

Taxonomy Five species of lamprey have been described from the upper Klamath basin: *L. tridentata* (dwarf Pacific lamprey), *L. lethophaga* (Pit-Klamath brook lamprey), *L. minima* (Miller Lake lamprey), *L. folletti* (Modoc brook lamprey), and *L. similis*. The dwarf, landlocked Pacific lamprey is the presumptive ancestor of the others. The Pit-Klamath brook lamprey seems to be generally accepted as the standard non-predatory lamprey of the upper Klamath and Pit River drainages, and the Miller Lake lamprey is accepted as an unusually small predatory species. The other forms are more controversial. The Modoc brook lamprey was described as a nonpredatory species (1) but has not been widely accepted as distinct (4, 5, 6). In contrast, the Klamath River lamprey is distinct not only morphologically but also biochemically (4, 7).

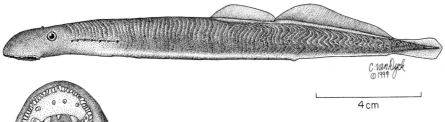

Figure 34. Klamath River lamprey adult and disc, 16.7 cm TL, Seiad Creek, Siskiyou County.

4 cm

1 cm

Names The name *similis* comes from the close resemblance of this species to dwarf Pacific lamprey. Other names are as for the Pacific lamprey.

Distribution This species was described from the upper Klamath River and Upper Klamath Lake in southern Oregon. However, it appears to be widespread in the lower Klamath and Trinity Rivers and tributaries (8). The predatory lamprey population in Copco Reservoir, Siskiyou County, is presumably this species (3). Lamprey ammocoetes identified as *L. similis* have been collected from the Merced River in the Central Valley, but their biochemical similarity to *L. tridentata* (4) suggests they are not *L. similis*.

Life History There is no specific information on the biology of this species, although adults seem to live in the Klamath River itself, as well as in lakes and reservoirs, where they prey on the native suckers and cyprinids.

Status ID. This designation is based on their limited range in rivers that have been severely modified by dams, diversions, and pollution. There is a real need for regular, systematic surveys of the upper Klamath basin for lampreys and other native fishes, to determine their status more accurately. A particular need for this species is to determine the habitat required for spawning and for ammocoetes.

References 1. Vladykov and Kott 1976b. 2. Vladykov and Kott 1979. 3. Coots 1955. 4. Docker et al. 1999. 5. Robins et al. 1991. 6. C. Bond and T. Kan, unpubl. ms. 1991. 7. Lorion et al. 2000. 8. J. Boyce, Humboldt State University, pers. comm. 2001.

River Lamprey, *Lampetra ayresi* (Gunther)

Identification The river lamprey is small (average TL of spawning adults about 17 cm) and predaceous, with an oral disc generally at least as wide as the head. The horny plates (teeth) of the oral disc are well developed (1, 2) but become progressively blunter in spawning individuals. The middle cusp of the transverse lingual lamina is well developed. There are 3 large lateral plates (circumorals) on each side, the outer 2 bicuspid, the middle one tricuspid. The supraoral plate has only 2 cusps that often appear as separate teeth, whereas the infraoral plate has 7–10 cusps. The eye is large compared with that of other California lampreys, the diameter being 1–1.5 times the distance from the posterior edge of the eye to the anterior edge of the first branchial opening. The number of trunk myomeres is high, averaging 68 in adults, 67 (range, 65–70) in ammocoetes. Adult river lampreys are dark on the back and sides, silvery to yellow on the belly. The tail is darkly pigmented. As they become sexually mature, the gut degenerates and the two dorsal fins grow closer together, eventually joining. Ammocoetes can be recognized by their pale heads (especially around the gill openings), a prominent line behind the eye spot, and a tail with a lightly pigmented center (3).

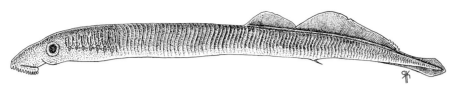

Figure 35. River lamprey. Adult, unknown locality, from Lee et al. (1980).

Taxonomy In 1855, William O. Ayres described the river lamprey as *Petromyzon plumbeus* from a single specimen collected in San Francisco Bay. Unfortunately, that name had already been given to a European lamprey. So in 1870 A. Gunther renamed it *P. ayresi.* In 1911 C. T. Regan decided this species and the European river lamprey, *Lampetra fluviatilis,* were identical. This diagnosis was accepted until 1958, when careful redescription of the river lamprey by V. D. Vladykov and W. I. Follett showed that it is indeed a distinct species, *L. ayresi* (1).

Names *Ayresi* is after William O. Ayres, who first recognized it as a species. Ayres was a San Francisco physician who was the first to describe a number of California's freshwater fishes. Other names are as for the Pacific lamprey.

Distribution River lampreys have been collected from large coastal streams from 20 km north of Juneau, Alaska, to San Francisco Bay (1, 2). In California most records are for the lower Sacramento–San Joaquin River system, but they have not really been looked for in most other streams. They are present in the Napa River, Sonoma Creek, and Alameda Creek, tributaries to San Francisco Bay (11), and in the lower Sacramento and San Joaquin Rivers, especially the Stanislaus and Tuolumne Rivers. A number are captured every year in the fish rescue facilities in the south Delta. They also appear to be regular spawners in Salmon Creek and in tributaries to the lower Russian River (Sonoma County) (12). In the Eel River a single adult female was collected at Cape Horn Dam. Outside California, they also apparently exist as widely scattered, isolated populations. In Oregon they are known only from the Columbia and Yaquina Rivers (separated by 182 km) (4). Likewise, they are known only from two large river systems in British Columbia, in the center of their range (10).

Life History The biology of river lampreys has not been studied in California, so information in this account is based on studies in British Columbia (5, 6), where the timing of life history events may not be the same owing to colder water or other factors.

Ammocoetes begin transformation into adults at about 12 cm TL, during summer. The process of metamorphosis takes 9–10 months, the longest known for any lamprey (6). Lampreys in final stages congregate immediately upriver from salt water and enter the ocean in late spring. Adults apparently spend only 3–4 months in salt water, where they grow rapidly to 25–31 cm TL.

River lampreys feed on a variety of fishes that are 10–30 cm TL, but most commonly herring and salmon (5, 7, 8). Unlike other lampreys in California, river lampreys typically attach to the back of the host fish, above the lateral line, where they feed on muscle tissue. Feeding continues even after the death of the prey. The effect of river lamprey predation on prey populations can be significant; in Canada, it is considered to be a major source of salmon mortality (8). River lampreys can apparently feed in fresh water, and a landlocked population may exist in upper Sonoma Creek, Sonoma County (9).

Adults migrate back into fresh water in autumn. The extent and timing of migrations in California are poorly known, although a mature adult found at Cape Horn Dam (25 May 1992) on the Eel River must have moved at least 250 km upstream. They spawn during February through May in tributary streams. While maturing, river lampreys shrink about 20 percent in length (5). They dig saucer-shaped depressions in gravelly riffles for spawning. Fecundity estimates from two females from Cache Creek (Yolo County) were 37,300 eggs (17.5 cm TL) and 11,400 eggs (23 cm TL) (1). Adults die after spawning. Ammocoetes remain in silty backwaters and eddies to feed on algae and microorganisms. The length of the ammocoete stage is not known, but it is probably 3–5 years, so total life span is likely to be 6–7 years.

River lampreys are capable of hybridizing with western brook lampreys under artificial conditions, but hybrids have not been observed in the wild. Apparently, a major barrier to hybridization is the slightly larger size of river lampreys (10).

Status ID. Trends in populations of river lamprey are unknown for the southern end of its range, but it is likely that the species has declined, along with the decline of suitable spawning and rearing habitat in the lower reaches of larger rivers. However, river lamprey are easy to overlook, so the species may be more abundant than indicated. It is abundant in British Columbia, but there are relatively few records from California. Its distribution, abundance, life history, and habitat requirements should be investigated in California.

References 1. Vladykov and Follett 1958. 2. Wydoski and Whitney 1979. 3. Richards et al. 1982. 4. C. Bond, Oregon State University, pers. comm. 5. Beamish 1980. 6. Beamish and Youson 1987. 7. Roos et al. 1973. 8. Beamish and Neville 1995. 9. Wang 1986. 10. Beamish and Neville 1992. 11. R. Leidy, USEPA, pers. comm. 1999. 12. M. Fawcett, pers. comm. 1998.

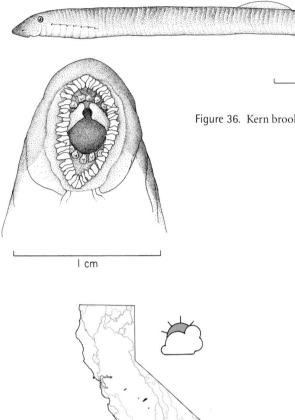

Figure 36. Kern brook lamprey adult and disc, 11.7 cm TL, Merced River, Merced County.

Kern Brook Lamprey,
Lampetra hubbsi (Vladykov and Kott)

Description The Kern brook lamprey is nonpredatory, with poorly developed plates (teeth) on its oral disc. Adults are 8–14 cm TL, ammocoetes 11–15 cm. The number of trunk myomeres is 51–57 (1, 2) with a mean of 54. The supraoral plate typically has 2 cusps. Between 3 and 4 (usually 4) lateral teeth are visible on each side of the disc, each with a single cusp (1). The disc is narrower than the head. The sides and dorsal region are gray-brown, and the ventral area is white. The dorsal fins are unpigmented, but there is some black pigmentation restricted to the area around the notochord in the caudal fin.

Taxonomy This brook lamprey was thought to be derived from the Pacific lamprey, based on its dentition (1). Biochemical evidence, however, indicates that it is most closely related to the river lamprey, as are most other brook lampreys (4).

Names Hubbsi is after Carl L. Hubbs, one of the great ichthyologists of the 20th century, and the description is published in a festschrift volume in his honor (1). Other names are as for the Pacific lamprey.

Distribution This species is endemic to the east side of the San Joaquin Valley. Kern brook lampreys were first collected from the Friant-Kern Canal but have since been found in the lower Merced, Kaweah, Kings, and San Joaquin Rivers (2). Ammocoetes found in the San Joaquin River between Millerton Reservoir and Kerckoff Dam probably also belong to this species (3), as do those collected in the Kings River above Pine Flat Dam (Fresno County). In 1988 ammocoetes and adults were collected from the siphons of the Friant-Kern canal when they were poisoned as part of an effort to eradicate white bass from the system.

Life History Principal habitats of the Kern brook lamprey are silty backwaters of rivers emerging from the Sierra foothills (mean elevation 135 m, range 30–327 m). Ammocoetes are usually in shallow pools and along edges of runs where flows are slight. They favor substrates that are a mixture of sand and mud at depths of 30–110 cm, where summer temperatures rarely exceed 25°C (2). This habitat also characterizes the lightless siphons of the Friant-Kern Canal, where ammocoetes are abundant at times. Presumably siphon populations do not contribute to the survival of the species, because adults derived from them wind up in the aqueduct itself. Adults in natural environments seek riffles with gravel for spawning and rubble for cover.

Judging from the times at which adults are collected, this lamprey undergoes metamorphosis in fall and spawns in spring, dying after spawning. Other aspects of its life history are not known but are presumably similar to those of the western brook lamprey.

Status IC. Relatively few unequivocal collections of this species have been made since it was first discovered in 1976. This is because most collections are of ammocoetes that cannot be reliably distinguished from those of the western brook lamprey. Probable populations are thinly scattered throughout the San Joaquin drainage and isolated from one another (2). This fragmented distribution makes local extirpations likely, without hope of recolonization, followed by eventual extinction. The probability of local extirpation is increased by the fact that all known populations but one are below dams, where stream flows are regulated without regard to the needs of lampreys and where fluctuations or sudden drops in flow may isolate or desiccate ammocoetes. Channelization or other work on the banks may eliminate backwater areas required by ammocoetes. Gravel beds needed for spawning may be eliminated or compacted, so they cannot be used by adults. Ammocoetes may also be carried to "sink" habitats such as the Friant-Kern siphons. Clearly, if this species is going to persist, flows and habitats of lower reaches of rivers of the San Joaquin drainage should be managed so as to consider its needs.

References 1. Vladykov and Kott 1976a. 2. Brown and Moyle 1993. 3. Wang 1986. 4. Docker et al. 1999.

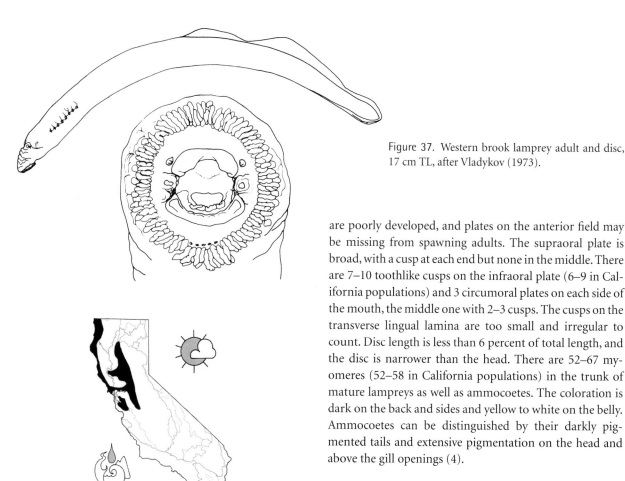

Figure 37. Western brook lamprey adult and disc, 17 cm TL, after Vladykov (1973).

Western Brook Lamprey,
Lampetra richardsoni Vladykov and Follett

Identification Western brook lampreys are small (up to 18 cm TL) and nonpredaceous. Tooth plates on the oral disc are poorly developed, and plates on the anterior field may be missing from spawning adults. The supraoral plate is broad, with a cusp at each end but none in the middle. There are 7–10 toothlike cusps on the infraoral plate (6–9 in California populations) and 3 circumoral plates on each side of the mouth, the middle one with 2–3 cusps. The cusps on the transverse lingual lamina are too small and irregular to count. Disc length is less than 6 percent of total length, and the disc is narrower than the head. There are 52–67 myomeres (52–58 in California populations) in the trunk of mature lampreys as well as ammocoetes. The coloration is dark on the back and sides and yellow to white on the belly. Ammocoetes can be distinguished by their darkly pigmented tails and extensive pigmentation on the head and above the gill openings (4).

Taxonomy The western brook lamprey was separated from the European brook lamprey, *L. planeri,* in 1965 (1). Populations in Oregon and California were subsequently described as *L. pacifica* (2). I follow Bond and Kan (3) and Robins et al. (11) in not recognizing *L. pacifica* and await a more complete study of brook lamprey systematics for definitive assignment. Bond (5) found that differences in myomere counts thought to distinguish *L. pacifica* and *L. richardsoni* ceased being definitive when populations were examined from the entire range of both types. Even *L. richardsoni* may not fit standard species definitions well because it is derived from the anadromous river lamprey, to which it is very similar biochemically (17). The presence of brook lampreys in coastal streams most likely represents many independent evolutionary events, rather than a single separation from river lampreys followed by dispersal of the nonpredatory form. Neither adults nor larvae of brook lamprey seem capable of entering salt water or of long-distance movement, although Beamish (12) has recorded at least one population that contains both predatory and nonpredatory adults, the predatory form capable of moving to the sea. This situation may be equivalent to that of threespine stickleback and rainbow trout, with repeated speciation of resident freshwater forms from anadromous forms. Brook and river lampreys will hybridize in the laboratory, but hybrids have never been observed in the wild (14).

Names Richardsoni refers to J. Richardson, a naturalist in the employ of the Hudson Bay Company, who wrote the first extensive account of the fish fauna of the Pacific Northwest in Volume 3 of his *Fauna Boreali-Americana* (1836). Other names are as for the Pacific lamprey.

Distribution Western brook lampreys are known from coastal streams from southeastern Alaska south to California, with major inland distributions in the Columbia and Sacramento–San Joaquin drainages (2, 9). In California populations have been identified mainly from the Sacramento drainage, including remote areas such as Kelsey Creek above Clear Lake (Lake County) (20). However, they are present above Pillsbury Reservoir in the Eel River (Mendocino County) (18) and in Mark West Creek (Sonoma County), a tributary to the Russian River (15). Spawning adults were collected in the Navarro River (Mendocino County) in 1999 (19). Ammocoetes collected from streams in the Los Angeles River basin may also be of this species (16, 21), although this population is now extirpated (6). It is likely that they occur in many streams along the California coast, especially in large rivers or their tributaries.

Life History Because western brook lampreys are difficult to collect and easy to overlook, little work has been done on their biology in California. Except for an early study by Hubbs (13), most information comes from studies in Washington (7, 8). Ammocoetes are most abundant in backwaters and pools of streams where silt and sand are mixed and populations can be as dense as 170 per square meter (7). Ammocoetes live 4–5 years in British Columbia and 3–4 years in Washington and California (7, 10, 13). Fastest growth and largest size (13–18 cm) are achieved in California (7) on a diet of algae (especially diatoms) and organic matter (10). Ammocoetes begin transforming in the fall and are mature in spring.

Spawners move into gravel riffles for spawning, where they construct nests slightly shorter than adult lengths. In Mark West Creek, Sonoma County (April 1994), brook lampreys constructed nests 15–20 cm in diameter in a gravelly riffle about 15 cm deep (15). Each nest pit was occupied by 2–4 individuals, although the largest lamprey (assumed to be female) did most of the excavating (15). Spawning begins when water temperatures exceed 10°C (7). Spawning behavior is similar to that described for Pacific lamprey (7, 9). The spawning season is apparently fairly short (March–April) in Coyote Creek, Alameda County (13), but it lasts as long as 6 months where flow conditions are more constant in Washington (7). Females produce 1,100–3,700 eggs, which hatch in about 10 days (10).

Status ID. Western brook lampreys are probably more common than records indicate because special effort has to be made to collect them and to separate ammocoetes from those of other species. However, it is unlikely that they can withstand severe pollution or habitat changes, so they are probably now restricted to less disturbed sections of streams. Systematics of the various populations assigned to this species merit investigation, because a number of the more isolated ones may deserve species status.

References 1. Vladykov and Follett 1962. 2. Vladykov 1973. 3. C. Bond and T. Kan, unpubl. ms. 1991. 4. Richards et al. 1982. 5. C. Bond, Oregon State University, pers. comm. 1998. 6. Swift et al. 1993. 7. Schultz 1930. 8. McIntyre 1969. 9. Morrow 1980. 10. Wydoski and Whitney 1979. 11. Robins et al. 1991. 12. Beamish 1987. 13. Hubbs 1925. 14. Beamish and Neville 1992. 15. M. H. Fawcett, pers. comm. 1998. 16. Culver and Hubbs 1917. 17. Docker et al. 1999. 18. Brown and Moyle 1996. 19. J. B. Feliciano, University of California, Davis, pers. comm. 1999. 20. T. L. Taylor, pers. comm. 1973. 21. C. L. Hubbs, pers. comm. 1974.

Sturgeons, Acipenseridae

Sturgeons are among the largest and most ancient of bony fishes. They are placed, along with paddlefishes and numerous fossil groups, in the infraclass Chondrostei, which also contains the ancestors of all other bony fishes. The sturgeons themselves are not ancestral to modern bony fishes but are a highly specialized and successful offshoot of ancestral chondrosteans, retaining such ancestral features as a heterocercal tail, fin structure, jaw structure, and spiracle. They have replaced a bony skeleton with one of cartilage and possess a few large, bony plates instead of scales. Sturgeons are highly adapted for preying on bottom animals, which they detect with a row of extremely sensitive barbels on the underside of their snouts. They protrude their extraordinarily long and flexible "lips" to suck up food.

Sturgeons are confined to temperate waters of the Northern Hemisphere. Only 8 of 25 species are found in North America, 2 in California. Most live primarily in salt water, moving up rivers only to spawn, but a few species live exclusively in fresh water. The anadromous forms are the largest fish in fresh water. The giant beluga sturgeon (*Huso huso*), which spawns in the Volga River of Eurasia, grows to 8.5 m (26 ft) and 1,297 kg (2,860 lb). White sturgeon are the largest freshwater fish in North America, apparently growing as large as 630 kg (1,400 lb) and more than 6 m (20 ft) long.

The history of sturgeon fisheries throughout most of the world has been one of overexploitation resulting in severe population reduction. The large size and sluggish nature of sturgeon make them vulnerable to netting and snagging, and their valuable caviar, isinglass, and flesh have made such fisheries very lucrative—while they last. Of equal importance, they live or spawn in large rivers, which have been almost universally dammed, diverted, and polluted. As a consequence, most species are threatened with extinction (Rochard et al. 1990; Birstein et al. 1997a). Proper management can restore overfished sturgeon populations, provided their spawning areas are not destroyed by pollution and competing uses of the water. Sturgeon culture is also starting to become an important segment of the aquaculture industry and raising sturgeon in hatcheries is a new tool for their conservation.

White Sturgeon, *Acipenser transmontanus* Richardson

Identification Adults have blunt, rounded snouts, with four barbels in a transverse row on the underside. The barbels are closer to the tip of the snout than the mouth. Their mouths have highly protrusible lips but lack teeth. Each fish has 5 widely separated rows of bony scutes (plates) on the body. The dorsal row has 11–14 scutes, the two lateral rows have 38–48 each, and the two bottom rows have 9–12 each, with 4–8 between the pelvic and anal fins. Large ventral scutes are absent behind the dorsal fin and anal fin, although tiny remnants (fulcra) may be present. The dorsal fin has one spine and 44–48 rays, while the anal fin has 28–31 rays. There are 34–36 gill rakers on the first gill arch. The ventral body surface is white, shading to gray brown on the back above the lateral scutes. The fins are gray and the viscera black. Young-of-year white sturgeon may be distinguished from green sturgeon by their 42 or more dorsal fin rays (greens have 35–40), more than 35 lateral scutes (greens have 30 or fewer), and 23 or more gill rakers on the first arch (greens have 15–19).

Taxonomy There is little controversy over the taxonomy of this species, which is most closely related to the green sturgeon. Populations from major river systems show some genetic differentiation, but not enough to warrant subspecies designations (16, 17).

Figure 38. White sturgeon, "about 700 lbs," Columbia River. Drawing by Paul Vecsei.

Names Just where the *white* in white sturgeon comes from is a bit of a mystery, because they are gray in color, but it probably refers to the pale color of their flesh compared with that of green sturgeon. *Acipenser* is Latin for sturgeon, while *trans-montanus* means "across the mountains," a reference to their wide distribution in the Columbia River system or to their presence west of the continental divide.

Distribution White sturgeon range in salt water from Ensenada, Mexico, north to the Gulf of Alaska, but they spawn only in large rivers from the Sacramento–San Joaquin system northward. At present, self-sustaining spawning populations apparently exist only in the Sacramento, Columbia (Washington), and Fraser (British Columbia) Rivers. Landlocked populations exist in the Columbia River basin above major dams (1, 18). In California white sturgeon are most abundant in the San Francisco estuary. This population spawns mainly in the Sacramento and Feather Rivers but may spawn in the San Joaquin River when flows and water quality permit (23).

Prior to the construction of Shasta Dam in the 1940s, the lower Pit River may have been an important spawning area (28). After Shasta Dam was built, trapping young sturgeon behind it, a landlocked population became established. This population reproduced for a while, maintaining a small fishery, but reproduction ceased following the construction of dams on the Pit River, which blocked access to historical spawning areas (28). White sturgeon are still occasionally caught in Shasta Reservoir, both long-lived residual fish and individuals from limited stocking attempts, especially in the 1980s. Historically, there may have been small runs in the Russian, Klamath, and Trinity Rivers as well. White sturgeon were once introduced into the Colorado River (19),

but there is no evidence the introduction was successful. They are now widely cultivated in California, and young are sold in aquarium stores, so individuals may be expected from other reservoirs and ponds. They have been planted in a number of reservoirs in southern California and the San Francisco Bay area, and occasional large fish are taken by anglers (e.g., a 21-kg sturgeon from Lafayette Reservoir, Contra Costa County) (22).

Life History White sturgeon spend most of their lives in estuaries of large rivers, moving into fresh water to spawn. They are usually most abundant in brackish portions of estuaries and move in response to salinity changes (9). A few make extensive movements in the ocean, and sturgeon tagged in the San Francisco estuary have been recaptured in the lower Columbia River and other estuaries between (2, 9). One tagged sturgeon was later recovered more than 1,000 km up the Columbia River. In estuaries adults tend to concentrate in deep areas with soft bottoms, although they may move into intertidal areas to feed at high tides.

The food of white sturgeon is taken on or close to the bottom. Young sturgeon (around 20 cm FL) feed mostly on crustaceans, especially amphipods (*Corophium* spp.) and opossum shrimp (*Neomysis mercedis*) (3, 4, 26). As they grow, their diet becomes more varied, although it still consists mostly of bottom-dwelling estuarine invertebrates, mainly clams, crabs, and shrimp. In the San Francisco Estuary most of these are introduced species, reflecting the ability of sturgeon to forage on whatever benthic prey are most readily available. In recent years a major item in the diet has been the overbite clam, *Potamocorbula amurensis*, which became extraordinarily abundant in Suisun Bay following its invasion in the 1980s. Fish assume increasing importance in the diets of larger sturgeon, especially herring, anchovy, striped bass, starry flounder, and smelt. When Pacific herring move into estuaries to spawn, white sturgeon may feed heavily on the eggs (6), as they do on eulachon eggs in the Columbia River (26). Other items recorded from the stomachs of large sturgeon in California include onions, wheat, Pacific lampreys, crayfish, frogs, salmon, trout, striped bass, carp, squawfish, suckers, and, in one case, a domestic cat (7). In captivity juvenile white sturgeon can adjust to artificial diets and grow rapidly when consuming food equivalent to 1.5 to 2.0 percent of their body weight per day, at 18°C (5).

Young white sturgeon grow rapidly in the San Francisco Estuary, reaching 18–30 cm FL by the end of their first year (9). Growth gradually slows as they become older, but they can reach 102 cm TL (40 in.) by their seventh or eighth year. In subsequent years they add 2–6 cm per year. Just how large they can grow is a matter of some dispute, because the largest fish were taken prior to 1900 and were subject to inaccurate measurements and exaggerated reporting. They may have achieved 6 m FL and 820 kg (1,800 lb), although the largest authentic record was of a specimen weighing 630 kg (22). Such large fish were probably more than 100 years old and were the largest fish in fresh water in North America. The largest white sturgeon taken in recent years, a 3.2-m FL fish from Oregon, was 82 years old (7). The largest recent record from California is of a female, 2.8 m TL, 210 kg, aged 47, accidentally caught in a fish trap. In 1963, however, CDFG recorded a dead sturgeon from Shasta Reservoir that measured 2.9 m TL, had an estimated weight of 225 kg, and was at least 67 years old (28). Sturgeon longer than 2 m and older than 27 years are rare (8). Age is determined by taking cross sections of fin rays and counting the number of visible rings, on the assumption that a new ring is laid down every year (8, 15).

Male white sturgeon are at least 10–12 years old and 75–105 cm FL before sexual maturity; females do not mature until they are 12–16 years old and 95–135 cm (9, 20). In captivity females may mature in 5 years and males in 3–4 (10). Maturation in adult sturgeon is apparently regulated by both photoperiod and temperature (21). When ready to spawn, sturgeon migrate upstream, although some movement to the lower reaches of rivers may take place in winter months prior to spawning. Spawning takes place between late February and early June when water temeratures range from 8 to 19°C, generally peaking around 14°C (18). Mature fish apparently start moving upstream in response to increases in flow, and spawning seems to be triggered by a pulse of high flow (23). Only a small fraction of the adult population spawns each year. In the Sacramento River most spawning apparently takes place between Knight's Landing (river mile 145) and Colusa (river mile 231) (23). Some fish may spawn on occasion in the Feather and San Joaquin Rivers (9, 11). White sturgeon presumably spawn either over deep gravel riffles or in deep holes with swift currents and rock bottoms, although substrates are gravel in the major Sacramento River spawning area. The adhesive eggs have been collected on the bottom at 10 m (10). In the Columbia River they spawn over bottoms of cobble and boulder, at depths of 3–23 m and bottom water velocities of 0.6–2.4 m/sec (18). When spawning is completed they move back down to the estuary. Males may spawn every 1–2 years, but females apparently have a 2- to 4-year wait between spawns. Longer intervals are also possible, especially if conditions are unfavorable.

Female Sacramento River white sturgeon are highly fecund, averaging 5,648 eggs per kilogram body weight (20). A "typical" female (1.5 m FL) will thus contain more than 200,000 eggs. The eggs are adhesive after fertilization and stick to the substrate. Larvae hatch in 4–12 days, depending on temperature (10). New larvae are about 11 mm long and at first swim in a vertical position, which presumably causes them to drift downstream toward the estuary (10). The yolk sac is absorbed in 7–10 days, and the larvae then begin swimming horizontally, actively feeding from the bottom. Juvenile sturgeon apparently have a greater tendency to live in the upper reaches of the estuary than do adults, indicating that the ability to adjust to salt water increases with size and age (12).

Spawning success varies from year to year, so the population in the San Francisco estuary tends to be dominated by a few strong year classes. Large year classes are associated with high outflows through the estuary in spring (9, 25). This relationship may result from larval sturgeon being moved quickly downstream to suitable rearing areas (27), where food is abundant and the probability of being sucked into diversions is low. Higher river flows may also stimulate larger numbers of sturgeon to spawn (9).

Because successful year classes may occur at wide and irregular intervals, the number of adult fish can vary widely. CDFG (9) estimated that in 1954 only 11,000 adult (>1-m) sturgeon existed in the estuary, but by 1967 the number had increased to 115,000. Numbers decreased to an estimated 74,000 adult fish in 1979, increased to 128,000 by 1984, declined to about 60,000 by 1990, but then reached record numbers (142,000) in 1997 (25). A decline in the adult population through the early 21st century is predicted, based on poor spawning success during the 1987–1992 drought with an increase again as the result of successful spawning in wet years starting in 1993 (25). The annual survival rate ranges from 74 to 90 percent, including fishing mortality that varies from 9 to 11.5 percent (9). In recent years improved angling techniques have gradually increased catch rate, but exploitation rates are still reasonable (25). To protect the most fecund females, maximum size limits (183 cm TL) have been imposed for the fishery, and this regulation, given current exploitation rates, seems sufficient to protect the population (25).

Status IE. White sturgeon support valuable commercial and sport fisheries in Canada, Oregon, and Washington (14, 24). In California they are taken in small numbers in the Native American fishery in the Klamath River and support a major sport fishery in the San Francisco Estuary.

White sturgeon in the San Francisco Estuary are a classic case of a valuable fish resource nearly wiped out by overfishing but restored through proper management (13). The large size and late age of maturity of sturgeons make them extremely vulnerable to overfishing, so it is not surprising that they were decimated by a commercial fishery that started in the 1860s and lasted until 1901. The peak catch was 1,660,000 lb taken in 1885. By 1895 the catch was down to 300,000 lb and declining annually. The fishery was closed in 1901 after a catch of less than 200,000 lb. Low catches in 1909, 1916, and 1917, when the fishery was reopened, indicated that the population had not recovered, so the commercial fishery was closed for good in 1917. In 1954 a year-round sport fishery was legalized, with a minimum size of 102 cm and a bag limit of one fish per day per fisherman. It was an immediate success, and large numbers were caught, mostly by snagging from party boats. Because snagging was considered unsportsmanlike, the method was outlawed in 1956. However, no other effective method had been found to catch sturgeon on hook and line, so the catch by anglers declined. Most sturgeon caught were taken by fishermen angling for other species, especially striped bass (13). In 1964 it was discovered that grass shrimp worked well as bait, and the sport fishery again intensified. In the 1980s additional pressure was exerted because fishing techniques had become more sophisticated (e.g., the use of sonic "fish finders"). Because of concern that harvest rates were too high, CDFG imposed new maximum (183 cm TL) and minimum (117 cm TL) size limits in 1991.

The value of managing this fishery is clearly indicated by the fact that present-day sturgeon catches are only slightly less per year than *average* commercial catches from 1875 to 1899, when the fishery was in decline. The unregulated commercial fishery nearly wiped out the population in a short time, whereas the present managed sport fishery promises to yield continuous returns for years to come. Even large sturgeon once again appear in the catch. In April 1973 a 190-kg, 2.8-m FL sturgeon was caught in the Sacramento River, a hook and line record.

Continued success of white sturgeon in the San Francisco Estuary is remarkable because almost all other species of fish have suffered major population declines in recent decades. The success can be attributed to good management coupled with the long life and high fecundity of the fish. These make it possible to maintain populations with a relatively small number of good spawning years. The sturgeon also have flexibility in their feeding habits; for example, they are now feeding on the abundant introduced overbite clam that is otherwise considered a disaster for the estuary. All this does not mean that we can afford to be sanguine about the white sturgeon's future. Continued alteration of the estuary and the Sacramento River is making successful spawning and rearing increasingly difficult. The long life span of sturgeon also allows for accumulation of contaminants such as PCBs, which may inhibit growth and reproduction (14). One concern over the abundance of overbite clams in the diet is that selenium and other toxic materials accumulated at high levels by the clams may be passed on to the fish.

Because white sturgeon are now successfully cultured, there is a tendency to think that reduced natural reproduction can be made up for by stocking hatchery-reared fish. In the long run, as the history of chinook and coho salmon in California has shown, reliance on hatcheries can create as many problems as it solves, or even more. If anything, we should be working toward improving spawning and rearing conditions in the wild for white sturgeon, recognizing that such efforts would benefit many other species as well. One place where a hatchery program for white sturgeon would seem to be justified is Shasta Reservoir. The sturgeon fishery that once existed there all but disappeared once dams denied fish access to historic spawning grounds in the Pit River. Planting juvenile sturgeon in the reservoir could at least restore a fishery for a native species.

References 1. C. Brown 1971. 2. L. Miller 1972a,b. 3. Radtke 1966. 4. Muir et al. 1988. 5. Hung et al. 1989. 6. McKechnie and Fenner 1971. 7. Carlander 1969. 8. Kohlhorst et al. 1980. 9. Kohlhorst et al. 1991. 10. Wang 1986. 11. Kohlhorst 1976. 12. McEnroe and Cech 1987. 13. Skinner 1962. 14. Emmett et al. 1991. 15. Brennan and Cailliet 1991. 16. Bartley et al. 1985. 17. Brown et al. 1992a. 18. McCabe and Tracy 1994. 19. Minckley 1973. 20. Chapman et al. 1996. 21. Doroshov et al. 1997. 22. C. Swift, pers. comm. 1999. 23. Schaffter 1997a,b. 24. Wydoski and Whitney 1979. 25. Schaffter and Kohlhorst 1999. 26. McCabe et al. 1993. 27. Stevens and Miller 1970. 28. T. Healey, CDFG, pers. comm. 2001.

Green Sturgeon, *Acipsenser medirostris* Ayres

Identification Green sturgeon are similar in appearance to white sturgeon, except the barbels are closer to the mouth than the tip of the long, narrow snout. The dorsal row of bony plates numbers 8–11, lateral rows, 23–30, and bottom rows, 7–10; there is one large scute behind the dorsal fin as well as behind the anal fin (both lacking in white sturgeon). The scutes also tend to be sharper and more pointed than in white sturgeon. The dorsal fin has 33–36 rays, the anal fin, 22–28. The body color is olive green with an olivaceous stripe on each side; the scutes are paler than the body.

Taxonomy Although there is no question as to the validity of this species, its geographic variation has received little attention. It is likely that Asiatic populations (Sakhalin sturgeon) belong to a different species, although they are similar morphologically to those in North America, even

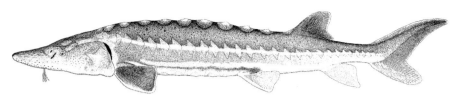

Figure 39. Green sturgeon, 160 cm TL. Drawing by Paul Vecsei.

sharing some unusual parasites (27). The Japanese population was described as *Acipenser mikadoi* based on one poorly preserved specimen (1), but the name is sometimes applied to the Asian form (the Sakhalin sturgeon in the Russian literature). The Asian form has about twice the DNA content of the North American form, and other molecular analyses indicate that the two forms are distinct (17).

Names In 1854 W. O. Ayres (2) described three species of sturgeon from San Francisco Bay, differentiated in part by the length of their snouts and named accordingly (*A. acutirostris, A. medirostris, A. brachyrhynchus*). The long- ("acute") and short-snouted forms were later identified as white sturgeon, leaving green sturgeon with an anomalous scientific name that translates as "middle snout." The common name is apt, because green sturgeon frequently have a distinctly green cast. Other names are as for the white sturgeon.

Distribution Green sturgeon are recorded from Mexico, the United States, Canada, Russia (Commonwealth of Independent States), Japan, and Korea, but the Asian records are those of the closely related Sakhalin sturgeon (now confined to the Tumnin River, Russia). As a general rule, these two species are rarely found below the 30th parallel, and their greatest abundance is between the 40th and 60th parallels. In North America green sturgeon range in the Pacific from the Bering Sea to Ensenada, Mexico. They are found in rivers only from British Columbia south to the Sacramento River. There is no evidence of green sturgeon spawning in Canada or Alaska, although small numbers are caught in the Fraser and Skeena Rivers (4). They are particularly abun-

dant in the Columbia River estuary, and individuals have been observed 225 km inland in the river; presently they are found almost exclusively in the lower 60 km and not upstream of Bonneville Dam. There is no evidence of spawning in the Columbia River or other rivers in Washington. In Oregon juvenile green sturgeon have been caught in several of the coastal rivers (6), but spawning has been confirmed only in the Rogue River (22).

In California the abundance of green sturgeon gradually increases northward of Point Conception. They are occasionally caught in Monterey Bay, but the southernmost spawning population is in the Sacramento River. They are occasionally captured in ocean waters off northern California, especially in bays, but spawning populations apparently existed historically only in the Eel River and in the Klamath-Trinity River system. The Eel River apparently no longer sustains a spawning run, although large sturgeon are occasionally observed in the lower river. The Klamath and Trinity Rivers remain as their principal spawning streams.

Life History The ecology and life history of green sturgeon have received little study, evidently because of the generally low abundance, limited spawning distribution, and low commercial and sport fishing value of the species. Green sturgeon are the most marine species of sturgeon, coming into rivers mainly to spawn, although early life stages in fresh water may last as long as 2 years.

Juveniles and adults are benthic feeders, and they may also take small fish. Juveniles in the San Francisco Estuary feed on opossum shrimp and amphipods (7). Adults caught in Washington had been feeding mainly on sand lances (*Ammodytes hexapterus*) and callianassid shrimp (27). In the Columbia River estuary green sturgeon are known to feed on anchovies and clams (5).

Green sturgeon migrate up the Klamath River between late February and late July. The spawning period is March–July, with a peak from mid-April to mid-June (6). Spawning times in the Sacramento river are probably similar because adult sturgeon are in the river, presumably spawning, when temperatures are 8–14°C. Spawning takes place in deep, fast water. In the Klamath River a pool known as the Sturgeon Hole (1.5 km upstream from Orleans, Humboldt County) is apparently a major spawning site, because leaping and other behavior indicative of courtship are often observed there during spring and early summer. In the Sacramento drainage capture of larval green sturgeon in salmon outmigrant traps indicates that the lower Feather River may be a principal spawning area. Indirect evidence indicates that green sturgeon may also spawn in the mainstem Sacramento River. Adults have been reported from as far upstream as Red Bluff, Tehama County (river km 383) and young from a number of places downstream (14, 15). Some spawning may also take place (or once did) in the lower San Joaquin River, because young green sturgeon have been taken at Santa Clara Shoal, Brannan Island State Recreational Area, Sacramento County (7). Preferred spawning substrate is likely large cobble, but it can range from clean sand to bedrock. Eggs are broadcast and externally fertilized in relatively fast water and probably in depths greater than 3 m (6). The importance of water quality is uncertain, but a small amount of silt is known to prevent the eggs from adhering to each other, thus increasing survival.

Female green sturgeon produce 60,000–140,000 eggs, about 3.8 mm in diameter. Based on the presumed similarity to white sturgeon, green sturgeon eggs probably hatch around 200 hr (at 12.7°C) after spawning; the larvae should be 8–19 mm long and the juveniles 2–150 cm TL (6). The juveniles seem to migrate out to sea before the end of their second year, primarily during summer and fall (6). In the Klamath River juvenile sturgeon outmigrate at 30–66 cm TL, when they are 1–3 years old, although many leave as yearlings (18, 24). They apparently remain near estuaries at first, but they migrate considerable distances as they grow (6). Fish between 70 and 120 cm TL are marine, so males spend 3–9 years at sea and females 3–13 years before returning (24). Individuals tagged by CDFG in the San Francisco Estuary have been recaptured off Santa Cruz, California; in Winchester Bay on the southern Oregon coast; at the mouth of the Columbia River; and in Gray's Harbor, Washington (9, 10). Most tags for green sturgeon from the San Francisco Bay system have been returned from outside that estuary (23).

Males and females grow at about the same rate, approximately 7 cm per year until they reach maturity at 130–150 cm TL, at which point growth slows (18, 24). Thus a 10-year-old sturgeon is about 105 cm TL; a 20-year-old, 160 cm TL; a 30-year-old, 195 cm TL; and a 40-year-old, 200 cm TL. However, males mature at younger ages than females and do not grow as large. The maximum length recorded in recent years from the Klamath River is about 270 cm TL (175 kg), and all fish over 200 cm TL are females (18, 20). Adults over 2 m TL and 90 kg are unusual (8). Mature fish are typically 15–20 years old. The largest fish have been aged at 42 years (24), but this is probably an underestimate (18), and maximum ages of 60–70 years or more are likely (6).

Status IC. Because of its low numbers and low culinary reputation, little attention has been paid to green sturgeon until recently. For example, Jordan and Evermann (11, p. 7) expressed what had been the most common attitude: "As a food-fish, it is of very inferior rank; indeed, it is commonly believed to be poisonous, but this belief is without warrant. Its flesh, however, is dark, has a strong, disagreeable taste, and an unpleasant odor, and is regarded as inferior to that of the white sturgeon." Even the roe has been rejected as unfit for caviar. In fact, the bad reputation of green sturgeon probably stems mostly from the dark color of the flesh because, properly prepared, it is quite tasty. As a consequence, a substantial fishery has developed in recent years. The following are reasons for being concerned about the status of green sturgeon in California and, consequently, in the world (15):

1. Green sturgeon and Sakhalin sturgeon appear to be in trouble throughout their ranges (21). Rochard et al. (12, p. 131) state, in their review of the status of sturgeons worldwide, that "Those [species of sturgeon] which do not have particular interest to fishermen (*A. medirostris, Pseudoscaphirhynchus* spp.) are paradoxically most at risk, for we know so little about them." In Japan Sakhalin sturgeon have apparently been extinct for 40 or more years (28), even though they once had spawning runs in rivers of Hokkaido. In Russia Sakhalin sturgeon is listed as a Category 4 species (probably endangered but with insufficient information to be classified as such). Borodin (3), however, indicates that it is highly endangered. Fishing for Sakhalin sturgeon is now officially forbidden in Russia. In Canada green sturgeon have been given "rare" status (1987) by the Committee on the Status of Endangered Wildlife in Canada, based on a general lack of biological information and uncommonness (4).

2. A number of presumed spawning populations (Eel River, South Fork Trinity River, San Joaquin River) have apparently been lost in the past 25–30 years, and the only known spawning now takes place in the Sacramento, Klamath, and Rogue (Oregon) Rivers, all of which are affected by water projects and intensive use of the watersheds. It is quite likely that these are the *only* spawning populations in North America.

3. The principal non–Native American fisheries for green sturgeon have been in Washington and the nearby Columbia River estuary, yet there is no evidence of sturgeon spawning in that region. It is highly probable that these fisheries depended on sturgeon from California attracted to the area for an unknown reason, perhaps owing to the abundance of food. The targeted green sturgeon fishery has now been halted, but considerable numbers of green sturgeon are still taken in the salmon gill net fishery in the lower river (19).

4. The Yurok tribe and other Native American tribes fish the annual run of green sturgeon in the Klamath River. The Yurok portion of the fishery is closely monitored; the annual catch declined from 389 fish in 1980–1988 to 256 fish in 1989–1997 (20). The average length of sturgeon in the catch, however, did not decline.

The various fisheries are harvesting at least 6,000–11,000 green sturgeon per year. Although there is no direct evidence of decline, the statistics are incomplete. It is possible that the fisheries discussed in the following paragraphs are "mining" a stock of large, old fish that cannot renew itself at present harvest rates.

A majority of the green sturgeon harvest has historically taken place in the *Columbia River region,* where they are caught by commercial fishermen, anglers, and Native American gill netters. There is little or no evidence of spawning in rivers of this region, and it is likely that fish harvested here migrated from Oregon or California, as indicated by limited recaptures of tagged sturgeon. Further evidence of lack of local recruitment into the fishery is that few juvenile sturgeon (<1.3 m) are caught (6). Commercial catch in the Columbia River region has fluctuated considerably. Between 1941 and 1951, catches averaged 200–500 fish per year, while between 1951 and 1971 catch averaged 1,400 fish per year (4). Between 1971 and 1989, an average of 21 tons of green sturgeon (ca. 2,000–4,000 fish) were harvested commercially each year (6). There have also been some notably high catches. In 1986 about 5,000 were harvested in the Columbia River estuary alone during a four-day sturgeon fishing season (6). When sport and Native American gill net catches are added in, the combined fisheries during this period were taking between 4,000 and 9,000 fish per year (26).

Concern over these high catches led to a ban on commercial fisheries targeted on green sturgeon in 1989, in both Oregon and Washington (26). However, fishermen gill-netting for salmon, fishing for white sturgeon, or trawling for other species can still keep green sturgeon caught incidentally, provided the fish are within a 48–66 inch TL slot limit. Between 1995 and 1999, the total catch (including sport and tribal fisheries in California) averaged about 2,000 fish, with the sport fishery taking about 500 of these.

The second largest fishery is probably in the *Klamath and Trinity Rivers.* A small number are taken in the sport fishery, but the main harvest is by the Native American gill net fishery. This fishery targets fish as they move up river to spawn during spring and again as they return seaward through the estuary, during June–August. It is mainly adults (>130 cm TL) that are captured (24, 25). Data on this fishery exist only since 1980, and the available harvest estimates are probably low because some of the green sturgeon harvest occurs prior to the annual monitoring activities of the

USFWS (25). In addition, the USFWS monitors only the sturgeon harvest on the Yurok Indian Reservation; catches by the Karuk and Hoopa tribal fishermen in the Klamath River basin are undetermined but are probably low (25). With that in mind, the adult harvest from the Klamath system has been between 100 and 800 fish per year. There seems to be, as yet, no indication of any recent decline.

Green sturgeon in the *Sacramento–San Joaquin drainage* are caught primarily by anglers fishing for white sturgeon. If we assume that green sturgeon longer than 102 cm (the legal size) are harvested in proportion to their numbers and at the same rate relative to white sturgeon, then exploitation rates have gradually increased since 1954 (13). Recent annual harvest rates for white sturgeon have been 9–11.5 percent per year. Presumably regulations adopted to reduce the catch of white sturgeon will also benefit green sturgeon, although the 183 cm TL maximum size still allows the largest female green sturgeon to be harvested.

The following is a description of the status of green sturgeon in the various drainages within California.

Sacramento–San Joaquin drainage. White sturgeon are the most abundant sturgeon in this system, and green sturgeon have always been uncommon. CDFG measured and identified 13,982 sturgeon of both species between 1954 and 1987. Based on these data, a 1:5 ratio of green sturgeon to white sturgeon is derived for fish less than 101 cm FL, and a 1:78 ratio for fish 101 cm or more FL (23). If we assume that sturgeon over 101 cm FL are adults, that green sturgeon and white sturgeon are equally vulnerable to capture, that their populations fluctuate in a similar manner, and that CDFG population estimates of white sturgeon (11,000–128,000, depending on the year) are accurate (13), then adult green sturgeon numbers in the estuary range from 140 to 1,600 fish. Numbers of juveniles are presumably even more variable, depending on episodic reproduction.

Eel River. Green sturgeon are the species usually caught in rivers, estuaries, and bays on the north coast from Tomales Bay to the Smith River. However, most early references regarding sturgeon from this area fail to distinguish the species. As a result, confusion has ensued as to their relative abundance in this region. Between the Sacramento and Klamath Rivers, only the Eel River has apparently supported spawning green sturgeon in the past. Historical accounts from 19th-century newspapers provide the earliest evidence of sturgeon in the Eel River. At this time sturgeon were reported from the mainstem, South Fork, and Van Duzen River (15). While not confirmatory, the lengths and weights given in these newspaper accounts would be consistent with adult green sturgeon. In the 1950s two young were collected in the mainstem Eel River, and large sturgeon were observed jumping in tidewater (16). Two additional young were taken from the Eel River in 1967. There are no confirmed records in the Eel River since then. However, adults are commonly collected in Humboldt Bay, a short distance to the north (15).

Klamath and Trinity Rivers. The largest spawning population of green sturgeon in California is in the Klamath River. Both green and white sturgeon have been found in the Klamath River estuary, but white sturgeon are taken infrequently. A investigation initiated in 1979 by USFWS found almost all sturgeon occurring above the estuary to be green sturgeon. The sturgeon spawn primarily in the mainstem Klamath River and mainstem Trinity River, but they have also been seen in the lower portion of the Salmon River (a Klamath tributary). In the Klamath the apparent upstream limit for spawning is Ishi Pishi Falls, upriver from Somes Bar, Siskiyou County (approximately river km 113). The Trinity River enters the Klamath at Weitchpec (river km 70), and spawning migrants penetrate the mainstem to about Grays Falls, Trinity County (river km 72).

Because of its limited distribution and our limited information about it, the green sturgeon deserves status as a species of special concern; it requires study to determine its population dynamics and ecological requirements. At least one population has been lost in California, and it is likely that the two existing spawning populations are smaller than they once were.

References 1. Jordan and Snyder 1906. 2. Ayres 1854. 3. Borodin 1984. 4. Houston 1988. 5. Wydoski and Whitney 1979. 6. Emmett et al. 1991. 7. Radtke 1966. 8. Skinner 1972. 9. Chadwick 1959. 10. Miller 1972b. 11. Jordan and Evermann 1923. 12. Rochard et al. 1990. 13. Kohlhorst et al. 1991. 14. Fry 1973. 15. Moyle et al. 1995. 16. Murphy and DeWitt 1951. 17. Birstein et al. 1997b. 18. USFWS 1979. 19. ODFW 1991. 20. D. Gale and D. Williams, Yorok Tribal Fisheries Program, pers. comm. 1998. 21. Birstein et al. 1997a. 22. A. Smith, ODFW, unpubl. obs. 23. D. Kohlhorst, CDFG, pers. comm. 24. Nakamoto et al. 1995. 25. T. T. Kisanuki, USFWS, pers. comm. 1994. 26. Oregon Department of Fish and Wildlife, unpubl. rpt. April 2000. 27. P. Foley, pers. comm. 1992. 28. K. Amaoka, pers. comm. 1990.

Herrings, Clupeidae

If sheer number of individuals is the criterion for success, herrings are one of the most successful families of fishes in the world. Early in the history of teleost evolution, they achieved plankton-feeding specializations that have allowed them to remain abundant. Herrings have highly protractile jaws and long, fine gill rakers for picking and filtering plankton. Their scales are cycloid, deciduous, and silvery, reflecting light like miniature mirrors to confuse predators. Their bodies are muscular yet deepened by a sharp keel on the belly. The keel eliminates the faint belly shadow most fishes have when seen from below, thus increasing the difficulty predators have in picking out individuals from a shoal. Indeed, most morphological specializations of the Clupeidae enable them to function in the huge schools in which they are typically found.

Although usually thought of as marine, herrings are also successful as anadromous and freshwater fishes. Thanks to humans, the ranges of some forms have been greatly extended, especially in North America, with results ranging from beneficial to disastrous. Small freshwater shads of the genus *Dorosoma* have long been regarded as ideal forage fish in reservoirs and large lakes, and consequently have been distributed throughout the United States. Unfortunately, most of these introductions took place before it was realized that plankton-feeding fishes can alter lake ecosystems by changing the zooplankton community, causing clear lakes to become green with algae and reducing the amount of food available to the young of game fishes. Often shad introduced as forage fish cause declines in the very predatory fishes whose populations they were supposed to enhance (DeVries and Stein 1990).

Only two clupeid species are regularly found in California's fresh waters, threadfin shad and American shad. Both were introduced with greater success than was perhaps ever imagined. In addition, Pacific herring (*Clupea harengeus*) occasionally wander into fresh water when they move into estuaries to spawn. Northern anchovy (*Engraulis mordax*), in the closely related family Engraulidae, wander into brackish water on occasion, usually as juveniles.

Threadfin Shad, *Dorosoma petenense* (Günther)

Identification Threadfin shad are small (rarely longer than 10 cm TL in California) with the typical deciduous scales, flattened bodies, and sawtooth bellies of most herrings. They are distinguished by the long, threadlike final ray of the dorsal fin and by the single dark spot behind the operculum. The mouth is oblique, small, and toothless. The upper jaw is longer than the lower. The dorsal fin (11–17 rays, usually 14–15) is falcate. Anal fin rays number 17–27; scales in the lateral series, 40–48; belly scutes (scales), 15–18 before the bases of the pelvic fins and 8–12 behind them. The intestine is long and convoluted, with a gizzardlike stomach. The gill covers are smooth or have a few faint striations. The overall color is silvery, although the back frequently has a black or bluish hue.

Taxonomy The subspecies of threadfin shad introduced into California is *D. p. atchafalayae* (after the Atchafalaya River, Louisiana), although subspecies designations may have little validity (1).

Names Doro-soma means lance-body, referring to the eel-like larvae. *Petenense* is after Lake Petén, Guatemala, from which the first specimens were described. *Threadfin* refers to the distinctive dorsal fin. *Shad* is apparently derived from the ancient Celtic name for herring.

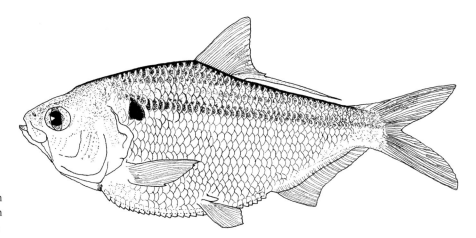

Figure 40. Threadfin shad, 10 cm SL, Sacramento–San Joaquin Delta. Drawing by A. Marciochi.

Distribution Threadfin shad are native to streams flowing into the Gulf of Mexico, south to Belize. In the Mississippi River and its tributaries they are found as far north as southern Indiana and Illinois (1). Shad from the Tennessee River at Watts Bar, Tennessee, were introduced into ponds in San Diego County in 1953 by CDFG (2). In 1954 fish from these ponds were introduced into San Vincente Reservoir, San Diego County, and Havasu Reservoir on the Colorado River. About 1,000 shad were planted in Havasu, and within a year they numbered in the millions and quickly spread downstream (2). In subsequent years they were planted by CDFG in reservoirs throughout the state, with the Sacramento–San Joaquin drainage planted in 1959 (3). From these transplants they have become established in the Sacramento–San Joaquin River system and its estuary, as well as in most of the lower Colorado River and the canals and drains of the Salton Sea region. Unauthorized plants have further expanded the range to more isolated lakes and reservoirs, such as Pillsbury Reservoir and Clear Lake, Lake County. They are occasionally taken in salt water from Long Beach to Yaquina Bay, Oregon (1, 4). Besides California, threadfin shad have been planted successfully in suitable waters throughout much of the United States, including Hawaii.

Life History Threadfin shad inhabit open waters of reservoirs, lakes, and large ponds as well as sluggish backwaters of rivers. In reservoirs they often congregate near inlets of small streams or along steep surfaces of dams. They prefer well-lighted surface waters and are seldom found below depths of 18 m (3). The best growth and survival occur in waters in which summer temperatures exceed 22–24°C and that do not become colder than 7–9°C in winter. Threadfin shad apparently cannot withstand water colder than 4°C for long (5). A sudden drop in temperature also causes high mortalities, as do prolonged periods of cold water. The population in the Sacramento–San Joaquin Delta experiences heavy die-offs every winter when the water cools to 6–8°C (6). In Clear Lake threadfin shad became abundant in 1985, but they were apparently extirpated during the exceptionally cold winter of 1990–1991. They reappeared in 1997, perhaps as the result of an illegal reintroduction (18).

Threadfin shad live mainly in fresh water and become progressively less abundant as salinity increases. Nevertheless, they can survive and grow in sea water (4). Salt water apparently inhibits reproduction: shad in the Salton Sea have failed to reproduce despite continuous recruitment from inflowing canals (7).

Threadfin shad form schools segregated by size and hence by age. Although shad concentrate in surface waters, young-of-year tend to be found in deeper water than adults, especially at night (8). When attacked by predatory fish such as striped bass, a school of shad will close together and hug the water surface, with some individuals leaping from the water at each attack. At such times they also are vulnerable to terns and other fish-eating birds.

Like all clupeids, threadfin shad are plankton feeders. They use their gill rakers to strain small (<1 mm) zooplankton, phytoplankton, and detritus particles from the water but feed individually on larger organisms, mostly zooplankton (9). This ability to feed by both filtering and picking allows for broad diets. Planktonic organisms often occur in their digestive tracts in roughly the same proportion as in the water during the day. At night, however, phytoplankton and detritus predominate in the stomachs, because low light levels restrict the ability of shad to feed by picking (9). Although shad can grow and reproduce on filtered prey alone, larger zooplankters, such as cladocerans and copepods, appear to be preferred (19). Large shad populations can virtually eliminate larger zooplankton species from lakes (10).

Threadfin shad are fast growing but short lived. Under optimal conditions they can increase in length 1–3 cm per month during the first summer of life, reaching 10–13 cm TL by the end of summer. Normally they reach only 4–6 cm TL by the end of their first year and 6–10 cm TL by the end of their second. Few live longer than 2 years or achieve more than 10 cm TL, although rare individuals may live as long as 4 years (8) and achieve lengths of 33 cm TL (11). The largest threadfin shad recorded from California was 22 cm TL, from the Salton Sea (7). Fish from non-reproducing saltwater populations frequently achieve larger sizes than are normal for fresh water (8). In fresh water one of the main factors limiting growth seems to be interspecific competition for food and space (8). Thus shad newly established in reservoirs tend to grow larger during their first year than their counterparts in more established populations.

Threadfin shad may spawn at the end of their first summer but usually wait until their second (8). Spawning takes place in California in April through August, peaking in June and July when water temperatures exceed 20°C (12). Spawning, however, has been observed at 14–18°C (13). Despite a protracted spawning period, each shad apparently spawns once a summer (8). Spawning is most often at dawn and centers around floating or partially submerged objects, such as logs, brush, aquatic plants, and gill nets. Small, compact groups of shad swimming close to the surface approach such objects rapidly, turning away just prior to collision. As they turn, eggs and sperm are released (14). The fertilized eggs stick to surfaces. Spawning is usually accompanied by splashing and leaping from the water. Females produce 900 to 21,000 eggs, the number increasing sharply with the size of the fish (8). The ability of threadfin shad to deposit eggs on floating objects may help account for their success in reservoirs, in which fluctuating water levels frequently expose embryos attached to fixed objects (12).

The embryos hatch in 3–6 days and larvae immediately assume a planktonic existence. They are weak swimmers and so are susceptible to entrainment in water diversions and power plant intakes. However, they are capable of some vertical migration, preferring surface waters during the day and deeper waters at night (12). The length of the planktonic life stage is not known exactly, but it is probably 2–3 weeks depending on temperature. Larvae metamorphose into juveniles at about 2 cm TL. Juveniles form dense schools and, in estuaries, are found in water of all salinities, although they are most abundant in fresh water (12).

Status IIE. Threadfin shad were brought into California in 1953 on the optimistic assumption that they were ideal forage fish for reservoirs that should greatly improve the growth rates of game fishes. Their desirability stemmed from their small size, high reproductive rate, and ability to occupy open waters presumed to be unexploited fish habitat. In many reservoirs managed for trout, especially those receiving large plants of catchable-size fish, the growth of trout larger than 28 cm FL can be extremely rapid when they feed on shad. The success of striped bass in Millerton Reservoir, Fresno and Madera Counties, is probably due largely to their diet of shad. In other reservoirs, large-size largemouth bass, black and white crappie, and white catfish also utilize threadfin shad. Unfortunately, shad are largely unavailable to small warmwater game fishes. Because the young of many centrarchids live in open water for extended periods of time, feeding on plankton, shad may actually compete with them by reducing plankton populations. In particular, threadfin shad may eliminate large species of planktonic crustaceans important in the diets of larval fishes (10). Thus, in some California reservoirs, growth and survival of young centrarchids, including largemouth bass, decreased after introduction of shad (15). In Clear Lake the establishment of shad was followed by the crash of *Daphnia* populations, decreased survival of juvenile largemouth bass, and huge increases in the numbers of piscivorous birds (e.g., western grebes, double-crested cormorants, white pelicans) (18). When the shad population collapsed in 1990, bass, bird, and zooplankton populations returned to preintroduction levels. When shad became reestablished (1998), *Daphnia* collapsed again and the birds returned (18). The increase was also accompanied by a huge die-off of shad in 1999, littering beaches with dead fish.

Mixed success with shad introductions has been experienced elsewhere, leading to attempts to eradicate them from some waters (16, 17). Unfortunately, although fisheries managers may have adopted a more cautious attitude toward threadfin shad introductions, enthusiasm for this fish still persists among some anglers, resulting in unauthorized introductions.

The ability of shad populations to increase explosively has allowed them to spread rapidly by "natural" means far beyond original introduction sites. The thousand shad introduced into Havasu Reservoir managed to provide enough offspring to colonize the entire lower Colorado River and Salton Sea basin in less than 18 months (3). In the Sacramento–San Joaquin drainage they quickly established populations downstream from reservoirs and then spread throughout the California Aqueduct system. In most areas they colonized, their effect—especially on native fishes with planktonic larvae and on young centrarchids—is unknown. In the Sacramento–San Joaquin Delta shad are a major item in the diet of striped bass and other piscivorous fishes, but their role in the ecosystem is poorly understood. Numbers in the Delta have gradually declined since the late 1970s, reflecting a general decline of planktonic fishes in the estuary.

References 1. Lee et al. 1980. 2. Dill and Cordone 1997. 3. Burns 1966. 4. D. Miller and Lea 1972. 5. Griffith 1978. 6. J. Turner 1966d. 7. Hendricks 1961. 8. Johnson 1970, 1971. 9. Holanov and Tash 1978. 10. Ziebell et al. 1986. 11. Carlander 1969. 12. Wang 1986. 13. Rawstron 1964. 14. Lambou 1965. 15. Von Geldern and Mitchill 1975. 16. DeVries and Stein 1990. 17. DeVries et al. 1991. 18. A. E. Colwell, Lake County Vector Control District, pers. comm. 1999. 19. Kjelson 1971.

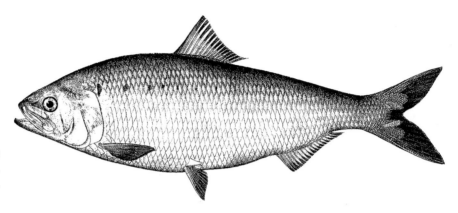

Figure 41. American shad, male, 38 cm SL, Norfolk, Virginia, 1878. USNM 25096. Drawing by H. L. Todd.

American Shad, *Alosa sapidissima* (Wilson)

Identification American shad are large (to 75 cm FL) clupeids with thin, deciduous scales, compressed bodies, and a sawtooth keel on their bellies. The mouth is terminal, and the upper and lower jaws are about equal in size, the lower fitting into the central notch of the upper (but sometimes projecting slightly). The dorsal fin (15–19 rays, usually 17–18) is short and straight edged, without the greatly elongated last ray characteristic of threadfin shad. Anal fin rays number 18–24 (usually 20–22); scales in the lateral series, about 50–55 (the lateral line is poorly developed); belly scutes (scales), 19–23 (usually 22–23) before the bases of the pelvic fins and 12–19 (usually 15–17) behind them (1). Gill rakers are long and slender, with 59–73 below the sharp bend on the first arch. The opercula have coarse, fanlike striations on their surfaces. Live fish tend to be steely blue on the back and silvery on the sides. They are distinguished by a row of 4–6 black spots that starts on the back just behind the operculum. The first spot is larger than the rest.

Taxonomy American shad show biochemical, meristic, and life history differences among watersheds in their native range along the Atlantic coast, but these differences are clinal, so no subspecies are recognized (10, 11, 24).

Names Alosa is derived from the old Saxon name for European shad (*Alosa alosa*); *sapidissima* means "most delicious," a fitting name for the most edible of the North American shads. American shad are often referred to as common or white shad.

Distribution American shad are native to the Atlantic coast from Labrador to the St. Johns River, Florida (1, 2). Between 1871 and 1881, more than 800,000 fry were caught in New York and planted in the Sacramento River (3, 18). By 1879 they were abundant. As a result of the California plants, they are now found from Todos Santos Bay in Mexico to Cook Inlet, Alaska, and a population has established itself as well in the Kamchatka Peninsula in Russia. Plants of shad were also made in the Columbia River in 1885 and 1886, although shad from California had already colonized the area by that time (18). They spawn in major rivers from the Sacramento drainage north to British Columbia.

The main shad runs in California are in the Sacramento River up to Red Bluff and in the lower reaches of its major tributaries (particularly the American, Feather, and Yuba Rivers), with smaller runs in the Mokelumne, Cosumnes, and Stanislaus Rivers as well as the north Delta and Old River in the south Delta. Small runs enter the Klamath, Russian River, and Eel Rivers. Apparently the only land-locked population of American shad in existence is in Millerton Reservoir, Fresno and Madera Counties, where they were accidentally introduced with striped bass between 1955 and 1957 (4, 5).

Life History Because of the importance of the species as a sport fish, American shad have been the subject of investigations by CDFG. Unless otherwise indicated, this account is based on summaries provided by Skinner (3), Stevens (6), Painter et al. (7), and Stevens et al. (8).

American shad, with the exception of the Millerton Reservoir population, are found in fresh water only when adults move up into rivers to spawn and when juveniles use rivers as nursery areas for the first year or two of life. What happens during the 3–5 years between emigration to sea and return for spawning is largely unknown, although shad tagged in the Sacramento River have been recovered from Monterey to Eureka. In their native Atlantic they make extensive movements up and down the coast from Florida to New Brunswick (9). The wide distribution of shad along the Pacific coast indicates that similar movements take place here as well, albeit mainly in years when oceanic conditions are favorable.

The first mature shad of each year's run appear in autumn in the lower portions of the estuaries, where they gradually adjust to low salinities. They do not move into fresh water until March–May, when water temperatures exceed 14°C. Peak runs and spawning usually occur at higher temperatures, 17–24°C in the Sacramento River. This means the first shad, usually unripe males, appear in late March or early April, but large runs are not seen until late May or early June. The runs become smaller again when water temperatures exceed 20°C, and few adults are seen after the first week of July. In the landlocked Millerton Reservoir population, mature shad start appearing in the San Joaquin River in late April, but some spawning occurs as late as early September (5); peak spawning occurs from mid-June to mid-July at water temperatures of 11–17°C (5). In their native range homing of shad to their natal streams is well documented (10). Shad have remarkable abilities to navigate and to detect minor changes in their environment (11). In the Sacramento River and its tributaries homing is generally assumed, although there is some evidence that numbers of fish spawning in major tributaries are proportional to flows of each river at the time the shad arrive. They are also capable of adjusting the timing of their runs to the timing of river outflows (24). However, spawning fish tagged in one year are most likely to turn up

in the same river in following years if they are repeat spawners (19).

In the Sacramento River, most males spawn for the first time at 3–4 years, and most females spawn at 4–5. Thereafter, fish may spawn every year, as shown by checks on their scales, with the oldest being 7 years. However, 70 percent of the run is made up of fish that have not spawned before.

Adult shad in the Delta feed in fresh water, unlike their counterparts on the Atlantic coast. This is likely a result of the abundance of large zooplankters in the Delta, as opposed to their scarcity and small size in most Atlantic spawning streams. Even in the Delta, however, the percentage of shad with empty stomachs is high, and feeding virtually ceases once main rivers are entered. The most abundant organism in their stomachs while they are in the Delta is mysid shrimp, followed by copepods, cladocerans, amphipods, clams, and fish larvae. On occasion they will eat other fishes. In Millerton Reservoir they prey on threadfin shad (5), and in the Sacramento River below Red Bluff Diversion Dam they occasionally prey on small chinook salmon (25).

Spawning takes place mostly in main channels of rivers over a wide variety of substrates, although sand and gravel are usual. Depth of the water ranges from 1 to 10 m, usually less than 3 m (1). Currents are typically 31–91 cm/sec (13), but shad at Millerton Reservoir spawn in currents of 20–60 cm/sec (5). Dissolved oxygen levels must be above 5 mg/liter (13, 21).

Spawning is a mass affair that peaks after dark and seems stimulated by falling light levels (1). In the San Joaquin River above Millerton Reservoir, all spawning takes place between 2100 and 0700 hr, with peaks between 2300 and 0400 hr (5). Each act is initiated when one or more males press alongside a female. The fish swim rapidly in a circle, side by side, releasing eggs and sperm. Other groups of shad usually spawn at the same time. During the spawning act, dorsal and upper caudal fins of spawning groups often break the surface, making distinctive splashing sounds. The higher the intensity of splashing, the more eggs are found in the water column (5). Each fish spawns repeatedly over a number of days, until the females have released all their eggs. Numbers of eggs produced by California shad have been determined only for the Millerton Reservoir population, where shad fecundities range from 98,600 to 225,600 (5). On the East Coast, average fecundity varies with geographic locality, and shad from New York (the source of California shad) produce 116,000–468,000 eggs, depending on the size of the female, with a mean around 250,000 (1). One reason for such high fecundities is that fertilization rates may be exceptionally low. In Millerton Reservoir only 2 percent of eggs collected had been fertilized (5).

Shad embryos are only slightly heavier than water, so they stay suspended in the current, gradually drifting downstream. They can be found at almost any depth during peak spawning season but are most numerous near the bottom. Hatching takes 8–12 days at 11–15°C, 6–8 days at 17°C (considered to be the optimal developmental temperature), and 3 days at 24°C (1). More rapid development at higher temperatures seems to be countered by lower survival rates of embryos.

Newly hatched shad are 6–10 mm TL and 9–12 mm TL when the yolk sac is finally absorbed and feeding begins (1, 12). At this point, if larvae do not feed within 2 days their survival may be severely depressed (19), demonstrating the need to reach a food-rich area at the appropriate time. Larval shad are planktonic for about 4 weeks. At this stage they cannot survive in salt water, although once they metamorphose into actively swimming juveniles (>25 mm TL) they can tolerate an abrupt switch to sea water (22). The first several months are usually spent in fresh water, but small shad can live in salinities of up to 20 ppt (20). They seem to prefer temperatures of 17–25°C (23). In the Sacramento River the main summer nursery areas are lower Feather River, Sacramento River from Colusa to the north Delta, and, to a lesser extent, the south Delta (8). During this period, they are extremely vulnerable to entrainment in agricultural diversions in the Delta, power plant cooling water diversions, and especially pumps of the State Water Project and Central Valley Project in the south Delta. Abundance indexes of juvenile shad have strong positive correlations with Delta outflows as a result of complex interactions among entrainment, larval feeding, and timing of spawning.

As the season progresses, the center of juvenile shad abundance moves closer to salt water. Entry into salt water takes place in September, October, and November, but it may start as early as late June, especially in wet years when outflows are high. Outmigrating shad are 5–15 cm FL. Most go out to sea, but some remain in the estuary for 1–2 years (8). In the Eel River juvenile shad spend their first summer in large schools in deep pools of the lower river. What they feed on is a mystery, but they may serve as food for adult shad trapped in the pools after spring flows drop.

While in the San Francisco Estuary, young shad feed on zooplankton, especially mysid shrimp, copepods, and amphipods. Although they feed primarily in the water column, they are opportunistic and will also take abundant bottom organisms (such as chironomid midge larvae) and surface insects (14). Most feeding takes place during the day, reflecting a reliance on vision for prey capture. The presence of summer concentrations of young shad in dead-end sloughs, where zooplankton are abundant, indicates that they will seek concentrations of food off migration pathways.

Growth of American shad varies with rearing environment and is probably related to a combination of temperature and availability of food. In Millerton Reservoir, where appropriate foods are relatively scarce, growth is slow compared with that in populations that go out to sea. At the end of the first year they average about 8 cm FL; second year, 16 cm; third year, 24 cm; fourth year, 30 cm; fifth year, 37 cm; sixth year, 39 cm; and seventh year, 42 cm (5). This is roughly comparable to the growth rates of males from the Yuba River, but females typically reach 42 cm FL by the end of their third year and 48 cm during their seventh year (5, 15). The largest fish in any year class are invariably virgin spawners, because repeat spawners have less energy to devote to somatic growth. There are no records of shad from California living longer than 7 years, although in their native range they live 11 years and reach 58 cm FL (1).

Status IID. The introduction of American shad into California was extraordinarily successful. By 1879, 8 years after the first introduction, a commercial fishery had developed (3). The fishery peaked in 1917 when 5.7 million lb were landed, representing about 2 million fish. Between 1918 and 1945 the catch ranged annually from 0.8 to 4.1 million lb. Much of this catch was canned. Between 1945 and 1957, the catch exceeded 1 million lb only once. Despite their reputation on the East Coast as an excellent food fish, American shad have never been particularly popular on the West Coast. Thus the commercial fishery was never very valuable, despite its size. Female shad, from which roe was removed for shipment to Asia, fetched $0.06–0.08 a pound in 1957; males brought less than $0.01 a pound.

In 1957 the inland commercial fishery was banned in favor of the sport fishery, mainly as a means to protect striped bass caught incidentally in gill nets set for shad (18). As a result, when shad migrate upstream today, especially in the Yuba, Feather, and American Rivers, anglers line the stream banks, often standing shoulder to shoulder when fishing is good. Spawning shad readily take a fly, jig, or small spinner and put up spectacular fights on light tackle. Ripe males are also caught at night by "bumping," a technique in which a long-handled dipnet is held vertically from the stern or side of a slowly moving boat. When a shad hits the net, it is twisted and lifted, and the shad is flipped into the boat. Some shad are also caught by dipnetting from banks. The present fishery seems to be mostly for sport, and a good many of the fish caught are either discarded or returned to the water.

Despite their popularity, American shad have been declining. In 1976 and 1977 the total spawner population was estimated at around 3 million fish, perhaps one-third the number that existed 60 years earlier (8). Although population estimates have not been made since 1977, indexes of juvenile shad abundance have declined steadily since then (16), and sport fishing catches appear down. The major cause in recent years is most likely the increased diversion of water from the rivers and the Delta, combined with changing conditions in the ocean, although pesticide affects on larvae and other factors may also be contributing. The shortage of adequate attraction flows in major spawning tributaries, such as the American River, may also have played a role in the decline (17).

References 1. MacKenzie et al. 1985. 2. Lee et al. 1980. 3. Skinner 1962. 4. Von Geldern 1965. 5. Ecological Analysts 1982. 6. Stevens 1966a. 7. Painter et al. 1979. 8. Stevens et al. 1987. 9. Talbot and Sykes 1958. 10. Leggett and Carscadden 1978. 11. Leggett 1973. 12. Wang 1986. 13. Emmett et al. 1991. 14. Levesque and Reed 1972. 15. Sanford 1975. 16. Herbold et al. 1992. 17. Snider and Gerstung 1986. 18. Dill and Cordone 1997. 19. Johnson and Dropkin 1995. 20. Limburg and Ross 1995. 21. Jenkins and Burkhead 1994. 22. Zydlewski and McCormick 1997. 23. Stier and Crance 1985. 24. Quinn and Adams 1996. 25. B. Vondracek, pers. comm.

Minnows, Cyprinidae

True minnows are one of the most abundant and widely distributed groups of freshwater fishes in the world, dominating streams of North America, Eurasia, and Africa. There are more than 250 species in North America alone, including introduced carp and goldfish. They range in length as adults from a few centimeters to more than 1 m.

The typical native North American minnow is a small, silvery fish. In California this description applies mainly to juveniles because the adults are often large (20 cm or more). The body is elongate and often has a dark band running along the side. The caudal fin is forked and the dorsal fin short, located just above the pelvic fins. True spines are absent from the fins, although carp, goldfish, and spinedaces have rays that are hardened and resemble spines. There are never teeth on the jaws, but pharyngeal teeth are well developed and often highly specialized. Scales are cycloid and typically are evenly distributed over the body but absent from the head. Socially most minnows are shoaling fish, schooling in many situations. During the breeding season, however, males of many species (e.g., fathead minnow) stake out territories and defend them from other fishes. Breeding males usually develop small, hard tubercles on their bodies and fins, particularly around the snout. The more conspicuous the tubercles, the more likely it is that the species builds nests and defends territories. The tubercles are inconspicuous on most native California minnows.

Many factors contribute to the success of the Cyprinidae. Perhaps most important are a well-developed sense of hearing, a fear substance they release when injured, the presence of pharyngeal teeth, and high fecundity. Their hearing is acute because they possess a series of small bones (Weberian ossicles) that connect the anterior lobe of the swim bladder to the inner ear. The swim bladder, being filled with gas, intercepts sound waves passing through the water (and the body of the fish). The vibrations are then carried to the inner ear by the ossicles, much as the bones

in the middle ear of mammals carry sound from the eardrum to the inner ear. This auditory system allows minnows to detect a much wider range of frequencies than most other fishes. Although the primary functions of such acute hearing are detection of predators and conspecifics and food finding, the auditory system is also used during breeding; the males of a number of species make sounds during courtship and territorial defense.

The sense of smell is also well developed in minnows and important in helping to avoid predators. If a minnow is injured so that the skin is broken, a special chemical present in the skin (fear substance) is released. The olfactory organs of minnows are highly sensitive to this substance. When it is detected, minnows immediately go into self-protective behaviors, fleeing or hiding. This mechanism is particularly valuable in weedy or turbid waters, where predators are difficult to see.

Pharyngeal teeth contribute to the success of minnows in much the same way that specialized jaw teeth contribute to the success of mammals on land. They allow minnows to specialize in feeding habits and to break up foods taken in through the toothless mouth. The pharyngeal teeth, located in the "throat" behind the last gill arch on each side, grind food against a hard plate on the roof of the buccal cavity. Minnows with different feeding habits tend to have different shapes, sizes, and numbers of pharyngeal teeth. Sacramento pikeminnows have pointed, knifelike teeth that point backward down the throat; these are well suited for retaining and cutting up the fish and large invertebrates they eat. Adult hardhead, which live with pikeminnows, have pharyngeal teeth that are flattened on the ends; they are suited for crushing algae and small invertebrates. Young hardhead, which feed primarily on aquatic invertebrates, have more knifelike teeth, which become flatter as the fish grow older.

Because the teeth are so distinctive, they can be used to distinguish species. The number and arrangement of the pharyngeal teeth are particularly useful characteristics, and tooth formulas frequently accompany descriptions of minnow species. Most minnows native to California have two rows of teeth on each side, so a typical formula reads 1,4-4,1, indicating one tooth on an inside row and four teeth on an outside row on each side. It is not unusual for the number of teeth on each side to differ slightly, owing to natural variation.

Despite all these advantages, many native minnow species are declining. Thicktail chub, Clear Lake splittail, Colorado pikeminnow, and bonytail have all become extinct in California within the past 40 years; together they represent 27 percent of the native cyprinids. Another presumably extinct species, the Clear Lake minnow, *Endemichthys grandipinnis* Hopkirk 1973, may be a hybrid and so will not be treated further.

California has (or had) 15 species of native minnows and 7 introduced species. The native species, together with other cyprinids native to rivers west of the Rocky Mountains, form an evolutionary group (clade) separate from other North American species (Cobern and Cavender 1992; Simons and Mayden 1998). Most are distinct enough to be placed in genera found only in western North America. In addition, four widely distributed native species in California have among them at least 27 putative subspecies: tui chub (10), hitch (3), California roach (8), and speckled dace (6). Many of these are poorly defined or undescribed; some may represent species. Unfortunately, some are likely to become extinct before they achieve formal taxonomic recognition.

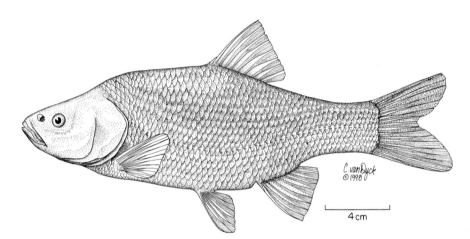

4 cm

Figure 42. Tui chub, 22 cm SL, Goose Lake, Modoc County.

Tui Chub, *Siphateles bicolor* (Girard)

Identification Tui chubs are typically chunky, large-scaled (41–64 scales along the lateral line) fishes with small, terminal, and slightly oblique mouths, stubby gill rakers, and a decurved lateral line. Gill rakers number 8–24, the left gill

arch usually bearing a few more than the right. The gap between the gill rakers is wider than the base of the gill rakers themselves. Both dorsal and anal fin rays number 7–9 (usually 8). All fins are rounded and short. The head becomes larger relative to the rest of the body in older fish and is usually convex in profile. A distinct hump may develop behind the head. The single-rowed pharyngeal teeth (0-5,5-0 or 0-4,4-0) are slightly hooked with narrow grinding surfaces. Live fish tend to be dusky olive, brown, or brassy on the back and white to silver on the belly. The younger the fish, the more silvery the body color. Adult size is highly variable; in springs they may only reach 10–12 cm SL, whereas those in large lakes may reach 30–40 cm SL or more.

The *pectinifer* form differs from the foregoing description in that the gill rakers are more numerous (29–40) as well as being long and slender. Distances between gill rakers are usually less than the width of the gill rakers themselves. The mouth is more oblique than that of typical tui chubs, and the profile of the head is slightly concave. The overall color at all sizes is silvery.

Taxonomy In most recent studies, the tui chub is placed in the genus *Gila* along with a number of other similar-appearing species from the western United States (46). However, biochemical evidence indicates that tui chubs (and a couple of other species) are more closely related to other endemic California minnows than to species of *Gila* (41). Therefore the generic name *Siphateles*, first applied by J. O. Snyder and widely used thereafter, has been resurrected and used here.

The tui chub is a highly successful species that presents fascinating problems in systematics. Almost every isolated or partially isolated drainage system in California, Nevada, and Oregon supports at least one distinctive form. J. O. Snyder, one of California's early ichthyologists, was so impressed by differences among the various forms that he described many as separate species. Today most of Snyder's species have been reduced to subspecies, but the taxonomic diversity reflected in his work still has not been satisfactorily resolved. Ten subspecies are recognized in California, but the number and taxonomic status of these forms are likely to change as we learn more about them.

Klamath tui chub. This subspecies (*S. b. bicolor*) of the Klamath system was the original recipient of the epithet *bicolor*, which created considerable confusion (see Names). It was originally considered a distinct species in the genus *Leuciscus* or *Tigoma* (3).

Cowhead lake tui chub. *S. b. vaccaceps* was described from a playa lake system in extreme northeastern California in 1980 (1). It is probably closely related to the Goose Lake tui chub and other chubs of the Oregon desert.

Goose Lake tui chub. This subspecies, endemic to Goose Lake on the California-Oregon border, was originally described by Cope in 1883 (2) as *Myoleucus thalassinus*. It was later assigned to *S. b. bicolor* (3), *S. b. formosa* (4, 5), and *S. b. thallassina* (6, 7). *S. b. bicolor* is now reserved for the Klamath tui chub. The name *S. b. formosa* was the old name originally applied to tui chubs supposed to have lived in the Sacramento–San Joaquin Valley. Because only a few poorly preserved specimens are known, the subspecies may be based on a mislabeled collection (47). Thus the appropriate name for the Goose Lake subspecies is *S. b. thallassina*.

Pit River tui chub. Hubbs et al. (8) listed this chub as an undescribed subspecies. Its relationships to other subspecies, especially the Goose Lake tui chub, need clarification.

Lahontan tui chubs. Perhaps the most intriguing systematic problem among tui chubs is the relationship between two forms in the Lahontan drainage, usually listed as the Lahontan creek tui chub, *S. b. obesa*, and the Lahontan lake tui chub, *S. b. pectinifer* (9, 10). The two are different enough in morphology that J. O. Snyder (4) placed them in

separate genera (*Siphateles* and *Leuciscus,* respectively). The differences in the gill rakers, much finer and more numerous in *pectinifer* than in *obesa*, are particularly striking. R. G. Miller (11) found that differences in gill rakers as well as slight morphological variations reflected differences in niche; *S. b. obesa* occurs in streams and lakes as a shallow-water bottom feeder, whereas *S. b. pectinifer* feeds on zooplankton in the open water of lakes. Studies in Pyramid Lake, Nevada, confirm that the two forms segregate by diet, distribution, and breeding times and places (12, 43).

Eagle Lake tui chub. This form is undescribed in part because it has long been considered a "hybrid" between *S. b. obesa* and *S. b. pectinfer*, based on the bimodal distribution of gill raker numbers (14). However, the isolated nature of Eagle Lake and its unusual limnological characteristics make it highly likely that its tui chub is distinct.

High Rock Springs tui chub. High Rock Springs is a spring system in Lassen County, an unusual and extremely isolated environment for tui chubs (45). Unfortunately, the form inhabiting this spring was driven to extinction in 1989 before it could be formally described. It deserves at least a posthumous description.

Owens tui chub. R. R. Miller (13) differentiated this subspecies (*S. b. snyderi*) from other tui chubs largely on the basis of number of radii on the scales, a character of questionable significance. Electrophoretic studies indicate that they are fairly distinct, and there is evidence of genetic differentiation within the Owens drainage as well (49). This form has been isolated in the Owens Valley for a long time, so it would be surprising if it were *not* different from other populations.

Mohave tui chub. *S. b. mohavensis*, originally native to the Mohave River, is biochemically one of the most distinct subspecies (48) and may warrant specific status (50).

Names The name tui chub is derived from the Paiute name for the species, *tui-pagwi*. *Pagwi* seems to be the Paiute word for minnow (17). Chub is an old English name of unknown origin, originally applied to a heavy-bodied European cyprinid, *Leuciscus cephalus*. *Bicolor* means two-colored. The scientific name for this species has a complex history (15). In most of the literature the name used is *Siphateles bicolor* or else *Siphateles* in combination with one of the names now used to designate subspecies. When *Siphateles* was merged into the genus *Gila*, the name became *Gila bicolor* (16). Unfortunately, the blue chub of the Klamath River system already had the name *Gila bicolor*, so the early synonym *coerulea* was adopted for it (16). Thus *G. bicolor* in the literature prior to 1964 is *G. coerulea*, whereas the tui chub is *S. bicolor*.

Distribution In California tui chubs are native mostly to interior drainages, except the Central Valley, and absent from all coastal drainages, except where introduced. Because they are hardy and used (illegally) as bait fish, they can be expected almost anywhere. In the Great Basin they are (were) present in many isolated springs and sloughs, including Cowhead Lake Slough (Modoc County) and High Rock Springs (Lassen County). They are abundant in Eagle and Honey Lakes and associated streams (Lassen County), in Lake Tahoe, and in the Truckee, Walker and Carson River drainages (where they are abundant in reservoirs). Tui chubs occur in much of the Owens River drainage, including Crowley Reservoir, isolated spring systems in Owens Valley, and Owens River gorge. The endangered Mohave tui chub was originally found throughout the Mohave River drainage but is now found only in San Bernardino County ponds isolated from its native river, mainly at Soda Springs (three ponds) and Lark Seep Lagoon (China Lake Naval Weapons Center). In the upper Klamath River basin Klamath tui chubs are found in lakes, sloughs, rivers, and reservoirs, downstream as far as Iron Gate Reservoir (although individuals have been collected downstream to the mouth of the Klamath River). In the Sacramento–San Joaquin drainage tui chubs are native only to Pit River downstream at least as far as Hat Creek and Britton Reservoir and to Goose Lake (Modoc County and Lake County, Oregon), although they have been introduced into some reservoirs (e.g., Almanor) and ponds in various locations (e.g., Point Reyes). Outside California they are found in a number of interior basins of Oregon (Catlow, Harney) and Nevada and are widespread in the Columbia River system in Washington and Oregon.

Life History Tui chubs occur in many habitats: isolated springs, large desert lakes, sloughs, meadow streams, sluggish rivers, and backwaters of swift creeks. The key feature of "typical" tui chub habitat is quiet water with well-developed beds of aquatic plants and bottoms of sand or other fine materials (18). Waters containing abundant tui chubs usually have summer temperatures in excess of 20°C and are alkaline. However, tui chubs do well under many limnological conditions—from the cold, clear, oligotrophic water of Lake Tahoe to the cool, productive waters of Pyramid Lake, Nevada, where the total dissolved solids are greater than 4,700 ppm, approximately 75 percent sodium chloride. Mohave tui chubs, the southernmost representative of the species, can survive temperatures from 2° to 36°C, but optimal temperatures are between 15° and 30°C (19, 20). This range of temperature tolerance is surprisingly narrow for a "desert" fish but may be typical for the species. The range of alkalinities tolerated is considerably greater, however, because tui chubs are regularly found at pH values greater than 9 and can tolerate pH levels of around 11 (21). Tui chubs are also tolerant of low dissolved oxygen levels. In Pyramid Lake they are regularly found at oxygen levels less than 50 percent saturation, and, when the water is cold, they will survive at less than 25 percent saturation (i.e., less than 4 mg/liter) (22).

During summer, in large, deep lakes, adult tui chubs tend to move into deep water during the day and return to shallow or surface waters at night (11, 23). In Lake Tahoe the pelagic form (*pectinifer*) schools well off the bottom, whereas the benthic form (*obesa*) shoals close to it. Thus the benthic chubs more commonly fall prey to lake trout, a deep-water benthic predator. Young-of-year chubs of both types remain in shallow water most of the summer, in large shoals, although strong wave action will drive chubs into deeper water among beds of aquatic plants. Larval tui chubs are planktonic (24), and the benthic and planktonic forms begin to segregate by diet and habitat at about 25 mm TL (12).

In shallow lakes with heavy growths of aquatic vegetation, such as Tule Lake, Modoc County, shoaling is less noticeable. Chubs tend to be dispersed among the aquatic plants in small groups, presumably as protection against predatory birds that are attracted to large aggregations. In autumn, in all types of lakes, the chubs seek out deep water in which to spend the winter, presumably on the bottom in an inactive state. In Pyramid Lake they concentrate at depths greater than 61 m, where both temperatures and oxygen concentrations are low (22). The spring reappearance of the chubs, at least in Eagle Lake, Pyramid Lake, and Lake Tahoe, is both sudden and spectacular, usually coming in mid-May (4, 11, 14). J. O. Snyder (4, pp. 66–67) described the spring return in Pyramid Lake vividly:

> On May 20 the weather suddenly settled and became warm. . . . About 2 o'clock the following morning there was heard a vigorous lapping of the water, which in the quiet air appeared entirely without cause until it was found to accompany the leaping of vast numbers of fishes. Far out and up and down the shores the surface of the water fairly boiled. Spring had come, and with it, in the dim light of early morning, myriads of fishes from the depths of the lake. Daylight revealed them everywhere, along the shore, among the boulders, and in the algae, hovering in enormous schools over the bars and moving about in the clear water of the sheltered bays.

Tui chubs are opportunistic omnivores with long intestines. Usually the majority of the gut contents consists of detritus, unidentified organic matter, and plant fragments. Given their abundance in many lakes, they may play an important role in nutrient cycling. However, it is hard to quantify detritus, so it is usually underreported or omitted from dietary studies, although it may be quite important nutritionally. In Eagle Lake 82 percent by volume of the gut contents of large tui chubs was detritus, 2 percent was algae, and the remainder was invertebrates (26). This is probably similar to the diet of most chubs over 10 cm SL that are reported to be feeding mainly on invertebrates. Thus chubs from ponds and springs were reported as feeding on aquatic insect larvae (especially chironomid midges) and benthic crustaceans (27, 28), and those from Big Sage Reservoir, Modoc County, fed on a mixture of plant material, plankton, insect larvae, and small tui chubs (29). In Lake Tahoe the food of benthic tui chubs was reported as 89 percent benthic invertebrates, 5 percent fish and fish eggs, 3 percent plankton, and 3 percent plants (11). The invertebrates consisted mostly of snails, small clams, caddisfly larvae, midge larvae, and crayfish. Benthic chubs in Pyramid Lake move into shallow areas at night to feed on insects, algae, and plant material (4). Detritus is presumably not important in the diet of pelagic (*pectinifer*) tui chubs, which feed, using their long gill rakers, almost exclusively (over 90%) on zooplankton (11, 39). Larval tui chubs feed on planktonic crustaceans and rotifers (30). Pelagic tui chubs continue to feed on zooplankton as they grow larger and as their gill rakers increase in number, whereas benthic tui chubs gradually switch to feeding on small benthic invertebrates (12, 25). In Eagle Lake young-of-year chubs feed on a mixture of benthic invertebrates, zooplankton, and small terrestrial insects blown in from the surrounding forest (26).

Tui chubs are long lived, although ages of large individuals have been consistently underestimated through the use of scales to age fish. When opercular bones are used for aging in place of scales (which show signs of partial resorption in older fish), large adults (30–40 cm SL) in Eagle Lake are aged at 12–33 years (26, 30, 42). Using scales, all such fish were aged at 6–7 years, the age at which they become sexually mature and growth slows (14, 42). In ponds scales indicate life spans of 3–4 years (31), whereas opercular bones indicate life spans of 6–7 years (28). For the first 2–3 years of life, scales, opercular bones, and length-frequency distributions tend to agree with one another for aging the fish. Thus tui chubs reach 5–10 cm SL in their first year, 6–18 cm in their second year, and 13–22 cm in their third year (14, 26, 27, 31, 32, 42). Growth slows at maturity, usually in the second to fourth year. In ponds and springs tui chubs rarely grow longer than 20 cm SL, but in large lakes fish measuring 30–40 cm SL are common. The largest tui chubs recorded from Eagle and Pyramid Lakes are around 42 cm SL (14, 26, 33).

Most spawning takes place between late April and early July, although in Lake Tahoe spawning apparently continues until the end of July (11). In springs and warm ponds spawning may occur from February through late August (28, 32). Multiple spawning by a single female is probably common, because all eggs do not ripen at the same time and larval tui chubs can be found well into August (25). Fecundities are high. A female from Eagle Lake measuring 28 cm FL contained 11,200 ripe eggs (14); females from an Oregon population measuring 15–28 cm TL contained 4,140–25,000 eggs (34); and Mohave tui chubs measuring 10–22 cm SL contained 3,800–50,000 eggs (31). Spawning in most places occurs at temperatures between 13 and 17°C (14, 24, 31), although Mohave tui chubs have been recorded spawning at 26°C (31). Tui chubs spawn in water less than 1.5 m deep, usually over beds of aquatic vegetation or algae-covered rocks and gravel, although in Lake Tahoe they spawn over sandy bottoms or in the mouths of streams (11, 14, 24). Spawning involves large, swirling aggregations, apparently with several males attending each female. In Pyramid Lake *obesa* and *pectinfer* forms seem to spawn at different times and places, reducing the potential for hybridization (43).

Newly fertilized eggs are 1.5–1.9 mm in diameter and adhere to aquatic plants or bottom (44). Embryos hatch in 3–6 days, and larvae start feeding soon after hatching. Although the larvae are mainly planktonic, in Eagle Lake they remain among aquatic plants until they reach about 2 cm TL, when they move into shallow water along the shore (14). In Lake Tahoe larvae also seem to concentrate in shallow, weedy nursery areas; as they grow, they spread out along the shore over both rocky and sandy areas (11). Scale formation starts at 20–25 mm SL.

Status IA–IE. Tui chubs are abundant and widely distributed, and so are not in trouble as a species. However, a number of the fascinating and ecologically diverse subspecies are in serious trouble and need special management.

Klamath tui chub. IE. This subspecies is still common in the Klamath basin in Oregon and California, although its numbers may be locally depleted owing to pollution of the larger lakes.

Cowhead Lake tui chub. IB. Endangered (proposed for federal listing, 1998). Cowhead Lake, Modoc County, was an alkaline lake drained to create pasture, although it probably dried up naturally on occasion. Today chubs survive only in a slough and ditches that drain the lake bed. During wet years, the slough may be a narrow channel as much as 6.4 km long for the fish, but during dry years this water may be reduced to a few pools, especially because inflowing streams are diverted for use in local ranches. Much of the slough is on private land, although some is on public (BLM) land. Most of the perennial water and deep pools, however, are on private land upstream of public land. The slough is attractive to cattle, so riparian vegetation that could provide cover for fish is largely missing and banks are heavily trampled. The drought years of 1986–1992 were especially hard on this subspecies, and only a small number of individuals likely survived the summer of 1992 (35).

Goose Lake tui chub. IB. In the summer of 1992 Goose Lake dried up. As lake levels dropped and the water became increasingly alkaline, large numbers of chubs were observed attempting to enter tributary streams, attracting thousands of white pelicans and other fish-eating birds to feast on them. The chubs and other endemic fishes apparently found temporary refuges in spring-fed pools in the streams, as well as in some reservoirs in the Thomas Creek drainage in Oregon, in which populations have been established for some time. Out of concern for the long-term survival of Goose Lake fishes, a Goose Lake Fishes Working Group was formed to develop management plans; it drew members from among agency biologists, private landowners, environmental groups, and other interested parties. It is hoped that implementation of voluntary management measures on both public and private land will forestall formal listing of the chub and other Goose Lake fishes as endangered species (35).

Pit River tui chub. ID. This chub is common in reservoirs and some streams in the Pit River basin, but its populations are scattered and status uncertain.

Lahontan lake tui chub. IC, apparently in low numbers. The *pectinifer* chub is abundant in Pyramid Lake, Nevada, and is at least present in Lake Tahoe. The chubs with long gill rakers and planktivorous diets in Stampede Reservoir on the Little Truckee River (39) may also belong to this subspecies, but no reservoir population can be regarded as secure. Concern for this form stems from the presence of kokanee and opossum shrimp (*Mysis relicta*) introduced into Lake Tahoe, which have depleted the zooplankton on which the chubs feed (35). A more recent threat has been the establishment of largemouth bass, which may prey on juvenile chubs in their inshore rearing areas. Their future is probably more secure in Pyramid Lake. Through most of the 20th century, levels of this lake fell steadily as the result of agricultural and urban diversions, but in recent years lake levels have risen as the result of increased inflows to protect cutthroat trout and cui-ui.

Lahontan stream tui chub. IE. The *obesa* chub is abundant and widely distributed in many habitats in watersheds of the eastern Sierra Nevada.

Eagle Lake tui chub. ID. Eagle Lake is a large, terminal lake that enjoys special management to protect a trophy fishery for endemic Eagle Lake trout. Nevertheless, it does not pay to be complacent about the future of Eagle Lake and its native fishes, including tui chub. The lake is growing in popularity as a tourist destination, and there is likely to be increased demand to "improve" the fishery, especially by introducing additional species that might prey on tui chub, compete with it, or spread diseases or parasites to it. Unfortunately, it is all too easy for irresponsible anglers to make unofficial introductions into the lake. There is therefore a need for a publicity campaign on the value of the native fishes, especially to ospreys and the other fish-eating birds that are abundant on the lake, and on the potential detrimental effects of introduced species. Perhaps one approach would be to promote a fishery for chubs themselves, similar to the one that once existed on the lake (14). The chubs are

large in size and tasty if properly prepared. Indeed, they were once a major food source for indigenous peoples throughout the Great Basin, especially the smaller ones that occurred in huge numbers in shallow water (40).

High Rock Springs tui chub. IA. This undescribed, dwarf tui chub quietly went out of existence in 1989, the victim of an unsuccessful attempt to farm fish in the effluent of a desert spring. High Rock Springs, Lassen County, is a warm spring system located on private land. In 1983 the rancher was issued an aquaculture permit by CDFG and introduced 1,000 Mozambique tilapia into a facility below the spring. The tilapia quickly colonized the spring system, and tui chubs disappeared within 6 years, presumably as a result of predation on their eggs and larvae (35).

Owens tui chub. IB? This subspecies is listed as endangered by both state and federal governments. Tui chubs are abundant in the Owens River drainage, especially in Crowley Reservoir. However, R. R. Miller, who described the subspecies, concluded that the fish in the main river were in fact introduced Lahontan creek tui chubs that had displaced the native chubs (13). "Pure" Owens chubs were then assumed to exist only in isolated springs, such as Hot Creek head springs, in the Owens River gorge (below Crowley Reservoir), and in Owens Valley Native Fish Sanctuary. The sanctuary was created specifically to protect endangered Owens pupfish and Owens tui chub. Because morphological differences between Owens and Lahontan chubs are small, an electrophoretic study was conducted to ensure that the isolated populations assumed to be Owens tui chubs had not introgressed with Lahontan tui chubs (36). This study could not discriminate the isozyme patterns of the two subspecies, although it did show that each of the isolated populations had some minor genetic differences that separated them from each other as well from the population in the main river. These ambiguous results suggest that more studies are needed, using more sensitive techniques, to determine the relationships among the various chub populations.

Mohave tui chub. IB. This is another subspecies listed as endangered by both state and federal governments. The Mohave tui chub is the only fish native to the Mohave River, San Bernardino County. In the 1930s arroyo chubs were introduced, presumably as bait, into reservoirs in the headwaters. They replaced tui chubs throughout the drainage through a combination of hybridization and superior ability to resist high flows (20, 37). A single population of Mohave tui chub persisted in isolated ponds at Soda Springs, which they presumably colonized in a major flood of the Mohave River. The largest of these ponds was converted into an ornamental lake to benefit customers of a resort, and the chubs thrived in part on bread thrown to them (31). The old resort is now a field station of the California State University system, and its ponds and springs are now managed largely to benefit tui chubs. A number of attempts have been made to establish Mohave tui chubs in other locations, but so far such attempts have been successful only in a pond at the China Lake Naval Weapons Center and in a small artificial pond near Hinkley, California. A recovery plan has been developed; the fish will qualify for upgrading to threatened status when six self-sustaining populations of at least 500 fish each are established (38).

References 1. Bills and Bond 1980. 2. Cope 1883. 3. Evermann and Clark 1931. 4. Snyder 1918. 5. Bond 1973. 6. Snyder 1908a. 7. Hubbs and Miller 1948. 8. Hubbs et al. 1979. 9. Hubbs 1961. 10. Hubbs et al. 1974. 11. R. G. Miller 1951. 12. Galat and Vucinich 1983a. 13. R. R. Miller 1973. 14. Kimsey 1954b. 15. La Rivers 1962. 16. Bailey and Uyeno 1964. 17. Loud 1929. 18. Bond et al. 1988. 19. McClanahan et al. 1986. 20. Castleberry and Cech 1986. 21. Falter and Cech 1990. 22. Vigg 1980. 23. Vigg 1978. 24. Cooper 1982. 25. Galat and Vucinich 1983b. 26. P. B. Moyle and students, unpubl. data. 27. Kimsey and Bell 1956. 28. McEwan 1988. 29. Kimsey and Bell 1955. 30. Scoppetone 1988. 31. Vicker 1973. 32. Cooper 1985. 33. Sigler and Sigler 1987. 34. Bond 1948. 35. Moyle et al. 1995. 36. W. Berg and P. B. Moyle, unpubl. data 1991. 37. Hubbs and Miller 1943. 38. USFWS 1984a. 39. Marrin and Erman 1982. 40. Butler 1996. 41. Simons and Mayden 1998. 42. Crain and Corcoran 2000. 43. G. Scoppetone, USGS, pers. comm. 1996. 44. Harry 1951. 45. Mills 1979. 46. Uyeno 1966. 47. C. L. Hubbs, pers. comm. 1974. 48. B. May, University of California, Davis, pers. comm. 1997. 49. B. May, pers. comm. 1999. 50. P. Harris, University of Alabama, pers. comm. 1998.

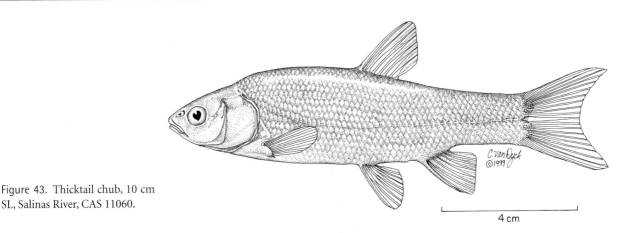

Figure 43. Thicktail chub, 10 cm SL, Salinas River, CAS 11060.

Thicktail Chub, *Gila crassicauda* (Baird and Girard)

Identification Thicktail chubs are heavy-bodied fish with short, deep, thick caudal peduncles; small, cone-shaped heads; 8–9 rays in both dorsal and anal fins; 16–20 rays in each pectoral fin; and 8–10 rays in each pelvic fin. The scales are large with 49–60 in the lateral line. The pharyngeal teeth (2,5-4,2) are sturdy and hooked. The 8–14 gill rakers (usually 10–12) are stubby and toothlike in appearance. The backs of living fish apparently ranged in color from green-ish brown to purplish black; the sides and belly were yellowish (1).

Taxonomy Although the thicktail chub is superficially similar to the tui chub, its double row of pharyngeal teeth suggests it is closer to members of the genus *Gila*. Its taxonomic position among the California minnows remains to be determined.

Names Thicktail is a reference to the wide caudal peduncle. Other common names include Sacramento chub and thicktail. *Crassicauda* means thicktail. For other names, see the accounts of tui chub and blue chub.

Distribution Thicktail chubs were once distributed throughout the Central Valley in lowland areas, in Clear Lake (Lake County), and in streams tributary to San Francisco Bay (1, 2), as well as in the Pajaro and Salinas Rivers (3, 7, 8). The species is extinct.

Life History Little is known of the habits of these once-abundant minnows because no one took an interest in them until they had become extremely rare. What little is known is summarized in Miller (1), who examined 101 fish ranging from 49 to 268 mm SL. He estimated that these comprised 98 percent of all specimens in scientific collections. Thicktail chubs were originally abundant in lowland lakes, sloughs, slow-moving stretches of river, and, during years of heavy runoff, surface waters of San Francisco Bay (2). The stubby gill rakers, short intestine, and stout, hooked pharyngeal teeth indicate that thicktail chubs were carnivorous, probably feeding on small fish and large aquatic invertebrates (1). They were part of the original valley floor fish assemblage that included hitch, Sacramento blackfish, Sacramento sucker, Sacramento perch, and tule perch (8). Thicktail chubs occasionally hybridized with hitch, and the hybrids were originally described in 1908 as a separate species (1).

Status IA. Bones of thicktail chub are among the most abundant fish remains in Native American middens along the Sacramento River (4) along tributary streams such as Putah Creek (2, 5) and in the Pajaro-Salinas drainage (3, 7, 8). In the 19th century the species was abundant enough to be common in the fish markets of San Francisco and to be served in saloons in Sacramento (6). By 1884, however, the

Sacramento Daily Record-Union (Feb. 9) reported that it was already "scarce in the river" and rarely appeared in the markets. Only a few were collected in the 20th century, with the last known specimen caught from the Sacramento River near Rio Vista in 1957, one of two specimens collected in that area since 1938 (2, 6). Extensive sampling of the Delta and lowland habitats of the Central Valley in recent years has failed to find any chubs.

Thicktail chubs most likely became extinct because they were unable to adapt to the extreme modification of valley floor habitats, particularly removal of tule beds, drainage of large, shallow lakes, reduction in stream flows, and modification of stream channels. However, equally or even more important was the introduction of alien predators, especially striped bass and largemouth bass. Thicktail chub may have been exceptionally vulnerable to predation, as indicated by their disappearance from Clear Lake, where habitat modifications were less severe than in the Central Valley.

References 1. R. R. Miller 1963. 2. Mills and Mamika 1980. 3. Gobalet 1990. 4. Schulz and Simons 1973. 5. P. D. Schulz, unpubl. rpt. 1995. 6. Schulz 1980. 7. Gobalet and Jones 1995. 8. Schulz 1995.

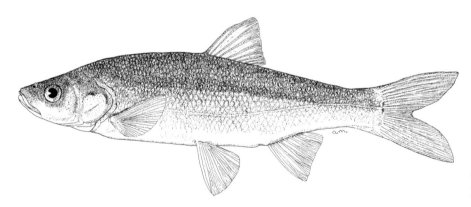

Figure 44. Blue chub, 15 cm SL, Tule Lake, Siskiyou County. Drawing by A. Marciochi.

Blue Chub, *Gila coerulea* (Girard)

Identification Blue chubs have moderately slender, compressed bodies, pointed snouts, relatively large eyes, and terminal mouths that extend back to the front of the eye. They have moderately fine scales (58–71 in the lateral line), 9 dorsal fin rays, 8–9 anal fin rays, and 14–17 rays in each pectoral fin. The two rows of pharyngeal teeth (2,5-5,2) are sharp and slightly hooked. The lateral line is decurved. They seldom exceed 35 cm SL and, alive, tend to be silvery on the sides and dusky on the back. Spawning males have blue snouts and are tinged with orange on the sides and fins.

Taxonomy The distinctiveness of this chub has been recognized ever since it was described by Charles Girard in 1856. Biochemical studies confirm its distinctiveness, even from other members of the genus *Gila* (10).

Names Gila is after the Gila River, New Mexico and Arizona, from which it was mistakenly assumed the first fish named to this genus had been collected; it actually came from the Zuni River, New Mexico (12). Blue (*coerulea*) chub is not very descriptive because they are no more or less blue than most California minnows, except for the blue snout of breeding males. For reasons explained in the account of tui chub, blue chubs were listed as *Gila bicolor* before 1964.

Distribution Blue chubs are widely distributed at lower elevations in the upper Klamath and Lost River systems of Oregon and California. In California they are found in Clear Lake Reservoir, Lost River, Lower Klamath Lake, and Tule Lake, as well as in canals and tributaries feeding them. Their native distribution was presumably above Klamath Falls, but they have now colonized Iron Gate and Copco Reservoirs downstream (in California) from the falls. They may have been introduced into other drainages in Oregon (3).

Life History Blue chubs are most abundant in warm (summer temperatures >20°C), quiet waters with mixed substrates (1). In the laboratory they lose equilibrium at temperatures of 28–33°C (mean, 31.5°C) (4), although they have been collected in the wild at temperatures as high as 32°C (11). Blue chubs are especially abundant in lakes, but they occur in a variety of habitats, from small streams and rivers to shallow reservoirs and deep lakes. In Boles Creek watershed, a tributary to Clear Lake Reservoir (Modoc County), they are common in permanent and intermittent sections, but most abundant in the small, shallow, weedy reservoirs on larger streams (5). In Upper Klamath Lake, Oregon, they are (or were) most numerous along rocky shores or in open water (2). They seem to avoid marshy shore areas. They have a high tolerance for low levels of dissolved oxygen, losing equilibrium at oxygen levels of 0.6–1.5 mg/liter at 20°C (4). Despite this tolerance, they are today largely excluded from deeper parts of Klamath Lake in summer because of oxygen depletion (2). As winter sets in and oxygen levels rise in deep areas, the chubs will move into them. In lakes blue chubs are often conspicuous as large schools moving in and out of shallow water.

Blue chubs are omnivorous. Twenty chubs from Willow Creek, Modoc County, in August 1972 (all 1 year old, 29–59 mm SL) had fed mostly (66% by volume) on chironomid midge larvae and pupae and on small numbers of water boatmen, water fleas, aquatic insect larvae, and various fly-

ing insects. Sixteen 2-year-old chubs (61–109 mm SL) had fed heavily on filamentous algae (68%) and aquatic and terrestrial insects. A similar diet was recorded for an Oregon population (6).

Like tui chubs, with which they are nearly always found, blue chubs grow fairly fast in their first 2–4 years of life, until they become mature at about 12–15 cm SL. After maturity, growth is slow, but the chubs are long lived and can reach at least 38 cm FL (7). A 34 cm FL chub was aged at 17 years (8).

Spawning occurs at any time from May through August, depending on locale and water temperatures (3). In Upper Klamath Lake, Oregon, spawning occurs in May and June over shallow gravelly or rocky areas at temperatures of 15–18°C. Spawning behavior in the lake was witnessed by C. R. Hazel (13):

> On the afternoon of May 4, 1966, I observed an estimated 200–300 blue chubs spawning at the shoreline on the northern end of Eagle Ridge. Spawning was taking place from near the surface to a depth of 0.3 to 0.5 m. The bottom was composed of large gravel and rubble of volcanic origin. The water was clear with a low concentration of blue-green algae (*Aphanizomenon*) . . . [and] the water temperature was 17°C. Two to several males would approach a female and exhibit rapid and violent agitations of the water, making it impossible to see exactly what was taking place. In some instances the female was pushed from the water onto dry land, and in a few situations, eggs were spawned outside the water. After these activities, egg masses were found attached to [submerged] rocks either on the sides or near the bottom edge. Many of the depositions were found along rocky edges at depths to 0.5 m.

Status IC. The blue chub was historically an extremely abundant fish within its limited range, and it remains a common fish in Upper Klamath and Agency Lakes, Oregon (9). However, its overall populations apparently declined in the 1980s and early 1990s as a result of multiple factors: drought, water diversions, pollution, and introduced species. The drought created additional stress in a system already stressed by the other factors. Diversions of water have dried up lowland habitats preferred by chubs or allowed organic pollutants to become so concentrated that upper and lower Klamath Lakes and Tule Lake are difficult for native fishes to inhabit. Lakes of the upper Klamath drainage are sumps for agricultural runoff, which carries fertilizers and animal waste, becoming increasingly eutrophic and less favorable to fish life, even when lake levels are high. In addition, alien fathead minnows, highly tolerant of polluted waters, have proliferated in recent years, with unknown effects on blue chubs and other native fishes (9). The best refuges appear to be the Boles Creek watershed and Clear Lake Reservoir in California, where blue chubs remain abundant (5, 8).

References 1. Bond et al. 1988. 2. Vincent 1968. 3. Lee et al. 1980. 4. Castleberry and Cech 1993. 5. Scoppetone et al. 1995. 6. Bird 1975. 7. Scoppetone 1988. 8. Buettner and Scoppetone 1991. 9. Simon and Markle 1997. 10. Simons and Mayden 1998. 11. D. Markle, Oregon State University, pers. comm. 1999. 12. R. R. Miller, pers. comm. 13. C. R. Hazel, pers. comm. 1974.

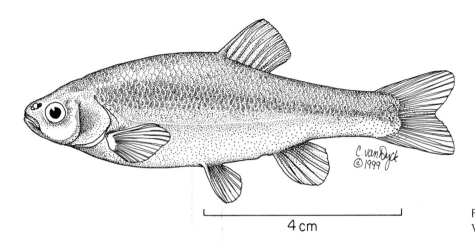

Figure 45. Arroyo chub, 8 cm SL, Ventura River, Ventura County.

Arroyo Chub, *Gila orcutti* (Eigenmann and Eigenmann)

Description Arroyo chubs are small, chunky fish that reach lengths of 120 mm SL; typical adult lengths are 70–100 mm. They have fairly deep bodies and caudal peduncles; large eyes (for a cyprinid); short, rounded snouts; and small, subterminal mouths. The pharyngeal teeth are hooked and closely spaced, with a formula of 2,5-4,2 (but counts may vary by 1–2 teeth). They have 7 anal fin rays and 8 dorsal fin rays. Gill rakers number 5–9. The lateral line has 48–62 scales, extends to the caudal peduncle, and is not decurved. Body color is silver or gray to olive green dorsally and white ventrally, usually connected with a dull gray lateral band. Males are distinguished from females by their larger fins and, when breeding, a prominent patch of breeding tubercles on the upper surface of each pectoral fin (1).

Taxonomy Miller (11) placed both *Gila orcutti* and *G. purpurea*, from Mexico and southeastern Arizona, in the subgenus *Temeculina*, indicating their distinctiveness. Analysis of mitochondrial DNA indicates a close relationship to other *Gila* from the Southwest, including the Colorado River (13).

Names Arroyo chubs are named for the gullies and small canyons (arroyos) of their native southern California. *Gila* is explained in the account of blue chub. *Orcutti* is for the botanist C. R. Orcutt, who in 1889 collected the first specimens, using a blanket as a seine (2).

Distribution Arroyo chubs are native to the Los Angeles, San Gabriel, San Luis Rey, Santa Ana, and Santa Margarita Rivers and to Malibu and San Juan Creeks (3). They have been successfully introduced into the Santa Ynez, Santa Maria, Cuyama, and Mojave River systems and other smaller coastal streams (e.g., Arroyo Grande Creek) (4). The most northern introduced population is in Chorro Creek, San Luis Obispo County. They are now extirpated from much of their native range, remaining abundant only in upper Santa Margarita River and its tributary De Luz Creek; Trabuco Creek below O'Neill Park; and San Juan Creek (San Juan Creek drainage), Malibu Creek (5), and West Fork of the upper San Gabriel River below Cogswell Reservoir (5). They also occur (but are scarce) in Big Tujunga Canyon; Pacoima Creek above Pacoima Reservoir; the Sepulveda Flood Control Basin, Los Angeles River drainage; and middle

Santa Ana River tributaries between Riverside and the Orange County line (5).

Life History Arroyo chubs are adapted to the warm, fluctuating streams of the Los Angeles Plain. Prior to the arrival of civilization and concrete, these streams were fluctuating, often muddy torrents in winter and clear brooks in summer, intermittent in some lower reaches. Arroyo chubs are most abundant in slow-moving or backwater sections of warm to cool (10–24°C) streams with muddy or sandy bottoms, but they are also found in fairly fast-moving (velocities of 80 cm/sec or more) sections of stream with coarse bottoms (12). They prefer depths greater than 40 cm (3, 12). Laboratory studies indicate that arroyo chub are physiologically adapted to survive the hypoxic conditions and wide temperature fluctuations common in coastal streams (6). In these habitats the chubs were originally associated with Santa Ana suckers, speckled dace, brook lampreys, threespine sticklebacks, and, in headwaters, rainbow trout.

They are omnivorous, feeding on algae, insects, and small crustaceans. However, in warmwater streams, most (60–80%) stomach contents consist of algae (7). They are also known to feed extensively on nematode-infested roots of floating water fern (*Azolla*). Invertebrates increase in the diet in number and variety during spring and are least abundant during winter (8). In a coolwater stream arroyo chubs fed largely on benthos, especially molluscs and caddisfly larvae, while sympatric rainbow trout fed largely on drifting invertebrates (9).

Arroyo chubs in the Santa Clara River reach about 60 mm SL in their first year, 70–75 mm in their second year, 75–80 mm in their third year, and 80–90 mm in their fourth year (1). Females first reproduce at 1 year of age. After their second year, females grow larger than males. Arroyo chubs rarely live beyond 4 years.

They are fractional spawners that breed more or less continuously from February through August, although most spawning is in June and July, in pools or in quiet edge water at temperatures of 14–22°C (1). During spawning, males follow a ripe female while actively rubbing their upper snouts below the female's pelvic fins. Rubbing and chasing lead to egg release, and eggs may be fertilized by more than one male (1). Embryos adhere to the bottom or to plants and hatch in 4 days at 24°C. The fry spend a few days after hatching clinging to the substrate but rise to the surface once the yolk sac has been absorbed (1). The next 3–4 months are spent in quiet water in the water column and usually among vegetation or other flooded cover. Arroyo chubs readily hybridize with California roach (7, 8) and Mojave tui chubs (10). As noted in the tui chub account, Mojave tui chubs have been completely eliminated from the Mojave River by arroyo chubs (6, 10).

Status IC. Arroyo chubs are presently common at only four places within their native range (5). They are scarce within their native range because the low-gradient streams in which they do best have largely disappeared (5). During 1986–1990, low-water conditions in the West Fork of the San Gabriel River were favorable to the chubs, allowing a temporary increase in numbers. The chubs became scarce again after the 1991–1992 rains but were common in 1993. Arroyo chubs are common in some streams where they have been introduced, especially the Santa Clara River, but such introduced populations have a history of hybridization with other cyprinids (although not in the Santa Clara River) (15) and cannot be regarded as secure (or genetically pure) (5).

If arroyo chubs were not abundant in a number of waters outside their native range and had they not thrived in those waters, they would qualify for listing as a threatened species. Their native range, like that of the sympatric Santa Ana sucker, is largely coincident with the Los Angeles metropolitan area, where most streams are degraded and fish populations reduced and fragmented, especially the low-gradient reaches that were optimal habitat (5). Populations in the Cuyama and Mojave Rivers are hybridized with California roach and Mojave chub, respectively (7, 10). Recently red shiners have been introduced into arroyo chub streams, and they may be excluding chubs from many areas (14). Chubs generally decline when red shiners and other exotics become abundant (15). In the Santa Margarita River a dramatic increase in arroyo chub abundance was noted after extreme high-flow events in 1997–1998 reduced the abundance of green sunfish, largemouth bass, redeye bass, and black bullhead (14). The potential effects of introduced species, combined with the continued degradation of urbanized streams, mean that this species is not secure, despite its fairly wide range.

Because of the uncertain status of most populations, annual surveys are needed for this species in its native range; these should be performed every five years at all known sites. Streams should be managed to favor arroyo chubs and other native fishes of the region. The strongest candidate for a native fish refuge is the West Fork of the San Gabriel River. In regulated streams releases that mimic the natural flow regime should favor arroyo chubs and other native fishes.

References 1. Tres 1992. 2. Eigenmann and Eigenmann 1893. 3. Wells and Diana 1975. 4. R. R. Miller 1968. 5. Swift et al. 1993. 6. Castleberry and Cech 1986. 7. Greenfield and Deckert 1973. 8. Greenfield and Greenfield 1972. 9. Richards and Soltz 1986. 10. Hubbs and Miller 1943. 11. R. R. Miller 1945a. 12. Bell 1978. 13. Simons and Mayden 1998. 14. C. C. Swift, pers. comm. 1998, 1999. 15. T. R. Haglund, University of California, Los Angeles, pers. comm. 1998.

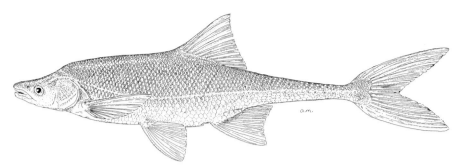

Figure 46. Bonytail, 30 cm SL, Green River, Wyoming. Drawing by A. Marciochi.

Bonytail, *Gila elegans* Baird and Girard

Identification Bonytails are readily recognized by their extremely narrow caudal peduncle and deeply forked tail; fine, embedded scales (75–99 along the lateral line); and small, flattened heads with small, elliptical eyes and terminal mouth. There is usually a conspicuous hump behind the head. Scales may be lacking on the dorsal and ventral surfaces as well as on the caudal peduncle. Dorsal and anal fin rays usually number 10–11; pelvic fin rays number 9–10. The pharyngeal teeth (2,5-4,2) are closely spaced, compressed, and hooked. The color of the back and sides ranges from dusky green to metallic blue with fine speckling; the belly is silvery to white. Breeding males become reddish orange on the head and sides below the lateral line and on the base of the anal and pectoral fins. Young fish lack the exaggerated morphology of the adults and bear a fairly close resemblance to young Colorado pikeminnow and other Colorado River *Gila* species.

Taxonomy The bonytail is one of four closely related *Gila* species in the Colorado River system. Miller (1) thought that this *Gila* "complex" could best be divided into two species: *G. cypha*, the bizarre humpback chub of the Grand Canyon, and *G. robusta*, the Colorado chub with four subspecies: *G. r. robusta, G. r. elegans, G. r. seminuda,* and *G. r. intermedia*. There now seems little doubt (2, 12) that the bonytail deserves recognition as a full species, as recom-

mended originally by Minckley and Deacon (3). Studies indicating that *G. elegans, G. cypha, G. intermedia,* and *G. robusta* are ecologically and reproductively segregated support this conclusion (2, 5, 14). The different forms of *Gila* presumably evolved to meet special ecological conditions in the Colorado River's varied waterways: *G. cypha* in the swift and turbulent water of the Grand Canyon and similar habitats in the Green and upper Colorado Rivers; *G. robusta* in the quieter pools and slower-moving waters of the main tributaries; *G. intermedia* for conditions in tributaries to the Gila River, Arizona, and *G. elegans* in the fast waters of the main river (14).

The taxonomy of these forms has long been uncertain because of hybridization among them (2, 4). However, Dowling and DeMarais (5) demonstrate that past hybridization is probably responsible for providing the genetic diversity necessary for development of the extreme morphological and ecological diversity among them. Indeed, a species (*G. seminuda*) endemic to the Virgin River, Utah, arose as a hybrid between *G. elegans* and *G. robusta* (5). In any case, the bonytail is the most distinctive of the forms from both a morphological and a genetic perspective (14).

Names Elegans means elegant. Members of the Colorado *Gila* complex are commonly referred to as Colorado chubs. An old common name for bonytail is Gila trout. For other names, see the account of tui chub.

Distribution Bonytails were originally widely distributed in the mainstem Colorado River and its tributaries in Wyoming, Utah, Colorado, Arizona, California, and New Mexico, as well as in Mexico. In California they were found only in the Colorado River where it borders the state. Today the principal wild population is in Mohave Reservoir, upstream from California, and this population is maintained primarily by stocking fish from the Dexter National Fish Hatchery (New Mexico) (4, 6). If any wild fish still exist in California, they would be large, old individuals in Havasu Reservoir. A few such individuals may also persist in the upper Colorado River. Fish from the Dexter Hatchery, however, were planted in Havasu Reservoir and the upper Colorado and Green Rivers starting in 1998 (13).

Life History Bonytails are usually considered to be primarily inhabitants of swifter waters of the large rivers of the Colorado system. This conclusion is based on their streamlined morphology, consisting of a slim, elongated body; fine, deeply embedded scales; wide pectoral fins; narrow caudal peduncle; and nuchal hump. According to Minckley (6), the limited information available on bonytail habits indicates they actually lived in flowing water in the less turbulent moving parts of the river, especially in areas with sandy bottoms. They apparently maintained themselves in the water column, where they could feed on insects and other food drifting in the current. Their odd morphology would not only help them maintain their position in such conditions, but also presumably help them persist through high-flow events or escape predators by moving through swift water. The water in which they were found was presumably often very turbid. In reservoirs they are a midwater species, aggregating over shoals 5–10 m deep a short distance from shore (6).

Vanicek and Kramer (6) found large bonytail (>20 cm TL) to be omnivorous surface feeders, taking terrestrial insects, filamentous algae, and plant debris such as leaves, stems, seeds, and horsetail stems. In reservoirs they will feed on zooplankton, algae, insects, and organic debris (6). Small fish (<3 cm TL) feed mostly on aquatic insect larvae; they become more dependent on drifting food as they grow larger (7).

The one study of a natural riverine population of bonytail indicates that they may grow to about 5.5 cm TL and 1 g in their first year, 10 cm TL and 8 g in their second year, and 16 cm TL and 31 g in their third (6). Breeding size (30–40 cm TL) was probably reached in 4–5 years. However, bonytail have the capacity to reach large sizes quickly. Under artifical conditions, they may grow over 30 cm in their first year (7). The largest fish known is about 64 cm TL, but most adults are 40–60 cm TL (4, 6). Growth slows drastically once reproduction begins, but adults may reach ages of 34–49 years (7).

Spawning apparently took place historically in May and June over gravel riffles or rubble-bottomed eddies at water temperatures of 15–20°C (6, 11), but bonytail have also been observed to spawn in reservoirs and in muddy-bottomed ponds at the Dexter National Fish Hatchery. Breeding behavior was observed in Mohave Reservoir, Nevada, in May (8). About 500 bonytail congregated over a gravel-covered shelf 9 m deep. As is typical of such cyprinid spawning groups, the males outnumbered the females by 2 to 1, and each spawning female was attended by 3–5 males. Eggs were broadcast over the gravel, to which they adhered. The spawning areas were not defended, and common carp were observed in the area, apparently feeding on the spawn.

The spawners were 28–36 cm TL. A female of 31 cm TL contained about 10,000 eggs.

Young fish are apparently planktonic for a short time after they hatch, but they are soon found in the quiet, shallow waters of the river's edge. In this habitat they are extremely vulnerable to predation by nonnative fishes that also aggregate there.

Status IA. Extinct in California as a naturally spawning, self-sustaining population, although individuals may be present as the result of planting programs. They are likewise extinct in almost all of their former range. The only remaining population of any size is in Mohave Reservoir, Nevada, and even this population is maintained by plants of fish from the Dexter National Fish Hatchery. A similar population exists in the upper Colorado and Green Rivers.

The effective extinction of bonytail is a legacy of the extreme development of the system for human use. The original river was seasonally warm and muddy with large annual fluctuations in flow. It has been replaced by cold, clear sections of river with regulated flows, by huge reservoirs of quiet water, and, in California, by a warm, depleted river that is polluted with salts and toxic chemicals. When big dams on the Colorado were being built, native species such as bonytail were considered trash fish that might interfere with the development of reservoir fisheries. Therefore, in 1962–1963, the largest deliberate fish poisoning operation ever attempted was carried out in the upper basin, mainly in the Green River and its tributaries (9). Over 715 km of river were poisoned, and millions of fish were killed. The kill was far from complete, but the native fishes never really recovered, although changes to the river caused by dams were probably mostly responsible for this outcome. It is nevertheless possible that, if the operation had not occurred, remnant populations of bonytail and other native fishes would persist in the upper basin reservoirs, as they do in Mohave Reservoir (9).

Persistence of bonytail in Mohave Reservoir indicates that adults can adapt to reservoir conditions. Indeed, they were once apparently among the most abundant fishes in Lake Cahuilla, an immense Pleistocene lake that existed periodically in the basin now occupied by the Salton Sea (10). Unfortunately, they do not seem to be able to complete their life cycle successfully in reservoirs. There is some evidence that they can survive and spawn in the modified riverine habitats, even those with reduced temperatures (11). Thus the most important proximate reason for their decline seems to be predation on embryos and young by alien fishes, such as common carp, threadfin shad, red shiner, channel catfish, green sunfish, and other species that thrive in reservoirs and backwater habitats. The ultimate reason for their decline, however, is the extreme modification of flows, habitats, and water quality of the Colorado River, because such conditions favor alien fishes. Thus the long-term survival of this species in the wild is questionable, because it probably will depend on continued hatchery propagation of young fish. Hatchery populations are subject to a variety of ills, from inbreeding (although so far this does not seem to be a problem) (4), to disease epidemics, to loss of funding for hatchery operations.

References 1. R. R. Miller 1946. 2. Holden and Stalnaker 1970. 3. Minckley and Deacon 1968. 4. Minckley et al. 1989. 5. Dowling and DeMarais 1993. 6. Vanicek and Kramer 1969. 7. Minckley 1991b. 8. Jonez and Sumner 1954. 9. Holden 1991. 10. Gobalet 1994. 11. Marsh 1985. 12. Simons and Mayden 1998. 13. Newsletter, Upper Colorado River Endangered Fish Recovery Program, 1998. 14. Minckley and De Marais 2000.

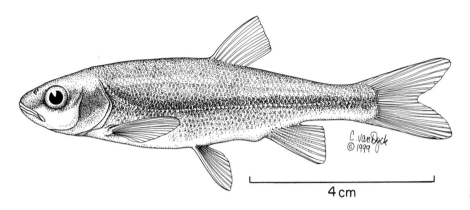

Figure 47. Lahontan redside, 8 cm SL, Willow Creek, Lassen County.

Lahontan Redside, *Richardsonius egregius* (Girard)

Identification Lahontan redsides are rather small and slender minnows (body depth divisible about 4 times into SL) with large eyes, terminal mouths, and deeply forked caudal fins. They are easiest to identify when in their spectacular breeding colors: a scarlet stripe in a field of yellow on each side, a shiny olivaceous back, and a silvery belly. In nonbreeding fish, the red color is greatly reduced or absent, but the stripe is still visible as a lateral band. The mouth is slightly oblique, the maxillary barely reaching the front edge of the eye. There are 7–8 (usually 8) dorsal fin rays, 8–10 (usually 9) anal fin rays, and 52–63 scales in the lateral line. Scales on the back behind the head tend to be crowded before the dorsal fin. Pharyngeal teeth (2,5-4,2) are strongly

hooked, and gill rakers are stubby, tending to expand toward the tips. The intestine is S-shaped, shorter in length than the body.

During spawning both sexes develop breeding tubercles on the body and head, but those on the males are larger and more numerous, and also occur on the pectoral fins. Males tend to be darker in color with a more intensely red stripe. When pressed down, the pectoral fins of males usually reach the base of the pelvic fins; those of females do not (1).

Taxonomy The Lahontan redside is closely related to a similar species (*R. balteatus*) in the Columbia River drainage and somewhat more distantly related to eastern minnows of the genus *Clinostomus,* also sometimes included in the genus *Richardsonius* (15). More distantly, it is related to the various species of *Gila.* Lahontan redsides hybridize with tui chubs and speckled dace (12, 13).

Names A variety of unofficial common names have been applied to the Lahontan redside, all referring to its breeding colors: Lahontan redshiner, Lahontan redside shiner, redside minnow, redside bream, red-striped shiner. *Richardsonius* is after Sir John Richardson (1787–1865), an English naturalist who described the only other species assigned to this genus, *R. balteatus. Egregius* means surprising. Just what surprised Charles Girard when he described this species in 1858 from a single specimen is not known. The complex history of its scientific nomenclature is given in La Rivers (2).

Distribution Lahontan redsides are native to streams and lakes of the old Lake Lahontan basin in northern Nevada and northeastern California. In California they are native to the following Great Basin drainages: Eagle Lake, Susan River, Truckee River, Walker River, and Carson River. They have been introduced into the Sacramento River system in several watersheds, so may be present in unexpected places outside their native range. Kimsey (3) reported a population in Mill Creek at the headwaters of the Rubicon River. This population, with those of other Lahontan fishes, may have been the result of a bait bucket introduction, although only a low divide separates the two drainages. More certain bait bucket introductions are in Bucks Lake, which drains into the North Fork Feather River; Loon Lake, which drains into the American River (16); various headwaters of the North Fork of the Mokelumne River around Bear Valley Reservoir; and Frenchman Reservoir and nearby streams (Frenchman, Little Last Chance, and Ramelli Creeks) in the upper Middle Fork Feather River drainage (17). Rutter (4) found redsides and other Lahontan fishes in Warner Creek, a tributary to the North Fork Feather River, but it is not known if they are still present there. An additional introduced population is present in Saddlebag Lake, Mono County (14). In theory, these last fish have access to southern California reservoirs by way of the Owens Aqueduct system.

Life History The habitat of Lahontan redsides was well described by Snyder (5, p. 54):

> This beautiful little fish is almost universally distributed throughout the brooks, rivers, and lakes of the region. It is found not only in the lower courses of the rivers where the water is deep and quiet, but it also stems the swift currents of the high mountain tributaries, following closely in the wake of the smallest trout. . . . It delights in the slow riffles and the quiet, shallow pools where large numbers may be seen swimming lazily about over the submerged bars, occasionally turning their silvery sides to the bright sun. In the lakes it congregates in large schools, swimming about submerged logs, tops of fallen trees, wharves, and other sheltered places.

In small streams, redsides prefer deep pools, where they shoal near the surface. Adults aggregate in higher-velocity water at the heads of pools, while juveniles prefer quieter water along edges or in backwaters (6). Their abundance in streams seems to be negatively affected by high winter flows (7) and by high densities of piscivorous brown trout. Redsides have shown considerable capacity to colonize reservoirs and may reinvade lower reaches of impounded streams in large numbers (7).

In lakes redsides are a shoaling littoral zone species that can live in a wide variety of conditions, from the cold waters of Lake Tahoe, to the alkaline waters of Eagle Lake (pH 9 or more), to fluctuating reservoirs. Typically they swim about in large schools close to the surface, generally staying over areas that have rocky bottoms. During the winter months, after water temperatures drop below 10°C, redsides disappear from shallows, presumably spending the cold months relatively inactive on rocky bottoms in deep water (1).

As their hooked pharyngeal teeth, short gill rakers, short intestine, and oblique mouth suggest, redsides are opportunistic feeders on invertebrates. In Lake Tahoe their diet consists about equally of surface insects, bottom-living insect larvae, and planktonic crustaceans (8); the predominant items in their stomachs vary with the area from which they have been feeding as well as with time of day. Thus in one study the percentage of bottom organisms in different samples ranged from 9 to 99 percent; the percentage of surface organisms, from 1 to 87 percent; and the percentage of planktonic forms, from 0 to 92 percent (8). Redsides in Tahoe feed at any time of day or night, but flying insects seem to be favored in evening and night; bottom and planktonic forms are favored during the day. In Eagle Lake (Lassen County) redsides feed mainly on planktonic cladocerans, caddisfly larvae, and amphipods (9). Individuals feed predominantly on one or another of the three, rarely mixing prey.

In small streams, redsides feed mainly on drifting insects, especially during daylight hours, but they will also feed on benthic insects and algae (6, 9). In Willow Creek, Lassen County, their diet is predominantly benthic invertebrates, especially caddisfly larvae and snails, which are taken mainly at night. The reason for night feeding seems to be exceptionally heavy predation pressure from aquatic birds (e.g., egrets, herons, kingfishers, pelicans), which forces the fish to be less active and remain in deeper cover during the day (9). Redsides in streams will also feed on the eggs of spawning Tahoe suckers (5, 8), and in some instances egg predation may limit sucker populations (7).

Studies on the age and growth of redsides indicate that growth rates are similar in streams, lakes, and reservoirs, although they are somewhat slower in colder streams and lakes, such as Lake Tahoe. Redsides average 34–55 mm SL (1–2 g) at the end of the first year, 51–63 mm (2–5 g) in the second year, 65–73 mm (7–9 g) in the third year, and 75–80 mm (9–11 g) in the fourth year (1, 10). Occasional fish will reach 14–17 cm SL. A single fish measuring 16 cm TL was 5 years old (16).

Based on size, most redsides become mature in their third or fourth summer. A few may mature in their second summer. The average number of eggs in 16 females from Lake Tahoe was 1,125 (1). The right ovary contains the majority of eggs. Spawning takes place at any time from late May through August, but most spawning seems to occur in the last 2 weeks of June at water temperatures of 13–24°C. In Lake Tahoe redsides either migrate up tributaries, such as Taylor Creek, to spawn over sand and gravel at the downstream end of pools, or spawn in shallows (<1 m) over gravel or small rocks (1, 11).

Spawning, according to Miller (8), "provides a scene of excitement, urgency, and confusion from which the observer despairs of any constructive outcome." Groups of 20–100 spawning fish swim about in a tight, swirling school close to the bottom. Release of sex products occurs when a small cluster of fish drops to the bottom and presses against the rocks. The fertilized eggs sink into crevices and adhere to surfaces.

After hatching, young fish leave for quiet, shallow water, often near the mouths of spawning streams. These areas usually have a protective cover of floating debris or overhanging bushes. Frequently small redsides shoal with young of other cyprinid species.

Status IE. Lahontan redsides are still abundant in most of their native range. Although they have been eliminated from a few streams by diversions, they have also successfully colonized a number of reservoirs, as well as streams, lakes, and reservoirs outside their native range. In some small streams their numbers may be limited by predation by nonnative brown trout.

References 1. Evans 1969. 2. La Rivers 1962. 3. Kimsey 1950. 4. Rutter 1908. 5. Snyder 1918. 6. Moyle and Vondracek 1985. 7. Erman 1986. 8. R. G. Miller 1951. 9. P. B. Moyle and students, unpubl. rpts. on Eagle Lake. 10. P. B. Moyle, unpubl. data. 11. Taylor 1990. 12. Hopkirk and Behnke 1966. 13. Calhoun 1940. 14. S. Parmenter, CDFG, pers. comm. 1998. 15. Simons and Mayden 1998. 16. Wang 1986. 17. R. Decoto, CDFG, pers. comm. 1999.

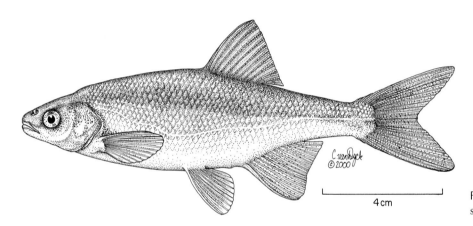

Figure 48. Hitch, 12 cm SL, Cosumnes River, Sacramento County.

Hitch, *Lavinia exilicauda* Baird and Girard

Identification Hitch have deep, laterally compressed bodies, small heads with upward-pointing mouths, moderately large scales, and decurved lateral lines. The body tapers to a narrow caudal peduncle, which supports a large forked tail. Hitch lack a sharp keel on the belly. They can reach lengths in excess of 35 cm SL, with the body becoming deeper as length increases. The long anal fin (11–14 rays) separates the species from most other California minnows; the origin of the dorsal fin (10–13 rays) is behind that of the pelvic fins. There are 54–62 lateral line scales and 17–26 gill rakers. The pharyngeal teeth (0-4 or 5-0) are long and narrow, slightly hooked, yet with fairly broad grinding surfaces. When small, hitch are silvery with a black spot at the base of the tail. Older fish lose the spot and become darker, with the largest fish approaching brownish yellow on the back.

Taxonomy Hitch are most closely related to California roach, with which they hybridize to produce fertile offspring (1). Hitch-roach hybrids are common in some larger tributaries of the Pajaro River and in the lower portions of Coyote and Alameda Creeks. Hitch also hybridize with Sacramento blackfish, although the hybrids are apparently sterile (2). They hybridized in the past with thicktail chub (4, 20).

The Clear Lake subspecies, *L. e. chi,* was described by Hopkirk (3) as a lake-adapted form. Another subspecies, *Lavinia e. harengus* from the Pajaro and Salinas Rivers, was described by Miller (4), based on its greater body depth and lower fin ray counts as compared with *L. e. exilicauda,* the Central Valley subspecies. However, *L. e. exilicauda* exhibits sexual dimorphism based on body depth, and there is considerable variability in body size and proportions among populations, so *L. e. harengus* should be reexamined (3). There is also a need to examine variation within Central Valley populations to see if other distinctive forms exist.

Names The name hitch is derived from the Pomo Indian name for this fish, as is the related name chi (5). However, Hopkirk (3) indicates that the name may have originally applied to Clear Lake splittail. *Lavinia* is a Latin feminine name whose application to hitch is somewhat of a mystery. The narrow caudal peduncle inspired *exili-cauda* (slender tail).

Distribution Hitch are native to the Sacramento–San Joaquin, Clear Lake, Russian River, and Pajaro-Salinas drainages. They have scattered populations throughout the Central Valley, from the Tulare Lake basin in the southern San Joaquin River drainage (7) to Shasta Reservoir in the northern Sacramento River drainage. In the San Francisco Bay region they are found in Coyote Creek, Alameda Creek, and other creeks draining Santa Clara, Contra Costa and Alameda Counties, as well as Suisun Creek, Napa County (8), and in the Delta. In the Monterey Bay region they are present in the Pajaro and Salinas Rivers and larger tributaries. There is also a small population in the Russian River. They are found throughout Clear Lake, Lake County, and in associated lakes, such as Lampson Pond, Thurston Lake, and lower Blue Lake, spawning in tributaries to the lakes. The current major spawning streams are, in roughly decreasing order of importance, Kelsey, Adobe, Seigler Canyon, Middle, Scotts, Manning, and Cole Creeks (29).

Hitch have been introduced into a few upstream reservoirs within their native range, such as Beardsley Reservoir (Tuolumne County) and Bass Lake (Fresno County). They have apparently been carried via the California Aqueduct to San Luis Reservoir, Merced County, and Pyramid and Silverwood reservoirs, Los Angeles County; they may have become established there, as well as in Aliso Canyon, a tributary to the Santa Clara River (9).

Life History Hitch are widespread in warm, low-elevation lakes, sloughs, and slow-moving stretches of river, and in clear, low-gradient streams. Their quiet water habitat is reflected in their rather deep, laterally compressed body shape. However, they can also be abundant in cool, clear, sandy-bottomed streams, such as Fresno River, Fresno County, or Putah Creek below Solano Diversion Dam (7, 8, 10, 11). In such streams smaller fish are often associated with run habitat where scattered beds of aquatic or emer-

gent vegetation serve as cover, while larger fish are found in deep pools associated with heavy cover and overhanging trees. In urban areas hitch may be found in low numbers in channelized streams with silty bottoms and turbid water (8). They can survive in such areas because they have the highest temperature tolerances among the native fishes of the Central Valley. In the laboratory juvenile fish acclimated to 30°C can withstand temperatures of nearly 38°C (critical thermal maximum) for short periods of time, although they will actively select temperatures of 27–29°C (acute preferred temperature) (27). Hitch can also withstand moderate salinities; in Suisun Marsh they have been found in salinities of 7–8 ppt, and in Salinas River lagoon, at salinities as high as 9 ppt (28).

In lakes adult hitch are usually pelagic. In Clear Lake juveniles are found in inshore shallow-water habitat and move into deeper offshore areas after approximately 80 days, when they are between 40 and 50 mm SL (12). While in shallow water, larvae and small juveniles require vegetation, such as tule beds, as refuge from predators. During the reproductive season, adult Clear Lake hitch migrate into the lower reaches of low-gradient tributary streams to spawn in gravel-bottomed sections that dry up during the summer (6, 12). Because hitch are not aggressive swimmers, their runs are easily blocked by small dams and other structures that impede upstream migration.

Before modern-day habitat alterations, hitch were associated with such fishes as Sacramento perch, Sacramento blackfish, thicktail chub, and splittail. Today their most common associates are introduced species, especially catfishes, centrarchids, and mosquitofish, although Sacramento blackfish, Sacramento sucker, and Sacramento pikeminnow are common associates in less disturbed habitats (8, 10).

The deep body; small, upturned mouth; long, slender gill rakers; and high but flat-topped pharyngeal teeth indicate that hitch are omnivorous open-water feeders. In Putah Creek hitch feed in summer on a mixture of filamentous algae, aquatic insects, and terrestrial insects (13). Small schools of hitch measuring 50–75 mm SL can be observed feeding, like trout, on drift at the heads of summer pools. In Clear Lake limnetic hitch greater than 50 mm SL feed primarily on *Daphnia* and other zooplankton (14), although insects may be taken on the surface when abundant (15). Juveniles (<50 mm SL) in the near-shore environment feed primarily on larvae and pupae of chironomid midges and other insects, as well as on small planktonic crustaceans (12, 15). Hitch feed primarily during the day (12).

Growth rates appear to be directly related to the productivity and summer temperatures of the environments in which they live. Clear Lake hitch grow much more rapidly than Sacramento hitch from high-elevation Beardsley Reservoir (5, 12, 16). In Clear Lake hitch reach 40–50 mm

FL within 3 months and measure 110–170 mm FL by the end of their first year, and 150–300 mm by the end of the second year; subsequent increases are 20–50 mm/year, with a maximum size of around 350 mm. Growth rates in San Luis Reservoir, Merced County, are apparently similar (26). Hitch in Beardsley Reservoir, in contrast, are only 40–50 mm FL by the end of the first year and 9–11 cm FL by the end of their second, with subsequent increments of 20–40 mm/year (16). In Putah Creek they average about 65 mm FL at the end of their first year (13) and reach 200–250 mm in 3–4 years. Females grow faster and larger than males. Scale analysis indicates that hitch live 4–6 years, but it is likely that analysis of the bony structures of large fish would yield greater ages.

Females usually mature in their second or third year; males mature in their first, second, or third year (16, 17). In the Pajaro River both sexes can mature during their second summer (age 1 year or more) when only 49–54 mm SL, and most fish longer than 70 mm are mature. Hitch are rather prolific: females from Beardsley Reservoir contained 3,000–26,000 eggs, with a mean of 9,000 (16). In Clear Lake average fecundity is 36,000 eggs, with a range of 9,000–63,000 (in a fish measuring 312 mm SL); their length-fecundity relationship is $F = 504[SL_{mm}] - 30,384$ (14).

Spawning takes place mainly in riffles of streams tributary to lakes, rivers, and sloughs, after flows increase in response to spring rains. They seem to require clean, fine to medium gravel and water temperatures of 14–18°C (5, 17), although the spawning requirements of the species are in need of further documentation. Smith (11), for example, observed spawning in the Pajaro River at 18–26°C in May–July, after low summer flows had been established. Hitch are also capable of reproducing in ponds and reservoirs. When they are present in ponds and reservoirs with Sacramento blackfish, the two species will hybridize, presumably because they are forced to share spawning areas. Likewise, hitch-blackfish hybrids were common in the Pajaro River when flowing water habitats were scarce during the 1976–1977 drought (28).

At Clear Lake spawning migrations usually take place from mid-March through May and occasionally into June. In 1992 the hitch runs started in mid-February and persisted until the streams dried in May–June (29). In the words of R. Macedo (6, p. 2), "As spectacular as any salmon run on the Pacific coast, hitch mass by the thousands and ascend the . . . streams. . . . The tumultuous splashing . . . and the appearance of herons, osprey, egrets, and bald eagles . . . signify . . . that the hitch are in. Along stream banks, raccoons, mink, otter, and even bears join the birds to feed on hitch." The hitch will also ascend and spawn opportunistically in various unnamed tributaries and drainage ditches. One year they even were observed spawning in a flooded meadow after swimming up a small ditch and across a flooded parking lot (30). Some may spawn in the shallow waters of Clear Lake itself, over clean gravel where there was wave action (17).

Spawning is a mass affair accompanied by vigorous splashing. A ripe female is closely followed by 1–5 males, who apparently fertilize eggs immediately after their release. There is no territoriality. Fertilized eggs are not adhesive but sink into interstices of the gravel before absorbing water and swelling to about 4 times their initial size (5). Swelling lodges embryos in the gravel, although large numbers of viable embryos can be observed at times drifting downstream, and dead embryos may accumulate in large numbers in pools and backwaters.

Hatching takes place in 3–7 days at 15–22°C, and larvae take another 3–4 days to become free-swimming (5, 18, 31). At about 25 mm TL, fry of Clear Lake hitch quickly move down into the lake (5). This behavior contributes to the success of hitch, because it permits reproduction in steams that dry up in summer (19). Small hitch spend the next 2 months shoaling in the lake's littoral region, usually among emergent tules, before moving out into open water, at about 50 mm TL. In permanent streams and in ponds, larval and postlarval hitch aggregate around aquatic plants or other complex cover in shallow water. They are most active during the day.

Status ID for all forms except Clear Lake, which is IC. The Clear Lake hitch is listed as a species of special concern (21) and appears to be in decline (6). Hitch were once abundant throughout their native range and an important food for Native Americans (6, 22, 23, 24). They are still commercially harvested on occasion from Clear Lake and may be the dominant fish in some streams (e.g., Auburn Ravine, Sacramento County [25], and sections of the Pajaro River, Santa Cruz and Monterey Counties [18]). However, today they are uncommon relative to other fishes in most places, and the scattered populations are increasingly isolated from one another. Hitch are becoming increasingly scarce, and some populations in streams flowing into the San Joaquin Valley have apparently gone extinct in recent years (7). The causes of the decline are uncertain, but it is presumably due to a combination of factors: the loss of adequate spawning flows in spring months (because of dams and diversions) and of summer rearing and holding habitat, as well as pollution and predation by nonnative fishes.

The principal threats to Clear Lake hitch are loss of spawning habitat and loss of nursery areas, factors that contributed strongly to the extinction of the Clear Lake splittail. The lower reaches of all their spawning streams dry up annually and probably did so naturally. However, these streams now go dry earlier in the season owing to stream diversions (6), and the result is spawning failures, especially during dry years. Clear Lake splittail formerly spawned somewhat later than did hitch, and early drying up of streams undoubtedly contributed to the demise of that species. This progressively earlier drying up of streams, if it proceeds unchecked, may seriously affect hitch as well. In streams such as Adobe and Kelsey Creeks upstream areas that were once used for spawning are now blocked by roads and other obstructions. Gravel mining on Kelsey, Scotts, and Middle Creeks has lowered the level of streambeds and the water table as much as 15 ft in some places; structures (mainly on Kelsey Creek) intended to aggrade gravel and raise the streambed present barriers to fish migration, especially during periods of low flow (6, 29). Fish passage facilities must be constructed specifically for hitch and other native cyprinids, which have slower critical swimming velocities than the salmonids for which most fish ladders are designed.

Hitch that make it over barriers and reach their spawning areas are unprotected and vulnerable in shallow water, where they are destroyed by local people by various means. (In the "sport" of "hitching," the fish are clubbed and thrown up on shore.) Recently increased levels of protection by CDFG and the implementation of educational activities for schoolchildren may lessen the extent of the destruction, although as of this writing it continues. An additional problem is that many of the marshy areas that once

ringed Clear Lake are now gone, limiting habitat available to larval hitch; such habitat loss is ongoing. A more recent threat has been the establishment of threadfin shad, which eliminate *Daphnia,* a principal food of hitch, from the plankton.

Overall, a thorough review of the abundance, distribution, status, and systematics of hitch is needed so that conservation strategies can be developed. Particular attention must be paid to Clear Lake hitch, which may deserve threatened status in the near future, and to hitch populations in the Russian River and in San Joaquin River tributaries.

References 1. Avise et al. 1975. 2. Moyle and Massingill 1981. 3. Hopkirk 1973. 4. R. R. Miller 1945b. 5. Murphy 1948b. 6. Macedo 1994. 7. Brown and Moyle 1993. 8. Leidy 1984. 9. Swift et al. 1993. 10. Moyle and Nichols 1973. 11. J. Smith 1982. 12. Geary 1978. 13. M. Dege, University of California, Davis, unpubl. data 1996. 14. Geary and Moyle 1980. 15. Lindquist et al. 1943. 16. Nicola 1974. 17. Kimsey 1960. 18. Swift 1965. 19. Cook et al. 1966. 20. R. R. Miller 1963. 21. Moyle et al. 1995. 22. Schulz and Simons 1973. 23. Gobalet 1990. 24. Broughton 1994. 25. Zimmerman 1995. 26. S. R. Johnson, unpubl. rpt. 1976. 27. Knight 1985. 28. J. J. Smith, California State University, San Jose, pers. comm. 1999. 29. R. Macedo, CDFG, pers. comm. 1996, 1999. 30. S. Hill, CDFG, pers. comm. 31. Wang 1986.

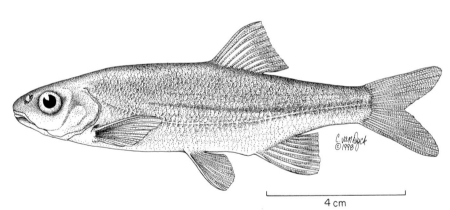

Figure 49. California roach, 10 cm SL, North Fork Tule River, Tulare County.

4 cm

California Roach,
Lavinia symmetricus (Baird and Girard)

Identification Adult California roach are small (usually less than 100 mm TL) and chunky bodied, with a narrow caudal peduncle. The eyes and head are relatively large; the mouth is small and slanted at a downward angle (subterminal). Some populations develop a distinctive "chisel lip," with a cartilaginous plate on the lower jaw. The dorsal fin is short (7–9 rays) and set behind the insertion of the pelvic fins. There are 6–8 anal fin rays. Fish with more dorsal and anal fin rays are probably hybrids with hitch. The scales are small, numbering 47–63 along the lateral line and 32–38 before the dorsal fin. The pharyngeal teeth are 0,5-4,0 and, although narrow and slightly hooked, appear to be adapted for grinding. The upper half of the roach is usually dark, ranging from dusky gray to steel blue. The lower half is usually a dull silver. During the breeding season, patches of red orange appear on the chin, operculum, and bases of the paired and anal fins. Males may develop numerous tiny breeding tubercles on the head at this time (38). Subspecies are distinguished by various distinctive subsets of these characters. Probably the most distinctive is the Red Hills roach, which has a dorsoventrally flattened body, small fins, and a chisel lip (1).

Taxonomy The California roach was first described as *Rutilus symmetricus* (Baird and Girard), from the San Joaquin River near Friant. It was subsequently reassigned to its own genus, *Hesperoleucus*, by Snyder (2) who described the following six species based on locality and morphological differences:

1. *Hesperoleucus symmetricus* from the Sacramento–San Joaquin Valley.

2. *Hesperoleucus subditus* from the Pajaro River system.

3. *Hesperoleucus venustus* from the San Francisco Bay system and the Russian River and Tomales Bay drainages.

4. *Hesperoleucus parvipinnis* from the Gualala River, Sonoma County.

5. *Hesperoleucus navarroensis* from the Navarro River, Mendocino County.

6. *Hesperoleucus mitrulus* from the Pit River system and Goose Lake, Modoc County.

Murphy (3) reanalyzed Snyder's data along with his own from coastal streams and concluded that the species should be relegated to subspecies status, as had been suggested by R. R. Miller a few years earlier (4). This diagnosis was accepted by most subsequent workers (e.g., 5, 6), even though Murphy's study was never published. Hopkirk (6) examined roach from coastal drainages and concluded that Murphy was correct in placing all roach in one species. However, he differed in his conclusions as to what populations should be recognized as subspecies. He considered *H. s. symmetricus,* *H. s. subditus,* and *H. s. parvipinnis* to be morphologically distinct subspecies, whereas *H. s. venustus* was not different from *H. s. symmetricus* (6). *Hesperoleucus s. navarroensis* was considered distinct, but included roach from the Russian River and tributaries to Tomales Bay (*H. venustus* in part), although the Tomales roach was thought to be

distinct enough to be recognized as a separate subspecies. Hopkirk (6) cautioned that his *H. s. symmetricus* possibly consisted of several subspecies, noting that a collection he examined from the Cosumnes River had some distinctive characters. Brown et al. (1) examined roach populations throughout the San Joaquin drainage and found that populations from more isolated tributaries (e.g., Kaweah and Tule Rivers) could be distinguished by multivariate analyses of morphometric data. The Kaweah River population was particularly distinctive because a high percentage had the "chisel lip" feature. A population originally discovered by B. Quelvog (CDFG) in small creeks near Sonora is so different that it undoubtedly merits subspecies status (1).

When DNA fingerprinting techniques were used to compare populations from four adjacent Sacramento Valley streams, there was evidence of fairly long isolation of the populations from one another (32), suggesting that more distant populations should be even more distinct. The California roach "complex" is in need of taxonomic reevaluation using biochemical techniques. Such a reevaluation may turn up new subspecies or even species, and perhaps merge presently recognized forms. Until then, I suggest that we either go back to recognizing Snyder's six species of roach or else recognize the following forms, based on a combination of morphology, meristics, and zoogeography. I prefer the latter course of action, and so recognize

1. Sacramento–San Joaquin roach, *L. s. symmetricus.* Sacramento and San Joaquin River drainages, except Pit River, as well as tributaries to San Francisco Bay. This is a complex of forms isolated in watersheds throughout Central California. Many of them are distinguishable from one another by morphology, genetics, or both, but the interrelationships are complex and poorly understood.

2. Clear Lake–Russian River roach, *L. s.* ssp. Clear Lake drainage and the Russian River. Morphologically these roach are similar to Sacramento roach, but they show a genetic relationship to each other and seem to represent a separate evolutionary line or lines (39).

3. Monterey roach, *L. s. subditus.* Tributaries to Monterey Bay, specifically Salinas, Pajaro, and San Lorenzo drainages.

4. Navarro roach, *L. s. navarroensis.* Navarro River.

5. Tomales roach, *L. s.* ssp. Walker Creek and other tributaries to Tomales Bay.

6. Gualala roach, *L. s. parvipinnis.* Gualala River.

7. Pit Roach, *L. s. mitrulus.* Upper Pit River and tributaries and tributaries to Goose Lake. Roach found in Oregon presumably belongs to this subspecies.

8. Red Hills roach, *L. s.* ssp., from Horton Creek and other small streams near Sonora, San Joaquin drainage.

An analysis of the mitochondrial and nuclear DNA of roach from the foregoing groups shows that many roach populations have very complex evolutionary histories, including (in some forms) past hybridization with hitch (39). Populations in most large watersheds show evidence of long isolation from one another, although there are distinct geographic groups as well. Although it is difficult to apply the Linnaean species concept to such an evolutionarily complex and dynamic group of fish, it is clear that the foregoing list of subspecies is, if anything, conservative in terms of providing recognition to roach diversity. It is likely that one or more of the above forms (or others yet to be recognized) will eventually be granted species status again.

The generic name *Lavinia* is preferred to *Hesperoleucus* because hitch (the only other species in the genus) and roach are interfertile, and the two species are closely related genetically (6, 7, 8, 9). The name *Lavinia* (Girard 1854) has precedence over the name *Hesperoleucus* (Snyder 1913). Roach hybridize extensively with hitch in tributaries to the Pajaro and Salinas Rivers and in Alameda and Coyote Creeks (4, 6) and with arroyo chubs in the Cuyama River (28).

Names The common name of California roach is derived from their superficial resemblance to one of the common minnows of Europe, the roach (*Rutilus rutilus*). Other names used in the past are western roach and Venus roach. For *Lavinia*, see the account of hitch. *Symmetricus* means symmetrical.

Distribution California roach are found throughout the Sacramento–San Joaquin River drainage, including the Pit River and tributaries to Goose Lake in Oregon. In coastal drainages, they are native to the Navarro, Gualala, and Russian Rivers; streams tributary to Tomales Bay; Pescadero Creek (San Mateo County); and, in the Monterey Bay drainage, San Lorenzo, Pajaro, and Salinas Rivers. At least three additional populations have resulted from introductions: Eel River (in the 1970s) in northwestern California (10); Soquel Creek, Santa Cruz County; and Cuyama River, San Luis Obispo and Santa Barbara Counties. The Cuyama River population may actually be native (28).

Life History Given their wide distribution, it is not surprising that California roach are found in a wide variety of habitats, although they appear to be excluded from many waters by piscivorous fishes, especially nonnative ones. California roach are generally found in small warm streams, and dense populations are frequently sighted in isolated pools in intermittent streams (11). They are most abundant in mid-elevation streams in the Sierra foothills and in the lower

reaches of some coastal streams. Roach are tolerant of relatively high temperatures (30–35°C) and low oxygen levels (1–2 ppm), a characteristic that enables them to survive in conditions too extreme for other fishes (12, 13, 14, 15). However, they also thrive in cold, clear, well-aerated "trout" streams (12, 17), in heavily modified habitats (16, 17), and in the main channels of rivers, such as the Russian and Tuolumne.

Within a watershed, roach can be found in a diversity of habitats, from cool headwater streams to the warmwater lower reaches. Their abundance in streams of the Clear Lake basin is positively correlated with temperature, conductivity, gradient, and coarse substrates and negatively correlated with depth, cover, canopy, and fast water (12). In streams tributary to San Francisco Bay, in contrast, they are most abundant in shady pools with sand, gravel, and bedrock bottoms and beds of aquatic plants. In the Pit River system roach are also characteristic of deep rock-bottomed pools in second- or third-order streams and in the Pit River itself (16). Most such habitat is characterized by low flow, moderate gradients, warm temperatures, and edge mats of duckweed and water ferns.

Although roach are characteristic of streams supporting assemblages of native fishes, they tend to be most abundant when found by themselves or with only one or two other species (15, 16, 18, 19). By themselves, roach will occupy the open waters of large pools; in the presence of predatory pikeminnows, roach are mostly confined to the edges of pools and to riffles and other shallow-water habitats (20, 21). In complex assemblages they concentrate in low-velocity (<40 cm/sec), shallow (<50 cm) water where fine substrates predominate (22). Nonnative green sunfish, however, can completely exclude roach from some streams, although the two species can coexist in large pools. For example, in Dye Creek, Tehama County, green sunfish have almost completely replaced roach in intermittent sections of the south fork, but roach dominate all habitats in the cooler, more permanent north fork that sunfish have been unable to invade; in the mainstem below the union of the forks, the two species coexist, but roach are largely absent from pools.

The ability of roach to survive in small tributaries has also led, through erosional captures of interior headwater streams, to their colonization of coastal streams where other cyprinids are absent, such as Navarro and Gualala Rivers. Such colonization could not have taken place through salt water because they are unable to tolerate very saline water. In August 1973 healthy roach were collected in Navarro River at salinities of 3 ppt, but those trapped downstream by the incoming tide died before salinities reached 9–10 ppt.

California roach feed largely by browsing on the bottom, but in the Tuolumne River (below Preston Falls) and in the Clavey River I have observed large roach feeding on drift organisms, including terrestrial insects, in fairly fast current.

They are omnivores. In small warm streams, filamentous algae typically dominates the diet, but aquatic insects and small crustaceans often make up 25–30 percent of their stomach contents by volume (24, 25, 26). In larger streams, such as the North Fork Stanislaus River, aquatic insects may dominate the diet at all times of the year (17). Crustaceans and small chironomid midge larvae are especially important to small roach. In adult roach the aquatic insects consumed reflect availability in benthos and drift. Small midge, mayfly, caddisfly, and stonefly larvae, along with elmid beetles, aquatic bugs, and amphipods, are taken roughly in proportion to their abundance on the bottom (17, 24). One roach from the Navarro River contained three larval lampreys. Because roach pick most food from silty bottoms, their stomachs usually contain considerable amounts of detritus and fine debris. Laboratory experiments suggest that retention of such fine material is facilitated by mucus secreted by epithelial cells and by gill rakers, so it presumably possesses nutritional value (36).

Growth is highly seasonal. Roach typically grow most rapidly in early summer (23, 26). In some streams they may take 2 years to reach 45 mm SL (23, 26). However, in permanent streams (e.g., Coyote Creek and the Stanislaus, Russian, and Navarro Rivers) roach frequently exceed 40 mm SL in their first summer, reach 60–75 mm in the second summer, and reach 80–95 mm in the third summer (17, 26, 27). Few exceed 120 mm SL. The oldest roach on record is a 6-year-old specimen from San Anselmo Creek, Marin County (26), but few live longer than 3 years.

Roach usually become mature after they reach 45–60 mm SL at 2 or sometimes 3 years of age (26). Fecundity ranges from 250 to 2,000 eggs per female depending on size (17, 23). Spawning is from March through early July, depending on water temperature (17, 26), although spawning activity has been observed in late July in the Russian River. Spawning usually takes place when temperatures exceed 16°C. The fish move up from pools into shallow, flowing areas where the bottom is covered with small rocks 3–5 cm in diameter. The fish spawn in large groups, each female repeatedly depositing eggs a few at a time in crevices between rocks (26). They are immediately fertilized by one or more males following close behind. A spawning aggregation was observed by J. Feliciano (pers. comm.) over several days in late May 2001 in the Navarro River:

> I observed a dense swarm of about 500 adult California roach crowded along the righthand margin of the stream. Roach were continually swimming in and out of the swarm from the surrounding pool. In the swarm, I observed some fishes jamming themselves head or tail first into the substrate, with their other ends clear of the water. Most of the splashing activity came from fish crowding around those individuals. The roach were quite active and oblivious; I clearly heard their splashing from ca. 20 m

downstream and was able to observe them from a point bar only 2 m away. The swarm progressed slowly upstream along the bank as fish moved in and out of the group.

This activity clears silt and sand from the interstices of the gravel, improving habitat for the fertilized eggs, which are adhesive and stick to the rocks. They hatch in 2–3 days, and larvae remain in crevices until large enough to swim actively around. Larval development is described by Fry (26). The population of roach in Bear Creek, Colusa County, apparently spawns in emergent vegetation, and newly hatched larvae remained among the plants for some time (23). Once the yolk sac is absorbed, larval roach feed mainly on diatoms and small crustaceans (26).

Status IA–E, depending on subspecies or population. Many populations of California roach are threatened to some degree because they are located in small streams that are vulnerable to human disturbance (especially diversion) and introduced predatory fishes (such as green sunfish), to which roach seem exceptionally vulnerable (1). The following are accounts by region.

Sacramento–San Joaquin roach. ID. Assuming this widely distributed form is indeed just a single taxon (which is unlikely), it is abundant in a large number of streams. Nevertheless, it is now absent from many streams and stream reaches where it once occurred (e.g., 15), and most populations are isolated by downstream barriers, such as dams, diversions, or polluted water containing predatory introduced fishes (32). Extirpations without recolonization can therefore be expected. Surveys by Moyle and Nichols (18) that were repeated by Brown and Moyle (19, 20) indicate that, in the San Joaquin drainage, the species has been eliminated from many streams since 1970, and from entire watersheds (e.g., the Fresno River) since the 19th century. The problems of conserving the many distinct evolutionary units of California roach are discussed by Brown et al. (1). Populations are increasingly being isolated from one another by artificial barriers. Much of their habitat is on private land, which is subject to development or intense grazing pressure. As a result many streams dry up more frequently or more completely than usual because of diversions and pumping from aquifers that feed them. Predatory fishes, such as largemouth bass and green sunfish, are often introduced into remaining deep pools to provide recreational fishing; such predators typically eliminate roach. However, the introduced Eel River population represents a major expansion of the range of this form (although the exact origin of the invaders has not been determined).

Clear Lake–Russian River roach. ID. These roach are abundant and widely distributed in both watersheds, but this situation could change rapidly with land and water use changes, especially in the Russian River.

Red Hills roach. IB. This highly distinctive form is found in a few small streams in an area partly administered by the BLM and characterized by serpentine soils and stunted vegetation. The largest population, of several hundred individuals, exists in Horton Creek, and smaller numbers occur in Amber and Roach Creeks (37). The limited area of serpentine soil in which this form occurs is subject to intense grazing, mining, and recreational use by off-road vehicles, which together significantly degrade the habitat. Activities causing streamside soil disturbance at the site of the main Horton Creek population pose a particularly serious threat (37), causing limited pool habitat to become shallower and warmer and reducing riparian cover.

Monterey roach. ID. Smith (29) found this roach widespread in the Pajaro and San Benito drainages, but somewhat less widely distributed than formerly. Since Snyder's (2) collections in 1908, they have disappeared from at least four sites owing to habitat alteration, including lowered water quality (increased turbidity, low dissolved oxygen) (29). Streams in the Monterey Bay drainages have been channelized, polluted, diverted, and otherwise altered by a combination of intensive agriculture and grazing, housing development, road building, and other human activities. Dams have reduced flood flows, resulting in upstream expansion of hitch; hybridization and competition with hitch have subsequently eliminated some roach populations (33). Recent losses of roach populations occurred when droughts eliminated isolated populations and dams or other human-made barriers prevented recolonization (33). Most original habitats of Monterey roach are on private land where there is little formal protection for aquatic organisms. Many populations share habitat with steelhead, and the listing of steelhead as a threatened species should help to provide protection for roach as well.

Navarro roach. IE. This form remains abundant in the Navarro River, where they may have benefited from opening of the canopy and warming of the water as the result of logging and agriculture. Even they, however, cannot survive the drying up of the river in some sections by diversion of both surface and ground water.

Tomales roach. IE. Most streams in the Tomales drainage (Marin County) have been heavily modified, but roach are nevertheless abundant in many areas. Their distribution is rather restricted, and most are on private lands that are heavily grazed. Thus siltation, bank erosion, and loss of riparian cover are constant problems. Equally important, the streams (e.g., Walker Creek) are dammed and diverted, regulating and reducing flows as well as creating conditions in which nonnative species are more likely to invade. Although the Tomales roach seems to be holding its own at the present time, its populations should nonetheless be monitored.

Gualala roach. IE. This form is common in the Gualala River (31) and is dominant in some headwater areas I have

examined. Its numbers may actually have increased temporarily as the result of warmer water associated with habitat degradation (34). Like the Tomales roach, this form has a rather restricted distribution within a watershed subjected to many insults (e.g., logging, road building, diversion of water by wineries) in recent years.

Pit roach. IB. This roach has disappeared from much of its former range in the upper Pit River drainage (16) and is confined to a few scattered populations either in small, isolated streams or in some regulated sections of the Pit River. Oregon stocks are classified as "sensitive-peripheral"; the only known Oregon population seems to be in Drews Creek, a tributary to Goose Lake (35). Presumably, each population is threatened by different factors, but the principal ones seem to be habitat loss (from, e.g., heavy grazing in riparian areas, road and housing construction, water diversions) and introduced predators, such as largemouth bass and green sunfish. Because populations are now widely scattered, local extinctions due to natural factors can also occur, but without hope of natural recolonization. As a result, the number of populations can be expected to dwindle over the years.

Overall, the California roach is in need of a comprehensive study of its status, systematics, and distribution. An analysis of the systematics of the species is especially required in view of the discovery of the Red Hills roach and indications that a number of undescribed forms may exist around the state (1). Immediate needs are to find streams in the Pit and San Joaquin River drainages that can be managed as refuges for local populations. All known stream habitats of the Red Hills roach should be protected and managed to benefit the species (and other native organisms in this unusual area); measures would include restrictions on mining, off-road vehicle use, and grazing. The Tomales, Navarro, and Gualala roach would benefit from watershed management practices that improve instream and riparian habitats. In absolute terms, most subspecies of California roach are still abundant, but there is growing evidence that local populations are disappearing one at a time (1, 16, 18, 20, 21). It would be prudent to at least stabilize populations of all taxa at their present levels of abundance in all major watersheds in which they occur. As a minimum measure, a system of Aquatic Diversity Management Areas (ADMAs) should be established; these should include special units for each distinctive population of roach in all geographic regions where they are native (31). A system of protected waters would protect not only this species, but entire biotic communities as well. In the meantime, populations should be monitored to ascertain that each form is holding its own.

References 1. Brown et al. 1992. 2. Snyder 1908c, 1913, 1916. 3. Murphy 1948c. 4. R. R. Miller 1945b. 5. Hubbs et al. 1979.

6. Hopkirk 1973. 7. Avise et al. 1975. 8. Avise and Ayala 1976. 9. Moyle and Massingill 1981. 10. Brown and Moyle 1996. 11. Moyle et al. 1982. 12. Taylor et al. 1982. 13. Knight 1985. 14. Cech et al. 1990. 15. Leidy 1984. 16. Moyle and Daniels 1982. 17. Roscoe 1993. 18. Moyle and Nichols 1973, 1974. 19. Brown and Moyle 1993. 20. Brown and Moyle 1991. 21. Brown and Brasher 1995. 22. Moyle and Baltz 1985. 23. Barnes 1957. 24. Fite 1973. 25. Greenfield and Deckert 1973. 26. Fry 1936. 27. K. Harsberger and C. Staley, University of California, Davis, unpubl. rpt. 1972. 28. Greenfield and Greenfield 1972. 29. Smith 1982. 30. CDFG 1991. 31. Moyle and Yoshiyama 1994. 32. Loggins 1997. 33. J. J. Smith, San Jose State University, pers. comm. 1999. 34. E. Gerstung, CDFG, pers. comm. 1998. 35. G. Sato, BLM, pers. comm., 1996. 36. Sanderson et al. 1991. 37. B. Quelvog, CDFG, pers. comm. 1998. 38. Murphy 1943. 39. J. Jones, University of California, Santa Cruz, pers. comm. 2001.

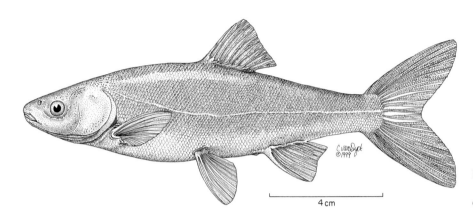

Figure 50. Sacramento blackfish, 13 cm SL, Putah Creek, Yolo County.

Sacramento Blackfish, *Orthodon microlepidotus* (Ayres)

Identification Sacramento blackfish can be readily recognized by their tiny scales (90–114 in lateral line); cone-shaped head with a flat, sloping forehead; round, elongated body; small eyes (adults); and narrow caudal peduncle. The mouth is terminal and slightly upturned, with narrow, only slightly protractile "lips." There are 9–11 rays in the dorsal fin, which has its origin above or slightly anterior to that of the pelvic fins. The anal fin has 8–9 rays, the pelvic fins 10. The pharyngeal teeth (0,6-6,0 or 0,6-5,0) are long, straight, and knifelike, with a narrow grinding surface on the dorsal side of each tooth. The 30–38 gill rakers are long and densely packed, and have a broomlike fringe at their tips. The color of small fish is silvery. Larger fish become progressively darker, especially on the back. The combination of fine scales and dark color gives large fish a dull, olivaceous sheen.

Taxonomy Sacramento blackfish are one of the most distinctive cyprinids in California, most closely related to hitch, with which they hybridize on occasion (1, 2). They have also hybridized with tui chubs (24). There is little evidence of geographic variation (3).

Names The common names usually refer to the shiny dark coloration of adults, hence "Sacramento blackfish," "greaser blackfish," or just "blackfish." In the older literature they are sometimes referred to as "hardhead," a name now reserved for *Mylopharodon conocephalus. Orthodon* means straight teeth; *microlepidotus,* small scales.

Distribution Sacramento blackfish are native to low-elevation reaches of the Sacramento and San Joaquin Rivers and their major tributaries, as well as to Clear Lake (Lake County) and the Pajaro and Salinas Rivers. They are present in the Russian River, but it is not certain if they are native there (3). They are present in a few central California reservoirs (e.g., Shasta, Hennessy, Lagoon Valley), but the extent of their distribution in these systems is poorly documented. In the San Francisco Bay region they are present in the Delta and in Coyote, Alameda, and Walnut Creeks (4). They have been transported by the California Aqueduct to San Luis Reservoir (Merced County), where they are common, and they can be expected in any reservoir in southern

California that contains water from the aqueduct. Since about 1986, they have been present in the lower Santa Ana River below Prado Reservoir (5). Blackfish were introduced into Lahontan Reservoir, Churchill County, Nevada, around 1964 and have spread to lakes in Stillwater Marsh and the Humboldt River drainage in Nevada (6, 23).

Life History Sacramento blackfish are most abundant in warm, usually turbid, waters of the Central Valley floor, often occurring in highly modified habitats otherwise dominated by nonnative fishes (7, 8, 9). Similarly, they are one of the most abundant fishes in Clear Lake, Lake County, and once were abundant in the large lakes (now drained) of the San Joaquin Valley. They are now common in oxbow lakes near rivers and in sloughs of the Sacramento–San Joaquin Delta (7). In streams, such as the Pajaro River or Putah Creek, blackfish are found in deep, turbid pools with soft (mud, clay) bottoms (10). They also thrive in fluctuating waters of reservoirs such as San Luis (Merced County) and Hennessy (Napa County) Reservoirs. One of their more extreme environments is Lagoon Valley Reservoir (Solano County), a warm, shallow, highly turbid recreational lake that becomes very alkaline in summer (pH 9–10). This flooded playa otherwise supports only Sacramento perch, fathead minnows, and mosquitofish. In Salinas River lagoon, they are common in areas where the salinity is around 7 ppt and have been collected at salinities of 9 ppt (27).

Blackfish show extraordinary physiological adaptations for surviving in extreme environments (12, 13). Adults are commonly found in waters where summer temperatures exceed 30°C and where dissolved oxygen levels may be very low. Optimal temperatures are 22–28°C (10), although growth is reduced and metabolic rates are increased at temperatures above 25°C (12). However, in the laboratory, juvenile blackfish can survive temperatures up to 37°C (11). This finding suggests that blackfish thrive under relatively moderate lakelike conditions but are adapted for surviving the periods of extreme conditions that occur during times of drought or low flow.

The feeding habits of blackfish are unusual for a North American cyprinid; they are primarily suspension feeders on planktonic algae and zooplankton, including rotifers, cladocerans, copepods, insect larvae, and suspended detritus (8, 9, 14, 15, 16, 17). Small (<2 cm SL) blackfish feed largely on zooplankton and insects by picking them from the water column or bottom (14, 17, 18). As they grow larger, the pumping of suspended material into the oral cavity becomes increasingly important. By opening and closing their mouths rapidly, blackfish suck in large amounts of water containing small food items (16, 17). The food material is carried to the roof of the mouth, where it is collected by mucus secreted by a palatal organ; clumps of mucus and food are then swallowed (19). The dense, broomlike gill rakers are not constructed to act as filters for food but work to direct the flow of water and food particles past the palatal organ, where the food is removed (19). Although this method of feeding would seem rather unselective, blackfish in ponds seem able to feed selectively on diatoms (8), and those in lakes, on larger algae and zooplankton species (6, 9, 16). When blackfish densities are high, their selective removal of algae-grazing zooplankton may result in blooms of algae, increased nutrient levels, and other major changes to lake ecosystems (6). The ability of adults to live on a diet of largely organic matter and algae is also reflected in their long, convoluted intestine, which is 4–7 times body length; the intestine is longest relative to body length in the largest fish (9, 20). Blackfish are not exclusively planktonic feeders but may also feed on soft, flocculent material, rich in organic matter and small invertebrates, from the bottom of lakes and ponds.

Growth of Sacramento blackfish is rapid in the first year. In Clear Lake they measure about 10 cm FL and weigh about 39 g at the end of their first year, growing rapidly to 25–26 cm and 230 g during their second year. During the third year growth differences between males and females usually become evident, with Clear Lake males reaching 34–35 cm (625 g) and females, 36–37 cm (710 g). Growth is slower in the following years. Growth rates for blackfish in ponds in the San Joaquin Valley (8), San Luis Reservoir (21), and Stone Lakes (Sacramento County) are similar to those of Clear Lake fish, except during the first year, when they tend to be faster (9). In Stone Lakes, male-female differences in growth rates were not found (9). Blackfish rarely exceed 50 cm FL and 1.5 kg (14). The maximum age as determined by scales is 5 years, but it is likely that the largest fish are at least 7–9 years old because scales do not give accurate readings for large cyprinids (22).

Sacramento blackfish of either sex become mature for the first time in their first, second, third, or fourth year, depending on how well the environment promotes growth (8, 14). Most mature in their second or third year. Males are more likely to mature at a younger age than females. Fecundities depend on body size, with a female measuring 171 mm FL producing about 14,700 eggs, a female measuring 350 mm FL producing 78,500 eggs, and a female measuring 466 mm FL producing 346,500 eggs (8). However, there is considerable uncertainty in these fecundity estimates, in part because individual blackfish may spawn over a fairly wide time span (8).

Mature males grow tiny breeding tubercles and seem darker than females during the breeding season. In Clear Lake spawning occurs between April and July at water temperatures of 12–24°C in shallow areas with heavy growths of aquatic plants. Spawning conditions are presumably similar elsewhere, although some blackfish may spawn as early as March (8). Because of turbid water, observations of spawning are few and incomplete. Murphy (14) observed spawning activity by a small school over a bed of aquatic vegetation in 90 cm of water. Males followed females closely, apparently fertilizing eggs as they were extruded onto plants. Similar behavior has been observed over rocks in water less than 18 cm deep (15) and in experimental ponds, where the fish spawned on strips of plastic. Spawning seems physiologically hard on blackfish. They develop spawning checks (interruptions of growth rings) on the scales (8, 14), indicating partial resorption to provide a last-minute supply of nutrients for developing gonads. In Clear Lake few fish manage to survive their second spawning, a fact that may account for the summer die-offs noted there. However, in Stone Lakes, blackfish regularly spawn 2–4 times (9).

Fertilized eggs stick to the substrate, and larvae are often concentrated in shallow water, especially near or in beds of aquatic plants (24). Larvae may also be found in open water (24). Juvenile blackfish are typically found in large schools in shallow water, often near cover. They can live on plant materials alone, but they grow fastest where animal prey is abundant (25, 26).

Status IE. The herbivorous filter-feeding habits of blackfish, coupled with their ability to survive in warm, turbid waters, allow them to succeed despite changes in their environment. Nevertheless, they are probably less abundant than formerly in native lowland habitats. They have been spread (and are still spreading) through introductions and aqueducts to a number of reservoirs, and their impacts on other fishes in these reservoirs and associated streams are not known. However, in Lahontan Reservoir, Nevada, blackfish apparently replaced native tui chub as the most abundant species. This observation suggests that their further spread to other watersheds and reservoirs should be prevented if possible.

They are important commercial fish, sold live in Asian fish markets in many California cities. They are prized for their culinary qualities and so are (or have been) harvested in large numbers from Clear Lake, San Luis Reservoir, and Lahontan Reservoir. Because of their ability to feed on algae and organic matter and to thrive in small ponds, they have considerable potential for warmwater aquaculture.

References 1. Moyle and Massingill 1981. 2. Avise et al. 1975. 3. Hopkirk 1973. 4. Leidy 1984. 5. Swift et al. 1993. 6. Byers and Vinyard 1990. 7. Turner 1966c. 8. Monaco et al. 1981. 9. Staley 1980. 10. Smith 1977, 1982. 11. Knight 1985. 12. Cech et al. 1979. 13. Campagna and Cech 1981. 14. Murphy 1950. 15. Cook et al. 1964. 16. Johnson and Vinyard 1987. 17. Sanderson and Cech 1992, 1995. 18. Cech and Linden 1987. 19. Sanderson et al. 1991. 20. Kline 1978. 21. Johnson 1976. 22. Scoppetone 1988. 23. La Rivers 1962. 24. Wang 1986. 25. Cech at al. 1982. 26. Fox 1988. 27. J. J. Smith, San Jose State University, pers. comm. 1998.

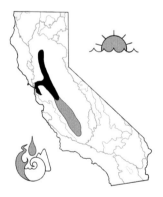

Sacramento Splittail, *Pogonichthys macrolepidotus* (Ayres)

Identification This large (to over 40 cm SL) cyprinid is readily recognized by the enlarged upper lobe of the tail, tiny barbels (sometimes absent) at the corners of the slightly subterminal mouth, and small head (head length divisible into body length less than 4.5 times) on an elongate body. The dorsal fin rays number 9–10; pectoral fin rays, 16–19; pelvic fin rays, 8–9; anal fin rays, 7–9; lateral line scales, 57–64 (usually 60–62); and gill rakers, 14–18 (usually 15–17). The pharyngeal teeth, usually 2,5-5,2, are hooked and have narrow grinding surfaces. The inner tooth rows are very small. Live fish are silvery on the sides, but become duller in color as they grow larger. The back is usually dusky olive gray. Adults develop a distinct nuchal hump on the back. During the breeding season, paired, dorsal, anal, and caudal fins are tinged with red-orange, and males become darker colored, developing tiny white tubercles on their heads and on the bases of the fins (28).

Taxonomy This species was first described in 1854 by W. O. Ayres as *Leuciscus macrolepidotus,* then by S. F. Baird and C. Girard as *Pogonichthys inaeqilobus.* Ayres's species description has priority as the official one, but *Pogonichthys* was accepted in recognition of its distinctive characteristics (1). The splittail is considered by some taxonomists to be more closely allied to cyprinids of Asia than to any North American species (2), but most evidence suggests it is allied with other endemic cyprinids of western North America. The genus comprises two species, *P. ciscoides* Hopkirk and *P. macrolepidotus* (1).

Names Splittail refers to the distinctive tail. *Pogon-ichthys* means bearded fish, referring to the small barbels, unusual

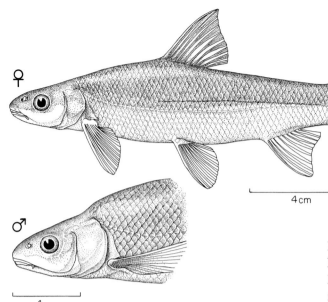

Figure 51. Sacramento splittail, 13.5 cm SL (head of large adult, 26 cm SL), Suisun Marsh, Solano County.

♀

♂

4 cm

4 cm

in North American cyprinids; *macro-lepidotus* means large-scaled.

Distribution Sacramento splittail are endemic to California, mainly to sloughs, lakes and rivers of the Central Valley. In the Sacramento Valley they were found in early surveys as far up the Sacramento River as Redding (below the Battle Creek Fish Hatchery in Shasta County), up the Feather River as high as Oroville, and in the American River to Folsom (3). Today they are largely absent from the upper parts of their distribution, although in wet years they may migrate up the Sacramento River as far as Red Bluff Diversion Dam (river km 391, Tehama County), and into the lower Feather and American Rivers (4, 5, 31). The Sutter and Yolo Bypasses, along the Sacramento River, are apparently important spawning areas today (5). In the San Joaquin River they were once found as far south as Friant (3). Archaeological evidence indicates that populations were present in the large lakes of the San Joaquin Valley floor, lakes Tulare and Buena Vista, where they were harvested by native peoples (6, 13). Recent surveys indicate that during wet years splittail may ascend the San Joaquin River as high as Salt Slough (river km 218) (8, 9, 10, 31). Successful spawning has been recorded in the lower Tuolumne River during wet years in the 1980s and 1990s, with both adults and juveniles observed at Modesto, 11 km upstream from the river mouth (10).

In the San Francisco Bay area Snyder (7) reported catches of splittail from southern San Francisco Bay and at the mouth of Coyote Creek in Santa Clara County, but they are now very rare there (11). During most years, except when they are spawning, splittail are largely confined to the Delta, Suisun Bay, Suisun Marsh, the lower Napa River, the lower Petaluma River, and other parts of the San Francisco Estuary (12, 30). In the Delta they are most abundant in the north and west portions when populations are low but are more evenly distributed throughout following years of successful reproduction (5, 14). An area with a particularly high concentration of splittail is Suisun Marsh, Solano County, especially during drier years (12). Occasionally, splittail are caught in San Luis Reservoir (15), which stores water pumped from the Delta, and a single specimen has been reported from Silverwood Reservoir, at the southern end of the California Aqueduct (16).

Life History Splittail are adapted for living in estuarine waters with fluctuating conditions, as well as in severe conditions that once occurred in alkaline lakes and sloughs on the floor of the Central Valley during droughts. They are remarkably tolerant of high salinities for a cyprinid and are regularly found at salinities of 10–18 ppt, although lower salinities seem preferred (12). Salinity tolerance increases with size, and adult splittail can tolerate salinities up to 29 ppt for short periods of time (17). Temperatures at which splittail are found are typically between 5 and 24°C, but fish acclimated to high temperatures can survive rapid changes and temperatures of 29–33°C for short periods (17). Splittail of all sizes can also survive low dissolved oxygen levels (<1 mg O_2/liter). These tolerances make them well suited to slow-moving sections of rivers and sloughs (18, 19). In Suisun Marsh, trawl catches are highest in summer, when salinities are 6–10 ppt and temperatures 15–23°C (18), although this observatioon reflects in part the annual influx of young-of-year fish from upstream. Young-of-year and yearling splittail in general are most abundant in shallow (<2 m) water (12) and show considerable capacity to swim against strong river and tidal currents (17).

Adult splittail show gradual upstream movement during the winter and spring months to forage and spawn in flooded areas (5, 15). During wet years, these upstream movements can be more directed and take the fish long distances upstream. Year class success of splittail is positively correlated with wet years, high Delta outflow, and flood plain inundation (5, 12, 15), presumably because adults are able to move upstream to suitable spawning areas and to find flooded vegetation for spawning, which also provides cover for larvae and young (5, 15).

Their small subterminal mouths, maxillary barbels, large upper tail lobes, and generalized pharyngeal teeth reflect the adaptation of splittail for feeding on bottom invertebrates in areas of low to moderate current. However, detrital material typically makes up a high percentage (50–60% by volume) of their stomach contents. In Suisun Marsh splittail foraged extensively on opossum shrimp (mainly *Neomysis mercedis*), benthic amphipods (*Corophium*), and harpactacoid copepods (19). After *N. mercedis* populations collapsed, mysid shrimp ceased being important in the diet (32). In the Delta they feed opportunistically on clams, crustaceans, insect larvae, and other invertebrates. When water levels rise in February and March, splittail often move into flooded areas to feed on earthworms (15). Rutter (3) reported large numbers of splittail feeding on loose eggs in areas where salmon were spawning, although overlap of these two species is rare today. They are largely diurnal feeders, with most intense feeding in early morning (15). Splittail are preyed on by striped bass and other piscivorous fishes. The desirability of the species as prey for striped bass has long been recognized by anglers, who fish for splittail in order to use them for bait.

Splittail are relatively long lived; analysis of scales indicates life spans of 5–7 years (19), but analysis of hard parts indicates that larger fish may be 8 or more years old (20). They reach about 110 mm SL in their first year, 170 mm in their second year, and 215 mm in their third year, growing about 35 mm/year thereafter. Both males and females can become mature by the end of their second year (19), although occasionally males mature in their first year and females may not mature until their third year (15). The sex ratio among mature individuals is 1:1 (15), but the largest and oldest fish are mostly females (20). Females are highly fecund; the largest may produce over 100,000 eggs (19, 29). A relationship between fecundity and length is $F = 0.0004(SL_{mm}^{3.40})$ (29), with larger females producing more eggs per millimeter, although this relationship may vary among years. A 1974 study found an average of about 165 ova per millimeter SL; a 1982 study, 600; a 1994 study, 151; and a 1996 study, 261 (19, 29, 34). The cause of this wide variation is uncertain, but it may be related to food availability (29).

Spawning can apparently take place any time from late February to early July (21, 28), with older fish reproducing first (15). Generally, gonadal development is initiated by the advent of autumn, with a concomitant decrease in somatic growth (19). In state and federal fish rescue facilities in the south Delta, adults are captured most frequently in January through April, when they are presumably engaged in spawning movements. The onset of spawning seems to be associated with rising water levels, increasing water temperature (to 14–19°C), and increasing day length. As they become ready to spawn, the fish move into flooded vegetation. Spawning is most frequent in March and April, and splittail appear to be fractional spawners, with individuals spawning over a period of several months (28). However, in some years spawning may take place within a limited period of time; in 1995, a year of extraordinarily successful reproduction, most splittail spawned over a short period in April (22). The fertilized eggs are adhesive and stick to submerged vegetation and debris until hatching. In captivity splittail will spawn on the sides and bottoms of net pens and in tanks (34).

Embryos hatch in 3–7 days, depending on temperature (34). Swim bladder inflation and active swimming and feeding begin 5–7 days later. Most larvae remain in shallow, weedy areas near spawning sites for 10–14 days before beginning to move into deeper offshore habitat as swimming ability increases (5, 21). Early larval stages may live in flooded vegetation because small prey (rotifers and microcrustaceans) are abundant there. Thereafter they focus on benthic crustaceans. A stock recruitment relationship in splittail is weak, indicating that under favorable environmental conditions a small number of large females can produce many young (5, 12). Young-of-year are caught in South Delta pumping plants in greatest numbers in April–August, presumably when moving downstream into the estuary (12).

Status IB. Sacramento splittail were listed by the USFWS as a threatened species in February 1999 because of the historic reduction in range and because of the large reduction in numbers during the severe drought of 1987–1993 (23, 24, 33). Their astonishing ability to recover under favorable conditions (5, 31), such as existed in 1995 and 1998, has alleviated fears of immediate extinction and ignited controversy over their actual status. However, given their history and distribution, their long-term survival is not assured.

Splittail have disappeared as permanent residents from portions of the Sacramento and San Joaquin Valleys because dams, diversions, channelization, and agricultural drainage have either eliminated or drastically altered much of the lowland habitat they once occupied or else made it inaccessible except during wet years. They are rare or seasonal in occurrence more than 10–20 km upstream of the Delta, except following years of unusually high reproductive success. In the San Joaquin Valley they seem to move into the lower river only during wet years; movements into the Sacramento and tributaries may be more frequent. As a result of these changes, today most are resident in the San Francisco Estuary, especially in the Delta and Suisun Marsh. Their abundance is strongly tied to outflow and the extent of flooded areas, especially the Yolo Bypass, presumably because spawning occurs over flooded vegetation. Thus when outflows are high at the right time of year (March–April), reproductive success is high, but when outflows are low, reproductive success is very low.

Within the present limited range, splittail have been estimated during most years to be only 35–60 percent as numerous as they were in 1940 (25). CDFG midwater trawl data indicate a decline from the mid-1960s to the late 1970s followed by a resurgence (with fluctuations) through the mid-1980s. From the mid-1980s through 1994, splittail numbers declined, with small increases in some years. In 1995 and 1998 the population increased dramatically, and the estuary and lower river habitats were flooded with juveniles. The 1995 and 1998 "boom years" demonstrated how splittail recruitment success fluctuates widely from year to year and over long periods of time (5). Large pulses of young fish were observed in the wet years 1982, 1983, and 1986, but recruitment was exceptionally low in 1980, 1984, 1985, and 1987–1994, which were mostly dry years (31). Not all wet years result in large splittail year classes, however. In 1996, for example, most high flows in the rivers occurred in December and January, before splittail were ready to spawn. Adult numbers tend not to show such dramatic fluctuations (5) because they are so long lived, with presumably high survival rates.

The long-term decline can be attributed to a variety of interacting factors (26), in the following approximate order of importance: (1) reduction in valley floor habitats, (2) modification of spawning habitat, (3) changed estuarine hydraulics, especially reduced outflows, (4) climatic variation, (5) toxic substances, (6) introduced species, and (7) exploitation.

Reduction in valley floor habitats. The Sacramento and San Joaquin valleys once had vast flood plains, with myriad sloughs and backwaters left from old river meanders, as well as a few large lakes, such as Lake Tulare. These quiet water habitats were presumably home to resident splittail because they resemble present-day habitats in the Delta and Suisun Marsh. They would have provided abundant food as well as necessary hydraulic connections for spawning. These habitats have now been almost entirely lost through drainage and diking for agriculture. Likewise, vast marshes of the Delta once provided extensive quiet water habitats that are scarce today. Elimination of these habitats eliminated the fish that lived in them, leaving the estuary to support splittail.

Modification of spawning habitat. Splittail spawn on terrestrial vegetation and debris on floodplains that are inundated by spring high flows, typically at depths between 0.5 and 2 m. An increase in the amount of flooded area presumably contributes to year class success in wet years both because of the increase in spawning habitat and because of the increase in larval rearing habitat. The longer residence time of water on floodplains during wet years also allows large "blooms" of zooplankton to occur, providing food for the larvae and juveniles. The decrease in the extent of floodplains in recent decades is consequently likely to be a major contributor to a decline in splittail numbers and to increased variability in recruitment success. Although these losses may be partially compensated for by the Yolo and Sutter Bypasses, the bypasses are not flooded with fish in mind. Yolo Bypass, for example, floods when water from the Sacramento River tops a concrete wall (the Fremont Weir) and spills over the Sacramento Weir. When the river drops, the bypass quickly drains. The bypasses are good habitat for splittail spawning when they are flooded for several weeks in March and April (35). If they flood and drain too early (e.g., December–January, as in 1996), they are not used for spawning.

Changed estuarine hydraulics. In the past 30 or so years, hydraulic conditions in the Delta have changed dramatically (12), but it is not clear if there is a cause-and-effect relationship between these changes and splittail abundance. It is possible that direct entrainment of larvae and juveniles in pumps of the South Delta may be part of the problem, although numbers entrained are directly related to abundance (5). More young are entrained during years with strong year classes, when, arguably, there are fish to spare. The increased movement of young-of-year into the central Delta as the result of changed hydraulics may lead to increased within-Delta entrainment and place small fish in conditions less favorable for growth and survival.

Climatic variation. Recent decades have seen some of the most extreme environmental conditions the estuary has experienced since the arrival of Euro-Americans. There were eight years of continuous drought, broken only by huge outflows in February 1986 and followed by exceptionally high precipitation in 1995 and 1997–1999. The prolonged drought had two major interacting effects: a natural decrease in outflow and an increase in the proportion of inflowing water being diverted. A natural decline in the numbers of splittail would be expected from reduced outflow, because of reduced availability of spawning and larval rearing habitat. However, the increase in diversions apparently decreased survival of splittail further through reduction in habitat, especially in the lower Delta and Suisun Marsh, and increased entrainment of larvae, juveniles, and adults. It is important to recognize that extreme floods and droughts have occurred in the past and that splittail have managed to persist through them. However, they did not historically experience the added stresses of reduced spawning and rearing habitat and increased diversion of water, making recovery from natural disasters much more difficult. Likewise, adult splittail in the past were not confined just to the estuary but presumably existed as several populations. Nevertheless, banner years for reproduction, such as the ones experienced in 1995 and 1998, can sustain the population through many subsequent years of less favorable conditions.

Toxic substances. The effects of pesticides and other toxic substances on splittail are not known, but there is considerable potential for negative interactions, especially when larvae are found in the Delta (26). This subject requires further investigation.

Introduced species. Introduced species are a perpetual problem in the San Francisco Estuary, and the problem worsens as new species are introduced through the dumping of ballast water from ships. The most recent problem introductions have been of several species of planktonic copepods and the overbite clam, *Potamocorbula amurensis.* The copepods seem to be replacing *Eurytemora affinis,* a copepod that has been the favored food of juvenile fish and of opossum shrimp. Opossum shrimp are in turn the favored prey of splittail. The native opossum shrimp has itself been replaced in much of the estuary by several slightly smaller similar species, *Acanthomysis* spp. The overbite clam, which invaded in 1986, may have an indirect effect on splittail because it has become extremely abundant in Suisun Bay, from which it is filtering out planktonic algae and small invertebrates, which constitute the base of the food web leading to mysid shrimp and splittail (27). Mysid shrimp, formerly the most important invertebrate in splittail diets, have become scarce in their diets, reflecting both changes in the species and reduced abundance overall (32). One possible consequence is reduced fecundity (29). If either white bass or northern pike become established in the estuary, which seems likely, their voracious predation will pose an additional threat.

Exploitation. A specialized fishery for splittail, prized as food by Asian Americans, has existed for a long time, as has the capture of splittail for bait by sport anglers. There is no evidence that this exploitation has contributed to a decline in the numbers of splittail. However, the food fishery concentrates on fish moving up into spawning areas, and large numbers are caught in some years; it should therefore be monitored and tightly regulated. A fishery that caught many large female splittail during a time when populations were already low could significantly affect the resilience of the species.

In short, splittail have tremendous ability to recover from events that cause major reductions in numbers. At the same time, their population has been stressed in an unprecedented manner by human activity. Therefore, their long-term persistence as a viable member of the fish fauna of California will depend on active management of the estuarine habitats, floodplains, and inflowing waters in ways that favor their reproduction and survival.

References 1. Hopkirk 1973. 2. Howes 1984. 3. Rutter 1908. 4. Baxter et al. 1996. 5. Sommer et al. 1997. 6. Gobalet and Fenenga 1993. 7. Snyder 1905. 8. Saiki 1984. 9. Brown and Moyle 1993. 10. T. Ford, Turlock Irrigation District, pers. comm. 1998. 11. Leidy 1984. 12. Meng and Moyle 1995. 13. Hartzell 1992. 14. Turner 1966a. 15. Caywood 1974. 16. Swift et al. 1993. 17. Young and Cech 1996. 18. Moyle et al. 1982. 19. Daniels and Moyle 1983. 20. L. Grimaldo, pers. comm. 1998. 21. Wang 1986. 22. Wang 1996. 23. Moyle et al. 1995. 24. USFWS 1996. 25. CDFG 1992b. 26. Bennett and Moyle 1996. 27. Nichols et al. 1990. 28. Wang 1995. 29. Feyrer and Baxter 1998. 30. Meng et al. 1994. 31. Baxter 1999, 2000. 32. Feyrer 1999. 33. USFWS 1999. 34. Bailey et al. 2000. 35. Sommer et al. 2001a.

Clear Lake Splittail, *Pogonichthys ciscoides* Hopkirk

Identification Similar in most respects to the Sacramento splittail, Clear Lake splittail differ in the following ways: more gill rakers (18–23, usually 21–23); more lateral line scales (60–69, usually 62–65); smaller fins; terminal mouth with absent or poorly developed barbels; small nuchal hump in adults; tail fin more or less symmetrical; and well-developed nuptial tubercles on the head and sides of breeding males (1).

Taxonomy This species was not described until 1973 despite its distinctive morphology and ecology (1). It is the only other member of the genus *Pogonichthys.*

Names Ciscoides means ciscolike, referring to its superficial resemblance to ciscoes (family Salmonidae) of the Great Lakes and elsewhere (1). Other names are as for Sacramento splittail.

Distribution Clear Lake splittail were endemic to Clear Lake, Lake County, and its tributary streams when spawn-

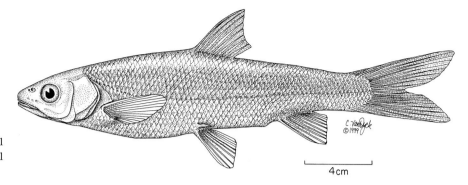

Figure 52. Clear Lake splittail, 21 cm SL, Clear Lake, Lake County, 31 March 1970.

ing. A single specimen is known from Cache Creek, the outlet of Clear Lake (1).

Life History Not much is known about Clear Lake splittail because there was little interest in them until after they became extinct. Their most distinctive features are adaptations for lake living. They once apparently schooled in large numbers over most of the lake, concentrating in littoral areas. Summer die-off of large splittail and other Clear Lake minnows seems to have been an annual event, although its exact cause is not known. Clear Lake splittail were more pelagic in feeding habitats than Sacramento splittail. They were observed eating ovipositing gnats and gnat egg rafts on the surface, as well as bottom-living gnat larvae and emerging pupae (2, 3). Of the diet of 22 splittail examined by Cook (7), 76 percent was zooplankton; the rest was insects or detritus.

Clear Lake splittail spawned in inlet streams in April and May, frequently migrating several kilometers upstream to suitable gravel riffles or areas with flooded vegetation. It is not known how long newly hatched splittail remained in the streams before returning to the lake, but it was probably at least three weeks. Once in the lake they apparently spent the first few months in the littoral zone.

Status IA. The species is globally extinct. Following a major, precipitous decline in the early 1940s (4), Clear Lake splittail managed to hang on until the mid-1970s. The most likely cause of their decline was diversion of streams during spawning and rearing seasons. Splittail apparently spawned

later than hitch (which have managed to maintain populations in the lake) and seem to have reared longer in the streams. Likewise, pikeminnows also spawned later in the season (April) than hitch and are now largely absent from the lake (although they persist in tributary streams).

It is possible that channelization of lower reaches of most tributaries was a major contributer to the decline by eliminating flooded areas needed by splittail for spawning and larval rearing. These aspects of their life history may have been particularly critical in dry years, when sudden reduction in water flows either trapped spawning adults or prevented young fish from moving into the lake (5). Other factors contributing to extinction may have been predation, competition, or diseases from introduced fishes. Although splittail managed to coexist with nonnative fishes for about 100 years, negative interactions may have acted synergistically with poor spawning success. It may be significant that splittail were still fairly easy to collect in Clear Lake in the early 1960s (1) and that their disappearance followed the explosive establishment of inland silversides in 1967. Silversides completely dominate the littoral zone of the lake, once the main habitat of juvenile splittail. Ironically, the huge schools of minnows once present in the shallow waters of the lake were referred to by early residents as "silversides" (6).

References 1. Hopkirk 1973. 2. Lindquist et al. 1943. 3. Cook et al. 1964. 4. Cook et al. 1966. 5. Murphy 1951. 6. Coleman 1930. 7. S. F. Cook, unpubl. data.

Hardhead, *Mylopharodon conocephalus* (Baird and Girard)

Identification Hardhead are large cyprinids, occasionally exceeding 60 cm SL, that resemble Sacramento pikeminnow, except that the head is not as pointed, the body is slightly deeper and heavier, the maxillary bone does not reach past the front margin of the eye, and a small bridge of skin

(frenum) connects the premaxillary bone (upper "lip") to the head. They have 8 dorsal fin rays, 8–9 anal fin rays, and 69–81 scales along the lateral line. The pharyngeal teeth (2,5-4,2) are large and molariform in adults, slender and hooklike in young fish. Young fish are silvery, gradually turning brown to dusky bronze on the back as they mature. Breeding males develop small white tubercles that cover the snout and extend in a narrow band along the side to the base of the caudal fin.

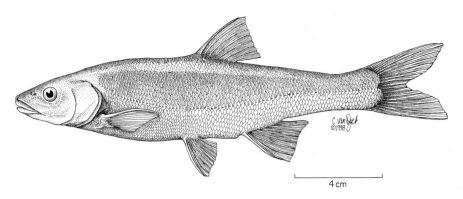

Figure 53. Hardhead, 33 cm SL, Deer Creek, Tehama County.

4 cm

Taxonomy *Mylopharodon conocephalus* was first described as *Gila conocephala* Baird and Girard (1) from a single specimen from the "Rio San Joaquin." In 1855 Ayres (2) redescribed the species as *Mylopharodon robustus*. Girard (3) then reclassified *G. conocephala* as *Mylopharodon conocephalus* and placed *M. robustus* as a closely allied second species. Jordan (4) united both forms as *Mylopharodon conocephalus* (5). There appears to be little morphological variation among hardhead populations (6). Although it is related to the four species of pikeminnow (*Ptychocheilus*), it is different enough to be retained in a separate genus (7, 8). Fossil evidence indicates that the genus has existed since at least the Miocene period (7, 38).

Names The origin of the name hardhead is obscure, particularly because it was applied to Sacramento blackfish, Sacramento pikeminnow, and other large minnows in the early literature. *Mylo-pharo-don* means mill-throat-teeth, referring to the molariform pharyngeal teeth; *conocephalus* means cone-shaped head, which is mildly descriptive.

Distribution Hardhead are widely distributed in low- to midelevation streams in the main Sacramento–San Joaquin drainage. They are also present in the Russian River (11). Their range extends from the Kern River, Kern County, in the south to the Pit River (south of the Goose Lake drainage), Modoc County, in the north (12, 13). In the San Joaquin drainage, the species is scattered in tributary streams and absent from valley reaches of the San Joaquin River (8, 9, 10). In the Sacramento drainage, the hardhead is present in most larger tributary streams as well as in the Sacramento River. It is absent from San Francisco Bay streams except the Napa River.

Life History Hardhead are typically found in undisturbed areas of larger low- to midelevation streams (8, 13), although they are also found in the mainstem Sacramento River at low elevations and in its tributaries to about 1,500 m (14). Most streams in which they occur have summer temperatures in excess of 20°C, and optimal temperatures for hardhead (as determined by laboratory choice experiments) appear to be 24–28°C (15). In a natural thermal plume in the Pit River, hardhead generally selected temperatures of 17–21°C, which were the warmest available (16). At higher temperatures hardhead are relatively intolerant of low oxygen levels, a factor that may limit their distribution to well-oxygenated streams and to surface water of reservoirs (17). They prefer clear, deep (>80 cm) pools and runs with sand-gravel-boulder substrates and slow velocities (20–40 cm/sec) (8, 12, 15, 18, 40). In streams adults often remain in the lower half of the water column (15, 18), although in reservoirs they can occasionally be seen hovering close to the surface (19, 20). Hardhead are always found in association with Sacramento pikeminnow and usually with Sacramento sucker. They tend to be absent from streams where introduced species, especially centrarchids, predominate (8, 13) and from streams that have been severely altered by human activity (21), although they can persist below dams under certain conditions. Their relatively poor swimming ability at low temperatures may keep them from moving up streams with natural or human-made velocity barriers that permit the passage of salmonids (39).

Hardhead are abundant in a few midelevation reservoirs used largely for hydroelectric power generation, such as Redinger and Kerkhoff Reservoirs on the San Joaquin River (Fresno County) and Britton Reservoir on the Pit River (Shasta County). They are most abundant in the upstream half of Britton Reservoir, where habitat is more riverine, and are less abundant in the more lacustrine habitat downstream, where introduced centrarchid basses are abundant (22). They are largely absent today from most warmwater reservoirs with high annual fluctuations in volume, although they can survive in such reservoirs in the absence of large populations of introduced predatory fishes.

In streams hardhead smaller than 150 cm SL often cruise about pools or slow runs during the day in small groups, rising to take insects from the surface, holding in areas of swifter current to eat insects and algae in the water column, or dropping to the bottom to browse (40). They are sedentary in streams, rarely moving more than a kilometer from home pools (23). Most movements away from home pools are presumably related to reproduction (23). Including such movements, the average home range of adult hardhead in a small foothill steam was estimated to be about 850 m (23). In Britton Reservoir large hardhead concentrate on warm summer days in surface waters (<1 m) and can often be seen remaining motionless close to the surface (19). This behavior makes them an important prey for bald eagles that nest in the area (20). In contrast, in streams adults will aggregate during the day in the deepest parts of pools or cruise about slowly well below the surface (40). They are most active in the early morning and evening when feeding.

Hardhead are omnivores that forage for benthic invertebrates and aquatic plant material on the bottom but also eat drifting insects and algae (40). In reservoirs they feed on zooplankton (24). Smaller fish (<20 cm SL) consume primarily mayfly larvae, caddisfly larvae, and small snails (14), whereas larger fish feed more on aquatic plants (especially filamentous algae), crayfish, and other large invertebrates. The ontogenetic changes in tooth structure are consistent with this dietary switch; juveniles have hooked teeth, characteristic of insectivores, whereas adults have large molariform teeth, needed for grinding hard prey and plants (14).

Hardhead typically reach 6–8 cm SL by the end of their first growing season, 10–12 cm in their second, and 16–17 cm in their third (14, 22, 25, 28). In the American River they can reach 30 cm SL in 4 years (14); in the Pit and Feather Rivers, it takes 5–6 years to reach that length (22, 25). In small streams resident hardhead rarely exceed 28 cm SL (28). Feather River fish measuring 44–46 cm SL were aged (using scales) at 9–10 years, but older and larger (to at least 60 cm SL) fish no doubt exist. If the older records are accurate, hardhead are capable of reaching up to 1 m TL (29).

Hardhead mature in their third year and spawn mainly in April and May (14, 23). Juvenile recruitment patterns suggest that spawning may extend into August in some foothill streams (26). Fish from larger rivers or reservoirs may migrate 30–75 km or more upstream in April and May, usually into tributary streams (24, 27). In small streams hardhead may move only a short distance from their home pools for spawning, either upstream or downstream (23). In

Pine Creek (Tehama County) resident hardhead aggregate during spawning season in nearby pools; spawning hardhead from the Sacramento River move into downstream reaches that dry in summer (23).

Spawning behavior has not been documented, but large aggregations of fish found during the spawning season suggest that it is similar to that of hitch or pikeminnow, with fertilized eggs deposited on beds of gravel in riffles, runs, or the heads of pools. Females, depending on size, can produce 7,000–24,000 eggs per year (23, 28). Grant and Maslin (23) noted that there were small undeveloped eggs in each ovary along with mature eggs, indicating that eggs may take 2 years to mature.

The early life history of hardhead is poorly known (26). After hatching, the larval and postlarval fish presumably remain along stream edges in dense cover of flooded vegetation or fallen tree branches. As they grow they move into deeper habitats, where those spawned in intermittent streams are swept down into main rivers, perhaps concentrating in low-velocity areas near the mouth. In Deer Creek (Tehama County) I have observed large aggregations of small juveniles (2–5 cm SL) in shallow backwaters. In the Kern River small juveniles concentrate along edges among large cobbles and boulders (41). Hardhead measuring 5–2 cm SL select habitats similar to those of adult fish. In Deer Creek this means pools or runs that are 40–140 cm deep, with water column velocities of 0–30 cm/sec (18). Such pools invariably contain Sacramento pikeminnows and Sacramento suckers.

Status ID, but IC in the San Joaquin drainage. Historically hardhead have been regarded as widespread and abundant in central California (2, 14, 29, 30, 31, 32, 33, 34). They are still widely distributed in foothill streams, but their populations are increasingly isolated from one another, making them vulnerable to localized extinctions. As a consequence they are much less abundant than they once were, especially in the southern half of their range. Reeves (14) summarized historical records and noted that they were found in most streams in the San Joaquin drainage; but in the early 1970s I found them in only 9 percent of sites sampled (8). Resampling many of the same sites about 15 years later indicated that a number of the populations had disappeared (10). They have a discontinuous distribution in the Pit River drainage, being present mainly in canyon sections of the main river and in hydroelectric reservoirs (13, 36). They are apparently still fairly common in the mainstem Sacramento River, in the lower reaches of the American and Feather Rivers, in some smaller tributary streams (e.g., Deer, Pine, Clear Creeks), and in some river reaches above foothill reservoirs. They have become extremely rare in the Napa River (11) and are uncommon in the Russian River.

Hardhead were abundant enough in Central Valley

reservoirs in the past to be regarded as a problem species, under the assumption they competed with trout and other game fishes for food. However, most reservoir populations proved to be temporary and were most likely the result of colonization by juvenile hardhead before introduced predators became abundant. Populations in Shasta Reservoir, Shasta County, declined dramatically within 2 years (14), although hardhead are still present there in small numbers (35). Similar crashes of large reservoir populations have been reported from Pardee Reservoir on the Mokelumne River, Amador/Calaveras County; Millerton Reservoir on the San Joaquin River, Fresno County; Berryessa Reservoir, Napa County; Don Pedro Reservoir, Tuolumne County; and Folsom Reservoir, El Dorado County (14).

The cause of hardhead declines appears to be habitat loss and predation by nonnative fishes. Hardhead require large to medium-size, cool- to warmwater streams with deep pools for their long-term survival. Such streams are increasingly dammed and diverted, eliminating habitat, isolating upstream areas, and creating temperature and flow regimes unsuitable for hardhead. Consequently populations are gradually declining or disappearing throughout the range of the species. A particular problem seems to be predation by smallmouth bass and other centrarchid basses. Hardhead disappeared from the upper Kings River when the reach was invaded by smallmouth bass (10). In the South Yuba River hardhead are common only above a natural barrier for smallmouth bass; only large adult hardhead are found below the barrier (37). The few reservoirs in which they are abundant today are those in which water level fluctuations (such as those for power-generating flows) prevent bass from reproducing in large numbers. However, either stabilization of water levels or increasing the amount of the drawdown in these reservoirs (which ex-

pose small hardhead to predation) can result in increased populations of centrarchid basses and decreased hardhead populations.

Although hardhead are still fairly common, their general long-term decline matches declines shown by other California native fishes. It would be prudent to stabilize hardhead populations while they are still at moderate levels. The best way to protect them would be to establish a number of Aquatic Diversity Management Areas in midelevation canyon areas in which normal flow regimes and high water quality would be maintained. Because hardhead are good indicators of relatively undisturbed conditions, a system of such managed waters would protect not only the species but also the entire biotic community of which it is a part. In the meantime, stream populations should be monitored to make sure that the species is holding its own. Particular attention should be paid to Napa and Russian River populations and to those in the San Joaquin drainage, which have the potential for extirpation in the near future.

References 1. Girard 1854. 2. Ayres 1855. 3. Girard 1856a. 4. Jordan 1879. 5. Jordan and Gilbert 1882. 6. Hopkirk 1973. 7. Avise and Ayala 1976. 8. Mayden et al. 1991. 9. Saiki 1984. 10. Brown and Moyle 1993. 11. Leidy 1984 and pers. comm. 12. Cooper 1983. 13. Moyle and Daniels 1982. 14. Reeves 1964. 15. Knight 1985. 16. Baltz et al. 1987. 17. Cech et al. 1990. 18. Moyle and Baltz 1985. 19. Vondracek et al. 1988a. 20. Hunt et al. 1988. 21. Baltz and Moyle 1993. 22. PG&E 1985. 23. Grant and Maslin 1997. 24. Wales 1946. 25. Moyle et al. 1983. 26. Wang 1986. 27. Moyle et al. 1995. 28. Grant 1992. 29. Jordan and Evermann 1896. 30. Evermann 1905. 31. Rutter 1903. 32. Follett 1937. 33. Murphy 1947. 34. Soule 1951. 35. J. M. Hayes, CDFG, pers. comm 1999. 36. Herbold and Moyle 1986. 37. Gard 1994. 38. Smith 1981. 39. Myrick 1996. 40. Alley 1977a,b. 41. L. Brown, USGS, pers. comm. 1999.

Sacramento Pikeminnow, *Ptychocheilus grandis* (Ayres)

Identification Sacramento pikeminnows are large (potentially over 1 m TL) cyprinids with elongate bodies; flattened, tapered (pikelike) heads; and deeply forked tails. The mouth is large, the maxilla extending behind the front margin of the eye, and is without teeth. The pharyngeal teeth (2,5-4,2) are long and knifelike. There are 8 rays in the anal fin, 8 rays in the dorsal fin, 15–18 pectoral rays, 9 pelvic fin rays, 65–78 scales along the lateral line, 38–44 predorsal scales on the back of the head, and 12–15 scale rows above the lateral line. Large fish are generally a dark, brownish olive on the back and gold-yellow on the belly. Small fish tend to be silvery on the sides and belly and have a dark spot at the base of the

tail. Fins of breeding adults are tinged with reddish orange. Spawning males develop tiny breeding tubercles on the head and a row of tubercles on the side that can extend to the base of the tail.

Taxonomy Despite its wide distribution in California, no distinctive regional forms of Sacramento pikeminnow have been noted, presumably because it is a highly mobile species favoring large streams. The Sacramento pikeminnow is one of four species of *Ptychocheilus*. Others are *P. lucius* in the Colorado River, *P. umquae* from rivers in west-central Oregon, and *P. oregonensis* in the Columbia River basin (1, 2). Within this group Sacramento pikeminnow appears to be most closely related to Colorado pikeminnow. The hardhead is closely related to pike-

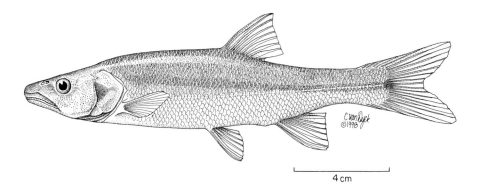

Figure 54. Sacramento pikeminnow, 15 cm SL, Eel River,
Mendocino County.

minnows but is distinct enough to be placed in its own
genus (*Mylopharodon*) (2).

Names Pikeminnow, adopted in 1998 by the American
Fisheries Society, is a replacement for the widely used name
"squawfish." Squawfish is a derogatory name conferred by
early settlers because pikeminnow was a common food fish
of Native Americans and therefore regarded as inferior. Because the name insults Native Americans (and indirectly a
fine fish), its replacement by pikeminnow as the official
common name is highly appropriate. Many other names
have also been applied to the species: Sacramento pike,
chub, whitefish, hardhead, chappaul, bigmouth, boxhead,
and yellowbelly. *Ptychocheilus* means folded lip, "the skin of
the mouth behind the jaws being folded" (3, p. 224); *grandis* means large.

Distribution Sacramento pikeminnows are found in creeks
and rivers throughout the main Sacramento–San Joaquin
River system, Pajaro and Salinas Rivers, Russian River, Clear
Lake basin, and upper Pit River. Sometime before 1975 they
became established in Chorro and Los Osos Creeks (San
Luis Obispo County), tributaries to Morro Bay (4), presumably via an aqueduct connecting these streams with the
upper Salinas River. They have also been transferred via the
California Aqueduct into reservoirs in southern California
(4). In about 1979 they were introduced into Pillsbury
Reservoir in the Eel River and have since spread throughout
the drainage (5, 7).

Life History Sacramento pikeminnows are widespread in
clear rivers and creeks of central California and present in
small numbers in the Sacramento–San Joaquin Delta. They
are largely absent from habitats that are highly turbid or
polluted and tend to be found in low numbers (mainly as
large adults) in lakes and reservoirs that contain centrarchid
basses. They are most characteristic of low- to midelevation
streams with deep pools, slow runs, undercut banks, and
overhanging vegetation. Although they are fairly secretive,
in large pools adults can be observed cruising about during
the day. They are most abundant in lightly disturbed, tree-
lined reaches that also contain other native fishes, especially
Sacramento sucker and hardhead (6). The smaller the
stream, the more likely pikeminnows are to be found only
in pools. Typically during low-flow periods during the day,
pikeminnow greater than 12 cm SL are found in water
deeper than 1 m with mean water column velocity of less
than 40 cm/sec, while smaller fish concentrate in shallower
areas with lower velocities, presumably in part to avoid predation by larger individuals (8, 9, 10, 11).

They generally live in waters with summer temperatures
of 18–28°C (7, 12, 13). Within this range pikeminnows often seek warmer temperatures if other aspects of the habitat are appropriate (12, 13). The maximum (acute) preferred temperature is around 26°C; temperatures above
38°C are invariably lethal (9). Temperatures lower than
38°C may also be lethal if the fish were previously living in
cooler water. Metabolic rates of pikeminnows increase with
temperature (14), although sustained swimming speeds
cannot exceed 2–2.5 body lengths per second (15). While
basically freshwater fish, Sacramento pikeminnows have
been found in Suisun Marsh in salinities as high as 8 ppt, although they are rarely found at salinities higher than 5 ppt.

Juvenile pikeminnows are typically found in small
schools, often mixed with other native cyprinids. The depth
a school selects is related to the size of the fish, because of
the dual threats of heron predation in shallow water and fish
predation in deeper water, although large pikeminnows
rarely pursue small fish during the day. Thus the smallest
fish (<30 mm) are typically found in the shallowest water at
stream edges. Larger fish may also school with other fishes;

I have observed mixed schools of pikeminnow and rainbow trout, all about 20–25 cm long, swimming about in tight formation in the Eel River. Schools of 15- to 25-cm pikeminnows in the Eel can contain several hundred individuals. Large pikeminnows typically cruise about in pools during the day in loose groups of 5–10 fish, although very large individuals may be solitary (11, 16). Often by midday, they become relatively inactive and return to cover (11, 17), although there are generally some still cruising about, feeding on surface insects or benthos (17). The largest fish emerge from cover and begin foraging as darkness falls, entering runs and shallow riffles to forage on small fish (40). Individual fish can move over 500 m during the night before returning to their "home" pools (40). Juveniles, in contrast, will forage actively during the day. The behavior of pikeminnows during colder months is not known, but they apparently seek deep cover (e.g., under submerged trees) that can serve as velocity refuges during high flows (16). Harvey and Nakamoto (40) found that individuals would move downstream 2–23 km to find suitable overwintering habitat but then would move back to their original pools, or to pools nearby, for the summer.

Pikeminnows are capable both of living a sedentary life style and of migrating long distances. In small streams adult pikeminnows may rarely leave a single pool or complex of pools (16, 17). Taft and Murphy (18) observed a tagged pikeminnow in the same pool for 3 years. However, in the Sacramento River pikeminnows move upstream past Red Bluff Diversion Dam during all months of the year; peak numbers (up to 10,000 per month) were typically observed in March, April, and May, when the fish were migrating to spawn (19). Some were tagged in the Delta, indicating an ability to migrate at least 400 km (20, 39). In the Eel River, although most adult fish are sedentary, individuals can move long distances; one radio-tagged pikeminnow was followed for 92 km, moving upstream (40). Most movement takes place at night.

As their pikelike appearance and sharp pharyngeal teeth suggest, pikeminnows are predators on large prey. Before the introduction of other predatory fishes such as largemouth bass, large pikeminnows were undoubtedly at the top of the aquatic food chain throughout the Central Valley. They are opportunists, taking prey on the bottom, at the surface, or in between, depending on type, abundance, and time of day. The size and kind of prey depend on the size of the fish. Pikeminnows under 10 cm SL feed predominantly on aquatic insects, switching to fish and crayfish between 10 and 20 cm (5, 17, 18, 19, 20). In the regulated lower American River, juvenile pikeminnows feed on small aquatic insects, especially corixids (water boatmen) and chironomid midge larvae; they also feed on larval suckers when they are abundant (38). Fish larger than 20 cm SL feed almost ex-

clusively on fish and crayfish, but large stoneflies, frogs, and small rodents have been found in their diets. In small streams the switch to fish may occur at a smaller size if potential prey (including smaller pikeminnows) are abundant (17). In the Eel River in the late 1980s, large pikeminnows fed on novel prey (lamprey ammocoetes, frogs), presumably because they were recent invaders to the system and were finding naïve prey (5). In order to avoid predation by large pikeminnows, California roach, Sacramento suckers, and rainbow trout seek out shallower or faster water than they would in the absence of pikeminnows (7, 21). However, large pikeminnows move into these habitats to forage at night. Curiously, threespine sticklebacks seem to have a hard time changing behavior in the presence of pikeminnows and are likely to co-occur with them only if the stream contains large amounts of dense cover (7, 27).

Pikeminnows in the Eel River forage on outmigrating juvenile salmon in spring, predation also characteristic of large pikeminnows holding below Red Bluff Diversion Dam on the Sacramento River (20). Although pikeminnows may consume large numbers of juvenile salmon, they are likely to have significant impact on salmon populations only where humans have created situations in which the natural ability of salmon to avoid predation is reduced, such as below dams (22) or in locations where pikeminnows are introduced, such as the Eel River (5). At Red Bluff heavy predation on salmon occurs mainly when the dam gates are closed, aggregating pikeminnows and disorienting small salmon in turbulent flows (39). In the Columbia River northern pikeminnow predation below dams is regarded as a major factor contributing to salmon declines, and considerable effort is spent on pikeminnow control, although dams and not pikeminnows per se are the ultimate cause of the problems (23, 24). Under natural conditions pikeminnows feed largely on nonsalmonid fishes such as sculpins (25, 39). The fact that large pikeminnows have low metabolic and digestive rates and that they feed infrequently, especially at low temperatures, also reduces their ability to affect salmonid populations during migrations (26).

Peak feeding usually occurs in early morning (small pikeminnows) or at night (large pikeminnows) (11, 17, 19). Nighttime predation rates at Red Bluff Diversion Dam were apparently enhanced when lights on the dam made prey more visible (20).

Pikeminnows are long lived and slow growing, well adapted to persist through periods of extended drought when reproductive success is low. Growth is usually continuous during the warmer months of the year (17), although it may temporarily cease during periods of drought or in streams that become intermittent (18). For the most part, determining the age of pikeminnows by reading scales is unreliable for older fish, although specimens have been aged at up to 12 years old by this method (28). Using opercular bones, pikeminnows measuring 66 cm SL from the Russian River have been aged at 16 years, suggesting that even older fish may not be unusual (29). Most populations of pikeminnows from rivers and reservoirs show fairly consistent growth rates for their first 5 years or so of life, reaching 50–85 mm SL at the end of their first year, 100–150 mm at the end of their second year, 170–250 mm at the end of their third year, 240–270 mm at the end of their fourth year, and 260–350 mm at the end of their fifth year (5, 16, 17, 28, 30, 39). Growth rates tend to be slowest in small streams and fastest in large, warm rivers. The highest growth rates on record are for the lower Sacramento River: 1.2–1.5 times higher than growth rates elsewhere after the first year (17, 39). There appear to be no differences in growth rates between the sexes. The largest Sacramento pikeminnow known, measuring 115 cm SL and weighing 14.5 kg, was caught in Avocado Lake, Fresno County, in an abandoned gravel pit just off the Kings River.

Sacramento pikeminnows typically become sexually mature at the end of their third or fourth year at 22–25 cm SL; males mature a year earlier than females. They may spawn annually thereafter, but they will not spawn in years when conditions are unfavorable (16, 28). Ripe fish move upstream during April and May (16, 18, 28), although larvae have been collected into July (31). Males usually arrive in the spawning area (gravel riffles or shallow flowing areas at the base of pools) first, when water temperatures rise to 15–20°C. Fish from large rivers or reservoirs usually move into small tributaries to spawn, whereas fish resident in small to medium-size streams typically just move into the nearest riffle (16, 18, 28).

The spawning behavior of pikeminnow has not been recorded in detail, presumably because they spawn largely at night (28). However, it is undoubtedly similar to that of other native cyprinids as well as northern pikeminnow (32). Males congregate in favorable spawning areas and wait for females (28). Any female swimming past a swarm of males is immediately pursued by one to six males. Spawning occurs when a female dips close to the bottom and releases a small number of eggs, which are simultaneously fertilized by one or more males swimming close behind her (32). Fertilized eggs sink to the bottom and adhere to rocks and gravel (31).

Fecundity is high (15,000–40,000 eggs per female, for fish measuring 31–65 cm SL) and related to size, although there is considerable variation in the estimates (16, 28, 33). In northern pikeminnow, the eggs hatch in 4–7 days at 18°C, and fry begin shoaling in another 7 days (33). These events are probably similar for Sacramento pikeminnow because, soon after spawning occurs, shoals of larvae or postlarvae can be observed in shallow pool edges or backwaters, often in association with larvae of other native fishes (31). As the

small fish become more active swimmers, they enter deeper water, especially in runs and along riffles in cover. Juvenile pikeminnows can disperse widely in their first year of life, colonizing stream reaches that have been dried up by drought (27) or made available to them through introduction (5). Young-of-year typically disperse downstream, whereas yearlings are more likely to move upstream (41).

Status IE. Sacramento pikeminnows are still common in central California and have expanded their range into the Eel River basin and creeks flowing into Morro Bay. Although they have become much less abundant in lowland habitats where they were once dominant predators, they have maintained large populations in the Sacramento River, foothill streams, and many regulated streams. When large reservoirs were created by damming Central Valley tributaries, pikeminnows and hardhead colonized the new reservoirs in high enough numbers to be considered a major management problem (34). However, after 10–15 years, the "rough fish problem" quietly went away on its own, presumably because of predation by centrarchid basses on naïve juveniles. Nevertheless, small populations of pikeminnows are still present in many reservoirs dominated by nonnative fishes, such as Pine Flat Reservoir (Fresno County), Anderson Reservoir (Santa Clara County), or Shasta Reservoir (Shasta County). They seem to persist by spawning in tributary streams, where juveniles remain during the vulnerable first 1–2 years of life. Pikeminnows still maintain large populations in hydropower reservoirs, which behave like giant riverine pools and are not drawn down annually (35).

As indicated previously, the ability of Sacramento pikeminnows to be significant predators on juvenile salmon is limited to unusual locations, such as those below Red Bluff Diversion Dam or in the Eel River (5, 22, 39). The degree of predation at Red Bluff Diversion Dam was greatly overestimated (20), resulting in a number of efforts to control pikeminnows. All—including annual "fish-outs" by anglers—failed. At one point an electrocution device, activated by a person viewing through a television camera, was installed in the fish ladder passing over the dam. The idea was to electrocute pikeminnows passing over the dam in order to reduce their population. The device worked for a short while, killing a number of pikeminnows, but then the pulse of migrants abruptly stopped. Apparently, the shocked fish had released fear substance, characteristic of cyprinids, which served to deter fish below the dam from proceeding. The migration was halted for several days, compounding whatever predation problem may have existed, because large fish then accumulated below the dam. The electrocution device was subsequently abandoned (36). The "problem" at Red Bluff Diversion Dam largely disappeared when gates were left open to allow safe salmon passage through the dam, coincidentally allowing pikeminnows to complete their spawning migration as well.

If the predatory nature of Sacramento pikeminnows gives them an undeservedly bad reputation, it also confers on them sporting qualities (18, 33, 37) recognized by every angler who hooks one (until he or she discovers that the struggling fish is not a trout or a bass). The culinary qualities of large pikeminnows are also underappreciated, although they fetch a good price in oriental markets and, like common carp, are excellent eating when properly prepared. More importantly, pikeminnows are a key component in many stream ecosystems and are fascinating to watch, cruising elegantly about their summer pools.

References 1. Carney and Page 1990. 2. Mayden et al. 1991. 3. Jordan and Evermann 1896. 4. Swift et al. 1993. 5. Brown and Moyle 1996. 6. Moyle and Nichols 1973. 7. Brown and Moyle 1993. 8. Moyle and Baltz 1985. 9. Knight 1985. 10. Brown and Moyle 1981. 11. Alley 1977a,b. 12. Baltz et al. 1987. 13. Dettman 1976. 14. Myrick 1996. 15. Cech et al. 1990. 16. Grant 1992. 17. Brown 1990. 18. Taft and Murphy 1950. 19. Bureau of Reclamation 1983. 20. B. Vondracek, J. A. Hanson, and P. B. Moyle, unpubl. ms. 1983. 21. Brown and Brasher 1995. 22. Brown and Moyle 1981. 23. Reiman and Beamesderfer 1990. 24. Beamesderfer et al. 1996. 25. Buchanan et al. 1981. 26. Vondracek 1987. 27. Smith 1982. 28. Mulligan 1975. 29. Scoppetone 1988. 30. Moyle et al. 1983. 31. Wang 1986. 32. Patten and Rodman 1969. 33. Burns 1966. 34. Dill 1938. 35. Vondracek et al. 1988a. 36. K. Marine, pers. comm. 37. Jordan and Evermann 1923. 38. Merz and Vanicek 1996. 39. Tucker et al. 1998. 40. Harvey and Nakamoto 1999. 41. J. J. Smith, San Jose State University, pers. comm. 1998

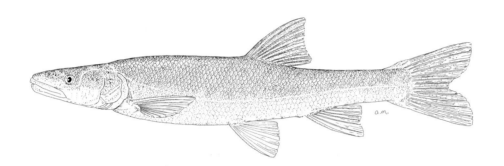

Figure 55. Colorado pikeminnnow, 35 cm SL, Green River, Wyoming. Drawing by A. Marciochi.

Colorado Pikeminnow, *Ptychocheilus lucius* Girard

Identification Colorado pikeminnow are large (up to 2 m), small-scaled cyprinids with elongate bodies, flattened, tapered (pikelike) heads, and deeply forked tails. Their scales are embedded, and there are usually more than 80 in the lateral line (76–97) and 18–23 rows above the lateral line. The toothless mouth is large and horizontal, the maxilla extending behind the front margin of the eye. The pharyngeal teeth (2,5-4,2) are long and knifelike. There are 9 rays in the anal fin and 9 in the dorsal fin, 14–16 pectoral fin rays, and 8–10 pelvic fin rays. The body tends to be silvery, but larger fish become dark on the back and white to yellow on the sides and belly. Juveniles are bright silvery on the sides and belly and have a dark spot at the base of the tail. Breeding adults are silvery on the sides, flecked with gold, and creamy on the belly. Spawning males develop tiny breeding tubercles on the head and a row of tubercles on the side that can extend to the tail.

Taxonomy See the account of Sacramento pikeminnow.

Names The trivial name *lucius* means pike, referring to the superficial resemblance of pikeminnow to true freshwater pikes (Esocidae). Jordan and Evermann (1) listed its common name as "white salmon of the Colorado" or "whitefish." Other names, including the replacement of the common name "squawfish" with "pikeminnow," are discussed in the account of Sacramento pikeminnow.

Distribution Colorado pikeminnows were once common in the Colorado River and its major tributaries from Wyoming (Green River), through Utah, Colorado, Arizona, New Mexico, Nevada, California, and Mexico. Today they are absent from the lower Colorado River (unless introduced from hatcheries) below Powell Reservoir and are largely confined to the Green River and its tributaries (especially the Yampa River), the upper mainstem Colorado River from Powell Reservoir to Palisades, Colorado, and the San Juan River below Navajo Reservoir, New Mexico (2).

Life History The Colorado pikeminnow is a big-river species. Large adults are (or were) found in turbid, silt-laden waters of the Colorado River and in large pools of its tributaries (3). Construction of a series of large impoundments destroyed much of this habitat and put the Colorado pikeminnow in danger of extinction. These events led to the fish being intensively studied—an expensive proposition considering how few are left. The results of these studies are summarized by Tyus (2) and Minckley (3), and these are the principal sources of information used in this account.

Adult Colorado pikeminnows move about actively in fairly large reaches of river (at least 5 km for a home range) but tend to spend much of their time near shore or in backwaters, where currents are slower and prey are abundant. Smaller (<40 cm SL) fish also frequent quiet waters at the river's edge or shallow pools with sand or silt bottoms (4), with the smallest most likely found in quiet backwaters. However, juveniles will move in and out of backwaters to other shallow-water habitats in response to rising and falling temperatures; they seem to prefer backwaters when they are warmer than the river itself (5). When food or habitat quality is poor, juveniles will move considerable distances (10 km or more) upstream to new areas (16). While Colorado pikeminnows naturally encounter seasonal temperature ranges of perhaps 4–30°C, optimal temperatures for swimming are 20–26°C, and those for growth are around 25°C (6). Colorado pikeminnows have moderately high salinity tolerance, surviving levels up to 12–14 ppt (about one-third the salinity of seawater) (7). Historically, low temperatures and high salinities were rarely limiting factors, but they probably are of major importance today; releases from dams have made much of the upper river colder and clearer, and irrigation return water has increased the salinity of the lower river.

Colorado pikeminnows were once the top aquatic carnivores at all life stages. Fish measuring less than 50 mm TL feed mostly on cladocerans, copepods, and chironomid midge larvae (4, 8). Aquatic insect larvae are the major food of fish measuring 50–100 mm TL; fish, especially other minnows, become increasingly important in the diet for individuals larger than 100 mm TL (4). Pikeminnow larger than 200 mm TL feed almost exclusively on other fish, but they will consume anything else that moves in or on the water, from large terrestrial insects (e.g., Mormon crickets) to small birds (2). Originally, their principal prey was the various species of suckers and chubs (*Gila*) that lived with them. Today they also consume abundant alien fishes (16) and occasionally get catfish, spines erected, lodged in their throats—with fatal consequences for both predator and prey. Feeding is sporadic. Vanicek and Kramer (4) found that 39 percent of the foreguts of large pikeminnows they examined were empty.

The Colorado pikeminnow is the largest cyprinid in North America, but the maximum size is open to debate. It is usually given as 1.8 m TL and 45 kg (2, 3, 9), which sounds considerably more precise than the "measurements" of 6 ft and 100 lb in earlier accounts (8). Accounts of pikeminnow over 1.1 m TL are old and anecdotal, although fish estimated as large as 1.5 m TL are known from an archaeological site (3). The largest fish caught in recent years of intensive sampling measured 96 cm TL (about 10 kg) (10). Adults typically measure 55–65 cm TL.

Colorado pikeminnows take a long time to reach large sizes. Individuals in the population studied by Vanicek and Kramer (4) averaged 44 mm TL at the end of the first year, 95 mm at the end of the second, 162 mm at the end of the third, 238 mm at the end of the fourth, 320 mm at the end of the fifth, and 391 mm at the end of the sixth, after which the fish mature. Using scales, a pikeminnow measuring 61 cm TL was determined to be 11 years old (4), a result that fits with the ages of fish of similar size raised in captivity. Using otoliths, pikeminnows have been aged up to 30 years. The largest and oldest fish are presumably females.

Colorado pikeminnows mature at 43–50 cm TL. They can make long migrations (over 200 km) to spawn in the same areas year after year. Migration in the upper Colorado begins in early summer, presumably in response to falling water levels, and spawning takes place in late June to early August after temperatures exceed 18°C, usually 20–22°C. Preferred spawning grounds are swift rapids in deep canyons, perhaps because potential egg predators are fewer there (2). Spawning success is highest in years when there are high spring flows, resulting in strong year classes that may dominate a population for years (15).

Spawning fish rest in pools or side eddies and then move abruptly into fast water to release eggs and sperm, with many males surrounding each female. Fertilized eggs adhere to rocks and gravel and hatch in 3–6 days. Larvae drift quickly and wind up in suitable rearing habitats 100–250 km downstream. After spawning, adults often follow the young downstream, returning to their original home ranges.

Young pikeminnows inhabit shallow edge habitats and small backwaters left behind by receding waters of summer, where they grow rapidly in response to abundant food and warm temperatures. Unfortunately, these same habitats are favored by alien fishes such as red shiner, which prey on the larvae (11).

Status IA. Extirpated in California and reduced to about 25 percent of its native range elsewhere, and to a small fraction of its original numbers (2). The Colorado pikeminnow is listed as endangered by both federal (1967) and California (1971) governments. Pikeminnows were once abundant in the lower Colorado River but by the early 1960s were probably extinct there. The last pikeminnow below Glen Canyon Dam (Arizona) was recorded in 1975 (3). Their disappearance from the lower river and rarity in the upper river are largely the result of drastic changes caused by large dams built in recent years (12). Neither the extensive reservoirs behind dams nor the cold, clear water flowing from them provides habitat appropriate for pikeminnow. In addition, dams block spawning migrations, curtailing reproduction. In habitats that remain suitable for Colorado pikeminnow, abundant alien fishes now prey on larvae and juveniles, and possibly compete with them for food. Some of these (e.g., catfish) may be unsuitable as prey for adult pikeminnow. A recovery plan has been written, revised, and implemented for Colorado pikeminnow. A key part of the recovery effort has been a major research program to determine limiting factors in order to ascertain which habitats need to be protected and enhanced. Because it is unlikely that any major dams on the Colorado River will be torn down, various experimental flow release programs are being tried to improve habitats in interdam reaches. Long-term survival, however, will probably depend on maintenance of relatively natural flow regimes in major upstream tributaries, such as the Green and Yampa Rivers (3, 16).

Part of the recovery program has been the breeding of Colorado pikeminnows in captivity at the Dexter National Fish Hatchery (New Mexico) and the release of thousands of juveniles into the watershed, including rivers from which they have been extirpated. Some fish have survived; yet the ability of introduced populations to become self sustaining is problematic, unless habitats are substantially improved and alien fishes removed. In addition, the ability of pikeminnows to reestablish the complex movement patterns needed for completion of their life history is questionable (2). Even with an extensive hatchery program and good intentions, it is unlikely that breeding populations of pikeminnow will become reestablished in the California portion of the Colorado River as long as poor habitat conditions persist and alien fishes that prey on larvae are present.

The story of the decline of Colorado pikeminnow is filled with irony, in particular the contrast of its status with that of northern pikeminnow. The northern pikeminnow has thrived in altered conditions created by dams on the Columbia River and is now subject to major "control" programs to reduce predation on juvenile salmon. One program has paid millions of dollars in bounties to anglers to kill large pikeminnows (13). The Colorado pikeminnow was itself subject to an eradication effort when, in 1962, 715 km of the Green River and its tributaries were poisoned with rotenone to eradicate "nongame" fishes that might have had adverse effects on trout fisheries in the soon-to-be filled Flaming Gorge Reservoir. The effects of the rotenone actually extended considerably farther downstream than intended, killing fish in the waters of Dinosaur National Monument. The operation was largely unsuccessful, but it apparently did eliminate some populations of native fishes above the reservoir (14). The controversy ignited by the huge operation created the first public awareness that native fishes of the Colorado River were in a serious state of decline and helped set the stage for future conservation efforts. A final irony: one of the items commonly fed captive brood stock of Colorado pikeminnow is hatchery-reared rainbow trout.

References 1. Jordan and Evermann 1923. 2. Tyus 1991a. 3. Minckley 1991a,b. 4. Vanicek and Kramer 1969. 5. Tyus 1991b. 6. Black and Bulkley 1985. 7. Nelson and Flickinger 1992. 8. Muth and Snyder 1995. 9. Jordan and Evermann 1896. 10. Young 1991. 11. Ruppert et al. 1993. 12. Ohmart et al. 1988. 13. Beamesderfer et al. 1996. 14. Holden 1991. 15. Osmundson and Burnham 1998. 16. Osmundson et al. 1998.

Speckled Dace, *Rhinichthys osculus* (Girard)

Identification The speckled dace is a small (usually less than 8 cm SL, occasionally to 11 cm SL), highly variable species distinguished by a thick caudal peduncle, a small subterminal mouth, a pointed snout, and small scales (47–89 in lateral line). The origin of the dorsal fin (6–9 rays, usually 8) is well behind that of the pelvic fins. The anal fin normally has 7 rays (6–8). The pharyngeal teeth (1,4-4,1 or 2,4-4,2) are strongly hooked and have only a slight grinding surface. Usually there is a tiny barbel at the end of each maxilla, and a small frenum (bridge of skin) often attaches the snout to the middle of the upper lip (premaxilla). Color is highly variable, but most fish over 3 cm have dark speckles on the back and sides, dark blotches on the side that often coalesce to resemble a dark lateral band, a spot at the base of the caudal peduncle, and a stripe on the head that runs through the snout. The background color on the back and sides is dusky yellow to dark olive, with the belly yellowish to whitish. The bases of the fins of both sexes turn orange to red during breeding, and males often have red snouts and lips as well. Males usually develop tubercles on the pectoral fins and head.

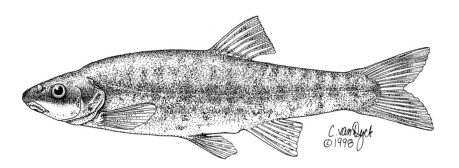

Figure 56. Speckled dace, 6.7 cm
SL, Johnson Creek, Modoc County.

Taxonomy The genus *Rhinichthys* is found in almost every drainage of North America, yet contains only eight recognized species, often in abundance. Their wide distribution reflects their ability to colonize new areas through headwaters, as well as their ability to adapt to new environments. Most species are highly variable and may represent complexes of species that are not yet recognized.

In the western United States no native fish species is as widely distributed or occupies such a wide variety of habitats as the speckled dace. Its adaptability is reflected in the variability of its body shape. Springs and slow streams may support small, chunky forms, whereas fast-moving streams support large, streamlined forms. The degree to which these distinctive morphological characters are fixed genetically or are plastic, capable of changing with the developmental environment, is unknown. Many different forms were described by early taxonomists and then later abandoned as the variable nature of the species became known. Jordan and Evermann (1), for example, divided this complex into 12 species, which have subsequently been reduced to one (2). Until modern molecular and morphometric techniques can resolve the relationships among the various forms, subspecies will continue to be recognized as a rule according to the regions in which they occur, although there are undoubtedly isolated populations within regions that also merit special taxonomic designation. There is at least some support for considering the following forms in California to be at least subspecies.

Lahontan speckled dace. *R. o. robustus* is widely distributed in streams and lakes of the northeastern Sierra, including the Walker, Carson, Truckee, Honey Lake, and Eagle Lake drainages. It conforms well to the general description of speckled dace, as do the following two forms.

Klamath speckled dace. *R. o. klamathensis* is found throughout the Klamath drainage, mainly in streams.

Sacramento speckled dace. The name *R. o. carringtoni*, applied to this dace, actually belongs to a form from the Snake River in Utah (38). Currently without formal description, it is found in streams throughout the Sacramento drainage and, historically, the western San Joaquin drainage as well. Dace in the Pit River and in streams tributary to Monterey Bay are also placed within this subspecies until definitive studies can be done.

Owens speckled dace. This dace and the Long Valley speckled dace are undescribed subspecies in the Owens drainage that are recognized by D. Sada on the basis of morphological and genetic analyses (3, 4, 5). The Owens speckled dace is found in the Owens River and its tributaries and seems distinct from the nearest other population in the Amargosa River.

Long Valley speckled dace. This small form is found only in Whitmore Spring and Little Alkali Lake in Long Valley in the Owens drainage. Like the Owens speckled dace, its closest relatives are dace found in Death Valley, which in turn are derived from dace in the Colorado River drainage (5).

Amargosa speckled dace. Gilbert (6) described *R. nevadensis* from Ash Meadows, Nevada, but the subspecific name *R. o. nevadensis* has been assigned to speckled dace in the Amargosa River canyon and Owens Valley as well (7). Dace from the Amargosa River in California differ somewhat morphologically from those in Ash Meadows. The former are characterized by a comparatively shallower head depth, a shorter snout-to-nostril length, a longer anal-caudal length, more pectoral fin rays, and fewer vertebrae, indicating that the two populations may be distinct (8, 9). However, genetic evidence for their separation is ambiguous, so I follow Sada (5) in referring to them all, including California populations, as *R. o. nevadensis*.

Santa Ana speckled dace. Morphological analyses (10) suggest that dace in southern California streams warrant subspecies status (11). Preliminary electrophoretic studies seem to confirm that Santa Ana speckled dace are distinctive (39), but the subspecies has yet to be formally described. These studies also indicate that this dace appears to be more closely related to dace of the Colorado River drainage than to northern populations.

Other forms. There are isolated populations of speckled dace in a number of places in California (e.g., the Cowhead Lake drainage and Surprise Valley, Modoc County) that may also merit special recognition but have simply not yet been examined closely. The population of dace in San Luis Obispo Creek in south-central California has been listed as a distinct taxon (12), but it may well have resulted from an introduction.

Names Speckled dace have a variety of unofficial common names, all of which include the word *dace:* western dace, Pacific dace, spring dace, dusky dace, and so on. The word *dace* is derived from the same Middle English word that gave rise to *dart* and was originally applied only to *Leuciscus leuciscus,* a lively European cyprinid. *Rhin-ichthys* means snout-fish; *osculus,* kissing, refers to the small flexible mouth. The history of scientific nomenclature for *R. osculus* is complicated. However, the generic name used in older literature is most often *Agosia* or *Apocope;* the species name is usually a variant of one of the names now used to designate subspecies (7, 10).

Distribution Speckled dace are the only fish native to all major Western drainage systems from Canada south to Sonora, Mexico. In California they are native to the Amargosa River (Death Valley); Owens Valley; eastern Sierra drainages from the Walker River north to Eagle Lake; the Surprise Valley and Cowhead Lake drainages; the Klamath-Trinity basin; the Pit River drainage, including the Goose Lake watershed; the Sacramento drainage as far south as the Mokelumne River; the San Lorenzo, Pajaro, and Salinas Rivers; San Luis Obispo, Pismo, and Arroyo Grande Creeks; the Morro Bay drainage; and the San Gabriel and Los Angeles river basins (12). They may also be present in headwaters on the west side of the San Joaquin Valley (e.g., Los Gatos Creek), but their presence there has not been confirmed. They are absent from the Clear Lake basin, the Russian River, and most small coastal drainages, as well as from the San Joaquin drainage. They are currently missing from the Cosumnes River drainage although present in watersheds on both sides of it. They are also absent from the lower Colorado River, although a single larval dace has been reported (13, 41). In the mid-1980s speckled dace were introduced by persons unknown into the Van Duzen River, a tributary to the Eel River, and it is likely they will

eventually spread throughout the watershed (14). An introduced population, presumably of Lahontan speckled dace, is also present in the headwaters of the North Fork Mokelumne River, from which it may eventually spread and contact the native Sacramento dace, which is present at lower elevations.

In some watersheds speckled dace may be limited to only small reaches of suitable habitat. Thus in the Pajaro River drainage they are presently found only in the San Benito River (15). Such limited distributions make populations prone to extinction, as happened to the only known populations in the San Francisco Bay drainage (16) and is happening to populations in southern California and the Owens Valley. Thus the species was likely more widely distributed in the past in California than it is today.

Life History Speckled dace occupy an extraordinary array of habitats: small springs, rushing brooks, pools in intermittent streams, large rivers, and deep lakes. Yet most of these habitats have a number of characteristics in common: clear, well-oxygenated water; abundant deep cover (rocks, submerged aquatic plants, overhanging vegetation, woody debris); and moving water from stream currents, wave action, or spring outflows (15, 17, 18, 19, 20, 21). Dace are generally small-stream (second- and third-order) specialists. They thrive in shallow (<60 cm), rocky riffles and runs, where they actively browse among rocks and plants. Their numbers may actually increase in streams that have been channelized or reduced in flow because of an increase in the extent of the shallow riffle habitat they prefer (15, 18). In some streams their ability to use their preferred riffle habitat is restricted by sculpins, which are also benthic insect feeders and compete for space (20). They often are most abundant in streams where sculpins are absent.

In lakes they live among the rocks, mostly in the zone stirred up by wave action (<1 m deep), although in Lake Tahoe they are common down to 8 m and have been taken there as deep as 61 m (22). In Eagle Lake, Lassen County, they are found among the rocks during the day but are common along sandy beaches at night (23). Dace adapted to warm water are tolerant of fairly high temperatures. In Owens Valley and in the Amargosa River they will live in water to 28–29°C (17). In intermittent streams in Arizona dace survive temperatures as high as 31°C and daily fluctuations of 10–15°C (24). On a seasonal basis dace may live in even more extreme conditions. Smoke Creek (Lassen County) supports a dace population that experiences water temperatures ranging from 0°C in winter (with anchor ice covering up to 95% of the substrate) to 29°C in summer (35). In the laboratory Klamath speckled dace can survive temperatures of 28–34°C and dissolved oxygen levels as low as 1 mg/liter (36). If conditions become too extreme and local populations are eliminated or greatly depressed by floods, droughts, or winter freezing, dace have remarkable abilities to recolonize or repopulate areas (34, 35).

Speckled dace are seldom found singly, yet they avoid forming conspicuous shoals except during breeding season. Typically small groups forage among the rocks as loose units. In Lake Tahoe and Eagle Lake they are most active at night, spending the day quietly among rocks or vegetation or in deep water (23, 25). In the Trinity River they have been reported as being most active both at night (26) and during the day (27). My own studies on streams in Lassen and Modoc Counties indicate that their nocturnal habits are strongly related to their vulnerability to bird predation. In Ash and Pine Creeks, where avian predators are scarce, dace are most active during the day (23, 28), whereas in Willow Creek, where a wide variety of avian predators are active because of the creek's proximity to Eagle Lake, dace are strongly nocturnal (23). Lake Tahoe dace become inactive in winter, although they do remain in shallow, rocky areas (22). In streams, however, they may be active all year if temperatures do not become too low (<4°C).

In general, speckled dace can be characterized as bottom browsers on small invertebrates, especially those taxa found in riffles, such as the larvae of hydropsychid caddisflies, baetid mayflies, and chironomid and simuliid midges (20, 23, 27, 28, 29). This feeding preference is reflected in their subterminal mouth, pharyngeal tooth structure, and short intestine. However, in lakes they feed opportunistically on large flying insects at the water's surface and on zooplankton (23, 25). Diet changes with season, reflecting prey availability. In the Trinity River in winter, the dominant food was chironomid larvae, with occasional mayfly and stonefly nymphs (30). The nymphs became dominant in the spring, yielding to emerging insects in summer. In the fall filamentous algae was important. A similar pattern was observed in Ash Creek, Lassen County (28), and Willow Creek, Humboldt County (27).

Age and growth have been determined primarily from length frequency analyses. Dace reach 20–30 mm SL by the end of their first summer (23, 27, 30), and in subsequent years they add, on average, 10–15 mm/year to their length, females growing slightly faster than males. However, growth can be reduced by many factors, especially severe environmental conditions, high population densities, or limited food availability (35). In most streams few fish survive more than 3 years or exceed 85 mm FL (23, 30). The largest dace I have encountered were 111 mm SL from Blue Creek, tributary to the Trinity River. In Lake Tahoe the largest fish recorded is 85 mm FL, but there seem to be five or six age classes (22).

Dace usually mature in their second summer. Fecundities of 11 dace (45–59 mm SL, mean 54 mm) from Pine Creek, Lassen County, ranged from 192 to 790 eggs, with a mean of 441 (23). Six dace of similar size (mean, 54 mm

SL) from nearby Willow Creek had a mean fecundity of 265 eggs (range, 195-370). Speckled dace can spawn throughout summer, but most such activity occurs in June and July, probably induced by rising water temperatures (30). In intermittent streams spawning may be induced by high-flow events (31). In lakes shoals of dace seek out shallow areas of gravel for spawning, or else migrate a short distance up inlet streams, where spawning occurs primarily on the gravel edges of riffles. Males congregate in a small area, from which they remove algae and detritus, leaving a bare patch of rocks and gravel. When a female enters she is immediately surrounded by a knot of males. The female wriggles the rear portion of her body underneath a rock or close to the gravel surface and releases a few eggs, while the males release sperm (31). The eggs sink into interstices and adhere to rocks. Embryos hatch in about 6 days (at 18–19°C), and larval fish remain in the gravel for 7–8 days (31). Speckled dace hybridize with Lahontan redside (32), presumably because both occasionally spawn at the same time and place.

After emerging, fry tend to concentrate in warm shallows, especially in channels between large rocks or among emergent vegetation. In Lake Tahoe, fry along with those of other cyprinids, move into shallow nursery areas, usually quiet swampy coves with an accumulation of floating debris. Scales first appear at 13 mm FL (30).

Status IB–E. Variable depending on subspecies or population. Widely distributed forms in major drainages are not in trouble, but most forms with limited distributions in arid areas are in danger of extinction, as are isolated populations of widely distributed subspecies. Speckled dace persist in an area as long as it has cool, flowing water; permanent pools; and a shortage of nonnative predators.

Lahontan speckled dace. IE. Abundant and widely distributed, although its populations can be depressed or eliminated by predatory alien brown trout (33).

Klamath speckled dace. IE. Abundant and widely distributed.

Sacramento speckled dace. IE. Abundant and widely distributed in the Sacramento and Pit Rivers. Its distribution is limited in the Pajaro and Salinas drainages, but it is common in the San Lorenzo River. It has apparently been extirpated from San Joaquin Valley streams and the Cosumnes River but its historic distribution is poorly known. The Salinas River population may be the source of fish present in San Luis Obispo Creek and Cuyama River, perhaps through introductions (37, 40).

Owens speckled dace. IB. The Owens dace has had its range greatly restricted by the introduction of alien trouts and water development. It is currently found in only a few scattered localities, including some irrigation ditches (3, 4, 5) and is in danger of extinction.

Long Valley speckled dace. IB. This newly discovered form is in danger of extinction because of an extremely limited habitat in a small part of Owens Valley (3, 4, 5).

Amargosa speckled dace. IB. Because it is confined to a few miles of desert stream (the Amargosa River in Amargosa Canyon, plus its tributary Willow Creek), this speckled dace is threatened with extinction by withdrawal of water from aquifers that feed the river. This water is being used to meet the needs of the ever-thirsty city of Las Vegas, as well as of local farms and towns. The extinction of this species will indicate that another unique desert aquatic ecosystem has been irretrievably lost (17).

Santa Ana speckled dace. IB. This form was petitioned for listing as a federal endangered species in 1994, but the petition was denied because it had not yet been formally described. Its range has been dramatically diminished (to a few headwaters of the San Gabriel, Los Angeles, and Santa Ana Rivers) by urban spread in the Los Angeles re-gion. Its extinction is likely unless it receives special protection (12, 17).

References 1. Jordan and Evermann 1896. 2. Hubbs et al. 1974. 3. Sada 1989. 4. Sada et al. 1993. 5. Sada et al. 1995. 6. Gilbert 1893. 7. La Rivers 1962. 8. C. Williams et al. 1982. 9. Deacon and Williams 1984. 10. Cornelius 1969. 11. Hubbs et al. 1979. 12. Swift et al. 1993. 13. Minckley 1973. 14. Brown and Moyle 1996. 15. Smith 1982. 16. Leidy 1984. 17. Moyle et al. 1995. 18. Moyle and Daniels 1982. 19. Moyle and Baltz 1985. 20. Baltz et al. 1982. 21. Moyle and Vondracek 1985. 22. Baker 1967. 23. P. B. Moyle and students, unpubl. rpts. 24. John 1964. 25. R. G. Miller 1951. 26. Moffett and Smith 1950. 27. Hiss 1984. 28. Li and Moyle 1976. 29. Moyle et al. 1991. 30. Jhingran 1948. 31. John 1963. 32. Calhoun 1940. 33. Strange et al. 1992. 34. Pearsons et al. 1992. 35. Sada 1990. 36. Castleberry and Cech 1993. 37. C. C. Swift, pers. comm. 1999. 38. C. L. Hubbs, pers. comm. 1974. 39. T. R. Haglund, University of California, Los Angeles, pers. comm. 1996. 40. Jordan 1894. 41. Winn and Miller 1954.

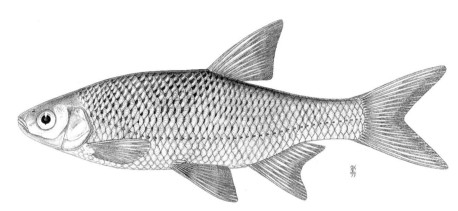

Figure 57. Golden shiner, 101 mm SL, Maryland. From Lee et al. (1980).

Golden Shiner, *Notemigonus crysoleucas* (Mitchill)

Identification Golden shiners are readily recognized by a deeply compressed body, a small head with a pointed snout and upward-pointing mouth, a strongly decurved lateral line, large deciduous scales, and a sharp, scaleless keel on the belly between the pelvic fins and the anus. The lateral line curves downward from the head (decurved) and has 44–54 scales. There are 7–9 (usually 8) rays in the dorsal fin, which has its origin behind that of the pelvic fins. The anal fin usually has 11–14 rays (range, 8–19); the pelvic fins, 9; and the pectoral fins, 15. The pharyngeal tooth formula is 0,5-5,0. Golden shiners usually measure less than 15 cm SL but occasionally reach 30 cm SL. They typically have a golden sheen to their scales, although a silvery color is also common, especially in smaller fish. The fins are generally colorless and lack dark basal spots. A faint dark stripe appears on the sides of live fish. Breeding males have a red-orange tinge on the pelvic and anal fins.

Taxonomy The golden shiner is one of the most distinctive small minnows in North America from a biochemical perspective, and it is often used as an "outgroup" for comparative taxonomic studies of North American cyprinids. Its relationship to other cyprinids is obscure, but it may be more closely related to Eurasian cyprinids than to other North American species.

Names *Notemigonus* means angled back, referring to the fish's angular body shape; *crysoleucas* is a combination of the Greek words for gold and white.

Distribution Golden shiners are native to most of eastern North America, including the Mississippi River system. They occur as far north as Quebec and as far south as Texas and Florida. Shiners from Illinois were introduced into Cuyamaca Reservoir (San Diego County) and the Feather River in 1891, and the Cuyamaca population was subsequently used by the California Fish Commission as a source of fish to plant as forage in other localities around the state (1). However, they did not become widely distributed until after 1955, when it became legal to use and raise golden shiners commercially as bait. Scattered populations exist in many waters throughout the state, and where they are not established individuals can often be found where careless anglers have dumped leftover bait. It is often difficult to tell if the few individuals found in many areas represent permanent or temporary populations.

Life History Golden shiners live primarily in warm, shallow ponds, lakes, and sloughs, where they are associated with beds of aquatic vegetation (2). In the Pit River, for example, they are abundant in a sluggish, highly turbid, muddy-bottomed reach in Big Valley but are rare or absent in faster-flowing sections above and below (11). They can tolerate temperatures up to 36–37°C and dissolved oxygen concentrations of <1 mg/liter (3). Often they are most abundant in low-elevation reservoirs and sloughs with other introduced fishes, such as largemouth bass, various species of sunfish, and mosquitofish (4). They are abundant in Ruth Reservoir (Trinity County), where mean summer temperatures are around 22°C and dissolved oxygen levels are fairly high; this reservoir also supports an assortment of other warmwater fishes, plus planted rainbow trout (5). Golden shiners occasionally become established in coldwater lakes (e.g., Dutch Lake, Fresno County, at 2774 m elevation), but they are likely to persist only if there are warm, shallow areas for breeding and rearing of young.

As the compressed body shape, deeply forked tail, and upturned mouth indicate, golden shiners are active fish that feed mostly on the surface or in midwater (6). Their triangular pelvic and pectoral fins give them considerable maneuverability, enabling capture of small swimming organisms with some precision. In lakes, golden shiners can switch rapidly between individually picking large zooplankters, such as *Daphnia,* to filter feeding on small zooplankters (16). This flexibility allows them to exploit a wider range of prey than many fishes of similar size. Zooplankton, particularly cladocerans like *Daphnia,* are the most important food for golden shiners of all sizes, followed closely by small flying insects taken at the water's surface. For example, *Daphnia* were the principal prey of golden shiners in Davis Reservoir, Plumas County (7). In a small coldwater lake (Castle Lake, Siskiyou County) shiners fed primarily on aquatic insects in shallow water when predation risk from rainbow trout was high, but fed more on *Daphnia* in open water when trout populations were reduced (15). Larger individuals occasionally take small fish, molluscs, and aquatic insect larvae. When animal food is in short supply, filamentous algae can be found abundantly in their stomachs.

Golden shiners are sight feeders and so are usually most active during the day. They are shoaling fish that form tight schools in littoral or pelagic areas when predator avoidance is a high priority. In such situations they may become nocturnal feeders, moving offshore after dark (12, 15). Golden shiner numbers in lakes may be regulated by piscivorous fishes (10).

Golden shiners grow faster in warm waters than in cold (13). In lowland California ponds they can reach 76 mm TL in one year; in higher, colder waters they reach only 36–46 mm TL. By the end of their second year they can reach 140 mm TL, after which growth slows down somewhat. In Ruth Reservoir, Trinity County, they averaged 56 mm FL at the end of their first year, 93 mm at the end of their second year, 116 mm at the end of their third year, 127 mm at the end of their fourth year, 140 mm at the end of their fifth year, 154 mm at the end of their sixth year, and 163 mm at the end of their seventh year (13). Females generally grow faster and achieve larger size than males, although in some situations their growth rates are similar (13). The maximum age recorded for golden shiners is 9 years, and the maximum length is about 260 mm SL (8).

The spawning season for golden shiners lasts from March through September in California, the exact time depending on water temperature. In a coolwater reservoir, shiners spawned from early June through early September, peaking in early July (13). Spawning usually begins when water temperature reaches about 20°C, although it has been recorded in water as low as 14°C; it rarely occurs above 27°C (2). Shiners spawn in shoals early in the morning. They are fractional spawners with fecundities at the beginning of spawning of 2,700–4,700 eggs or more (2). Each female deposits her adhesive eggs on submerged vegetation and bottom debris, where they are fertilized immediately by one or more males trailing close behind (2). Occasionally, active nests of largemouth bass are selected as spawning sites. Survival of eggs and larvae may actually be higher in this situation, presumably because the adult bass protects the nest (8). Embryos hatch in 4–5 days at 24–27°C (2). Newly emerged fry school in large numbers close to shore, often in association with aquatic plants. Initially, larvae feed primarily on small rotifers and epiphytic algae (especially diatoms), but they gradually switch to small crustaceans (14).

Status IIE. Golden shiners are extensively propagated as a baitfish in California. Consequently, they are introduced throughout the state, with unknown effects on native fish and fisheries. In coldwater lakes, they can reduce zooplankton populations and thus reduce growth and survival of trout. Of the three legal bait minnows in California (golden shiner, fathead minnow, and red shiner), golden shiners seem least able to establish large, permanent populations in streams and natural lakes, although they do so readily in reservoirs. In natural situations their populations seem to be largely eliminated by predatory fishes with which they co-occur in California reservoirs (9). Unfortunately, it is difficult to predict situations in which golden shiner popu-

lations will become established and pose a problem. Therefore, bait fishing with golden shiners and other minnows should ideally be banned in California. At the very least, golden shiners used as bait should be restricted to fish raised in the state. This would prevent the introduction of the rudd (*Scardinius erythrophthalmus*), a European minnow similar to golden shiner that is sometimes sold in the eastern United States and may be found in bait shipments of golden shiners. The rudd is in the process of becoming widely distributed in the eastern United States.

References 1. Dill and Cordone 1997. 2. Becker 1983. 3. Smale and Rabeni 1995. 4. Moyle and Nichols 1973. 5. Vigg and Hassler 1982. 6. Keast and Webb 1966. 7. Erman et al. 1983. 8. Carlander 1969. 9. Kramer and Smith 1960. 10. Johannes et al. 1989. 11. Moyle and Daniels 1982. 12. Hall et al. 1979. 13. Kisanuki 1980. 14. Hatch 1988. 15. Elser et al. 1995. 16. Ehlinger 1989.

Figure 58. Fathead minnow, breeding male, 55 mm SL, Maryland. From Lee et al. (1980).

Fathead Minnow, *Pimephales promelas* Rafinesque

Identification Chunky fish seldom exceeding 85 mm TL, fathead minnows can be distinguished by their thickened first dorsal fin ray; small, slightly oblique mouth; and crowding of scales behind the head. The head is short, blunt, and broad on top. The lateral line seldom extends beyond the anterior half of the body. There are 44–54 scales in the

lateral series. Dorsal rays are 8; pelvic fin rays, 8; anal fin rays, 7; and pharyngeal teeth 0,4-4,0, with oblique grinding surfaces. The intestine is 2–3 times the body length, and the peritoneum is black. The back is usually dark, tending toward brown or olive, with scales outlined by pigment; the sides are dull and dusky, often with the black peritoneum showing through. Small fish or individuals from turbid waters may be pale whitish to silvery. Breeding males have conspicuous tubercles on the snout (usually 16, in three rows), chin, and pectoral fins and a spongy pad on the back of the head; they turn nearly black (particularly on the head), with two wide, pale vertical bands on their sides.

Taxonomy Fathead minnows in California have multiple origins and continue to be brought in from Arkansas and other states. Occasionally, pink-colored fathead minnows are brought into or reared in the state.

Names The word *minnow* is an Old English word of possible Latin origin. In Great Britain it is applied primarily to the cyprinid *Phoxinus phoxinus,* but use of the term in America has been broadened to include all small cyprinids. *Pime-phales* means fat helmet ornament; *pro-melas,* before black. Both terms refer to the head of spawning males, which is dark colored and swollen.

Distribution Fathead minnows are native to most of the eastern and midwestern United States and Canada as well as to parts of northern Mexico, except for the Atlantic slope and the Gulf states east of the Mississippi River. Their use as bait and forage fish has resulted in introductions throughout the West. They first came into California as bait in the Colorado River fishery in the early 1950s and were subsequently reared in central California by both commercial breeders and CDFG (1). CDFG then introduced them widely as forage and allowed them to be extensively propagated for bait, resulting in establishment in many areas, including southern California (1). They are now widely established in the Sacramento–San Joaquin basin, upper Klamath basin, Colorado River, and many coastal drainages. They can be expected in any watershed where conditions are appropriate for their survival, thanks to irresponsible bait anglers.

Life History Fathead minnows can survive in a wide variety of habitats, but they do best in pools of small, muddy streams and in ponds, where other fish are scarce. They can be characterized as pioneers, first to invade and last to disappear from intermittent streams and other fluctuating aquatic environments (2). They are capable of tolerating alkalinities of more than 2,100 mg/liter (3) as well as low dissolved oxygen levels (<1 mg/liter), high levels of organic pollution and turbidity, and temperatures up to 33°C (4, 5). They prefer temperatures of 22–23°C (4). With their high reproductive rates and parental care, they "explode" in temporary aquatic habitats. For example, Olcutt Lake, Solano County, is a large vernal pool that in years of heavy rain connects to small sloughs of the Sacramento–San Joaquin Delta; during these years fathead minnows invade and become abundant. The minnows pose a threat to the rare vernal pool invertebrates, but fortunately the lake dries up, eliminating the minnows. In more stable environments fathead minnows seem poor competitors with other species, especially other cyprinids. When they do occur with other species, they are generally found in association with beds of aquatic vegetation.

Like other cyprinids (presumably), fathead minnows avoid predators in part through their keen sense of smell. Most obviously, they avoid areas where fear substance from other minnows has been released, because it signifies a recent attack by a predator (12). They will also avoid, by smell, habitats or areas where fear substance has been detected on a regular basis, even after the substance has long dissipated.

This avoidance behavior can be passed on to conspecifics, including those that have not directly experienced an association with fear substance (13). Just as remarkable, fathead minnows learn to recognize odors of predatory fish and avoid areas where the odors are strong (14). To counter this ability, northern pike, at least, have special areas where they defecate, to reduce the problem of continually releasing distinctive substances into the water (15).

Despite their terminal mouths, fathead minnows are opportunistic bottom browsers on filamentous algae, diatoms, small invertebrates, and organic matter (6). This diet is indicated by their grinding pharyngeal teeth and long intestine. It is likely, absent other fishes, that they feed on whatever small organisms are most abundant on the bottom, in midwater, or among aquatic plants. They obtain nutrition from organic debris but grow on such a diet only if it is mixed with a small proportion of invertebrates (11).

Growth rates of fathead minnows are highly variable, influenced by factors such as temperature, food availability, and population size. Growth normally ceases at low temperatures (<7°C), but this may be the result of low food availability. At the end of their first growing season (age 0) they measure 25–64 mm TL, and they may reach 84 mm TL in their second season (age I fish). Few fish reach ages II or III or approach the maximum recorded length of 109 mm TL (4, 7). Size also depends on sex, because males grow larger than females.

The age of sexual maturity is variable: first spawnings have been recorded by fish just a few months old, by yearlings, and by 2-year-olds (4, 7). This variability has undoubtedly contributed to the success of fathead minnows in fluctuating environments. In the warm waters of California it is likely that spawning in the first summer of life is common. Another factor contributing to the success of the species is the ability to spawn repeatedly throughout the summer once water temperature exceeds 15–16°C, although reproduction becomes less frequent at high temperatures and ceases at 32°C (4, 8). Thus, although a female can carry anywhere from 600 to 2,300 eggs, usually fewer than a third will be ripe at any one time. Total egg production per female, especially in a newly established population with a low density of fish, may greatly exceed the number of eggs each female contains at one time. A single female spawned 12 times in 11 weeks, producing 4,144 eggs (8). Most fish usually die 30–60 days after the onset of spawning (4, 8).

Breeding males are highly territorial, accounting for their larger size, dark coloration, and well-developed breeding tubercles. The center of each territory is usually a flat stone, board, or branch at a depth of 30–90 cm that serves as an egg-laying site. Root masses, water lilies, old tires, and vertical stakes may also be used (2, 4). Males defend their nests from other males with such vigor that occasional injuries result, especially to eyes, presumably from contacts with breeding tubercles. Males improve a nest site by enlarging a hollow underneath the rock or stick and by removing small pieces of debris. Because the sticky eggs are usually laid on the undersurface, males clean it off by rubbing with the head pad. The pad is also used for tending the developing embryos; mucous secretions from it rub off onto embryos and may increase their survival rate (4, 9). Males also nibble at embryo masses to remove dead and foreign material.

While males defend their territories, females swim nearby in loose schools. When ready to spawn, one approaches a male, who then goes through a courtship display that culminates in his leading her into the nest, egg laying, and fertilization. Males spawn with several females over an extended period of time, and nests have been found containing more than 12,000 eggs in various stages of development (4). The eggs, about 1.3 mm in diameter, hatch in 4–6 days at temperatures around 25°C (4, 8). Newly hatched larvae measure about 4.8 mm TL and remain in the nest for a few days after hatching.

Status IIE. Fathead minnows, along with golden and red shiners, are legal bait minnows in California, and this means they have been widely distributed in the state by anglers and bait dealers. They have established populations in many areas but are usually only locally abundant. However, they have become extremely abundant and may be displacing native cyprinids such as blue chub in Upper and Lower Klamath lakes in Oregon and California and in Tule Lake, California (5, 10). Ironically, fathead minnows in the Klamath lakes may have come from the release of animals used for pollution bioassays (10) rather than from bait buckets. Although it can be argued that it is already too late, their use as bait minnows in California should be banned to safeguard native fishes, especially California roach, that live in intermittent stream habitats favored by fathead minnows. Ideally, bait fishing with live minnows should be banned in general, because anglers are prone to release their leftover bait wherever they are fishing, creating the potential for establishment of new populations. In addition, an essential part of any protocol calling for the use of freshwater fish in bioassays should be to rear and keep them in escape-proof systems and then destroy them when each project ends (10).

References 1. Dill and Cordone 1997. 2. Cross 1967. 3. McCarraher 1972. 4. Becker 1983. 5. Castleberry and Cech 1993. 6. Keast 1966. 7. Carlander 1969. 8. Dobie et al. 1956. 9. Wynne-Edwards 1932. 10. Simon and Markle 1997. 11. Lemke and Bowen 1998. 12. Mathias and Smith 1992. 13. Chivers and Smith 1995a. 14. Chivers andSmith 1995b. 15. Brown et al. 1995.

Red Shiner, *Cyprinella lutrensis* **(Baird and Girard)**

Identification Red shiners are small (usually <70 mm TL) minnows with deep, compressed bodies and terminal mouths. The lateral line is decurved, with 33–36 scales. There are 8 rays in the dorsal fin, 8–9 in the anal fin, 13–15 in each pectoral fin, and 8 in each pelvic fin. The pharyngeal teeth are 0,4-4,0 or 1,4-4,1, with narrow grinding surfaces. Nonbreeding fish are buff to steely blue on the back, silver on the sides (sometimes with a faint dark lateral band), and white on the belly. Breeding males have numerous tubercles on the head, sides, and fins; they have red to orange caudal, anal, pelvic, and pectoral fins and steely blue sides. Their heads are red on top and pinkish on the sides, with conspicuous purplish crescents immediately behind the opercles. Red shiners are best distinguished from juveniles of native minnows by their body shape, the absence of a spot on the caudal peduncle, and outlined scales on the back and upper sides.

Taxonomy Red shiners were formerly placed in the genus *Notropis*, which has been divided into a number of separate genera (1). The subspecies to which the fish in California belong is uncertain.

Names *Cyprinella* means small carp (genus *Cyprinus*). *Lutrensis* means otter, referring to Otter Creek, Arkansas, from which the first specimens were collected. The term *shiner* is widely applied to small, silvery minnows in North America.

Distribution Red shiners are native to streams of Western and Central states that drain into the Mississippi River and Rio Grande. Use as bait led to their establishment in the

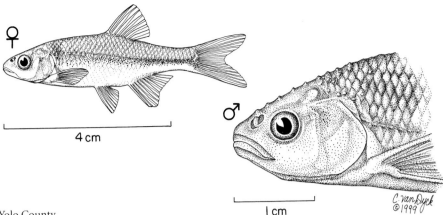

Figure 59. Red shiner, Putah Creek, Yolo County.

Colorado River between 1950 and 1953 (2) and in freshwater ditches around the Salton Sea. It is likely that these fish are descended from shiners that escaped from an Arizona bait farm, which had brought them originally from Texas. In 1954 shiners were taken by CDFG to the Sacramento–San Joaquin drainage and planted in Lake County ponds, but there is no evidence this introduction succeeded (3). However, after it was adopted as a bait minnow it became widely distributed in southern California (4) and the San Joaquin Valley (5). Red shiners became established in Coyote Creek (Santa Clara County) in 1986 (23). As of this writing they are colonizing Sacramento Valley streams (e.g., Cache Creek, Yolo County) and coastal streams. They are also establishing themselves in southern California and are present at least in San Juan and Aliso Creeks, Orange County, and Big Tijunga Creek, Los Angeles County (24). Red shiners can be expected anywhere in the state, despite the fact that it is illegal to use them as bait north of the San Joaquin Valley (3).

Life History Red shiners thrive in unstable environments, such as intermittent streams, as well as highly disturbed or polluted environments, such as drainage ditches and some reservoirs (e.g., Millerton Reservoir, Fresno County). In the San Joaquin Valley they are most abundant in turbid, alkaline, shallow, slow-flowing water (5). In the laboratory, red shiners can tolerate pH values of 4–11, salinities of up to 10 ppt, dissolved oxygen levels as low as 1.5 mg/liter, and sudden changes in temperature of 10–21°C, although they will avoid extreme conditions (including clear, cool water) when given the chance (6). They are extremely tolerant of high temperatures and have been collected from water as warm as 39.5°C (7), although they prefer summer temperatures around 25–30°C. In the Colorado River they seem most common in backwaters and sloughs, avoiding areas of strong current. In general, largest numbers are found in water less than 30 cm deep, with velocities of 10–50 cm/sec, over silt or other fine substrates and near instream cover (8).

Red shiners characteristically swim about in large schools, feeding on whatever organisms are most abundant, especially small crustaceans, aquatic insect larvae, surface insects, and, when necessary, algae (9, 10). They also feed on larval fish seasonally abundant in backwater habitats (11). Morphologically they seem best adapted for taking small invertebrates in midwater or from aquatic plants in quiet water. Most feeding is during daylight, although there may be a peak of activity at dawn (12).

Growth is most rapid during the first summer, when they reach 25–30 mm SL. In subsequent years they can grow 5–15 mm/year, achieving a maximum length of 80 mm SL and a maximum age of 2.5–3.0 years.

Red shiners mature in their second summer of life, and only a few live to spawn in their third summer (13). Females vary in fecundity because they are fractional spawners; in unspawned females, eggs appear in three distinct size classes, and the number of mature ova ranges from 485 to 1,200 (9, 15). Spawning occurs at water temperatures between 15 and 30°C, permitting a long breeding season. In their native range they can spawn from May to October, but most spawn in June and July (14). The presence of fish in spawning colors in Cache Creek (Yolo County) in late June and in Millerton Reservoir in June and July indicates that spawning times may be similar in central California. Some may cease spawning during severe conditions in midsummer, but resume again in the fall (13). Red shiners spawn in slow-flowing water, and embryos stick to a variety of substrates, including aquatic plants, gravel and sand, tree roots, logs, and other submerged debris. Active sunfish nests are also used (9). Apparently, red shiners can spawn either in groups or on territories held by individual males. Nonterritorial males court females by swimming closely beside them with erect fins. A chase for a meter or so usually follows, often resulting in one or more fish leaping from the water. Spawning occurs when male and female swim side by side, fins erect, over suitable substrate (16). The numerous breeding tubercles of males are used for contacting females during courtship and holding them during spawning (17).

Little has been published about the early life history of the red shiner, although larval development has been described (15, 18).

Status IIE. The red shiner is a true weedy species, spreading rapidly once established and displacing native cyprinids wherever it goes. Its initial success in the Colorado River was unexpected, reflecting the poor knowledge of its biology in the 1950s (19). The species spread rapidly through the Colorado River and its tributary streams. It has been implicated as a predator on larvae of Colorado River native fishes and is therefore a major obstacle to recovery (11). In the Moapa River, Nevada, establishment of red shiner and other alien species was associated with the decline of native fishes (20). In the Virgin River, Arizona, Nevada, and Utah, red shiners were recorded in 1972 as displacing Virgin River spinedace (21).

In 1976, in the first edition of this book (p. 204), I wrote: "Because red shiners have potential for becoming established in the warm intermittent streams of California where they would compete with endemic fishes, their use as bait fish outside the Colorado River system should be discour-

aged." In 1979 the Citizen's Nongame Advisory Committee, appointed by CDFG, and of which I was a member, recommended that red shiner be banned as a bait fish outside the Colorado River. A CDFG staff review of the recommendation agreed (22), but the state Fish and Game Commission capitulated to the bait-fishing industry's protests and permitted red shiner to continue to be used for bait (3). As a direct result, it may now be threatening native cyprinids in southern and central California, although there are no studies available to document this. Given the circumstances, it would seem appropriate for CDFG, through special assessment of the bait-fishing industry, to fund a major study of the red shiner and its effects on native fishes to determine if any control strategies are possible. Despite its wide distribution and abundance, the red shiner should still be banned as a bait fish, to prevent further expansion of its range.

References 1. Mayden 1989. 2. Hubbs 1954. 3. Dill and Cordone 1997. 4. Swift et al. 1993. 5. Jennings and Saiki 1990. 6. Matthews and Hill 1977, 1979. 7. Carlander 1969. 8. Peters et al. 1989. 9. Becker 1983. 10. Minckley 1982. 11. Ruppert et al. 1993. 12. Harwood 1972. 13. Farringer et al. 1979. 14. Cross 1967. 15. Wang 1986. 16. Minckley 1959. 17. Koehn 1965. 18. Saksena 1962. 19. Miller 1952. 20. Deacon and Bradley 1972. 21. Deacon 1988. 22. Gleason 1982b. 23. J. J. Smith, San Jose State University, pers. comm. 1999. 24. C. C. Swift, pers. comm. 1999.

Goldfish, *Carassius auratus* (Linnaeus)

Identification Goldfish in the wild can be as variable in color and body shape as those in pet stores. However, in wild populations there is strong selection (presumably by predatory birds and fish) for more protectively colored wild phenotypes: usually olive on the back, silvery to shiny bronze on the sides, white to yellow on the belly, and dusky on the fins. Like common carp, goldfish are heavy bodied and possess stout, serrated spines at the beginning of the dorsal and anal fins. Unlike carp, they lack barbels at the corners of

their thin-lipped, terminal mouths. Goldfish also tend to be deeper bodied and have a more rounded belly than carp. Counting the spine (actually a hardened ray) and the two smaller spines next it, they have 15–21 rays in the long dorsal fin and 5–6 rays in the anal fin. There are 25–31 large scales along the lateral line; the pharyngeal teeth (0,4-4,0) are blunt and comblike. Breeding males develop small tubercles on the sides of the head and pectoral fins.

Taxonomy Goldfish will hybridize with common carp. The hybrids, when bred and raised in captivity, are known as "silver carp" and sold in Asian markets for food. Some hybridization may take place in the wild in California as well.

Names Carassius is the Latinized common name (French, carassin; German, Karausche) of the closely related Crucian carp (*Carassius carassius*), a native of western Europe. *Auratus* means gilded or golden.

Distribution Wild goldfish originally ranged from eastern Europe to China. They are now established worldwide in suitable waters. In California they may have been established in the wild as early as the 1860s (1). They are spread

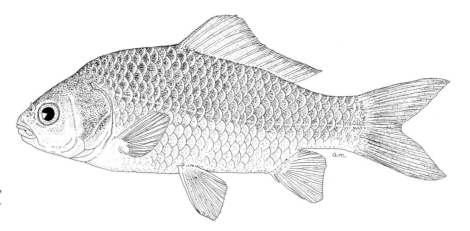

Figure 60. Goldfish, 16 cm SL, Putah Creek, Yolo County. Drawing by A. Marciochi.

by aquarists and bait fishermen. Large established populations are present in some southern California reservoirs and in canals, sloughs, and reservoirs of the Central Valley, as well as in Clear Lake (Lake County). Individuals from recent releases and from natural spawnings are likely to be found almost anywhere in the state where water is sufficiently warm.

Life History Although goldfish are known to survive water temperatures from 0 to 41°C, populations generally become established only in warm (27–37°C), often oxygen-deficient water in areas where winters are mild (2). They can be found in many habitats but seem especially well suited to fertile farm ponds, small backyard ponds, warmwater reservoirs, and sloughs with heavy growths of aquatic vegetation. They do well in highly disturbed and polluted habitats dominated by other alien fishes (11). Goldfish can become established in cold, oligotrophic lakes provided there is a littoral area large and warm enough for breeding. They rarely establish permanent populations in streams, although they are sometimes abundant in reaches below reservoirs containing reproducing populations (3). In clear streams they are strongly associated with deep pools with dense cover, whereas in turbid streams they are associated with deep pools (3). They may, however, move up into riffles and runs to graze on algae.

Goldfish are omnivores that feed heavily on algae, as their long intestine and closely spaced gill rakers suggest. They also consume zooplankton, large amounts of organic detritus, and aquatic macrophytes, indicative of feeding on the bottom as well as in midwater. Adult goldfish collected in November from sloughs of the San Joaquin River (Fresno County) were feeding mostly (58% by volume) on planktonic diatoms, together with a few strands of filamentous algae. The rest of their diet was organic detritus with a few fragments of higher plants. The diet of 71 goldfish from sloughs of the Sacramento River in November and April was similar, except that the April fish had also eaten chironomid

larvae and cladocerans (45%). In Clear Lake goldfish feed mainly on algae and aquatic macrophytes, mixed with zooplankton (9). Goldfish will also occasionally take insects and small fish (4). Young-of-year feed on zooplankton and small aquatic insect larvae (4, 9).

Growth rates in goldfish are highly variable, depending on environmental conditions. Overcrowding particularly stunts growth. Thus at the end of the first growing season they may range in length from 15 to 105 mm SL (2, 5). In California young typically reach 50–90 mm in their first year (9). In Sacramento River sloughs and in Clear Lake, normal growth in subsequent years is 15–25 mm/year, the amount decreasing with age. Thus goldfish in their fourth year from the Sacramento River measured 117–161 mm SL, although similarly aged fish from the San Joaquin River and from Clear Lake measured 161–215 mm SL (9). Goldfish may reach 41 cm TL and weigh 1.5 kg, but fish more than 25 cm TL are uncommon. In Clear Lake, however, shoals of goldfish measuring 22–30 cm SL may be encountered on occasion. Using scales, these fish have been aged at 5–10 years (9). Goldfish more than 40 cm SL are most likely goldfish-carp hybrids or simply misidentified carp. Females generally grow larger and live longer than males. As a result the male:female sex ratio changes from 1:1 in small fish to 13–16:100 in fish measuring more than 15 cm TL (5). Although fish in the wild rarely live longer than 6–8 years, maximum ages of 30 years have been recorded in aquaria (2, 6, 7).

Wild goldfish mature by their third or fourth year, males almost always maturing during the second or third year. Goldfish are serial spawners, so the number of eggs per female is highly variable. The number of eggs also varies with the size and health of the fish. Nine fish (average length, 135 mm SL) from the Sacramento River contained an average of 19,900 mature eggs, the numbers ranging from 8,000 in one fish measuring 121 mm SL to 29,000 in one measuring 168 mm SL. In Clear Lake fecundity estimates for individuals ranged from 9,000 eggs in a fish measuring 24 cm FL to nearly 72,000 in one measuring 28 cm FL (9). However, absolute fecundities seem to be in the range of 160,000–380,000 eggs per female (2). Spawning requires temperatures of 16–26°C (8). At higher or lower temperatures gonads do not develop completely, and eggs laid may not develop successfully. Overcrowding will also inhibit spawning. Under normal conditions goldfish spawn several times per season, laying 2,000–4,000 eggs each time (4). In California the first spawning takes place in April or May. Spawning usually occurs at sunrise on sunny days, over aquatic vegetation, flooded grass, roots, leaves, and other submerged objects. The spawning act is similar to that of carp, a male following close behind the female and fertilizing the eggs immediately after their release. The fertilized eggs are highly adhesive and hatch in 5–7 days. Larvae and small juveniles seek heavy cover among aquatic vegetation (8).

Status IID. Although goldfish are widely distributed in California, their ecological role is not well understood. For the most part, they are not very abundant except in severely disturbed habitats. In mud-bottomed ponds their feeding activities may eliminate aquatic plants and greatly increase turbidity (10). Occasionally they become so abundant in reservoirs that control measures are desirable (1). Unfortunately, the control of pet and occasional illegal bait releases, although highly desirable, seems impossible. In some reservoirs large goldfish are harvested and sold live as food in oriental markets.

References 1. Dill and Cordone 1997. 2. Becker 1983. 3. Smith 1982. 4. Dobie et al. 1956. 5. Breder and Rosen 1966. 6. Trautman 1957. 7. Carlander 1969. 8. Wang 1986. 9. University of California, Davis, unpubl. studies. 10. Richardson et al. 1995. 11. L. Brown 2000.

Figure 61. Common carp, 36 cm SL, Suisun Marsh, Solano County. Fish print by Christopher M. Dewees.

Common carp, *Cyprinus carpio* Linnaeus

Identification Common carp are large-scaled, heavy-bodied cyprinids with two barbels on the upper lip on each side of subterminal mouths. The rear barbel is longer than

the front. The dorsal fin is long, with 17–21 rays preceded by a stout, serrated spine plus 2 small spines (all actually hard rays). The anal fin also has a spine (plus 2 small spines), followed by 5–6 rays. The pelvic fins contain 5–7 rays; the caudal fin usually has 19, 17 of which are branched. There are 32–38 scales along the lateral line in most wild carp, although there are varieties that lack scales completely (leather carp) or have only a few patches of large, irregular scales (mirror carp). The pharyngeal teeth (3,1,1-1,1,3) are large and molariform. Adult carp are gold-green to bronze in color, with red-tinged pectoral, pelvic, and anal fins. Juveniles tend to be brown to gray, with terminal mouths and tiny barbels.

Taxonomy Common carp in California (and North America generally) are descended from domesticated carp from Germany and perhaps Japan. Balon (1, p. 9) indicates that feral carp in North America still resemble heavy-bodied domestic varieties as much as the ancestral carp of the Danube River, which is a "powerful, elongated, and torpedo-shaped animal with large regular scales and a golden (yellow-brown) color." Koi are brightly colored domestic carp originating in Japan.

Names The word *carp,* and its relative *carpio,* is an ancient one; forms of it were used by the Roman and Celtic peoples of Europe, and similar words are present in most European languages (1). The generic name *Cyprinus,* first used by Linnaeus in 1758, seems to be an indirect reference to its great fecundity because the name is probably derived from Cyprus, the island home of Venus.

Distribution Common carp have been introduced into suitable waters worldwide, a practice probably started in Europe by the Romans, who cultured them. Although common carp is widely regarded as having been first cultivated in China and then somehow brought to Europe, Balon (1) presents convincing evidence that it evolved in the Caspian–Black Sea region, from where it spread naturally to the Danube River. The Romans apparently got their fish from the Danube. Carp were then spread throughout medieval Europe for culture in the ponds of monasteries and became very popular as food fish. Because of the high esteem in which they were (and are still) held in Europe as food and sport fish, they were brought to California in 1872 by Julius A. Poppe, who stocked a pond in Sonoma Valley with five carp from Germany. He sold their progeny widely throughout the West (2). In 1879 the California Fish Commission started raising carp with broodstock provided by the U.S. Fish Commission. From these sources and new imports from the eastern United States, carp were planted all over California and the western United States. By 1896 they were widely distributed, but their disadvantages were starting to become so apparent that official stocking was halted. Today common carp are found in rivers, lakes, and reservoirs throughout North America.

In California carp are present in the Sacramento–San Joaquin drainage, the Salinas and Pajaro basins, the Russian River, Clear Lake, the Colorado River, some Lahontan drainage reservoirs and rivers, and the Owens River, as well as along coastal southern California. To the best of my knowledge they are absent from the Klamath River basin, all North Coast watersheds, the Pit River, Eagle Lake and other isolated Great Basin watersheds, and the Death Valley region. However, it would not be surprising to find them in any of these places.

Life History Common carp are most abundant in warm, turbid water, especially reservoirs, at low elevations, but they also manage to live in some trout streams and a few coldwater reservoirs at high altitudes, such as Shaver Lake, Fresno County (1,320 m). They are generally most abundant in eutrophic lakes, reservoirs, and sloughs with silty bottoms and growths of submergent and emergent aquatic vegetation. In streams they are associated with turbid water; deep, permanent pools; high alkalinity; and soft bottoms (3, 17). Cover, such as submerged tree branches, becomes more important as water becomes clearer. Juveniles also prefer deep pools, but they will move into shallow water if there are dense beds of aquatic vegetation for cover (3). Carp are active at water temperatures of 4–24°C, although the optimum temperature for growth seems to be around 24°C (4). One of the main reasons carp have succeeded so well in the West is their ability to survive under adverse conditions. They can withstand exceptionally high turbidity, sudden temperature changes, high temperatures (31–36°C, depending on acclimation temperatures), and low oxygen concentrations (0.5–3.0 ppm) (4, 5). They can survive in deoxygenated water by gulping air at the surface (13) and pumping an air-water mixture across the gills. Carp can inhabit estuaries as well as freshwater environments, although they apparently must spawn in fresh (or nearly fresh) water (6). They can survive salinities up to 16 ppt (5) and are regularly found in the San Francisco Estuary at salinities of 10–12 ppt.

In lakes and reservoirs carp seldom occur deeper than 30 m. They usually overwinter, however, in deeper waters of lakes and streams, moving into shallow water to feed and breed as the water warms up in spring. If preferred feeding areas are exceptionally shallow, they will move in to feed only during early morning and evening. They also move into flooded fields to feed and breed in the spring.

In general carp are omnivorous bottom feeders, although animal food (particularly aquatic insect larvae and small molluscs) seems to be more important in their diet than plants (4, 5, 7). Their diet changes with their age. Newly hatched carp feed on both zooplankton (e.g., rotifers and copepods) and phytoplankton (algae). As they increase in size, they begin to feed on benthic insect larvae. By the end of their first summer they are eating most available bottom invertebrates. Adults will feed heavily on aquatic plants and on algae, which might be expected given their long gut (3–4 times body length) and molariform teeth. However, small animals associated with plants may be as important nutritionally as the plants themselves. The preferred animal foods are aquatic insect larvae, especially midge larvae (Chironomidae), followed by aquatic crustaceans, molluscs, and annelid worms. Fish, probably dead before eaten, and fish larvae and eggs, including carp eggs, have been found in their diets (5).

Carp typically root around on silty bottoms, stirring up aquatic insects, which they then pick from the water. They frequently take silt into their mouths and then spit it out, picking out organisms thus suspended. The effect of this behavior in shallow lakes and ponds is to uproot aquatic plants that provide cover and food for other fishes and waterfowl and to greatly increase turbidity, cutting down on sunlight available for plant growth (14, 15). Although turbidity created by carp can be responsible for the disappearance of game fish from an area, more often than not carp were not the creators of adverse conditions but rather moved into an area already disturbed (8). The ability of carp to colonize new areas or reinhabit streams and lakes that have dried up and then refilled is legendary. In lowland streams they are typically the first fish to return to streams following drought, their backs cutting the water as they splash through riffles. Their ability to move long distances is well documented (5, 9). A tagged carp in the Missouri River moved nearly 1,100 km upstream in just over 2 years.

Growth of carp varies considerably according to summer water temperatures, length of growing season, quality of water, and food availability (4, 5). During their first summer of life they may reach 7–36 cm SL, averaging 10–15 cm SL. During their second year they can double in length and add 10–12 cm in each following year, although growth tends to slow down after the fourth or fifth year. Increase in weight follows a similar pattern, although it too can be highly variable. In the wild carp seldom live longer than 12–15 years or exceed 80 cm SL and 4.5 kg. However, they have been recorded as living as long as 47 years in captivity. The largest carp ever caught (from South Africa) weighed 37.9 kg; the largest one caught in North America (from Mississippi) weighed 37.2 kg (10). The largest carp recorded for California was caught in Lake Nacimiento, San Luis Obispo County, and weighed 26.3 kg (16).

Spawning takes place in spring and early summer when water temperatures start to exceed 15°C, with highest activity at 19–23°C (5). The first indication of spawning is large shoals of carp swimming slowly about in open water near beds of aquatic plants, usually close to shore, their dorsal fins and backs frequently breaking the surface. Soon they separate into smaller groups, which move into shallow, weedy areas, preferably recently flooded, and quickly begin to spawn, accompanied by splashing. Usually, each female is closely pressed by two or three smaller males. Spawning occurs at any time of day or night, but it seems to peak in late evening and early morning.

A female lays about 500 eggs at a time and, depending on size, will deposit 50,000–2,000,000 eggs during a season (4, 5). Eggs are adhesive and stick to plants, tree roots, and bottom debris (6). Embryos hatch in 3–6 days, and newly hatched larvae measure 3–7 mm TL. These quickly drop to the bottom or attach to vegetation, where they live on the contents of their yolk sac for a few days. Soon they start feeding on zooplankton and become increasingly active swimmers as their fins develop, occasionally moving up into the water column. By the end of their first week, most carp fry have moved into beds of emergent or submerged vegetation. They seldom leave protective cover until they have attained 7–10 cm TL and are fairly secure from predation.

Status IIE. In the California watersheds into which they have been introduced, common carp have reached the maximum extent of their range. Despite the great disdain in which it is held by anglers and managers, the fish is increasingly popular as koi, an ornamental pond fish. Koi are carp nevertheless, and if they escape into the wild they are capable of establishing wild populations, much like goldfish. Thus, under present regulations, it seems likely that carp will eventually become established in watersheds, such as the upper Klamath basin, from which they are now fortuitously absent.

The introduction of common carp to North America is now widely regarded as a serious mistake, although the decision was a very popular one in the 1870s (11). Congressmen scrambled to have carp raised by the U.S. Fish Commission planted in their districts, an action facilitated by the rapidly developing network of railroads (11). Carp have probably displaced or reduced populations of native fishes in some areas and have been responsible for destruction of shallow waterfowl habitat in various parts of the country (8). However, their ecological role in California streams and reservoirs is poorly understood because they are so characteristic of disturbed and polluted habitats. It is possible that, through their foraging behavior, they decrease local water clarity and prevent dense beds of aquatic plants from growing, but there is no direct evidence for this in California.

Carp have low value as forage for piscivorous fishes because the most vulnerable stages of their life history are spent well hidden. However, they do have virtues as a food and game fish—virtues that are slowly being rediscovered in California and elsewhere (12). They grow rapidly and achieve large size in polluted water that supports few other fish. They can provide good sport, because they are wary, large, and often surprisingly difficult to catch, and put up a good fight when hooked. Carp fishing tournaments are becoming increasingly popular, even catch-and-release tournaments. Common carp can be a real culinary treat when properly prepared, and are highly appreciated by diverse ethnic groups in California. A commercial fishery exists for them in Clear Lake (Lake County) and in some reservoirs.

Controlling carp is both difficult and expensive. Probably the most effective means are intensive commercial fishing in large bodies of water and the use of fish poisons in small bodies of water. Efforts should certainly be made to exclude carp from waters that do not now contain them. Serious consideration should be given to banning the sale or keeping of koi in watersheds from which carp are now absent.

References 1. Balon 1995. 2. Dill and Cordone 1997. 3. Smith 1982. 4. McCrimmon 1968. 5. Becker 1983. 6. Wang 1986. 7. Minckley 1982. 8. J. Moyle and Kuehn 1964. 9. Sigler 1958. 10. Panek 1987. 11. Moyle 1984. 12. AFS 1987. 13. Nakamura 1994. 14. Lougheed et al. 1997. 15. Wilcox and Hornbach 1991. 16. Files, CDFG, Region 4. 17. Brown 2000.

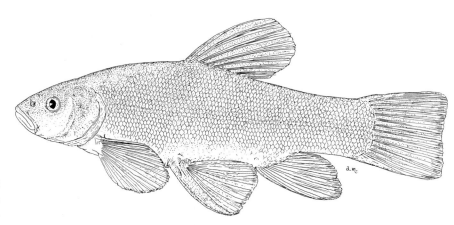

Figure 62. Tench, 23 cm SL, pond near Lobitas Creek, San Mateo County. Drawing by A. Marciochi.

Tench, *Tinca tinca* (Linnaeus)

Identification Tench are deep and thick-bodied; covered with tiny, deeply embedded scales (90–115 in the lateral line); and very slimy. The mouth is small and terminal with a single barbel at the end of each maxilla. The caudal fin is squared and the dorsal and anal fins are well rounded, each with 8–9 rays. The pharyngeal teeth are in a single row, usually 5-4. In California tench may reach sizes of 60–80 cm TL and 2–3 kg. Mature males possess a thick ray on the leading edge of each pelvic fin. The color of the back varies from dark green to black, becoming bronze on the sides and belly. Some individuals may be a gold-bronze color overall. The fins are dark.

Names *Tinca* is the Latin word for tench, and the Old English name tench is derived from it.

Distribution Tench are native to most of Europe except northern Scandinavia. In 1922 12–24 fish measuring 10–15 cm TL were brought to California from Italy and intro-

duced as sport and food fish into a private reservoir near Lobitas Creek (San Mateo County) by an Italian-American rancher (1, 2). They were still present in this reservoir in 1973. They were subsequently spread to other ponds and reservoirs in Santa Cruz and San Mateo Counties, and it is not known if any of these populations persist. A population was established in the 1950s in a pond in Humboldt County, near the Trinity River, but it was eradicated in 1976 (2).

Life History No work has been done on tench biology in California, but it has been studied in Europe (3, 4) and Tasmania (5), where the fish have also been introduced.

Tench are fish of warm, quiet waters that do best in farm ponds, oxbow lakes, sloughs, castle moats, and deep, slow-moving stretches of rivers. They are generally associated with muddy bottoms and heavy growths of aquatic macrophytes. Tench can survive water temperatures as high as 30–35°C, oxygen concentrations under 1 ppm, and salinities up to 12 ppt. Although tench from northern Europe can apparently withstand temperatures close to freezing, California tench, descended from southern European populations, may not be able to withstand such low temperatures. The optimum temperature for growth seems to be between 12 and 30°C.

Tench are rather sluggish and are not very aggressive toward other tench or other fishes, earning them the reputation of the "Physician of Fishes" (6, p. 134). They are usually solitary and strongly nonmigratory. During hot summer months they tend to congregate in deep holes and shady areas, seeking cooler water. They tend to forage during the night and move into heavy cover, such as deep cattail stands, during the day (7).

Invertebrates that live on the bottom or on aquatic plants are their main food. Tench 6–12 cm TL feed primarily on aquatic insect larvae, especially those of mayflies, damselflies, chironomid midges, and caddisflies. Larger fish depend on whatever large invertebrates are most abundant. Thus large tench from one pond fed mostly on pulmonate snails; those from another, on oligochaete worms; and those from another, on insect larvae, especially chironomids (5, 7). Algae and aquatic plants become important only when overcrowding in a pond reduces invertebrate populations. Tench are probably not able to survive on a purely vegetarian diet. Tench measuring less than 6 cm TL feed on small crustaceans among aquatic plants, especially cladocerans, copepods, and amphipods. Small chironomid larvae and water mites may also be taken. Newly hatched fry take mostly small crustaceans, especially nauplii, along with rotifers and diatoms.

The growth of tench is slow for a large cyprinid, averaging about 3 cm/year for the first 4 years and becoming progressively slower thereafter. A fish measuring 30 cm TL will probably be at least 9 years old. In Tasmania tench grew fastest in farm ponds, slowest in a large lake, and moderately well in a sluggish river. In Europe they commonly reach 64 cm TL and weigh 2 kg, although fish weighing nearly 4 kg have been caught. In California tench may reach 2–3 kg (1).

Tench mature during their third or fourth year, males usually maturing a year before females the same age.

Spawning is in summer (May–August in Europe), after water has reached 18°C. Tench aggregate for spawning in areas of heavy plant growth, each female laying around 500,000 eggs per kilogram body weight (4). The adhesive green eggs, each about 1.2 mm in diameter, stick to aquatic plants. They hatch in 6–8 days, and the 2- to 3-mm-long fry begin feeding a day or so later.

Status IIB. Tench were an unauthorized introduction into California. Fortunately, their slow growth, confinement to isolated ponds in small coastal drainages, and generally low desirability have kept them from spreading. However, their hardiness in and out of water and their high fecundity do facilitate their spread into other river systems. Although they seem to be innocuous compared with carp, their potential for offering competition for food, especially to native cyprinids, is high enough that introduction into other waters should be prevented. Because they are presently found in only a few small ponds without public access, their further spread seems unlikely, especially because local ranchers seem to have lost interest in them. A thorough survey of their populations is needed; if possible, eradication should be attempted.

References 1. Shapovalov 1944. 2. Dill and Cordone 1997. 3. Wheeler 1969. 4. Varley 1967. 5. Weatherley 1959. 6. Walton 1653. 7. Perrow et al. 1996.

Grass carp, *Ctenopharyngodon idella* Steindacher

Description Grass carp are solid, moderately slender fish with a wide, scaleless head and a terminal mouth. They may reach over 1 m SL. The scales are large (34–45 in the lateral series) and outlined in black, most with a dark spot at the base. The dorsal fin is short (8 rays) and spineless, with its origin in front of that of the pelvic fin. It has 9 anal fin rays, 18–20 pectoral fin rays, 8 pelvic fin rays, and 15–16 gill rakers. The pharyngeal teeth are 2,5-4,2 with rough, elongate

grinding surfaces (1). The fish generally have a silvery-white appearance, although the back and sides may be olivaceous, the head gray, the belly white to yellow, and the fins dark.

Taxonomy The varieties of grass carp planted in California are mostly triploid males or hybrids with bighead carp, *Hypopthalmichthys molotrix,* both of which are supposed to be sterile.

Names Cteno means comb, while *pharyngo-don* means pharyngeal teeth, referring to the rough, comblike surfaces of the teeth. *Idella* seems to be a combination of the Greek and Latin words for small, but the reference is obscure. Grass carp have been called white amur in an attempt to whitewash fears of many biologists that they might become another common carp in terms of habitat destruction.

Distribution Grass carp are native to large rivers of central-east Asia, from the Amur River of China and Siberia to Thailand (2). They were brought into Arkansas in 1963, cultured, and released into an Arkansas lake in 1970 (5) and the Arkansas River in 1971 (1). From there they quickly spread

Figure 63. Grass carp. Painting by J. Tomelleri.

through the Mississippi drainage and established reproducing populations, despite opinions that they would not be able to do so. Fish dealers in Arkansas also marketed live fish for aquatic weed control, so they can now be found in most states and in Mexico, with a number of reproducing populations. Although grass carp are officially prohibited from most of California, fish have been illegally imported from Arkansas a number of times and planted in ponds; when such populations were found CDFG eradicated them (e.g., from golf course ponds in the Carmel Valley in the 1980s). Grass carp have been legally introduced into canals in the Coachella and Imperial Valleys in southeastern California for weed control. Since 1979 sterile triploid grass carp have been released experimentally, and they are now fairly common in the region. Triploid grass carp are expensive and, given the weed control mythology associated with grass carp, it is likely that wild, self-sustaining populations will eventually be established in the state, most likely in the Colorado and San Joaquin Rivers.

Life History Grass carp are native to large, temperate river systems, where they forage in backwaters and shallow areas. They seem capable of living in a wide variety of conditions, including ponds, irrigation canals, and lakes. They survive in waters with near-freezing temperatures in winter and are likely to die in summer only when temperatures reach 38–39°C (3). Optimal temperatures for growth are around 25°C, but they will feed at temperatures ranging from 3° to 33°C. Grass carp can survive oxygen levels of less than 1 mg/liter and salinities of 17 g/liter (perhaps higher for short periods) (3). Adults can thus invade estuaries as well as freshwater environments.

Grass carp are restless fish that can move hundreds of kilometers in rivers within short periods of time. They feed constantly, and if they find good feeding conditions (beds of aquatic plants) they will stay in one area for an extended period (3). This behavior resulted in their colonizing much of the Mississippi and Missouri River systems in the 30 years following their introduction in Arkansas (4). It also allows them to quickly locate beds of aquatic plants in lakes and other large bodies of water.

Adult grass carp are omnivores, with a strong bias toward plants. The biggest fish are the most herbivorous. Their herbivory is surprising considering that their intestine is short (only 2–3 times the body length); other herbivorous fishes have much longer digestive tracts, to provide the surface area needed for breaking down plant material. Their need to consume large amounts of plant material to compensate for a short digestive tract is presumably one reason they are so effective at plant control. Their digestive efficiency is increased by the powerful pharyngeal teeth, which break open plant cells. Juveniles feed largely on aquatic invertebrates, mainly benthic but occasionally planktonic, and begin switching to plant material at 3–4 cm TL (3). Adults consume almost any kind of plant given the opportunity (including terrestrial vegetation hanging over water) but seem to prefer submerged macrophytes, especially such relatively "soft" forms as the exotic weed *Hydrilla verticilla*. Less preferred plants (such as water hyacinth) are likely eaten only after more palatable plants have been consumed. Omnivory in grass carp asserts itself once they have depleted beds of aquatic plants and they switch to diets of benthic invertebrates, such as crayfish and clams (3).

Grass carp can grow rapidly and reach large sizes. In their native Amur River, they grow 9–10 cm/year in their first 4 years of life, after which they become mature and growth slows to 6–7 cm/year for the next 3 years and 2–5 cm/year thereafter (3). However, larger fish may show weight increases disproportionate to length increases. Faster growth occurs in warmer waters, and some individuals reach over 5 kg within 2 years. They apparently reach lengths of 1–1.5 m TL, with weights of 30–36 kg. As would be expected of such large fish, they are long lived, with life spans in excess of 15 years.

Grass carp will become mature within 2 years in warm climates, 4–5 years in temperate climates, and 8–10 years in colder climates, usually at lengths of 60–70 cm TL and weights of 4–5 kg. Females produce, depending on size, 237,000 (a 68-cm female) to 1.7 million eggs (a 1-m female); fecundities average about 618 eggs per gram of ovary weight (3). Spawning is initiated by a rise in water temperature (above 18°C, optimal 20–25°C) and a rise in water level. Spawning grass carp seek out open riverine areas with moderate currents, because fertilized eggs are semipelagic and must be suspended for several days before hatching. Spawning behavior is typical of cyprinids, with each female pursued closely by two or more males (3). The larvae are apparently pelagic for a period before transforming into juveniles that inhabit shallow water.

Status IIB. So far as is known, there are no self-reproducing populations of grass carp in California as of this writing. Sterile, triploid grass carp are widely used for weed control in southern California irrigation canals, however, so the species is likely to be encountered in many places. Illegal introductions of normal grass carp can occur because grass carp are easy to obtain from out-of-state dealers despite prohibitions. It would therefore not be surprising if they became established in California rivers, which seem to have all the conditions grass carp need for successful reproduction.

The use of grass carp for aquatic weed control is controversial, but the following statements about them are widely accepted (3, 5):

1. They can be very effective at reducing and occasionally eliminating beds of submerged aquatic plants and as such are an alternative to herbicides.

2. They are most effective in controlling weeds in confined situations (such as ponds, canals, and small lakes) or in situations in which they can be stocked at high densities.

3. All aquatic plants are not equally palatable to grass carp, and their selective feeding can actually result in an increase in undesirable aquatic weeds in some situations.

4. They have the potential to become established in most river systems of the United States and Mexico.

5. Their ability to eliminate beds of aquatic plants means they can drastically change aquatic ecosystems by reducing the amount of cover available to small fish of other species (including predatory fish), by increasing algae blooms (decreasing water clarity), and by changing the distribution and abundance of aquatic invertebrates.

6. The effects of feral grass carp on large ecosystems (e.g., rivers of the Midwest) are not known, but they must be assumed to be negative until proven otherwise. However, obvious large-scale effects on riverine ecosystems have not been observed (5).

7. Triploid grass carp are a fairly safe method of weed control because a vast majority are sterile and because, despite their long lives, they represent a reversible management action. The biggest problem with their use is the likelihood of cheating, with the much cheaper and more easily obtainable normal grass carp being substituted for triploid individuals.

8. Grass carp should be used for weed control only after careful consideration of alternatives; once introduced, their populations should be carefully monitored.

Clearly the use of grass carp for aquatic weed control should be tightly regulated. The ban on their use, even of triploid forms, north of the Tehachapi Mountains should be continued because the agencies responsible for their regulation do not have adequate staff to monitor introductions.

References 1. Etnier and Starnes 1993. 2. Dill and Cordone 1997. 3. Chilton and Moeneke 1992. 4. Lever 1996. 5. Leslie et al. 1996.

Suckers, Catostomidae

Suckers are a highly successful group even though they lack the diversity of species of the minnows (Cyprinidae), with which they share the order Cypriniformes. With the exception of a few plankton-feeding forms, they are bottom browsers, sucking up small invertebrates, algae, and organic matter with their fleshy, protrusible lips. Their comblike pharyngeal teeth serve to break up items entering the long, coiled intestine. The ability of suckers to thrive on abundant food little exploited by other fishes, combined with the mobility conferred by their solid, muscular bodies, has permitted a small number of species to become abundant in a wide variety of habitats, including mountain and foothill streams, reservoirs and lakes, tidal sloughs, and large rivers. In addition, they possess the characteristics that have made cyprinids so successful, such as a well-developed sense of hearing, fear substance, and high fecundity. Specializations include an enlarged Weberian apparatus (for hearing), a complex mouth structure (for vacuum cleaner–like suction feeding), and tetraploidy (G. Smith 1992). Like the large cyprinids so characteristic of California, most suckers have a life history that combines large size with long life and high fecundity, enabling them to persist through long periods of unfavorable environmental conditions.

Suckers are an ancient family, with fossils dating back to the early Cenozoic (Paleozoic). Ancestral suckers, large deep-bodied forms, were once found throughout Asia and North America. The closest living relatives to suckers in the Cypriniformes are likely to be various Asiatic groups (Smith 1992). However, Asia today supports only two sucker species, one ancient relict (*Myxocyprinus asiaticus*) in China and one recent invader from North America (*Catostomus catostomus*). Thus the sucker success story is primarily a North American one, especially the evolution of the "standard" stream suckers (*Catostomus, Moxostoma*). There are three basic ecological types of suckers: *(1)* deep-bodied suckers, most with terminal mouths, inhabiting open waters of large lakes and sluggish rivers; *(2)* small mountain suckers, with horny plates on their lower lips for scraping algae and invertebrates from rocks in fast-moving streams; and *(3)* typical suckers, which occupy a wide range of habitats but are mainly stream dwellers. The specialization of the lake and mountain suckers allows two or more species to coexist in waters that presumably would otherwise support only one.

The success of suckers has given them a bad reputation among anglers, who frequently accuse them of competing with game fish for food and space. This accusation is rarely justified. Too often the presence of suckers and the absence of game fishes are considered to be part of a cause-and-effect relationship when, in fact, the lack of game fishes (especially trout) may be due to poor habitat, low water quality, or overfishing. Suckers may even be beneficial to game fish populations as forage fish that utilize food (algae and detritus) largely unavailable to predatory fishes. They also have some importance as commercial and sport fish: they reach large sizes, put up a good fight on light tackle, and are quite edible. They were an important source of food for Native Americans (Lindstrom 1996). Hubbs and Wallis (1948) pointed out that those in Yosemite Valley preferred Sacramento suckers to trout as food.

Ten species of suckers are included in this book as part of the California fish fauna, but a case can be made for adding three others to the list. Flannelmouth suckers (*Catostomus latipinnis*) were historically part of the lower Colorado River fish fauna but disappeared from the California portion in the late 19th century, for unknown reasons, although they were probably always uncommon. However, in 1976 they were reintroduced into the tailwaters of Davis Dam in Nevada, where they became established in about 25 km of river (G. Mueller, U.S. Bureau of Reclamation, pers. comm. 2000). A few have been subsequently captured in California, although the open, sandy-bottomed habitat is largely unsuitable for them. They have the distinction of being the first native fish to be extirpated from California's waters, and then successfully reintroduced. The tenuous nature of their presence in California nevertheless removes them from further consideration here. An undescribed sucker (*Catostomus* sp.) lives in Wall Canyon Creek on the Nevada side of Surprise Valley. These suckers may wash into alkaline Surprise Lake, covering the valley floor in Modoc County, during times of high runoff (C. L. Hubbs,

pers. comm. 1974). Bigmouth buffalo (*Ictiobus cyprinellus*) were once established in reservoirs of southern California and in the lower Colorado River, but there are no recent records of them.

It is a sad comment on the state of California's native fish fauna that six of ten native sucker species are rare, endangered, or potentially endangered. On the other hand, three species (Sacramento, Tahoe, and Owens suckers) are doing quite well in reservoirs and other human-altered habitats.

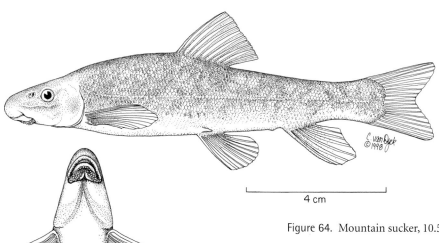

4 cm

Figure 64. Mountain sucker, 10.5 cm SL, Martis Creek, Placer County.

Mountain sucker, *Catostomus platyrhynchus* (Cope)

Identification Mountain suckers are small, sleek suckers (typically 12–20 cm TL as adults) with a subterminal mouth and fleshy, protrusible lips covered with numerous large papillae. The lips have deep lateral notches at the juncture of the upper and lower lips and a shallow, median cleft on the lower lip. On the lower lip there are two semicircular bare areas on the inner margin, next to which a round car-

tilaginous plate (for scraping) is often visible. The front of the upper lip is also without papillae. There are 23–37 gill rakers on the first gill arch, 75–92 lateral line scales, and 8–13 (usually 10) rays in the dorsal fin. The pelvic fins have 9 rays and a well-developed axillary process at the base. The intestine is long (4.5–6 times the body length), and the peritoneum is black. Fish are brown to olive green dorsally and laterally and white to yellow ventrally. A lateral band or a series of blotches is usually present along the sides. Mature males have a dark, red-orange lateral band above a black-green band. The fins also take on a red-orange color during spawning season. Breeding males develop tubercles over the entire body and all fins (except for the dorsal fin), with the tubercles on the enlarged anal fin being especially prominent. In females tubercles are restricted to the dorsal and lateral areas of the head and body.

Taxonomy The mountain sucker was described in 1874 as *Minomus platyrhynchus* from specimens collected in Utah (1). The genus was subsequently changed to *Pantosteus* (2), which was applied to several other forms, most importantly *Pantosteus lahontan* from the Lahontan basin of California and Nevada (3) and *Pantosteus jordani* from the Columbia and upper Missouri Rivers (4). However, G. R. Smith (5) concluded that all small suckers with a cartilaginous plate in the lower lip in the Great Basin and Columbia River drainage were one species. He further concluded that differences among mountain-type suckers and other "standard" suckers were not sufficient to merit generic distinction, although *Pantosteus* was maintained as a subgenus (17). His extensive review led to designation of most forms as *Catostomus platyrhynchus,* including mountain suckers in California. Nevertheless, given the long isolation of various populations from one another, a reevaluation of their taxonomic status using modern statistical and molecular techniques is merited. It would not be surprising if a number of distinct taxa, including the Lahontan form, reemerged from such an analysis.

Names Cato-stomus means inferior (down) mouth; *platyrhynchus* means flat-snout, although the snout is, if anything, rounder than that in most other sucker species. The name mountain sucker is used because the species often lives in cool mountain streams.

Distribution As presently recognized, the mountain sucker has an extraordinarily wide distribution in western North America (6). In Canada it is found in an Arctic drainage (Saskatchewan River) and various watersheds in Saskatchewan, Alberta, and British Columbia. In the United States it is present on both sides of the Rocky Mountains, including in streams in the upper Missouri River drainage in Montana, South Dakota, and Wyoming. Other Western states in which it is found include Colorado, Utah, Idaho, Washington, Oregon, and Nevada. The major Western watersheds or zoogeographic regions it inhabits include the Columbia, Bonneville, Lahontan, and Colorado. In California the mountain sucker is native to Lahontan drainage river basins: Walker, Carson, Truckee, and Susan. It is absent from the Eagle Lake basin. In Nevada it is also found in the Quinn, Humboldt, and Reese Rivers (7). Today the mountain sucker is found in the North Fork of the Feather River, in the Sacramento drainage, especially in Red Clover Creek, a tributary to the North Fork. The Feather River population presumably resulted from an irrigation diversion from the Little Truckee River that carries water across the divide (16). In addition, the California Academy of Sciences has at least one specimen taken in the lower Sacramento River, indicating that this sucker could spread further in the drainage.

Life History The characteristic habitat of mountain suckers is clear streams with moderate gradients, 3–15 m wide and less than 2 m deep, with rubble, sand, or boulder bottoms. However, they also live in a variety of other waters, such as large rivers and turbid streams. They are occasionally found in lakes and reservoirs but are notably absent from large lakes, such as Tahoe, Eagle, and Pyramid Lakes. Within their

entire range they have been recorded at elevations as high as 2,800 m and at temperatures of 1–28°C (5). Within streams they are usually found in pools, especially those containing aquatic macrophytes, logs, or deeply undercut banks. In swifter water they are typically found in velocity refuges behind rocks or under logs. In Lahontan streams the abundance of mountain sucker is positively correlated with pools and negatively correlated with riffles (8, 9). The suckers typically select areas with mean water column velocities of 0.1–0.5 m/sec and depths of 0.5–1.8 m (9). Within these areas they are most abundant in dense cover, especially around rootwads (9).

Mountain suckers form exclusive shoals and segregate from other catostomids in much of their range (10), yet this is not the case for California populations, which form mixed aggregations with Tahoe suckers (9). There is a positive correlation between the abundance of mountain suckers and that of Tahoe suckers and speckled dace (8). They are also common associates (and prey) of various native and introduced trout.

Mountain suckers feed mostly on algae and diatoms as well as on small quantities of aquatic insects and other invertebrates (5, 11). They feed by scraping food from the substrate, and this strategy results in sand and grit also being ingested. The importance of algae in their diet is indicated by the movement of suckers into areas coincident with "blooms" of algae on the rocks (9). The diet of juveniles (<30 mm TL) contains a higher proportion of insects (11).

In Montana mountain suckers reach 60–65 mm TL in their first year and 90-100 mm TL by the second year (11). Average growth rates are greatest during the first year and decrease gradually through the third year, after which growth is slow and constant. Individuals rarely exceed 17 cm TL but occasionally reach 23 cm TL (11). Given the length distributions of suckers observed in California streams, this pattern of growth is probably true here as well. Females are larger than males, live longer (7–9 years versus 7 years for males), and mature later (in their third or fourth year at 9–17 cm TL) (5, 11). Males mature in their second or third year at 6–14 cm TL (5, 11). Fecundity is variable, females producing between 990 (for a specimen measuring 13 cm TL) and 3,710 (for a specimen measuring 18 cm TL) eggs (11).

Mountain suckers are fairly unusual for a stream-dwelling fish in western North America in that they spawn in midsummer (June to early August) rather than in spring (8, 9, 10). They move into small streams in late July for spawning and for feeding on algae on rocks (15). Spawning takes place in gravelly riffles immediately upstream of deep pools and is probably nocturnal. The fertilized eggs are adhesive and stick to the gravel. Temperatures at times of spawning are 11–19°C (8, 9, 10), although fish in breeding condition were noted in Sagehen Creek at temperatures of 9–12°C (9). Larval and juvenile suckers are found on stream edges and in beds of aquatic plants in or near pools (10).

Status ID. The mountain sucker is present in scattered populations in California and Nevada, which show high variability in numbers (8, 9, 12). However, its populations in California seem to be in a general decline (8, 9), with the exception of the introduced population in Red Clover Creek and the population in East Fork Carson River and its tributary, Hot Springs Creek (12). The decline is tied to stream alterations and modifications, especially construction of dams and reservoirs that isolate populations. Mountain sucker populations have a hard time persisting in reservoirs. Because their favored habitats are the lower reaches of streams, now flooded by reservoirs, the remaining habitat supports only small populations that are vulnerable to extirpation. In contrast, in East Fork Carson River, a stream without a major reservoir on the mainstem, sucker populations in 1988 were estimated to range from 1,000 to 44,000 per kilometer

of stream (13). High densities of mountain suckers may also exist in the lower Truckee River above Reno.

Streams in which mountain suckers have had sharp declines have also seen declines of Lahontan speckled dace and mountain whitefish (14). Thus the decline of mountain suckers is probably a good indicator that the native fish and invertebrate assemblages of many Lahontan drainage streams in California are in some trouble. It is therefore important that a number of streams in the basin be identified as targets for management—specifically for maintaining the integrity of the native biotic community, which includes mountain sucker.

References 1. Cope 1874. 2. Cope and Yarrow 1875. 3. Rutter 1903. 4. Evermann 1893. 5. G. River Smith 1966. 6. Lee et al. 1980. 7. La Rivers 1962. 8. Olson and Erman 1987. 9. Decker 1989. 10. Hauser 1969. 11. Marrin 1980. 12. Erman 1986. 13. J. Deinstadt, CDFG, unpubl. data 1996. 14. Olson 1988. 15. Decker and Erman 1992. 16. D. Erman, University of California, Davis, pers. comm. 1998. 17. G. R. Smith 1992.

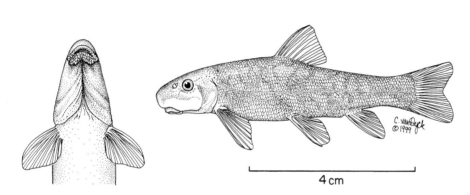

Figure 65. Santa Ana sucker, 6 cm SL, San Gabriel River, Los Angeles County.

Santa Ana Sucker, *Catostomus santaanae* (Snyder)

Identification Santa Ana suckers resemble mountain suckers, to which they are closely related. They are small (usually <16 cm SL) and have deep notches at the junctions of the upper and lower lips, with a shallow median notch in the

lower lip that allows 3–4 rows of papillae to cross it. Papillae are large on the lower lip and distributed in a convex arc on the anterior portion. The jaws have cartilaginous scraping edges inside the lips. The fontanelle beneath the skin on the top of the head is closed in fish larger than 7 cm SL. There are 21–28 gill rakers on the external row of the first arch and 27–36 on the internal row. There are 67–86 lateral line scales, 9–11 (usually 10) dorsal fin rays, and 8–10 pelvic fin rays. The axillary process at the base of the pelvic fins is simple. The caudal peduncle is deep, measuring 8–11 percent of SL. The intestine is long, with up to 8 coils, and the peritoneum is black. Color in living fish is silvery white on the belly and dark gray on the sides and back, with irregular dorsal blotches on the sides and faint patterns of pigmentation arranged in lateral stripes (1). The membrane between the rays of the caudal fin is pigmented, whereas the anal and pelvic fins usually lack pigment. Breeding males have tubercles on most parts of the body, although they are heaviest on the anal fin, caudal fin, and lower half of the caudal peduncle. Females grow tubercles on the caudal peduncle and fin.

Taxonomy *Catostomus santaanae* was originally described as *Pantosteus santa-anae* by Snyder (2) from the Santa Ana River, Riverside County. In a subsequent revision of the nomenclature (1) the hyphen was omitted from the specific name and the genus reduced to a subgenus of *Catostomus*. Santa Ana suckers exhibit higher variability in anatomical characteristics than other members of the subgenus *Pantosteus* (1), such as the number of papillae on the anterolateral corners of the lower lip, pigmentation of the caudal interradial membrane, and development of the axillary process. Within the species, however, there is little differentiation among populations from the three adjacent but isolated rivers (1), and individual populations show limited genetic variation (3). Santa Ana suckers hybridize with introduced Owens sucker in the Santa Clara River (3).

Names Both common and trivial names are after the Santa Ana River, from which the first specimens were collected.

Distribution Santa Ana suckers are native to the Los Angeles, San Gabriel, Santa Ana, and Santa Clara river systems of southern California (1). In the Los Angeles and San Gabriel River drainages they once occurred downstream to the mouths (8) but are now restricted to the larger stream sections that still exist in headwater areas (4). In the Santa Ana River they survive only in the lower portions, mainly in reaches with flows enhanced by waste water (Mt. Roubidoux downstream to a few kilometers below Imperial Highway). They have been extirpated from the upper Santa Ana River drainage, where they were once present in Fish and Santiago Canyons and in Cajon and City Creeks (4). In the Santa Clara River, Santa Ana suckers were first collected in the 1930s and are therefore often regarded as introduced (1, 5). However, it is possible they are native (3). They are widespread in the drainage, occurring downstream to near the mouth (8). They hybridize with the Owens sucker in the vicinity of Fillmore (4). Fish upstream in the Soledad Canyon area are pure Santa Ana suckers (3).

Life History Santa Ana suckers live in small to medium-size (<7 m wide) permanent streams in water ranging in depth from a few centimeters to a meter or more (1, 6). They require cool (<22°C), flowing water, with flows ranging from slight to swift. Although Santa Ana suckers are usually found in clear water, they tolerate seasonal turbidity. Preferred substrates are generally gravel, rubble, and boulder, but occasionally they are found on sand or mud substrates.

Santa Ana suckers are often associated with algae, but not with macrophytes.

The best description of present-day Santa Ana sucker habitat is provided by Deinstadt et al. (6) for the West Fork of the San Gabriel River. The West Fork is a small (typical summer flow 0.1 m^3/sec, width 5–8 m, depths mostly 15–30 cm), permanent stream that flows through a steep, rocky canyon with chaparral-covered walls. Overhanging riparian plants, mainly alders and sedges, provide cover for fish. Santa Ana suckers use all areas and do not require streamside cover when larger, deeper holes and riffles are present. In the Santa Ana River suckers concentrate in tributaries or in sections of river that are fed by high-quality effluent from sewage treatment plants. Greenfield et al. (7) recorded Santa Ana suckers entering the Santa Clara River from a recreational lake. However, they probably do not usually inhabit reservoirs, because they are not known from Piru, Morris and San Gabriel Reservoirs (8).

Streams in southern California are subject to periodic, severe flooding that results in drastic decreases in sucker populations (7). Santa Ana suckers, however, are adapted for living in such unpredictable environments and quickly repopulate following floods. Such adaptations include short generation time (early maturity), high fecundity, and a relatively prolonged spawning period. These characteristics enable Santa Ana suckers to recolonize streams rapidly by producing more young over a longer time span. The short generation time allows Santa Ana suckers to reproduce early in life, as the probability of adult mortality is high. The small size also probably enables individuals to utilize a greater range of instream refuges than would be available to larger fish during high flows.

Like mountain suckers, Santa Ana suckers feed mostly on algae (especially diatoms) and detritus, which they scrape from rocks and other surfaces. In the Santa Clara River 98 percent of their diet consists of algae and detritus, although small numbers of aquatic insect larvae are also taken (7). Larger fish generally feed more on insects than do smaller fish.

Age and growth studies are difficult because Santa Ana suckers lack strong annuli on the scales. Nevertheless, by examining otolith and length frequency distributions, Greenfield et al. (7) found that *(1)* at the end of their first 6 months of life, Santa Ana suckers from the Santa Clara River averaged 33 mm SL; *(2)* they matured during their second summer and usually died at the end of their third summer at 75–110 mm SL; *(3)* a few suckers lived through a fourth summer (age III+), reaching 140–160 mm SL; and *(4)* males and females grew at the same rate.

Spawning is from mid-March to early July, with peak activity usually in April. Fecundity appears to be exceptionally high for a small sucker species, ranging from 4,423 eggs in a female measuring 78 mm SL to 16,151 eggs in a female measuring 158 mm SL (7), although the high counts may be based on immature eggs (8). Fecundity appears to increase with body weight in a linear fashion (7).

Spawning takes place over gravelly riffles, and spawning behavior is presumably similar to that of other stream catostomids. Fertilized eggs are demersal and adhesive and hatch within 36 hr (at 13°C). The development of embryos and larvae is described by Greenfield et al. (7). The mouth becomes subterminal in position when larvae reach 16 mm SL.

Status IB. In January 1999 the USFWS determined that the Santa Ana sucker merited listing as a threatened species, citing massive habitat change and introduced species as causes of its decline (11). The native range of the species is largely coincident with the Los Angeles metropolitan area, so it is not surprising that most populations have declined or been extirpated. The status of the Santa Ana sucker in each of its drainages is as follows (8, 9, 10):

Los Angeles River. Once widespread in this drainage, Santa Ana suckers have been found in recent years only in lower Big Tujunga Creek, in 20–30 km of stream below Big Tujunga Dam. The population appears to be hanging on, although it shows wide fluctuations in numbers (8).

San Gabriel River. The Santa Ana sucker is still fairly common in this drainage, although the population numbers fewer than 5,000 fish in most years. They inhabit about 40 km of the contiguous West, North, and East Forks of the San Gabriel River, but the North Fork population is very small. The West Fork population exists mainly below Cogswell Reservoir, where it is subject to the vagaries of regulated flows. The San Gabriel River population is mostly found in Los Angeles National Forest, but it is likely to persist only under appropriate land management.

Santa Ana River. A population of a few hundred to a few thousand fish exists in the seminatural stretch of river between Prado Dam (a flood control structure) and a concrete drop structure at Weir Canyon Road, Yorba Linda (8, 11). Below this area, the river channel is cleared and channelized, providing little habitat. Upstream of Prado Dam, another smaller population exists in about 6 km of stream between Norco and Riverside, mainly in effluent from sewage treatment plants (8, 11). Much of the bottom is sand, so the limited gravel-bottomed areas near Riverside (which are separated from downstream areas by impassable drop structures) are presumably crucial to the survival of the population as spawning areas (8). Most water is diverted into settling ponds between the two reaches, and suckers do not survive in the ponds (8). Water quality is constantly threatened by many and various local inputs. The fluctuations in sucker numbers, combined with water quality and other problems associated with urbanization, indicate that the Santa Ana River population is not secure.

Santa Clara River. Santa Ana suckers are still present in the lower part of the main river from the estuary to a few kilometers upstream from the mouth of Sespe Creek (7, 11). They are also present in Sespe Creek and in the Soledad Canyon reach of the main river. The biggest population appears to be that in Sespe Creek, where hybridization with Owens suckers has occurred. The most secure population is that in Soledad Canyon, although numbers were greatly reduced during the 1985–1992 drought.

The Santa Ana sucker is threatened by elimination or alteration of its stream habitats, reduction or alteration of stream flows, pollution, and introduced species. It is adapted for surviving extreme environmental perturbations, so populations can recover from disasters provided there is a permanent refuge for a core population. The fact that this fish is in such trouble is indicative of the poor state of streams in the Los Angeles Basin.

In lowland areas virtually all of the habitats once used by this species have been channelized, frozen in concrete, dewatered, or otherwise altered. In upland areas most streams either have been dammed and diverted or are continually threatened by mass erosion of destabilized hillsides (from road building, offroad vehicle use, gravel extraction, forest fires, and development), by gold dredging and other mining activities, and by grazing and other heavy uses of riparian areas. For example, mining activity has increased in recent years in Cattle Canyon, a tributary of East Fork San Gabriel River, resulting in the apparent elimination of sucker populations in the canyon.

A number of the remaining populations of Santa Ana sucker live below dams or in sections of stream dependent on waste water from sewage treatment plants. The flows of Big Tujunga Creek below Big Tujunga Dam vary so much that an artificially enhanced trout population cannot maintain itself, and all native fishes are subject to extirpation, as almost happened to the sucker around 1989 or 1990. The population in West Fork San Gabriel River is constantly threatened by accidental high-water releases (with heavy sediment loads) from Cogswell Reservoir, which have devastated this stream several times in the past. In the Santa Ana River, the main population depends on adequate releases of water from sewage plants in Riverside. The water passes over a series of drop structures in the riverbed, which allow only downstream movements of fish. Upstream of Riverside dams and diversions have eliminated the sucker and its habitat.

Where habitats are suitable, introduced species are a constant threat. For example, the sucker formerly inhabited the upper Santa Ana River in the San Bernardino Mountains but seems to have been eliminated by predation from alien brown trout. Large numbers of Santa Ana suckers exist in the Soledad Canyon area of the upper Santa Clara River, but the potential exists for hybridization with introduced Owens suckers that inhabit the lower river (7). Other populations are continually threatened by introduced species, such as red shiner (a potential competitor and egg predator) and green sunfish (a potential predator).

In the long run, this species will persist only if several streams in its range are managed for native fishes. Immediate steps should be taken to protect their habitats in all drainages, including assurance of adequate flows. Studies on the life history requirements of the species should also be undertaken. As an immediate conservation measure, the East and West Forks of the San Gabriel River should be given status as native fish management areas or refuges, to protect not only the sucker but also other native fishes. Protection of native fishes should have priority over use of the stream for other purposes, including maintenance of the wild trout fishery, gold dredging, and recreation.

References 1. Smith 1966. 2. Snyder 1908b. 3. Buth and Crabtree 1982. 4. Swift et al. 1993. 5. Bell 1978. 6. Deinstadt et al. 1988. 7. Greenfield et al. 1970. 8. C. C. Swift, pers. comm. 1998, 1999. 9. R. N. Fisher, San Diego State University, pers. comm. 1998. 10. T. R. Haglund, University of California, Los Angeles, pers. comm. 1997. 11. Federal Register 64(16): 3915–3923 (1999).

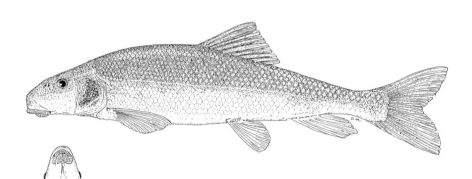

Figure 66. Sacramento sucker, 14 cm SL, Ash Creek, Modoc County. Drawing by A. Marciochi.

Sacramento Sucker, *Catostomus occidentalis* Ayres

Identification Sacramento suckers are "typical" suckers, with subterminal mouths and large fleshy lips covered with papillae (4–6 rows on the upper lip). The lower lip is evenly joined to the upper on both sides and has a deep median indentation with just one row of papillae bridging the two sides. The dorsal fin (11–15 rays, usually 12 or more) is slightly longer than it is high, its origin usually closer to the base of the caudal fin than to the tip of the snout. The anal fin has 7 rays (occasionally 6 or 8). There are 56–75 scales along the lateral line, with 10–17 scale rows above it and 8–10 below it. Adult suckers tend to be greenish to brown on the back and dusky yellow-gold to white on the belly. Spawning fish develop a dark stripe on the sides, which is lined with or is entirely dark red, especially on spawning males. Spawning males (and often females as well) also have numerous tubercles on the pelvic, anal, and caudal fins. Young suckers are gray all over, slightly darker on top, with 3–4 poorly defined splotches on the sides.

Taxonomy The Sacramento sucker is a highly variable species both within and among populations (1, 2). W. O. Ayres described it in 1854, from specimens purchased in a San Francisco fish market (3). Subsequently three other forms were described (4, 5, 6), which became recognized as subspecies. The validity of two of the subspecies has been questioned owing to the lack of strong differences in morphometric and morphological traits (2). However, given the isolation of the four subspecies from one another, I recommend maintaining the subspecies designations until a genetic analysis has also been performed throughout the range of the species. A form to include in such an analysis would be the sucker from the upper Kern River basin, which co-occurs naturally with the distinctive golden trouts.

Catostomus occidentalis occidentalis is the typical form found in the Sacramento and San Joaquin Rivers and tributaries, as well as in the Russian River, Clear Lake, and streams tributary to San Francisco Bay. *C. o. mniotiltus* (Pajaro sucker) is a coarser-scaled form (60–64 lateral line scales versus more than 64 lateral line scales in other forms) found in the Pajaro and Salinas Rivers; it is arguably the most distinctive of the four forms (2, 5). *C. o. humboldtianus* (Humboldt sucker) is confined to the Eel, Bear, and Mad Rivers of Humboldt County (4). *C. o. lacusanserinus* (Goose Lake sucker) is isolated in the Goose Lake watershed; it was originally described on the basis of one specimen (6).

Names Western sucker is a frequently used but unofficial common name. *Cato-stomus* means inferior (down) mouth; *occidentalis* means western.

Distribution The Sacramento sucker is a common, widely distributed species in central and northern California. In the Sacramento–San Joaquin drainage it occurs in streams and reservoirs from the upper Goose Lake basin in Oregon to the upper Kern River in the San Joaquin–Tulare drainage. On the coast it occurs in the Mad, Bear, Eel, Navarro, Russian, Pajaro, and Salinas Rivers as well as Lagunitas Creek (Tomales Bay). It has been moved through water transfers into Cayucos Creek, Whale Rock Reservoir, Chorro Creek, and Morro Creek, all in the Morro Bay drainage (San Luis Obispo County), but the status of these populations is uncertain (7). Sacramento sucker can be expected to be found in southern California reservoirs, after being transferred there via the California Aqueduct.

Life History Sacramento suckers are found in a wide variety of waters from cold, rapidly flowing streams to warm sloughs to low-salinity sections of the San Francisco Estuary. They are most abundant in clear, cool streams and rivers (8, 9) and in lakes and reservoirs at moderate elevations (200–600 m). Adults are most numerous in larger streams; juveniles are often most abundant in tributary streams or shallow reaches of large rivers where adults have spawned. They are typically associated with native minnows (especially Sacramento pikeminnow, hardhead, and California roach), but it is common to find them in waters otherwise dominated by alien species. Different sizes are found in different microhabitats (10, 11, 12, 28). Larval suckers (<14 mm SL) concentrate over detritus bottoms or among emergent vegetation in warm, protected stream margins. Juvenile suckers (<50 mm SL) stay on or close to the bottom, foraging in shallow (20–60 cm), slowly flowing (<10 cm/sec) water along stream margins. Smaller fish seek the shallowest water. In the absence of predators such as pikeminnow, juvenile suckers use deeper water (13). During the day subadult and adult suckers are usually found in deep water of pools and runs or beneath undercut banks near riffles. Large suckers seek areas where they are relatively safe from avian predators (herons, osprey) and where stream velocities are less than 40 cm/sec. In clear streams large suckers are mostly found either in deep cover or in deep pools during the day (28).

Sacramento suckers are not particularly fussy when it comes to choosing water temperatures (10). They can be found in streams where temperatures rarely exceed 15–16°C and in streams where temperatures may reach 29–30°C (14). Preferred temperatures seem to be around 20–25 °C, which may be optimal for growth (15). In the laboratory 36°C is the upper lethal temperature for suckers acclimated to warm water (15). Suckers also seem to have fairly high salinity tolerances; large adults have been found in Suisun Marsh, living in salinities exceeding 13 ppt.

Suckers often occur in small, loose groups of foraging fish. Feeding can be an almost continuous activity, but usually suckers are most active at night. In streams adults spend the day browsing or resting on the bottom of deep pools or in flowing areas with strong surface turbulence (28), moving up into riffles to forage in the evening. In lakes they spend daylight hours in fairly deep water, moving into shallows to feed at night. Feeding activity is greatly reduced during the colder months of the year. Then dense aggregations of large suckers are sometimes found underneath ledges and logs in deep pools of large rivers. Sacramento suckers can colonize new habitats rapidly; sections of stream that were dry during a severe drought were reinhabited by suckers within a year of return to normal flows (28).

The food of Sacramento suckers is much like that of other suckers: algae, detritus, and small invertebrates associated with the bottom. In Thomes Creek, Tehama County, the digestible portion (i.e., that excluding sand) of the gut contents of adults ranged from 50 percent (by volume) invertebrates in winter to 1–12 percent invertebrates at other times of the year; the remainder was detritus and algae (16). In Hat Creek, Shasta County, in September, suckers more than 40 cm long had algae (mostly diatoms) making up 40 percent of the gut contents (17). The bulk of the remaining portion consisted of invertebrates, especially chironomid and caddisfly larvae. In smaller suckers (11–22 cm FL) hydracarinid mites and blackfly larvae were also important. In suckers less than 9 cm long cladocerans were most important (17). In the Russian River (August 1973) postlarval suckers with their terminal mouths and short digestive tracts were surface and midwater feeders on early instars of aquatic insects (18). As they transformed into juveniles, with subterminal mouths and long intestines, their food consisted mostly of diatoms, filamentous algae, and protozoans. Small juveniles (24–38 mm SL) ate a wide variety of small organisms, as well as indigestible items such as sand grains, suggesting development of the bottom-browsing habits of adults. The bulk of their diet, and that of large adults, consisted of filamentous algae, diatoms, and detritus. Invertebrates made up less than 20 percent of the diet.

Growth in Sacramento suckers is as variable as their habitats. In the upper Merced River (Yosemite National Park), where the water is cold year round, yearling fish averaged only 47 mm SL (19). Suckers of comparable age in the lower Merced River and Hat Creek averaged 80 mm SL (17, 20). In a coldwater section of North Fork Feather River, yearling suckers averaged 74 mm FL, while in a warmer section they averaged 145 mm FL (21). In Ruth Reservoir, Trinity County, they averaged 94 mm FL (29). Record growth apparently occurred in a newly established population in Whale Rock Reservoir, San Luis Obispo County, where suckers reached 174 cm FL in their first year (26). Thereafter annual increments in length range from 12 to 87 mm, averaging around 40 mm, although the rate slows down in older fish. Fish at first maturity (usually ages 4–6) range from 200 to 320 mm FL, depending on the stream or reservoir (16). In the North Fork Feather River, where it was unlikely (because of a fish poisoning operation) that any suckers were older than 10 years, 7- to 10-year-old fish measured 350–420 mm FL (21), an age-size relationship consistent with that reported from other studies (16, 29). However, a sucker measuring 560 mm FL, from the cold waters of Crystal Springs, Shasta County, was determined to be 30 years old using opercular bones, and it is quite likely that many suckers over 400 mm are considerably older than 10 years (22). Male and female suckers grow at the same rates, but the largest and oldest fish are generally female (16).

It is fairly typical for a sucker population to have a nonuniform age structure, with strong year classes indicating that spawning or survival of young is not completely successful every year. Reproductive success is highest during wet years, when high flows provide increased access to spawning habitat and flood shallow areas favored by larvae and small juveniles, reducing predation, especially by alien fishes. When sucker populations are hit with a major disaster, such as attempts to eliminate populations by poisoning, the few survivors may have extremely successful reproduction in succeeding years and flood the environment with young; one or two strong age classes may march through the population, inhibiting reproduction through competition for food and space and eventually creating a dense population of large suckers, often in poor condition (17, 22).

Spawning first occurs during the fourth, fifth, or sixth year and is often preceded by a migration to a spawning stream, typically a tributary to a large river or reservoir. Ripe suckers in lakes and reservoirs often congregate at the mouths of streams prior to migration, and they may start moving into spawning streams as early as late December. The immediate trigger for spawning runs from Pine Flat Reservoir–Kings River, Fresno County, seemed to be sudden warming of inflowing creeks after a series of warm days (24), although increases in flow were also implicated (23). During a 5-year period, spawning runs began at temperatures ranging from 5.6 to 10.6°C. A sudden cooling spell halted migration until the water warmed up again. Once a migration has started, suckers may move considerable distances (more than 50 km) upstream. In some streams the number of spawning migrants can be in the thousands; in Thomes Creek, as many as 240,000 suckers from the Sacramento River have been estimated to be present in spawning reaches (16). Most spawning takes place over gravel riffles between late February and early June, although peak spawning usually occurs in March and April (16, 23). Goose Lake suckers apparently spawn mainly in late April and May (27). However, the presence of larval suckers in mid-August in the Russian River and other coolwater streams indicates that spawning can take place as late as early August (25). Spawning temperatures are usually 12–18°C (25).

Limited spawning also takes place in lakes and reservoirs. I have observed shoreline spawning in Pine Flat Reservoir, where temporary streams flow into the lake. In creeks with resident populations of suckers, spawning adults will move from pools to riffles for spawning in response to increases in flows. In lower Putah Creek, Yolo County, spawning can occur as soon as early February if flows increase sufficiently to cover spawning gravels to a depth of 30 cm or more. When flows drop, spawning ceases, but it resumes when flows increase again. However, when flows are continuously high, spawning can be initiated in Putah Creek with no obvious flow or temperature cues.

Sacramento sucker spawning behavior is like that of other catostomids. Large numbers congregate in a spawning area, with each spawning female accompanied by 2–7 males. Vigorous splashing during spawning by the female and closely attending males creates a slight depression in the gravel. In February 1999 I observed spawning in Putah Creek, Yolo County. A sinuous line of five or six fish, accentuated by the dark stripe on their sides and their orangish tails, weaved about, headed upstream. The lead fish (the female) started to tremble and dipped downward, compressed between 3 or 4 of the following males. When the group hit bottom, eggs and sperm were released, and the water around the suckers suddenly became cloudy with a puff of brown silt, which drifted downstream. At one point at least five groups of suckers spawned simultaneously, and the entire spawning area erupted in splotches of drifting brown.

The fertilized eggs either adhere to gravel or bits of debris or else bounce along the bottom until they are caught in the gravel or washed to a small backwater (16). Although spawning can occur at any time of the day or night (16), it frequently seems to peak in early morning. Most females presumably release their eggs over a fairly short period of time once spawning has started. Individual females may spawn in as many as 7 years (16).

Fecundity is highly variable, with only a weak positive relationship to the size of the female. In Alpine Lake, Marin County, females measuring 28–38 cm FL contained 4,700–11,000 eggs (20), whereas in Thomes Creek females measuring 32–48 cm FL contained 10,300–32,300 eggs (16).

The embryos hatch in 2–4 weeks, and the larvae remain in or among the gravel. The postlarvae emerge and are soon washed into warm shallows or among flooded vegetation, where they often occur in large aggregations. In Thomes Creek there was a mass exit of postlarval and small juvenile suckers (10–30 mm FL) in a 3-week period in May, but larger juveniles (59–90 mm FL) moved down to the Sacramento River in small numbers continuously as long as flows were high enough to permit it (16). Most of the larger juveniles were presumably holdovers from the previous summer. In some spawning streams juveniles will spend 2–3 years in the stream before finally moving down to a large river or reservoir during high flows. In streams with resident populations of suckers, juveniles stay in shallow, dense cover as long as possible, as a haven from predators, especially centrarchid basses.

Status IF, except Goose Lake sucker, ID. The Sacramento sucker is one of the few species of native fishes that has thrived despite massive changes to California's waterways. Although it is scarce or absent from many lowland habitats where it once occurred, it has expanded its populations in many areas by taking advantage of reservoirs and regulated streams. It has also persisted in many now-isolated streams because of its ability to both withstand adverse environmental conditions and flood the environment with young when favorable conditions return. Its extraordinary ability to recover from disasters is reflected in the general failure of fisheries agencies to eliminate it from streams where it was perceived to be competing with trout for food and space. After a major poisoning operation, the population is usually back to its former abundance (or higher) within 6–9 years (21). An exception to this "rule" is lower Hat Creek, Shasta County, which was poisoned in 1968 to eradicate suckers that dominated the fish fauna (30), and where suckers are at present a minor part of a fish fauna dominated by rainbow trout. The suckers did not return in part because a barrier was constructed downstream before the poisoning operation; it prevents use of the creek by suckers moving up from Britton Reservoir (30). Other factors may also have contributed to continued suppression of sucker populations, such as maintenance of large populations of predatory trout.

In fact, the idea that suckers cause rainbow trout populations to collapse is largely a myth. When trout populations are small and sucker populations are large, the circumstance is usually due to a combination of factors, such as water temperatures or flow regimes that are marginal for trout, high mortality of trout from disease and stress, and removal of large trout by anglers. In natural situations the two species show strong segregation in use of resources (10). In addition, small trout can be observed following big suckers around, feeding on invertebrates stirred up by their browsing. Young-of-year suckers are prey for trout on occasion. More important, larval suckers are typically abundant when fall-run juvenile chinook salmon are moving downstream and are often heavily fed upon by the salmon (32, 33, 34). They may thus contribute to rapid growth and increased survival rates of juvenile chinook.

The ecological role of Sacramento suckers in streams is poorly understood and worthy of study. Among potential major roles for the species are the following: *(1)* keystone species affecting the composition of invertebrate communities through grazing (directly and indirectly), *(2)* rearing substrate for parasitic glochidia larvae of increasingly rare native mussels, *(3)* high-energy food resource for juvenile salmon and trout, and *(4)* prey for otters, ospreys, bald eagles, herons, and other predators. Spawning aggregations of large suckers are especially important for eagles just before or during their nesting season, as an easy source of energy (31, 35).

The Goose Lake sucker is listed as a state Species of Special Concern because the Goose Lake basin has been altered by agriculture, especially diversions that block spawning migrations and may cause Goose Lake to dry up more rapidly under conditions of severe drought (27). Riparian grazing has also reduced cover, pool depth, and other factors important to suckers in streams. On the other hand, reservoirs created by the diversions also serve as drought refugia for suckers. After a crisis created by drought dried up the lake 1992, cooperative efforts between agencies and landowners resulted in a better understanding of the distribution of suckers (e.g., its presence in ranch reservoirs) and in other efforts to improve conditions for all fish in the basin (27).

References 1. Martin 1967. 2. Ward and Fritzsche 1987. 3. Ayres 1854. 4. Snyder 1908c. 5. Snyder 1913. 6. Fowler 1913. 7. Swift et al. 1993. 8. Moyle and Nichols 1973. 9. Brown and Moyle 1993. 10. Baltz and Moyle 1984. 11. Moyle and Baltz 1985. 12. Baltz et al. 1987. 13. Brown and Moyle 1991. 14. Cech et al. 1990. 15. Knight 1985. 16. Villa 1985. 17. Brauer 1971. 18. J. Norton and J. White, unpubl. data 1974. 19. Hubbs and Wallis 1948. 20. Burns 1966. 21. Moyle et al. 1983. 22. Scoppetone 1988. 23. Mulligan 1975. 24. B. Cates, unpubl. ms. 1971. 25. Wang 1986. 26. Underwood 1983. 27. Moyle et al. 1995. 28. G. E. Smith 1982. 29. Kisanuki 1980. 30. Smith et al. 1999. 31. Jackman et al. 1999. 32. Merz and Vanicek 1996. 33. Merz 1998. 34. P. B. Moyle, unpubl. data. 35. Hunt et al. 1988.

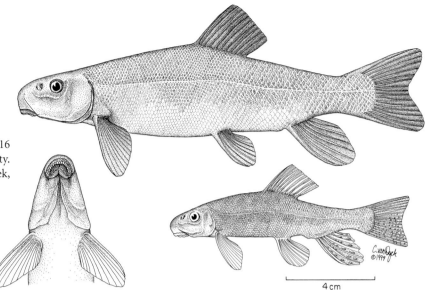

Figure 67. Modoc sucker. Top: Female, 16 cm SL, Johnson Creek, Modoc County. Bottom: Male, 9 cm SL, Washington Creek, Modoc County.

Modoc Sucker, *Catostomus microps* Rutter

Identification Modoc suckers are small (usually <16 cm SL), with short heads (head length divisible into standard length 4–5 times), small eyes (orbit width less than 6% of SL), and small scales (79–89 [usually >81] in the lateral line). Scales below the lateral line number 9–12; scales above the lateral line number 14–19 (usually 15) (1, 2, 3, 4). There are 10–11 rays in the dorsal fin, 7 in the anal fin, and 8–10 in the pelvic fins. Recent studies indicate that ranges of meristic counts may be somewhat broader than those given here (13). The axillary process is absent from the pelvics. The lower lip has a deep medial notch, with only one of 5–6 rows of papillae connecting the two halves. The upper lip has 2–4 (usually 2) continuous rows of papillae. The fontanelle beneath the skin on top of the head is usually closed, or nearly so. Alive, they are deep gray to greenish brown above, changing to yellow or white on the belly. Breeding males develop an orange-red lateral band, orange fins, and breeding tubercles on the fins and body. Breeding females are less colorful and have few or no tubercles.

Taxonomy The Modoc sucker presents an interesting taxonomic and zoogeographic puzzle. A study of its meristic and morphometric characters suggested that its closest relatives are two species of suckers (*C. wigginsi, C. leopoldi*) that occur in north-central Mexico, and the Klamath smallscale sucker (12), leading to speculation that these species were derived from an ancestral species once widespread throughout the Southwest (12). However, genetic studies indicate that it is most closely related to the Sacramento sucker, with which it co-occurs (13). The distribution of Modoc suckers into a small number of isolated populations suggests that there may be interesting intraspecific taxonomic patterns as well, representing either two separate invasions of suckers from the Sacramento River or repeated evolution of the Modoc sucker phenotype. Although some hybridization takes place where the two species co-occur, it is apparently rare and insufficient to create problems for the Modoc sucker (13).

Names Modoc suckers occur primarily in Modoc County, which was named for the Modoc Indians, who otherwise have a history of being treated very badly by our civilization. "Modoc" is the English mangling of the name of the Modoc people for themselves, People of Tule Lake, roughly "Moatakni maklaks" (5). *Cato-stomus* means inferior (down) mouth. *Microps* means small eye.

Distribution Modoc suckers were originally described from Rush Creek, Modoc County, a major tributary to Ash Creek, and were thought to be confined to that watershed (1). At present populations are known from two small watersheds in the upper Pit River watershed and tributaries to Goose Lake in Oregon. The Pit River populations are in the Ash Creek and Turner Creek watersheds. In the Ash Creek watershed fish are found in Johnson Creek (tributary to Rush Creek), Rush Creek (probable), Dutch Flat Creek, and Willow Creek (apparently a hybrid population) (6, 7). In the early 1970s small numbers were still present in upper Ash Creek, Lassen County (8), but it is unlikely that any exist there today. In the Turner Creek watershed fish are found in Turner, Hulbert, and Washington Creeks, as well as Coffee Mill Gulch (6, 7) and Garden Gulch (13). Modoc-like suckers are present in Cedar Creek, formerly a tributary to the South Fork Pit River, which now flows into Moon Reservoir (3, 15), although it appears they are most closely related to Tahoe suckers and so may represent a separate species or invasion (13). In Oregon Modoc suckers were originally reported from Bauers Creek (Lake County) in the Goose Lake drainage in the 1930s (14), a record that was largely ignored or forgotten despite the presence of museum specimens (13). They were rediscovered in 1997 by Stewart Reid of the USFWS, who also found records of additional fish from nearby Thomas Creek (13). The Bauers Creek population was confirmed to be present in 2001 (13).

Life History Modoc suckers are pool dwellers in a few small, often intermittent, headwater streams flowing through meadows and dry forests (elevation 1,286–1,567 m). Sections of stream in which Modoc suckers live are characterized by moderate gradients, low summer flows, and high spring flows fed by local snowmelt. They are most abundant in reaches dominated by large mud- and rock-bottomed pools partially shaded by overhanging trees and shrubs and containing cool (<25°C), moderately clear water. Deep (1–2 m) pools may be essential as drought refuges. Within pools there is some segregation by size, with the smallest fish occurring among rocks in shallow water and larger fish in the deepest areas (0.5–2.0 m), near or under overhanging tree roots and plants. They are largely absent from sections dominated by riffles, including channelized sections (8).

Even in areas where they are most abundant, Modoc suckers seldom dominate the fish fauna; they usually make up less than 20 percent of the fish present. They are commonly associated with speckled dace, rainbow (redband) trout, Pit-Klamath brook lamprey, California roach, and Pit sculpin, and occasionally with Sacramento sucker, brown trout, and Sacramento pikeminnow. The abundance of the latter four species is negatively correlated with the abundance of Modoc suckers (8). This is partly the result of different habitat preferences (sculpin), but predation may play a role in the cases of brown trout and pikeminnow. Modoc suckers have been found in the stomachs of brown trout, which is known to reduce sucker populations in other streams.

The feeding habits of Modoc suckers are like those of other sucker species: more than 75 percent detritus and algae, the rest aquatic insect larvae and crustaceans that live in or on muddy substrates or among filamentous algae. Chironomid midge larvae seem to be particularly important (8). However, the jaw structure of these suckers indicates that they may be better adapted to scraping algae from rocks than most other sucker species (2).

Modoc sucker growth rates for the first 4 years of life are similar to those of other suckers in small streams. Thus they average 7 cm SL at 1 year, 11 cm SL at 2 years, 14 cm SL at 3 years, and 18 cm SL at 4 years. However, unlike most other sucker species, including Sacramento sucker, they apparently seldom grow larger than 15 cm SL (8). The largest and oldest Modoc sucker known was 28 cm SL and 5 years old (as determined from scales). The typical small size of Modoc suckers may be a response to small, cool streams rather than the result of an intrinsic limit on maximum size. The largest collection of Modoc suckers measuring more than 15 cm SL was taken from a warm irrigation ditch along Rush Creek that had deep (2–3 m), permanent pools. The largest and oldest fish are typically females.

The small size and short lives of Modoc suckers are partially compensated for by maturation at an early age. Most males and females mature in their third year at about 12 cm SL. A few males mature during their second year. Spawning takes place over fine gravel in the lower end of pools or in riffles between mid-April and early June. When stream flows increase, the suckers move upstream, typically into small tributaries, for spawning. Spawning has been observed from midmorning to late afternoon, at temperatures of 13–16°C (9), in water around 15 cm deep. Spawning behavior is similar to that of other suckers: Males enter the spawning grounds first and wait for females. When a ripe female enters, 2–3 males quickly assume positions on each side, and milt and eggs are released simultaneously (9). Fertilized eggs drop into interstices in the gravel. Females have a fairly high fecundity for their size, compared with other suckers: two females, measuring 162 and 165 mm SL, contained 6,395 and 12,590 eggs, respectively (8).

Status IB. The Modoc sucker is formally listed as an endangered species by both state and federal governments. When Alan Marciochi and I studied these fish in the early 1970s, their situation was regarded as perilous, because of the poor condition of their California streams. These fish barely managed to persist through two periods of severe drought, but today their situation is secure enough that upgrading the species to threatened status can be seriously considered. The rediscovery of the Oregon populations lends additional credence to this suggestion, although the systematic interrelationships of the various populations must be worked out first. The interacting factors that contributed to the endangered status of the suckers were *(1)* isolation, *(2)* stream channelization, *(3)* grazing, *(4)* water diversions, and *(5)* brown trout predation.

Isolation. Modoc suckers are known from only two widely separated watersheds of the Pit River, Ash Creek and Turner Creek, and from two streams in the upper Goose Lake basin. The natural isolation of the watersheds from one another does provide some security for the species in that a disaster affecting one system is not likely to affect the others. All have large enough drainages that extinction due to natural causes is unlikely if the watersheds remain in good condition. However, in Ash Creek the once widespread sucker is now found in populations in isolated small headwater streams, for which local extinction is always a threat (7).

Stream channelization. Until fairly recently, it was common practice in the Pit River region to straighten out and dredge streams flowing through meadows, to reduce flooding and increase grazing time in spring. The long-term consequences of such channelization are often negative even in terms of grazing benefits (e.g., dry meadows become invaded by junipers and sagebrush), but they are disastrous for Modoc suckers. Channelization eliminates pools, so long sections become unsuitable habitat (10).

Grazing. The watersheds supporting Modoc suckers have been subjected to grazing by cattle and sheep for over 100 years, on both private and public land. Loss of riparian vegetation from grazing—combined with roading, logging, and other practices that cause water to run off the land faster—has resulted in severe downcutting of stream channels in the Washington Creek watershed, although enough deep pools still remain to support suckers. Everywhere, concentration of cattle on stream banks has led to reduction of cover through slumping banks, elimination of overhanging plants, and sedimentation of pools. Conditions have typically been worst on private land, where cattle and sheep have been kept in fenced pastures next to streams that have often been channelized as well.

Water diversions. In 1977 there were at least 26 diversions on Modoc sucker streams in California (3), and many of them still exist today. Water in Modoc sucker streams is diverted for two main reasons: irrigation of pastures and Christmas tree farms, and elimination of meadows too soggy for grazing. Irrigation diversions have been a particular problem in Ash and Rush Creeks, where diversions reduce flows and dams probably reduce or eliminate movement of fish. Dutch Flat Creek originally flowed through a wet meadow but was diverted to flow along one edge of the meadow, resulting in downcutting to bedrock (and drying up of the meadow, now a juniper flat). In both situations habitat for Modoc sucker has been reduced.

Brown trout predation. Brown trout are the most piscivorous of nonnative trouts and have a well-deserved reputation for reducing or eliminating populations of other fishes. When large trout are found in pools in Modoc sucker areas, suckers are scarce and tend to be found in the stomachs of the trout.

With all these simultaneous threats, it is a bit of a miracle that the Modoc sucker did not become extinct before interest in its conservation developed. Many active measures have been taken by the USFS, BLM, CDFG, and other agencies to restore Modoc sucker populations by improving their habitat and reducing the likelihood of further invasions of unfriendly species. Livestock have been fenced out of riparian zones along many streams, substantially improving cover for suckers (11). Barriers to upstream movement of Sacramento suckers have been constructed on lower Johnson and Turner Creeks. Stream improvement measures, including deepening of pools and attempts at bank stabilization, have been instituted in Turner Creek. In 1990 an effort to eliminate green sunfish, bluegill, and largemouth bass from Washington Creek through poisoning was unsuccessful, but it may nevertheless have given the sucker populations a boost. Key habitat on private land in Dutch Flat Creek was purchased, and fencing, bank stabilization, and other measures were taken to improve the habitat. These measures and others have made the future of the Modoc sucker much brighter. Much more, of course, still needs to be done, such as rehabilitating other stream reaches on private land, eliminating brown trout and other alien fishes, and restoring the suckers, where feasible, to other tributaries to Ash Creek.

References 1. Rutter 1908. 2. Martin 1967, 1972. 3. Ford 1977. 4. Cooper 1983. 5. Pease 1965. 6. Mills 1980. 7. Scoppetone et al. 1992. 8. Moyle and Marciochi 1975. 9. Boccone and Mills 1979. 10. Moyle 1976. 11. Yamagiwa 1996. 12. G. R. Smith 1992. 13. S. Reid, USFWS, pers. comm. 2001. 14. Schultz and DeLacey 1935. 15. P. B. Moyle, unpubl. obs.

Tahoe Sucker, *Catostomus tahoensis* Gill and Jordan

Identification Tahoe suckers have large heads (head length divisible 4 times into SL), long snouts (half of head length), and fine scales (82–95 in the lateral line, 16–19 rows above it, 12–15 rows below it). The caudal peduncle is thick, the least depth divisible 12 times into SL. Their subterminal mouths are large, with papillose lower lips so deeply incised that usually only one row of papillae crosses completely. The upper lip has 2–4 rows of papillae. The frontoparietal fontanel (on top of the skull, beneath the skin) is usually open. The dorsal fin has 9–11 rays; the anal fin, 7; and the pectoral fins, 14–16. Live fish tend to be dark olive on the back and upper half of the sides, the dark contrasting sharply with the yellow or white of the belly and lower half of the sides. There is typically a well-defined lateral band. In

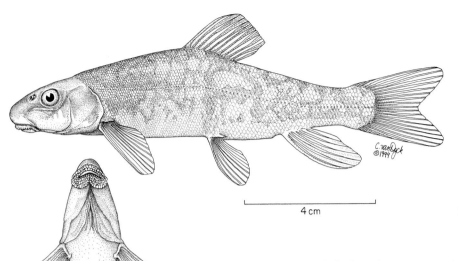

Figure 68. Tahoe sucker, 11 cm SL, Martis Creek, Placer County.

4 cm

breeding males this band becomes a bright red stripe running across brassy-colored sides. Breeding tubercles are well developed on the anal and caudal fins. Larvae are described by D. Snyder (21).

Taxonomy There seems to be little controversy over the taxonomy of this species, which is related to other "standard" suckers in Western drainages (20). The sandbar sucker, *C. arenarius* (1), has proved to be the same species as *C. tahoensis* (2). A complete taxonomic history is given in La Rivers (3).

Names Named for Lake Tahoe, this species has also been called Nevada sucker and red sucker.

Distribution Tahoe suckers are native to the Lahontan drainage system of California and west-central Nevada, including the Carson, Walker, Truckee, and Susan Rivers, and Eagle Lake. They are abundant in natural lakes (e.g., Eagle, Tahoe, Pyramid [Nevada], Independence, Webber) and in reservoirs. Either through canals or thanks to introduction by fishermen, Tahoe suckers have apparently become estab-

lished in the upper reaches of the Feather and Rubicon Rivers in the Sacramento system (4).

Life History Tahoe suckers are the "typical" suckers of the northeast side of the Sierra Nevada in California and Nevada. They occur in a wide variety of habitats but achieve their greatest sizes and numbers in large lakes, such as Lake Tahoe and Pyramid Lake, and perhaps in reservoirs as well. In Lake Tahoe and Pyramid Lake adults tend to be found at depths less than 15 m (5, 6). Miller reported "herds" of large Tahoe suckers in Lake Tahoe at depths of 10–13 m "moving over the bottom and feeding in a manner suggesting the grazing of sheep" (5, p. 155). Occasional suckers can be found as deep as 300 m in Lake Tahoe (7), but they are rare below 60 m in Pyramid Lake (6). In alpine (2,067 m) Webber Lake, Sierra County, small groups of adult suckers roam around at all depths (to 18 m) during the day but show a strong inshore movement to forage in shallow water (<1 m) at night (8). Juvenile suckers (65–140 mm SL) tend to stay in shallow water at all times, forming tight shoals (8). Small (<50 mm SL) suckers are found mainly in spawning streams but occasionally occur in shallow water of lakes as well.

In small streams small suckers select shallow (<40 cm) areas with slow currents (<20 cm/sec), whereas adults select pools and runs deeper than 60 cm, typically in association with heavy cover (9, 10). During the day, adults may rest closely packed together under overhanging banks and other cover, emerging to feed at night (10). Tahoe suckers can also show seasonal movements in and out of streams, especially from reservoirs. In lower Sagehen Creek adult suckers move in from Stampede Reservoir in June through August, partly for spawning and partly for foraging (10, 11). Although Tahoe suckers are found in many waters that rarely exceed 15–16°C in summer, they are also common where summer temperatures may exceed 25°C. They are well adapted for growing and feeding in the fluctuating temperature regimes of small streams (12).

Tahoe suckers are omnivorous bottom feeders. In streams adults ingest a wide variety of organisms, as well as inorganic and detrital material, but mostly algae and small benthic invertebrates (especially larvae of chironomid midges and caddisflies). Juveniles feed mainly on cladocerans and other animal material associated with aquatic plants and beds of algae (9). Invertebrates seem to make up a larger part of their diet in natural lakes (>60%), although detritus, which has some nutritional value, is always significant. Lake suckers consume whatever small forms are abundant in the benthos (8, 13). In Pyramid Lake algae, midge larvae, and small crustaceans found in algal mats are the principal foods of adults (3). In Lake Tahoe midge larvae, amphipods, and annelid worms "in a bulky matrix of sand" are dominant (5, p. 53). In Eagle Lake their summer diet is predominately Tricoptera larvae and pupae and amphipods (14). In Webber Lake molluscs, midge larvae, and amphipods are dominant, although in Stampede Reservoir aquatic macrophytes and midge larvae make up the bulk of the diet (13). Postlarval suckers (<4 cm FL) feed mostly on zooplankton, chironomid larvae, and small terrestrial insects. As they grow larger, they become increasingly bottom oriented. As a result, the variety of organisms taken increases until, at about 13 cm FL, the diet is the same as that of adult suckers (7). In Webber Lake juvenile suckers feed largely on midge larvae and pupae (characteristic of open areas), whereas adults feed more on amphipods (characteristic of aquatic vegetation) (13).

Growth in Tahoe suckers varies, presumably in response to food availability and water temperature (13). Growth is fastest during the first year of life, suckers averaging 40–70 mm SL, and continues to be fairly steady (20–40 mm/year) until maturity is reached at ages 3–6. In streams 5-year-old fish are likely to measure 120–130 mm SL, in lakes 140–160 mm (13, 14). The annual increment is less in older suckers (5–10 mm). The fastest-growing populations studied have been those in Eagle Lake, Lassen County, where 7-year-old fish measure 20–30 cm SL and 10-year-old fish measure 30–35 cm SL (14). Fish larger than 15 cm or older than 7 years are rare in streams, but fish in natural lakes may reach over 60 cm SL and ages of 27 years or more (14, 15). Males and females have similar rates of growth.

Fecundity varies with size and age of female (7, 16). In Tahoe and Pyramid Lakes the number of eggs ranges from 2,400 in a female measuring 15 cm FL to 59,300 in a female measuring 43 cm FL. Mean fecundity is around 20,550 eggs (typical of about a 31-cm female) (16).

Spawning takes place in March–August, the time of year depending on altitude and water temperature, although typically it occurs in March–May. In Tahoe, Pyramid, and Eagle Lakes there appear to be two spawning populations: one that spawns in streams, consisting of fish less than 25 cm SL, and another that spawns in the lakes, containing larger fish as well (7, 16). Lake spawning usually takes place when temperatures are 12–23°C (16). Spawning sites typically have rock and gravel bottoms at depths of 5–18 m, although some spawning may also take place in shallower areas. In streams the preferred spawning grounds are gravel riffles with few large rocks. Stream spawning is generally preceded by nighttime upstream migrations when water temperatures reach 11–14°C; an increase in flows may also stimulate the movement. Historically, Tahoe suckers migrated more than 80 km up the Truckee River from Pyramid Lake in roughly a week (3).

Spawning is described by Snyder (1, p. 43):

The males appear first on the spawning beds and are always represented there in large numbers, each female being attended by from two to eight or more. Twenty-five males were seen attending one female in a pool. Occasionally another female would enter the pool from below, when she would be met and inspected by a school of males and then allowed to pass without further notice. Several of these passing females proved on examination not to be ripe. On account of the presence of so many males nothing definite can be observed of the spawning act, more than that the eggs are extruded and shaken down in the gravel by the female while the males struggle over and under her, churning the water to foam by their activities.

During spawning season, males space themselves evenly on spawning riffles but do not seem to be territorial or even aggressive. When a female approaches they will leave their stations, spawn, and then resume them again (7). Intense spawning activity may result in the creation of shallow, nestlike depressions in sand or gravel (1), although none were observed in runs of suckers from Lake Tahoe (7). The vigorous spawning act, however, does seem to ensure that most of the adhesive yellow eggs get buried in gravel.

Spawning success varies considerably from year to year. Large numbers of young-of-year typically appear during years when there are sustained high flows during spawning (17, 18). This presumably is the result of flooded vegetation, which provides habitat for larval and postlarval fish. This habitat has abundant food (small invertebrates), warm temperatures, and shelter from both predators and high stream velocities. However, when brown trout are abundant in a stream, their predation on small suckers as they emerge from protected habitats may keep populations small (17, 18).

Status IF. Tahoe suckers are common throughout their range and are typically one of the most abundant fish where they occur. Their role in stream and lake food webs is not well understood, but they are often a major prey of large trout (5). There is no evidence that they have a negative influence on game fish populations, and in some streams their numbers may be greatly reduced by brown trout predation (17, 18). They were once an important food source for Native Americans (19). Although no fishery exists for them at the present time, both Snyder (1) and La Rivers (3) reported them to be excellent eating.

References 1. Snyder 1918. 2. Hubbs and Miller 1951. 3. La Rivers 1962. 4. Kimsey 1950. 5. R. G. Miller 1951. 6. Vigg 1980. 7. Willsrud 1971. 8. Marrin 1983. 9. Moyle and Vondracek 1985. 10. Decker 1989. 11. Decker and Erman 1992. 12. Vondracek et al. 1982, 1989. 13. Marrin et al. 1984. 14. P. B. Moyle and students, unpubl. studies. 15. Scoppetone 1988. 16. Kennedy and Kucera 1978. 17. Strange et al. 1992. 18. Strange 1995. 19. Lindstrom 1996. 20. G. R. Smith 1992. 21. D. E. Snyder 1983.

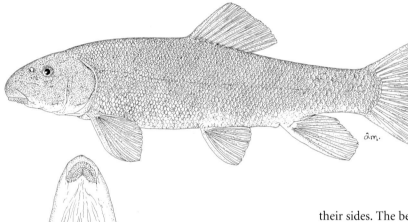

Figure 69. Owens sucker, 28 cm SL, Crowley Reservoir, Mono County. Drawing by A. Marciochi.

Owens Sucker, *Catostomus fumeiventris* Miller

Identification Owens suckers are similar to Tahoe suckers in having large heads, long snouts, coarse scales, and thick caudal peduncles. Their subterminal mouth is large, and the papillose lower lip is deeply incised. Lateral line scales number 66–85 (usually fewer than 80), with 13–16 rows above the lateral line and 9–11 below. Pectoral fins have 16–19 rays; dorsal fin, 10 rays, and pelvic fins, 9–10 rays. Adults are slate-colored dorsally, and this coloration occasionally becomes very dark; they often have weak, blue iridescence on their sides. The belly is a dusky or smoky color. Reproductive adults develop a dark, red-tinged stripe on the sides. The paired fins may be faintly tinged with reddish-amber as well.

Taxonomy Owens suckers were first described as a population of *C. arenarius* (1), a species now merged with *C. tahoensis*. Although they were subsequently recognized as being distinctive by C. L. Hubbs in 1938 (2), they were not formally described until 1973 by R. R. Miller (3). They are a close relative of the Tahoe sucker (11). The introduced population in the Santa Clara River hybridizes with the Santa Ana sucker (4, 5, 6).

Names *Cato-stomus* means inferior (down) mouth; *fumeiventris* means smoky belly, referring to the characteristic color pattern.

Distribution The Owens sucker is endemic to the Owens River watershed in southeastern California and is widely distributed in streams of Owens Valley, including Owens River and Bishop Creek. It is most abundant in Crowley Reservoir, Mono County. Other populations exist in Convict Lake, Mono County, and Lake Sabrina, Inyo County. There is also an introduced population in June Lake in the Mono Lake Basin. A population became established in the 1930s in the Santa Clara River, Los Angeles County, via Owens Aqueduct. It is apparently present in lower Sespe Creek of this drainage, the outflow from Fillmore Trout Hatchery, and Piru Creek and Reservoir (7). Adults have been observed spawning in the outflow from Fillmore Trout

Hatchery in large numbers, although the numbers seem to have declined in recent years (7).

Life History In the Owens River and two of its tributaries, lower Rock Creek and lower Hot Creek, Owens suckers are most abundant in sections with long runs and few riffles (8). The substrate in these sections consists mostly of fine material, with lesser amounts of gravel and rubble. Water temperatures are typically 7–13°C with pH 7.9–8.0. Adults also thrive in lakes and reservoirs, such as Convict Lake (native) and Crowley Reservoir, where they seem to occur on the bottom at all depths.

The life history of Owens suckers is undoubtedly similar to that of the closely related Tahoe sucker (3). They seem to be nocturnal feeders that ingest aquatic insects, algae, detritus, and inorganic matter sucked off the bottom. Age and growth have not been studied, but they rarely exceed 50 cm SL. They spawn from early May to early July. The population in Crowley Reservoir spawns in springs and gravel patches along the lake shore as well as in tributary streams, sometimes in large numbers (10). On 17 May 1975, 500–1,000 adults were observed spawning in a 200-m section of Hilton Creek (10). At about the same time, small numbers were observed spawning in the reservoir at depths of 1–2 m (10). Larvae transform into juveniles when they reach 19–22 mm TL and are usually found in quiet, sedge-dominated margins and backwater areas (3).

Status ID. Owens suckers have adapted to the damming of the Owens River and the creation of Crowley Reservoir, so they still have large populations in a good portion of their native range. Successful introductions into June Lake and the Santa Clara River have also been made. They have showed some capacity to adjust to the presence of non-native fishes; they were once the only fish in Convict Lake, which they now share with alien trout species. However, their total range is limited, and the bulk of their population seems to depend on reservoirs that are dominated by introduced game fishes, so their populations do need to be monitored. Because of their abundance and wide distribution in Owens Valley, they have not yet been introduced into Owens Valley Native Fish Sanctuary north of Bishop, although this facility is available to strengthen their protection should this ever be necessary (9).

References 1. Snyder 1919a. 2. Shapovalov 1941. 3. R. R. Miller 1973. 4. Hubbs et al. 1943. 5. Crabtree and Buth 1981. 6. Buth and Crabtree 1982. 7. Swift et al. 1993. 8. Deinstadt et al. 1986. 9. E. P. Pister, Desert Fishes Council, pers. comm. 1998. 10. C. C. Swift, pers. comm. 1999. 11. G. R. Smith 1992.

Klamath Largescale Sucker, *Catostomus snyderi* Gilbert

Identification This species is similar to other *Catostomus* species, with its short head, subterminal mouth with papillose lips, large scales, and solid body, although the caudal peduncle is thicker than that in most other species. There is a deep medial incision on the lower lip, resulting in only one row of papillae extending across the lip. The upper lip is narrow and has 4–5 complete rows of papillae. The dorsal fin is short (11 [occasionally 10 or 12] rays), with a basal length equal to or shorter than that of the longest dorsal ray. The dorsal fin insertion is closer to the snout than to the caudal fin. There are 7 anal fin rays. Scales are large, 67-81 along the lateral line, 11–14 scale rows above it, and 8–12 rows below it. Gill rakers on the first gill arch number 30–35 (usually more than 32) in adults and 25–28 in juveniles. The rakers in adults have well-defined processes (bony bumps). The dorsal surface of adult fish is greenish, and the ventral surface is yellow-gold. The coloration of reproductive adults has not been described.

Taxonomy Catostomus snyderi was first described by C. H. Gilbert from Upper Klamath Lake (1). It most closely resembles other large "standard" suckers of western drainages, such as the Tahoe sucker and largescale sucker (*C. macrocheilus*) of the Columbia River basin, and is presumably derived from a common ancestor (10). However, today it is closely tied genetically to the Lost River, shortnose, and Klamath smallscale suckers with which it shares the Klamath Basin (13). In fact, it shares a gene pool with the other three suckers as the result of recent hybridization events, even though each of these species was the result of millions of years of independent evolution. Nevertheless, the species maintain their morphological identities, although hard-to-identify hybrid individuals also exist.

This type of complex genetic interaction is fairly common among western fishes (e.g., Colorado River *Gila* species) and seems to be one way of dealing, in an evolutionary

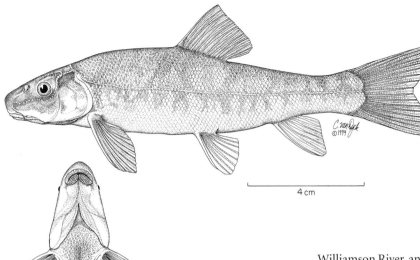

Figure 70. Klamath largescale sucker, 12 cm SL, Sprague River, Oregon. OSU 013739.

Williamson River, and the Williamson River above Klamath Marsh (2). In California they are or have been found mainly in the Lost River drainage and in the Klamath River downstream to Iron Gate Reservoir (11). Individuals may be present in the lower Klamath River.

Life History Of the three large suckers endemic to the upper Klamath Basin, the least is known about the Klamath largescale sucker. It seems to be the least lake-dependent of the three species, although it is (or was) found in natural and unnatural lakes of the basin. There is some evidence that it needs fairly high water quality because it is largely absent from highly eutrophic Upper and Lower Klamath Lakes, except in some bays where inflowing streams improve water quality (3). Although these suckers can apparently withstand, for short periods, temperatures as high as 32°C, dissolved oxygen levels of 1 mg/liter, and pH levels in excess of 10 (3, 4, 5), conditions in polluted lakes may exceed even these limits. Tributary streams that support largescale suckers rarely exceed 25°C. It is likely that historically the majority of the large adults were found in lakes, especially in deep water, whereas juveniles lived either in streams or in shallow areas of lakes. However, there are reproducing populations in a number of the larger streams (3).

Klamath largescale suckers are presumably benthic omnivores, as are other large *Catostomus* species. In Upper Klamath Lake juveniles were found to feed mainly on zooplankton (9).

Growth rates of this fish have not been documented, but it is likely that they become mature in 4–6 years at lengths of 20–30 cm FL. A male measuring 31 cm FL was aged as 7 years old (8). The maximum recorded age is 31 years from a fish measuring 46 cm FL (6).

Spawning migrations from Upper Klamath Lake, Oregon, occur from March through early May, depending on flows and temperatures of the Sprague River, although temperatures can range from 5.5 to 19°C during this period (7). Upstream migrations can be fairly lengthy, and the ability of

sense, with harsh, rapidly changing environments, in which high genetic diversity and morphological flexibility are advantageous for responding to unpredictable change (14). When environmental conditions stabilize, natural selection works strongly to produce divergent phenotypes, the kinds of fish we humans recognize as species.

In the Klamath basin, the naturally flexible arrangement among the sucker species has been exacerbated by severe human-caused change to the lakes and streams. The fact that the different forms have persisted despite these changes is a tribute both to their evolutionary flexibility and to the general fitness of the basic phenotypes, the result of eons of selection. For the most part, the species (as described in this book) segregate in their reproduction and ecology. Therefore there seems little reason not to continue to recognize the four species as such. (See further discussion in the account of shortnose sucker.)

To make matters a bit more complicated, it seems that the largescale suckers in the Sprague River (Oregon), a tributary to upper Klamath Lake, are genetically distinct from all other suckers (13). Their status awaits resolution.

Names *C. snyderi* is named for J. O. Snyder, the California ichthyologist who first recognized that Klamath largescale suckers were undescribed and brought this fact to the attention of C. H. Gilbert, his teacher at Stanford University (1).

Distribution The Klamath largescale sucker is native to the Klamath River and Lost River–Clear Lake systems of Oregon and California. They are known from Upper Klamath Lake, the Clear Lake–Lost River system, the entire Sprague River, the lower 20 km of the Sycan River, the lower largescale suckers to use fish ladders to pass barriers has contributed to their survival (3). Males tend to migrate earlier than females, and spawning can continue through May (2). Fecundity was estimated for three females at 39,697 (353 mm SL), 64,477 (405 mm SL), and 63,905 eggs (421 mm SL) (2).

Status IC. Listed as a species of Special Concern in California, the Klamath largescale sucker is a poorly known native to waters that have been highly modified by dams, diversions, grazing, and pollution. California populations are on the edge of its rather limited range. The Lost River drainage in California has been highly altered and contains large populations of introduced predatory fishes, such as yellow perch and Sacramento perch. The largescale sucker hybridizes with the Lost River sucker and the shortnose sucker, both of which have been formally listed as endangered by both the USFWS and CDFG. All this evidence indicates that Klamath largescale sucker may be on its way to becoming a threatened species, especially in California. It is possible that it is already absent from the state. However, in Oregon the populations of the fish are in good shape compared with the status of the other two sucker species, because they do have some stream populations (which may be distinct), they do not depend on lakes for rearing, and they are able to ascend barriers, especially if fish ladders are present (3). The Klamath largescale sucker should be regarded as a key member of the evolutionarily complex group of Klamath basin suckers and protected in order to maintain the genetic and ecological diversity of the group.

It is clear that more work needs to be done on the systematics, distribution, and ecology of this species. Even more clearly, there is a major need to rehabilitate stream and lake habitats for all native fishes in the upper Klamath basin.

References 1. Gilbert 1898. 2. Andreasen 1975. 3. Scoppetone and Vinyard 1991. 4. Castleberry and Cech 1993. 5. Falter and Cech 1991. 6. Scoppetone 1988. 7. Stubbs and White 1993. 8. Buettner and Scoppetone 1991. 9. Scoppetone et al. 1995. 10. Smith 1992. 11. M. Beuttner, USFWS, unpubl. data 1997. 12. G. Scoppetone, USFWS, pers. comm. 1998. 13. Tranah 2001. 14. Dowling and Secor 1997.

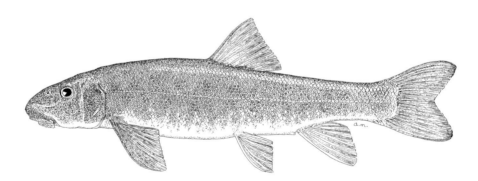

Figure 71. Klamath smallscale sucker, 18 cm SL, Scott River, Siskiyou County. Drawing by A. Marciochi.

Klamath Smallscale Sucker, *Catostomus rimiculus* Gilbert and Snyder

Identification The Klamath smallscale sucker is a typical sucker with a subterminal mouth, fine scales (81–93 in the lateral line, 15–18 above the lateral line, and 11–13 below the lateral line), 10–12 dorsal fin rays, 6–7 anal fin rays, and 16–18 pectoral fin rays. The eyes are small and the fontanel on top of the skull is either narrow or closed. The lips are large with large papillae. The upper lip has 5–6 complete rows of papillae while the lower lip has 4 rows, at least 2 of which go completely across, leaving a shallow median cleft. Klamath smallscale suckers are dusky olive brown on the back and sides, and yellow to white on the belly.

Taxonomy This species resembles the Modoc sucker (6) but apparently is not closely related to it (see the Taxonomy section in the account of Modoc sucker). Within the species there are meristic and genetic differences between the populations in the Klamath and Rogue Rivers (1). Emerging genetic evidence suggests that the two forms may eventually merit treatment as separate species (7, 8). The genotype of the Klamath form reflects limited past hybridization with suckers from the upper Klamath Basin (see the account of Klamath largescale sucker), increasing their genetic diversity (2, 7). There is also a population of dwarfed fish in the isolated upstream areas of Jenny Creek, a tributary to the Klamath River in Oregon, although it is genetically similar to the fish in the main river.

Names *Rimi-culus* means split-small, referring to the shallow cleft of the lower lip. Other names are as for Sacramento sucker.

Distribution Klamath smallscale suckers are widely distributed in the Trinity River and its larger tributaries, the Klamath River and tributaries below Klamath Falls, Oregon, and the Rogue River in Oregon. There are also a few records from the upper Klamath basin (8).

Life History Despite its wide distribution in three river systems, the life history and ecology of Klamath smallscale sucker have not been extensively studied, although it is unlikely that it differs in any major respect from the life histories of other typical suckers. Klamath smallscale suckers seem to be most abundant in deep, quiet pools of main rivers and in slower-moving stretches of tributaries, but they can be found in faster-flowing habitats when feeding or breeding. Moffett and Smith reported that "it is common to see large schools feeding along the bottom of the pool areas any time of the year" (3, p. 19). They are also common in reservoirs, such as Copco Reservoir on the Klamath. In

tributary streams they are typically associated with speckled dace, prickly (or marbled) sculpin, and juvenile steelhead.

Klamath smallscale suckers migrate up tributary streams to spawn in spring; spawning in tributaries to Copco Reservoir has been observed from mid-March to late April (4). Juvenile suckers are most abundant in small streams used for spawning. Fecundities of three fish were 15,300 (38 cm SL), 20,000 (38 cm SL), and 16,400 eggs (35 cm SL) (4).

Klamath smallscale suckers seldom achieve large sizes, and 50 cm FL would be exceptionally large. A sample of suckers ($N = 38$) from the Klamath River below Iron Gate Reservoir, which I aged using scales, indicates that they reach about 11 cm FL in their second year, 15 cm FL in their third year, 20 cm in their fourth year, 23 cm in their fifth year, 26 cm in their sixth year, 31 cm in their seventh year, 34 cm in their eighth year, and 35 cm in their ninth year. Using opercular bones, fish measuring 45 cm SL have been aged at 15 years (5).

Status IE. The Klamath smallscale sucker is a common species in the Trinity, Klamath, and Rogue River watersheds. If anything, dams and diversions have increased its habitat by providing more quiet, warm waters. The species is overdue for an extensive study of its life history, ecology, and taxonomy.

References 1. Snyder 1908a. 2. Miller and Smith 1981. 3. Moffett and Smith 1950. 4. Knudsen and Mills 1980. 5. Scoppetone 1988. 6. G. R. Smith 1992. 7. Tranah 2001. 8. D. Markle, Oregon State University, pers. comm. 2001.

Lost River Sucker, *Catostomus luxatus* (Cope)

Identification Lost River suckers are large as adults (up to 1 m TL and 4.5 kg) and can be distinguished by their long, narrow head; small, subterminal (almost terminal) mouth; thin upper lip with only a moderate number (2–5 rows) of

large papillae; deeply notched lower lip with 1–3 rows of papillae; short, widely spaced, triangular gill rakers (27–31 on the first arch) with no processes; and small eyes. The dorsal fin has its origin located only slightly in front of the origins of the pelvic fins and has 10–12 rays. There are 7–8 anal fin rays, 10 pelvic fin rays, and 82–113 scales along the lateral line, with 13–16 above it and 8–12 below. They are light brown to black dorsally, often brassy on the sides, fading to white or yellow on the belly.

Taxonomy The Lost River sucker has both unique features and features that tie it to other distinctive suckers. The confusion this caused has led to it being placed in both *Chasmistes* and *Catostomus,* as well as its own monospecific genus, *Deltistes.* The sucker was originally described as *Chasmistes luxatus* (1), then as *Catostomus rex* (2), followed by *D. luxatus* (3). The last placement was largely on the basis of the unusual gill rakers. R. R. Miller (4) thought the gill

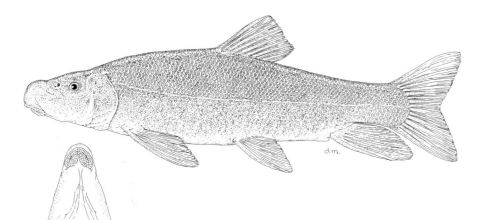

Figure 72. Lost River sucker, 38 cm SL, Clear Lake Reservoir, Modoc County. Drawing by A. Marciochi.

rakers were not necessarily diagnostic and placed the fish back in *Catostomus*. However, he reconsidered this opinion after a study of fossil fishes, which indicated that *Deltistes* was closest to the *Chasmistes* suckers (5). Its distinctiveness was validated by the phylogenetic study of G. R. Smith (19). There is indeed no other species quite like it. Recent genetic studies, however, indicate that it shares a common gene pool with other Klamath basin suckers, suggesting that the genus *Catostomus* is most appropriate. As discussed in the account of Klamath largescale sucker, the Lost River sucker is both part of a fascinating evolutionary phenomenon and a morphological species.

Names Luxatus means dislocated or "put out of joint," referring to the crumpled appearance of the snout of many large fish. In the Klamath lake area, Lost River suckers are known to anglers as mullets. The Klamath Indians called them tswam (1).

Distribution Lost River suckers are native to the Lost River and upper Klamath river systems, especially large lakes in these systems (Tule Lake, Upper Klamath Lake, Lower Klamath Lake). In the Klamath system they are found upstream (when spawning) in the Williamson and Sprague Rivers of Oregon and downstream (as a result of recent colonization) as far as Copco Reservoir and probably Iron Gate Reservoir. In the Lost River system they are found in the main river and in Clear Lake Reservoir, including upstream a few kilometers into Willow and Boles Creeks (6). A small nonreproducing population exists in Tule Lake (17). They are apparently absent from Lower Klamath Lake.

Life History Lost River suckers seem to be adapted for lake living, although they move out of their lakes to spawn in tributary streams, formerly in huge numbers (8). Their optimal habitat is defined by conditions that existed in the large lakes prior to their degradation. The lakes were shallow (<12 m) but fairly clear (Secchi depths typically >1 m), cool (summer temperatures 16–24°C), and moderately alkaline (pH 7.2–9.2) (9, 11). The water was well mixed by summer winds, and so was oxygenated from top to bottom (6–10 mg/liter). These conditions allowed the growth of large beds of submerged aquatic plants and extensive marshes along the edges (9), providing plenty of invertebrate food for adults and dense cover for larvae and juveniles. Today Clear Lake (a natural lake converted into a reservoir by a dam) comes closest to meeting these conditions, although it is highly turbid and does not support large beds of aquatic plants. Suckers are found throughout the reservoir, mainly at depths of less than 1.5 m (10, 11).

The Klamath lakes have become highly eutrophic, and suckers are now largely confined to regions near inflowing tributaries or other areas where water quality is not inhospitable to fish life (10). In particular, low oxygen levels may explain why suckers are largely absent from much of the Klamath lakes and Copco Reservoir (11). Die-offs of suckers begin when dissolved oxygen levels become lower than 1.58 mg/liter, and major die-offs occur when oxygen levels drop to 1.05 mg/liter (16). In late summer, during times of severe oxygen stress, the suckers may move up into the rivers for refuge (10). pH levels higher than 10 and temperatures higher than 31–32°C are also likely to be lethal, at least for juveniles (18). In Clear Lake Reservoir suckers apparently concentrate in deep areas in winter, but in summer they are more widely distributed (11). Native fish associates of the Lost River sucker are or were the shortnose sucker, Klamath largescale sucker, tui chub, blue chub, and rainbow trout, as well as various lampreys and sculpins.

Adult and juvenile suckers are bottom feeders, as indicated by the large quantities of detritus in their guts, along with chironomid midge larvae and amphipods; small amounts of zooplankton are also consumed (7, 11). The distinctive morphology of the mouth, head, and gill rakers may be an adaptation for "grazing" on invertebrates and diatoms growing on aquatic plants, a food source now much less abundant in the lakes.

Lost River suckers grow rapidly in their first 5–6 years, reaching 35–50 cm FL in this period (11). When individuals reach maturity at ages 5–9, growth slows down, and mature fish measuring 50–60 cm FL may be 7–15 years old. Larger fish may be over 20 years old; a fish from Upper Klamath Lake measuring 74 cm FL was aged at 43 years (12). Early reports mentioned fish as large as 1 m long (1), but fish over 65 cm are rare today (11). The largest and oldest fish are females. Recruitment has been limited at times, resulting in populations of mostly large, old individuals; in 1986 95 percent of the suckers captured in Upper Klamath Lake were 19–30 years old (7). An exception appears to be Clear Lake Reservoir, where smaller and younger fish are usually common (11).

Spawning occurs in larger tributaries to the lakes: Sprague and Williamson Rivers (Upper Klamath Lake), Sheepy Creek (Sheepy Lake, Oregon), Lost River (Tule Lake), and Willow and Boles Creeks (Clear Lake). In the Lost River, however, the embryos do not survive (17). Apparently only part of the population of adult suckers spawns in any given year (13). Fish begin making short migrations up into streams when discharge increases at any time from early February through early April, although March is probably the most frequent month of movement (7, 10, 11, 13). In Willow Creek radio-tagged suckers were found to migrate only 3–6 km and remain on spawning grounds for 2–3 weeks (13). Temperature per se does not appear to be a critical factor in stimulating migration, although a rise in temperature is often associated with an increase in stream flows into the lake. Migration has been observed while ice was still on Clear Lake Reservoir and inflowing temperatures were 4–7°C (11, 13). Spawning has been observed at temperatures of 4–19°C (10), but it is apparently less frequent or may stop above 12°C (13). Spawning may also take place in large springs in the lakes (7), and springs may be used for spawning in streams as well (13).

For spawning, fish select riffle or run areas with rocky or gravel substrates (>1.25 cm in diameter), depths of 21–128 cm, and water velocities of 1–84 cm/sec (7, 13). Patches of gravel introduced into a spawning area will be used heavily by spawning fish if flow and depth conditions are correct (10). Spawning behavior is similar to that of other suckers, in which one female spawns with several males and fertilized eggs drop into interstices in the rocks. Each female produces 102,000–236,000 eggs (7).

Larvae emerge and spend at best only a short time in shallow water along stream edges before moving into lakes. Larval downstream movement occurs mostly at night over about a 6-week period from late March to early June; the timing of outmigration depends on spawning time (10, 11, 13). Juvenile suckers aggregate along the shoreline in water

less than 50 cm deep, preferably in openings in beds of emergent plants or near beds of submerged macrophytes (10). They also select areas with fairly high water quality (dissolved oxygen levels > 4.5 mg/liter, pH < 9) (10). As they increase in size, the suckers become increasingly bottom oriented. Presumably this habitat selection behavior, which probably exposes the small suckers to high predation rates, was highly adaptive when deltas of the inflowing streams supported large expanses of marsh, which would provide cover for small fish. These marshes are now largely absent.

Status IB. State (1974) and federally (1988) listed as Endangered. Lost River suckers were once an important food resource in the Klamath and Lost River basins. In 1879 Cope (1) reported that Lost River suckers, fresh and dried, were a staple food of the Modoc and Klamath tribes. The suckers were still abundant in 1894 when Gilbert found them to be "the most important food fish in the Klamath Lake region" (8, p. 6). He also mentioned that attempts had been made to can suckers commercially, as well as to render them for oil. Prior to 1924 large numbers were taken annually from Sheepy Creek for consumption by people and hogs (14). In more recent years a major snag fishery for "mullet" developed in the Williamson and Sprague Rivers, with over 10,000 fish being harvested in 1968; levels declined thereafter, to 687 in 1985 (7, 16).

Lost River suckers and their principal habitats have been subjected to just about every environmental insult possible, with no end in sight. The suckers are gone from Lower Klamath and Sheepy Lakes, uncommon in Upper Klamath and Tule Lakes, and common only in Clear Lake Reservoir. That a few thousand fish manage to hang on in various lakes is a tribute to their longevity, fecundity, and persistence in spawning. The catalogue of insults below is largely derived from the USFWS Recovery Plan (10).

Drainage. Some of the earliest attempts to eliminate the lacustrine habitats involved draining various lakes to create farmland. After 1924 most of Sheepy Lake, Lower Klamath Lake, and Tule Lake were drained. Although the lakes partially reflooded after farming attempts failed, sucker populations never became reestablished in Sheepy and Lower Klamath Lakes and remain small in the miniaturized Tule Lake. Equally important, starting in 1905, the Bureau of Reclamation began draining and diking the vast marshes around and near the lakes and rivers, eliminating over 100,000 acres of wetlands. This included many kilometers of critical native fish rearing marshes along the edges of lakes. This is most likely one of the most important causes of decline, because low survival of juvenile suckers is clearly a major problem. Drainage of marshes presumably also greatly increased direct input of agricultural pollutants into lakes, both from drained marshlands and from surrounding areas, because the lakes were no longer buffered by nutrient-absorbing wetlands.

Dams. Dams have had two major direct effects: blocking spawning migrations and flooding marshlands. Chiloquin Dam (built 1914–1918) on the Sprague River (Oregon) partially blocked access of suckers to much of their historical spawning habitat, increased their vulnerability to fisheries when they aggregated below the dam, reduced recruitment of gravel to downstream spawning areas, and perhaps increased the likelihood of hybridization with other sucker species. Anderson-Rose Dam on the Lost River may have had similar effects on fish migrating up from Tule Lake. On the other hand, by increasing the size of Clear Lake Reservoir, this dam may have increased the amount of habitat for fish during most years, and it may be the long-term best hope for the species. Ironically, the large, shallow reservoir was created as a means for evaporating large quantities of water in order to reduce the amount of water flowing to the Tule Lake region. However, in the drought years of 1991 and 1992 Clear Lake Reservoir was drawn down so low (maximum depth 1.2 m) by the Bureau of Reclamation to supply water to farmers that many fish were lost downstream (17). Concern over the survival of the remaining fish in the face of winter freezing was sufficiently strong that some were captured and sent to Dexter National Fish Hatchery in New Mexico, as potential brood stock.

Diversions. Diversions reduce stream flows, lower lake levels, and send juvenile suckers down canal systems from which they cannot return on their own. Diversion of water from natural lakes decreases their ability to dilute pollutants, increases their alkalinity, and reduces shoreline marsh habitat. Studies on Upper Klamath Lake indicate that water quality becomes significantly worse when diversions greatly reduce lake level, increasing the blooms of noxious algae. Diversion of water from streams can reduce flows and deepen habitats to the point that they can no longer support suckers.

Organic pollutants. The lakes in the upper Klamath basin are surrounded by agriculture, and excess fertilizers; wastes from cows, humans, and other animals; and various other contaminants are dumped into them via irrigation return water. When these pollutants are added to the increase in phosphorus from "natural" sources, the lakes become more and more nutrient rich (hypereutrophic). Inpouring of nutrients has converted naturally productive lakes into green soups of bluegreen algae, with wide daily fluctuations in oxygen and carbon dioxide levels. In the lakes and rivers during summer, dissolved oxygen concentrations are frequently low, while temperatures, turbidity, pH, algae, and bacteria are all at high levels. Low levels of dissolved oxygen are associated with fish die-offs (16). The environment is thus increasingly hostile to native fish life. The major exception to this set of conditions is Clear Lake Reservoir, which is not highly polluted; despite its name, it is naturally very turbid (17).

Chemical contaminants. Watersheds feeding the lakes are heavily contaminated with pesticides and other chemicals used in agriculture, mosquito control, and forestry, including persistent organochlorines (e.g., DDT) that were banned long ago. The low populations of benthic organisms and low amphibian populations in Tule Lake may be an indication that this is a growing problem. The lack of benthic food organisms may particularly be a problem for Lost River suckers. However, effects of such contaminants are probably dwarfed by other problems in the lakes.

Grazing. Where regular agriculture is lacking, grazing of livestock is a major use of the land. Riparian areas are largely unfenced, so riparian vegetation is often gone, resulting in bank destabilization, loss of cover for fish, and increased sedimentation. This is particularly a problem in spawning streams, where suitable substrates for sucker spawning may be buried or imbedded and cover for larvae eliminated. In Modoc National Forest, much of the stream habitat above Clear Lake Reservoir has been fenced to exclude cattle.

Logging. Sedimentation is a result of large-scale logging and road building in the upper watersheds. The volcanic soils in the area are highly vulnerable to erosion and, when mobilized through erosion, fill streams and lakes with nutrient-rich sediments. This process further exacerbates the eutrophication of the lakes and buries spawning areas.

Exploitation. Lost River suckers have a long history of being exploited, mainly because of their large size and accessibility while spawning. Such large, late-maturing fish are easy to overfish, especially if the fishery targets large, old, high-fecundity females that have high reproductive potential. It is therefore possible that the sport fishery in the 1960s and 1970s helped accelerate the decline of this species. Fishing is now banned.

Introduced species. Fishes introduced into these watersheds include predatory bass, yellow perch, and Sacramento perch, which inhabit both streams and lakes. More recently fathead minnows have become extraordinarily abundant in the large lakes (except Clear Lake Reservoir) and may offer competition to small suckers just by their sheer abundance. The effects of these introductions, and others, are not known; they thrive in part because habitats have been altered in ways that favor them. Many of these species (but not fathead minnow) coexisted with the suckers before water quality became such a severe problem, so they may be able to coexist in recovered systems. The impact of invaders is lowest in Clear Lake Reservoir, where only Sacramento perch are abundant (17).

As usual, all these problems interact with natural environmental fluctuations. Severe droughts exacerbate all human-caused problems and create severe competition for limited water between humans and fish. This was seen dramatically in 1991 and 1992, when the Bureau of Reclamation drained Clear Lake Reservoir to keep downstream farmers from failing. This was done despite the fact that the reservoir is the best remaining habitat for both Lost River and short-nose suckers and supports the most viable populations. Ironically, during the severe drought of 2001, enormous controversy was generated when the remaining water in the reservoir was reserved for fish, rather than farms. The fish were widely blamed for creating economic problems that in reality had a long and complex history. These crises would have been much less controversial if agricultural practices in the basin had not already badly polluted Upper Klamath Lake, making it an untenable habitat for the suckers. Major fish kills (such as those in 1996 and 1997) occur there, so the suckers are under continuous threat of extirpation (17).

Dozens of efforts, big and small, now seek ways to allow the suckers to persist, including culturing of fish by the Klamath tribe. Part of the motivation for these efforts has been to restore the lakes to a condition more conducive to fishing and recreation and to improve the quality of water sent downstream into salmon habitat. However, the most pow-erful motivation has clearly been the Endangered Species Act, with its considerable powers of economic persuasion, especially its ability to limit the "take" of Lost River and shortnose suckers. Beyond the ongoing actions, many others are needed to make rivers and lakes more habitable for suckers and native life in general (10). An economic analysis indicates that measures taken to restore wild sucker populations could, in the long run, have a net benefit to both local and national economies and greatly improve the quality of life in the upper Klamath basin (15).

References 1. Cope 1879. 2. Eigenmann 1891. 3. Seale 1896. 4. R. R. Miller 1959. 5. R. R. Miller and Smith 1981. 6. Buettner and Scoppetone 1991. 7. Scoppetone and Vinyard 1991. 8. Gilbert 1898. 9. Hazel 1969. 10. Stubbs and White 1993. 11. Scoppetone et al. 1995. 12. Scoppetone 1988. 13. Perkins and Scoppetone 1996. 14. Coots 1965. 15. ECO Northwest 1994. 16. Martin and Saiki 1999. 17. G. Scopptone, USFWS, pers. comm. 1999. 18. Saiki et al. 1999. 19. Smith 1992.

Shortnose Sucker, *Chasmistes brevirostris* Cope

Identification Adult shortnose suckers are distinguished by large heads with oblique, terminal mouths; thin (for a sucker), striated lips; and deeply notched lower lips. Lip papillae are minute and few (0–5 rows). The snout is blunt, frequently with a small hump; the body is nearly cylindrical. The gill rakers, numbering 32–41, are slender, triangular, and densely tufted at the ends (although some end in knobs instead); lateral line scales number 67–92, with 12–20 scale rows above the lateral line and 9–13 rows below it; dorsal fin rays number 10–13, and anal fin rays number 7 (1). Shortnose suckers show considerable variability in morphology and meristics, especially in the Lost River watershed; for example, some may have subterminal mouths and fairly wide lips. Suckers from the Upper Klamath Lake basin differ somewhat from those in the Lost River basin, tending to have wider and deeper heads, terminal mouths with thin lips without papillae, and more (39 versus 35) gill rakers (1). Juveniles are best distinguished from other suckers by the separated lobes of the lower lip, often with a ridge between them; the caudal peduncle is short and deep; and the anal fin, when depressed, will reach past the beginning of the caudal fin (2). Live fish are dark on the back and sides, ranging from silvery to white on the belly and lower lip region of the head. Spawning fish have a reddish cast to their scales in a lateral band.

Taxonomy The shortnose sucker is one of the distinctive lake suckers of the genus *Chasmistes* that are native to large lakes of the western United States and that have a fossil history going back millions of years (17). It was described as *Chasmistes brevirostris* in 1879 by E. D. Cope (3). Two additional species of *Chasmistes*, *C. stomias* and *C. copei*, were described from Upper Klamath Lake a few years later (4, 5). The former "species" seems to represent large, spawned-out specimens of *C. brevirostris*, each with a well-developed hump on the snout (or, perhaps, hybrids with the Lost River sucker), whereas the latter "species" seems to represent large specimens without the snout hump. The confusion in identifying shortnose suckers experienced by early workers continues to this day because of the sucker's highly variable morphology. The variability may be partly the result of hybridization with other sucker species in the region (6, 18). The hybridization is largely the result of the high degree of alteration of the aquatic environments of the basin in recent decades, but it also represents an evolutionarily positive phenomenon in the sense that higher genetic and morpho-

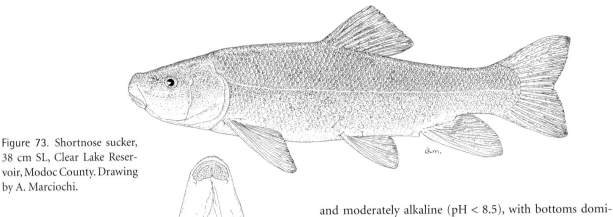

Figure 73. Shortnose sucker, 38 cm SL, Clear Lake Reservoir, Modoc County. Drawing by A. Marciochi.

logical diversity may increase the ability of the sucker to persist through periods of rapid change. Shortnose suckers do not seem to have a hard time recognizing one another, even if we do.

The now-extinct population of *Chasmistes* in Lake of the Woods, Oregon, has been referred to *C. stomias* (6), and its taxonomic position relative to the shortnose sucker is not certain.

Names Chasmistes means "one who yawns," referring to the large, flexible mouth; *brevi-rostris* translates as short-snouted. The Klamath Indians called the shortnose sucker xoöptu (3).

Distribution Shortnose suckers are native to the upper Klamath River and Lost River basins in Oregon and California. The foci of their original distribution were Upper Klamath, Lower Klamath, Sheepy, Tule, and Clear Lakes, as well as larger, deeper sections of river, although they were found in tributary streams while spawning. In the Lost River drainage today they are found in Clear Lake Reservoir, the main river below the reservoir, and the Boles Creek and Willow Creek drainage above the reservoir, with populations present in small reservoirs scattered along the creeks (1). A small population is also present in Tule Lake, the terminus of the Lost River (2). In the Klamath River basin they are present today in Upper Klamath Lake (Oregon) and its major tributaries (Williamson, Sprague, and Wood Rivers) and perhaps the lower reaches of smaller tributary streams as well. Small numbers are present in Iron Gate and Copco Reservoirs (California) on the Klamath River (2).

Life History Shortnose suckers are adapted for life in large, shallow lakes that are cool (summer temperatures rarely higher than 25°C), clear, well oxygenated (4–9 mg/liter),

and moderately alkaline (pH < 8.5), with bottoms dominated by large beds of aquatic vegetation. The only lake today that approaches these conditions, although it is fairly turbid, is Clear Lake Reservoir. Here the fish are generally found in water greater than 1.5 m deep. The other lakes in its range have mostly turned into shallow, eutrophic, anoxic, warm lakes that have heavy blooms of planktonic algae (mainly *Aphanizomenon)* encouraged by high levels of phosphorus entering the lake from outside sources. As a result, in hypereutrophic Upper Klamath Lake, the remaining suckers huddle around inflowing water from streams and springs, where water quality is less noxious than that in the lake. To a certain extent they can withstand adverse conditions; they can tolerate temperatures up to 31–33°C, oxygen levels near 1–2 mg/liter, and pH levels of around 10 (2, 7, 8, 16), but it is clear they prefer more moderate conditions. Presumably a tolerance for adverse conditions allowed them to survive through natural periods of drought, when lake levels were low. However, it has not allowed them to adjust to the extreme conditions that exist more or less continuously at present. Occasionally shortnose suckers will inhabit pools in streams; both juveniles and adults were observed using a 1.7-km stretch of Willow Creek, favoring areas with undercut banks and depths greater than 50 cm (11). Native fishes associated with shortnose suckers are or were the Lost River sucker, Klamath largescale sucker, tui chub, blue chub, and rainbow trout, as well as the various lampreys and sculpins.

Adult shortnose suckers, like other *Chasmistes* species, feed primarily on zooplankton, especially cladocerans (9), although the guts of only a few adults have been examined. The presence of detritus in fish from Clear Lake Reservoir indicates that they may also feed close to the bottom. Juvenile suckers apparently are more benthically oriented, feeding mainly on chironomid larvae and other insects (9).

Shortnose suckers can grow to about 50 cm FL, although fish over 45 cm FL are unusual today. The oldest fish recorded was 33 years old (48.5 cm FL; from Copco Reservoir), but in Clear Lake Reservoir fish measuring 35–40 cm FL are 10–15 years old (9, 10). Growth is highly variable among individuals and is most rapid in the first 5 years of life, by which time most fish are around 30 cm FL; thereafter growth takes place at a typical rate of less than 1 cm/year (9). Females apparently grow faster and reach larger sizes than males. Maturity usually sets in at ages 5 or 6, although some males may mature at age 4 (15).

Spawning takes place in tributary streams or, occasionally, in springs in lakes (2). Movement upstream to spawning areas can begin as early as late February and take place as late as early May. In large rivers in Oregon spawning migration typically begins in late March or early April, but in Willow and Boles Creeks, California, it typically begins as soon as stream flows start to rise in response to melting snows, often in February and March. Increase in stream flow seems to be the most important trigger for spawning migrations, because spawning has been observed at temperatures from 5.5 to 19°C (2). Reproductive suckers can move at least 46 km upstream in Boles Creek for spawning, although most do not move that far (9). They move similar distances up the Sprague and Williamson Rivers, Oregon. Spawning shortnose suckers select moderately fast-flowing areas of stream (velocities of 18–125 cm/sec, typically 66–120 cm/sec), with large gravel to cobble substrates and depths of 11–130 cm (usually 30–90 cm) (2, 11). Like other suckers, shortnose suckers spawn in small groups, usually a female followed by several males. The fertilized eggs are scattered over the substrate and sink into interstices in the gravel. Each female can produce large numbers of eggs; two females measuring 49 cm FL contained 36,763 and 56,217 eggs (12). The average fecundity of females is estimated to be around 38,000 eggs (11).

The embryos develop for several weeks (the exact time depends on temperature) before hatching. Larvae begin moving downstream almost as soon as they can swim, but most emigration takes place between late April and early June (2, 11). Larvae move mainly at night, with peak emigration times around midnight (11). During the day, they hang out in the water column in shallow (<50 cm) water over hard substrates (2). Once in lakes, larvae transform into benthically oriented juveniles, which prefer to aggregate along the edges of beds of aquatic plants, in shallow, clear water with high levels of dissolved oxygen.

Status IB. The shortnose sucker has been officially declared Endangered by both federal (1988) and state (1974) govern-

ments. In the Klamath River shortnose suckers were once so abundant that they, along with Lost River and Klamath largescale suckers, were a major food source. Spawning runs of the three suckers were later fished commercially in Oregon and then became subject to a snag fishery for "mullet." The causes of the decline are many, but drainage of lakes and marshes, dams on spawning streams, and heavy organic pollution of lakes are the main culprits. These are discussed in detail in the account of the Lost River sucker, which shares the same habitats. In addition, the population in Lake of the Woods, Oregon (perhaps a separate species), was eliminated by "rough fish" control measures to improve trout fishing, mainly treatment of the lake with a fish poison (13).

In the Klamath River basin there are apparently only a few thousand adult shortnose suckers left, mostly large, old individuals. Successful recruitment has been unusual, and only the long life and high fecundity of adults has allowed the species to persist. In the Lost River basin the population in Clear Lake Reservoir has been estimated at around 73,000 fish, and successful recruitment occurs on a regular basis (11). This relatively large population, however, is concentrated in a reservoir that can be drained during severe drought or made shallow enough that suckers are subject to intense bird predation. A similar situation exists with the small population in Gerber Reservoir in Oregon. The concentration of fish in reservoirs means that they are very vulnerable to major disasters, such as a pesticide spill or dam breakage. A recovery plan for the shortnose sucker (11) notes the importance of maintaining multiple populations, especially in recognition of the morphological, life history, and presumably genetic differences between populations in the Klamath River and Lost River basins. An evaluation of the economic consequences of improving conditions for native suckers in the two basins indicates that such an investment could have a net positive effect on regional economies (14).

References 1. Buettner and Scoppetone 1991. 2. Stubbs and White 1993. 3. Cope 1879. 4. Gilbert 1898. 5. Evermann and Meek 1897. 6. Andreasen 1975. 7. Castleberry and Cech 1993. 8. Falter and Cech 1991. 9. Scopettone et al. 1994. 10. Scoppetone 1988. 11. Perkins and Scoppetone 1996. 12. Coots 1965. 13. Bond 1966. 14. ECO Northwest 1994. 15. G. G. Scoppetone, USFWS, pers. comm. 1999. 16. Saiki et al. 1999. 17. Miller and Smith 1967, 1981. 18. Tranah 2001.

Razorback Sucker, *Xyrauchen texanus* (Abbott)

Identification Razorback suckers measuring more than 2 cm TL are distinguished by the sharp-edged keel on the back before the dorsal fin. They have a subterminal mouth with weakly papillose lips; the lower lip has a deep median

cleft, which completely separates the two halves. There are 68–87 lateral line scales, 13–16 dorsal fin rays, 7 anal fin rays, 36–50 gill rakers, and a well-developed fontanel on the top of the skull. Live fish are dusky to olivaceous on the back, brownish on the sides, and yellow-orange to white on the belly. Spawning males become nearly black on the back and

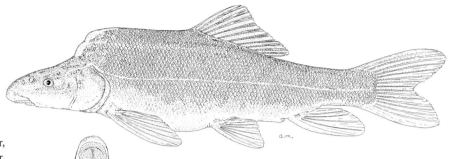

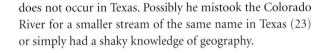

Figure 74. Razorback sucker, 36 cm SL, Green River, Wyoming. Drawing by A. Marciochi.

bright yellow on the belly, with an orange band on each side. Breeding tubercles are well developed on the anal, caudal, and pelvic fins, as well as on the caudal peduncle and posterior part of body (1, 5). Males in general are smaller than females, with slimmer bodies, larger fins, and a more pronounced keel on the back (5). Spawning females may also have tubercles, although they are less developed than those in males, and have similar spawning colors.

Taxonomy This species was described in 1861 as *Catostomus texanus* but placed in its own genus in 1889 (2, 5). The species has apparently been around for at least 5 million years (31), and its closest relatives are *Chasmistes* and *Deltistes* suckers, large lake species (29). There is some evidence of introgressive hybridization with the flannelmouth sucker, *C. latipinnis* (3, 4).

Names Razorback describes the sharp keel on the species well, much better than the older common name humpback sucker, which implies deformity (2). *Xyrauchen* translates as "razor nape." Just why C. C. Abbott used the trivial name *texanus* when he described the species is not known, for it

does not occur in Texas. Possibly he mistook the Colorado River for a smaller stream of the same name in Texas (23) or simply had a shaky knowledge of geography.

Distribution The razorback sucker was originally found throughout the Colorado River and its major tributaries in Wyoming, Utah, Colorado, New Mexico, Arizona, Nevada, and California (5). It was presumably once found as far south as the Colorado Delta in Mexico during periods of high flow. When the Salton Sea filled in 1904 and 1905, it was colonized by razorback suckers, repeating natural invasions that took place when the predecessor of the Salton Sea, Lake Cahuilla, existed (6). Today their distribution is limited to scattered individuals (5). In California there are probably no self-sustaining populations, but a few individuals have been found in the river as far south as Imperial County, especially in the Senator Wash area. Most fish are releases from fish hatcheries. As of 2000, nearly 60,000 suckers had been planted in the reach between Davis Dam (Nevada-Arizona) and Parker Dam–Havasu Reservoir (California-Arizona), but there is no evidence of natural recruitment (26, 30).

Life History The conspicuous, knife-edged hump of these fish is an adaptation for living in the swift, muddy waters of large rivers. The hump, together "with the long, flat, sloping head, undoubtedly steadies the fish against the bottom in currents where the water has a tendency to push down on the anterior portion of the body while the dorsal keel provides increased stability when faced into the current" (2, p. 360). However, this adaptation is presumably used primarily to move through swift water during floods or migrations, because most of the time razorback suckers concentrate in slower-moving sections of river, deep pools, backwaters, or oxbows (5). They will also readily inhabit reservoirs, gravel pits, and other habitats on the river. Historically it is likely that floodplains were important habitat for all life history stages, and even at the present time adults will selectively use permanent, weedy backwater habitats in the lower river (27). Overall they are associated with shallow (<2 m) areas with bottoms of sand and mud, where velocities are less than 50 cm/sec (5), although large, radio-tagged suckers have been found to use areas as deep as 3.4 m with velocities of up to 60 cm/sec, even in summer, when foraging on sand bars in midchannel (7). Juveniles primarily inhabit warm, shallow areas, which today often means ditches and irrigation canals (27).

Although razorbacks establish localized populations in river reaches or in reservoirs, there is evidence from tagged fish that many individuals move considerable distances. One fish moved 266 km upstream over a 4-year period but then moved back to its area of origin (5). The relationship between such movements and spawning is not known.

Adults generally swim about in small schools, often in water less than 1 m deep, feeding on the bottom. In reservoirs, however, they seem to be found mainly in water deeper than 1 m. Their usual food is algae and detritus, although aquatic insect larvae and zooplankton are also consumed (8, 9, 10). The presence of zooplankton in the diet—retained by long, complex gill rakers—indicates that the suckers are at least partially adapted for plankton feeding (10), which may help them persist in reservoirs. Larval suckers feed first on diatoms, detritus, and algae but quickly switch to rotifers and small planktonic crustaceans (11, 12). As the suckers become more benthic in orientation, chironomid larvae and other small insects become important in the diet.

Razorback suckers have highly variable growth rates. In ponds, where conditions for growth are nearly optimal, a few fish from a single brood may reach 40 cm TL in their first year, while others only reach about 5 cm TL; however, the disparity in size disappears after 5 years or so, at 55–60 cm TL (24). In captivity most females reach 30 cm or more and mature in 3–4 years, but in the wild growth to such size and maturity usually takes 7–9 years. Back-calculations of growth from otoliths indicate that wild males reach 10–11 cm SL in their first year and wild females about 14 cm SL. By the fifth year (after which growth tends to slow down significantly), males average about 20–22 cm SL and females 26–28 cm SL. Thereafter both sexes grow at a rate of about 7–8 mm/year (16). In Mohave Reservoir, once fish reach 60 cm or so, growth becomes so slow that it is not detectable. Razorbacks are reputed to reach 90 cm (7.3 kg) (13, 16), but there are no recent records of fish over 76 cm SL (5–6 kg), and fish longer than 50 cm have apparently always been uncommon. The largest and oldest fish are females. Most fish present in the lower river in the 1980s and 1990s were large (40 cm SL or more) individuals that were 24–45 years old (16), although smaller and younger fish of hatchery origin are also present today. There is evidence of only occasional natural recruitment, even though successful spawning is observed (5). In the Green River (Utah and Colorado) there has been enough natural recruitment from spawning during years of high flow to maintain a small, precarious population (22).

The first sign of spawning is the appearance of loose shoals of males near potential spawning sites, starting in November or December; females come in to join the shoals for spawning for short periods of time in the following two months. Individuals use multiple spawning sites but may return to the same areas year after year (13). Spawning takes place January through June, depending on water temperatures, but usually occurs before April in the lower river (e.g., Mohave Reservoir) (5, 20). Water temperatures recorded during spawning have ranged from 10.5°C to 21°C, but optimal temperatures seem to be around 15°C (5). However, optimal temperatures for embryo survival are 20–25°C (18). In the upper river spawning is usually associated with the increasing flows of spring runoff events (19). Spawning takes place over alluvial fans of inflowing streams, gravel bars of the main river, and shallow waters of reservoirs, where substrates are suitable (clean gravel and cobble), water is kept in motion by current or waves, and depths are less than 3.5 m (often less than 0.6 m) (5). Spawning behavior is similar to that of other suckers and can take place at any time of the day or night (5, 13). "One female is attended by 2 to 12 males, the group moving slowly in circles of three to five feet in diameter. Upon reaching a suitable spawning site, the female, closely pressed by the male on either side, settles on the bottom and starts to vibrate her body. When this act reaches a convulsive stage, the eggs and milt are simultaneously expelled. As this occurs, the three fish move forward and upward, leaving a cloud of silt and sand as spawning is consummated" (13, p. 107). The repeated spawning acts will clean a bottom area of silt and other fine debris and create shallow depressions 20 cm or more in depth (5). Each female spawns many times, releasing anywhere from 36,000 to over 140,000 eggs (average, 1,800–2,100 eggs/cm SL) over the spawning period (5).

The fertilized eggs adhere to the substrate and hatch in 1–2 weeks. Larvae emerge from the bottom, absorbing their yolk sac in 7–12 days depending on temperature (5). They then drift downstream into shallow backwaters or rise into the open waters of shallow bays of reservoirs (5, 11, 14). In such habitats they grow rapidly on plentiful food, but, outside isolated ponds, few or none reach the juvenile stage, because they are eaten by green sunfish, red shiners, and other alien predators (5, 11, 28).

Status IB. The razorback sucker is listed as Endangered by both state (1974) and federal (1991) governments. Razorback suckers were once one of the most abundant fishes in the lower Colorado River and served as a major food source for the Mojave people and other tribes that lived along the river. Commercial fisheries have existed for them at various times and places. When the Salton Sea filled with Colorado River water, razorback suckers were among the most abundant colonizers, both historically and prehistorically (6). By the 1950s, however, they were uncommon throughout their range, and today they are extremely rare, especially in the lower river (5, 27).

The decline of razorback sucker reflects how much the lower Colorado River has been altered from its original condition (15). The near extinction of this fish is all the more remarkable because it can live and spawn in reservoirs and ponds, is exceedingly long lived, and has very high fecundity—characteristics that should favor its survival. The ultimate cause of its decline has been the alteration of the natural flow regime of the river through the construction of numerous dams and diversions throughout the entire Colorado River basin, with concomitant changes in water quality, physical habitats, and frequency of flood events. Limited natural recruitment in the upper river has been weakly associated with unusual high-flow events (22). The changed river greatly favors an array of nonnative fishes—at least 25 species. These fishes in turn prey on embryos, larvae, and juveniles of razorback suckers, to the point that none survive (5). Particular problems have been the explosive spread of red shiners, which inhabit backwaters and prey on larvae (17), and large populations of channel and flathead catfish, which prey on juveniles (21). Although other factors no doubt contribute to the decline, successful rearing of razorback suckers in fish exclosures and in isolated ponds indicates the major importance of predation by alien fishes (5, 11).

There is recent evidence (mainly collection of larvae) of reproducing populations in Havasu Reservoir, in the river below Parker Dam (San Bernardino County), and in Senator Wash (Imperial County) (23, 25). However, recruitment of young appears to be slight, and populations are maintained mainly by hatchery introductions (26, 30). Recruitment is only likely to happen if larvae resulting from early spawning drift downstream and settle in habitat that is relatively predator free, such as a canal that was dry during the previous year. Some of these fish may grow fast enough to become too large (20–40 cm) to be prey for most piscivorous fish and hence survive into the following years (if their habitat is not destroyed or dried up). Other fish present there are either hatchery fish from stocking operations or, perhaps, exceedingly old individuals from previous eras. There is little hope that natural, self-sustaining populations of razorback suckers can be reestablished in California as long as conditions in the lower river remain unchanged. Their best hope for survival is in the upper river or in large tributaries, in reaches managed specifically for native fishes.

A major recovery program (adopted in 1998) is in effect for the razorback sucker. It focuses on rearing fish in hatcheries and using the resultant progeny in attempts to reestablish populations in the wild, primarily in the upper river basin, and especially the Green River (5). Part of the critical habitat is a 100-year floodplain in parts of the lower river (27). The program has met with only modest success, but three national fish hatcheries and at least four grow-out facilities are devoted to rearing Colorado River native fishes. In addition, experimental restoration of high-flow events and habitats is being tried. If pursued indefinitely, such efforts may result in partial recovery of razorback sucker populations, although probably not in California. A successful recovery plan for razorback sucker will thus involve restoration of flow regimes and floodplain habitats similar to the natural ones in a number of major tributaries to the Colorado River, augmentation of sucker populations with hatchery-reared fish, and careful genetic monitoring (23).

References 1. Minckley 1991b. 2 La Rivers 1962. 3. Hubbs and Miller 1952. 4. Buth et al. 1987. 5. Minckley et al. 1991a. 6. Gobalet 1992, 1994. 7. Tyus 1987. 8. Jonez and Sumner 1954. 9. Dill 1944. 10. Marsh 1987. 11. Marsh and Langhorst 1988. 12. Papoulias and Minckley 1992. 13. Sigler and Miller 1963. 14. Carter et al. 1986. 15. Ohmart et al. 1988. 16. McCarthy and Minckley 1987. 17. Ruppert et al. 1993. 18. Marsh 1985. 19. Tyus and Karp 1989. 20. Bozek et al. 1991. 21. Marsh and Brooks 1989. 22. Modde et al. 1996. 23. Modde et al. 1995. 24. R. R. Miller, Univ. Mich, pers. comm. 1974. 25. W. L. Minckley, Univ. Arizona, pers. comm. 1997, 1998. 26. G. Mueller, USGS, pers. comm. 2000. 27. Bradford and Gurtin 2000. 28. Marsh and Minckley 1989. 29. G. Smith 1992. 30. Mueller et al. 2000. 31. Hoetker and Gobalet 1999.

Bullhead Catfishes, Ictaluridae

The North American catfishes (Ictaluridae), with about 45 recognized species, are but a small part of the large catfish order (Siluriformes), which contains more than 2,200 species, most which live in fresh waters of the tropics. They are related to minnows (Cyprinidae), linked to them by Weberian ossicles, the small chain of bones used to transmit sound from the swim bladder to the inner ear.

Within the Ictaluridae there are three distinct groups: large, "typical" catfishes and bullheads (*Ameiurus, Ictalurus,* and *Pylodictis*), small madtoms (*Noturus*), and blind cave catfishes (*Satan, Trogloglanis,* and *Prietella*). All of these fishes have much in common: *(1)* nocturnal, bottom-feeding habits; *(2)* absence of scales; *(3)* 8 barbels (4 pairs): 2 on the snout, 2 on the end of the maxillae, and 4 on the chin; *(4)* a well-developed adipose fin; *(5)* hundreds of tiny teeth arranged in bands on the roof of the mouth; and *(6)* rays that have been modified into stout spines on the pectoral and dorsal fins. The spines are apparently the main reason why catfish are not taken as often by predatory fish as one would expect given their large numbers. The spines can be held or locked in an erect position, making each fish a large spiny mouthful; the sheath of skin over the spine contains poison that can cause considerable discomfort for the unlucky predator or angler. The locking mechanism of the spines also serves as a means for making a wide variety of sounds, through rubbing the base of the pectoral spines on a special part of the cleithrum, a bone in the pectoral girdle (Fine et al. 1997). Spines are used by fishery workers to determine the ages of catfish, because annual rings are visible in thin cross sections through them.

The ictalurid catfishes are native only to waters east of the Rocky Mountains, except in Mexico. The seven species found in California have all been introduced. However, fossil catfishes are known from California and other parts of the West. There is some uncertainty over the distribution of the three species of "square-tail" bullhead catfishes (black, brown, and yellow bullheads) in California because the species are easily confused. Therefore individuals of these species should be identified carefully.

The catfishes support major sport fisheries in warm waters of California. They have large, self-sustaining populations, attain large sizes, and are highly edible. Channel catfish are a major aquaculture species in California as well.

Black Bullhead, *Ameiurus melas* (Rafinesque)

Identification Black bullheads are stout-bodied catfish distinguished by the combination of a square-tipped, slightly notched tail; darkly pigmented membranes between light-colored rays of the short (19–23 rays, including rudimentary rays), rounded anal fin; dark chin barbels (always darker than the chin); pectoral fin spines that are smooth to rough but rarely with strong "teeth" on the rear edge; jaws equal in length (although the upper one may protrude slightly); and a pale vertical bar usually found at the base of the caudal fin. There are 15–21 gill rakers on the first arch. Their colors are solid (not blotched) and are typically dark on the back and pale on the belly. Live adult black bullheads are often bright gold-yellow on the sides and belly. The young are black to dusky in color with white bellies.

Taxonomy Black bullheads were formerly placed in the genus *Ictalurus*, but bullhead catfishes (including the white catfish) are now recognized as constituting a distinct evolutionary lineage from other catfishes and have been placed back in their own genus (1, 2).

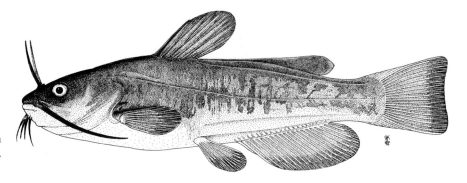

Figure 75. Black bullhead, 10 cm SL, Maryland. From Lee et al. (1980).

Names *A-mei-urus* translates as "without less tail," presumably referring to the lack of a fork in the caudal fin; *melas* means black. The name bullhead was originally applied in England to freshwater sculpins. It was apparently transferred to various catfishes in North America because of similarities in head shape. An old name for the bullhead catfish is horned pout. Black bullhead is a bit of a misnomer because the fish are typically bright yellow; the name apparently originated from the solid black anal fin, used as a distinguishing feature.

Distribution Black bullheads are native to much of the United States east of the Rocky Mountains, except the eastern seaboard, penetrating into southern Canada (2). Their range has been greatly filled in and expanded through introductions and now includes most Western states. Their exact year of introduction into California is not known, because most early introductions were simply recorded as "catfish" or "bullheads," and these are reported as early as 1874. However, the earliest confirmed record is dated 1942, from the Colorado River (3), and it is likely that earlier records of "bullheads" were all brown bullheads. They are now widespread and common in the Central Valley (including the San Francisco Estuary). They are the most common catfish in coastal drainages from San Luis Obispo County south to the Mexican border (16). They are present in scattered localities in the streams tributary to Monterey Bay (e.g., Llagas Creek) but are likely to become more widespread there. They are widespread but uncommon in the lower Colorado River. A limited population is found in the Lost River drainage (Antelope Creek) (4), so it is likely they will eventually spread throughout the upper Klamath drainage. In the Owens River drainage they are common in ponds and streams on valley floors. They are also apparently common in the lower reaches of the Carson, Walker, and Truckee Rivers in Nevada (13, 14), but I am not aware of records from California for these watersheds. On the North Coast they are found in the Russian River drainage but are absent from watersheds farther north to the Oregon border, including the lower Klamath River. They appear to be absent from the upper Pit River and Goose Lake. However, given the ease with which they can be moved around by anglers, they can be expected anywhere in the state, especially in ponds. In fact, I have the suspicion that they have replaced brown bullhead as the most common bullhead catfish in California, if not the most widely distributed, and are frequently mistaken for brown bullhead or yellow bullhead as a consequence.

Life History Little work has been done on black bullheads in California, and this summary is largely based on work done in other states (5, 6, 7). Their preferred habitats are ponds, small lakes, river backwaters, and sloughs and pools of low-gradient streams with muddy bottoms, slow currents, and warm, turbid water. In California these habitats are typically farm ponds, sloughs, reservoirs, and the highly altered lower reaches of rivers. They are capable of surviving water temperatures up to 35°C (38°C under laboratory conditions) (15). They are quick to invade new areas and are often abundant in ditches, intermittent streams, and other temporary habitats, including waters in which dissolved oxygen concentrations drop to 1–2 mg/liter (15). They are also capable of living in the brackish waters of estuaries; in Suisun Marsh I have found them living in salinities as high as 13 ppt. In California they are typically associated with other alien species that favor highly altered environments, especially bluegill, green sunfish, inland silverside, carp, red shiner, fathead minnow, goldfish, channel catfish, and threadfin shad (17).

Black bullheads are highly social and are usually found in loose shoals. Adults tend to be in physical contact with each other during the day, when they are deep in beds of aquatic plants or under some other cover. They come out to forage actively at night. Young-of-year bullheads, in contrast, swim about during the day in tight schools. Despite their diurnal habits, young black bullheads feed mostly at dawn and dusk, although adults presumably feed continuously throughout the night (8). Black bullheads of all sizes are omnivorous bottom feeders. They feed extensively on aquatic insects, crustaceans, and molluscs and occasionally take live fish and scavenge on dead ones. In sloughs of the Sacramento–San Joaquin Delta they feed on, in order of importance, fish, amphipods, isopods, snails, and other invertebrates (9). In reservoirs they eat earthworms and terrestrial insects as water levels rise over previously dry areas, and they move out into open water to feed when planktonic midge larvae and pupae are abundant (10). In ponds and small lakes chironomid midge larvae typically dominate their diet (18). Their stomachs almost invariably contain substantial detritus, algae, and pieces of aquatic plants, although the nutritional value of this material to bullheads has not been determined.

The growth of black bullheads is highly variable and depends on conditions such as temperature, food availability, and degree of overcrowding. Under optimal conditions with artificial feeding, they reach 30 cm TL (500 g) in a year. However, in the wild they need 3–9 years to reach a similar size. In Putah Creek, Yolo County, bullheads measuring 20–25 cm TL are 2–3 years old. The maximum size is apparently around 61 cm TL (3.6 kg), reported for a fish from New York (7). In ponds black bullheads often form stunted populations with small sizes (17–23 cm TL) at maturity.

Fecundity varies from 1,000 to 7,000 eggs per female but is typically 2,500–3,000 eggs (11). Black bullheads spawn in June and July, usually after water temperatures exceed 20°C. A sudden rise in temperature may trigger spawning (11). Before spawning, the female of each pair constructs a shallow depression by fanning away fine materials with her pec-toral fins and pushing out larger objects with her snout. As the nest nears completion, the male frequently touches the female with his barbels or rubs up against her. When they are ready to spawn, they line up head to tail, and the male wraps his tail fin over the head of the female. The female quivers and releases a number of eggs, which the male fertilizes (12). The fertilized eggs stick to each other, forming a yellow mass in the nest, which parental fish stir with their pelvic and anal fins, keeping the embryos well oxygenated. Once they hatch (5–10 days depending on temperature), the young, laden with yolk sacs, stay in the nest until they can swim freely, usually another 4–5 days (7).

The young stay together for 2–3 weeks in a tight ball that seems to be in continuous motion. The ball of young is guarded by one or both parents until the young reach about 25 mm TL, at which point they disperse into shallow water, sometimes swimming about in small shoals. In general they are gregarious fish during the day but more dispersed at night.

Status IID. Black bullheads seem to be expanding their range in California and becoming increasingly abundant in highly disturbed lowland aquatic environments. Their impact on other species, native and nonnative, is not known. In ponds and small streams they may form populations consisting of individuals too small for harvest, but in many places they do support small fisheries. Their use as a pond or reservoir fish in general should be discouraged in favor of other catfishes—or even better, native non-catfish species

References 1. Lundberg 1982. 2. Etnier and Starnes 1993. 3. Dill and Cordone 1997. 4. Buettner and Scoppetone 1991. 5. Carlander 1969. 6. Minckley 1973. 7. Becker 1983. 8. Darnell and Meierotto 1965. 9. Turner 1966a. 10. Applegate and Mullan 1967. 11. Dennison and Bulkley 1972. 12. Wallace 1967. 13. La Rivers 1962. 14. Sigler and Sigler 1987. 15. Smale and Rabeni 1995. 16. Swift et al. 1993. 17. L. Brown 2000. 18. Campbell and Branson 1978.

Brown Bullhead, *Ameiurus nebulosus* (Lesueur)

Identification Brown bullheads are plain, heavy-bodied catfish with square tails and blunt snouts supporting 8 dark (usually) barbels, including a long one at each corner of the wide, terminal mouth. The anal fin is short (21–24 rays, including rudimentary rays, but easily countable rays number 19–22), with membranes that are the same color as the rays. The pectoral and dorsal fin spines have 5–9 sawlike teeth on their posterior edges and feel very rough. Dorsal fin rays number 6–7. There are 11–15 gill rakers on the first gill arch. Adults are plain yellow-brown on their sides with dark mot-tling. Their bellies are white to yellow, and they lack a pale band at the base of the caudal fin.

Taxonomy See the account of black bullhead.

Names Nebulosus means clouded, referring to the mottled sides. This was the original "square-tail catfish" of California introductions. Other names are as for black bullhead.

Distribution The native range of brown bullhead included most of the United States east of the Great Plains and southeastern Canada, although its exact historical distribution is

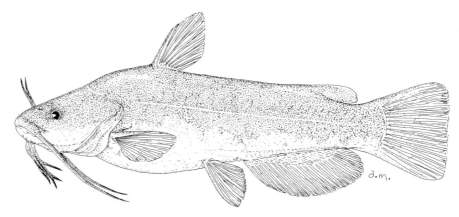

Figure 76. Brown bullhead, 18 cm SL, Clear Lake, Lake County. Drawing by A. Marciochi.

obscure because of misidentifications and introductions. It has been widely introduced throughout the western United States and southwestern Canada and is present in every major river system in the West.

The brown bullhead arrived in California in 1874, when 70 fish from Lake Champlain, Vermont, were planted in ponds and sloughs in Sacramento County (1). This apparently was the only introduction into the state, but the species quickly became abundant and was soon widely distributed throughout California. By 1890 the California Fish Commission claimed to have planted catfish (mostly brown bullheads) in every county of the state (1). They were an important part of the commercial fishery of the Delta by the early 1890s. They are established today in the following locations: *(1)* most larger coastal drainages from Southern California to the Klamath River, including the Eel and Mad Rivers; *(2)* the upper Klamath basin, including the Lost River; *(3)* the entire Sacramento–San Joaquin system, including the Pit River and possibly Goose Lake; *(4)* the lower reaches of the Truckee, Walker, and Carson Rivers in Nevada (so potentially in California as well); and *(5)* the Owens River. In the Colorado River most, or perhaps all, records may be misidentifications (13), and their presence should be confirmed with voucher specimens. Lack of records from the remaining, mostly arid, watersheds probably reflects my ignorance of their existence in stock ponds and other wa-

ters on private lands. In short, they can be expected just about anywhere in the state, although great care must be taken not to confuse them with black or yellow bullheads.

Life History Brown bullheads are the most widely distributed bullhead in California, both because of their early introduction and because they can adapt to a wide variety of habitats, from warm, turbid sloughs to clear mountain lakes. They are most abundant, however, in larger bodies of water, such as sloughs of the Sacramento–San Joaquin Delta, Clear Lake (Lake County), and foothill reservoirs, where they are usually associated with the deep end of the littoral zone (2–5 m), beds of aquatic plants, and muddy bottoms. In rivers brown bullheads are found mainly in sluggish, low-gradient reaches in association with deep pools, high turbidity, beds of aquatic plants, and soft substrates (2).

They can live at temperatures from nearly 0 to 37°C, but optimum temperatures for growth seem to be 20–33°C (3). It is not unusual to find them in "trout" streams, especially in the Sierra, although such streams typically reach 20°C or more in the summer and have sluggish, soft-bottomed reaches (e.g., Grizzly Creek, Plumas County). When temperatures are low, they burrow into loose substrates and become torpid (3, 15), and this behavior may explain their ability to persist in cold streams. However, feeding has been observed at temperatures as low as 4°C (4). In the Eel River small populations of brown bullhead live in warmer reaches of the mainstem, where small fish can be found in both riffles and pools. However, it is not certain whether or not these populations would exist without periodic infusions from Pillsbury Reservoir upstream (5).

Brown bullheads tolerate a wide range of salinities and alkalinities. I have found individuals in Suisun Marsh at salinities in excess of 13 ppt. A population was established in alkaline Eagle Lake (Lassen County) in the early 1900s when lake levels were high and alkalinities moderate (pH presumably around 8). It persisted until the early 1980s, by which time pH values regularly exceeded 9. Low oxygen levels

(<1 mg/liter) can also be survived by brown bullheads, either by becoming torpid when they coincide with low temperatures or through gulping air when temperatures are high (3).

Their social behavior is similar to that of black bullhead. They are most active at night and will form feeding aggregations. However, size-based feeding hierarchies determined by aggressive behavior are also known (6). Although they concentrate in favorable habitats, such as weedy bays or sloughs, they are also capable of extended movement. Tagged fish in a California reservoir have been recaptured as far as 26 km from the point of release (7).

Foraging brown bullheads swim along the bottom at an angle, their barbels just touching the substrate. The barbels locate food through a combination of taste and feel. When a fish detects something edible, it quickly turns around and snaps it up, often taking in detritus or algae at the same time (8). When small (<60 mm TL), they feed mostly on chironomid midge larvae and small crustaceans, but they consume larger insect larvae and fish as they increase in size. Brown bullheads from the Sacramento–San Joaquin Delta eat mostly amphipods, isopods, crayfish, dragonfly larvae, and snails (9). In Clear Lake they feed extensively on inland silversides and other small fish that are abundant inshore (14). However, they are opportunistic, omnivorous scavengers, so that almost any organism of suitable size can be expected in their diet (3, 7).

The growth of brown bullheads is fairly rapid in California. They reach 7–10 cm TL in their first year, 10–14 cm in their second year, 14–20 cm in their third year, and 19–28 cm in their fourth year (7, 14). Much faster and slower rates have been recorded (10). The maximum length is about 53 cm TL and the maximum weight about 2.2 kg, but fish more than 30 cm TL and 450 g are uncommon. The oldest fish on record was captured from Clear Lake 8 years after it had been tagged. Because it already measured 25 cm TL when tagged, it was probably at least 10 years old (11).

Brown bullheads normally first breed in their third year. The spawning season in California is typically May through mid-July and usually begins when water temperatures exceed 21°C (3). The first step in spawning is nest construction. Nests are typically depressions in sand or gravel in close proximity to cover, such as beds of aquatic plants or a fallen tree. They are constructed by females, with some assistance from males. Nesting may occasionally occur in a burrow in a bank, hollow log, or pile of rocks (3). Spawning is preceded by courtship behavior in which the male and female swirl around the nest, head to tail. Spawning occurs when the two fish are side by side, facing in opposite directions (3). The eggs are laid in batches until the female has deposited her entire clutch. Females lay 2,000–14,000 eggs, depending on size (10), in a gelatinous mass. Usually both sexes tend the developing embryos by chasing away predators and other bullheads and by stirring the embryos with their fins and mouths. After hatching (6–9 days, depending on temperature), yolk sac–fry remain in the nest for another week or so, closely guarded by the parents. When the young become free swimming, they leave the nest in a tight, swirling ball, which is guarded by the parents until the young begin to disperse at about 50 mm TL (3). The primary guard of the nest and young is the male, although survival of the young is significantly higher when the male is assisted by the female (12).

Status IID. The brown bullhead is abundant and widely distributed in California and is an important contributor to recreational fisheries. For example, it is the most important catfish in the fishery of Clear Lake, Lake County (1). It is most abundant in habitats altered by human activity, such as reservoirs, and its impact on native fishes and aquatic ecosystems is unknown. It is reared in small numbers in California's aquaculture industry, primarily for planting in ponds for fee fishing.

References 1. Dill and Cordone 1997. 2. Moyle and Daniels 1982. 3. Becker 1983. 4. Keast 1968a. 5. Brown and Moyle 1996. 6. Keen 1982. 7. Emig 1966. 8. Keast and Webb 1966. 9. Turner 1966a. 10. Carlander 1969. 11. McCammon and Seeley 1961. 12. Blumer 1985. 13. W. L. Minckley, Arizona State University, pers. comm. 1999. 14. E. Bianchi et al., University of California, Davis, unpubl. ms. 1978. 15. Loeb 1964.

Yellow Bullhead, *Ameiurus natalis* (Lesueur)

Identification The yellow bullhead is a stout-bodied catfish, with a straight-edged to slightly rounded tail and a wide, terminal mouth lined with dense rows of small, sharp teeth. It is distinguished by a slightly rounded anal fin with 24–27 rays (including rudimentary rays) that are the same color as the membranes between the rays and are all nearly the same length. The 6 chin barbels are pale yellow to white, with a long barbel at each corner of the mouth. The pectoral and dorsal fin spines are stout and sawtoothed on their hindmost surfaces (but not as sharply so as in the brown bullhead). There are 13–15 gill rakers on the first gill arch. Typical adult coloration is yellow-brown on the sides, without mottling, with a white chin and belly. The fins are dusky. There is no pale vertical bar

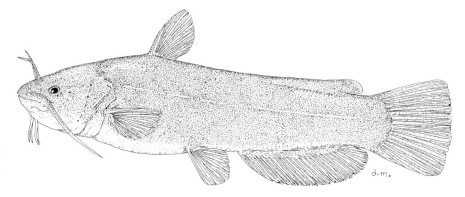

Figure 77. Yellow bullhead, 21 cm SL, Colorado River. Drawing by A. Marciochi.

on the caudal fin base, but the anal fin often has a dark stripe running across it.

Taxonomy See the account of black bullhead.

Names *Natalis* translates as "having large buttocks," a reference to the obese specimens originally described by the French naturalist C. A. Lesueur in 1819. Other names are as for black bullhead.

Distribution Yellow bullheads are native to most of the United States east of the Rocky Mountains, from the Great Lakes to Texas and Mexico. They have not been as widely introduced in the western United States as black and brown bullheads and have only a few populations established outside their native range. Their exact date of introduction into California is not known, but the first authenticated records are from the Colorado River in 1942, by which time the species was already widespread in the lower river (1). They are fairly common in the lower Colorado River, especially in backwaters, well into Mexico (2). Outside the Colorado River and associated canals, records are few. In Southern California they have been recorded from Evans Reservoir, Riverside County, and from the Santa Ana River, Orange County (3), and so may be present in other reservoirs in the region as well. There are occasional reports of this fish from the Sacramento–San Joaquin Delta and elsewhere in northern California (4), but it is highly likely that the records are misidentified brown bullheads. Extensive sampling of the Delta by CFG biologists over many years did not produce any yellow bullheads (10, 11). Thus all records of yellow bullheads from northern California should be regarded as suspect unless confirmed by expert examination of specimens. I regard them as established only south of the Tehachapi Mountains.

Life History Yellow bullheads are usually found in clear, warm water in low-gradient streams with permanent flows and rocky bottoms, or in shallow weedy bays of lakes (5, 6, 7). Lack of suitable habitat may explain why they are so uncommon in California. In the Colorado River they occur mainly in clear, weedy backwaters and canals, where dissolved oxygen levels can be low and temperatures high. They have the ability to withstand slightly lower dissolved oxygen levels and the same high temperatures as black bullhead (9). Compared with the information available on black and brown bullheads, little is known about yellow bullheads, especially in California. However, their life history does not seem to be strikingly different from that of the other two species, even though there are habitat differences (6).

Yellow bullheads are nocturnal and omnivorous, but they consume less plant material than other bullheads. Their diet consists mainly of aquatic insects, molluscs, crustaceans (especially crayfish), and fish (5).

Growth is similar to that of brown bullheads; in their native range they can achieve lengths of more than 47 cm TL and weights of up to 1.4 kg (6, 8). Breeding age, behavior, and time are also similar to those for brown bullheads (6), as are fecundities (8).

Status IIC. Yellow bullheads are uncommon in California and are likely to remain so. There is little reason to attempt to establish them where they are not already present because they do not seem to have any outstanding advantages for fisheries over other species of catfish. Their effects on native fishes in the lower Colorado River are not known, but they are part of the suite of predators that prevent reestablishment.

References 1. Dill and Cordone 1997. 2. Ruiz-Campos 1995. 3. Swift et al. 1993. 4. Contreras 1973. 5. E. Miller 1966. 6. Becker 1983. 7. Minckley 1973. 8. Carlander 1969. 9. Smale and Rabeni 1995. 10. Turner 1966a. 11. D. Kohlhorst, CDFG, pers. comm. 1999.

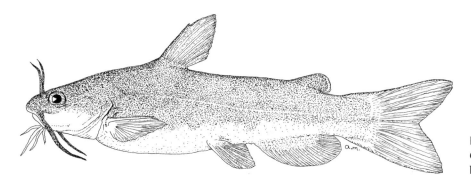

Figure 78. White catfish, 11 cm SL, Clear Lake, Lake County. Drawing by A. Marciochi.

White Catfish, *Ameiurus catus* (Linnaeus)

Identification White catfish are stout-bodied bullheads with forked tails. The fork of the tail is not as deep as in channel and blue catfish, and the tail lobes are rounded, the upper often slightly longer than the lower. There are 22–25 rays in the rounded anal fin (including 2–3 rudimentary rays), 5–6 soft rays in the dorsal fin, and 8–9 soft rays in each pectoral fin. The spine on each pectoral fin has 11–15 sharp teeth on its rear edge. There are 18–21 gill rakers on the first gill arch. The mouth is terminal, with long, dark maxillary barbels; the chin barbels are white. The head becomes disproportionately large in individuals measuring more than 40 cm TL. The body is usually gray-blue to blue-black on the back and sides and white on the belly. Some fish may have a mottled appearance, and those taken from extremely turbid water may be very pale.

Taxonomy See the account of black bullhead.

Names Catus means catlike. White catfish are the "fork-tailed catfish" of much of the early California fish literature. Other names are as for black bullhead.

Distribution White catfish were originally found in the lower reaches of coastal streams from the Hudson River, New York, south into Florida, including a few streams entering the Gulf of Mexico. California populations are apparently derived from either 56 or 74 fish imported in 1874 from the Raritan River, New Jersey, and planted in the San Joaquin River near Stockton (1). They spread rapidly through the Central Valley and were planted in many reservoirs and lakes, including Clear Lake (Lake County). They apparently had been introduced into San Diego County reservoirs by the 1940s, but by the 1980s they had colonized a number of southern California reservoirs by way of the California Aqueduct. They are found in large reservoirs on streams tributary to Monterey Bay. On the North Coast they are found in the Russian River (including Mendocino Reservoir) and Ruth Reservoir on the Mad River. They became established in the lower Eel River around 1990 (2).

They are notably absent from the Klamath River basin, from the Colorado River, and from the Pit River and Goose Lake. There are no records from Great Basin drainages in California, but white catfish are apparently present in Nevada in the Carson and Truckee River basins (3), so they could be found in some waters of eastern California.

Life History White catfish evolved in the sluggish lower reaches of large coastal streams of the Atlantic coast, so it is not surprising to find them abundant in the Sacramento and San Joaquin Rivers and in the San Francisco Estuary. They avoid the deep, swift channels favored by channel catfish and are most abundant in slow-current areas—such as Frank's Tract (a submerged island), the east and central Delta, and the south Delta around Old River—which they share with other warmwater fishes (4). During the day they tend to avoid heavy beds of aquatic plants and water less than 2 m deep, but they move into shallower water (<50 cm) at night (18). They can live in salinities as high as 11–14.5 ppt (5, 6, 17), although they disappear from Suisun Marsh when salinities exceed 8 ppt. They are also very successful in Clear Lake, Lake County, numerous reservoirs, and some farm ponds in California. They are usually found in water that exceeds 20°C in summer and can survive temperatures of 29–31°C (7). In reservoirs they concentrate at depths of 3–10 m during late spring and early summer. They tend to disperse in summer, although the bulk of the population is located below 10 m. If the reservoir stratifies, depth distribution is modified and catfish seek out temperatures greater than 21°C. In winter they are found mostly at depths of 17–30 m (8). Tagging studies in lakes and reservoirs indicate that white catfish wander about but that there are no regular seasonal migrations (9, 10). However, in the Delta most angler recaptures of tagged fish take place near the site of release, as do recaptures of fish in subsequent years as part of CDFG studies (18). Delta white catfish also seem to aggregate in the deepest part of sloughs and channels in winter and then disperse more widely in the warmer months (18).

White catfish are carnivorous bottom feeders but occasionally swim into surface waters of reservoirs to prey on plankton-feeding fishes. On the bottom they eat whatever organisms are most available, smaller fish eating smaller organisms. Thus young-of-year catfish (4–10 cm FL) in the Delta feed mostly on amphipods, opossum shrimp, and chironomid midge larvae. As they grow larger, their diet becomes more diversified and includes fish and large invertebrates, but amphipods and opossum shrimp are still the most important items (4). This may explain why adult white catfish in the Delta are much slower growing than other populations, in which the adults feed primarily on fish (11). In Putah Creek, Yolo County, juvenile white catfish feed mainly on aquatic insects, especially baetid mayfly larvae, while older fish feed on a range of items from algae to crayfish to bullfrog tadpoles. In reservoirs threadfin shad are particularly important, although in Clear Lake a wide variety of fishes (but especially inland silversides) are eaten (12, 19). White catfish also commonly feed on carrion; parts of dead birds and mammals have been found in their stomachs, as have parts of American shad that died after spawning (12, 13).

Growth rates of California white catfish vary widely. One of the slowest-growing populations known anywhere is that in the south and central Delta, which for the first 8 years of life averages 125, 163, 192, 214, 229, 243, 258, and 272 mm

FL, respectively (11). Growth in the north and west Delta is slightly faster. In the Sacramento River growth rates are higher than those of Delta fish after the third year of life (following a switch to a piscivorous diet); by ages 7 and 8, these catfish are 9–10 cm longer than their Delta counterparts (11). The fastest-growing fish in California live in Clear Lake, where white catfish for the first 8 years average 119, 183, 244, 297, 343, 372, 396, and 407 mm FL, respectively (11, 12). Males grow faster and become larger than females (11). Growth rates in Putah Creek appear to be similar for the first 3–4 years, although a 7-year-old fish measured only about 35 cm FL. In their native habitats white catfish can attain lengths of over 60 cm TL and weights of 3 kg, but fish over 2 kg are unusual. The official state angling record, however, is a fish from a pond in William Land Park weighing 10 kg (assuming this was not a misidentified channel catfish).

White catfish mature at 20–21 cm FL, which means that in California they are usually 3–4 years old. Spawning takes place in June and July when water temperatures exceed 21°C (12), but it can occur into September (17). The male builds a nest on sand or gravel, near cover, or in cavelike situations among rocks. Reproductive and parental behavior is similar to that of the bullheads, although care of young is only by the male (14). Each female lays 2,000–3,000 eggs, which hatch in about a week at 24–29°C.

Status IID. White catfish are the most important catfish for sport fishing in the Sacramento–San Joaquin Delta and support important sport fisheries in many lakes and reservoirs. Their popularity is due to their abundance, accessibility, and size. In the Delta 81,000–460,000 kg of white catfish were harvested annually by commercial fishers until 1953, when the fishery was banned (15). At the time it was thought that the commercial fishery was overexploiting catfish and reducing the catch of anglers. Mortality rates of larger catfish declined after closure of the fishery, so much so that white catfish are now regarded as underexploited; all catch limits for the sport fishery were removed in 1988 (15). In other California waters they are also underharvested and so are likely to remain a popular sport fish. Efforts to expand their range, however, should not be allowed, because adults are piscivores and therefore most likely to change ecosystems into which they have been introduced (16). Their introduction into Clear Lake, for example, was associated with the decline of native cyprinids in the lake (1).

References 1. Dill and Cordone 1997. 2. Brown and Moyle 1996. 3. Sigler and Sigler 1987. 4. Turner 1966a. 5. Ganssle 1966. 6. Perry and Avault 1969. 7. Kendall and Schwartz 1968. 8. Von Geldern 1964. 9. McCammon and Seeley 1961. 10. Rawstron 1967. 11. Schaffter 1997a. 12. E. Miller 1966. 13. Borgeson and

McCammon 1967. 14. Breder and Rosen 1966. 15. Schaffter and Kohlhorst 1997. 16. Moyle and Light 1996. 17. Wang 1986. 18. D. Kohlhorst, CDFG, pers. comm. 1999. 19. E. Bianchi et al., University of California, Davis, unpubl. obs. 1978.

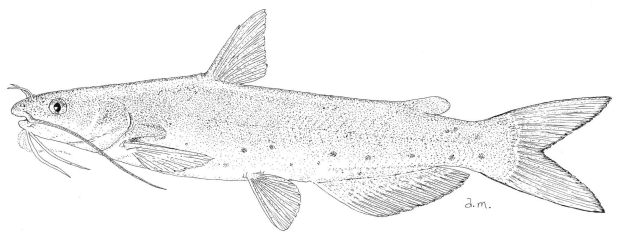

Figure 79. Channel catfish. Top: Adult, 24 cm SL, Clear Lake, Lake County, by A. Marciochi. Bottom: Juvenile, 13 cm SL, Maryland, from Lee et al. (1980).

rays, the pectoral fins, 4–5 rays. The barbels are dusky to white in color and the maxillary barbels are black and longer than the head. The normal color of adults is gray-blue on the sides, often with an olive gold tinge, fading to white on the belly. Young-of-year often have black-tipped fins. Spawning males become dark, with enlarged heads, thickened lips, fatty pads behind and above their eyes, and thickened fin membranes. Males have a distinct urogenital papilla that extends toward the tail, so males have just one opening behind the vent, while females have two.

Taxonomy See the account of black bullhead.

Names Icta-lurus means fish-cat; *punctatus* means spotted. Spawning males have been called chuckle-headed catfish or mistaken for blue catfish. Channel catfish have also been called spotted catfish.

Channel Catfish, *Ictalurus punctatus* (Rafinesque)

Identification Channel catfish are elongate, small-headed catfish distinguished by deeply forked tails with pointed lobes, rounded anal fins with 24–29 rays, upper jaws that project beyond their lower jaws, and conspicuous eyes (for a catfish). They can usually be recognized by the presence of tiny, conspicuous black spots scattered over the light-colored back and sides. The spots may be absent or few on very large or very small fish. The dorsal fins have 5–6 soft

Distribution Channel catfish were originally distributed throughout the Mississippi-Missouri River system southward into northeastern Mexico, but their range has been expanded through introductions to almost all parts of North America. The history of channel catfish in California is murky (1). They were first planted in 1891 in Cuyamaca Reservoir, San Diego County, and in the Feather River, but this transplant seems to have failed in both places. Additional plants in San Diego County took place in 1895 and 1922, but their success is also doubtful. Sometime between

1925 and 1930, a group of Roseville businessmen planted 65 channel catfish in the American River; it is assumed that Central Valley channel catfish populations were subsequently derived from this plant, although there are no confirmed records of their presence until 1942 (1). Likewise, in the Colorado River channel catfish seem to have become established in the 1920s, probably from plants in Arizona or Nevada as early as 1912. In southern California it is most likely that channel catfish plants in reservoirs in the 1920s and 1930s led to their establishment (1). From the 1960s onward, public and private fish hatcheries have reared channel catfish, resulting in their distribution throughout the state in public waters and private ponds. They can thus be expected almost anywhere where conditions are suitable, although self-sustaining populations occur mainly in warmwater reservoirs, lakes (e.g., Clear Lake, Lake County), and sluggish riverine areas (e.g., upper Pit River, Sutter Bypass and Delta channels on the Sacramento River, lower Colorado River). They are absent from North Coast watersheds north of the Russian River, including the Klamath River. However, a single adult channel catfish was captured in an irrigation canal in 1996 in the Klamath basin, so they may be present there (12).

Life History As their streamlined bodies and deeply forked tails indicate, channel catfish are adapted for living in main channels of large streams. In rivers adults typically spend days in pools or beneath logjams or undercut banks, moving into faster water to feed at night. Young-of-year, however, will live full time in riffles, taking advantage of rocks that break the current. Optimal habitat for channel catfish of all sizes is supposedly clear warmwater streams with sand, gravel, or rubble bottoms (2, 3). Thus in the Platte River, Nebraska, adult channel catfish were found in areas of dense cover, where depths were mostly greater than 60 cm and velocities were less than 40 cm/sec, that were close to areas with much more rapid flows (4). Juveniles (<21 cm TL), in contrast, were found at depths of 10–70 cm, usually in fast water (10–80 cm/sec) over sandy bottoms (4). However, they grow well in a wide variety of water bodies, from farm ponds to reservoirs to turbid, muddy-bottomed rivers like the lower Colorado. This tolerance is one of the main reasons they are the most commonly cultured North American catfish. Although they can live in waters with oxygen concentrations as low as 1–2 mg/liter, they grow best above 3 mg/liter and at temperatures of 24–30°C. They can withstand temperatures of 36–38°C, with 39°C being lethal (5). Despite their tolerance for moderate salinities, channel catfish in the Delta are not common in brackish water (6); I have not collected them in Suisun Marsh at salinities greater than 10 ppt.

Channel catfish are reputed to be omnivorous, but detritus and plant material frequently found in their stomachs may be the result of accidental ingestion with invertebrates and fish. They are not fussy eaters, however, because they adjust readily to living on commercial catfish chow in captivity and consume a wide variety of organisms in the wild. For small channel catfish (<20 cm FL), the main food is crustaceans (amphipods in the Delta) and the larvae of aquatic insects. As the catfish grow larger, fish and crayfish become increasingly important, although catfish of all sizes will consume aquatic insects. Usually fish measuring more than 30–38 cm TL are piscivorous, but any organism of appropriate size, including small mammals, is eaten (2, 3, 6).

Channel catfish are on average a fast-growing species, but there is considerable variation in growth from population to population. In good habitat they typically will reach 7–10 cm TL in the first year, 12–20 cm in the second year, 20–35 mm in the third year, 30–40 cm in the fourth year, and 35–45 in the fifth year (7). In California channel catfish in Lake Havasu Reservoir on the Colorado River grow considerably more slowly than the population in the main river (8). Fish from the river reach 53 cm FL in their seventh year; those in the reservoir take 12 years to reach the same size. In their native range channel catfish have been reported reaching more than 1 m TL, weighing over 26 kg (South Carolina), and living for nearly 40 years (Quebec). In California fish measuring more than 53 cm TL (2.5 kg) or more than 10 years old are unusual. The largest California fish on record, from Santa Ana River "lakes," Orange County (caught in 1994), weighed 23.9 kg.

The age and size of channel catfish at first spawning are highly variable; ages from 2 to 8 years have been recorded, as have lengths of 18–56 cm TL (7). Channel catfish typically must grow to 30 cm TL and be 3 or more years old before spawning. Spawning requires temperatures of 21–29°C, with 26–28°C being optimum (3, 10). In California this means they spawn in April through August, depending on the region (11). A few fish may spawn more than once in a season (7, 9). However, it is not unusual for planted populations not to reproduce at all, especially in reservoirs and ponds, so these populations must be maintained by continuous stocking.

Probably the main reason some populations of channel catfish fail to reproduce naturally is lack of suitable spawning sites. They require cavelike sites for nests, preferring old muskrat burrows, undercut banks, logjams or riprap made up of large rocks. In ponds they will use old barrels or similar containers for nest sites. The first signs of spawning are darkened males cleaning and defending nest sites, which are often in short supply. Females pair off with males well before spawning but do not assist in nest cleaning (3). The head-to-tail spawning behavior is similar to that of black bullheads (3, 10). Spawning occurs repeatedly over 4–6 hours until the female has deposited all her eggs. The fertilized eggs adhere to each other. Each female lays

2,000–70,000 eggs, depending on size (7), about 8,800 per kilogram of body weight (3). The male tends the developing embryos by aerating them with vigorous movements of his body. The embryos hatch in 5–10 days (usually 6–7), and new larvae measure 10–12 mm TL (11). The young start actively swimming about 1–2 days after hatching and leave the nest after about 7 days. Usually the male ceases guarding them at this point or shortly thereafter. Young may stay in a shoal of siblings for a week or two before dispersing into shallow, flowing water at about 25 mm TL. Some juveniles also enter the water column, resulting in their collection in ichthyoplankton surveys (11).

Status IID. Channel catfish are a popular and widely distributed sport fish in California because they are easy to raise in hatcheries, have fairly fast growth rates, and are capable of reaching large sizes. In recreational fishing ponds they are usually most successful in association with largemouth bass and bluegill. Their ready availability from private fish farms has resulted in their spread throughout the state. The impact of channel catfish on native fishes, amphibians, and invertebrates is not known, but given their predatory nature, it is unlikely to have been positive.

Channel catfish are an important aquaculture species throughout the United States, and large numbers are raised in catfish farms in the Central Valley, along the Colorado River, and elsewhere. In 1997 about 6 million pounds of channel catfish were reared in California, with a market value of about $11.8 million. Most of these catfish were sold to specialty food markets that sell live fish. Most processed channel catfish sold in supermarkets in California are imported from farms in the southeastern United States (13).

References 1. Dill and Cordone 1997. 2. E. Miller 1966. 3. Becker 1983. 4. Peters et al. 1989. 5. Allen and Strawn 1968. 6. Turner 1966a. 7. Carlander 1969. 8. Kimsey et al. 1956. 9. Dill 1944. 10. Clemens and Sneed 1957. 11. Wang 1986. 12. D. Markle, Oregon State University, pers. comm. 1998. 13. F. Conte, University of California, Davis, pers. comm. 1998.

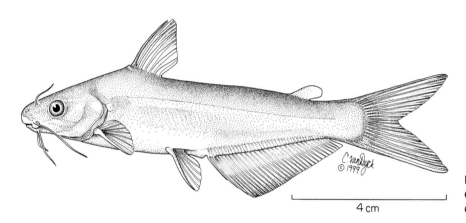

Figure 80. Blue catfish, 9 cm SL, Clifton Court Forebay, Contra Costa County.

4 cm

Blue Catfish, *Ictalurus furcatus* (Lesueur)

Identification Blue catfish are long bodied with deeply forked tails, terminal mouths with projecting snouts, and a pale color; they resemble the much commoner channel catfish. However, the anal fin is long (30–35 rays) with a straight edge that tapers downward to the end of the caudal peduncle. Their color is pale blue to olive on the sides (white on the belly), and there are no black spots. The bodies of adults are steeply humped before the dorsal fin and can be fairly stout. Their eyes appear small in the head and the barbels are white; the maxillary barbels are just barely longer than the head. They can be sexed using characters given for channel catfish.

Taxonomy See the account of black bullhead.

Names In the midwestern United States blue catfish are often called white catfish, fulton, or Mississippi catfish. *Ictalurus* means fish-cat; *furcatus* means forked.

Distribution Blue catfish are native to the main channels of the Mississippi and Missouri Rivers and their principal tributaries from South Dakota southward, as well as Gulf coast rivers well into Mexico. In 1969 blue catfish from Arkansas were introduced into Jennings Reservoir, San Diego County (1). Subsequently they were introduced into other southern California reservoirs (including Sutherland, El Capitan, San Vicente, Irvine, Santee, and Matthews Reservoirs), as well as into ponds of a commercial fish breeder in Imperial County. In 1978 a single adult fish was caught in the Sacramento–San Joaquin Delta, and in 1984 and 1985 juveniles were collected, indicating that reproduction was taking place (2). They continue to be found in the Delta, but in very low numbers (11). Their introduction into central California was presumably due to escapees from catfish farms. Their use in aquaculture is likely to further their spread, especially into the Colorado River.

Life History Deep channels of big rivers are the original habitat for blue catfish, but they also do well in large reservoirs and fish farm ponds. In rivers they remain on the bottom during the day in deep (8–10 m) areas with moderate currents. They seem to avoid muddy-bottomed pools and backwaters, except in spring, when they spawn. At night they are often found feeding in rapids or other swift-flowing areas. In reservoirs they prefer deep water but may move into shallows to feed at night. They can survive a wide range of temperatures (0–37°C) and salinities (up to 22 ppt), although they seem to grow best at temperatures around 27°C and at salinities between less than 7 and 8 ppt (3, 4).

The feeding habits of blue catfish are similar to those of channel catfish, except they are more strongly piscivorous and nocturnal. They feed mostly on crustaceans and aquatic insects when young, but they will take fish when they are as small as 10 cm TL. Once they reach 20–30 cm, fish are their main source of food, although large invertebrates may also be eaten (5). They also seem to consume fish larger than those eaten by channel catfish. In southern California reservoirs they were introduced in part to feed on the abundant Asiatic clam *Corbicula*, but they do so to only a limited extent.

The growth rates of blue catfish seem to be about the same as, or slightly less than, those of channel catfish living in the same waters (6, 7). Limited data from southern California reservoirs indicate that their growth is decidedly slower than that of channel catfish (8), but the two species have similar growth rates in warmwater reservoirs in other states (4). Exceptional growth of blue catfish in California has been observed only in El Capitan Reservoir, which is deep and turbid. Unlike channel catfish, blue catfish can reach lengths of more than 1.6 m and weights of more than 45 kg, at least in their native big rivers. The largest blue catfish from California, caught in 1996 in Lower Otay Reservoir, Orange County, weighed 37.3 kg (9). Just how large they actually can grow is debatable, because most of the "record" catfish were in fact caught before reliable records were kept. However, blue catfish weighing 90–100 kg may once have been caught (10). None approaching such weights have been caught in the past hundred years, however. Just how old such monster catfish would be is also a matter of conjecture; 50 years or older would not seem unreasonable, although the oldest blue catfish on record are 21 years old.

Spawning takes place in early summer when water temperatures reach 21–25°C. Blue catfish use hole nests like channel catfish, so it can be assumed that their spawning and parental behavior are similar. The extent to which they spawn successfully in California is not known.

Status IIC. There seem to be three main reasons why blue catfish were introduced into California: commercial catfish farmers wanted to raise them, anglers thought they could provide a trophy sport fishery, and agencies thought they might be useful in control of nuisance clams (4). Given their ecological similarity to channel catfish, blue catfish add little to California's sport fisheries except another species that can grow fairly large. When planted in reservoirs they mostly just replace channel catfish in the fishery. Because blue catfish seem to grow more slowly and are harder to catch than channel catfish, their planting may actually decrease total catfish catch. They do not seem to have as much value for aquaculture compared with channel catfish and are reared in only small numbers in the state. Given their limited value and given that they have some potential to harm populations of other fish because of their predatory habits, their use in California beyond their present range should be discouraged.

References 1. Dill and Cordone 1997. 2. Raquel 1986. 3. Perry 1968. 4. Pelzman 1971. 5. Brown and Dendy 1961. 6. Carlander 1969. 7. Perry and Avault 1969. 8. M. Lembeck, unpubl. rpts. 9. Taucher 1987. 10. Cross 1967. 11. D. Kohlhorst, CDFG, pers. comm. 1999.

Flathead Catfish, *Pylodictis olivaris* (Rafinesque)

Identification Flathead catfish have an extremely large flat head with small eyes that are located toward the top. The mouth is terminal with a lower jaw projecting beyond the upper jaw. The caudal fin is slightly rounded and slightly indented in the middle with a yellow to white patch on the upper lobe (except in large adults). The anal fin is short (14–17

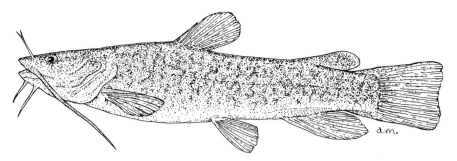

Figure 81. Flathead catfish, 14 cm SL, Rio Conchas, Mexico. Drawing by A. Marciochi.

rays, including 2 small rays that are hard to see) and rounded; the adipose fin is large and projecting. The spine in each pectoral fin is rough on both sides. The body is surprisingly slender in contrast to the head. Flathead catfish are black when young. As they increase in size they first become olive, mottled with brown on the sides and back, and then a plain olivaceous yellow-brown. Males can be distinguished from females by their distinct genital papilla with a small round opening at the tip. The genital papilla of females is smaller and recessed. The two urogenital openings of the female, however, appear as a longitudinal slit (1).

Names Pylodictis is a misspelling of *Pelodichthys,* meaning mud fish (2); *olivaris* refers to their greenish coloration. They are often called mud cats by anglers.

Distribution Flathead catfish are native to most of the Mississippi-Missouri drainage, as well as to the Rio Grande and rivers along the coast of the Gulf of Mexico to northeastern Mexico. They were introduced by the Arizona Game and Fish Department into Martinez Reservoir on the lower Colorado River in 1962 and were first recorded in California in 1966 (3). They have since spread into canals of the Imperial Valley (4) and into the Gila, Salt, and Verde Rivers of Arizona.

Life History Flathead catfish inhabit the turbid waters of large rivers and reservoirs. The adults prefer to live on the bottom of deep pools (1–2 m or deeper) or under rocks and large logs in areas with strong flow (5, 6). Areas with complex logjams are particularly favored. Juveniles live in riffles

and runs with rocks or other complex structure, preferring deep cover. Summer temperatures of rivers with flathead catfish are typically around 24–34°C. Adults are solitary for most of the year and seldom wander far from home pools (7). Like other catfishes, they are largely nocturnal in habit and move at night into shallow flowing areas to forage.

The feeding behavior of adult flathead catfish reflects their sedentary habits. They usually lie in wait in one place until suitable prey organisms come near enough to be inhaled by the sudden opening of their enormous mouths. Adults (25 cm TL or more) feed mostly on fish and crayfish, including native minnows and suckers in the Colorado River. In the California portion of the river small flatheads (<10 cm TL) feed largely on aquatic insect larvae, gradually becoming more piscivorous as they grow larger (5, 7). Large (18–81 cm TL) flathead catfish feed on aquatic insects, crayfish, and fish, including red shiner, channel catfish, common carp, and threadfin shad (8).

In their native range growth is fastest in large, muddy rivers with an abundance of small fish for prey. Not surprisingly, therefore, growth in the lower Colorado River is fast, with the fastest growth recorded in the Coachella Canal (9, 10). In the Colorado River (in contrast to the Coachella Canal) they reach around 11 cm TL (versus 12 cm TL) at age 1, 20 (24) at age 2, 38 (39) at age 3, 46 (48) at age 4, 59 (58) at age 5, 71 (69) at age 6, 79 (72) at age 7, 93 (80) at age 8, and 100 or more at age 9 (9, 10). They can live as long as 19 years and achieve lengths greater than 1.4 m and weights greater than 45 kg (11). It is not unusual to catch large (9–13 kg) flathead catfish from the lower Colorado River; the largest on record weighed about 27 kg (caught in 1992).

Male flathead catfish usually do not become mature until they are 3–5 years old and exceed 38 cm TL; females wait until they are 4–6 years old and in excess of 46 cm TL (11). Spawning takes place in early summer (presumably May through early July in the Colorado River) after temperatures reach 22–24°C (7). The fish form pairs, and both sexes either construct a nest depression or occupy and enlarge submerged holes in streambanks (12).

The male courts the female in the nest by rubbing repeatedly against her. When she is ready to spawn he wraps his tail around her head, and the female releases 30–50 eggs, which he swims over and fertilizes (12). Each female lays 4,000–59,000 eggs, depending on size (1). Once the female has laid all her eggs, the male chases her off the nest. He guards the embryos and keeps them stirred up with his mouth and fins. After hatching, the young form a tight shoal that stays in or around the nest for several days and is guarded by the male. The shoal gradually disperses as the young assume a solitary existence and move into riffles and runs.

Status IID. Flathead catfish are well established in the lower Colorado River. They are more abundant than angler catches indicate because catching them requires both knowledge of their habits and the ability to sit for long night hours, fishing on the bottom with large bait fish. The patience of the angler is also tried because flathead catfish normally take their time in swallowing bait, and it is all to easy to jerk the bait out of a fish's mouth.

Although flathead catfish were introduced after the major decline of native fishes in the lower Colorado River, they are now a major impediment to reintroduction of native minnows and suckers owing to their predatory habits (13). Because of their potential negative effects on native fishes and on fisheries for other species, any effort to introduce them into rivers and reservoirs beyond the Colorado River should be strongly opposed.

References 1. Turner and Summerfelt 1971. 2. Jordan and Evermann 1896. 3. Dill and Cordone 1997. 4. Swift et al. 1993. 5. Minckley and Deacon 1959. 6. Peters et al. 1989. 7. Becker 1983. 8. Minckley 1982. 9. Pisano et al. 1983. 10. Young and Marsh 1990. 11. Carlander 1969. 12. Breder and Rosen 1966. 13. Marsh and Brooks 1989.

Pikes, Esocidae

The pikes are a small (five species) but widespread family of voracious predators. They are readily recognized by their elongate bodies with dorsal and anal fins symmetrically placed far back on the body and by their long, pointed, and flattened snout containing a large mouth lined with sharp teeth. This morphology admirably serves them in their role as an ambush predator of large invertebrates, fish, frogs, ducklings, and anything else that moves in the water. The dorsal and anal fins support the forked tail in helping the pike's muscular, streamlined body to push against the water, producing the rapid acceleration needed for an ambush attack. The surprised prey are usually grabbed in midbody, turned, and swallowed head first. This basic predatory strategy has served the pikes well since at least the Paleocene period, over 60 million years ago (Nelson 1994). Today they are often the most abundant large predators in north temperate lakes and slow-moving streams.

Because of their popularity as a game fish, a number of attempts have been made to introduce pike into California. In the 19th century, some effort was made to establish northern pike (*Esox lucius*), muskellunge (*E. masquinongy*), and grass pickerel (*E. americanus*), but the introductions failed (Dill and Cordone 1997). In recent decades introductions of esocids have been opposed by CDFG because of their potentially devastating impacts on salmon and trout fisheries and on native fishes. Unfortunately, some self-serving anglers successfully introduced northern pike into reservoirs in the upper Feather River basin, where efforts to eradicate them have been both controversial and expensive.

Northern Pike, *Esox lucius* Linnaeus

Identification The northern pike is readily recognized by the typical pike morphology described in the family account (Fig. 20, in key, p. 83), with most fish measuring 15–60 cm TL. The duckbill-like snout is well over half the length of the head, with the maxilla of the upper jaw reaching past the middle of the eye. The lower jaw projects beyond the upper jaw and contains large canine-like teeth in the front and more peglike teeth in the back. Both the tongue and the roof of the mouth have patches of fine, backward-curving teeth. The elongate body, nearly cylindrical in cross section, is covered with fine cycloid scales and in living fish is quite slimy. The tail is forked. According to Etnier and Starnes (1), lateral line scales number 105–148; dorsal fin rays, 15–19; anal fin rays, 12–15; pectoral fin rays, 14–17; pelvic fin rays, 10–11; and branchiostegal rays, 14–16. Color is variable, but it is generally dark olive or gray on the back and sides and white to yellow on the belly. There are typically rows of light, oval spots on the sides and finer white spots on the head. Juvenile pike often have oblique dark bars running down the sides. Paired fins are usually tinged with orange, and the median fins contain dark blotches.

Taxonomy Despite its Holarctic distribution and ancient lineage, the northern pike has not been divided up into taxonomic subunits. All five species in the family Esocidae are in the genus *Esox*, and the species can hybridize.

Names When Linnaeus described this species from Europe in 1758, he gave it two old names for northern pike. *Esox* is a Latinized Gaulish word for large fish, used by the Roman naturalist Pliny, while *lucius* is the Latin name. Pike is a name commonly applied to long fish with pointed heads.

Distribution Northern pike are native to northern Eurasia (Russia, Ukraine, and most of Europe) and northern North

America, including all of Canada (except the Maritime provinces) and most of the eastern United States south to Nebraska, Missouri, Ohio, and West Virginia (1, 2, 3). Their range in the United States (and Europe) has been greatly expanded by introductions, especially in the West, where they are now present in both the upper Columbia and Colorado Rivers, and are apparently spreading downstream (4). They are present in a number of reservoirs in Nevada (5). In California they were illegally introduced into Frenchman Reservoir (Plumas County) in the Middle Fork Feather River basin in the 1980s (6). In 1991 and 1992 CDFG eradicated them from the reservoir, the Middle Fork Feather River below the reservoir, and nearby streams, using the toxicant rotenone. The massive operation included treatment of all streams and ponds in the Sierra Valley and resulted in the killing of thousands of pike (11). The operation was successful in part because drought had reduced the amount of water needing treatment and had dried up the river below the valley, keeping the pike from spreading farther (11). Unfortunately, around 1994, an ignorant, selfish angler illegally introduced them into Davis Reservoir, farther down the watershed (6). They are still present there, despite eradication efforts. A single pike has also been caught in Oroville Reservoir, about 200 km downstream. Unless they are eradicated from Davis Reservoir and further illegal introductions prevented, their establishment elsewhere in California is likely.

Life History In their native range northern pike are inhabitants of cool, clear lakes, sluggish streams, and river backwaters that possess abundant shallow water with extensive beds of aquatic plants. Davis Reservoir and inflowing Grizzly Creek fit this description well, as do many waters in California that do not have pike in them. Pike are typically found in water less than 4 m deep in close association with beds of aquatic macrophytes (7). Larger pike (>40 cm TL) are usually located at the edge of the beds in deeper water, while smaller pike tend to favor the interior of the plant beds in shallower water (7, 8). This habitat choice reflects both the cryptic coloration of pike (which matches the color and patterns of plant beds) and their use of plants as cover to ambush prey and hide from predators. For young-of-year pike, Casselman (8) developed a general rule that the preferred depth increased by 10 cm for every 10–12 cm of length. Young-of-year pike generally prefer to hide in beds of submerged aquatic plants as opposed to among the stems of emergent plants (8).

Not surprisingly, the optimal temperature for growth (generally assumed to be the preferred temperature) reflects the steady habitat shift with size in pike. Pike fry seem to do best when temperatures are 25–26°C, whereas older young-of-year prefer temperatures in the range 22–25°C and older fish prefer temperatures around 19°C (8). Nevertheless, pike can live and grow at both higher and lower temperatures.

The upper lethal temperatures are 28–30°C, depending on past thermal history, and winter temperatures of 1–4°C are commonly experienced (8). Pike also have considerable tolerance for low oxygen levels (0.5–2 mg/liter can be lethal, depending on temperature), but they grow best at levels near saturation (8). Northern pike are primarily freshwater fishes, but they can inhabit water with salinities of up to 10 ppt and breed in water measuring 7 ppt; salinities above 18 ppt are lethal (3). In the Baltic Sea they occur at salinities of 1–7 ppt but are most abundant in areas where the salinities are lowest (19). Because they are visual predators, pike are most abundant in fairly clear water, and growth rates increase as water clarity increases, all else being equal (8).

When not breeding, pike tend to be solitary. They are not territorial, but most individuals have home ranges in which they forage. Just outside these areas they have places used regularly for defecation, presumably to decrease the possibility of fear scents being released by cyprinid prey, which could warn fish away from foraging areas (9). Despite their tendency to stay in one place, adult pike can also move long distances. Pike dispersing down the Green River in Colorado and Utah moved an average of 75 km/year, and one individual moved 110 km in 13 months (4).

Northern pike are ambush predators, and prey size increases with body length and jaw dimensions (10). Their elongate, muscular bodies, forked tail, and rear-placed median fins give them the ability to accelerate rapidly from one place, assisted by jets of water squeezed from their bucchal cavity, through the opercular openings (2). Fry feed mainly on zooplankton, amphipods, and benthic insect larvae, the size of which gradually increases as the fish grow larger (7). Their prey is generally associated with aquatic vegetation. By the time pike measure 25 cm TL, they are preying largely on other fish, including other pike. Although they are opportunistic predators, soft-rayed fishes (minnows, suckers, salmonids) are preferred as prey. A wide array of organisms have been found in pike stomachs, including spiny-rayed fish (e.g., bluegill), frogs, aquatic snakes, and small mammals and birds (2). In the Green River small minnows and suckers were most commonly taken (4). In Davis Reservoir pike were found to consume other pike, pumpkinseed sunfish, brown bullhead, golden shiner, largemouth bass, rainbow trout, and signal crayfish, which were the main species available as prey (11, 12).

Growth rates of pike vary, depending on how close habitat conditions are to those described as optimal. Casselman (8) examined the growth rates of 82 populations within their native range and found the following fork lengths at the end of each year: 1: mean, 18 cm; range, 9–34 cm, 2: 32, 16–51, 3: 42, 23–64, 4: 49, 29–74, 5: 56, 33–81, 6: 62, 38–86, 7: 64, 43–89, 8: 68, 44–93, 9: 71, 47–97, and 10: 76, 55–110. In Davis Reservoir mean fork length at the end of the first year was 23 cm; that at the end of the second year, 39 cm;

and that at the end of the third year, 53 cm, suggesting that growth rates were close to the upper end of the range for pike (12). Growth rates in Frenchman Reservoir were similar (11). Pike rarely live longer than 10–12 years, and fish over 7 years old are uncommon (8). The biggest pike from California, from the Middle Fork Feather River below Frenchman Reservoir, was about 5 years old and weighed about 9 kg (11). The maximum age and size recorded from North America, for a pike from a cold Canadian lake, are 29 years and 110 cm FL (14.2 kg) (8).

Pike become mature in their second or third year of life, at sizes ranging from 30 to 50 cm FL; males usually mature at earlier ages and smaller sizes than females (13). Males of fast-growing pike such as those in Davis Reservoir can presumably mature at the beginning of their second year of life. Fecundities of female pike are variable but generally increase with size at a rate of two to five eggs per gram of body weight (8). They tend to be highest in pike from fast-growing populations. Thus 3-year-old females from Davis Reservoir, weighing 2.1–2.5 kg, could be expected to produce 30,000–80,000 eggs each in one season.

Pike are late winter–early spring spawners (February–April), with ripe fish seeking streams or shallow (<50 cm) areas in lakes where there are either dense beds of aquatic plants or flooded marsh vegetation (7, 13). Migration to the spawning areas is apparently triggered by a combination of increasing daylight, increasing temperatures, and increasing inflow from streams. Pike can move some distance (25–50 km) to spawn if necessary and will home to their natal areas and areas in which they spawned previously (14). Adults will move into the spawning areas at temperatures of 1–4°C, but spawning takes place only when temperatures are in the range 5–19°C; preferred temperatures are 6–14°C (7, 13). If inshore temperatures are too warm, pike will sometimes spawn over plant beds in water 2–7 m deep (7). The first fish to appear in the spawning areas are males, waiting for females to enter. Each entering female is then courted by one to three males, which swim alongside her in close contact (2, 13). The female releases 5–60 eggs during each spawning act, which are fertilized by the males as they are released; the spawning act takes place repeatedly over a 1- to 2-hr period over a wide area (2, 7). The fertilized eggs adhere to the plants, where they hatch in 12–14 days. For the next 4–15 days, the larval pike are attached to the plants by means of adhesive glands on the tops of their heads (7). When they reach 10–12 mm TL, they are released from the plants and begin actively feeding on zooplankton. Movement out of the spawning areas begins at about 20 mm TL, often triggered by falling water levels (7), although this is also the period when small pike can disperse to new areas, carried by increased stream flow.

Status IIC. I had hoped that, by the time this book was in production, northern pike would have been eradicated from California, so I would not have to write about them. Unfortunately they are still present. Northern pike are clearly a highly undesirable species for the state, with great potential for becoming significant predators on soft-rayed native fishes (such as juvenile chinook salmon, steelhead, and splittail) in natural waters of the Central Valley, including the San Francisco Estuary. Their voracious predation could disrupt salmon and trout fisheries in streams and reservoirs and further endanger various native fishes. For example, in Spain (a region with climate and regulated rivers similar to those of California) introduced northern pike have disrupted fish assemblages and reduced fish diversity in both streams and lakes (15, 16). Such impacts are generally predicted following the invasion of a top predatory fish (17). In the native range of northern pike, management of the species must frequently take into account their ability to affect the populations of other fish species (20). The potential adverse impacts of pike have long been recognized in California, and their import, possession, or transport is banned under state law—a statute that was willfully disobeyed in their introduction.

Pike are an excellent target for an eradication program for a number of reasons. They are known to be harmful to native fish and fisheries. They have the potential to spread (or be spread) widely in California. They are confined to just one reservoir and its inflowing stream. They have previously been successfully eradicated from a similar reservoir. An attempt by CDFG to eradicate pike from Davis Reservoir in 1997 using rotenone failed for a host of reasons—many of them related to the intense local opposition to the use of a fish poison in the reservoir, which was the water supply for the town of Portola. Opposition to the treatment led to rotenone being applied when the reservoir was too full and the water temperatures too low for it to be completely effective, although the treatment presumably reduced the ability of pike to disperse downstream when the reservoir spilled. By 1999 the pike were back in the lake in large numbers.

A management plan for the pike (21) recommended a series of measures to contain them, including a special screen on the dam to prevent emigration of small pike and a variety of means to reduce pike numbers in the reservoir, including various kinds of capture efforts. Eradication using rotenone was specifically excluded as a management measure. A model of pike population dynamics developed by T. Foin and students (18) indicates that any control measure short of eradication is likely to leave enough pike in the reservoir that their further spread is likely. Fortunately, the next big reservoir downstream, Oroville, is too deep and steep-sided to support pike reproduction, so it presumably will serve as at least a partial barrier to further dispersal into the Feather and Sacramento Rivers (18). Unfortunately, the most probable means of further dispersal will be by narrow-minded anglers who consider themselves above the law and put their own angling preferences above those of other people or above the need for conservation of native fishes, including chinook salmon. The sooner pike are eradicated from Davis Reservoir, the better for California fish and fisheries.

References 1. Etnier and Starnes 1993. 2. Becker 1983. 3. Jenkins and Burkhead 1994. 4. Tyus and Beard 1990. 5. Sigler and Sigler 1987. 6. Dill and Cordone 1997. 7. Grimm and Klinge 1996. 8. Casselman 1996. 9. Brown et al. 1995. 10. Hart and Hamrin 1988. 11. CDFG, unpubl. data. 12. D. J. Valle, Portola High School, unpubl. data 1997. 13. Billard 1996. 14. L. M. Miller et al. 2001. 15. Elvira et al. 1996. 16. Rincon et al. 1990. 17. Moyle and Light 1996. 18. Foin et al. 2001. 19. Lappalainen et al. 2000. 20. Paukert et al. 2001. 21. http://www.dfg.ca.gov/ northernpike/mgpike.htm.

Smelts, Osmeridae

Smelts are small silvery fishes with adipose fins that can occur in enormous numbers in inshore marine and fresh waters of the Northern Hemisphere. McAllister (1963, p. 4) characterized smelt as elegant and tasty, many of them having a "curious cucumber odor." McPhail and Lindsey (1970), however, noted that at least one writer had labeled this odor as being that of putrid cucumbers. The smell is definitely more pronounced in some species (e.g., delta smelt) than others. The smell is caused by the compound *trans*-2-*cis*-6-nonadienal, but its function, if any, is unknown (McDowall et al. 1993). Smelt are also characterized by having jaw teeth, no axillary processes, eight rays in the pelvic fins, and adhesive demersal eggs. Because each smelt species is rather variable but also similar to other species, considerable confusion exists in the literature as to the identity of many populations. McAllister's 1963 revision of the family cleared up much of the confusion, but also added a bit (see the account of delta smelt).

The origin of the word *smelt* is not known, but it is tempting to relate it either to *smell*, for cucumbery reasons, or to *smolt*, the silvery immature stage of salmon. Both relationships are doubtful because a Latin version of the word has been around since at least the eighth century, and similar words are present in a number of Germanic languages.

The tastiness of smelt has made them much sought after, and the tendency of spawning fish to congregate in large numbers in streams or along beaches has made the seeking easy. They are important economically both as sport fish and as commercial fish, at least in North America. They give much pleasure to masochistic fishermen who plunge into freezing surf and cold rivers to scoop them up with dipnets. Although best when eaten fresh, they were formerly dried in large numbers. Dried eulachon, which have high oil content, were burned as candles by the Yurok Indians.

Smelt are almost always associated with coastal regions in subarctic to temperate waters of the Northern Hemisphere. Species may be marine, freshwater, anadromous, or estuarine in habit. Not surprisingly, smelt are often important components of marine and freshwater ecosystems as major planktivores that are prey for fish, birds, and mammals.

Only 12 species of smelt are recognized. Yet seven of them occur in California (Moyle 1992) and four are covered in this book. Because the seven species occupy such a wide array of habitats in California and represent a wide array of relationships of fish to humans, it is worth comparing them all here.

1. Delta smelt, *Hypomesus transpacificus*. A California endemic, this species is confined to the San Francisco Estuary and is threatened with extinction.

2. Surf smelt, *H. pretiosus*. Surf smelt are an abundant marine species but are sometimes found in estuaries at salinities as low as 4 ppt. They are the most important commercial and sport fish among the California smelts, with around 225,000 kg harvested each year.

3. Wakasagi, *H. nipponensis*. Native to Japanese lakes, wakasagi were introduced into California reservoirs as a forage fish. They have invaded the San Francisco Estuary and are now regarded as a problem species.

4. Longfin smelt, *Spirinchus thaleichthys*. Longfin smelt is an estuarine-anadromous species that has declined from its major importance in estuarine ecosystems in California.

5. Night smelt, *S. starksi*. Like the surf smelt, this is an inshore marine species that is subject to sport and commercial fisheries, although it is not harvested in such large numbers.

6. Eulachon, *Thaleichthys pacificus*. The "candlefish" is anadromous and was once important in both Native American fisheries and a sport dipnet fishery. Like other anadromous fish in California, it has become scarce.

7. Whitebait smelt, *Allosmerus elongatus*. This is a rather uncommon marine species about which little is known.

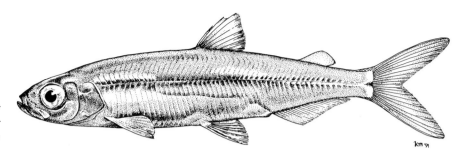

Figure 82. Delta smelt, Sacra-mento–San Joaquin estuary. Courtesy U.S. Fish and Wildlife Service.

Delta Smelt, *Hypomesus transpacificus* McAllister

Identification Delta smelt are slender-bodied fish that typically reach 60–70 mm SL, although a few may reach 120 mm SL. The mouth is small, with a maxilla that does not extend past the midpoint of the eye. The eyes are relatively large, with orbit width contained approximately 3.5–4 times within head length. Small, pointed teeth are present on the upper and lower jaws. The first gill arch has 27–33 gill rakers, and there are 7 branchiostegal rays. The gill covers lack strong concentric striations. The pectoral fins reach less than two-thirds of the way to the bases of the pelvic fins. There are 8–11 (usually 9–10) dorsal fin rays, 8 pelvic fin rays, 10–12 pectoral fin rays, and 15–19 anal fin rays. The lateral line is incomplete, with 53–60 scales. There are 4–5 pyloric caeca. Live fish are nearly translucent and have a steely blue sheen to their sides. Occasionally there may be one chromatophore (a small dark spot) between the mandibles, but usually there is none.

Taxonomy The delta smelt was originally considered a disjunct population of the widely distributed pond smelt, *Hypomesus olidus,* which occurs in fresh and brackish waters of Alaska, northwestern Canada, and coastal northeastern Asia. Hamada (1) recognized pond smelt and delta smelt as different species and renamed the pond smelt *H. sakhalinus,* retaining the name *H. olidus* for delta smelt and the similar-appearing wakasagi. McAllister (2) redescribed delta smelt as *H. transpacificus,* but with Japanese and California subspecies, *H. t. nipponensis* for wakasagi and *H. t. transpacifi-*

cus for delta smelt. *H. sakhalinus* went back to being *H. olidus.* Although the Japanese and California smelts are morphologically similar, it really made little sense to group them in the same species because they are separated from each other by the Pacific Ocean, with no populations in estuaries or lakes along thousands of kilometers of intervening coast. V. A. Klujanov, a Russian worker (3), recognized the species as distinct (as *H. transpacificus* and *H. nipponsensis*) based on skull structure, but his work was largely ignored until the cessation of the cold war. Similar conclusions were reached in a 1997 revision of the genus (31). Genetic studies (4, 5) have confirmed that delta smelt is indeed a well-defined species and indicate that it is most closely related to surf smelt, *H. pretiosus,* with only a comparatively distant relationship to wakasagi. Unfortunately, wakasagi introduced into California hybridize with delta smelt (5), although there is no evidence of introgression, suggesting that the hybrids are sterile. Genetic studies also show that delta smelt is just one intermixing population (5).

Names Delta smelt were formerly identified with pond smelt, and so the name pond smelt is used in the literature prior to 1970. Delta smelt was adopted by the American Fisheries Society (1970). *Hypo-mesus* means below-middle, referring to the position of the pelvic fins; *transpacificus* refers to the rather peculiar (and incorrect) notion that this species had populations on both sides of the Pacific Ocean and "to the friendship of Japanese and Canadian ichthyologists" (2, p. 36).

Distribution Delta smelt are endemic to the upper San Francisco Estuary, principally the Delta and Suisun Bay. They occur in the Delta primarily below Isleton on the Sacramento River side and below Mossdale on the San Joaquin River side. They are found seasonally throughout Suisun Bay and in small numbers in larger sloughs of Suisun Marsh. They move into sloughs and channels of the western Delta (e.g., Lindsey Slough) when spawning (mainly March–April). During high-outflow periods they may be washed into San Pablo Bay, but they do not establish permanent populations there. During periods of drought the center of delta smelt abundance has been the northwestern Delta in the channel of the Sacramento River. During years of average to high outflow they may concentrate anywhere from the Sacramento River around Decker Island to Suisun Bay prior to spawning movements.

Life History Delta smelt are euryhaline fish that typically rear in shallow (<3m), open waters of the estuary. They are mostly found within the salinity range of 2–7 ppt, but they can be found at salinities ranging from 0 to 18.4 ppt (6) and can tolerate salinities up to 19 ppt (7). In general delta smelt prefer to rear in or just above the region of the estuary where fresh water and brackish water mix and hydrodynamics are complex as a result of the meeting of tidal and riverine currents. This region is typically in Suisun Bay. During the 1987–1992 drought, the smelt were concentrated in deep areas in the lower Sacramento River around Decker Island, where the bottom salinity hovered around 2 ppt much of the year (8), apparently because the salt water–fresh water mixing zone was located in this region. However, smelt may also be common in this region during nondrought years, a finding that suggests they are attracted to favorable hydraulic conditions that allow them to maintain position. The idea that smelt do not consistently rely on the mixing zone for location is suggested by observations in such years as 1993, when the smelt continued to be common in Suisun Bay during the summer, even after the 2-ppt isohaline had retreated upstream (9). Smelt survival rates are generally highest in years when the area with salinities of 2 ppt or less includes shallow water in upper Suisun Bay during April–June, but the relationship is not consistent. Temperatures seem to have little effect on smelt distribution, because the smelt are found at temperatures ranging from 6 to 28°C, although 28°C is close to their lethal limit of 29°C (6, 7). Overall their behavior suggests a preference for low-salinity areas with tidal currents. The smelt are relatively poor swimmers (maximum swimming velocities of around 28 cm/sec). They tend to select portions of the water column that have relatively low velocities and to swim in short bursts (strokes) followed by a period of rest (glides) (21). This stroke-and-glide swimming behavior, combined with diel shifts in position in the water column (in response to tidal currents), presumably allows them to stay within relatively limited regions, where planktonic food organisms are also concentrated. These regions include main channels of the Delta and Suisun Marsh and open waters of Suisun Bay, where the waters are well oxygenated and temperatures are relatively cool (usually <20–22°C in summer).

Although nonbreeding delta smelt are usually aggregated together in limited areas, perhaps largely as a result of estuarine hydrodynamics, they do not appear to be strongly shoaling. Indeed the stroke-and-glide swimming behavior likely makes maintenance of coordinated schools difficult. Instead individual fish apparently hang out in the water col-umn and rely on their small size and transparency to hide them from predators in turbid water. The fact that they are rarely found in the stomachs of such predatory fish as striped bass, white catfish, and black crappie, even when they are abundant (10, 11), is a good indication that this strategy is successful.

Delta smelt feed primarily on planktonic copepods, cladocerans, amphipods, and, to a lesser extent, insect larva. Smelt measuring less than 12 mm FL feed almost exclusively on immature stages (copepodids and nauplii) of calanoid copepods, with adult copepods gradually becoming a more important component of the diet with increasing size (25, 26). Larger fish may also feed on the opossum shrimp, *Neomysis mercedis* or *Acanthomysis* sp., as well as other larger zooplankters (12, 25). Historically the most important food organism for all sizes was the euryhaline copepod, *Eurytemora affinis,* although in recent years invading species, *Pseudodiaptomus* spp. (mainly *P. forbesi*) have become a major part of the diet, especially in summer, when *P. forbesi* largely replaces *E. affinis* in the plankton (12, 25). However, *P. forbesi* is a somewhat larger species than *E. affinis,* and larval delta smelt show a strong preference for *E. affinis* as a consequence. The decreased abundance of *E. affinis,* both in absolute terms and relative to the abundance of other copepods, may be reducing the survival of larval smelt (25, 26).

Growth of delta smelt is rapid, and juvenile fish reach 40–50 mm FL by early August (12, 13, 14, 15). The period of most rapid growth occurs after they reach 30 mm and a greater variety of food becomes available to them (27). By this time young-of-year dominate the population, and adults become increasingly scarce. The smelt reach the typical adult length of 55–70 mm SL in 7–9 months (usually by September). Growth during the next 3 months slows down considerably (only 3–9 mm total), presumably because most of the available energy is channeled toward gonadal development (14, 15). Growth is apparently strongly tied to food supply. In recent years delta smelt have averaged 5–10 mm smaller at a given age than previously (16), an observation that is most likely the result of depletion of zooplankton in Suisun Bay by the invading overbite clam, *Potamocorbula amurensis.* The abrupt change from a population dominated by single-age adults during spawning in spring to a population dominated by juveniles in summer demonstrates that most adults die after they spawn. A small number either do not spawn in their first year or spawn a second time, some reaching lengths of 90–120 mm SL in their second year.

In September or October delta smelt begin a gradual, diffuse migration upstream toward spawning areas in the upper Delta. It may take several months for an individual to reach a spawning site. Spawning takes place between February and July, as inferred from collections of larvae (17,

18). However, most spawning takes place from early April through mid-May and occurs in sloughs and shallow edge waters of channels in the upper Delta and in the Sacramento River above Rio Vista. Spawning has also been recorded in Montezuma Slough near Suisun Bay (17) and may occur in Suisun Slough in Suisun Marsh in some years or in the Napa River "estuary." Most spawning seems to take place at 7–15°C (17). Ripe smelt and recently hatched larvae have been collected at 15–22°C, so it is likely that spawning can take place over the entire range of 7–22°C. Temperatures optimal for survival of embryos and larvae have not yet been determined, but it is likely that survival decreases as temperature increases beyond 18°C.

Spawning apparently takes place mainly at night, coincidental with new or full moons, presumably when the tide is low (29). This would keep embryos from being exposed by a drop in tide and ensure their hatching during the neap phase of the tidal cycle (29). In laboratory tanks spawning behavior begins when a female separates herself from other fish and takes up a position on or near the bottom (28). One or more males take up positions near the female, and one male repeatedly dips toward the tank floor. Spawning behavior, as described by J. Lindberg (28), is as follows: "The female occasionally came out into the current and pointed her head into the current and the male aligned himself so their flanks touched. The pair swam forward in this fashion about two centimeters off the floor of the tank, and a small cloud of milt was observed as they moved forward. This behavior was repeated and occasionally the primary male would chase off the two attending males." The eggs are broadcast over the bottom in a series of spawns, in a single event (19). Delta smelt eggs are demersal and adhesive, sticking by means of a tiny stalk to hard substrates (17). Although exact spawning areas have not been documented, it is likely that they contain gravel, sand, or other submerged material that is washed by gentle currents, close to main river channels.

Females (59–70 mm SL) lay 1,200–2,600 eggs (12), and the relationship between female size (FL) and fecundity has been determined (19) to be: Number of eggs = $0.266FL^{2.089}$. If this relationship holds true for the few large (90–110 mm SL) second-year smelt in the delta, such fish may contribute disproportionately to the egg supply and may be of critical importance in some years (29). Most eggs are produced in the left ovary, which is 6–15 times larger than the right ovary (19). A similar asymmetry is present in the testes of males.

At 14.8–16.5°C embryonic development to hatching takes 9–13 days, and feeding begins 4–5 days later (19). Newly hatched smelt have a large oil globule that makes them semibuoyant, allowing them to stay just off the bottom (19), where they can feed on rotifers and other microscopic prey. Because they are on or near the bottom, they are usually not swept downstream until the swim bladder and fins are fully developed several weeks later. Once the fins are well developed and the swim bladder is filled with gas, larvae become more buoyant and can rise up higher into the water column (19). At this stage (16–18 mm TL) most are presumably washed downstream until they reach the mixing zone or the area immediately above it. They appear to be strong enough swimmers to maintain themselves within the mixing zone by moving up and down in the water column, keeping themselves suspended and circulating with the abundant zooplankton that also occur in this zone (29).

In a near-annual fish like delta smelt, a strong relationship would be expected between the number of spawners present in one year and the number of recruits to the population in the following year. Instead the stock-recruitment relationship for delta smelt is weak or absent, accounting for about a quarter of the variability in recruitment (18). This relationship does indicate, however, that factors directly affecting numbers of spawning adults (e.g., entrainment, predation) can have a significant effect on delta smelt numbers during the following year. There is some indication that this lack of a strong stock-recruitment relationship is due to different sources of mortality being important in different years, especially if those sources affect different life stages.

It is interesting to speculate as to why the delta smelt has, until recently, been such a successful species in the estuary. The 1-year life cycle and low fecundity imply that, from the smelt's point of view (evolutionarily speaking), the estuary was a remarkably constant place, ensuring successful reproduction and larval survival every year. The tiny number of 2-year-old fish in the population would seem to be a meager insurance policy, at best, against environmental disaster. In the undisturbed estuary, an extensive, food-rich, brackish water mixing zone would have been present no matter how wet or dry the year. In wet years it would have been down in Suisun Bay or even San Pablo Bay, whereas in dry years it would have been in the undiked Delta, or perhaps even upstream in the Sacramento River and associated wetlands. Thus good conditions for survival would likely have been present in virtually every year. The natural tendency of delta smelt to "follow" good conditions in the estuary for both spawning and rearing may also help explain why in some years part of the spawning population shifts to the San Joaquin side of the estuary, apparently in response to increased outflow from the San Joaquin River. Their continued abundance following the diking and draining of the Delta and the gradual reduction of freshwater flows through the Delta presumably reflects the relatively benign climate of the period up to the 1970s, providing enough outflow to keep good rearing conditions available in Suisun Bay in most years. The increased variability in climate, combined with increasingly drastic human-caused changes to the estuary, resulted in a shift of the smelt's status from abundant to endangered.

Status IB. Delta smelt were officially listed as Threatened in 1993 by both the state and federal governments (16), although Endangered status is perhaps more fitting. They were once one of the most common pelagic fish in the upper San Francisco Estuary, as indicated by their abundance in CDFG trawl catches (13, 15, 20). Smelt populations have fluctuated greatly in the past, but between 1982 and 1992 they were consistently very low. The decline became precipitous starting in 1982 (12, 16). During the period 1982–1993 most of the population was confined to the Sacramento River channel between Collinsville and Rio Vista. From 1992 to 2000 numbers were mostly low, but abundance indexes were within historical (pre-1980) levels for 5 of 7 years, although lower than the historical average (35). In 1994–1998 delta smelt were found mainly in Suisun Bay. The true number of delta smelt is not known and is extraordinarily difficult to estimate, because handling a smelt is likely to kill it. However, their pelagic life style, short life span, spawning habits, and relatively low fecundity indicate that a substantial population is necessary to keep the species from becoming extinct.

The causes of decline of the delta smelt population are multiple and synergistic, and it is likely that different causes are important in different years.

Reduction in outflows. Distribution of smelt is related to amount of fresh water flowing through the estuary (outflow). During dry years they concentrate in the Sacramento River and upper Delta, while in wet years they concentrate in the lower river and Suisun Bay (16). This means that when outflows are low they are *(1)* more restricted in distribution, *(2)* concentrated in a less favorable environment (river channel), and *(3)* more vulnerable to diversion in the massive pumping plants of the federal Central Valley Project and the State Water Project. During most years increased diversion of water from the Sacramento and San Joaquin Rivers and tributaries has reduced outflows. The diversions, small and large, start in upstream portions of watersheds and end in the projects' pumping plants in the south Delta. In particular, spring (March–June) outflows created by snowmelt are diminished in all but wet years, so that the total amount of outflow is reduced, as is the number of weeks of high spring outflow. In many years high rates of diversion in the south Delta have created net reverse flows up the lower San Joaquin River, making delta smelt more vulnerable to entrainment. The overall effect is particularly severe in years when total water available from runoff is low or when smelt have spawned in the San Joaquin side of the estuary. For fish and most other Delta organisms, moderately high spring outflows are important because they cause the low-salinity zone of the estuary to be located in Suisun Bay. Suisun Bay is broad and shallow, so when the mixing zone is located there nutrients and algae can circulate in sunlit waters, allowing algae to grow and re-

produce rapidly. This provides plenty of food for zooplankton, which in turn serve as food for delta smelt and their larvae. Low outflows place the mixing zone in the deep, narrow channels of the Delta and Sacramento River, where the productivity of phytoplankton is lower because much of the water is beyond the reach of sunlight, meaning that fewer fish can be supported. The total area available for fish is also reduced. If the food supply is inadequate, fish presumably either starve to death or are subject to increased mortality from predation as a result of slower growth rates.

Despite the foregoing reasoning, a positive statistical relationship between outflows and delta smelt abundance does not exist. This is presumably because the effects of outflow on smelt are complex, affecting not only abundance but also distribution patterns and, perhaps, spawning times. The lowest smelt numbers generally occur in years of either low outflow or extremely high outflow, but outflow and smelt numbers show no relationship at intermediate outflows. Obviously other factors strongly affect smelt abundance.

Entrainment losses to water diversions. Water is pumped out of the estuary through numerous small diversions for farms of the Delta and, especially, the large diversions in the south Delta. Water is also pumped through power plants for cooling. It is possible that entrainment loss can be a factor limiting delta smelt populations in years when large numbers of larvae are pumped through Central Valley Project and State Water Project plants. This pumping also changes the hydraulics of the Delta, such that small fish wind up in Delta channels rather than down in Suisun Bay, where they are immune to diversion. Once in the channels delta smelt are more vulnerable to the hundreds of siphons and pumps that irrigate Delta islands, especially when irrigation return water from the islands keeps the water in the channels moving back and forth. The peculiar stroke-and-glide swimming behavior of smelt makes them particularly vulnerable to entrainment or, if a diversion is screened, to impingement on the screen (21). Larger fish entrained at Central Valley Project and State Water Project plants are trapped and trucked back to the Delta. Unfortunately, delta smelt are easily stressed and are very likely to die from such handling (7, 19, 21). Entrainment of adults is especially likely to have negative consequences if it occurs just before or during spawning season or if it selectively removes larger females.

High outflows. The years of major smelt decline have been characterized not only by unusually dry years with exceptionally low outflows (1987–1991) but also by unusually wet years with exceptionally high outflows (1982, 1986, 1996). High outflows presumably flush adult smelt out of the system along with much of the zooplankton. This means that not only is the spawning stock of smelt reduced, but its food supply is as well. Furthermore, depletion of established populations of invertebrates and fish may make it easier for alien species of copepods, clams, and fish to colonize the estuary, an outcome that may be detrimental to smelt.

Changes in food organisms. In recent years a number of alien species of copepods have invaded the estuary and increased in numbers, while the numbers of the historically dominant euryhaline copepod, *Eurytemora affinis,* have declined. Whether or not the decline of *E. affinis* is caused by competition with alien species, by selective predation, or by changes in estuarine conditions that favor the alien species is not known. One species, *Sinocalanus doerri,* swims faster and therefore avoids predation more easily than *E. affinis* (23). Another group of species, *Pseudodiaptomus* spp., are larger than *E. affinis* and so more difficult for small smelt to capture. The decreased abundance of favored copepods and increased difficulty of feeding on alien copepods may increase the likelihood of larval starvation or decrease growth rates.

Another potential indirect cause of larval starvation is the recent (1986–1987) invasion of the overbite clam, *Potamocorbula amurensis,* which is now abundant in Suisun Bay. This clam has reduced phytoplankton and zooplankton populations in the bay with its high filtration rates and dense populations. It has obviously not been responsible for the smelt decline, which began before the invasion of the overbite clam, but it may help keep smelt populations at low levels by reducing the availability of zooplankton for larval food. In recent years delta smelt have been growing more slowly and reaching smaller sizes as adults, an outcome that may be a response to the decreased abundance of prey in Suisun Bay (16).

Yet another complicating factor is the occasional rise in abundance of the diatom *Melosira* in the Delta, at times to the point that it is the most abundant species of phytoplankton. This diatom grows in long chains and is very difficult for zooplankton to graze on; thus the composition or abundance of zooplankton (or both) may also be tied to diatom blooms. The causes of the increase in chain diatom abundance are not known, but it may be related to an increase in water clarity experienced in recent years.

Toxic substances. The estuary receives a variety of toxic substances, including agricultural pesticides, heavy metals, and other products of our urbanized society. The effects of these toxic compounds on larval fishes and their food supply are poorly known, but there is growing evidence that larval fish are suffering direct mortality or additional stress from low concentrations of toxic substances. It is possible that the planktonic organisms on which smelt feed may on occasion be depleted by short aperiodic flushing of high concentrations of pesticides (e.g., carbofuran) through the system (30). It is not known if these substances also are affecting delta smelt directly or indirectly. There is some suspicion that, in July 1997, some kind of toxic event in Suisun Bay lowered survival of small smelt, but the evidence is weak (for example, water samples showed low general toxicity in the laboratory).

Disease, competition, and predation. There is as yet no direct evidence that disease, competition, or predation has caused delta smelt populations to decline, despite the abundance of introduced species in the estuary. However, the diseases and parasites of delta smelt have not been studied, and it is possible that survival rates may be lowered by parasites, such as a tapeworm that has appeared in smelt in fairly high frequency in recent years (29). The effects of predation or competition from introduced planktivores (such as threadfin shad, inland silverside, and wakasagi) have likewise not been studied. It is quite possible that, at low population levels, interactions with these species could prevent recovery. In particular, inland silversides are usually collected where delta smelt are spawning, and they could be major predators on eggs and larvae (24). They are known to be voracious predators on fish larvae, and it is likely that smelt larvae hang out around spawning areas for a few weeks before becoming pelagic and moving downstream. The invasion of the estuary by silversides also roughly coincides with the period of smelt decline.

In past years efforts to enhance striped bass populations by planting large numbers of juveniles from hatcheries or by rearing fish entrained at the pumping plants could have had a negative effect on other pelagic fishes. Enhanced predator populations, without a concomitant enhancement of the populations of prey, such as delta smelt, may result in excessive pressure on prey species. A particular problem had been posed by the planting of thousands of juvenile striped bass at Rio Vista, near areas where delta smelt have concentrated in recent years. Those fish may have affected smelt by depleting local zooplankton populations or by preying on smelt, even if predation rates were low. In 1992 planting of juvenile striped bass was halted by CDFG because of the potential effects of their predation on juvenile winter-run chinook salmon. However, a much smaller planting program was subsequently started, using only fish rescued from the pumping plants (and reared to a larger size), and CDFG plans for recovery of Delta fisheries include restoration of striped bass to historical levels—levels that may be incompatible with smelt and salmon restoration.

Loss of genetic integrity. Wakasagi were successfully introduced into reservoirs in the Sacramento drainage and have subsequently been collected from downstream areas. They have apparently been present in the Delta since the 1970s but did not become widespread until the 1990s (5). Their hybridization with delta smelt has the potential to cause loss of the genetic integrity of the latter species. Fortunately, it is likely that hybrids are sterile, given the large genetic differences between the two species (4). Nevertheless, interbreeding may cause the loss of valuable gametes of delta smelt and further reduce their ability to recover.

Ever since the delta smelt was listed as a threatened species, it has been studied to find ways to promote its recovery. The first step in the official recovery process was to describe the critical habitat of the delta smelt, which was defined to include all of the legal Delta (as defined by Section 12220 of California's Water Code of 1969), Suisun Bay, and Suisun Marsh (22).

The declaration of critical habitat means that all habitat-altering activities taking place within the region must be analyzed as to their effect on delta smelt and then modified if their effect is likely to be negative. Because critical habitat includes the entire Delta and much of the rest of the estuary, it is implicit that recovery of delta smelt requires recovery of natural processes in the estuary, including outflow (9). Finding ways to protect the smelt and other species, while not disrupting water supplies, was a major reason for the Bay-Delta Accord of 15 December 1994. The CALFED process that the accord set in motion is resulting in major efforts toward habitat recovery for smelt and other species, although the effectiveness of these measures remains to be seen (32). If they do not succeed in at least restoring delta smelt populations as defined in the 1996 recovery plan (22), then more drastic measures to improve estuarine functioning may have to be implemented.

As a backup measure, delta smelt culture techniques and facilities are being developed (19, 33). The smelt have now been reared through their entire life cycle in order to make experimental fish available without depleting wild populations (34). However, if hatchery propagation is to be successful, the fish must be released into an environment that provides ample food, low levels of toxic compounds, and low entrainment losses. Thus water management in the Delta will always be a key factor for smelt survival.

References 1. Hamada 1961. 2. McAllister 1963. 3. Kljukanov 1970. 4. Stanley et al. 1995. 5. Trenham et al. 1998. 6. Swanson and Cech 1995. 7. Swanson et al. 2000. 8. Herbold 1994. 9. Jassby et al. 1995. 10. Stevens 1966b. 11. Turner 1966b,c. 12. Moyle et al. 1992. 13. Erkkila et al. 1950. 14. Ganssle 1966. 15. Radtke 1966. 16. Sweetnam 1999. 17. Wang 1986. 18. Sweetnam and Stevens 1993. 19. Mager 1996. 20. Stevens and Miller 1983. 21. Swanson et al. 1998. 22. USFWS 1996. 23. Meng and Orsi 1991. 24. Bennett and Moyle 1996. 25. Lott 1998. 26. Nobriga 1998a,b. 27. Grimaldo et al. 1998. 28. J. L. Lindberg, pers. comm. 1992. 29. W. A. Bennett, University of California, Davis, pers. comm. 2000. 30. H. W. Bailey, University of California, Davis, pers. comm. 1997. 31. Saruwatari et al. 1997. 32. Moyle 2000. 33. Lindberg et al. 1999. 34. J. L. Lindberg, pers. comm. 2001. 35. Rockriver 2001.

Wakasagi, *Hypomesus nipponensis* McAllister

Identification Wakasagi are slender-bodied fish that typically reach 70–90 mm SL, although a few may reach 150 mm SL. Their mouth is small, with a maxilla that does not extend past the midpoint of the eye. Their eyes are relatively large, with orbit width contained approximately 3.5–4 times within head length. Small, pointed teeth are present on the upper and lower jaws. The first gill arch has 29–36 gill rakers, and there are 7 branchiostegal rays. Strong concentric striations are lacking on the gill covers. The pectoral fins reach less than two-thirds of the way to the bases of the pelvic fins. There are 8–11 dorsal fin rays, 8 pelvic fin rays,

11–14 pectoral fin rays, and 14–17 anal fin rays (12). The lateral line is incomplete, with 54–60 scales in the lateral series. There are 4–7 pyloric caeca. Live fish are nearly translucent and have a steely blue sheen to their sides. There are usually 5–77 chromatophores (small dark spots) on the isthmus between the mandibles, the best single character for distinguishing wakasagi from delta smelt (which have 0–1). The number of chromatophores increases with the size of the fish (11).

Taxonomy See the account of delta smelt. The closest relative of this species is presumably the widespread Asian marine smelt, *Hypomesus japonicus*.

Names Wakasagi is the Japanese common name for this smelt, which translates literally as icefish, a reference to its color. *Nipponensis* refers to Nippon, an alternate name for Japan. It is sometimes called freshwater smelt in California (1). Like the delta smelt it has also been referred to as pond smelt.

Distribution Wakasagi are native to lakes and estuaries of Hokkaido, Japan, but they have been introduced into lakes on Honshu and Kyushu (2). In 1959 CDFG imported 3.6 million fertilized eggs, many of them dead on arrival, attached to palm fiber mats (3). The embryos originated from an introduced population of smelt in Suwa Lake, east

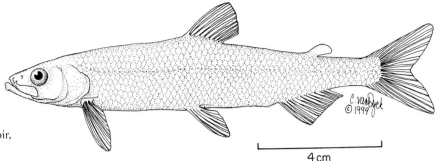

Figure 83. Wakasagi, Folsom Reservoir, Sacramento County.

4 cm

of Tokyo. The mats were placed in streams feeding into Freshwater Lagoon (Humboldt County) and Dodge (Lassen County), Shastina (Siskiyou County), Spaulding (Nevada County), Jenkinson–Sly Park (El Dorado County), and Big Bear (San Bernardino County) Reservoirs. The purpose of the introductions was to provide forage for rainbow trout and other salmonids in coldwater reservoirs. Populations became established in all but Dodge and Big Bear Reservoirs (15).

In 1972 and 1973 about 77,000 fish from Shastina Reservoir were moved to Almanor Reservoir (Plumas County), where they became established. By 1976 fish from Almanor Reservoir had moved about 100 km down the North Fork Feather River and colonized Oroville Reservoir. They may also have colonized Butt Valley Reservoir about this same time. Fish from Spaulding Reservoir on the Bear River colonized Rollins Reservoir, about 35 km downstream, by 1976 and Camp Far West Reservoir, another 40 km downstream, a few years later. Fish from Jenkinson Reservoir, on a tributary to the Cosumnes River, presumably invaded Folsom Reservoir via the canal maintained by the El Dorado Irrigation District, which dumps water into the American River. They were first detected in Folsom in 1989 (15). Wakasagi were present in the San Francisco Estuary as early as 1974, but they were not detected until about 1990, when they were found in the American River below Folsom (4). Small numbers were found throughout the estuary by 1998, including downstream into Suisun and San Pablo Bays (11). From the Delta, wakasagi have been transported via the California Aqueduct to San Luis Reservoir (11). Further spread to southern California reservoirs is therefore likely. Hybrids with delta smelt have also been found in the Delta (4). The most likely source of wakasagi in the Delta is Folsom Reservoir, because they are abundant in the reservoir and large amounts of water are periodically released from it to control salinity in the estuary. However, collection of wakasagi in the Cosumnes River in 2001 suggests they also could have moved down directly from Jenkinson Reservoir.

In short, wakasagi are widespread in the Sacramento River watershed, including the San Francisco Estuary, and have established populations in the Klamath Basin (Shastina Reservoir) and at one North Coastal locality (Freshwater Lagoon). Further spread is likely, especially to waters downstream of their present locations.

Life History The wakasagi has been reasonably well studied in Japan (5, 6, 7), where it is an important food fish, but the only life history study in California was carried out in Shastina Reservoir (8, 13). In Japan it can be either anadromous or resident in fresh water, and apparently some populations exhibit both life history strategies (7). It is not certain if the strategies have a genetic basis, but if they do, California populations should not be inclined to go out to sea. Nevertheless, California wakasagi tolerate a wide range (0–29 ppt) of salinities (14). They can also live in a wide range of temperatures. In the laboratory critical thermal maxima are 27–29°C and minima are 2–4.5°C (14). The range of 14–21°C is probably the optimal one for both growth and reproduction. Like delta smelt, wakasagi are pelagic plankton feeders, but they appear to be a schooling species (unlike delta smelt). They also are faster swimmers than delta smelt (maximum speed, circa 43 cm/sec for adults) (14).

Wakasagi are opportunistic planktivores. In Shastina Reservoir the most important items in the diets of fish measuring greater than 50 mm FL were chironomid midge pupae, cladocerans (mainly *Bosmina*), and copepods, although relative amounts varied with availability. Other food items included chironomid larvae, phantom midge larvae (Chaoboridae), and smelt eggs. Fish measuring less than 50 mm FL fed mainly on cladocerans (8). Diets of wakasagi in Japanese lakes are similar (6), although they apparently often take fish larvae as well (9); they have feeding peaks in early morning and evening, when light levels are low (6). In the Delta they apparently feed mainly on planktonic copepods (11).

In Japan most individuals from anadromous stocks live 1 year, spawn, and die. Some freshwater populations have a similar life history pattern (6), whereas others live up to 4 years (2, 7). However, wakasagi in Japan in any situation rarely live more than 2 years because of heavy exploitation (7). In Shastina Reservoir 20–35 percent of spawning fish were 2 years old, but none were older (8). A majority of 2-year-olds were females. The mean size of wakasagi in Shastina Reservoir was 73 mm FL at first spawning and 94 mm FL at second spawning. The largest fish were about 120 mm FL. In the San Francisco Estuary wakasagi over 100 mm have been collected, suggesting that some fish are living at least 2 years.

Wakasagi spawn in April and May, usually moving up rivers a short distance to deposit fertilized eggs in shallow areas of gravel or sand. In Japan they also spawn on sandy beaches (2). The presence of delta smelt–wakasagi hybrids suggests that both species select similar areas for spawning in the Delta. The free-swimming juveniles colonize open waters of lakes and reservoirs and grow quickly to adult size. In Japan the young of anadromous stocks move out to sea in July and August but return to fresh water in November and December (7). The few survivors of spawning return to the sea after spawning.

Status IIE. Wakasagi are clearly expanding their range in central California, and the consequences of this expansion are likely to be hard on native fishes, especially delta smelt. They threaten delta smelt because the two species hybridize and are likely to compete for food and spawning sites; wakasagi may also prey on delta smelt larvae. As of 2001 wakasagi had not become abundant in the San Francisco Estuary, which is surprising because they have broader temperature and salinity tolerances than delta smelt and can swim faster, presumably giving them greater ability to avoid being sucked into water diversions (14). It is possible that they are highly vulnerable to predators such as striped bass, perhaps because of their schooling behavior.

From the perspective of water managers, wakasagi are an enormous nuisance in the Delta, because small individuals are extremely difficult to distinguish from small delta smelt. For example, because the delta smelt is listed under federal law as a threatened species, the big State Water Project and Central Valley Project pumps in the south Delta can be forced to halt their pumping if they kill ("take") too many delta smelt. The inability of workers at the pumping plants to distinguish readily between the two species may mean that at times the plants are shut down while awaiting identification of entrained smelt by biochemical means. If wakasagi continue to expand at the expense of delta smelt, achieving recovery of delta smelt as mandated under the Endangered Species Act will become increasingly difficult, making it possible that restrictions on the use of Delta water (such as the timing and amount of pumping from the south Delta) to protect smelt will continue indefinitely.

It is ironic that the original introduction of wakasagi was regarded as a reversible experiment and that reservoirs selected for that introduction were ones that could be "chemically treated if it were found the smelt were undesirable" (3, p. 141). Once established the smelt were, of course, found to be desirable, even without proper evaluation of their effects, resulting in their transfer to other locations. One general effect of wakasagi introductions has been the collapse of kokanee fisheries in reservoirs into which they have been introduced (1). This outcome suggests that wakasagi deplete zooplankton populations in reservoirs, with negative effects on other fishes with life stages that depend on zooplankton (e.g., largemouth bass). Wakasagi in reservoirs do become important prey of rainbow trout and other salmonids introduced annually at near "catchable" sizes (13), and they apparently improve the growth rates of these fish as a consequence. However, they may also simply displace other planktivorous prey fishes (e.g., threadfin shad) (1) with no real net gain in the growth rates of predatory fishes. Introduction of wakasagi provides yet another example of the "Frankenstein effect," in which an introduction carried out for a narrow, albeit beneficial, reason (e.g., to improve reservoir fisheries) winds up having long-term negative consequences that were totally unforeseen (10).

In Japan wakasagi are a favored food fish, supporting a specialized fishery. Exploitation rates of mature fish are high and, to compensate for them, fry resulting from artificial propagation are planted in the millions annually (6). It was the long history of artificial propagation of wakasagi that made it so easy to bring them to California.

References 1. Dill and Cordone 1997. 2. McAllister 1963. 3. Wales 1962. 4. Trenham et al. 1998. 5. Hamada 1961. 6. Seki et al. 1981. 7. Utoh 1988. 8. Wigglesworth 1975. 9. Shiraishi 1957. 10. Moyle et al. 1986. 11. Aasen et al. 1998. 12. Sweetnam 1995. 13. Rogers 1984. 14. Swanson et al. 2000. 15. D. Lee, CDFG, pers. comm. 1998.

Longfin Smelt, *Spirinchus thaleichthys* (Ayres)

Identification Longfin smelt can be distinguished from other California smelt by their long pectoral fins (which reach or nearly reach the bases of the pelvic fins), incomplete lateral line, weak or absent striations on the opercular bones, low number of scales in the lateral series (54–65), and long maxillary bones (which in adults extend just short of the posterior margin of the eye). The lower jaw projects forward of the upper jaw when the mouth is closed. Small, fine teeth are present on both jaws, as well as on the tongue, vomer, and palatines. The number of dorsal rays is 8–10; anal rays, 15–22; pectoral rays, 10–12; gill rakers, 38–47; and pyloric caeca, 4–6. Orbit width is divisible into head length

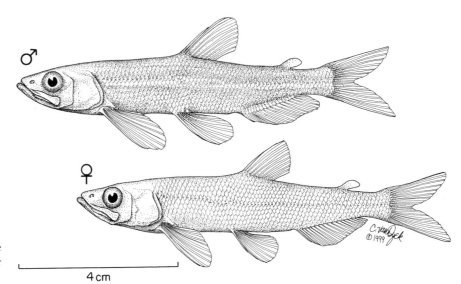

Figure 84. Longfin smelt, male and female, 8 cm SL, Russian River estuary, Sonoma County.

4 cm

3.6–4.5 times, and the longest anal rays are divisible 1.4–2.2 times into head length (1, 2). The lining of the gut cavity (peritoneum) is silvery with a few scattered speckles. The sides of living fish appear translucent silver, while the back has an olive to iridescent pinkish hue. Mature males are usually darker than females, with enlarged and stiffened dorsal and anal fins, a dilated lateral line region, and breeding tubercles on paired fins and scales (1).

Taxonomy Longfin smelt were at one time considered to be two species: Sacramento smelt (*S. thaleichthys*) in the San Francisco Estuary and longfin smelt (*S. dilatus*), accounting for the rest of the populations. McAllister (1) merged the two species because meristic characters separating Sacramento smelt from other populations represented the southern end of a north-south cline in characters rather than a discrete set. This analysis was confirmed by an electrophoretic study (3), which showed only minor differences in allele frequencies between smelt populations in Lake Washington (Washington) and San Francisco Bay. The differences were, however, sufficient to demonstrate no recent gene flow between the two populations. Longfin smelt in the San Francisco Estuary are isolated from other populations and are the southernmost of the species. They are similar in this respect to a recognized run of chinook salmon (e.g., winter-run chinook) and fit the definition of an Evolutionarily Significant Unit (ESU) established by NMFS (4). A population must satisfy two criteria to be considered an ESU: *(1)* it must be substantially reproductively isolated from other conspecific population units, and *(2)* it must represent an important component in the evolutionary legacy of the species. The longfin smelt in the San Francisco Estuary fills both of these criteria.

The closest relative of the longfin smelt is the night smelt, *Spirinchus starksi* (3). A third *Spirinchus* species, *S. lanceolatus,* occurs in northern Japanese waters and differs from *S. thaleichthys* in several morphological characters and in timing of spawning (1).

Names "Longfin" describes the pectoral fins well. *Spirinchus* means breath-beginning, referring to the conspicuous duct that connects the air bladder to the gut. *Thaleichthys* means rich fish, but is probably a reference to the related eulachon, *Thaleichthys pacificus.*

Distribution Populations of longfin smelt in California have historically been known from the San Francisco Estuary, Humboldt Bay, the Eel River estuary, and the Klamath River estuary (5). Spawning longfin smelt have been recorded from the Van Duzen River in the Eel River drainage, and a sample from there is in the fish collection at Humboldt State University. There are also recent records from the mouth of the Klamath River, so presumably a small population still exists there. They were collected from the Russian River estuary in 1996 and subsequent years, so this presumably is a resident population as well (19). In the San Francisco Estuary longfin smelt are rarely found upstream of Rio Vista or Medford Island in the Delta. Adults occur seasonally as far downstream as South Bay, but they are concentrated in Suisun, San Pablo, and North San Francisco Bays. They also occur on a regular basis in Gulf of the Farallones, just outside the Golden Gate, but are most likely to be found there in wet years (6, 21). The southernmost record of the species range is a single fish from Monterey Bay (7).

Outside California, longfin smelt are found in scattered bays and estuaries from Coos Bay, Oregon, to Prince William Sound, Alaska. They are found in Skagit Bay, Grays Harbor, and Willapa Bay in Washington; the lower Columbia River; and Yaquina and Coos Bays, Oregon (5). Landlocked populations occur in Lake Washington, Washington, and Harrison Lake, British Columbia (8).

Life History Longfin smelt in California are euryhaline, nektonic, and anadromous. Adults and juveniles can be found in the open waters of estuaries, mostly in the middle or at the bottom of the water column. They are found at salinities ranging from nearly pure seawater to completely fresh water, although most seem to prefer salinities in the range 15–30 ppt once past the early juvenile stages (21). They can occupy water as warm as 20°C in summer, but summer preferred temperatures seem to be around 16–18°C. The wide salinity and temperature preferences reflect the ability of smelt to occupy different portions of the estuary according to time of year and stage of life cycle. In the San Francisco Estuary, the center of their distribution gradually moves down the estuary during summer. They concentrate in most years in San Pablo Bay in April–June and become more dispersed (many moving into Central San Francisco Bay) in late summer. There is a gradual shift in population upstream in fall and winter, as yearlings begin to move upstream to spawn. The exact distribution pattern of longfin smelt varies from year to year. During winter months, when fish are moving upstream to spawn, high outflows may push many back into San Francisco Bay, whereas drought years may find them concentrating in Suisun Bay (20).

The landlocked smelt in Lake Washington, Washington, show daily vertical migrations, moving into deep water during the day and into the upper water column at night (5). This habit may explain why juvenile and adult longfin smelt are usually captured in trawls in the lower half of the water column in the San Francisco Estuary (21), where most sampling takes place during the day. Vertical migrations may be tracking diurnal movements of their principal prey, opossum shrimps, *Neomysis mercedis* and *Acanthomysis* spp. Copepods and other crustaceans are also important prey at times, especially to small fish. These are similar to their feeding habits in Lake Washington, where reductions in populations of mysid shrimp resulted in reduced growth rates of the smelt (8, 25). Longfin smelt are eaten by predatory fishes, birds, and marine mammals. They are a major prey of harbor seals, *Phoca vitulina,* in the Columbia River (5).

Growth in California populations is similar to that in more intensively studied Washington populations (8, 25). Most growth takes place in the first 9–10 months of life, when the smelt typically reach 6–7 cm SL. Growth rates level off during the first winter, but there is another period of growth during the second summer and fall, when they reach 9–11 cm SL. Weight gains may be considerable as gonads develop during this latter period. The largest smelt, presumably females in their third year of life, measure 12–15 cm SL.

Spawning in longfin smelt takes place in fresh water, over sandy or gravel substrates, rocks, and aquatic plants (5, 9). In the San Francisco Estuary, spawning appears to take place mainly below Medford Island in the San Joaquin River and below Rio Vista on the Sacramento River (9). The lower end of the spawning habitat seems to be upper Suisun Bay around Pittsburg and Montezuma Slough in Suisun Marsh (9). Sacramento longfin smelt have a rather protracted spawning period; adult movements indicate that some spawning may take place as early as November (21), whereas larval surveys indicate that spawning may occur into June (9). Most spawning is from February through April, because larval smelt are most abundant in this period and large smelt become rare after this time. Older and larger smelt apparently spawn later than smaller ones (9). Males evidently precede females in the spawning run upriver (10), and spawning occurs at night. Spawning in the San Francisco Estuary occurs at water temperatures of 7.0–14.5°C (5), although spawning has been observed at lower temperatures in other areas, such as Lake Washington (9). Each female lays 5,000–24,000 adhesive eggs (8), but the number can apparently vary considerably from year to year (11). The mean fecundity of smelt in the San Francisco Estuary in the early 1970s was 9,752, which is on the low end of the range and may not reflect the fecundity in other years. Most longfin smelt die after spawning. A few smelt, mostly females, live another year, although it is not certain whether or not they spawned previously.

The embryos hatch in 40 days at 7°C (8). Newly hatched larvae are 5–8 mm long (9) and buoyant. They quickly move into the upper part of the water column and are swept down into more brackish parts of the estuary. During years when periods of high outflows coincide with the presence of larval smelt (e.g., 1980, 1982, 1983, 1984, 1986), the larvae are mostly transported to Suisun and San Pablo Bays; in years of lower outflow, they are transported to the western Delta and Suisun Bay (6, 21). Larvae are strong enough swimmers that they can move up and down in the water column in order to maintain position within the mixing zone of the estuary. Metamorphosis into juveniles probably begins 30–60 days after hatching, depending on temperature (5). The distribution of young-of-year smelt largely coincides with that of the larvae.

There is a strong positive correlation between winter and spring Delta outflow and longfin smelt abundance the following year. There is also a strong correlation between juvenile survival (adult abundance) in the San Francisco Estuary and Delta outflow (12), as well as the position of the 2-ppt isohaline (13). The reason seems to be that flows increase the rate of transport into rearing habitat in Suisun and San Pablo Bays and reduce the probability of larvae being retained in the Delta, where they are exposed to entrainment, pesticides, and other adverse factors. High freshwater outflows also increase the volume of brackish water (salinity, 2–18 ppt) rearing habitat required by larval and juvenile smelt. It is likely that longfin smelt larvae, like striped bass larvae, have higher survival rates in brackish water (14).

Status IC. The longfin smelt is listed by CDFG as a Species of Special Concern because of its long-term decline in California, although it is secure in the more northern parts of its range. Longfin smelt were once one of the most abundant species in both the San Francisco Estuary and Humboldt Bay (16). It is likely that they were an important component of the smelt fishery in the estuaries in the late 19th century, although species were not recorded. In recent years longfin smelt populations have declined dramatically in the San Francisco Estuary and have become very small in the Eel River estuary and Humboldt Bay. The populations in the Klamath and Russian River estuaries are (and probably always have been) small.

In the San Francisco Estuary longfin smelt historically showed wide fluctuations in abundance, reflecting both population trends and their concentration in some years in areas that were not sampled. Numbers of longfin smelt typically reached their lowest levels during drought years, but quickly recovered when adequate winter and spring flows were once again present. Longfin smelt catches were exceptionally high in 1982, but numbers plummeted in following years (15) and remained at low levels subsequently. The number of smelt in a given year seems to be a function of the number of spawners and of outflow during the spawning and larval periods in the previous year (6). Despite reasonably good outflows in 1995–1999, smelt numbers remained fairly low when a strong upward trend might have been expected. The decline in longfin smelt abundance paralleled that for other fishes in the estuary, such as delta smelt, but it was, if anything, even more precipitous. Longfin smelt declined in rank abundance from first or second in most trawl surveys during the 1960s and 1970s to seventh or eighth at present (22).

In Humboldt Bay surveys in the early 1970s found longfin smelt to be one of the most abundant fish in the bay

and an important component of the food webs (16). However, only a few longfin smelt have been collected from Humboldt Bay in recent years despite extensive sampling (23, 26). Likewise, there are no recent records from the Eel River estuary (24). Longfin smelt are still present in the Klamath River estuary, but confirmed records are few; two spent males were collected in November 1992 (21). The status of the recently discovered population in the Russian River estuary is uncertain.

Because of the severe decline of longfin smelt in California, in 1992 the Natural Heritage Institute petitioned USFWS to list it as an endangered species. The petition was denied in 1993, largely on the basis of taxonomy and the fact that the smelt is not in trouble throughout its range. Had the smelt been a salmon, its California populations (ESUs) would probably have been listed.

The causes of the decline of longfin smelt from northern estuaries are not known, but they are probably similar to those of their decline in the San Francisco Estuary, which are multiple and synergistic.

Reduction in outflows. Reduction in outflows through water exports is probably the single biggest factor affecting longfin smelt abundance in the San Francisco Estuary. To demonstrate the effects of the state and federal pumping plants on the smelt, a regression equation has been calculated relating smelt numbers to Delta outflow (22). This equation predicts that mean spring (March–May) outflows much less than 3,400 cfs will result in reproductive failure of the smelt. Such flows for 2–3 years in a row would probably result in extinction of the longfin smelt in the estuary. During the period 1986–1994 outflows in most years were perilously close to that number, pushed there by the increase in diversions during long periods of drought, when smelt numbers would be naturally low. This circumstance resulted in abnormally low numbers of longfin smelt being produced. This amplification of normal drought effects has been compounded by the ability of upstream reservoirs to retain more winter-spring runoff because reservoirs have been below flood control limits. Later release of this water for export may have exacerbated the normal drought-year decline of this species beyond the impacts of reduction in annual totals. Thus even during wetter years (1995–1999) the abundance of smelt has been lower than would be predicted by the past relationship (12) between smelt abundance and outflow.

Entrainment losses to water diversions. One of the effects of decreased outflow is increased vulnerability of longfin smelt of all sizes to entrainment in State Water Project and federal Central Valley Project pumping plants, agricultural diversions within the Delta, and power plants. The effects of direct entrainment of smelt in the two pumping plants are not well understood because of limited information on what proportion of the population at each life stage is entrained and on survival rates of the fish that are salvaged and returned to the Delta. Although large numbers of adult longfin smelt are captured at the pumping plants, it is unlikely that many individuals of this species survive the experience. If they do, they are probably consumed by piscine and avian predators attracted to the predictable commotion of trucks releasing fish. In any case, the fact that rates of capture at the pumping plants have increased even as populations have been decreasing suggests that direct entrainment is a significant source of mortality.

Entrainment of larvae in agricultural diversions within the estuary is largely unquantified. Entrainment in Delta agricultural diversions was presumably a fairly constant source of mortality for 50–100 years, until flows across the Delta increased as the result of increased pumping by the State Water Project and Central Valley Project. These pumps not only remove more water than formerly, they also pump water earlier in the year, when longfin smelt are spawning and larval fish are present. The changed hydraulics increase the exposure of larval, juvenile, and adult smelt to in-Delta entrainment, predation, and other factors.

The importance of entrainment of longfin smelt, especially larval smelt, in the cooling water of power plants is not well known. The potential for entraining significant numbers of larvae is, however, considerable, especially when smelt populations are low.

Climatic variation. Variation in climate in California since 1982 has been the most extreme since the arrival of Europeans. The period 1985–1992 was one of continuous drought, broken only by the record outflows of February 1986; 1995–2000, in contrast, was a period of unusually high rainfall. The 1986 flood occurred during the peak spawning season for longfin smelt and quite likely washed a high percentage of spawning fish and their offspring far downstream, perhaps beyond the Golden Gate. This event was particularly unfortunate because the smelt were already showing signs of precipitous decline, and the washout may have exacerbated the problem. The prolonged drought had two major interacting effects: a natural decrease in outflow and an increase in the proportion of inflowing water being diverted. A natural decline in smelt numbers would be expected from reduced outflow because of reduced availability of brackish water habitat for larvae and juveniles. However, the increase in diversion exacerbated the decline in smelt survival through a combination of further reduction in brackish water habitat and increased entrainment of larvae, juveniles, and adults. It is important to recognize that extreme floods and droughts have occurred in the past and that smelt have managed to persist in large numbers. Unlike today, the smelt historically did not experience the extreme conditions caused by increased diversion of water, nor were their numbers depleted to the point that recovery from natural disasters becomes much more problematic. Thus six

consecutive wet years did not result in a dramatic recovery of smelt populations.

Toxic substances. Pollution is an insidious problem in the estuary because toxic compounds, especially pesticides, come from many sources, may be episodic in nature (and therefore hard to detect), and may affect mainly the early life history stages of fish, for which mortality is hard to observe. There is no evidence that toxic compounds have affected populations of longfin smelt over the long term. This is not surprising, because the smelt spawn early in the season, when few agricultural chemicals are being applied and flows for dilution are high. It is possible that episodic high concentrations of chemicals applied in winter (e.g., carbofuran) may have negative effects on smelt if episodes coincide with major spawning times. The short life span and plankton-feeding habits (resulting in a short food chain) of longfin smelt reduce the probability of accumulation of toxic materials in their tissues.

Predation. Predation is a poorly understood but potentially important factor affecting longfin smelt abundance. The principal piscivore in the estuary is striped bass, which was introduced over 100 years ago, replacing native piscivores such as Sacramento perch, thicktail chub, and steelhead. Longfin smelt remained abundant despite the explosion of striped bass numbers, and in recent years smelt decline has coincided with the decline of striped bass. It is therefore unlikely that striped bass predation per se is responsible for smelt decline. It is possible that concentrated striped bass predation in Clifton Court Forebay has some effect on longfin smelt populations. Smelt are drawn into this forebay before being entrained in State Water Project pumps. However, the smelt probably do not survive entrainment in any case, so whether they are eaten by bass or entrained by the pumps is a moot point.

Inland silversides are potential predators on eggs and larvae of longfin smelt, but their effects, if any, are not known. Speculation that they are important in this regard is based on (1) their invasion of the estuary in the late 1970s, roughly coincident with the decline of both delta and longfin smelt; (2) their concentration in shallow waters where smelt spawn; and (3) their known effectiveness as predators on larval fishes (17).

Introduced species. Invasions by alien species are a perpetual problem in the San Francisco Estuary, especially those introduced in the ballast water of ships. Among recent problem introductions have been those of several species of planktonic copepods and the overbite clam. The copepods are regarded as a problem because they have partially replaced *Eurytemora affinis,* a copepod that had been the favored food of larval fish. One of the alien copepod species (*Sinocalanus doerri*) is difficult for larval fish to capture, but it occurs mostly upstream of concentrations of smelt larvae; it may only be a problem if diversions keep smelt larvae upstream. Other alien copepods are readily eaten by smelt, but differences in size and behavior may make them less desirable. The overbite clam, in contrast, may have a direct effect on smelt populations because it has become extremely abundant in San Pablo and Suisun Bays, from which it appears to be filtering out most planktonic algae, the base of the food web on which smelt depend (18). The decline of opossum shrimp, a major food of smelt, is most likely related to the overbite clam invasion. The clam is not, however, a direct cause of the initial decline of longfin smelt, because it did not invade until after February 1986, when the estuary's biota had already been devastated by immense outflows (18). Nevertheless, its abundance is probably a factor in the failure of smelt populations to increase dramatically during the 1995–1999 wet period.

The Delta Native Fishes Recovery Team, appointed by USFWS, developed recovery criteria and recommendations for longfin smelt in the San Francisco Estuary. Essentially, recovery of longfin smelt depends on restoration of more natural estuarine hydraulics in all years. For Humboldt Bay, studies are needed to determine conditions for restoration of the smelt as a major player in the ecosystem. The nature and extent of smelt populations in the Eel, Klamath, Russian River estuaries should be determined, as well as the factors that limit their abundance.

References 1. McAllister 1963. 2. Morrow 1980. 3. Stanley et al. 1995. 4. Waples 1991a. 5. Emmett et al. 1991. 6. Baxter 2000. 7. Eschmeyer et al. 1983. 8. Dryfoos 1965. 9. Wang 1986, 1991. 10. Wydoski and Whitney 1979. 11. Chigbu and Sibley 1994. 12. Stevens and Miller 1983. 13. Jassby et al. 1995. 14. Hall 1991. 15. USFWS 1994. 16. Barnhart et al. 1992. 17. Bennett and Moyle 1996. 18. Nichols et al. 1990. 19. M. Fawcett, pers. comm. 1998. 20. Armour and Herrgesell 1985. 21. Baxter et al. 1999. 22. B. Herbold, USEPA, pers. comm. 1998. 23. R. Fritzsche, Humboldt State University, pers. comm. 1996. 24. L. Brown, USGS, pers. comm. 1996. 25. Chigbu and Sibley 1998. 26. D. Sweetnam, CDFG, pers. comm. 2001.

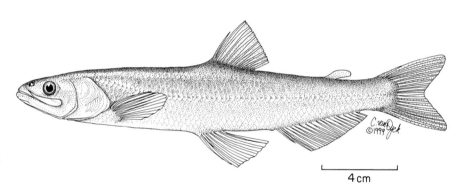

Figure 85. Eulachon, 20 cm SL, Oregon Coast off Garibaldi. OSU 003855.

4 cm

Eulachon, *Thaleichthys pacificus* (Richardson)

Identification Eulachon are the largest Pacific North American smelts, reaching 20–30 cm TL. They have compressed, elongate bodies and large, oblique mouths; the maxilla usually reaches just past the middle of the eye but can extend beyond the posterior margin of the eye in adults. Gill covers possess strong concentric striations, and the pectoral fins, when pressed against the body, reach about two-thirds of the way to the bases of the pelvic fins. Body depth is 15–20 percent of standard length; head length is 20–26 percent of standard length. The lateral line is complete, with 70–78 scales. There are 8–12 pyloric caeca, 10–13 dorsal fin rays, 8 pelvic fin rays, 10–12 pectoral fin rays, 18–23 anal fin rays, 17–23 slender gill rakers on the first arch, and 7–8 branchiostegal rays. Their jaws have small pointed teeth, which may be missing from spawning fish, especially males. Small teeth are also present on the tongue and palatines, and a pair of moderately large canines is present on the vomer. The lining of the gut cavity (peritoneum) is pale with dark speckles. Eulachon are brown to dark blue on the back and head with a silvery white belly and unmarked fins. Spawning males develop a distinct midlateral ridge and numerous distinct tubercles on the head, body and fins. Females may have a few poorly developed tubercles. The flesh is very oily.

Taxonomy The eulachon is the only species in the genus *Thaleichthys*. Although there are significant differences in meristic characters (e.g., number of vertebrae) among fish from different rivers (1), there have been no genetic studies documenting the discreteness of the disparate stocks.

Names Eulachon (pronounced something like oolak-on) is the name of the Chinook people for this fish. They are also known as candlefish because the flesh is so oily that dried eulachon were burned by the Yurok and other native peoples as candles after insertion of a wick. *Thale-ichthys* means rich fish, a reference to their oily flesh; *pacificus* refers to their exclusively Pacific distribution.

Distribution The main eulachon population in California is in the Klamath River, Del Norte County; runs are also found in the Mad River and Redwood Creek, Humboldt County. They have also been reported in small numbers in the Smith River and in tributaries to Humboldt Bay (15). The California populations are the southernmost of the species. To the north eulachon have been found spawning in coastal streams from Oregon to Bristol Bay, Alaska. Oceanic distribution parallels the coastal distribution, but eulachon have been collected westward in the Bering Sea to the Pribilof Islands (2) and as far south as Bodega Head, Sonoma County (3); San Francisco Bay (16); and Point Buchon, San Luis Obispo County (17). Most eulachon runs occur in larger rivers, such as the Fraser and Columbia, although smaller streams may also have runs. Rivers in Washington that have (or had) large spawning runs are the Columbia, Cowlitz, Grays, Kalama, Lewis, Sandy, and Nooksack (4). In Oregon eulachon apparently spawn only in the Umpqua River (2).

Life History Eulachon are anadromous, spending most of their life at sea and then spawning in the lower reaches of coastal rivers. In the ocean they occur in both deep and shallow water above the continental shelf, presumably close to their home streams, although their marine movements are poorly known. In the ocean they feed mainly on euphausid shrimps, copepods, and other crustaceans (2, 6). They do not feed while in fresh water.

Larvae washing out of streams transform into free-swimming juveniles at about 50–80 mm TL. They grow 40–50 mm/year for the next 3 years. Most eulachon mature during their third year, and adult length averages around 17 cm TL, with a range of 14–20 cm. Most move upstream to spawn in their third year and die after spawning, although a fraction live to spawn once again the following year (6). A few fish may live to 5 years (2).

Spawning migration occurs sometime between December and May (2, 4), usually peaking in February–March. It typically begins in mid-March in California and extends into May at the northern end of their range (5). Fish appear in the Klamath River in March and April. Spawning migration commences when river temperatures are above 4°C, but it slows or stops if the temperature drops below 4°C or exceeds 8°C (2, 5). Migrating fish seldom penetrate more than 10–12 km upstream. However, spawning has been recorded more than 160 km up the Columbia River.

During the spawning migration adults evidently travel along the bottom of estuarine and river channels (2) and also in shallows at the water's edge (11). There is some indication, based on meristic characters, that eulachon return to natal streams (1). Males arrive at the spawning grounds first and remain there longer than females; hence more males are taken by sport and commercial fisheries. Spawning occurs at 4–10°C where water velocities are moderate and bottoms consist of pea-size gravel or gravel mixed with sand, wood, and other debris (2). Spawning takes place en masse at night. Fertilization is external and females produce an average of 25,000 eggs. Range in fecundity has been variously reported as 17,300–60,000 (4) and 7,000–31,000 (2), but it depends on female size. Each egg is surrounded by two membranes. The outer membrane ruptures when the egg hits the bottom, and its adhesive edges stick to the substrate. The outer membrane is attached by a short stalk to the inner membrane, which still surrounds the egg. The embryo is thus anchored to the bottom until it hatches in 2–3 weeks (7). Hatching occurs in 19 days at 8.5–11.5°C and in 30–40 days at 4.4–7.2°C (2). The feeble, transparent larvae (4–7 mm TL) stay near the bottom and are quickly washed out to sea.

Eulachon constitute an important food source for marine and anadromous fishes. At times they are heavily preyed upon by halibut, cod, salmon, and sturgeon, as well as by marine mammals, including finback whales, porpoises, orcas, seals, and sea lions (2).

Status IC, possibly IB. Eulachon have declined in California but appear to be abundant in the northern parts of their range (2). The largest run in the United States is the Columbia River run, which has remained stable, with 1990 commercial landings of nearly 2.8 million pounds and recreational catch in some years equaling commercial catch (8).

Eulachon spawning runs in the Klamath River and other streams were at one time the basis of a sport dipnet fishery and a Native American subsistence fishery. The runs were regarded as a major event because the fish moved up in huge swarms, followed by large flocks of predatory seabirds. Heavy runs also occurred in both the Mad River and Redwood Creek up to the mid-1970s (11, 12), and there were incidental reports of fish, but no regular large runs, in the Smith River (12). In the Mad River spring runs of eulachon occurred in spurts, each run lasting a few days and the runs occurring perhaps 2 weeks apart. The fish would "pour" by, and at times several pounds of fish could be taken in a single dipnet haul (12). In Redwood Creek, "massive runs" of eulachon occurred during 1972-1974; the fish would form a "black mass" so thick they could be caught by hand (13). Spawning occurred primarily between the ocean and mouth of Prairie Creek (a major tributary) and up Prairie Creek about 0.5 km.

In the Klamath River eulachon have been scarce since the 1970s, with the exception of three years: they were plentiful in 1988 and moderately abundant again in 1989 and 1999. Native American fishers interviewed in 1992 likewise attested to a recent decline (11, 14). In 1996 only one individual was reported taken from the Klamath River (14). Since the 1970s eulachon have also not been seen in Prairie Creek. There have been no occurrences of eulachon reported by fishery biologists in Redwood Creek and the Mad River since at least the mid-1980s (11, 12, 13).

The factors responsible for the decline of California eulachon populations are unknown. Given the extended ocean life phase of eulachon and the apparently sporadic nature of their abundance in recent years, it is likely that oceanic conditions may be important determinants of the size of spawning runs. The relationship between El Niño/Southern Oscillation (ENSO) events and eulachon abundance must be investigated, because it is likely that warmer ocean temperatures and decreased upwelling affect eulachon growth and survival. However, eulachon are sensitive to a number of environmental factors, and their recent decline in California streams may be the result of changes in water quality or spawning habitat in the lower reaches of rivers. In Redwood Creek, for example, extensive modification of the creek mouth and lagoon area by levee construction has rendered much of the estuary unsuitable for juvenile salmon and steelhead (9), and it is likely that eulachon have been likewise affected.

In more northern regions, especially British Columbia, eulachon still support a small but valuable commercial fishery. They have been rated highly as gourmet food and have been described as "unsurpassed by any fish whatsoever in delicacy of the flesh, which is far superior to that of the trout" (10, p. 521). In California eulachon have never been particularly important commercially. In 1963, however, heavy runs in the Klamath and Mad Rivers and Redwood Creek resulted in a catch of 56,000 lb (3). In the past eulachon were important to Native Americans for food, fuel, and oil. The oil, described as having "a very attractive flavor" (10, p. 521), can be rendered down to a fatty substance that was highly valued by the Pacific Northwest tribes. Dried eulachon were burned as candles, providing light on short winter days. Eulachon are now less preferred by Native American fishermen than salmon and lampreys (11), but they are nevertheless clearly an important component of the cultural legacy of riparian peoples.

Given the lack of quantitative data on eulachon populations, an important first step in their management is to monitor year-to-year abundance in the Klamath River, Mad River, and Redwood Creek. It is not clear whether eulachon populations in California would be amenable to, or benefit from, active management. However, perhaps "rehabilitation" of selected streams (e.g., the lower reaches of Redwood Creek) to previous levels of environmental quality could improve their spawning success and early life stage survival.

References 1. Hart and McHugh 1944. 2. Emmett et al. 1991. 3. Odemar 1964. 4. Wydoski and Whitney 1979. 5. Morrow 1980. 6. Barraclough 1964. 7. Carl and Clemens 1953. 8. Oregon Department of Fish and Wildlife 1991. 9. Larson et al. 1983. 10. Jordan and Evermann 1896. 11. T. Kisanuki, USFWS, pers. comm. 1998. 12. J. Waldvogel, University of California Co-operative Extension, pers. comm. 1996. 13. S. Saunders, Prairie Creek Hatchery, Humboldt County, pers. comm. 1996. 14. D. Gale, Yurok Tribal Fishery Program, pers. comm. 1998, 1999. 15. Jennings 1996. 16. R. D. Baxter, CDFG, pers. comm. 1999. 17. R. Lea, CDFG, pers. comm. 1998.

Salmon and Trout, Salmonidae

No family of fish has excited as much interest through the centuries as the Salmonidae, at least in the Western world. The Salmonidae contains legendary fishes: salmon, trout, and char (subfamily Salmoninae); whitefish (subfamily Coregoninae); and grayling (subfamily Thymallinae), over which controversies over conservation and management continue to rage. All members (66 or more species, depending on who is counting) are native to the Northern Hemisphere, but species have been introduced into suitable waters all over the world. A salmonid fish is generally recognizable by its fusiform body, forked tail, adipose fin, and axillary scale (usually visible as a distinct process) at the base of each pelvic fin. Most juveniles have bands on the side (parr marks).

Because salmonids have a long history as major sport and commercial fishes, a delightful vocabulary has developed for various stages of salmonid life histories. Spawning adults construct a *redd* (nest depression) in which *alevins* (sac-fry) hatch from embryos. When *fry* develop vertical bars on their sides, they are called *parr.* In anadromous forms parr lose their parr marks and turn silvery as they start moving out to sea. They are then termed *smolts.* Males that have spent only a year at sea but have returned to spawn are *grilse,* although such males are more often called *jacks.* Similar females, which are rare, are called *jills. Kelt* is a rarely used term for spawned-out fish. A dedicated salmonid angler can be termed a *finatic.*

Salmonids have been around since the Eocene, and most species have probably been extant for millions of years (Stearley 1992). However, their present distribution and abundance have been strongly shaped by Pleistocene events. In northern and mountain areas they followed the advance and retreat of continental glaciers, rapidly colonizing new streams and lakes, whereas in southern areas their ranges expanded and contracted according to changes in rainfall and sea level associated with glaciation. Salmonids thrive in such dynamic environments as long as the water is fairly cool (<22°C maximum) and well oxygenated. In fresh water populations quickly adapt to local conditions, from tiny, clear, high mountain streams, to alkaline lakes, to large coastal rivers, to intermittent streams. Anadromous salmon and steelhead are major pelagic predators of northern seas and show an astonishing variety of life history adaptations, both within and between species, that maximize their ability to use freshwater habitats for reproduction and ocean habitats for feeding. Their ability to move long distances through the ocean has resulted in colonization of remote coastal streams from Baja California to the Aleutian Islands. The tendency of anadromous forms to give rise to land-locked forms has resulted in numerous distinct populations in lakes, headwaters, and other isolated environments.

Thus an evolutionary premium has been placed on fish that are adaptable in behavior and life history patterns, opportunistic in feeding, and capable of moving long distances through both fresh and salt water. Perhaps because they are tetraploid (have twice as much genetic material as most fishes), they show rapid evolutionary responses to new environments, developing distinctive forms often described as species or subspecies. For example, in isolated mountain streams, golden-colored trout have evolved independently in many areas (e.g., California golden trouts, Paiute cutthroat trout, redband trouts), probably within the past 10,000–50,000 years.

The evolution of many local forms, races, and runs of salmonids has always posed a challenge for taxonomists, as well as for anglers who desire to give distinctive-looking trout the taxonomic respectability of a Latin name. Early taxonomists, such as David Starr Jordan or John Otterbein Snyder, simply described every fish they caught as a new species, but it soon became evident that most of these "species" had characters that broadly overlapped those of other forms. In reaction, many forms were lumped together, and the fascinating diversity within species was largely ignored. Natural diversity was then confused by widespread planting of hatchery salmonids of mixed origin. In recent years the tools of molecular genetics have allowed for more positive definitions of distinct forms (regardless of physical appearance) and for determination of relationships among them—information that has been useful for conservation efforts. Molecular genetics has been particularly helpful in identifying runs of salmon and steelhead, which otherwise are distinguished mostly by life history traits. Today eight species native to California are recognized, including one char and one whitefish. Within the two native "trout" species (rainbow and cutthroat) numerous subspecies are recognized, although a number have not been formally described. Anadromous forms (salmon, steelhead, sea-run cutthroat) arguably have genetically distinct runs in every major stream, but they are increasingly divided up by life history strategy (run timing), genetic groupings, or some mixture of the two.

Despite the natural richness of California's salmonid fauna, numerous attempts have been made to establish still other species. Four species became established and widespread: brook trout, lake trout, brown trout, and kokanee. Four others were introduced but did not become established. Introductions of Bonneville cisco (*Prosopium gemmiferum*), lake whitefish (*Coregonus clupeaformis*), and Atlantic salmon (*Salmo salar*) failed immediately. Arctic grayling (*Thymallus arcticus*) occasionally established self-sustaining populations following their numerous introductions; the most recent such population was in Lobdell Reservoir (Mono County), and it lasted about 25 years. Grayling are now extinct in the state. The most pervasive salmonid introduction in California, however, has been that of "native" rainbow trout, which are now found in most suitable streams outside the original native range and in reaches upstream of natural barriers within the native range. California rainbow trout have also been planted in many places throughout the world, as have California chinook salmon (although much less successfully).

It is a reflection of the state of aquatic habitats in California that two species of native salmonids (bull trout, pink salmon) are extirpated and that most varieties of anadromous salmonids are in danger of extinction, as are endemic subspecies of rainbow and cutthroat trout. Considering that these fishes have historically supported major fisheries, native salmonid management in California must in general be regarded as a failure over the past century. Only recently have major efforts, many heroic, some successful, been made to restore or recover declining populations. A fundamental problem in restoration, especially for anadromous salmonids, is understanding the causes of declines, which are a synergism between natural and human-related factors. Natural shifts in production regimes in the ocean (the Pacific Decadal Oscillation and ENSO events) can cause major declines in salmon populations even if stream and river habitats are in good condition (Hare et al. 1999), as can droughts even if ocean conditions are favorable. Likewise, excessive ocean or stream harvesting can mask positive effects of stream and river restoration programs, and large releases of hatchery-reared fish can confound determination of causal factors underlying declines of wild fish. It is clear that restoration of the diversity and production of salmonid fishes requires an understanding of how all these factors interact—and a great deal of patience. Large investments in habitat restoration and better management practices for fisheries, hatcheries, and regulated rivers are clearly needed, even though returns on investments will take a long time to be realized.

Mountain Whitefish, *Prosopium williamsoni* (Girard)

Identification Mountain whitefish can be distinguished from other California fishes that possess an adipose fin by the combination of large scales (74–90 in the lateral line); a small, toothless, ventral mouth; a short dorsal fin (12–13 rays); and a slender body that is nearly cylindrical in cross section. The head is short (about 20% of TL), with a laterally compressed snout that overhangs the mouth. Gill rakers are short (19–26 on the first gill arch) and armed with small teeth. Branchiostegal rays number 7–10 per side; anal fin rays, 11–13; pelvic fin rays, 10–12; and pectoral fin rays, 14–18. The axillary process at the base of the pelvic fins is well developed. The tail is forked. The body is silvery and olive green to dusky on the back, and scales on the back may be outlined in dark pigment. Breeding males develop nuptial tubercles on the head and sides. Juveniles are silvery with 7–11 dark, oval parr marks on each side.

Taxonomy Mountain whitefish are widely distributed across the West with many disjunct populations, such as those found in streams in the Lahontan drainage in California and Nevada. It is highly likely that a thorough taxonomic analysis across their range will reveal a number of distinct units. The Lahontan population especially deserves close scrutiny because it is the one most isolated from other populations.

Names Prosopium is from the Greek word meaning face or mask, referring to the large bones in front of the eyes, which

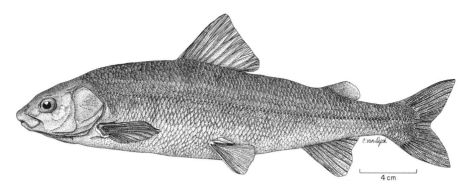

Figure 86. Mountain whitefish, 28 cm SL, Sagehen Creek, Placer County.

give the head its distinctive appearance; *williamsoni* is after Lt. R. S. Williamson, who commanded the Railroad Survey of California and Oregon, during which this species was first collected.

Distribution Mountain whitefish are one of the most widely distributed species in western North America. Outside California they are found throughout the Columbia River watershed (including Wyoming, Montana, Oregon, Washington, Idaho, British Columbia, and Alberta), the upper reaches of the Missouri and Colorado Rivers, the Bonneville drainage, and the Mackenzie and Hudson Bay drainages in the Arctic. In California they are found in streams and lakes (including Lake Tahoe) on the east slope of the Sierra Nevada, in the Truckee, Carson, and Walker River drainages. They are absent from Susan River and Eagle Lake.

Life History Mountain whitefish are most common in clear, cold streams with large pools that exceed 1 m in depth and in mountain lakes. Generally they live in waters with summer temperatures of 11–21°C (16). In California most of their populations are found at elevations of 1,400–2,300 m. In Lake Tahoe they generally live close to the bottom in fairly deep water, although they move into shallows during spawning season. They typically swim about in schools of 5–20 fish. Studies of whitefish in Sheep River, Alberta, indicate that some individuals stay in limited areas throughout

their lives, but most show complicated movements in relation to feeding, spawning, and overwintering (9).

As their subterminal mouths and body shape suggest, they are bottom-oriented predators on a wide variety of small aquatic insects (1, 2, 3). They feed in part by stirring up the bottom with their tail and pectoral fins and then turning to feed on exposed invertebrates (17). Small juveniles feed mainly on tiny chironomid midge, blackfly, and mayfly larvae (12), with their diet becoming more diverse with size. In Walker River adults feed mainly on larvae of mayflies (56% by volume) and caddisflies (34%) during summer (3). In Lake Tahoe they also feed on bottom-dwelling invertebrates: snails, dragonfly larvae, chironomid midge larvae, mayfly larvae, caddisfly larvae, crayfish, and amphipods (4). Small amounts (about 10%) of zooplankton and surface insects are also taken (4). Whitefish diets are strongly tied to abundance of prey, although fish measuring greater than 10 cm SL feed on a greater variety of organisms, including larger prey, than smaller fish. Most feeding takes place at dusk or after dark. However, in streams they will feed on drifting invertebrates, including terrestrial insects, during the day (10).

Growth is highly variable, depending on habitat, food availability, and temperature (18). Growth of fish from a small alpine lake (Upper Twin, Mono County) was similar to that of fish from high-elevation waters in other states: 11 cm SL at the end of year 1, 13.5 cm at year 2, 15 cm at year 3, 17 cm at year 4, and 20 cm at year 5 (1, 5). Fish from rivers at lower elevations seem to be 25–30 percent larger at any given age after the first year. Young reared in tributaries to Lake Tahoe were largest in the Truckee River (8.6 cm FL at 10 months) and smallest (7.3–7.8 cm) in small tributaries (4). Large individuals (25–50 cm SL) are probably 5–10 years old. The oldest fish on record (from Canada) is 17 years; the largest seems to be one measuring 51 cm FL and weighing 2.9 kg from Lake Tahoe (6). A standardized length-weight equation is $\log_{10} W_g = -5.086 + 3.036 \log_{10} TL_{mm}$ (11).

Spawning takes place in October through early December at water temperatures of 1–11°C (usually 2–6°C) (12). High embryo mortality is experienced at temperatures above 9°C (16). Spawning is preceded in streams by upstream or downstream movements to suitable spawning areas, possibly as the result of homing to historical spawning grounds (12). From large rivers whitefish may move upstream into smaller tributaries for spawning, but the behavior is variable (12). Movement is often associated with a fairly rapid drop in water temperature (9). From lakes whitefish migrate into tributaries to spawn, but some spawning may take place in shallow waters as well. Favored spawning areas are riffles (or wave-washed areas in lakes) where depths are greater than 75 cm and substrates are coarse gravel, cobble, and rocks less than 50 cm in diameter (10). Whitefish do not dig redds but scatter eggs over gravel and rocks, where they sink into interstices. The eggs are not adhesive. Little is known about spawning behavior, but they seem to spawn at dusk or at night, in groups of more than 20 fish (10). They become mature in their second through fourth year, although the exact timing depends on sex and size. Each female produces an average of 5,000 eggs, but fecundity varies with size, from 770 to over 24,000 (7, 8, 18) or around 11–12 eggs per gram of body weight (10). The embryos hatch in 6–10 weeks (or longer, depending on temperatures) in early spring. Newly hatched fish are carried downstream into shallow (5–20 cm) backwaters, where they spend their first few weeks. As fry grow larger, they gradually move into deeper and faster water, usually in areas with rock or boulder bottoms (12). Fry from lake populations move into the lake fairly soon after hatching and seek out deep cover, such as beds of aquatic plants.

Status ID. Mountain whitefish are still common in their limited California range, but their populations are fragmented. There is no question that they are less abundant than they were in the 19th century, when they were harvested in large numbers by Native Americans and then commercially harvested in Lake Tahoe (5, 14). There are still runs in tributaries to Lake Tahoe, but they are relatively small and poorly documented. Whitefish apparently were already reduced in numbers by the 1950s (4). They still seem to be fairly common in low-gradient reaches of the Truckee, East Fork Carson, East and West Walker, and Little Walker Rivers (15). Small populations are still found in Little Truckee River, Independence Lake, and some small streams, such as Wolf and Markleeville Creeks, tributaries to the East Carson River (15). Their populations in Sierra Nevada rivers and tributaries have been fragmented by dams and reservoirs, and whitefish are generally scarce in reservoirs. A severe decline in the abundance of whitefish in Sagehen and Prosser Creeks followed the construction of Stampede and Prosser Reservoirs, respectively (13, 15). These observations all suggest that they are less abundant and less widely distributed than formerly. A thorough survey of their distribution and abundance is needed, along with a study of their taxonomic status in relation to other populations of mountain whitefish.

Mountain whitefish are an underappreciated game fish because their cyprinid-like appearance belies (to most anglers) their culinary and sporting qualities. Ichthyologist-angler J. O. Snyder wrote in 1918 that "it rises to a fly . . . , is as game as trout, and by some is preferred as a game fish" (14, p. 69). Anglers have also held them in low regard because of their supposed competition with trout for food, an assumption for which there is no real evidence (3). Likewise, there is no evidence that they prey on trout eggs or fry (10). In fact it is possible that alien trout may limit whitefish populations by preying on their fry, recorded as an item in brook trout diets (3).

References 1. Sigler 1951. 2. Pontius and Parker 1973. 3. Ellison 1980. 4. R. G. Miller 1951. 5. McAfee 1966. 6. Cordone and Frantz 1966. 7. Scott and Crossman 1973. 8. Sigler and Sigler 1987. 9. Davies and Thompson 1976. 10. Thompson and Davies 1976. 11. Rogers et al. 1996. 12. Northcote and Ennis 1994. 13. Erman 1986. 14. Snyder 1918. 15. E. Gerstung, CDFG, pers. comm. 1999. 16. Ihnat and Bulkley 1984. 17. T. L. Taylor, Entrix, Inc., pers. comm. 1990. 18. Wydoski 2001.

Coho Salmon, *Oncorhynchus kisutch* (Walbaum)

Identification Coho are fairly large salmon, with spawning adults typically attaining 55–70 cm FL and weighing 3–6 kg. They have 9–12 dorsal fin rays, 12–17 anal fin rays, 13–16 pectoral fin rays, and 9–11 pelvic fin rays. Lateral line scales number 121–148, and the scales have single pores. There are 11–15 branchiostegal rays on either side of the jaw. Gill rakers are rough and widely spaced, with 12–16 on the lower half of the first arch.

Spawning males are typically intensely dark red on the sides, with the head and back dark green and the belly gray to black. Females are drabber and paler than males, often appearing a dull, dark pink on the sides. "Standard" spawning males are characterized by strongly hooked jaws and slightly humped backs. The jaw is less strongly hooked in jack males and is only slightly hooked in females. Both sexes have small black spots on the back, dorsal fin, and upper lobe of the caudal fin, with no spots on the lower lobe of the caudal fin. The gums of the lower jaw are gray, except the upper area at the base of the teeth, which is generally whitish. Parr have 8–12 narrow parr marks centered along the lateral line. The parr marks are narrow and widely spaced. The adipose fin of parr is finely speckled, imparting

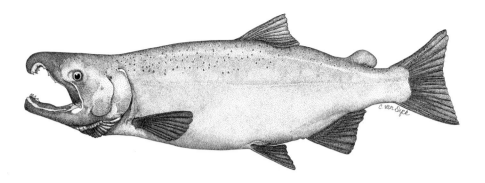

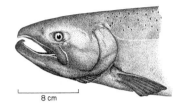

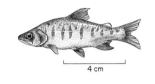

Figure 87. Coho salmon. Top: Spawning male, 51 cm SL, British Columbia. Bottom left: Spawning female, 50 cm SL, British Columbia. Bottom right: Parr, 7 cm SL, Scott Creek, Santa Cruz County.

to it a gray color, while their pelvic, pectoral, and dorsal fins lack spots and are often tinted orange.

Taxonomy The genus *Oncorhynchus* contains six species of "salmon," two species of "salmon-trout," and several species of "trout" (the number of which depends on who is counting). "Salmon" are those species (*O. kitsutch, keta, gorbuscha, tshawytscha, nerka, masou*) in which females (and usually males) die after spawning and are for the most part anadromous. "Salmon trout" are cutthroat trout (*O. clarki*) and rainbow trout (*O. mykiss*), which are either anadromous or resident in streams and which can spawn multiple times. "Trout" are species derived from anadromous forms but that are now completely landlocked (e.g., Mexican golden trout, *O. chrysogaster*). The six salmon species, however, form a group with a common ancestor (1). Within that group, coho and chinook salmon are more closely related to each other than to other salmon, a fact that may explain occasional human-induced hybridization (2).

Coho salmon have thousands of semi-isolated populations in coastal streams over a wide range. At the same time,

fish from different regions mix at sea, and individuals may "stray" into nonnatal streams for spawning. These two opposing and dynamic evolutionary forces keep coho salmon (and other salmon) surprisingly uniform in morphology and life history throughout their range, while producing runs that show strong, genetically based adaptations to local or regional environments. In California coho populations are the southernmost for the species, and they have adapted to the extreme conditions (for the species) of many coastal streams. Allozyme data indicate that California stocks are genetically differentiated from stocks in more northern areas. On the basis of such data, as well as other evidence such as life history attributes, NMFS has divided California coho populations into two ESUs, each representing groups of populations that interbreed more with each other than with other populations and that exhibit adaptations to regional environments (3). The *southern Oregon–Northern California coasts ESU* is composed of populations in streams from Cape Blanco in southern Oregon (just north of the Rogue River) to Punta Gorda (Mattole River, Humboldt County) in northern California. In the ocean these fish tend to be found mainly off California. The *central California coast ESU* extends from Punta Gorda to the San Lorenzo River, Santa Cruz County; it includes the southernmost populations of the species as well as those in San Francisco Bay.

Names Silver salmon is a common name often used in California, but coho salmon has gained wide usage and is the common name adopted by the American Fisheries Society. Coho is derived from a Native American dialect name for the species. *Oncorhynchus* means hooked snout; *kisutch* is J. J. Walbaum's Latinization of the vernacular name used in the Kamchatka Peninsula of Russia in the 16th century (4).

The name was as written down by Georg Wilhelm Steller, a German naturalist who participated in Russian exploration of the north Pacific coast of Asia in 1733–1744. He collected specimens of most species of Pacific salmon and took extensive notes on their biology, but he died on his return journey. In 1784 Thomas Pennant, an Englishman, described the salmon using Russian materials and attached to his descriptions common names recorded in an account of the expedition; the account he used had been written in Russian and translated into English. In 1792 Johann Julius Walbaum, another German naturalist in Russian employ, took the names Pennant used and converted them into Latin, as the species names for the salmon. Walbaum's names became the official scientific names for the species. If the species names of Pacific salmon do not seem to make any sense in any language, consider the transformations of words that must have taken place in the chain of events leading to their formal adoption!

Distribution In North America coho salmon historically spawned in most coastal streams from central California to the Kukpuk River near Point Hope, Alaska (5). In Asia they ranged historically from North Korea and northern Japan (Hokkaido) to the Anadyr River in Russia. In California spawning populations were once found in most coastal streams from the Smith River (Del Norte County) south to the San Lorenzo River (Santa Cruz County), with individual fish straying as far south as Big Sur River. Today the southernmost populations are found in Scott and Waddell Creeks (Santa Cruz County), although a small run is maintained in the San Lorenzo River by artificial propagation (36). There are historic records of the occurrence of coho in 582 California streams, but by 1991 about half these streams had lost their populations (6). Coho salmon once ascended Klamath River and its tributaries at least as high as Klamath Falls, Oregon, but are now blocked from the uppermost river by Iron Gate Dam, about 306 km from the mouth. Likewise, in Trinity River they can ascend only as high as Lewiston Dam, about 306 km from its mouth on the Klamath (6). In the Eel River system they formerly spawned in about 390 km of tributaries to South Fork Eel River, lower mainstem Eel River, and Van Duzen River (8). In the Sacramento drainage coho salmon were never common, but a small population probably once spawned in the McCloud and Upper Sacramento Rivers, as well as in some tributaries to San Francisco Bay (9). There was a small population using Corte Madera Creek in San Francisco Bay, although it most likely is now gone. As of 1998 coho were present in Scott and Waddell Creeks (Santa Cruz County), with no other populations present in coastal streams south of San Francisco Bay. J. J. Smith (37) has noted a few adults in San Vicente Creek (Santa Cruz County) and in Gazos and Pescadero Creeks (San Mateo County), but their origin is

uncertain. The first population north of San Francisco Bay is found in Redwood Creek, Marin County. Thereafter small populations are scattered in coastal streams and rivers. In the ocean most coho salmon spawned in California streams remain in waters off California or southern Oregon (5, 27). They have been caught in ocean waters as far south as Baja California, Mexico (5).

Life History The life history and habitat requirements of coho salmon have been well documented, from the classic study in Waddell Creek by Shapovalov and Taft (10) to more recent summaries (5, 11, 12, 13).

Juvenile coho are generally at highest densities in deep (≥1 m), cool pools with plenty of overhead cover, especially in summer, but they use a wide variety of habitats if cover, depths, temperature, and velocities are appropriate. They are typically associated with instream cover (such as undercut banks, logs, and other woody debris) close to areas that are productive for feeding. Juveniles show pronounced shifts in habitat with season, especially in California streams (14, 15, 42). In spring, when stream flows are moderate and fish are small, they are widely distributed through riffles, runs, and pools. As stream flows diminish in summer, they increasingly concentrate in pools or deeper runs. During winter, before emigration, they seek refuges from high-velocity flows generated by winter storms. Especially important are large off-channel pools with complex cover or small spring-fed tributary streams. Availability of overwintering habitat is one of the most important and least appreciated factors influencing the survival of juvenile coho in streams.

Juveniles prefer and presumably grow best at temperatures of 12–14°C. They do not persist in streams where summer temperatures reach 22–25°C for extended periods of time or where there are high fluctuations in temperature at the upper end of their range of tolerance (11, 13). In the Mattole River watershed (Humboldt and Mendocino Counties), coho were found to be absent from tributaries in which the maximum temperature exceeded 18°C for more than a week, suggesting that thresholds for persistence may be lower than once thought (43). Temperatures exceeding 25–26°C are invariably lethal. Preferred water velocities are 0.09–0.46 m/sec, depending on habitat, and juveniles actively seek refuges from high velocities (16). For fish to hold in fast-moving water, dissolved oxygen levels must be near saturation. Typical coho rearing streams are very clear. Even moderate silt loads can damage the gills of small coho and reduce growth rates; likewise, even short periods of high turbidity or silt loads can be detrimental to the emergence, feeding, and growth of young coho (11, 13). It is worth noting that some juveniles rear in the freshwater portions of estuaries and lagoons rather than in streams.

Studies by J. Nielsen (17, 22) indicate that habitat use by juvenile coho in some California streams is more complicated than is generally appreciated. She found four distinct types of juveniles, perhaps with a genetic basis, which she termed estuarine, margin, thalweg, and early pulse juveniles. *Estuarine* juveniles move downstream into estuaries soon after emergence and rear in intertidal areas. *Margin* juveniles remain in stream margins and backwaters during summer, where growth is typically slow, so that yearling fish move downstream at less than 70 mm SL. *Thalweg* juveniles are the "standard" juveniles that rear in deeper parts of the main channel, feeding and growing steadily all season long; they are around 100 mm SL when they smolt and head out to sea. *Early pulse* juveniles show two pulses of growth, one in spring and one in autumn, and transform into smolts at greater than 100 mm SL. Nielsen (17) characterizes their behavior as "trout-like" in that they hang out under deep cover during the day and forage on drifting invertebrates at dawn and dusk. These four types of juveniles presumably have different survival potentials under varying conditions. The early pulse juveniles in particular may have a strong advantage during times of drought because they can dominate pool habitats and remain in areas with cool seeps during the day (17). Estuarine juveniles are probably especially scarce today because most small estuaries in California are much shallower and warmer than they were historically (as a result of siltation from logging, road building, and agriculture) and so are much less suitable as rearing habitat. When large numbers of juvenile coho of hatchery origin are released into a stream, this delicate subdivision of habitat breaks down, along with social hierarchies, and survival of wild coho is reduced (22).

Another factor complicating habitat use by juvenile coho salmon is competitors and predators. Principal competitors for the food and space of juvenile coho are other salmonids, especially chinook salmon, steelhead, and cutthroat trout. Coho will segregate from steelhead of similar size, dominating pools, while steelhead occupy runs and riffles (25). Temperature plays an important role in segregating chinook salmon parr from coho parr: coho stay in cool tributaries; chinook live in warmer main rivers. Despite a considerable degree of habitat segregation among juvenile salmonids, interactions are common. The more aggressive coho typically dominate, causing other species to grow more slowly (18, 26). Juvenile coho also prey on other salmonids, and this may increase segregation and be a major cause of mortality for other species (27). However, when habitat conditions in California streams favor juvenile steelhead so that their densities are higher than those of coho, growth of coho may be suppressed through competition for food in crowded pools, especially when flows are low, and through aggressive interactions with large 1- to 2-year-old steelhead (19). Large juvenile steelhead and cutthroat trout can also be predators on coho juveniles in large pools. Reduction in such pools and elimination of large

trout as the result of logging and other human activities can actually cause a temporary increase in summer growth and survival of coho in some streams (20), although the positive effects are likely to be negated by loss of coho overwintering habitat and their increased vulnerability to bird and snake predation. Predators are one of the largest sources of direct mortality in streams, and juvenile coho are constantly having to adjust their behavior in order to balance the risk of predation with the need to forage (21).

The foraging behavior of juvenile coho is complex, but the fish are usually placed into three foraging categories: territorial, floater, and nonterritorial fish (21, 22). *Territorial* coho are typical thalweg juveniles that defend feeding territories in flowing water from other coho and salmonids. They usually are among the fastest-growing fish in the stream. *Floaters* are small, slow-growing coho that live in the same areas as territorial fish but either are constantly on the move, avoiding territorial fish, or occupy stream margins. *Nonterritorial* coho are found mostly in pools individually and in small shoals, often feeding in the water column at the upstream end. During winter territorial behavior largely disappears when fish aggregate in deep cover, move into side channels, or move up into small clear tributaries (5).

Emigration from streams in California takes place in March, April, and May and begins when groups of 10–50 fish abandon their deep cover or feeding territories and enter the mainstem of the river system (5, 11). Most of this movement takes place at night. Outmigration typically peaks from late April to mid-May, if conditions are favorable. Migratory behavior is tied to a combination of factors, such as rising or falling water levels, day length, water temperature, food densities, phase of the moon, and dissolved oxygen levels, although it is also clearly a "programmed" behavior. Downstream movements are not continuous, but are interspersed with periods of holding and feeding in areas of low current velocity (23). The outmigrants are mostly 1 year old and measure 10–13 cm FL, although a few larger 2-year-olds may also be present. Parr marks are still prominent in early migrants, but later migrants are silvery, having transformed into smolts. In the estuary smolts often linger for a period, moving up and down with tidal currents, suggesting that a period of estuarine residence is preferred for adjusting their osmoregulatory system to seawater (23).

After entering the ocean, immature salmon initially remain inshore, close to the parent stream, where they feed on pelagic marine invertebrates. They gradually move northward, staying over the continental shelf. Coho salmon can range widely in the north Pacific; the movements of California fish are poorly known but it appears that although some move as far north as Alaska (5), most stay in California and Oregon waters. Curiously, most coho caught off California in ocean fisheries were reared in coastal Oregon streams (natural and hatcheries). In 1990, for instance,

112,600 coho were caught in commercial and recreational ocean fisheries, a number that greatly exceeded the production capability of California populations alone (40). Oceanic coho tend to school together, but schools break apart when feeding occurs (5). Although it is not known if schools are mixtures of fish from different streams, fish from each region tend to be found in the same general area. Oceanic coho salmon become increasingly piscivorous as they increase in size, feeding voraciously and opportunistically on a wide variety of small pelagic marine fishes; however, shrimp, crabs, and other pelagic invertebrates continue to be important food in some areas (5). Presumably one reason California coho may not move far in the ocean is the productivity of the upwelling system off the California coast, which provides high densities of food and cold temperatures. During ENSO events, when productivity declines and temperatures increase, coho growth and survival decrease (25). There is also some evidence that growth and survival of oceanic coho may decrease when a region is flooded with large numbers of hatchery fish (26).

In streams juvenile coho can also be voracious feeders, ingesting any organism that moves or drifts over their holding area. Their diet is mainly aquatic insect larvae and terrestrial insects, but small fish are taken when available. The importance of different foods depends on the season and on the preferences of individual fish. During winter months—when temperatures in California streams are typically 5–10°C, flows are high, and water is turbid—coho feed infrequently and opportunistically (24). In Pudding Creek (Mendocino County) winter coho fed on flying insects and mayfly larvae when flows were low but on earthworms when flows were high (24). In spring and summer territory sizes decrease as food abundance increases and growth rates increase. When adults are spawning, loose eggs and fragments of decaying carcasses can be major foods for juvenile coho, improving growth and body condition during a period when other food is often scarce (38).

Peaks of feeding in streams are typically at dawn and dusk, when drifting insects are most available, but daytime feeding on abundant prey is also common. A similar pattern has been noted for feeding by juveniles in the ocean (39).

Growth in fresh water, as indicated previously, varies with a number of factors, but smolts leaving California streams as "yearlings" (12–15 months old) measure 8–15 cm FL. Some juveniles will achieve even larger sizes before emigration by staying 2 years in the stream. In Prairie Creek (Humboldt County) 20 percent of the emigrating smolts in 2000 were 2-year-old fish (42). Once they enter the productive marine environment, young coho grow 1.1–1.5 mm/day, reaching 40–50 cm FL in their first year at sea and returning to spawn after 16–18 months at sea at 60–80 cm FL (3–6 kg) (5, 10). Males that return as jacks, after 6 months at sea, are typically around 40 cm FL (10). The largest coho from California was caught in Lagunitas Creek (Marin County) in 1959 and weighed 10 kg.

California coho salmon have a fairly strict 3-year life cycle, with about half spent in fresh water and half spent in salt water. The main exception to the 3-year rule are jack males, which are essentially 2 years old. The combination of a 3-year life cycle and a strong homing instinct means that each stream has three distinct populations based on the timing of runs, which are isolated both temporally and spatially from one another. The jacks, however, keep runs from being genetically isolated from one another, as do rare precocial females.

Coho salmon migrate up and spawn mainly in streams that flow directly into the ocean or are tributaries of large rivers. Spawning migrations begin after heavy late fall or winter rains breach sand bars at the mouths of coastal streams, allowing fish to move into lagoons. Upstream migration typically occurs when stream flows are either rising or falling, not necessarily when streams are in full flood. The timing of return varies considerably, but in general they return earlier in the season in more northern areas and in larger river systems (27). In the Klamath River coho run between September and late December, peaking in October–November. Spawning itself occurs mainly in November and December (28). The early part of the run is dominated by males, with females returning in greater numbers during the latter part of the run. Coho move up the Eel River 4–6 weeks later; arrival in the upper reaches peaks in November–December (27). In short coastal streams of California, most coho return during mid-November through mid-January (27). In the southernmost populations in Scott and Waddell Creeks (Santa Cruz County), spawning migrations often do not occur until November or December (10), and spawning may extend into February or even early March (37). In Oregon streams spawning can occur as late as mid-March if drought conditions delay rains or runoff (5).

Females typically choose a spawning site near the head of a riffle, just below a pool, where water changes from smooth to turbulent flow and there is abundant medium to small gravel. The flow characteristics of redd locations usually ensure good aeration, and the circulation facilitates fry emergence from gravel. Each female builds a succession of redds in the same place, moving upstream as she does so and depositing a few hundred eggs in each (44). Thus spawning takes about a week to complete, during which time each female lays 1,400–3,000 eggs. There is a positive correlation between fecundity and size of females, but California coho have lower fecundities than fish from more northern populations (5). A dominant hooknose male accompanies a female during spawning, and one or more subordinate or jack males may also engage in spawning. In a tributary to Lagunitas Creek I once watched a large hooknose male defend a redd against a subordinate male and at least two jacks. At one point the hooknose male actually grabbed a jack between its jaws and lifted it out of the water with a shaking motion. Nevertheless, when the female decided to spawn, the jacks joined the hooknose male in the act. Both males and females die after spawning, although the female, recognizable by her visibly worn and whitened tail, may guard a nest for up to 2 weeks.

Embryos hatch after 8–12 weeks of incubation; the time is inversely related to water temperature but also has a genetic component (33). Hatchlings remain in the gravel until their yolk sacs have been absorbed, 4–10 weeks after hatching. Under optimum conditions, mortality during this period can be as low as 10 percent; under adverse conditions of high scouring flows or heavy siltation, mortality may be close to 100 percent (27). Upon emerging, alevins seek out shallow water along stream margins. Initially they form shoals, but as they grow bigger the shoals break up and many juveniles (parr) set up individual territories.

Status IB. The two coho salmon ESUs in California were listed (in 1996 and 1997) as Threatened by NMFS because of a 90–95 percent decline in abundance in the previous 50 years and evidence of continuing decline (41). The central California ESU in particular is in danger of extinction (34), and populations south of San Francisco Bay are listed as Endangered (1995) by the state. Calculating the exact extent of the decline is difficult because records of coho numbers are few, even for individual streams. Historical estimates of statewide coho salmon abundance are very rough, made by knowledgeable fisheries managers based on limited catch statistics, hatchery records, and personal observations of runs in various streams. In years of high ocean productivity, California streams may once have supported nearly a million spawners (35). Maximum estimates for number of coho spawning in the 1940s range from 200,000 to 500,000 (36). Coho numbers held at about 100,000 statewide in the 1960s (29, 35), with 40,000 in the Eel River alone, and then dropped to a statewide average of around 33,500 for the 1980s (6). Coho salmon numbers in California, including hatchery stocks, are presently less than 6 percent of the conservative estimates of their abundance during the 1940s, and there has probably been at least a 70 percent decline in numbers since the 1960s. In the drought years 1988–1990, about 31,000 adult coho salmon entered California streams each year (6). However, hatchery fish made up 57 percent of this total, and many populations in the Klamath and other rivers contained at least some fish of recent hatchery ancestry. Hatchery stocks, without exception, have in their ancestry fish from other river systems, often ones outside California (6).

Coho salmon are widely distributed in coastal streams of California. Their populations show large fluctuations, but the general trend has been downward, especially in wild populations of small coastal streams. Of 582 coastal streams that historically held coho salmon, recent records for 244 of them indicated that 40–50 percent had lost their coho runs (6). In Del Norte County 45 percent of streams for which there are reliable records have lost their coho, mainly in the Klamath-Trinity system. Corresponding percentages for other counties are as follows: Humboldt County, 31 percent; Mendocino County, 41 percent; Sonoma County, 86 percent. For the four counties farther south the value is 56 percent, but this number excludes streams in the Sacramento drainage and includes streams with extremely low populations that are enhanced by hatchery production. Big-river populations are presently maintained by hatcheries for the most part. The Sacramento drainage, specifically McCloud River, supported coho salmon in the 19th century (7, 30), but they were extirpated before any good records could be kept. Historical numbers of spawners in the Klamath River system have been estimated at 15,400–20,000, with 8,000 for Trinity River (30). Only 1,700 coho returned to Klamath Basin hatcheries in 1990 (40), and 3,100 returned in 1991 (31).

The largest concentration of wild fish (with little or no hatchery influence) remaining was thought to be in South Fork Eel River, estimated to have runs of about 1,300 fish. A 1990 survey, however, revealed a population one-half to one-third that size. Lagunitas Creek (Marin County) supports one of the more consistent small-stream coho runs. This stream and its tributaries historically supported 500–2,000 adult spawners yearly (36); in recent years, the numbers seem to be stable (32). A similar, if much smaller, self-sustaining run exists in nearby Redwood Creek. Brown et al. (6) considered 5,000–7,000 fish to be a realistic assessment of total naturally spawned adults returning to California streams each year since 1987, although this number includes some stocks that contain fish of recent hatchery derivation. Presently, there are considerably fewer than 5,000 wild coho salmon (no hatchery influence) spawning in California each year (9, 35). Many of these fish are in populations of fewer than 100 individuals. These small populations are likely to be below the minimum sizes required to preserve genetic diversity and to buffer them from natural environmental disasters. There is every reason, therefore, to think that California's coho populations are continuing to decline.

The reasons for the decline include poor land use practices (especially those related to logging and urbanization) that degrade streams, genetic and behavioral interactions of wild stocks with hatchery fish, introduced diseases, and overharvesting (6). These significant human factors are superimposed on natural factors—mainly floods, droughts, and ENSO conditions in the ocean—that naturally cause coho populations to fluctuate. Populations at natural lows may now not be able to recover because of damaged habitats. Although all salmon are affected by these factors, California coho are especially affected because virtually all wild females are 3 years old. Therefore a severe winter flood or summer drought, in conjunction with human-caused factors, can eliminate one or more year classes from a stream. There is good evidence that this has happened repeatedly in coastal drainages, where decline of coho is linked to poor stream and watershed management. In more northern streams (Mendocino to Del Norte Counties), most damage has been done by post–World War II logging practices that removed riparian vegetation and woody debris from channels, caused stream temperatures to increase, filled pools with silt and gravel, altered stream channels, and otherwise modified habitats. In more southern streams road construction, poor farming and grazing practices, and water diversions have been major causes of coho declines. At present populations are so low that moderate fishing pressure on wild coho may prevent recovery, even in places where stream habitats are adequate. Likewise, predation by seals and sea lions on returning fish when populations are low may prevent recovery. Prior to the declaration of coho as a threatened species, existing regulatory mechanisms—such as fishing regulations, forest practice rules, and stream alteration agreements—were demonstrably inadequate to protect the species in California, Oregon, and Washington, and populations declined steadily and precipitously as a result.

The key to stopping the decline of coho salmon is to protect their spawning and rearing streams and to restore damaged habitat. This is a difficult task because it means modifying or halting logging, farming, and road construction activities in dozens of coastal drainages and implementing habitat restoration plans in hundreds of streams. In many streams it means that major rehabilitation projects must be funded and completed. The continued closure or limitation of the fishery until population trends statewide are reversed is also a necessity. Given the large scale of coho problems, innovative approaches to stream restoration must be tried, working with landowners, timber companies, and gravel miners. For example, logging operations in sensitive drainages should be required to leave wide buffer zones along all streams (including fishless tributaries and seasonal streams) and to log and build roads in ways that add no silt to streams and do not increase the risk of landslides. In areas already degraded by logging, rootwads

and other large woody debris should be added to streams to create pools. Juvenile overwintering habitat must be identified in each stream and special protection or enhancement accorded to it.

Serious consideration should be given to eliminating all production hatchery programs, especially those that rely on nonnative stocks. This would reduce the effects of interbreeding of hatchery coho with wild coho and reduce the spread of hatchery diseases to wild fish. Where population augmentation is deemed necessary, small-scale, localized hatchery operations using local wild stock could be set up as temporary measures (but these must be used with extreme caution, with firm closure deadlines), as was done on Freshwater Creek, Humboldt County.

Monitoring populations is a necessity; spawning streams should be identified and populations censused annually. This would allow population trends to be followed and provide a focus for restoration efforts. The challenges of managing a resource as diffuse as coho salmon are considerable, but if declines are not reversed soon we are likely to lose most, if not all, of our wild populations, including the southernmost populations of the species.

References 1. Stearley and Smith 1993. 2. Bartley et al. 1990, 1992. 3. Federal Register 60(142): 38,011–38,030 (1995) and 61(212): 56,138–56,149 (1996). 4. Netboy 1974. 5. Sandercock 1991. 6. Brown et al. 1994. 7. Evermann and Clark 1931. 8. Mills 1983. 9. Leidy 1984. 10. Shapovalov and Taft 1954. 11. Hassler 1987. 12. Pearcy 1992. 13. Emmett et al. 1991. 14. Nickelson et al. 1992a,b. 15. Taylor 1988. 16. Fausch 1993. 17. Nielsen 1992a,b. 18. Taylor 1991. 19. Harvey and Nakamoto 1996. 20. Iwanaga and Hall 1973. 21. Martel 1996. 22. Nielsen 1994. 23. Moser et al. 1991. 24. Pert 1993. 25. Johnson 1988. 26. Emlen et al. 1990. 27. Baker and Reynolds 1986. 28. USFWS 1979. 29. California Advisory Committee on Salmon and Steelhead Trout 1988. 30. U.S. Commission for Fish and Fisheries 1892. 31. CDFG 1992a. 32. Trihey and Associates 1996b. 33. Konecki et al. 1995. 34. Adams et al. 1996. 35. Mills et al. 1997. 36. E. Gerstung, CDFG, pers. comm. 1999. 37. J. J. Smith, San Jose State University, pers. comm. 1998. 38. Bilby et al. 1998. 39. Brodour and Pearcy 1987. 40. A. Baracco, CDFG, pers. comm. 1994. 41. Weitkamp et al. 1995. 42. Bell et al. 2001. 43. Welsh et al. 2001. 44. Briggs 1953.

Chinook Salmon,
Oncorhynchus tshawytscha (Walbaum)

Identification Spawning adults of chinook salmon are olive brown to dark maroon without conspicuous streaking or blotches on the sides. Spawning males are darker than females and have hooked jaws and slightly humped backs.

There are numerous small black spots on the back, dorsal fin, and both lobes of the tail in both sexes. They can be distinguished from other spawning salmon by color pattern, particularly the spotting on the caudal fin and the black gums of the lower jaw. They have 10–14 major dorsal fin rays, 14–19 anal fin rays, 14–19 pectoral fin rays, 10–11 pelvic fin rays, 130–165 pored lateral line scales, and 13–19

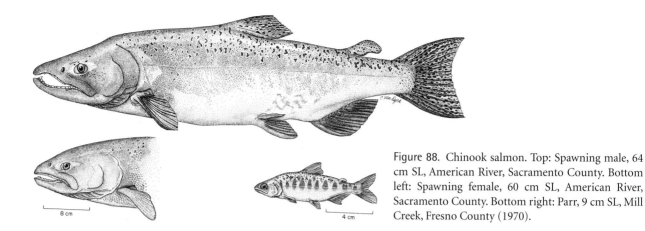

Figure 88. Chinook salmon. Top: Spawning male, 64 cm SL, American River, Sacramento County. Bottom left: Spawning female, 60 cm SL, American River, Sacramento County. Bottom right: Parr, 9 cm SL, Mill Creek, Fresno County (1970).

branchiostegal rays on each side of the jaw. More distinctively, they possess a larger number (>100) of pyloric cecae than other salmon. Gill rakers are rough and widely spaced, with 6–10 on the lower half of the first gill arch. Spawning adults are the largest Pacific salmon, typically 75–80 cm SL (9–10 kg), with lengths in excess of 140 cm (45 kg). The largest on record for California weighed 38.6 kg (1). Parr have 6–12 parr marks, each equal to or wider than the spaces between and most extending below the lateral line. The adipose fin of parr is pigmented on the upper edge but clear at its center and base; the dorsal fin occasionally has one or more spots on it, but other fins are clear.

Taxonomy Within the genus *Oncorhynchus,* chinook salmon are most closely related to coho salmon (2, 3), with which they occasionally hybridize (4). (See the account of coho salmon.) Within the species there are many distinct populations, usually recognized as "runs" or "stocks," that show genetically based adaptations to local and regional environments. In California, runs that are widely recognized by fisheries managers and others who work with chinook salmon are, from north to south: *(1)* Smith River fall run (and spring run), *(2)* Klamath-Trinity fall run, *(3)* Klamath-Trinity spring run, *(4)* Klamath late fall run, *(5)* Redwood Creek fall run, *(6)* Little River fall run, *(7)* Mad River fall

run, *(8)* Humboldt Bay tributary fall run, *(9)* Eel River fall run, *(10)* Bear River fall run, *(11)* Mattole River fall run, *(12)* Garcia River fall run, *(13)* Russian River fall run, *(14)* Central Valley fall run, *(15)* Central Valley late fall run, *(16)* Sacramento winter run, and *(17)* Central Valley spring run. The four major Central Valley runs can be distinguished by molecular techniques (58). In the Klamath-Trinity and Central Valley basins, stocks in major tributaries are often recognized independently as well, with considerable justification, based on small but important differences in genetics and life histories. Recognition of tributary runs increases the total by about 12 runs in the Central Valley and at least 6 in the Klamath-Trinity basin. The San Joaquin fall run is recognized as a distinct unit for management because it is the southernmost run of the species, other runs have been extirpated from the basin, and timing of migrations shows small differences from Sacramento runs. However, thanks to hatcheries, transplants, and straying of Sacramento River fish, it may no longer be genetically distinguishable from runs in the American River. A number of runs have been extirpated in the past hundred years, most famously the San Joaquin spring run, but some apparently disappeared before they were formally recognized (e.g., the Eel River spring run). NMFS (5) recognizes five ESUs of chinook salmon in California, based on genetic and life history similarities among geographically proximate populations:

1. *Southern Oregon and California coastal ESU.* This ESU covers fall-run chinook salmon in coastal streams from Cape Blanco in Oregon south to San Francisco Bay. It includes fall-run chinook in the lower Klamath River and some spring runs in Oregon as well.

2. *Upper Klamath and Trinity Rivers ESU.* This ESU includes all fall-, late-fall-, and spring-run chinook salmon in the Trinity River and in the Klamath River upstream of the mouth of Trinity River.

3. *Central Valley fall-run ESU.* This ESU covers fall-run

and late-fall-run salmon in the Sacramento and San Joaquin Rivers and tributaries.

4. *Central Valley spring-run ESU.* This ESU covers spring-run chinook salmon in both rivers and their tributaries, although it exists today only in the Sacramento River drainage.

5. *Sacramento River winter-run ESU.* This unique run originally spawned in cold waters of the McCloud, Pit, and upper Sacramento Rivers but is presently found only in the mainstem Sacramento River, below Keswick Dam.

Names King salmon is a widely used name in California, but chinook salmon is the official common name selected by the American Fisheries Society for use throughout the world. Other names occasionally applied are spring salmon, quinnat salmon (especially in New Zealand), and tyee (for large adults). Chinook is the name of a large tribe of Native Americans that lived along the Columbia River. *Tshawytscha* is derived from the name for these fish used by the natives of the Kamchatka Peninsula of Russia as it was written down by a German naturalist, translated into Russian, converted into English, and put into Latin by another German naturalist (see the account of coho salmon for details and for other names).

Distribution In North America chinook salmon occur in streams north to Kotzebue Sound, Alaska. In Asia they are found from northern Japan (Hokkaido) and the Amur River of Russia north to the Aanadr River of Russia (2, 8). Although chinook salmon are often caught in the ocean off southern California, the southernmost spawning runs have been in the Central Valley, specifically in the San Joaquin and Kings Rivers (Fresno County) (28). Chinook salmon are widely distributed in the pelagic zone of the north Pacific Ocean; the southern extent of their distribution depends on ocean temperatures. For the most part, they are rarely found south of 40° north latitude, except off California, where they are regularly found south of Monterey Bay (about 35° north latitude) (2).

Along the North Coast of California, spawning runs occur (or occurred) in larger coastal streams north of San Francisco Bay to the Oregon border, as indicated in the Taxonomy section of this account. The largest of these runs has been in the mainstem Eel River and its tributary, Tomki Creek. In the Central Valley spawning occurred in all major streams draining the Sierra Nevada and (in the north) the Cascades, although distribution of spawning fish has been severely truncated by dams blocking access to upstream areas (6). From south to north, these streams are as follows: Kings, San Joaquin, Merced, Tuolumne, Stanislaus, Calaveras (intermittent spawning), Mokelumne, Cosumnes,

American, Feather River, Yuba, and mainstem Sacramento Rivers; then Butte, Big Chico, Deer, Mill, Antelope, Battle, and Cow Creeks; then Little (upper) Sacramento, McCloud, and Pit Rivers (6). In the Pit River chinook salmon once ascended high enough to spawn in Hat Creek and the Fall River, Shasta County (6). On the west side of the Sacramento Valley, where water is less abundant and more seasonal than on the east side, runs of chinook salmon historically occurred in most years in the following creeks: Clear, Cottonwood, Stony, Thomes, Stillwater, Cache, and Putah (6).

In the Klamath River chinook salmon once ascended into Upper Klamath Lake, Oregon, to spawn in the major tributaries to the lake (Williamson, Sprague, and Wood Rivers), but access to this region was blocked by Copco Dam, built in 1917 (5, 7). Today they are known to spawn in a number of streams besides the mainstem Klamath River; starting upstream, these include Bogus Creek, Shasta River, Scott River, Indian Creek, Elk Creek, Clear Creek, Salmon River (including major forks and Wooley Creek), Bluff Creek, Blue Creek, and the lower reaches of some of the other smaller tributaries to the mainstem. In the Trinity River chinook salmon spawn in the mainstem (with their upstream distribution limited by Lewiston Dam), the north and south forks, Hayfork Creek, New River, and Canyon Creek.

CDFG plants "surplus" juvenile salmon to support fisheries in large reservoirs with some degree of success. Reservoirs in which they have been planted include Shasta, Almanor, and Berryessa. Adult salmon have been observed ascending streams tributary to Shasta and Almanor (55), but there is no evidence yet of spawning success.

Since 1872 many attempts have been made to establish chinook salmon elsewhere in the world, but the only successful transplants seem to have been made to New Zealand and the Laurentian Great Lakes (8, 9). The fish in New Zealand became established as the result of repeated plants of juveniles originating from embryos from uncertain sources in the upper Sacramento River drainage, but most likely from Battle Creek (57).

Life History Chinook salmon show a fascinating array of life history pattern adaptations that allow them to take advantage of diverse and variable riverine environments. Healey (2) divides the life history strategies into two basic types, stream-type and ocean-type, and notes that within these two broad categories there are local variations that are typically recognized as runs or stocks. *Stream-type chinook* have adults that run up streams before they have reached full maturity, in spring or summer, and juveniles that spend a long time (usually >1 year) in fresh water. *Ocean-type chinook* have adults that spawn soon after entering fresh water, in summer and fall, and juveniles that spend a relatively short time (3–12 months) rearing in fresh water. In California, where the salmon are at the southern end of their range, both types are present, with a wide array of variations on the themes. Because the amount of time some juveniles rear in fresh water depends in part on conditions in the river, the classification of some runs as having ocean- or stream-type life history categories may be ambiguous.

Variations on life history themes are named for the timing of spawning runs of adults (fall-run, late-fall-run, winter-run, and spring-run), but there are differences among them at all life history stages, as shown for Central Valley chinook runs (Table 11). Other river systems in California support or supported mainly fall-run and spring-run chinook, with timing of life history events similar to that of Central Valley chinook, although smaller runs have narrower migration time windows. For example, in the Klamath River the first fall-run chinook historically appeared in mid-July and the run was finished by early November, although in recent years run timing has become 1–4 weeks later, apparently as a result of the operation of a hatchery and fish ladder (10, 11). As might be expected from such an adaptable species, there are local versions of each life history variation. Thus Blue Creek, a fairly pristine tributary to the lower Klamath River, supports just a late fall run of salmon (12). In a tributary to the Smith River (Del Norte County) three distinct runs have been noted: mid-November to mid-December, late December to mid-January, and late January to mid-February (13). In the San Joaquin River, fall-run chinook arrive later than they do in the Sacramento River (Table 11). Clearly chinook salmon have enormous capacity to adapt to local conditions.

Fall-run chinook salmon are unambiguous ocean-type chinook adapted for spawning in lowland reaches of big rivers and their tributaries. They move up from the ocean in late summer and early fall in mature condition and typically spawn within a few days or weeks of arriving on the spawning grounds. Juveniles emerge from the gravel in spring and move downstream within a few months, to rear in mainstem rivers or estuaries before heading out to sea (14). The strategy allows salmon to take advantage of extensive high-quality spawning and rearing areas in valley reaches of rivers, which are often too warm to support salmon in summer. The success of this strategy is reflected in the fact that fall-run chinook have historically been the most abundant run in California. It has also made them ideal for rearing in hatcheries, almost to the exclusion of other runs. An interesting component of this strategy is a high rate of "straying" of adults from natal streams that allows them to take advantage in wet years of favorable conditions in streams not normally used for spawning or to colonize new spawning areas that develop as a result of fluvial processes. In recent years straying of fish of presumed hatchery origin has resulted in the establishment of spawning runs of 100–200 fish in Guadalupe River and Coyote Creek, tributaries to south San Francisco Bay. Smolts were found in Guadalupe River in 1998 (56).

Late-fall-run chinook salmon are mostly stream-type chinook found mainly in the Sacramento River today. They are the largest and most fecund salmon in California, commonly weighing 9–10 kg as adults, in part because they historically came in as mainly 4- and 5-year-old fish (15, 16). Adults typically hold in the river for 1–3 months before spawning. They are adapted for spawning and rearing in reaches of mainstem rivers, such as the upper Sacramento, that remain cold and deep enough in summer for rearing of juveniles. Juveniles grow rapidly in these reaches, so by the time they enter the ocean, after 7–13 months in fresh water, they measure 150–170 mm FL, with all the survival advantages of comparatively large size confers (16).

Winter-run chinook salmon are unique to the Sacramento River. They typically migrate upstream as immature silvery fish during winter and spring and then spawn several months later in early summer. As chinook salmon go, the adults tend to be small and have low fecundity because most return to spawn as 3-year-olds (16). Winter-run chinook were adapted for spawning and rearing in the clear, spring-fed rivers of the upper Sacramento basin, especially the McCloud River, where summer temperatures were typically 10–15°C. These conditions were created by glacial and snow melt water percolating through porous volcanic formations that surround Mt. Shasta and Mt. Lassen, and that cover much of northeastern California. Today Shasta Dam denies access to their historical habitats and they persist mainly because water released from Shasta Reservoir during summer has for the most part been cold. The residence time of juveniles in streams is less than a year (5–10 months), followed by an indeterminate time in the estuary. They are thus intermediate in characteristics between stream- and ocean-type chinook (2), a further indication of their uniqueness.

Spring-run chinook salmon enter rivers as immature fish in spring and early summer and exhibit a classic stream-type life history pattern, although the stay of some juveniles in fresh water may be less than a year. They historically migrated upstream as far as they could go in the larger tributaries to the Sacramento, San Joaquin, Klamath, and Eel Rivers, where they held for several months in deep, cold pools. They then spawned in early fall, and juveniles reared in the streams for 3–15 months, depending on flow conditions (15, 16). This strategy allowed salmon to take advantage of midelevation habitats inaccessible during summer and fall (owing to high temperature and low flows in lower reaches) and difficult to use during high-flow periods (when holding pools are scoured). As a result of this strategy, spring-run chinook were once nearly as abundant as fall-run chinook and were the dominant run in the San Joaquin watershed (17, 18). Today, however, access to most

Table 11

Generalized Life History Timing of Central Valley Chinook Salmon Runs

	Migration period	Peak migration	Spawning period	Peak spawning	Juvenile emergence period	Juvenile stream residency
Sacramento River basin						
Late fall run	October–April	December	Early January–April	February–March	April–June	7–13 months
Winter run	December–July	March	Late April–early August	May–June	July–October	5–10 months
Spring run	March–September	May–June	Late August–October	Mid-September	November–March	3–15 months
Fall run	June–December	September–October	Late September–December	October–November	December–March	1–7 months
San Joaquin River basin (Tuolumne River)						
Fall run	October–early January	November	Late October–January	November	December–April	1–5 months

Source: Yoshiyama et al. (1998).

of their historical spawning areas is blocked by dams, and they persist in just a few streams in the Sacramento and Klamath drainages.

Once juvenile salmon emerge from the gravel, they initially seek areas of shallow water and low velocities while they finish absorbing the yolk sac. Many, however, disperse downstream, especially if high-flow events correspond with emergence (2). Dispersal behavior shows variation among fry that emerge from a single redd, with larger individuals most likely to disperse (19). Movement occurs mostly at night and tends to cease after a couple of weeks, when fry settle down into rearing habitat in streams or estuaries. The social behavior of juveniles varies from schooling to territoriality. Stream-type juveniles are more likely to be territorial and behave aggressively toward one another than are ocean-type juveniles (20). However, I have observed juvenile chinook salmon (70–80 mm FL) in Deer Creek (Tehama County) foraging together in small groups in open areas among rocks in riffles and runs or at the tail end of pools. Such behavior may have been partially induced by interactions with larger, aggressive rainbow trout common in the area. Indeed interactions among species of salmonids in streams are complex and can result in displacement of individuals of one species by those of another; usually the pattern that emerges is one of segregation in use of microhabitats (21, 22, 23). Often segregation is strongly related to size of fish, regardless of species.

In general, there is a shift in microhabitat use by juvenile chinook to deeper and faster water as they grow larger. Microhabitat use and foraging behavior can be influenced,

however, by the presence of predators (other fish, birds), which may force fish to select areas of heavy cover and suppress foraging in more open areas (24). During the night, juvenile chinook may abandon their foraging areas in swift-moving water and retreat to quiet edge waters or pools (2), as an energy-conserving measure or as a way to avoid predation from pikeminnows, which often feed at night.

A major limiting factor for juvenile chinook salmon is temperature, which strongly affects growth and survival. For Central Valley fall-run chinook fry, optimal temperatures for growth and survival are 13–18°C (25), although throughout the range of chinook positive growth is experienced at temperatures of 5–19°C (65). Few fish can survive temperatures greater than 24°C for even short periods of time, and at around 22–23°C major mortality is experienced in wild populations (26, 65). At sublethal temperatures growth is reduced and predation rates may be increased as a consequence. Presumably, there are slight (1–2°C) differences in optimal and lethal temperatures of chinook salmon of different runs and stocks. Temperature can interact with turbidity to affect the survival of juveniles. At moderate levels, reduced water clarity reduces the ability of predators to find juvenile salmon, although as clarity decreases salmon have an increasingly hard time finding their own prey (24, 27). Not surprisingly, salmon fry tend to move downstream, and smolts emigrate to the ocean when freshets increase river flow, increase turbidity, and decrease temperatures. Peak periods of movement tend to be at night, further reducing predation risk of small salmon. At Red Bluff Diversion Dam on the Sacramento River, this adaptive behavior worked against the small salmon because bright lights on the dam enabled pikeminnows and other predators to feed effectively at night on fish coming over the dam spillway! When it was realized that enhanced predation may have been contributing to the decline of chinook salmon stocks, the lighting was changed.

Juvenile chinook salmon move downstream at a wide variety of sizes and conditions. In the Sacramento River juvenile salmon can be found moving downstream during all months, as both fry and smolts. Spring-run juveniles tend to peak in winter (January–February) and then again in spring (April) (29, 30). Fall-run juvenile outmigration peaks in spring (March–April). Winter-run juveniles move mainly in September–January, whereas emigrating late-fall fish seem to occur in spring through early fall (April–October). In general, stream-type juveniles move downstream and out to sea as smolts, at lengths of 80–150 mm FL, but ocean-type (fall-run) juveniles move downstream at small sizes (30–50 mm FL) to rear in the estuary. Spring-run chinook in Butte Creek, Butte County, however, move out as both small fry and smolts. Movement into the estuary may vary with year. In the Klamath River fall-run juvenile salmon apparently move into the estuary in larger numbers in years when river flows are low and temperatures high than they do when conditions for rearing are better in the river (29). In the Sacramento River fall-run juveniles move into freshwater portions of the estuary, including the Yolo Bypass, to rear in February and March, resulting in substantially higher growth and survival rates than in rivers (14, 62). This habit allows them to reach 80–100 mm FL before they turn into smolts and migrate out to sea. Juveniles from other runs apparently do not spend as much time in the estuary but pass through fairly rapidly on their way to sea. Whether or not this rapid passage is a recent phenomenon as the result of drastic changes in estuarine habitat or is the historical pattern is not clear.

Downstream movements of juveniles of all runs serve not only to disperse and move them toward the ocean, but also to provide access to temporary habitats in which slightly warmer temperatures and abundant food may encourage rapid growth. The tendency of juveniles in rivers to move toward shallow edges, especially during the day, puts them in heavy cover or among emergent vegetation, where invertebrates are abundant and where many predators have a hard time finding them. In the Central Valley during high-flow periods, these fish historically moved into the floodplain, where they could rear for several months. The Yolo Bypass, an artificial floodplain near Sacramento, apparently serves that function periodically today. During periods of flooding large numbers of juvenile salmon can be found there, where they have exceptionally high growth rates (62). Likewise, juveniles may move into lower reaches of intermittent tributaries for rearing during high-flow periods, presumably because such areas have abundant food and few large predators (31).

Once in the ocean, juvenile chinook from California rivers tend to stay along the California coast, although there may be a general northward movement of fish, leading to a few being found off Washington (2). Concentration of California salmon in marine waters off the state is not surprising considering their high productivity. This productivity is caused by upwelling generated by the complex phenomenon known as the California Current, a southward-moving current originating in the Gulf of Alaska. In these food-rich waters, juvenile salmon swim, presumably in schools, at depths that vary with season (0–100 m) but are typically deeper (20–45 m) than those of most other salmon (2). The importance of ocean productivity to chinook populations is indicated by a decline in ocean survival of salmon during years when the current does not flow as strongly and upwelling decreases (32). The ocean stage of the chinook life cycle lasts 1–5 years.

While in fresh water, juvenile chinook salmon are opportunistic drift feeders and eat a wide variety of terrestrial and aquatic insects. In the regulated Sacramento and lower American Rivers, small salmon (40–80 mm FL) feed mainly on larvae and pupae of chironomid midges, baetid mayfly larvae and adults, and hydropsychid caddisfly larvae, although a wide variety of other organisms are taken as well (33, 66, 67). Likewise, in the fluctuating flows of the lower Mokolumne River, juvenile salmon were found at different times feeding predominately on zooplankton, on chironomid larvae, and on the larvae of Sacramento suckers (34). Zooplankton was taken when large amounts of cladocerans and copepods were being flushed from an upstream reservoir, whereas suckers were consumed when they were abundant in shallow water. In regulated rivers, such as the Sacramento, prey size does not increase substantially with fish size, despite the increase in mouth gape, because of the low availability of large organisms (66). However, if food is abundant, size of prey seems to make little difference in growth rates (62, 68). In the Sacramento–San Joaquin Delta, terrestrial insects are by far the most important food, but crustaceans are also eaten (35). When juvenile chinook enter flooded areas during high water, they consume large amounts of the zooplankton and small insect larvae that favor such areas. In contrast, in Mattole River lagoon, juvenile chinook feed on aquatic (drift) and terrestrial insects, largely ignoring abundant zooplankton and benthic amphipods (61). Juvenile chinook feed mostly during the day, with peaks at dawn and during the afternoon, perhaps because their most common prey, chironomid larvae and pupae, drift mostly during the day (36).

After juveniles enter the ocean, they become voracious predators on small fish and crustaceans. Small fish feed heavily on invertebrates, such as crab larvae and amphipods (11). As they grow larger, fish increasingly dominate their diet. They typically feed on whatever pelagic planktivore is most abundant, usually herring, anchovies, juvenile rockfish, and sardines off California (2). Once the switch to fish is made, growth is rapid, amounting to 0.35–0.57 mm/day (2). Thus at age 2 Sacramento River fall-run chinook average about 55 cm FL; at age 3, about 70 cm FL; at age 4, about 90 cm FL; and at age 5, about 100 cm FL. Obviously there is considerable variation in length at different ages. Salmon with stream-type life histories are typically 5–15 cm smaller at a given age because they enter the ocean as relatively small 1-year-old fish (2). Because ocean growth *rates* are similar among different runs but *sizes* at ocean entry are different, the lengths of adults returning to spawn at a given age will differ among runs. Thus Sacramento late-fall-run fish seem to be bigger on average than salmon of other Sacramento runs because they spawn mainly as 4- or 5-year-olds, even though they have a stream-type life history, whereas winter-run fish are among the smallest of the spawning Sacramento salmon because they are stream-type salmon that spawn mostly at age 3.

Over the past several decades, strong selection for large fish by the commercial fishery has meant that small 3-year-old fish of all runs are increasingly abundant. In the case of San Joaquin River tributaries, a significant percentage (up to 67%) of the run in some years has been 2-year-old males (jacks or grilse) or, surprisingly, females (jills; 14% in 1996) (38). Usually, 2-year-olds are only males. Harvesting of older and larger fish not only selects for smaller adult salmon but also increases the variability in run sizes. When salmon return at multiple ages they can both reduce the impact of disasters that might wipe out the spawn of one year (e.g., a scouring flood) and ensure that fish spawned in different years are able to interbreed.

In the ocean chinook salmon home to their natal region over great distances. Various celestial orientation mechanisms have been proposed to explain how salmon find their way to the mouth of their natal river, but there is evidence that an internal compass (possibly geomagnetic) may be involved. Migrating salmon have been followed with sonic tags and found to move at a fairly steady pace, day and night, at depths (35–40 m) that make navigating by sun and stars unlikely (38). Once they reach the region of the stream mouth, many "landmarks" are available to guide them further, including geomagnetic anomalies, visual cues, and distinctive odors from their home streams (39). Upstream migration takes place mainly during the day, with fish apparently tracking stream odors on which they imprinted when small (2). While a majority of fish do home to the same stream in which they were hatched—behavior that accounts in part for the extraordinary adaptations to local conditions found in salmon—some also "stray" and wind up in the wrong stream. Straying is presumably also an adaptive mechanism, allowing salmon to colonize newly opened areas and to mix genetically with other runs, especially those in other streams close to the natal stream.

In rivers chinook salmon can migrate upstream more than 2,000 km (Yukon River, Alaska), although most migrate relatively short distances (<150 km). Historically in the Sacramento River system some migrated over 630 km to spawn in the Fall River. In the Klamath River some probably migrated around 450 km to their highest point in Oregon, although they can travel only 306 km (as far as Iron Gate Dam) today. In terms of elevation, they reach about 1,800 m in Mill Creek (Tehama County), one of the highest elevations known for spawning Pacific salmon.

Once they reach their home stream, salmon first select areas for holding, although fall chinook may spawn without any delays. Spring chinook select large deep (usually >2 m) pools, typically with bedrock bottoms and moderate velocities. In California spring chinook usually hold where mean water column velocities are 15–80 cm/sec: often under ledges, in deep pockets, or under the "bubble curtain" formed by water plunging into pools (15). The fish do not necessarily stay in the same pool all summer long, but move between pools, usually with a net upstream movement. Holding areas are near spawning areas, which may be the tails of holding pools. Chinook have been observed digging redds and spawning at depths from a few centimeters to several meters and at water velocities of 15–190 cm/sec, but most seem to spawn at depths between 25 and 100 cm and velocities of 30–80 cm/sec (2). Winter-run chinook salmon are a major exception to this generalization, because they usually spawn at depths of 1–7 m in the Sacramento River (40). Regardless of depth, the key to successful spawning is having an adequate flow of water around developing embryos, which means they have to be buried in coarse substrate (typically a mixture of gravel and small cobbles) with a low silt content. When each redd is dug, the female essentially cleans an area measuring 2–10 m^2, loosening gravel and mobilizing "fines" so that embryos will have access to a steady flow of oxygen-containing water (2). Redd sites are apparently chosen in good part by the presence of subsurface flow. This is one reason that redds from previous spawners are often desirable places for later fish to spawn (to the detriment of early embryos).

Spawning behavior is similar to that of coho salmon, including the presence of small jack males that spawn as streakers. In addition, mature 1-year-old males have been observed that have never gone to sea (2) and are assumed to spawn by sneaking into the nest of large adults. Some of these precocious parr, which have enormous testes (about 21% of body weight) may actually survive to spawn a second time (54). The combination of regular and irregular males ensures a high degree of fertilization of eggs—more than 90 percent. Each female produces 2,000–17,000 eggs. Although the number of eggs increases with body size, the relationship is not as strong as that in other salmon; it varies among populations and runs (2). For example, Sacramento fall-run chinook appear to have exceptionally high fecundity for a given size (2). The average fecundity of females in the Sacramento River in recent years has been estimated to be 3,700 for winter-run, 4,900 for spring-run, 5,500 for fall-run, and 5,800 for late-fall-run fish, with differences resulting from a combination of average body size and other factors (16).

For maximum embryo survival, water temperatures must be between 5 and 13°C and oxygen levels must be close to saturation (2, 30, 65). Under such conditions embryos hatch in 40–60 days and remain in gravel as alevins (sac-fry) for another 4–6 weeks, usually until the yolk sac is fully absorbed. After emerging from gravel, fry are typically washed downstream into back- or edge water areas, where velocities are low, cover is dense, and small food items are abundant. As they grow larger and more agile, they move into deeper and faster water.

Status IA–E. Chinook salmon are in long-term decline in California, although they are not, *as a species,* in immediate danger of extinction. Through the 21st century, however, our society will be making decisions that will determine whether we have just a few "museum runs" supported by hatcheries or whether we continue to have a diversity of chinook salmon runs in the major rivers of the northern part of the state that support fisheries. We are in danger of turning what was once a major cultural resource (63) into a curiosity known to only a few. We have already lost some major runs, most notably the southernmost run, the San Joaquin spring run, which once numbered in the hundreds of thousands. Other runs are barely hanging on. The following is an assessment of the status of the 17 runs commonly recognized in the state as of 1999: *(1)* Smith River fall and spring run; *(2)* lower Klamath River fall run; *(3)* Upper Klamath-Trinity fall run; *(4)* Klamath–Trinity spring run; *(5)* Klamath late fall run; *(6)* Redwood Creek fall run; *(7)* Little River fall run; *(8)* Mad River fall run; *(9)* Humboldt Bay tributaries fall run; *(10)* Eel River fall run; *(11)* Bear River fall run; *(12)* Mattole River fall run; *(13)* Russian River fall run; *(14)* Central Valley fall run; *(15)* Central Valley late fall run; *(16)* Sacramento winter run; and *(17)* Central Valley spring run.

Smith River fall-run chinook. ID? The Smith River, on the Oregon border, is a relatively unaltered stream that has never supported a particularly large chinook salmon population. Annual estimates are generally 15,000–30,000 fish. There is no evidence of a long-term decline in the fall run, but data are limited. There also seem to be a few (about 500) spring-run chinook in the system, but their past and pres-

ent status is poorly understood (17). NMFS (5) classifies all Smith River chinook as part of their southern Oregon–California coastal ESU (SOCC-ESU).

Lower Klamath River fall-run chinook. IC. NMFS (5) considers salmon in the lower Klamath River, below the mouth of the Trinity River, to be part of its SOCC-ESU, while those spawning in the upper Klamath and Trinity Rivers are considered a part of its upper Klamath and Trinity River ESU (UKTR-ESU). The lower Klamath River fish, including the run up Blue Creek, are tied to Smith River fish genetically and also share some details of life history (e.g., late spawning compared with upper-river fish) (10, 12). The differences between fish from the upper and lower rivers are probably becoming increasingly obscured by the large numbers of hatchery fish released into the upper Klamath and Trinity Rivers that undoubtedly "stray" into lower Klamath spawning areas. Regardless of genetics or origin, the Klamath River fall run (from both the upper and the lower river) was once one of the most numerous runs in California, totaling perhaps 500,000 annually at one time. From 1876 to 1933 the runs supported a large inriver commercial fishery, which in turn supported several canneries near the mouth of the river. The numbers caught during this period are not certain because coho salmon and various chinook salmon runs were reported together. But it is likely that peak catches (in 1915) were around 100,000, dropping to fewer than 20,000 fish 20 years later (7). The intense fishing pressure also resulted in average fish being small and mostly 3-year-olds. The last cannery shut down in 1933, and the commercial fishery switched to offshore troll fisheries, sport fisheries, and an inriver gill net fishery by the local Native Americans. In the 1960s and 1970s these fisheries harvested about 350,000 chinook of all types from the Klamath, a majority of them presumably originating from hatchery production. Wild populations of chinook continued to decline, however. The offshore commercial fishery was essentially shut down in the 1990s. The numbers of wild-spawning lower Klamath fall-run chinook of natural origin (such as those in Blue Creek) are poorly known, but they presumably total 2,000–3,000 fish in some years.

Upper Klamath–Trinity fall-run chinook. IC. The status of wild fish in these two rivers is hard to determine because millions of hatchery juveniles have been released over the years. If fish spawning in the mainstem and streams with heavy hatchery influence are excluded, annual numbers of wild fish are probably 20,000–40,000, although total numbers of fall-run fish have been as low as 11,000 (in 1991). With the upper reaches of the rivers cut off by dams and reaches below dams having reduced and altered flow regimes, the number of salmon using the remaining habitat is clearly but a small fraction of the original total. Unfortunately, salmon of the upper Klamath and Trinity Rivers were not counted independently of other runs before the 1930s.

However, fall-run salmon in the Shasta River (arguably a distinct run), tributary to the upper Klamath, historically numbered 20,000–80,000 or more fish per year alone; they now number a few hundred to a few thousand. The decline of the Shasta River chinook was caused by degradation of the watershed and stream habitat by agriculture, which accelerated following the increase in irrigation permitted by construction of Dwinnell Dam in 1926 (41, 42). Based on numbers like these, it is likely that the majority of the 500,000 or more fish that once entered the Klamath-Trinity basin were upper Klamath–Trinity fall-run chinook.

Klamath-Trinity spring-run chinook. IB. Spring-run chinook in the Klamath-Trinity system are on the verge of disappearing. They are lumped in by NMFS (5) with fall-run and late-fall-run fish in the UKTR-ESU because of genetic similarities, but substantial differences in life history traits merit separation of the runs for conservation purposes. In the Klamath drainage the principal remaining run is in the north and south forks of the Salmon River and in Wooley Creek, tributary to the Salmon River. The north and south forks of the Trinity River, and possibly New River, also support a few fish (43). The large run of spring chinook in the mainstem Trinity River is apparently maintained entirely by hatchery production.

The Klamath-Trinity system once supported spring-run chinook populations that totaled more than 100,000 fish. Even this estimate is probably low, because spring-run fish were apparently the main run of chinook salmon in the Klamath River in the 1800s, but it was depleted by the end of the century as the result of hydraulic mining and commercial fishing (11). In each of four upper Klamath tributaries alone, historical run sizes were estimated by CDFG (43) to be at least 5,000: Sprague River (Oregon), Williamson River (Oregon), Shasta River, and Scott River. Runs in the Sprague and Williamson Rivers were probably extirpated before 1900 as the result of dams constructed in Oregon; if any fish remained, they were eliminated with the construction of Copco Dam across the main river in California in 1917. The run in Shasta River, probably the largest tributary run in the Klamath drainage, disappeared in the early 1930s as a result of habitat degradation and increased summer water temperatures caused by Dwinnell Dam. The smaller Scott River run was extirpated in the early 1970s by a variety of causes.

In the Trinity River runs that once existed above Trinity Dam included an estimated 5,000 or more fish in mainstem Trinity River above Lewiston and 1,000–5,000 each in Stuart Fork Trinity River, East Fork Trinity River, and Coffee Creek (43). In the Salmon River drainage an estimated total of 500–1,500 adults used the north and south forks and Wooley Creek each year through the mid-1990s, but the number dropped to fewer than 200 in 1998 (59). In South Fork Trinity River numbers have ranged from 0 to 300 in recent years, down from the 7,000–11,000 fish that once held

in the stream (15). The low numbers now using the south fork are largely a response to the 1964 flood, which triggered landslides that filled in holding pools and covered spawning beds.

Klamath late-fall-run chinook. IA? This run, considered part of the UKTR-ESU by NMFS (5), has always had a shaky identity. However, the presence in mainstem rivers and large tributaries of large fresh salmon that seem to come up after the fall run is finished has been long noted (11, 12). It is quite likely that, even if late-fall-run chinook did once exist as a distinct run in the Klamath system, they no longer do because of changes in flow and temperatures in the mainstem rivers and because of the flooding of the environment with juvenile hatchery-reared fall-run fish.

Redwood Creek fall-run chinook; Little River fall-run chinook; Mad River fall-run chinook; Bear River fall-run chinook; Humboldt Bay tributaries fall-run chinook; Mattole River fall-run chinook. IB. These runs (and those in the Smith, Eel, and Russian Rivers) are considered by NMFS to be part of the SOCC-ESU, which is listed as Threatened. Most of the larger coastal tributary streams have chinook salmon entering them from time to time, but these streams appear to have had consistent annual runs, although even historically the runs were probably only a few hundred fish each. NMFS (5) lumps them all in the SOCC-ESU, but given the poor supporting evidence it is equally reasonable to put all coastal populations south of the Klamath River into a separate ESU. The small size of these populations has always made them vulnerable to extirpation from natural and unnatural causes, and not surprisingly their numbers have typically been less than 100 per stream in recent years, although CDFG (44) regards these populations as "viable."

Eel River fall-run chinook. IB. The Eel River basin historically had fall chinook runs that averaged about 93,000 fish, sometimes reaching nearly 600,000 fish—enough to support canneries on the lower river (53). By the 1950s the runs were averaging about 24,000 fish (14,500–38,000). The changes to channel spawning and rearing habitat wrought by the major floods of 1955 and 1964, in combination with overfishing and poor ocean conditions, caused a decline from which they never really recovered. Thus by the 1990s runs often numbered fewer than 5,000 fish, with numbers in upper reaches (e.g., Tomki Creek) dwindling to fewer than 50 fish in many years (53). The Eel River probably once also supported a small spring-run chinook population, but there are no records of its abundance.

Russian River fall-run chinook. IA. Early records of chinook salmon in the Russian River are scant, but given the habitats and flows once typical of the river, it is logical to assume that it supported a run (45). Unfortunately, the salmon disappeared with the advent of agriculture and water projects in the basin. Attempts to reestablish self-sustaining runs through annual planting of hatchery fish do not appear to have been successful, although some spawning has been observed in the basin (45). Hatchery fish have been of multiple origin, but mainly Sacramento River and Klamath River stocks (5).

Central Valley fall-run chinook. ID. This ESU is considered a candidate for Threatened status by NMFS (5), has always been the most abundant run in the Central Valley, and may have numbered over a million spawners in some years (46). It occurred in all major tributaries, each of which had a distinct run. In the period 1967–1997 average numbers in the Sacramento River, including hatchery fish, ranged from 107,300 to 381,000 fish, with an average of around 200,000 fish. In the San Joaquin system annual numbers have been smaller and more variable, ranging from 1,100 to 77,500, with half being fewer than 10,000 fish. The runs approached extinction during the drought-influenced years 1989–1992. In both cases runs are heavily supplemented with fish of hatchery origin (10–65%, depending on run, year, and who is counting) from large hatcheries on Battle Creek and the Feather, American, Mokelumne, and Merced Rivers. It is not certain to what extent naturally spawning salmon depend on hatcheries to maintain their populations or vice versa. Reduction in the ocean fishery combined with favorable ocean conditions resulted in increased returns of this run, in both hatchery and wild fish, in the late 1990s. In part because of the uncertainty of the role of hatchery fish in the ESU, NMFS (5) has considered listing it as a threatened species.

Central Valley late-fall-run chinook. IB. This run is listed by CDFG as a Species of Special Concern (15). Although NMFS (5) regards it as part of the fall-run ESU, it has such a distinctive life history pattern that it needs separate recognition and management. The historical abundance of late-fall-run chinook is not known because it was officially recognized as distinct from fall-run chinook only after Red Bluff Diversion Dam was constructed in 1966. In order to get past the dam, salmon ascended a fish ladder in which they could be counted with some accuracy. The four chinook salmon runs present in the river were revealed as peaks in the counts, although salmon passed over the dam during every month of the year. Like those of winter-run and spring-run chinook, their numbers have declined since counting began in 1967. In the first 10 years of counting (1967–1976) the run averaged about 22,000 fish; it declined to an average of about 10,000 fish through 1990 and then to about 6,700 fish in 1991–1994 (64). There have been no counts of 20,000 fish or more since 1975, although 16,000 fish were counted in 1987. After 1991 full counts were no longer made because the gates at Red Bluff Diversion Dam had been opened to allow free passage of winter-run chinook adults and smolts. As a result migrants no longer have to pass over the ladder. (This is a good thing, because delays below the dam, caused by the inability of fish to locate the ladders, were apparently a major source of mortality for salmon.) No reliable estimates of late-fall-run numbers have been available since 1994, so their actual status is not known. Given the fact that they are largely spawners in the mainstem, where many threats to their existence exist, there is little reason to be optimistic about their long-term survival. Some late-fall-run chinook are now reared at Battle Creek Hatchery to supplement wild stocks. It is likely that the San Joaquin River also once supported a late fall run, but it is now extinct.

Central Valley spring-run chinook. IB. Sacramento spring-run chinook were listed as Threatened by the California Fish and Game Commission in 1998 and by NMFS in 1999. Spring-run chinook salmon in the Sacramento–San Joaquin River system historically made up one of the largest sets of runs on the Pacific coast (46). Commercial fisheries for spring chinook caught in excess of 567,000 fish in 1883 alone (versus 213,000 fall-run fish). Runs in the upper San Joaquin River probably exceeded 200,000 fish at times, and it is likely that an equal number of fish were once produced by the combined spring runs in Merced, Tuolumne, and Stanislaus Rivers. However, early historical population levels were never measured. In 1955 CDFG estimated that with proper water management the San Joaquin drainage could still produce about 210,000 wild chinook salmon per year, with fall-run chinook (originally a minor portion of the San Joaquin salmon runs) replacing spring-run populations lost to dam construction. The last large spring run in the San Joaquin River occurred in 1945, when 56,000 fish made it up the river (47). When Friant Dam was completed in 1948 the remaining fish were cut off from their upstream habitats.

The impact of the dam and efforts to rescue San Joaquin spring-run chinook salmon were recorded by CDFG biologist George Warner (48, p. 62): "In 1948, disaster struck. Friant Dam . . . had been completed and the Bureau of Reclamation assumed control of the river . . . [and] bureau officials diverted water desperately needed by salmon down the Friant-Kern Canal to produce surplus potatoes and cotton in the lower San Joaquin Valley. Only enough water was released in the river to supply downstream canals and some of the pumps." CDFG crews succeeded in trapping 1,915 spring-run chinook and trucking them to the base of Friant Dam. The fish were able to hold in the cold releases through the summer and then spawn successfully in the fall. Unfortunately, when the juvenile salmon attempted to move out to sea, they ended up stranded in a dry stretch of river. In the words of Warner (48, p. 65): "The tragic conclusion to the history of the 1948 spring run was that the only beneficiaries of our efforts to salvage a valuable resource were the raccoons, herons, and egrets." Efforts to rescue the run in 1949 and 1950 also failed; thus San Joaquin spring-run chinook salmon became extinct.

The Sacramento River drainage as a whole is estimated to have supported spring-run chinook runs exceeding 100,000 fish in many years between the late 1800s and 1940s (6, 17), but these estimates may be low by a factor of 3 or 4 (60). As in the San Joaquin drainage, these chinook populations were also drastically reduced following the construction of barrier dams. Historical run sizes for tributaries to the Sacramento River were estimated to be 15,000 or more above Shasta Dam (McCloud River, Pit River, Little Sacramento River); 8,000–20,000 in the Feather River above Oroville Dam; 6,000–10,000 in the Yuba River above Englebright Dam; and 10,000 or more in the American River above Folsom Dam.

The decline of spring-run chinook in the Sacramento drainage began when streams were disrupted by gold mining and irrigation diversions, but it accelerated following the closure of Shasta Dam in 1945, denying spring-run salmon access to major spawning grounds in the McCloud, Pit, and upper Sacramento Rivers (a loss of at least 250 km of habitat, including the best holding, spawning, and rearing habitat in the Sacramento drainage). The principal habitats remaining open to spring-run chinook are Deer, Mill, and Butte Creeks (Tehama and Butte Counties), which historically were minor habitats for these salmon. Diking and draining of the sinks and channelization of Butte Creek ironically may have allowed salmon easier penetration into upstream areas, especially after flows were enhanced by transfer of water from the Feather River into Butte Creek for power generation. In 1969–1979 estimates of spawning fish in Deer and Mill Creeks averaged 2,300 and 1,200 fish, respectively. From 1980 to 1999 the estimates were 695 and 400 fish (64). Butte Creek supported runs that varied from 10 to 8,700 or more fish from 1951 to 1979, although the early records are poor and inconsistent (17, 64). In 1980–1989 the estimates averaged 530 fish, and the number jumped to 3,550 fish in the following decade, thanks to large runs in 1996 (7,480) and 1999 (20,260) (64). Spawning populations in other tributary streams, such as Antelope and Big Chico Creeks, are considerably smaller, with various small creeks supporting runs of fewer than 100 fish, and often none at all.

The mainstem Sacramento River also supports some putative spring-run chinook. Estimates of "spring-run" fish spawning in the river range from 3,700 to 21,000 fish between 1969 and 1997, with an average population of 6,700 fish per year overall and 2,500 since 1990 (64). However, most of these fish appear to be hybrids with fall-run fish. Such hybridization takes place both naturally and artificially. Natural hybridization takes place in cold waters emanating from Shasta Dam, in which early-migrating, immature fish hold through the summer; they are not spatially segregated from later-arriving fall-run fish, so interbreeding takes place. Many, and perhaps most, of the "spring-run"

fish also originate in Coleman (on Battle Creek) and Feather River hatcheries, where fish designated as spring-run are nearly indistinguishable from hatchery fall-run fish (46). These hybrid fish are able to sustain themselves in the hybrid conditions of a regulated river, but it is problematical to count them as spring-run.

Sacramento winter-run chinook. IB. This unique run is listed as Endangered by both state and federal governments. It is endangered because Shasta Dam completely cut the run off from its historical spawning grounds in cold tributary rivers. It survived only because releases from Shasta Reservoir were cold enough in most years for embryos and young to survive through summer. It now persists in about 70 km of river below Keswick Dam, where it declined to near extinction as the result of difficulty in passing Red Bluff Diversion dam and high mortality from excessively warm water during drought years. Run size dropped from nearly 120,000 fish in 1969 to 191–1,200 fish in recent years, with an average of 600 in 1990–1997 (64). Efforts to protect the run have included a cradle-to-grave hatchery program (49) and modification of release facilities at Shasta Dam to allow colder water to be placed in the river when the reservoir is low. A recovery plan for winter-run fish has been written as well (40), including innovative criteria for recovery that take into account the uncertainties of estimating their numbers (51). Probably the best action that can be taken to protect winter-run chinook is to restore runs to Battle Creek, where it should be possible for the fish to complete their life cycle without excessive manipulations of flows, temperatures, or fish.

Causes of Chinook Salmon Declines All chinook salmon runs in California have declined, some to extinction. The declines will continue, interrupted by times of high returns from fortuitous natural conditions, unless major restoration efforts are successful. These efforts must address the multiple causes of the decline and look beyond further technological fixes (52). Because chinook salmon are big-river fish, the state of their populations is a good reflection of the ecological health of California's major rivers.

The single biggest cause has been the construction of massive dams and diversions on all major rivers. In the Sacramento–San Joaquin system these dams have denied chinook salmon access to over half the stream reaches they once used and to over 80 percent of their historical holding and spawning habitat (6). Likewise, chinook salmon have been denied access to the upper reaches of the Klamath and Trinity Rivers and their larger tributaries. Most severely affected by dams have been stream-type chinook, which use upper stream reaches for holding, spawning, and rearing. Dams generally also render habitat below them less suitable for salmon by reducing flows, increasing temperatures, causing rocks in streambeds to become deeply embedded so they cannot be used for spawning, reducing cover for juveniles, and changing channel configurations. The massive reduction in salmon numbers has had far-reaching and little-understood consequences for aquatic ecosystems of which the salmon were once a part. The salmon were a major source of energy and nutrients for nutrient-starved rivers and riparian areas in the mountains, bringing the productivity of the California Current oceanic region far inland. No doubt the abundance of fish and wildlife in river corridors that once supported large populations of native peoples was partly tied to the annual fertilization effect of salmon runs.

Beyond dams, many other factors have contributed to the decline of chinook salmon, and each run has a litany of special problems associated with it. The most commonly mentioned general factors will merely be listed here (in no particular order) because they are covered in greater detail elsewhere in this book and in the literature cited at the end of this account.

1. *Fisheries,* both in the ocean and in streams, can easily overharvest salmon, especially when fish from already-depleted wild runs are mixed with fish of hatchery origin in the catch. Hatchery fish can sustain much higher catch rates than wild fish, but they cannot usually be harvested selectively before they reach the hatchery.

2. *Entrainment* of juveniles in diversions as they move downstream can cause major losses. The significance of the losses is often controversial, including losses to the two big pumping plants in the south Delta. The usual solution to the problem, screening diversions, can be expensive and is often ineffective. A more effective strategy is to combine screens with a reduction in diversions during periods of high outmigration.

3. Loss of floodplain and estuarine rearing habitat by *diking and draining* has had an unknown impact, although there is growing evidence that such habitat was once of major importance for the growth and survival of juvenile salmon.

4. Enhanced *predation* can be a major cause of death of juvenile salmon where removal of cover or unfavorable hydraulic conditions make them unusually vulnerable to piscivorous birds and fish. A particular problem is posed by nonnative predatory fish, such as striped bass, which can consume outmigrating salmon in large numbers. Artificial enhancement of striped bass numbers and the introduction of new predators (e.g., northern pike) are likely to increase these predation rates.

5. *Competition* from hatchery-reared juveniles for food and space in streams and from adults for spawning areas is just beginning to be taken seriously as a contributing cause of the decline of wild fish. Flooding an environment with hatchery fish can have a negative impact on everything from social structure to food supplies to the spawning success of wild fish. During years of low ocean productivity, even competition for limited food in the ocean is possible.

6. *Diseases,* native and introduced, can be propagated in hatchery fish and spread to wild fish.

7. *Pollution* is an ongoing and pervasive problem. Major sources of human sewage were cleaned up following passage of the Clean Water Act, but organic pollution, especially from livestock, can still be a problem in some areas. Although fish kills from heavy metals, pesticides, and other toxicants still occur on occasion, much more common and harder to deal with are the sublethal effects, which may reduce the ability of fish to deal with disease and other natural sources of stress (50). There are also ongoing threats of major fish kills due to toxic wastes originating from the abandoned Iron Mountain Mine near Redding and similar sources.

8. *Loss of riparian forests* can increase the temperatures of streams and decrease the amount of trees and other large woody debris that falls into streams. Cool temperatures and dense cover are important for salmon survival in streams.

9. *Siltation* of spawning areas from catastrophic and chronic sources reduces spawning success. The ultimate example of this was furnished by the hydraulic mining debris that flowed down many rivers following the Gold Rush, wiping out many runs temporarily. However, in more recent times landslides and chronic siltation from poor land management (especially related to road construction, logging, and grazing) often reduce the ability of spawning areas to support fish.

10. The effects of *introduced fish, invertebrates, and plants* can be many and various, if poorly documented. California's waters, but especially its estuaries, have a high rate of invasion of alien organisms (mainly from the ballast water of ships). New species can change the way ecosystems function or the availability of food and cover, to the detriment of juvenile salmon.

11. *Natural factors*—such as long periods of drought, extreme flooding events, and periods of low ocean productivity—can also decimate salmon populations. In fact, salmon populations naturally showed wide fluctuations in response to long periods of drought or ocean productivity. Most contemporary global

warming scenarios for California foresee an increase in climatic variability, meaning an increase in the natural variability of salmon populations. This prediction provides additional incentive to get other human-caused sources of population variability under control if we are to maintain salmon populations.

With all these factors harming salmon populations, the restoration of remaining runs to self-sustaining and harvestable levels is extremely difficult and expensive. Hundreds of small problems have to be dealt with, while bigger fixes involving the management of entire ecosystems are being implemented. Imaginative solutions are required, such as *(1)* removing dams and barriers on some streams (e.g., Battle Creek); *(2)* restoring runs to streams from which they have been eliminated, ranging from small streams (e.g., Putah Creek) to large rivers (e.g., the San Joaquin River); *(3)* restoring meander belts and floodplains; *(4)* creation of better rearing habitat in estuaries; *(5)* changing hatchery and harvest practices to favor wild fish; and *(6)* tying the restoration of salmon to the creation of a better environment for all creatures, including humans. As Michael Black (52, p. 70) states, "Instead of gleefully manipulating nature, we must use our considerable intellectual gifts to modify our own beliefs, behaviors, and cultures in keeping with a biologically healthy river. By emulating what wild anadromous fish require to thrive, salmon and steelhead may just be capable of showing us the way."

References 1. Fry 1973. 2. Healey 1991. 3. Stearley and Smith 1993. 4. Bartley et al. 1990. 5. Myers et al. 1998. 6. Yoshiyama et al. 2001. 7. Snyder 1931. 8. Scott and Crossman 1973. 9. Lever 1996. 10. Shaw et al. 1997. 11. Emmett et al. 1986. 12. Gale et al. 1998. 13. Waldvogel 1988. 14. Kjelson et al. 1982. 15. Moyle et al. 1995. 16. Fisher 1994. 17. Campbell and Moyle 1991. 18. CDFG 1998. 19. Bradford and Taylor 1997. 20. Taylor 1990. 21. Chapman and Bjornn 1969. 22. Everest and Chapman 1972. 23. Stein et al. 1972. 24. Gregory 1994. 25. Marine 1997. 26. Baker et al. 1995. 27. Gregory and Levings 1998. 28. Moyle 1970. 29. Wallace and Collins 1997. 30. Vogel and Marine 1991. 31. Maslin et al. 1997. 32. Kope and Botsford 1990. 33. Merz and Vanicek 1996. 34. Merz 1998. 35. Sasaki 1966. 36. Sagar and Glova 1988. 37. CDFG, unpubl. data 1988–1997. 38. Ogura and Ishida 1995. 39. Groves et al. 1968. 40. NMFS 1996a. 41. Wales 1951. 42. Ricker 1997. 43. CDFG 1990. 44. CDFG 1992a. 45. Steiner 1996. 46. Yoshiyama et al. 1998. 47. Fry 1961. 48. Warner 1991. 49. Arkush et al. 1997. 50. Dethloff 1998. 51. Botsford and Brittnacher 1998. 52. Black 1995. 53.Hill and Webber 1999. 54. Robertson 1957. 55. T. Cannon, pers. comm. 1998. 56. J. J. Smith, San Jose State University, pers. comm. 1998. 57. Quinn et al. 1996. 58. Banks et al. 1999. 59. E. R. Gerstung, CDFG, pers. comm. 1999. 60. F. Fisher, CDFG, pers. comm. 1998. 61. Busby and Barnhart 1995. 62. Sommer et al. 2001. 63.Yoshiyama 1999. 64. Yoshiyama et al. 2000. 65. McCullough 1999. 66. Petrusso and Hayes 2001b. 67. Martin and Saiki 2001. 68. Petrusso and Hayes 2001a.

Kokanee (Sockeye Salmon), *Oncorhynchus nerka* (Walbaum)

Identification With their solid bright red bodies and green heads, spawning kokanee are unmistakable. Males develop a distinct hump on the back, and the snout becomes long and hooked, with large teeth. Both sexes lack black spots on the back or caudal fin and have 19–27 long, slender, and rough gill rakers on the lower half of the first gill arch. Non-spawning kokanee can be distinguished from trout by the lack of spotting (except occasional vague spots on the dorsal fin and fine speckling on caudal fin), long anal fin (13–18 rays), 28–40 slender, closely spaced gill rakers, and slightly oblique mouth. They also have 11–26 complete dorsal rays, 11–21 pectoral rays, 9–11 pelvic rays, 120–150 lateral line scales, and 11–16 branchiostegal rays on each side. Parr have 8–14 oval parr marks centered on the lateral line that are narrower than the spaces between them. The backs of nonreproductive fish have a blue-green sheen, the sides are silvery, and the fins are without spotting.

Taxonomy Kokanee are nonanadromous sockeye salmon that are frequently referred to as a subspecies (*O. nerka kennerlyi*) of sockeye. However, landlocked populations of sockeye have evolved independently in many different places, and these populations differ from each other in their characteristics, so there seems little reason to maintain the subspecific name (1). Each population of kokanee is genetically distinct from the anadromous form that presumably gave rise to it, demonstrating reproductive isolation (2).

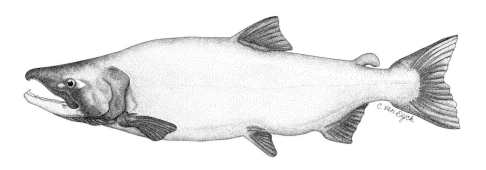

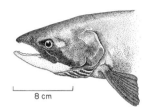

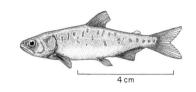

Figure 89. Sockeye salmon. Top: Spawning male, 49 cm SL, British Columbia. Bottom left: Spawning female, 46 cm SL, British Columbia. Bottom right: Parr, 8.5 SL, Alaska.

Hybridization can nevertheless take place, usually by small jack males of kokanee spawning with female sockeye (2). Kokanee populations in California, established through introductions, have many different origins (11).

Names Sockeye salmon are often called red or blueback salmon and kokanee are sometimes called redfish. Sockeye is an approximation of the name given it by the Native Americans who lived along the Fraser River in Canada, as is kokanee. *Nerka* is a Russian name for sockeye salmon. For other names see the account of coho salmon.

Distribution Spawning populations of sockeye salmon and their derivative kokanee populations are native to rivers and lakes of North America from the Columbia River north to the Yukon River in Alaska. Individual anadromous sockeye found in streams south of the Columbia system, including California, are probably nonspawning strays or fish from kokanee populations that decided to go out to sea. In Asia sockeye and kokanee occur from northern Japan to the Anadyr River in Russia (1). Kokanee have been successfully introduced into coldwater lakes throughout Canada and the northern and western United States. They have also been successfully introduced into New Zealand (12). Kokanee were brought into California in 1941 from Idaho and successfully established in Salt Springs Reservoir on the North Fork Mokelumne River (Amador County), in order to establish salmonid fisheries in fluctuating reservoirs. Progeny from this population, apparently mixed with other kokanee imported from outside the state, were planted in a number of natural lakes in the Tahoe region, including Lake Tahoe (11). In 1951 kokanee from Kootenay Lake, British Columbia, were planted in Shasta Reservoir to provide forage for rainbow trout of the Kamloops strain, also imported from British Columbia. Plants were subsequently made in many of California's lakes and reservoirs, with varying degrees of success (3). Today kokanee populations are present in many coldwater reservoirs and some lakes in the Sierra Nevada and in lakes and reservoirs scattered around the state. They are present in Lewiston and Trinity Reservoirs on the Trinity River, and occasional juvenile "sockeye" salmon collected in salmon outmigrant traps in the lower Trinity River presumably originated from these fish (15). They are also reared in California hatcheries to fingerling size and planted, by air, in about 30 lakes and reservoirs (4). In 1992 all hatchery fish originated from Bucks Lake, Plumas County (4).

Life History In California anadromous sockeye salmon occur only as rare strays mixed in with runs of chinook and coho salmon, so their life cycle will not be dealt with here. In any case, the life cycle of kokanee is similar to that of sockeye salmon except that kokanee mature in lakes rather than the ocean.

Kokanee prefer well-oxygenated open waters of lakes and reservoirs, where temperatures are 10–15°C. They inhabit surface (1–3 m) waters as long as temperatures remain in the preferred range or colder. As surface waters warm up in summer, they gradually move deeper. In Lake Tahoe they are found most of the year at depths of less than 4 m, but in July through September they concentrate at depths of 17–40 m (5). In large midelevation reservoirs (e.g., Shasta Reservoir) they stay in the hypolimnion during summer. Heavy kokanee mortality will occasionally occur when the hypolimnion becomes depleted of oxygen.

As the fine gill rakers of kokanee attest, their main food is zooplankton. In Lake Tahoe, prior to the introduction of opossum shrimp (*Mysis relicta*), they fed mostly on water-fleas (*Daphnia pulex*); copepods and midge larvae were minor items (5). After the establishment of shrimp, which depleted zooplankton (6), kokanee diets became dominated by midge pupae, copepods, and terrestrial insects (7). In Oroville Reservoir kokanee feed primarily on copepods and cladocerans, taking small fish and insects on occasion (8). Kokanee diets change little as fish grow larger, although newly emerged fry in streams subsist on aquatic insects for short periods of time. Dietary changes with season primarily reflect changes in available zooplankton (8). Feeding ceases just prior to spawning. In some lakes kokanee show daily movements up and down the water column, moving up into warmer waters to feed at night and then down into cooler waters to digest their meals during the day. Apparently, this strategy maximizes the efficiency of food use (13).

Growth in kokanee can be fairly rapid for a freshwater salmonid. They typically reach 10–25 cm TL in their first year, 18–31 cm in their second year, 22–44 cm in their third year, and 23–47 cm in their fourth year (9). In Oroville Reservoir they average 27 cm at the end of year 2, and 37 cm at the end of year 3 (8). In the 1960s Lake Tahoe kokanee tended to be at the upper end of each length group, and fish longer than 53 cm (1.4 kg) were present (5). Following the depletion of larger zooplankters by opossum shrimp, growth rates decreased dramatically, with spawning adults reaching only 24–35 cm (7, 14). In contrast, sea-run sockeye usually reach 65–80 cm TL (4–7 kg) before spawning. Males typically grow larger than females.

The size and age of spawning kokanee depend in part on growing conditions (e.g., food availability, light, and temperature regimes) and in part on the origin of the stock. Some populations complete their life cycle in 2 years; others (e.g., the one in Lake Pend Oreille, Idaho) take as long as 7 years. Most populations mature in 4 years, including those present in Lake Tahoe. In many instances fish from populations with 4-year cycles in their native lakes mature in 3 years in California (e.g., those in Oroville Reservoir), presumably as a result of better growing conditions (8). Some kokanee transplants into California, however, have been from populations with 2-year life cycles because these populations are alleged to have superior growth rates. Some plants of these fish have been made into lakes that already contained fish with 3- or 4-year life cycles, but it is not known if the two groups of fish coexist as they do naturally in some British Columbia lakes or if they interbreed. Most kokanee measure at least 20 cm TL before they spawn, but mature fish as small as 16 cm have been recorded.

Kokanee normally spawn between early August and early February, but spawning has been recorded in California as late as early April. The time of spawning is determined in part by genetic background and in part by lake and stream temperatures. In Taylor Creek, the principal spawning stream for Lake Tahoe kokanee, spawning occurs between late September and late November, although most takes place in early October (14). Most spawning takes place in streams in gravel riffles a short distance from the lake when temperatures are 6–13°C. Taylor Creek is enhanced as a spawning stream because USFS increases flows from an upstream dam (14). However, lake spawning in beds of gravel close to shore, usually at depths less than 18 m, is sometimes important in lakes and may be important in Lake Tahoe in some years (5).

The first sign that spawning is about to begin is congregation of kokanee near the mouths of streams or near lake spawning sites. Like other salmon, kokanee home to the stream in which they were hatched (or planted as fry) and locate the stream in part by its distinctive odor (10). The female builds the redd and defends the area from other females while her male partner defends the area from other males. Spawning behavior is similar to that of other salmon. Each female contains 200–1,800 eggs, larger fish containing more eggs. Fertilized eggs are buried beneath 5–15 cm of gravel by the female. Many females die before releasing all their eggs. Particularly low spawning success is found in Lake Tahoe kokanee, of which only 11 percent and 28 percent of the dead females examined in 1967 and 1968, respectively, were spawned out and 30 percent and 46 percent, respectively, had died without spawning at all (5). This low success rate is compensated for in part by the high survival rates of embryos. Spawning fish live 2–4 days in Taylor Creek, which is fairly typical for the species (14).

Fry emerge in April through June and move downstream immediately. Most movement takes place at night. Some may begin feeding while in the stream, but most do not start until they enter the lake.

Status IID. Kokanee were originally introduced into California to provide fisheries in reservoirs that would either be self-sustaining or only require periodic plants of small fish. They were also planted as forage for more piscivorous salmonids. Enthusiasm for kokanee led to their becoming widely established during the 1950s and 1960s in large, coldwater lakes and reservoirs. They have largely failed to provide a forage base for trout in reservoirs, although they are consumed in small numbers by larger trout. In small lakes they may actually depress trout growth and population size by competing for zooplankton. Angling for kokanee has become popular in recent years, resulting in expansion of the hatchery program to raise fingerlings, including bringing in fish from outside the state (4, 11). Even so, most populations are probably underexploited because specialized fishing techniques (e.g., trolling with small lures in deep water) are required during much of the season. The only time they seem particularly vulnerable to more conventional salmonid fishing techniques is when they congregate off stream mouths prior to spawning.

The introduction of kokanee into reservoirs is generally regarded as a success because they support a sport fishery at relatively low maintenance cost. When introduced into natural lakes they undoubtedly changed ecosystem processes because they are fairly efficient planktivores. Their impact on native fishes, such as *pectinifer* tui chubs in Lake Tahoe or Lahontan cutthroat trout in Independence Lake, is not known but is probably negative. Nevertheless the annual spawning run in Taylor Creek is now celebrated in an annual festival for humans and an annual feast for

bald eagles and other birds. Future hatchery programs should rely only on stocks already present in the state, to reduce the probability of disease being brought in with fish from out of state.

References 1. McPhail and Lindsey 1970. 2. Wood and Foote 1996. 3. Seeley and McCammon 1966. 4. Hendrickson and Hendrickson 1993. 5. Cordone et al. 1971. 6. Richards et al. 1991. 7. Frantz 1979–1981. 8. Hiscox 1979. 9. Carlander 1969. 10. Lorz and Northcote 1965. 11. Dill and Cordone 1997. 12. Lever 1996. 13. Bevelhimer and Adams 1993. 14. Beauchamp et al. 1994. 15. T. Kisanuki, USFWS, pers. comm. 1998.

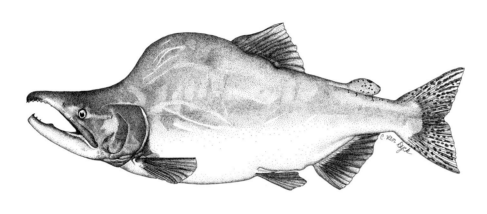

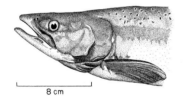

Figure 90. Pink salmon. Top: Spawning male, 46 cm SL, Alaska. Bottom left: Spawning female, 44 cm SL, Alaska. Bottom right: Parr, 6 cm SL, Alaska.

Pink Salmon, *Oncorhynchus gorbuscha* **(Walbaum)**

Identification Spawning male pink salmon have a pronounced purple hump behind the head, a greatly enlarged and hooked snout, and jagged teeth. Spawning females are troutlike in form and olive green on the sides, with long, dusky, vertical spots. Marine-phase fish are steely blue to blue-green dorsally, white ventrally, and silvery on the sides. The back and upper parts of the lateral surfaces have large black spots, which are also present on the adipose and caudal fin lobes Nonspawning pink salmon can be distinguished from other salmon by the combination of large black spots on the back, dark oval blotches on both tail lobes, and 16–21 gill rakers on the lower half of the first gill arch. The mouth is terminal and there are sharp teeth on both jaws, the vomer, the palatines, and the tongue. They have 10–16 complete rays in the dorsal fin, 13–19 in the anal fin, 14–18 in each pectoral fin, and 9–11 in each pelvic fin. The scales (147–198 in the lateral line) are deeply embedded in spawning fish. Branchiostegal rays number 10–15 on each side of the jaw. The maximum size recorded for pink salmon is 76 cm SL (6.3 kg), but fish over 60 cm (2.5 kg) are unusual. Young in fresh water are always small (less than 5 cm TL), silvery, and without parr marks or spots on the dorsal fin.

Taxonomy This species was first described in 1792 and has since been regarded as one of the most distinctive Pacific salmon. There are biochemical differences among pink salmon stocks in different river systems (1, 2) but, except for size, no major morphological differences. However, the fairly strict 2-year life cycle of pink salmon means that in each river system even- and odd-year runs are genetically distinct and may be more closely related to runs in other river systems of the same year type than to the run of a different year type in the same system (2). From Washington on south, pink salmon are predominately odd-year fish, including those that once spawned in California. NMFS (2) determined that there are only two distinct ESUs in the Pacific Northwest: the Snohomish River (Washington) even-year pink salmon and the southern odd-year pink salmon, occurring from southern British Columbia to Washington and presumably Oregon and California. However, no genetic data exist on the southernmost populations.

Pink salmon are most closely related to chum salmon, but "natural" hybrids have been noted between pink and chinook salmon in tributaries to the Laurentian Great Lakes (14).

Names Humpback salmon or simply "humpy" is a widely used common name. They are called pink salmon because of their pink flesh, which increases their value as commercial fish. *Gorbuscha* is derived from the Russian word for humpback. For other names see the account of coho salmon.

Distribution Spawning pink salmon ascend coastal streams of northern Asia, from Korea through Japan to Siberia (3). Along the Pacific coast of North America they range from the MacKenzie River in the Yukon Territory of Canada south to coastal streams of California. In the ocean they have been documented as far south as La Jolla (4). However, the largest runs on the southernmost end of their range are in streams tributary to Puget Sound (2). In California small numbers have been reported from the San Lorenzo River (5), the Sacramento River and tributaries (6), the Klamath River (7), and the Russian, Garcia, and Ten Mile Rivers (8). A pink salmon caught in the Mad River also was reported in the popular press (*Arcata Union*, Sept. 6, 1928; 16), which stated that this species had been frequently taken in the Mad River by net fishermen many years earlier. Pink salmon have been observed spawning in the Ten Mile and Garcia Rivers (8), and Fry (9) observed at least six pink salmon redds in the lower Russian River in 1955. Irregular occurrences of spawning in some Mendocino County streams have also been reported (10). During the 1800s

pink salmon were found in the Sacramento River, "which it [*sic*] ascends in tolerable numbers in October" (17, p. 54). In 1891 pink salmon trapped at Baird Station on the Mc-Cloud River were spawned artificially and the young released in the river (11). During the 1930s commercial fishermen on the Sacramento River reportedly captured a dozen or more pink salmon in some seasons (6). During the period 1949–1958, 38 pink salmon were taken in the Sacramento River watershed; this total included 12 fish from Coleman National Fish Hatchery, 4 in Mill Creek, and 3 at Nimbus Fish Hatchery on the American River (6). Limited spawning occurred in the Sacramento–San Joaquin river system during 1989, because seven pink salmon smolts were salvaged at the state's J. E. Skinner Fish Protective Facility near Tracy in March 1990 (15).

Life History Pink salmon are best known for their 2-year life cycle (2, 3), although occasionally 3-year-old fish are reported (12, 18). Adults move into fresh water between June and September and spawn from mid-July to late October, depending on geographic location. Spawning in California has been recorded only in September and October (9, 11). Most pink salmon spawn in the intertidal or lower reaches of streams and rivers. Upstream migrations of 100–700 km occur in some river systems, and there are records of fish 300–400 km up the Sacramento River system (11). Spawning streams are determined by the physiological requirements of the salmon, as follows (13). Optimal temperatures are 5.6–14.4°C; 0.0°C and 25.6°C are the lethal limits. Spawning generally occurs at temperatures of 7.2–12.8°C, with 4.4–13.3°C optimal for hatching. Embryos and alevins require fast-flowing (21–101 cm/sec) and well-oxygenated (>6 mg/liter) water for normal development and survival.

Spawning occurs in gravelly riffles with water depths of 20–60 cm. The six redds built by females in the lower Russian River were all situated along stream edges where the substrate was finer (9). No redds were found in the middle portion of the riffle where the substrate was composed of coarser gravel. Spawning males are aggressive and defend territories in riffles. They often inflict severe wounds on each other with their large jaw teeth during conflicts. While males fight, females dig redds. To dig, a female turns on her side and repeatedly cuts at the gravel with her tail, displacing gravel downstream. When the depression is approximately 90 cm long and 45 cm deep, the female signals the male of her readiness to spawn by sinking to the bottom until her anal fin touches the gravel. The male swims alongside the female and the two quiver and gape while simultaneously releasing eggs and sperm. Not all eggs are released in one spawning, so the female digs a new redd above the old one and buries the eggs from the preceding spawning while doing so. She may thus dig several redds in succession and spawn with more than one male.

A female usually lays 1,200–1,900 eggs over several days (2, 3). Both males and females die a few days to a few weeks after spawning. Embryos hatch in 4–6 months, presumably in February and March in California. Alevins emerge from the gravel in April or May, at which time the yolk sac has been absorbed. The fry, measuring 35 mm TL, immediately begin to migrate downstream into the estuary. Migration takes place at night, and fish usually reach the estuary in one night. Once in the estuary they form large schools and remain in inshore areas for several months before moving out to sea. Most juveniles do not remain in fresh water long enough to feed, although those that hatch from redds farther upstream will feed on aquatic insects. At sea juveniles feed on small crustaceans and other invertebrates. Maturing adults feed mostly on fish, squid, euphausids, amphipods, and copepods.

Pink salmon wander great distances in the ocean, and tagged fish have been captured 2,700 km (1,700 mi) from where they were tagged (10). However, they are fairly faithful to their parent streams and return there for spawning. The 2-year life span of pink salmon results in distinctive populations, which form odd- and even-year spawning runs. Some streams may support major runs of both (odd and even) years, whereas others may support major runs of one or the other year. Historically, the southernmost pink salmon fisheries in North America landed large numbers only in odd-numbered years.

Status IA. Pink salmon are extinct in California but abundant in the northern parts of their range. In Alaska and Canada they support major commercial fisheries. In an evaluation of the status of southern populations, NMFS concluded that some populations of odd-year pink salmon in Washington were in decline but that the ESU did not merit listing as threatened. Populations in Oregon and California were regarded as not persistent enough for consideration (2). This lack of persistence is probably a recent phe-

nomenon. In California they have never been common enough to excite much attention, although they occurred on a regular basis in the past. In the late 1880s they were included in the salmon catch sent from the North Coast to San Francisco markets (11). In recent times they were still common enough to be noticed. Taft (8, p. 198) cited reports that considerable numbers of pink salmon were running in northern California streams in 1937: "many quite large schools of them" in the Ten Mile River and "several hundreds" in the Garcia River, "spawning all over from the Red Bridge to the western boundary of the Indian Reservation, a distance of about two miles." They also were observed in the Russian River during that year (8).

Today, however, pink salmon are extremely rare in California. Most fish now recorded in the state are probably fish that strayed while at sea and followed other species of salmon upstream. Their occurrence in the Russian River in 1937 and evidence of limited spawning in 1955 (9) would indicate that this run was the southernmost for the species, except for occasional spawners in the Sacramento River. A run in the Russian River has not been recorded since 1955. Given the major changes that have taken place in the river since then—such as gravel mining, construction of Dry Creek Dam, and a number of major pollution events—it is not surprising that pink salmon no longer spawn there. Thus, although occasional pink salmon are observed in California, they must be regarded as one of the species extirpated from the state.

References 1. Beacham et al. 1985. 2. Hard et al. 1996. 3. Heard 1991. 4. Hubbs 1946. 5. Scofield 1916. 6. Hallock and Fry 1967. 7. Snyder 1931. 8. Taft 1938. 9. Fry 1967. 10. Roedel 1953. 11. U.S. Commission for Fish and Fisheries 1894. 12. Omel'chenko and Vyalova 1990. 13. Emmett et al. 1991. 14. Rosenfield 1998. 15. D. McEwan, CDFG, pers. comm. 1990. 16. S. Van Kirk, pers. comm. 1997. 17. California Fish Commission 1880. 18. Anas 1959.

Chum Salmon, *Oncorhynchus keta* (Walbaum)

Identification Spawning male chum salmon are heavy-bodied and slightly humped, and have a long, hooked snout with conspicuous canine-like teeth at the ends of the jaws. They are dark olive on the back and dark maroon with irregular greenish vertical bars on the sides. Females are similar in color, although the maroon is less well developed; they lack a hump, often have a distinct midlateral stripe, and have a jaw that is less hooked. In the ocean silvery chum salmon can be distinguished from other salmon by the absence of large black spots on the back and caudal fin (although black speckles may be present) and by the 18–28 short, smooth gill rakers on the first gill arch. They have

10–14 major rays in the dorsal fin, 13–17 in the anal fin, 14–16 in each pectoral fin, and 10–11 in each pelvic fin. The scales (124–153 in the lateral line) are deeply embedded in spawning fish. Branchiostegal rays number 12–16 on each side of the jaw. Their maximum size is about 1 m SL (15 kg), but they are typically less than 80 cm SL (6–7 kg). The parr have 6–14 small, pale parr marks, and the width of the light areas between the marks is greater than the width of the marks themselves. There is no spotting on the fins, and the back is mottled green, the sides silvery green.

Taxonomy Chum salmon are most closely related to pink and sockeye salmon, forming a subgroup within *Oncorhynchus* (1). Chum salmon show a strong homing ten-

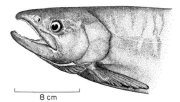

Figure 91. Chum salmon. Top: Spawning male, 60 cm SL, British Columbia. Bottom left: Spawning female, 48 cm SL, British Columbia. Bottom right: Parr, 7 cm SL, Klamath River, Del Norte County.

8 cm

4 cm

dency to their natal streams (2), resulting in genetic differentiation of spawners in different streams. There is some evidence that, even within a single river system, genetic differences associated with spatial separation of spawners may occur (2). Nothing is known about the relationship of California populations to other populations of chum salmon, but presumably they are linked to the closest large populations in Washington.

Names Chum salmon have been called dog salmon because they were the salmon Eskimos fed to their sled dogs. They are sometimes marketed as "silver-brite" salmon. *Keta* means "fish" in a dialect of the Amur people in southeastern Russia (2). See the account of coho salmon for details and for other names.

Distribution Chum salmon have the widest natural geographic distribution of the Pacific salmons, ranging from Korea up along the Arctic coast of Russia, and from the Mackenzie River on the Canadian Arctic coast through

Alaska and then southward into California. Historically, they were reported to occur in streams from San Francisco to the Bering Straits (3, 20), and were considered to be "abundant in the fall, from Sacramento northward" (4, p. 55). In the Pacific Northwest chum salmon are abundant in streams of British Columbia and Washington, including the Columbia River. They become progressively less abundant in the ocean and as spawners in Oregon streams south to Cape Blanco (5, 6). Only small populations exist from the Rogue River on south.

In California small runs of chum salmon were historically present in streams from the Sacramento River north. In the 1880s chum salmon constituted a minor portion of the salmon catch from the Humboldt County coast sent to San Francisco markets, and they also occurred in catches in the Sacramento River (7). Based on a 10-year (1949–1958) survey of the Sacramento River system, during which 68 chums were recorded, Hallock and Fry (8) concluded that a very small run was present. A few fish are still taken in the Sacramento drainage, but no spawning has been recorded in recent decades.

Today small runs of chum salmon still seem to maintain themselves in three California rivers: Smith, Klamath, and Trinity. In the Smith River drainage a small run is reported more or less annually in Mill Creek, a tributary to the estuary (9). Spawning behavior has been observed there as well (14). A few adult and juvenile chum salmon have also been observed annually in the South Fork Trinity River, the apparent remnant of a larger run that existed there prior to the 1964 flood (15). Evidence of successful spawning is the annual capture of small numbers of juveniles (38–58 mm FL) in a rotary screw trap set at river km 34 on the Trinity River (16). Evidence of spawning in the Klamath River is provided by collections of juvenile chum salmon dating back to at least 1944 (in the California Academy of Sciences). Like pink and sockeye salmon, individual chum salmon may also occur in North Coast rivers as strays that presumably move upstream with coho or chinook salmon. Chums have been found in ocean waters as far south as San Diego (10, 11), but the southernmost freshwater record has been the San Lorenzo River, Santa Cruz County (12).

Life History Chum salmon normally complete their life cycle in 3–5 years, although a few males may complete it in 2 and some females may live as long as 7 years. They are highly migratory and versatile in their use of fresh and marine waters. Their life history and habitat requirements have been well studied in British Columbia and Alaska, and most of the information presented here is taken from these studies (2, 5, 6). They can spawn in intertidal areas, but some populations in the Amur River of Russia and the Yukon River of Alaska and Canada spawn 2,500 km or more up-river. Normally chum salmon spawn within 200 km of the ocean. There are no natural, completely landlocked forms. They appear unable to hurdle waterfalls and other barriers that present few difficulties for the passage of other salmon species. In general chum salmon (like pink salmon) have a short freshwater and an extensive marine life stage, and they are especially dependent upon estuaries during the non-migratory juvenile stage. In North America there is a northern (early-run) stock that spawns from June through September and a southern (late-run) stock that spawns from August through January. In Washington, Oregon, and California all stocks are late run. Early-run fish generally spawn in mainstems of streams, whereas late-run fish spawn in smaller streams that have more favorable winter temperatures. In the Sacramento River they have been captured from early August to early February (8). In West Branch Mill Creek, a tributary of the Smith River, chums were observed entering during mid-December, when stream flows were high (9). No fish were seen in years lacking high December flows, although it is possible that chums spawned in mainstem Smith River or its other tributary streams during those years.

Chum salmon adults and maturing juveniles are epipelagic in the ocean, but all stages are bottom oriented in rivers and streams (2). Adults migrate upstream in water velocities up to 2.44 m/sec and spawn in velocities of 46–101 cm/sec. Upstream migration occurs in water between just above freezing and 21.1°C, with an optimum range of 8.3–15.6°C. Optimum spawning temperatures are 7.2–12.8°C, with oxygen levels greater than 80 percent of saturation, although short dips to 5 mg/liter can be survived. Spawning gravels are typically 1.3–10.2 cm in diameter, but eggs and alevins are found primarily in medium-size gravel (2–4 cm diameter). In the Columbia River drainage chum salmon redds were composed of 13 percent gravel greater than 15

cm, 81 percent less than or equal to 15 cm, and 6 percent silt or sand. In a survey of redds in Washington, 80 percent were located in depths of 13.4–49.7 cm, with a mean depth of 27 cm. Incubation temperatures are 4.4–13.3°C, although embryos can survive colder temperatures after they have developed for a period and become cold tolerant. Optimum outmigration river temperatures for fry are 6.7–13.3°C.

Adults show strong homing behavior to natal streams, in which they select spawning sites in which there are good intragravel flows (upwelling). Females are territorial and dig and spawn in a series of 4–6 redds, each one immediately upstream of the previous one. A decreasing number of eggs is laid in later redds. The combined set of redds averages 2.8 m^2 in size, and the female guards the last redd pocket until she dies. Males, which are sexually active for 10–14 days, may spawn with several females, and they are physically aggressive toward other males. Large dominant males defend females vigorously. Subdominant, or satellite, males may sneak spawn—that is, they will approach a spawning pair from downstream and attempt to fertilize some eggs. Large females can lay over 4,000 eggs, but the average fecundity is 2,400–3,100 eggs per female.

Fertilized eggs are 6.0–9.5 mm in diameter and hatch after about 2–6 months of incubation, usually from December to February. Alevins are 20–24 mm long at hatching and grow to 30–35 mm while in the gravel; they absorb their yolk sac in 30–50 days and then emerge from the gravel. Fry in streams measure 30–70 mm TL, depending on the distance they migrate from the spawning grounds to the estuary. Fry typically emerge from the gravel at night and immediately migrate downstream. Migration is mainly nocturnal in some river systems, but they may migrate during daylight in other areas. Fry do not school as strongly as do pink or sockeye fry, and they are attracted to the shade or darkness of aquatic vegetation.

Fry may not feed in fresh water if their downstream migration is short; if they are in fresh water for a lengthy period, they consume small crustaceans and insects, with chironomid larvae being of particular importance. In estuarine and nearshore marine areas they take epibenthic prey, such as harpacticoid copepods and gammarid amphipods. As they move into deeper water and grow larger, calanoid copepods, hyperiid amphipods, crustacean larvae, larvaceans, euphausids, pteropods, and fishes become part of the diet. The major prey of chum salmon in the ocean are gelatinous zooplankton (jellyfish, ctenophores, and salps), and they presumably are the reason chums have a large baglike stomach, unique among salmonids (13).

Status IB. Endangered in California. Chum salmon are the second most numerous salmon in the North Pacific region, but they are in long-term decline in their southern range. In California they are increasingly rare, although they have probably always been uncommon. Most populations, including the one in the Sacramento River, have been extirpated. Today they occur sporadically and in very low numbers. The three rivers in which there seems to be some evidence of annual or nearly annual spawning (Smith, Klamath, and Trinity) probably support populations of 10–50 fish each, and it is doubtful that they will be viable in the long run.

Nevertheless, chum salmon continue to appear in California's rivers. For example, in the 1980s and 1990s perhaps 20 chum salmon were seen at Nimbus Fish Hatchery on the American River or caught by fishermen in the upper American River (17). There are no recent records of chums observed during stream surveys in the northern Sacramento River drainage (18) or in the San Joaquin drainage (19).

The historical uncommonness of chum salmon in California makes it difficult to identify factors that have negatively affected their abundance. Chum salmon in general do not migrate far upriver in the southern part of their range (2), and the lower reaches of coastal California streams are often the most degraded reaches. Habitat deterioration of spawning areas from logging, road building, mining, and other factors has certainly contributed to population decreases.

If chum salmon are to exist in California, regular surveys of the South Fork Trinity, Klamath, and Smith Rivers are needed to determine the status of the few fish spawning. The exact timing and place of spawning must be determined. Suitable habitat, flow, and water quality should be maintained in order to protect and enhance *all* the imperiled salmonids (including summer steelhead) in those rivers. Once key spawning areas are known, specific plans for enhancing populations should be established.

References 1. Stearley and Smith 1993. 2. Salo 1991. 3. Jordan and Gilbert 1882. 4. Eigenmann 1890. 5. Pauley et al. 1988. 6. Monaco et al. 1991. 7. U.S. Commission for Fish and Fisheries 1892. 8. Hallock and Fry 1967. 9. J. Waldvogel 1988 and pers. comm. 10. Eschmeyer et al. 1983. 11. Messersmith 1965. 12. Scofield 1916. 13. Welch 1997. 14. P. Foley, pers. comm. 1996. 15. T. Mills, CDFG, pers. comm. 1996. 16. T. Kisanuki, USFWS, pers. comm. 1996. 17. R. Ducey, CDFG, pers. comm. 1996. 18. R. Painter, CDFG, pers. comm. 1996. 19. M. Paisano, CDFG, pers. comm. 1996. 20. Rogers 1974.

Rainbow Trout, *Oncorhynchus mykiss* (Walbaum)

Identification Rainbow trout are highly variable in color, body shape, and meristic characters. Nevertheless, adults can usually be recognized as silvery trout with numerous black spots on the tail, adipose fin, dorsal fin, and back (best developed anteriorly) and an iridescent pink to red lateral band. The spots on the tail are typically in radiating lines. The cheeks (opercula) are also pinkish, the back iridescent blue to nearly brown, the sides and belly silver, white, or yellowish. Resident stream forms are generally darker than lake or sea-run forms. The mouth is large, the maxillary bone usually extending behind the eye, with well-developed teeth on the upper and lower jaws, head and shaft of the vomer, palatines, and tongue. Basibranchial teeth are absent. The dorsal fin has 10–12 principal rays; the anal fin, 8–12 principal rays; the pelvic fins, 9–10 rays; and the pectoral fins, 11–17. The tail is slightly forked. There are 16–22 gill rakers on each arch and 9–13 branchiostegal rays. The scales are small, with 110–160 pored scales along the lateral line, 18–35 scale rows above the lateral line, and 14–29 scale rows below it.

The coloration of young is similar to that of adults except that they also have 5–13 widely spaced, oval parr marks centered on the lateral line; the interspaces are wider than the parr marks. Juveniles also possess 5–10 dark marks on the back between the head and dorsal fin, white to orange tips on the dorsal and anal fins, and few or no black spots on the tail. Adults from small streams may retain the color patterns of parr.

Taxonomy Rainbow trout are the most abundant and widespread native salmonid in western North America. They are successful because they have adapted to a wide variety of habitats (including fish hatcheries) and are flexible in life history patterns. As a result many local populations are distinctive and have been awarded taxonomic recognition. Variation, however, is often considerable. Distinctive characters, especially colors, are often in part phenotypic responses to local conditions and may be lost if fish are transferred to another habitat (1). Mixing of rainbow trout in hatcheries and indiscriminate planting have further blurred distinctions among populations, especially in California.

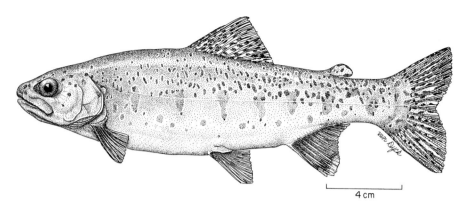

Figure 92. Rainbow trout, 21 cm SL, California.

Nevertheless, molecular techniques are increasing our ability to recognize common ancestry and evidence for the genetic basis of differences among populations. Major controversies center mainly on giving subspecies designations to various evolutionary groups in nonanadromous populations and in designating ESUs in anadromous populations (steelhead) and their resident derivatives. Such designations are, for better or worse, important for protecting distinctive trout and their habitats.

The complex nature of rainbow trout systematics is reflected in attempts to synthesize existing knowledge (2, 3). Rainbow trout and cutthroat trout have generally been thought to be more closely related to one another than either is to Pacific salmon species, because of similarities in appearance and life histories (60). However, rainbow trout are more closely related to salmon species than they are to cutthroat trout, which have more ancestral characteristics than rainbow trout or Pacific salmon (3, 66). Like cutthroat trout, rainbow trout are prone to becoming isolated in headwater areas or in streams distant from the ocean and rapidly evolving distinctive forms. Occasionally natural barriers break down and isolated forms reunite with the main gene pool, resulting in "hybrids" that confuse taxonomists. Of course, the process of barrier breakdown has been accelerated by human activity. Behnke (2) thinks that, prior to modern disruption of rainbow trout gene pools, there were three distinct groups: redband trout of the upper Columbia and Fraser River basins, redband trout of the Sacramento–San Joaquin River drainage, and coastal rainbow trout. Redband trout is the general designation given to native, mostly resident, forms in interior basins, whereas coastal rainbow trout is used to refer to all anadromous coastal forms and their recent resident derivatives, from Alaska to Baja California. All three groups are represented in California: Columbia redbands by hatchery introductions (Kamloops rainbow), native redbands by golden trout and a number of other forms, and coastal rainbows by steelhead and most rainbows in coastal and Central Valley streams.

There are many unresolved questions about the relationships and identity of native rainbow trouts in California (73), but the following classification seems to make sense from a conservation perspective. Probably the weakest part of this classification is the recognition of six distinct groups (ESUs) of steelhead in California, based on analysis by NMFS of genetic and life history data. Given the genetic diversity of steelhead populations (73), more groups are likely to be recognized in the future. However, recognition of these six groups as distinct stocks of steelhead or rainbow trout does help to conserve diversity in life history adaptations. All forms listed as "steelhead" have nonmigratory populations in their watersheds.

Coastal rainbow trout, *O. m. irideus*
    Klamath Mountains Province steelhead
        Klamath winter steelhead
        Klamath summer steelhead
    Northern California steelhead
        North Coast winter steelhead
        North Coast summer steelhead
    Central Valley steelhead
    Central Coast steelhead
    South/Central Coast steelhead
    Southern steelhead
Redband trout
    Upper Kern redband trout
        California golden trout, *O. m. aguabonita*
        Little Kern golden trout, *O. m. whitei*
        Kern river rainbow trout, *O. m. gilberti*
    Upper Sacramento redband trout
        McCloud River redband trout, *O. m. stonei*
        Goose Lake redband trout, *O. m. subsp.*
        Eagle Lake rainbow trout, *O. m. aquilarum*

Klamath Mountains Province steelhead. This is an ESU recognized by NMFS that includes coastal rainbow trout from the Elk River in Oregon through the Klamath and Trinity Rivers. Rainbow trout in this region are linked by common traits in their genetics and chromosomes (4) and also by the presence of an unusual life history form: half-pounders, immature steelhead that return from the sea to overwinter in fresh water (see Life History). There are two distinct spawning types: winter (ocean-maturing) steelhead and summer (stream-maturing) steelhead. Winter steelhead typically move upstream between November and April and spawn fairly soon after their arrival on the spawning grounds. Summer steelhead migrate between late April and June and spend summer in deep pools in canyons, spawning in December–April. They are distinguished from winter steelhead by time of migration (5), the immature state of the gonads at migration (6), and the location of spawning areas (5, 7). Attempts to distinguish juvenile summer and winter steelhead and resident juvenile rainbow trout using otolith nuclei widths, scale circuli densities, and visceral fat content have been only partially successful (8, 9). The temporal and spatial isolation of spawning fish from winter steelhead serves to maintain genetic differences (77), although summer steelhead in this ESU and in others are more closely related to winter steelhead in their watershed than to summer steelhead in other ESUs.

Northern California steelhead. These steelhead make up another ESU recognized by NMFS, which includes all trout in streams from Redwood Creek (Humboldt County) to the Gualala River (Sonoma County), including the Eel River. This ESU is closely related to the Klamath Mountains Province ESU and contains both winter and summer steelhead and, apparently, half-pounders as well. There are no summer steelhead south of Matolle River. The differences between summer and winter steelhead are the same as those discussed for Klamath Mountains Province steelhead.

Central Valley steelhead. Rivers of the Central Valley contain only winter steelhead, although summer steelhead may have been present prior to the construction of large dams (61). The winter run might be better termed the "fall run" because they start entering fresh water in August, with a peak in late September–October, after which they hold until flows are high enough in tributaries to enter for spawning (62, 79). This ESU does not include steelhead in streams tributary to San Francisco and San Pablo Bays. Nonmigratory populations of rainbow trout not of hatchery origin belong to this ESU, as would populations of presumably "residualized" steelhead that live in reservoirs above major dams and migrate into tributary streams.

Central California coast steelhead. This is an ESU that includes coastal populations of winter steelhead from the Russian River south to Aptos Creek (Santa Cruz County), including fish in streams tributary to San Francisco and San Pablo Bays.

South/Central coast steelhead. This ESU comprises winter steelhead populations found in three tributaries to Monterey Bay (Pajaro, Salinas, and Carmel Rivers), in the small streams of the Big Sur Coast, and small intermittent streams of San Luis Obispo County, south to Point Conception.

Southern California steelhead. The Southern California steelhead is one of the most distinctive steelhead ESUs in terms of both genetics and life history (10). It basically includes all populations south of Point Conception, historically into Baja California. Curiously, the ESU shows not only unique genetic characteristics but also high genetic diversity, suggesting that it developed from a population that survived in a Baja California refuge during the Pleistocene and that has recently come into contact with steelhead of more northern origin (76). Its high diversity may help to explain the remarkable capacity of this ESU to persist in seemingly unfavorable environments (76, 78).

The southernmost anadromous populations today appear to be in Malibu Creek (Los Angles County) and San Mateo Creek (San Diego County) (61). Southern steelhead are winter-run steelhead that persist in streams whose lower reaches flow through coastal plains, which present substantial migration passage problems to and from distant headwater spawning and rearing habitats. These reaches are essentially passable only when winter rains create substantial flows for short periods. Their occurrence in such a demanding environment requires distinctive ecological and physiological adaptations.

Resident rainbow trout. This is simply a catchall designation for hundreds of nonanadromous wild rainbow trout populations that exist throughout California and that are either derived naturally from steelhead (and would therefore be part of the appropriate ESU) or, more likely, are of mixed hatchery and native origin. It has no validity as a taxonomic unit except to indicate the presumed mongrel nature of most rainbow trout populations.

Upper Kern redband trout. This group of three subspecies is treated separately in the section of this book on golden trout. They are most closely related, among rainbow trout, to redband trout in the McCloud and Pit Rivers (65). The reasons for separate treatment are related to history, convenience, and interest, not taxonomy.

McCloud River redband trout. These trout are native rainbows with brick-red bands on their sides that live in the McCloud River watershed (Fig. 93, p. 283). They have a long and cloudy taxonomic history, confused by the introduction of nonnative strains of rainbow trout into the system and by past natural (if limited) connections to Sacramento River populations (73). The name *stonei* was used by D. S. Jordan to designate resident redband trout found in the upper Sacramento River drainage but mainly in the upper McCloud River (11), whereas Behnke (2) uses it as a name of convenience for all redband trout populations, of multiple origins, in the McCloud and Pit drainages. To complicate matters further, Sheepheaven Creek, a tiny tributary to the

McCloud River, contains a distinctive population of red-band trout (12, 13, 65) that may deserve separate subspecific designation on its own. Regardless, rainbow trout in the upper McCloud watershed are a collection of isolates that deserve special recognition and protection (65), reflecting evolutionary responses to a complex and changing environment. Probably there are one or more distinctive redbands in tributaries to the Pit River as well, but most populations have hybridized with coastal rainbows, so taxonomic recognition may not be possible.

Goose Lake redband trout. These are genetically distinct (13, 65) redband trout that persist in the tributaries to Goose Lake, as well as in the lake itself. Behnke (2) suggests that Goose Lake redbands may be most closely related to redband trout of nearby Warner Basin, Oregon. However, genetic studies indicate a closer relationship to McCloud River redbands (65) with some genetic connections to Sacramento River coastal rainbows (72). Regardless of their complex ties to other trout populations, there is adequate evidence to regard Goose Lake redband trout as a distinct ESU.

Eagle Lake rainbow trout. J. O. Snyder (14) described this trout as a subspecies of rainbow trout. Needham and Gard (1), pointing out that all other native Great Basin trout populations are cutthroat trout, suggested that Eagle Lake rainbow trout were descended from introduced or immigrant rainbow trout from the Feather or Pit River drainages. Busack et al. (15), in an extensive electrophoretic, karyotypic, and meristic analysis, found that, even though the Eagle Lake trout is electrophoretically close to both coastal and redband rainbow trout and is meristically most similar to coastal rainbow trout, its karyotype (of 58 chromosomes) is like that of redband trout. They suggested that Eagle Lake rainbow trout are derived from immigration or unrecorded introduction of "a rainbow trout with 58 chromosomes" (p. 423). However, molecular evidence demonstrates that they are most closely related to other California redband trout (65, 72). Rainbow trout presumably colonized Eagle Lake via upper Pine Creek and an ancient connection with a headwater tributary of the Pit River.

Hatchery strains. Hatchery strains of rainbow trout are typically of mixed origins because of intense selection for traits favorable for hatchery production, such as rapid growth under crowded conditions, resistance to disease, and high fecundity (21). Such strains are true domestic animals, as distinct from their wild ancestors as cows and horses are from their ancestors. Some hatchery strains, however, are maintained with little or no crossing with other strains, although they are still highly domesticated. Kamloops rainbow trout (a strain of *O. m. gairdneri*) from British Columbia were imported repeatedly from 1950 to 1986 and reared in California hatcheries because of their reputation for fast growth, large size, and ease of catch.

Kamloops rainbow are still widely planted in lakes and reservoirs, especially in high-elevation lakes, although some lines have been hybridized with California rainbows (21). Another hatchery strain usually kept relatively pure is one derived from Eagle Lake rainbow trout. This strain does especially well in alkaline lakes and reservoirs and can attain large size faster than other strains because it matures at a later age.

Other rainbow trout. Royal silver rainbow trout, supposedly native to Lake Tahoe and now extinct, present a zoogeographic puzzle similar to that of Eagle Lake rainbow trout. However, there is little reason to doubt that J. O. Snyder's 1912 description of "*Salmo regalis*" was based on large-size rainbow trout derived from fish introduced in the 1860s and 1870s. His royal silver rainbow trout differs little from rainbow trout of known origin that have grown to large sizes in other large lakes. A similar situation exists in regard to the legendary emerald trout "*Salmo smaragdus*" of Pyramid Lake, Nevada (2).

Names The scientific name of rainbow trout has a long and esoteric history that resulted in a sudden shift to *O. mykiss* after nearly 150 years of calling it *Salmo gairdneri* or *S. irideus* (16). The rainbow trout was first described from Kamchatka populations in 1792, by J. J. Walbaum, as *Salmo mykiss*. In 1836 J. Richardson described steelhead from the Columbia River as *S. gairdneri*. In 1855 Gibbons described juvenile steelhead from San Leandro Creek (now buried in Oakland) as *S. iridea*. Subsequently North American biologists and anglers tended to refer to resident rainbow trout as *S. irideus* and steelhead as *S. gairdneri*, until they gradually recognized that steelhead and resident rainbow trout were really the same species. Although *S. irideus* faded from use, the name persisted in the common name "rainbow trout" (*irideus* means rainbow). Resident rainbow trout were originally called "brook trout" but began being called rainbow trout after the introduction of "true" brook trout from the eastern United States. More recently Behnke (2) has resurrected *irideus* as the subspecific name for coastal rainbow trout.

Meanwhile North American scientists, after some initial confusion as to which fish the name *S. mykiss* applied to (2), generally refused to recognize Russian rainbow trout as belonging to the same species as the one here, presumably for patriotic reasons. With the end of the cold war, the oldest species name for rainbow trout, *mykiss*, was finally applied to North American forms. At the same time, systematic work indicted that rainbow trout were more closely related to Pacific salmon, *Oncorhynchus* spp., than to Atlantic salmon and trout, *Salmo* spp., hence the name *O. mykiss*. *Mykiss* is another one of the transformed Kamchatkan common names for salmonids (see the discussion in the account of coho salmon for this and other names).

Somewhat ironically, it is likely that rainbow trout originated in North America and invaded Eurasia during the Pleistocene period (2).

Distribution Rainbow trout were originally native to Pacific coast streams from the Kuskokwim River in Alaska down to streams in Baja California. The southernmost population is *O. m. nelsoni,* a redband trout isolated in the Rio Santo Domingo in the mountains of Baja California (67). In the Columbia River drainage they were found throughout Oregon, Washington, and British Columbia and penetrated as far inland as major falls on the Snake River in Idaho. The easternmost populations are found in the Athabasca River in Alberta, the result of a stream capture from the Fraser River system of British Columbia (2). In Asia rainbow trout are native mainly to the north Pacific coast south of the Kamchatka Peninsula. In salt water, steelhead are found throughout the North Pacific ocean.

In California coastal rainbow were originally found in all permanent streams from San Diego County north to the Klamath River drainage. It is uncertain if anadromous fish once present in the upper Klamath basin were coastal rainbows or anadromous redband trout (2); prior to 1917 steelhead spawned in the tributaries to Upper Klamath Lake, Oregon (24). Coastal rainbows are also native to the Sacramento–San Joaquin system, including Pit and lower McCloud Rivers, where they hybridized naturally with redband trout. Most Central Valley streams probably originally contained steelhead in reaches readily accessible from the ocean and resident populations above barriers or in less accessible streams, such as those that historically emptied into Tulare and Buena Vista Lakes in the San Joaquin Valley.

The distribution of various steelhead ESUs is given under Taxonomy. Of special interest is the distribution of the life history variety known as *summer steelhead* (17)—although it might more accurately be called "stream-maturing steelhead" (61). Summer steelhead runs have been recorded from the Middle Fork Eel, mainstem Eel, Van Duzen (tributary to the Eel), Mattole, Mad, North Fork Trinity, New (tributary to the Trinity), and South Fork Trinity Rivers, as well as Canyon Creek (in the Trinity system), the Klamath River drainage (Dillon, Elk, Indian, Red Cap, Bluff, and Clear Creeks), the Salmon River, Wooley Creek (tributary to the Salmon), Redwood Creek, and the Smith River (5). Up to 50 percent of California summer steelhead are concentrated in the Middle Fork Eel River. Other Eel River populations (North Fork Eel, Black Butte River, Woodum Creek, and Larabee Creek) are now gone (17).

Redband trouts, including three golden trouts and Eagle Lake trout, occur in isolated places at the edges of the coastal rainbow range in the Sacramento–San Joaquin basin. Their distribution, as indicated under Taxonomy, has been fragmented and confused by the introduction of hatchery rainbow trout. *Goose Lake redband trout* are endemic to Goose Lake and its major tributaries (Lassen and Willow Creeks in California and the extensive Thomas Creek system and Crane Creek in Oregon) as well as to smaller streams, such as Cottonwood and Pine Creeks in California and Augur, Bauer, Camp, Cox, Drews, Shingle Mill, Snyder Meadow, and Warner Creeks in Oregon. Joseph, Parker, and East Creeks, tributaries of the upper Pit River in California, also contain trout genetically similar to Goose Lake redband (13). *McCloud River redband trout* have been reported from creeks tributary to the McCloud River, such as Sheepheaven, Tate, Edson, and Moosehead Creeks (2, 12, 13, 18) and from the McCloud River above Middle Falls (2, 12, 13, 17, 18). Redband trout from Sheepheaven Creek were transplanted into nearby Swamp Creek in 1972 and 1974 and into Trout Creek in 1977 (68). They are now established in both streams. *Eagle Lake rainbow trout* are endemic to Eagle Lake, Lassen County, and its main tributaries, Pine and Papoose Creeks (15, 17). They have been planted in numerous waters throughout California from hatchery stocks originating from trout captured annually at the Pine Creek egg collecting station and from domestic brood stock. The trout have also been exported to other states and to Canada. It is unlikely that naturally reproducing populations of pure Eagle Lake trout are present in any of these planted waters.

Rainbow trout have been introduced into coldwater streams throughout most of the world. They are now present in South and Central America, Africa, Asia (including India), Europe, New Zealand, Australia, Papua New Guinea, Tasmania, Hawaii, and Réunion (in the Indian Ocean) (19). In North America they are found in every state and province that has cold waters. Likewise, they are probably the most widely distributed fish in California; their natural distribution has been greatly expanded by transplants into most coldwater streams and lakes, including many waters that were originally fishless. The California rainbow trout gene pool has likewise been expanded by introductions of trout from British Columbia and elsewhere (22). They are the principal species raised in California trout hatcheries and are widely stocked even where they cannot reproduce. Many, if not most, wild rainbow trout populations around the world had their origins in California. Supposedly, most of these fish originated from the lower McCloud River, where Baird Hatchery in the late 1800s produced large numbers of fertilized rainbow trout eggs for export; these fish were apparently hybrids of steelhead and resident fish. However, in New Zealand at least, most of the trout apparently originated from Sonoma Creek, Sonoma County (20).

Life History Few, if any, fishes have been as intensively studied as rainbow trout. There are many reasons for this phenomenon, but the most prominent are *(1)* their worldwide distribution in cold waters; *(2)* their ease of culture, thanks to which they are readily available as experimental animals; *(3)* their significant value for aquaculture and fisheries; *(4)* their diversity of life history strategies; and *(5)* their mystique among anglers, who support studies or become fish biologists themselves. This account is far from comprehensive, focuses on California populations, and relies heavily on my personal experiences of working with this fish.

The life history patterns of California rainbows are both variable and flexible. Two basic patterns are migratory life history and resident life history; both types often exist in the same population, but dominance of one or another is frequently a defining trait for a population. Migratory rainbows are either sea-run (anadromous), lake-run (limnodromous), or within-river (potadromous) migrators. In California most lake-run fish are derived from steelhead in reservoirs behind impassable dams. However, Eagle Lake trout migrate into the headwaters of Pine Creek, and Goose Lake redbands migrate into tributaries of Goose Lake.

Steelhead have two basic life history patterns, winter and summer. *Winter steelhead* enter streams from the ocean when winter rains provide large amounts of cold water for migration and spawning. They typically spawn in tributaries to mainstem rivers, often ascending long distances. They return to the ocean after spawning, if possible. *Summer steelhead* (also known as spring-run steelhead) typically enter rivers as immature fish during receding flows of spring and migrate to headwater reaches containing deep pools. They spend summer in these pools, where they mature to spawn in winter or spring.

In the ocean the distribution of different steelhead stocks is poorly known, but it is likely that most California fish, especially those from southern California, do not wander far from the California coast. Some populations of steelhead have an additional variant in their life history pattern, the *half-pounder*. These are immature fish, measuring 25–35 cm FL, that overwinter in fresh water after spending a summer in the ocean (23). In large rivers some steelhead, mainly small males, move only as far as the river but return to tributary streams for spawning. In contrast, resident rainbow trout often spend their entire lives in a few hundred meters of stream, although some may migrate considerable distances within a stream system to find suitable spawning grounds. It is likely that most resident populations of trout produce individuals that are prone to wander more than others, helping to maintain gene flow among populations and reestablishing populations that have become extinct. For example, when Goose Lake (Modoc County) dried up in 1992, runs of large redband trout that lived in the lake disappeared. The only Goose Lake redband trout remaining were resident in small alpine tributaries that flowed into the lake and were above the reach of lake spawners. After the lake refilled the migratory lake population reestablished itself, presumably from fish dispersing from upstream resident populations.

Regardless of life history strategy, for the first year or two of life rainbow trout are found in cool, clear, fast-flowing permanent streams and rivers where riffles predominate over pools, where there is ample cover from riparian vegetation or undercut banks, and where invertebrate life is diverse and abundant. In streams, there are strong shifts in habitats with size and season: the smallest fish are most often found in riffles; intermediate size fish, in runs; and large fish, in pools (74, 75). In smaller streams larger trout often migrate to large rivers, lakes, or the ocean. A key characteristic of all these habitats is cool temperatures. Rainbows are found where daytime temperatures range from nearly 0°C in winter to 26–27°C in summer, although extremely low (<4°C) or extremely high (>23°C) temperatures can be lethal if the fish have not previously been gradually acclimated. Even when acclimation temperatures are high, temperatures of 24–27°C are invariably lethal to trout, except for very short exposures (25, 26). Thus juvenile steelhead disappeared from a section of Big Sulphur Creek (Mendocino County) when hot springs caused summer temperatures to rise above 26°C for extended periods (27). For large trout, especially adult steelhead, lethal temperatures are usually around 23–24°C.

When temperatures become stressful in streams, survival requires trade-offs. Juvenile steelhead, faced with the increased energetic costs of living at high temperatures, will move into fast riffles to feed because food is most abundant there, even though there are additional costs associated with maintaining position in fast water (28). In Sespe Creek, Ventura County, where summer temperatures regularly exceed 27°C, trout seek out the bottoms of pools where springs keep temperatures lower (17–21°C) during the day; however, these same areas may have low, potentially lethal, levels of dissolved oxygen (29). In the Eel River, mass mortality of juvenile steelhead was observed after large numbers sought out relatively small spring areas to avoid lethal temperatures and presumably depleted the oxygen (69). At high temperatures rainbows are also much more vulnerable to unusual stress (e.g., being caught by an angler) and likely to die as a consequence. When temperatures are high for trout but optimal for a coexisting fish species, interactions may reduce trout growth (63).

The optimal temperatures for growth of rainbow trout are around 15–18°C, a range that corresponds to temperatures selected in the field when possible (30). Thus in a section of the Pit River containing a thermal plume from an inflowing cold tributary, rainbow trout selected temperatures of 16–18°C (30). However, many factors affect choice of temperatures by trout (if they have a choice), including the availability of food. Under the fluctuating conditions present in most streams in summer, the mean daily temperatures optimal for growth are likely to be 2–3°C lower than those under more constant conditions (25). The optimal temperatures for fry may also be somewhat lower than those for juveniles.

At low temperatures rainbow trout survive oxygen concentrations as low as 1.5–2.0 mg/liter, but normally concentrations close to saturation are required for growth. Activity is reduced as oxygen concentration drops, even at low temperatures (31). Their tolerance of varying chemical conditions of water is also broad. They can live at pH values from 5.8 to 9.6. All other factors being equal, their best growth seems to be achieved in slightly alkaline waters (pH 7–8), although Eagle Lake trout have adapted to highly alkaline waters (pH 8.4–9.6).

In streams different sizes of rainbow trout show distinct preferences for different microhabitats as defined by depth, velocity, substrate, and cover (32, 33, 34, 74, 75). Fry (<50 mm SL) typically concentrate in shallow (<50 cm) water along stream edges, where water column velocities are low (1–25 cm/sec). Juveniles (50–120 mm SL) occur in deeper (50–100 cm) and faster (10–30 cm/sec) water, usually among rocks or other cover. Larger fish seek out a wide variety of deeper habitats (often including "pockets" behind rocks, runs, or pools) but typically stay close to fast water capable of delivering drifting invertebrates to them, such as inflowing water at the head of pools. Adult trout increase their foraging efficiency by moving into high-velocity water only to feed and then quickly returning to low-velocity areas for holding.

Predators have a strong effect on microhabitats selected by rainbow trout. Small trout select places to live based largely on proximity to cover in order to hide from both avian predators (kingfishers, mergansers, herons) and predatory fish. Birds are a threat primarily either in shallow water or near the surface, whereas predatory fish (including large trout) approach from deep water. In the Eel River, for example, the invasion of predatory pikeminnows resulted in a dramatic shift from juvenile trout being present at a wide range of depths to being present mainly in riffles too shallow for large pikeminnow foraging (37). Thus mean depth dropped from 70 to 39 cm and mean water column velocity increased from 19 to 44 cm/sec.

Even though rainbow trout are the only fish species found in many California streams, more often than not they occur with other salmonids (especially juvenile coho and chinook salmon in coastal streams and brown trout in interior streams), as well as with sculpins, suckers, and one or two species of minnows, such as speckled dace or California roach. It is unusual, however, to find more than three to four other species in abundance in streams where rainbow trout are common. They interact successfully with other species, rarely competing with nonsalmonids but often dominating other salmonids. Thus a study of interactions between Sacramento suckers and rainbow trout produced no evidence of competition (32). To the contrary, juvenile trout will follow large suckers around as they browse on the bottom and pick up invertebrates disturbed by sucker feeding. In coastal streams juvenile steelhead interact with juvenile coho and chinook salmon, with the result that each species selects different microhabitats in complex ways (35, 36). Juvenile steelhead possess more cylindrical bodies, shorter median fins, and larger paired fins than other salmonids with which they co-occur, giving them an advantage when holding or swimming in fast water (58). However, competition among difference size classes of steelhead in different habitats may result in reduced growth of one size class at the expense of another. In degraded streams, where shallow water predominates, abundant small steelhead may suppress the growth of larger ones (64). When alien brown trout and rainbow trout are found in the same stream, adult brown trout tend to select slower areas with undercut banks and other cover, pushing rainbow trout into faster, more open water, where they are more vulnerable to anglers and predators.

One of the main reasons rainbow trout are such successful competitors is that they are highly aggressive and often defend feeding territories in streams (38). Other salmonids recognize aggressive displays of rainbow trout (e.g., rigid swimming, flared operculae, nipping at the caudal peduncle of invading fish) and usually react either by fleeing or by challenging the trout with similar displays, perhaps driving it off its position. The winners of such interspecific contests are determined by a number of factors, but relative size and habitat preferences play leading roles. Aggressive displays are also important in interactions among rainbow trout at a site. Individual trout may set up feeding territories, which they then defend from each other. The number of territories depends on many factors, but probably the most important are size of fish, speed of current, water temperature, and availability of cover. Superimposed on this territorial mosaic, however, is a dominance hierarchy in which large fish are dominant over small fish and hold much larger territories within which small fish are tolerated (38). In pools, where feeding takes place mainly at the inflow, the social structure is much looser, and trout of similar size may shoal when not feeding. In the Eel River I have observed steelhead measuring 20–25 cm FL schooling in big pools with pikeminnows of similar size.

Stream-dwelling rainbow trout feed mostly on drifting aquatic organisms and terrestrial insects, but they will also take active bottom invertebrates. Thus stomachs from a sample of trout taken from one stream at the same time are likely to contain a hodgepodge of terrestrial insects, adult and emergent aquatic insects, aquatic insect larvae, amphipods, snails, and occasional small fish. Individual trout, however, tend to specialize in the organisms on which they feed, even over a long period of time, and do not take the whole range of foods available (39). Diet also changes with size; larger fish tend to take larger prey. Rainbow trout are nevertheless very opportunistic; for example, steelhead juveniles in the Trinity River during an April study were feeding largely on ants (40). In the lower American River, where fluctuating flows from dam releases limit the diversity of benthic organisms, small steelhead fed largely on adults and larvae of small mayflies and chironomid midges (41). Feeding was reduced during a year when water level fluctuations were more extreme.

When water is turbid from sediment, drift feeding is reduced (59). In the McCloud River, which is slightly turbid owing to suspended glacial material, rainbow trout feed mainly on the bottom, and the classic evening "rise" to feed on drifting and terrestrial organisms is often not seen (42). Rainbow trout can feed at any time of day or night, but there are typically feeding peaks at dawn and dusk, when drift levels are still high and there is enough light to see drifting organisms, as well as terrestrial insects that are more active at night. In winter feeding is considerably reduced from summer levels, and trout feed mostly on bottom-dwelling invertebrates.

In lakes feeding varies with the availability of prey. Although benthic invertebrates and zooplankton seem to be preferred, terrestrial insects are eaten when other foods are scarce. In Eagle Lake, Lassen County, in June and July, even large trout (30–50 cm SL) will often be found with stomachs full of zooplankton, although others will be filled with leeches, caddisflies, or amphipods. Later in the season they may switch to feeding on small fish, especially the abundant tui chubs. In general, rainbow trout in lakes eat more fish than do stream-dwelling rainbows, although fish normally do not become an important part of the diet until the trout reach 30–35 cm TL. In reservoirs rainbow trout achieve rapid growth on planktivorous fishes such as threadfin shad and wakasagi. As in streams, feeding is most intense during summer but can continue throughout winter at temperatures as low as 1°C (43).

After steelhead leave their home streams, they feed on estuarine invertebrates and marine krill, but as they increase in size fish gradually become more important to their diet. The large size and rapid growth achieved by steelhead can be attributed in large part to their diet of fish, squid, and crustaceans taken in ocean surface waters (44). In streams adult steelhead feed opportunistically, but most caught by anglers have empty stomachs. However, 95 percent of adult fish in Deer and Mill Creeks, Tehama County, were found to contain food, mainly caddisfly larvae and salmon eggs (80).

Growth rates in nonmigratory rainbow trout depend on temperature, food availability, flow, and trout densities (71). In small, high-gradient streams California rainbow trout typically reach 75 mm FL at the end of their first year, 140 mm at the end of their second year, 190 mm at the end of their third year, and 235 mm at the end of their fourth year (45). In warm, low-gradient streams they may reach 90–100 mm FL in year 1, 150–210 mm in year 2, 210–300 mm in year 3, and 300 mm or more in year 4, although fish older than 3 years are rare (especially in heavily fished populations). In large productive streams, such as the upper Sacramento River, fish may reach 140–150 mm FL in year 1 and measure 380–400 mm by year 4, growing 30–50 mm/year in subsequent years as they feed increasingly on fish (45). In the neighboring McCloud River, which is cooler, growth rates are similar to those in small headwater streams, but fish may reach 30–35 cm FL by living 6 or 7 years (46).

Growth of steelhead in fresh water is also highly variable, but sizes of 10–12 cm FL at the end of year 1 and 16–17 cm at end of year 2 are fairly typical in larger streams where food is abundant. In small California streams with low summer flows, steelhead usually measure 5–9 cm FL at the end of their first summer and 10–16 cm at the end of their second summer. If summer flows are higher and food is abundant, they may reach 10–20 cm FL in their first year. An additional spurt of growth may occur in spring, just prior to smolting (70), giving smolts age 1 and above an additional size advantage. Steelhead smolts migrate out to sea at 1–3 years of age, at 10–25 cm FL. After 1–2 years at sea they return at 35–65 cm (1.4–5.4 kg) (47).

In alpine lakes and reservoirs rainbow trout reach 10–16 cm FL in their first year, 13–20 cm in their second, and 19–22 cm in their third. In such lakes they seldom live longer than 6 years or grow to more than 40 cm FL. In Eagle Lake trout are raised in a hatchery for 18 months until they reach 30–40 cm FL and then planted in the lake. Trout measuring 43–46 cm TL are 2 years old, and those 46–56 cm are 3 years old. Similar growth is achieved by fish planted as fingerlings in some reservoirs (e.g., Crowley Reservoir, Mono County), but generally it is somewhat slower, especially after the first year. The largest known nonsteelhead rainbow trout (from Jewel Lake, British Columbia) weighed 23.9 kg (48), although the largest one caught by angling (from Lake Pend Oreille, Idaho) weighed 16.8 kg. The largest such fish from California (Feather River) weighed 9.6 kg; the largest California steelhead known (Smith River) weighed 12.4 kg (caught in 1976). The largest steelhead on record, from Alaska, weighed 19.1 kg (48). The oldest rainbow trout known are those from Eagle Lake; they once reached 11 years before the population became supported by hatchery fish. Steelhead occasionally live for 9 years, but in general rainbow trout 6–7 years old are unusual.

Most nonanadromous rainbow trout mature in their second or third year, but the time of first maturity varies from the first to the fifth year. Mature fish can be of any size from about 13 cm FL on up. Most wild rainbow trout are spring spawners, from February to June, but low temperatures in high mountain areas may delay spawning until July or August. In some streams in the Bay-Delta region, such as Putah Creek below Monticello Dam, spawning takes place in December.

For steelhead, age at maturity depends on the combination of years in fresh water (1–3 years) plus years at sea (1–4 years). In their classic study of steelhead in Waddell Creek, Santa Cruz County, Shapovalov and Taft (6) identified 32 different freshwater-saltwater combinations, but most fish were of four types: 2/1 (30%), 2/2 (27%), 3/1 (11%), and 1/2 (8%). The relative abundance of these types varies from river to river. In the lower Klamath River over half the spawners are 2/2, the percentage increasing in tributaries, with the added wrinkle that most return to fresh water as half-pounders as well (47). In addition, in the Klamath and a few other North Coast drainages, there are runs of both winter and summer steelhead, with the latter fish coming in while still immature and delaying spawning for 8–10 months (5). To make matters even more complicated, small precocial jack males that may have spent only a few months at sea, or not gone to sea at all, are present in most steelhead populations. This variability in life history strategy presumably allows steelhead to maintain their abundance and genetic diversity in the face of high variability in both ocean and stream conditions, and allows them to use a wide variety of stream habitats.

California winter steelhead enter coastal streams after rains increase flows, which in turn breach sandbars on mouth lagoons and permit passage through lower reaches. Fish may move upstream any time during the period December–March, although the peaks for such activity are typically in January and February. Summer steelhead seem to enter streams as flows taper off in spring and spawn the following winter. Steelhead and other rainbow trout have well-developed homing abilities and usually spawn in the same stream and area in which they had lived as fry. This means that races and runs of trout develop that are adapted to local conditions. Summer steelhead, for example, prefer holding in deep (3 m or more), cold (10–15°C) pools during summer, but they sometimes persist even when temperatures reach 25–27°C for short periods of time. These fish are also capable of spawning in tributaries that dry up during summer, because fry emigrate soon after hatching.

As in most other salmonids, the female digs a redd with her tail, usually in the coarse (1–13 cm diameter) gravel of the tail of a pool or in a riffle. Water velocities over redds are typically 20–155 cm/sec, and depths are 10–150 cm. Mating behavior between a pair of large adult fish is similar to that of other salmonids but is complicated by the presence of other males, which sneak in to spawn along with the mated pair. In steelhead, the sneaker males can range from small parr (15–20 cm FL) that have probably never been to sea, to jacks, to slightly smaller subordinate sea-run males, kept at

bay by the aggressive attacks of the dominant male. Mature parr can spawn with females even if a large male is absent (6), a strong indication that this is indeed a successful alternative way to be a male. Both resident rainbows and steelhead can spawn annually, but it is not unusual for fish to skip a year between spawns. After spawning spent steelhead often move gradually downstream and hang out in pools for periods of time during the downstream migration. In Waddell Creek females seem to move downstream fairly quickly after spawning while males tend to linger for the chance of additional spawning; as a result the weight loss of both sexes is similar by the time they return to the ocean (70). Steelhead can spawn up to four times, but mortality rates between fish of succeeding ages are high, typically 50–75 percent, so that very few fish spawn so often.

The number of eggs laid per female depends on size and origin but ranges from 200 to 12,000 eggs. Rainbow trout measuring under 30 cm TL typically contain fewer than 1,000 eggs; steelhead contain about 2,000 eggs per kilogram of body weight.

The eggs hatch in 3–4 weeks (at 10–15°C), and fry emerge from the gravel 2–3 weeks later. The fry initially live in quiet waters close to shore and exhibit little aggressive behavior for several weeks.

Status IA–E, IID. It is ironic that whereas rainbow trout are probably the most widely distributed fish in California, many of their distinctive populations are in danger of extinction. Their wide distribution is largely a result of two factors: *(1)* the ease with which they are raised in hatcheries and then planted to support fisheries and *(2)* planting of rainbow trout in coldwater streams outside their native range. When fish of hatchery origin dominate, native strains tend to disappear. This section therefore deals first with hatchery trout and then with native forms.

Hatchery rainbow trout. The term *hatchery rainbow trout* refers to any resident trout that has spent part of its life cycle in a hatchery; most such fish are thoroughly domesticated, having been the result of 50–100 generations of selection for life in hatcheries. Hatcheries were developed to support trout fisheries because rainbow trout are the most popular game fish in California. After World War II the perception of fisheries agencies was that demand for trout was far beyond the natural reproductive capacities of wild populations, especially when so many trout streams had been altered by dams. Therefore CDFG began devoting a considerable portion of its fishing license revenues to rearing domestic trout for planting on a put-and-take basis, a practice that still continues. (Put-and-take trout are raised to be caught as quickly as possible.) Most trout planted measure 18–30 cm TL and are caught within 2 weeks of planting (49). This is fortunate because hatchery-raised rainbows are ill adapted for survival in streams and are likely to die of starvation or stress within a few weeks if not caught. If large numbers are planted in streams with wild trout populations, their sheer numbers are likely to disrupt established hierarchies, making wild fish more vulnerable to angling. Such streams generally must be continually planted if any sort of trout fishery is to be sustained because neither wild nor domestic trout can maintain themselves very easily. The increasing popularity of catch-and-release fishing has resulted in many anglers questioning the value of planting domestic trout in streams. One response has been to plant fish mainly in roadside streams subject to heavy angling use or in streams that cannot sustain wild trout.

In lakes the survival rates of planted catchable-size fish are much higher than those in streams because a comparatively low expenditure of energy is required to stay alive (and become adjusted to the environment) in the absence of current. In addition, the trout are less vulnerable to angling and predators. In some reservoirs, such as Crowley Reservoir, food is so abundant and available to hatchery rainbows that they grow very quickly and to fairly large sizes. In hundreds of once-fishless alpine lakes, rainbow trout are planted, usually by airplane, as fingerlings. Despite the low productivity of these lakes, enough trout survive and grow to support back country fisheries. A hidden cost of these aerial transplants has been changes in lake ecosystems, represented by extirpation of some invertebrates and elimination of many populations of the mountain yellow-legged frog (*Rana muscosa*) (55). In many lightly fished lakes planted on a routine basis, trout have established self-sustaining populations, so stocking is not really necessary to sustain fisheries (50). It is clear that the policy of aerial planting of trout in wilderness lakes needs careful evaluation (see also the discussion as part of the account of brook trout).

Despite generally low survival rates of planted trout in streams, a few apparently survive and occasionally interbreed with wild trout. This was probably more generally true in the past, when plants were frequently made of trout of wild origin. Thus indiscriminate planting of rainbow trout has led to loss through hybridization of many populations of rainbow, redband, and golden trout, as well as of cutthroat trout. Only in the past few decades have the aesthetic values of distinctive local populations been officially recognized and large-scale efforts made to conserve them.

Hatchery steelhead. Steelhead have been propagated in California hatcheries since the 1870s based on the idea that hatcheries would greatly increase steelhead numbers. The idea gained in popularity after dams and diversions cut off access to much of their historical upstream habitats (51). Hatchery steelhead have been a mixed blessing, at best. Although they probably maintained the steelhead fishery in the Sacramento River and a few other places, they also allowed habitat protection and restoration to be largely ig-

nored until recently. Often native populations (e.g., that in the Sacramento River) declined despite hatcheries. It is likely that hatcheries often removed more native fish as brood stock from the river than they were able to produce. In response hatcheries imported stock better adapted to hatchery conditions from elsewhere (e.g., Eel River stock for the Nimbus Hatchery on the American River). Even though domestic steelhead typically have lower survival rates once released than wild steelhead (52), they can be produced in large numbers to make up for it. Hatchery steelhead, both as juveniles and as adults, then have negative genetic and behavioral interactions with remaining wild steelhead in streams, continuing the downward spiral of wild populations.

Coastal rainbow trout (resident). IE. Coastal rainbow trout are the common wild rainbow trout in most of California, either as natural populations or through introductions into other areas. Although the genetic identities of distinct local populations have been lost in many instances as a result of planting hatchery fish, wild strains adapted to local environmental conditions may persist (53). Some resident fish present above dams may represent landlocked versions of the original steelhead populations.

Klamath Mountains Province steelhead. Winter-run steelhead ID; summer steelhead IB. Listed as a candidate for Threatened status by NMFS in 1998, steelhead in the Klamath-Trinity basin have had their range reduced by the construction of major dams on the Klamath, Trinity, and Shasta Rivers, with further declines caused by downstream changes to channels and water temperatures from decreased flows. Poor watershed management (connected with such practices as grazing, logging, and road building) has contributed to declines as well, especially as a result of siltation of holding pools and spawning riffles and increases in water temperatures due to loss of shading. Interactions with hatchery steelhead have contributed to further declines of wild populations, as may have fisheries, including catch of steelhead in gill nets on the high seas. Winter-run steelhead are nevertheless still widely distributed and fairly common in the basin, although much less abundant than formerly. Summer steelhead, however, are in danger of extinction because, in addition to all the usual causes of decline, they are exceptionally vulnerable to poaching when oversummering in pools. As a consequence, during the 1990s there were perhaps 1,000–1,500 adults divided among eight populations—less than 10 percent of their former abundance (17).

Northern California steelhead. Winter-run steelhead IC; summer steelhead IB. Steelhead in North Coast streams were listed as Threatened by NMFS in 2000. They are still widespread, but their numbers continue to decline for the same reasons as in the Klamath-Trinity region, with each river having its own suite of problems. The continuation of logging practices destructive to steelhead streams was one of the main reasons given for the listing of this ESU by NMFS. In the Eel River introduced pikeminnows have, through predation and competition, decreased the capacity of the mainstem river to grow juvenile steelhead. The Middle Fork Eel River is also one of principal remaining strongholds for genetically distinct summer steelhead (77), although their numbers have declined from their historical abundance; the river supports an annual run of 400–1,700 fish, which are vulnerable to poaching (17).

Central Valley steelhead. IB. This ESU was listed as Threatened by NMFS in 1998. Winter steelhead were once widely distributed in the Sacramento and San Joaquin drainages, but construction of dams on most of its tributaries separated them from historical spawning and rearing areas. The principal remaining wild populations are a few hundred fish that spawn annually in Deer and Mill Creeks, Tehama County, and a population of unknown size in the lower Yuba River. Apparent wild steelhead are found elsewhere in the Sacramento system, mainly in the cold tailwaters of dams, but their identity is confused by the presence of hatchery fish (of Eel River origin in the American and Mokelumne Rivers) and by the presence of various strains of rainbow trout of hatchery origin in rivers. With the possible exception of a small population in the lower Stanislaus River, steelhead appear to have been extirpated from the San Joaquin basin (54).

Central coast steelhead. IB. This ESU was listed as Threatened by NMFS in 1997, based on an estimated 85 percent decline in abundance between 1960 and 1997. Even in 1960 its numbers must have been reduced from historical levels, given urbanization, intensive grazing, agricultural conversion, dams and diversions, and many other insults to watersheds that have been accumulating for 150 years. The Russian River was once the third most productive watershed for wild steelhead in California (after the Sacramento and Klamath Rivers), with annual runs of 20,000–60,000 fish supporting an important instream fishery (57). In recent years numbers have been highly variable, in the 500–10,000 range, with most fish of hatchery origin. The only major hatchery is Warm Springs Hatchery in the Russian River drainage, which was built to compensate for the impact of a major dam on Dry Creek; since the dam was built, steelhead numbers have declined, on average, "only" by a factor of seven (55). In small streams of coastal San Mateo and Santa Cruz Counties, steelhead numbers seem to have been relatively stable, if low, since the 1970s. Most water diversions and severe logging occurred earlier, as did major declines. In tributaries to San Pablo and San Francisco Bays, steelhead have been nearly eliminated from lowland streams by dams, diversions, urbanization, and flood control projects. However, many upstream habitats have healthy resident trout populations, derived from steelhead.

South-Central coast steelhead. IB. These steelhead were also listed as Threatened in 1997 by NMFS. In the five largest streams in the region (Pajaro, Salinas, Carmel, Big Sur, and Little Sur Rivers), a total of fewer than 500 fish spawned annually during years of drought (1987–1991), down from around 5,000 in the early 1960s (4). Numbers in small creeks in San Luis Obispo County were also very low during the drought. Numbers have partially rebounded to predrought levels, although these levels were depressed from historical levels. For example, in a study that crossed ESU boundaries, steelhead were noted to have been present historically in 72 streams in San Mateo, Santa Cruz, and Monterey Counties. In the early 1990s they were still at historical levels in 17 streams, in decline in 32, extirpated from 1, and of unknown status in 26 (56). Causes of decline in the region are multiple but are directly related to dams and diversions, siltation of spawning areas, and blocking of access to spawning areas by culverts and other barriers (56). Populations in small, permanent streams of the Big Sur Coast have remained in good condition.

Southern steelhead. IB. This distinctive ESU was listed as Endangered by NMFS in 1997, reflecting the fact that its distribution largely coincides with the concentration of human populations in southern California. Most of its streams have been dammed, diverted, and urbanized to one degree or another. Of 92 streams in which it historically spawned in the six south coastal counties, it is now absent from 39, including all streams south of Ventura County, except Malibu Creek (Los Angeles County) and San Mateo Creek (San Diego County) (61). The total stream miles in which juveniles now rear amount to less than 1 percent of the historical number (17). In all larger streams its numbers are a small fraction of its original abundance. Although there is considerable interest in at least maintaining the remaining anadromous populations of this fish, restoration will be difficult given increasing population pressures combined with the effects of global warming, which is likely to make both streams and the ocean off southern California less habitable for southern steelhead. Nevertheless, restoration of this unusual ESU is possible if adequate flows are provided, habitats are restored on a watershed scale, and access is provided to historical spawning and rearing areas (72, 76, 78).

McCloud River redband trout. IB. Because of questions about the exact distribution of "pure" forms of this trout, its status is poorly known, although even the most generous definition of the form would still make it a species of concern because of its limited distribution. Its populations center in a few small creeks, most notably Sheepheaven Creek, that have until recently been subjected to abuse from logging and grazing (17). The upper McCloud River has been subjected to annual plants of hatchery rainbow trout and heavy fishing (halted in 1995), so the mainstem fish are genetically closest to coastal rainbows (65). An interagency conservation agreement was signed in 1998 that limits the stocking of trout above Middle Falls to trout other than rainbow trout. In 1998 brook trout were planted (62). However, redband trout in tributaries would be even more secure (e.g., free from the threat of disease) if all stocking of trout was halted, regardless of species.

Goose Lake redband trout. IB. The long-term persistence of this fish depends largely on the health of populations in the headwaters of streams flowing into Goose Lake in Oregon and California, even though much of the conservation attention has focused on large fish in the lake itself. The extirpation of the lake population during a drought and its subsequent partial recovery indicate the probable importance of downstream colonization of the lake from headwater populations. Because of the high level of concern over extirpation of Goose Lake redbands (and other native fishes) from Goose Lake when it dried up, conservation efforts have been under way in the watershed, by both agencies and private landowners, to restore streams (e.g., by changing grazing practices) and to remove or alter migration barriers. The future of this fish is much more promising now than it once was (17).

Eagle Lake rainbow trout. IB. This would seem to be a form that is secure, because it supports a major trophy fishery in Eagle Lake and is reared in large numbers in hatcheries by CDFG for planting in the lake. The fishery is supported by two types of Eagle Lake trout: those that are reared from "wild" fish collected annually at a weir on the mouth of Pine Creek and those that are from domesticated brood stock of Eagle Lake trout. It is assumed that domestic fish do not mix with fish of wild origin returning to Pine Creek, the historical spawning stream for Eagle Lake trout, which seems unlikely. The Eagle Lake trout depends on hatcheries for its existence because Pine Creek became inaccessible for spawning in most years after roads, railroads, grazing, and logging changed its basic hydrology. In 1950 a small number of fish were taken from the creek by CDFG and spawned in the hatchery. The 600 trout that resulted then became brood stock, and some were replanted in the lake. In 1959 the annual program of trapping fish for hatchery spawning began (with 16 fish), and little or no natural reproduction was allowed to take place (62). These actions saved the Eagle Lake trout from extinction.

Unfortunately, 50 years of reliance on hatchery production has probably altered the genome of the fish, and this change may increase the difficulty of reestablishing a natural population. A major effort is now under way to restore the Pine Creek watershed, using a Coordinated Resource Management Process that involves dozens of agencies, ranchers, and interest groups. The goal is to restore a naturally spawning population of Eagle Lake trout. Restoration problems are being solved one by one, as the stream is fenced, old channels restored, barriers created by roads and a railroad breached, and land use patterns changed. One major problem that still exists is the presence of a large population of alien brook trout in the prime spawning and rearing area for the trout, 45 km upstream from Eagle Lake. This population will probably have to be eradicated if large numbers of Eagle Lake trout are ever to be produced naturally (17). It is possible, however, that reduction of the brook trout population followed by heavy planting or reproduction of Eagle Lake rainbows could eventually eliminate brook trout, given the fairly warm summer temperatures of upper Pine Creek. Weaning of the Eagle Lake trout and its fishery from dependence on hatcheries is a necessary but long-term process, as is restoration of most wild rainbow trout populations.

References 1. Needham and Gard 1959. 2. Behnke 1992. 3. Stearley and Smith 1993. 4. NMFS 1996a. 5. Roelofs 1983. 6. Shapovalov and Taft 1954. 7. Everest 1973. 8. Rybock et al. 1975. 9. Winter 1987. 10. Nielsen et al. 1997. 11. Jordan and Evermann 1923. 12. Gold 1977. 13. Berg 1987. 14. Snyder 1918. 15. Busack et al. 1980. 16. Smith and Stearley 1989. 17. Moyle et al. 1995. 18. Hoopaugh 1974. 19. Lever 1996. 20. Scott et al. 1978. 21. Busack and Gall 1980. 22. Dill and Cordone 1997. 23. Kesner and Barnhart 1972. 24. Fortune et al. 1966. 25. Hokanson et al. 1977. 26. Bjornn and Reiser 1991. 27. Kubicek and Price 1976. 28. Smith and Li 1983. 29. Matthews and Berg 1997. 30. Baltz et al. 1987. 31. Cech et al. 1990. 32. Baltz and Moyle 1984. 33. Moyle and Baltz 1985. 34. Smith and Aceituno 1987. 35. Hartman 1965. 36. Everest and Chapman 1972. 37. Brown and Moyle 1991. 38. Jenkins 1969. 39. Bryan and Larkin 1972. 40. Boles 1990. 41. Merz and Vanicek 1996. 42. Tippets and Moyle 1978. 43. Elliott and Jenkins 1972. 44. Barnhart 1986. 45. Snider and Linden 1981. 46. Sturgess and Moyle 1978. 47. Hopelain 1998. 48. Hart 1973. 49. Butler and Borgeson 1965. 50. Knapp and Matthews 2000. 51. Hallock et al. 1961. 52. Berejikian 1995. 53. Gard and Seegrist 1965. 54. R. M. Yoshiyama, unpubl. data 1999. 55. Knapp et al. 2001. 56. Titus et al. 1994. 57. Steiner 1996. 58. Bisson et al. 1988. 59. Barrett et al. 1992. 60. Allendorf and Waples 1996. 61. D. McKuen, CDFG, pers. comm. 1999. 62. E. R. Gerstung, CDFG, pers. comm. 1999. 63. Reeves et al. 1987. 64. Harvey and Nakamoto 1997. 65. Nielsen et al. 1999. 66. Stearley 1992. 67. Nielsen et al. 1998. 68. J. F. Hayes, CDFG, pers. comm. 69. L. R. Brown, USGS, pers. comm. 1990. 70. J. J. Smith, San Jose State University, pers. comm. 1999. 71. Railsback and Rose 1999. 72. Douglas 1995. 73. Bagley and Gall 1998. 74. Vondracek and Longanecker 1993. 75. Baltz et al. 1991. 76. Nielsen 1999. 77. Nielsen and Fountain 1999. 78. Carpanzano 1996. 79. McEwan 2001. 80. Burns 1974.

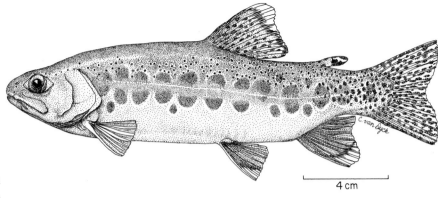

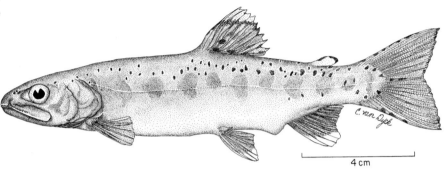

Figure 93. Top: California golden trout, 18 cm SL, Clarence Lake, Fresno County. Bottom: McCloud River redband trout, 19 cm SL, Sheepheaven Creek, Shasta County.

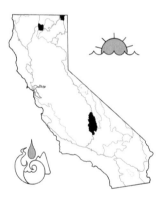

Upper locations are for redband trout.

Golden trout, *Oncorhynchus mykiss* subspp.

Identification Golden trout are rainbow trout (see the account of this species), so the basic rainbow trout characteristics apply to this subspecies. However, the coloration of golden trout is spectacularly bright: the belly, cheeks, and branchiostegals are bright red to red-orange; lower sides bright gold; central lateral band red orange; and back deep olive green. About 10 parr marks are usually present, even in adults, centered on the lateral line. Spots are large and concentrated on the dorsal and caudal fins. Body spotting is highly variable, but spots are usually scattered across the back with a few below the lateral line; the *gilberti* form is heavily spotted. The pectoral, pelvic, and anal fins are orange, the latter fins having white to yellow tips preceded by

a black band. The dorsal fin also has a white to orange tip. The following meristic characters usually fit golden trout as defined below: scales in lateral series, 150–210; scales above lateral line, 34–45; pelvic fin rays, 8–10; gill rakers, 17–21; pyloric ceca, 20–40; vertebrae, 58–61 (1, 2). Basibranchial teeth are present in some *gilberti* and *whitei* trout.

Taxonomy Golden trout are three subspecies of rainbow trout native to the upper Kern River basin: California golden trout, *O. m. aguabonita*; Little Kern golden trout, *O. m. whitei*; and Kern River rainbow trout, *O. m. gilberti*. They are here treated separately from the rest of the rainbow trout subspecies because *(1)* historically the two classic golden trouts (*aguabonita, whitei*) have been treated together as a separate species (which they decidely are not), *(2)* the California golden trout is the official state fish of California, and *(3)* there is a great deal of interest in their taxonomy, biology, and conservation.

The three golden trouts are part of the redband-rainbow trout complex found in isolated areas of California and Oregon and apparently represent remnants of the first invasion of rainbow trout into the region. They were subsequently replaced in lowland areas by coastal rainbows, leaving the isolated populations to go in their own evolutionary directions. The three trouts of the upper Kern basin have color patterns that make them very distinctive to the human eye but are otherwise similar to other types of rainbow trout. As a consequence they have been subject to much taxonomic confusion and controversy (1, 2, 3, 4). Three *species* of golden trout were originally described: *Salmo aguabonita* from the South Fork Kern River (Volcano Creek), *S. whitei* from the Little Kern River, and *S. roosevelti* from Golden Trout Creek, along with Kern River rainbow trout as a subspecies of rainbow trout (*S. gairdneri gilberti*) (5). However, the first two forms were eventually recognized as subspecies of *S. aguabonita*: *S. a. aguabonita* and *S. a. whitei*, and *S. roosevelti* became a color variant of *S. a. aguabonita*. More recently studies of meristic variation demonstrated that the two "classic" golden trouts and Kern River rainbow trout are all rainbow trout, rather than separate species (1, 2). This status was actually recognized as far back as 1893 by D. S. Jordan, who described golden trout as an offshoot of Kern River trout and made both of them subspecies of rainbow trout (19). Genetic evidence confirms both the relatedness and distinctiveness of the three forms (3, 6). Although "pure" populations of the three forms can be distinguished biochemically (6), the identity of most Kern River rainbows is cloudy. The present-day population of wild rainbow trout in the Kern River that are heavily spotted like original Kern River rainbow appear to be hybrids between California golden trout and coastal rainbow trout, with golden trout alleles becoming more prevalent as one moves upstream (13, 27).

Names See the account of rainbow trout for genus and species names. *Aguabonita* means beautiful water and "is the name of a cascade on Volcano Creek, near which this trout abounds" (7, p. 504). *Gilberti* is after Charles Gilbert, a taxonomist who described many species of western fishes, while *whitei* honors Stewart Edward White, a naturalist who wrote about golden trout. California golden trout are occasionally referred to as Volcano Creek golden trout.

Distribution Golden trout, as defined here, are native only to the upper Kern River basin, Tulare and Kern Counties.

Kern River rainbow trout were once widely distributed in the system; in the mainstem they probably existed downstream as far as Keyesville (below where Isabella Dam is today) and in the South Fork downstream as far as Onyx. Today populations defined as Kern River rainbow trout live in the Kern River from Durrwood Creek upstream to Junction Meadow. Populations were established through transplantation in Rattlesnake and Osa Creeks, and possibly upper Peppermint Creek and others (23). Much of their remaining habitat is in Sequoia National Forest (about 29 km) and Sequoia National Park (about 40 km).

Little Kern golden trout are native to and are still found in the Little Kern River above the falls on the lower river, although some of these populations show signs of introgression with coastal rainbow trout (23).

California golden trout are native to Golden Trout Creek (of which Volcano Creek is a small tributary) and the South Fork Kern River in the upper Kern River basin (2, 3). However, this fish has been translocated into many other waters within and outside California. Even before they had been formally described, they were being moved by enthusiastic fishermen to other drainages in the Sierra Nevada! One early transplant was into Cottonwood Lakes not far from Golden Trout Creek. The lakes have served as a source of golden trout eggs for stocking other waters. As a result they were introduced into more than 300 high mountain lakes and streams outside their native range in California alone. About 100 of these lakes have since lost their golden trout populations (23). In any case, it appears that most transplanted golden trout populations in California, including those in Cottonwood Lakes, are hybridized with coastal rainbow trout of hatchery origin (24).

California golden trout have become established in mountain waters in other Western states and provinces, especially the Rocky Mountain states, as a result of trades among hatcheries of various trout stocks in the 1920s and 1930s. Most of these populations have also apparently hybridized with either coastal rainbows or cutthroat trout (2).

Life History Most studies on golden trout have been on populations of California golden trout, so this summary is largely of their biology. Presumably the biology of the other two forms is similar.

Golden trout are largely native to streams of the Kern Plateau at elevations above 2,300 m. Because the valleys of the plateau were not subjected to Pleistocene glaciation, they are broad, flat, and filled with alluvium, creating wide meadows through which streams meander. These streams, the principal habitat of golden trout, are wide, shallow, and exposed, with limited riparian vegetation to provide cover. The bottoms consist largely of sand, gravel, and some cobble. The water is clear and usually cold, although summer temperatures can fluctuate from 3 to 22°C on a daily basis (8). Preferred habitats of the trout are pools and areas associated with undercut banks, aquatic vegetation, and clumps of sedges (21, 22). The exposed, downcut nature of the streams today is largely the result of heavy grazing of livestock, which began in the 1860s, causing compaction and accelerated erosion of loose alluvial deposits (9, 20). The trout are also found in higher-gradient streams above the valleys, in more conventional pool-riffle-cascade habitats, but many of these populations may have been the result of early transplants above barriers. Outside their native range golden trout occur in a wide range of habitats, from mountain lakes to small cold streams. The principal characteristic of the high-elevation waters in which golden trout have established self-sustaining "pure" populations is the absence of other trout species, although they do coexist naturally with Sacramento suckers in part of their native Kern

River basin. Kern River rainbows probably also coexisted with Sacramento pikeminnow and hardhead, where their elevational ranges overlapped.

The high mountain habitat of golden trout is tied to their brilliant colors. Although the colors may fade dramatically in golden trout kept in hatcheries or planted in lakes, without doubt they have a genetic basis (10). It is clear, therefore, that the bright colors have an adaptive significance, particularly because similar brilliant coloration has evolved independently in other Western trout from high mountain areas, such as Paiute cutthroat trout, Gila trout (*O. gilae*), Apache trout (*O. apache*), and Mexican golden trout (*O. chrysogaster*). The usual explanation given is that bright colors make fish less visible to predators in clear streams with bottoms of bright, rust-colored volcanic rocks. Although this may be a partial explanation of the phenomenon, especially in smaller streams of the upper Kern River basin, the bottoms of streams to which golden-colored trout are native are not consistently brightly colored, especially in areas where decomposed granite makes up much of the substrate. In addition, birds and mammals likely to prey on trout are infrequent in high mountain areas. An alternative explanation is behavioral. Most stream-dwelling trout species, especially males, assume bright colors during the breeding season. It is advantageous for male trout to temporarily sacrifice some of their cryptic coloration to increase their chances of reproductive success. Brightly colored males tend to be most attractive to females and to have the greatest success in defending breeding territories. Stream-dwelling trout often defend feeding areas from other trout when they are not spawning, but the advantages of being brightly colored are outweighed by the disadvantages of being more visible to predators. Because golden trout evolved in an area where predators are scarce, it would be advantageous for them to retain brilliant colors even when not spawning. The most brightly colored fish would have the greatest success in defending feeding territories and be able to grow faster, increasing reproductive success by achieving larger sizes and, perhaps, maturity at younger ages.

One indication of the shortage of natural predators in golden trout habitat is that golden trout are active throughout the day and night, although they do prefer to hold near cover and are most likely to be found in open water at night (21). They have home ranges in small streams, typically measuring around 16–18 m (22), but they rarely move more than 5 m in a day (21, 22). Long-distance movements seem to take place mainly at night (21).

Golden trout feed at all times of day and at temperatures as low as 2°C (21). Their food is essentially every invertebrate that lives in or falls into their waters. In streams these are primarily larval and adult aquatic insects, plus a few terrestrial forms. In lakes golden trout eat mainly caddisfly larvae, chironomid midge larvae, and planktonic crustaceans (11). The stomachs of golden trout in a Sierra lake in July 1970 contained large caddisfly larvae, with cases, and hundreds of tiny seed shrimp (Ostracoda). The latter organisms were swarming among beds of rushes that grew close to shore. The ability of golden trout to feed on such microcrustaceans has undoubtedly contributed to their success in mountain lakes.

In small streams in their native range golden trout have slow growth rates, reflecting the low productivity, the short growing season, and (in some areas) the high densities of trout (8). They can live up to 9 years, which is remarkably long for a stream-dwelling trout. In streams they typically attain 3–4 cm SL by the end of their first summer of life, 7–8 cm by the end of their second summer, and 10–11 cm by the end of their third summer, and they grow 1–2 cm/year thereafter, reaching a maximum size of 19–20 cm SL (8). Introduced populations in lakes grow somewhat faster; they reach lengths of 4–5 cm FL during the first year, 10–15 cm by the second, 13–23 cm during the third, and 21–28 cm by the fourth (11, 12). In lightly fished lakes golden trout reach 35–43 cm FL by the seventh year. The largest golden trout from California weighed 4.5 kg and was taken from Virginia Lake, Madera County; the largest on record, from Wyoming, weighed nearly 5 kg and measured 71 cm TL (11). Quite likely these fish were hybrids with coastal rainbows.

Golden trout become mature in their third or fourth year and spawn when water temperatures reach 10–15°C and high spring flows decline, usually by mid-May through June (15, 16). Mature females (>95 mm) dig wide, shallow redds in riffles with surprisingly small substrates (4- to 12-mm gravel), shallow depths (5–20 cm), and water velocities of 30–70 cm/sec (15, 16). Spawning activity is highest during midafternoon, when water temperatures are highest (15). Although spawning has been observed in lakes, it is rarely, if ever, successful, and attempts to establish golden trout in lakes without inlets or outlets suitable for spawning have mostly failed (14). Each female lays 300–2,300 eggs, the number depending on the size of fish according to the formula $N = 10.44FL_{cm} - 1290$ (11).

The embryos hatch in about 20 days at 14°C. The fry, measuring 25 mm TL, emerge from the gravel 2–3 weeks after hatching. Fry from lake populations move into the lake at around 45–50 mm TL (11).

Status IB. Golden trout are the official freshwater fish of California and have been accorded high priority for preservation and management. Their continued existence, especially in their native range, requires intensive management and continuous monitoring.

Little Kern golden trout were listed by USFWS as Threatened in 1977 after surveys determined that unhybridized populations existed in only six small streams in the Little Kern basin, about 10 percent of their original

160 km of stream, and in a nearby stream (Coyote Creek) established from a transplant made in the 1880s (17). The immediate cause of decline was competition with brook trout and hybridization with coastal rainbow trout, although habitat degradation from logging and grazing also reduced the ability of some streams to support trout. To eliminate alien trout, stream by stream, section by section, alien and hybrid trout were chemically treated by state and federal agencies. This work was completed in 1998. Unfortunately, hybrid Little Kern golden trout reappeared in the mainstem and some tributaries, as a result of any or all of the following: illegal planting of rainbow trout; stocking of genetically contaminated, hatchery-reared golden trout; or possible treatment failures (23). Up to 70 percent of the watershed may have to be treated again to remove hybrid fish (23).

Kern River rainbow trout are listed as a Species of Special Concern by CDFG. The form was thought to have disappeared through introgression with nonnative rainbow trout (18), although it may have originated as a result of natural invasion of coastal rainbows into golden trout streams (13). Genetic studies in the 1980s suggested that this fish was still extant in some of its native range (3), but more recent studies indicate that genetically distinctive fish that can be assigned to this taxon no longer exist in most areas (27). However, the continued presence of heavily spotted, brightly colored fish that look like original Kern River rainbow trout has encouraged management efforts to maintain this phenotype. Primary threats to remaining populations are continued introgression with nonnative rainbow trout and habitat losses from poor watershed management (connected with such practices as grazing, logging, and road building), combined with such stochastic events as floods, drought, and fire. For example, some of the present habitat of the fish suffered from the Flat Fire of 1976 and subsequent landslides that filled in pools and deposited silt in spawning areas. In addition introduced beaver have significantly altered the river in Kern Canyon in Sequoia National Park, flooding meadows and increasing braiding and meandering in the channel (25), thus reducing habitat available for trout.

Efforts are being made to identify those streams still retaining the "best" Kern River rainbow trout. A management plan for the upper Kern River basin (above Isabella Reservoir) has been drafted, and it contains recommendations for enhancing the native trout populations. Problems addressed in the plan include grazing in riparian areas and heavy recreational use of the basin. In order to reestablish populations of Kern River rainbow trout, CDFG biologists have recommended that anglers be allowed to keep only two fish in the upper basin, with a maximum length of 10 inches (25 cm). There are currently plans to replace nonnative rainbow trout stocked in tributary streams with catchable-size fish identified as Kern River rainbow trout; if this program does not succeed, the entire stocking program in the basin will be reevaluated (26). Surveys to monitor trout populations and identify habitats in need of improvement are scheduled at 5-year intervals.

California golden trout are listed as a Species of Special Concern by CDFG. Until about 1980, they did not arouse much conservation concern because they had been so widely transplanted and because they seemed to be doing well in their native range. However, transplanted populations either did not persist or hybridized with other trout. In their native range they became threatened by a combination of invasion of nonnative trout and habitat degradation. Alien brown trout are a continuous threat as predators and competitors, even though they were largely eradicated from golden trout streams in the early 1980s and barriers were constructed to prevent their reinvasion. Unfortunately, most barriers are temporary. In the South Fork Kern River the two artificial barriers need frequent repair, especially after high-flow events. The uppermost barrier (at Templeton Meadows) was recently reconstructed and appears to be effective in excluding brown trout. The lower barrier (at Monache Meadows) has not been very effective, and in 1993 CDFG biologists found a reproducing population of brown trout above it. It is most likely that the trout ascended the barrier during high flows, when the water drop was less than 1 m, but it would also have been relatively easy for anglers to have moved fish over the barrier (23).

Another threat to California golden trout is degradation of their streams from livestock grazing, which continues (legally) even though the streams are now located in the Golden Trout Wilderness Area (Inyo National Forest). Some reaches of stream from which livestock are excluded have higher populations than reaches to which livestock are allowed access, trampling banks, eating riparian plants, and polluting the water (20). Other fenced reaches, however, show little improvement because grazing upstream still causes sedimentation and affects processes that create the deep, narrow channels needed by trout (23). Despite grazing, golden trout densities (1.3–2.7 fish per square meter) and biomass (16–21 g/m^2) are among the highest recorded for trout streams anywhere (20). Unfortunately, the wide and shallow stream morphology created by grazing favors small trout, so few fish exceed 150 mm TL.

In the long run the survival of golden trout in their native habitats will depend on considering the upper Kern Basin as the truly special place it is. As the only major unglaciated watershed in the Sierra Nevada, it contains other unusual or endemic plants and animals as well. The fragile meadow systems through which classic golden trout streams flow must be treated with special care through the elimination of grazing, most roads, harmful recreational practices (e.g., offroad vehicle use), logging, and other degrading factors. As much as possible, alien fishes must be eliminated from the basin above natural barriers. Where aliens cannot be eliminated entirely, artificial barriers should be constructed to protect upstream areas, recognizing that they are bound to fail periodically and that expensive reclamation projects will have to be repeated. Angling regulations (preferably catch-and-release only for native trout and keep-all-you-catch for nonnative trout) should be strictly enforced, and educational programs should be put in place to discourage anglers from moving nonnative trout into native trout waters.

References 1. Schreck and Behnke 1971. 2. Behnke 1992. 3. Berg 1987. 4. Gold and Gall 1975, 1981. 5. Jordan 1893. 6. Nielsen et al. 1999. 7. Jordan and Evermann 1896. 8. Knapp and Dudley 1990. 9. Odion et al. 1988. 10. Needham and Gard 1959. 11. Curtis 1934. 12. Needham and Vestal 1938. 13. Bagley and Gall 1998. 14. McAfee 1966. 15. Knapp and Vredenburg 1996. 16. Stefferud 1993. 17. Christenson 1986. 18. Gerstung 1980. 19. Jordan 1894b. 20. Knapp and Matthews 1996. 21. Matthews 1996a. 22. Matthews 1996b. 23. E. R. Gerstung, CDFG, pers. comm. 1999. 24. R. Leary, University of Montana, unpubl. rpt. to CDFG 1998. 25. D. Lentz, CDFG, pers. comm. 1996. 26. D. Christenson and S. J. Stephens, pers. comm. 1996. 27. M. Bagley and B. P. May, University of California, Davis, pers. comm. 1999.

Figure 94. Top: Lahontan cutthroat trout, 25 cm SL, Granite Lake. Bottom: Paiute cutthroat trout, 16 cm SL, Silver King Creek, Alpine County.

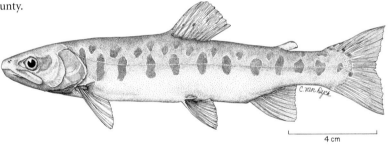

Cutthroat Trout, *Oncorhynchus clarki* **(Richardson)**

Identification Cutthroat trout are usually recognized by numerous black body spots and yellow to red slashes of pigment under each side of the lower jaw. However, cutthroat slash marks are faint or absent in young (<8 cm TL) or sea-run trout and are sometimes present on rainbow trout. More definitively, they possess basibranchial teeth, which can be detected by feeling the floor of the "throat" with one finger. Coastal cutthroat are similar to rainbow trout in overall body color, but spotting is heavier, particularly below the lateral line and on the posterior half of the body. Spots are also frequently present on the anal and paired fins, which otherwise are uniform in color. Lahontan cutthroat are similar to coast cutthroat, except that the body tends to be dark olive to reddish yellow and the spots are larger and fewer. Paiute cutthroat trout have coppery, greenish, or yellowish bodies with parr marks retained by the adults; their bodies and caudal fins are largely without black spots, although there are generally a few on the dorsal and adipose fins. Cutthroat trout in general have larger mouths (longer maxillary bones) and more slender bodies than rainbow trout. The teeth are well developed on the upper and lower jaws, head and shaft of the vomer, palatines, tongue, and basibranchial bones. The dorsal fin has 9–11 major rays; the anal fin, 8–12 major rays; the pelvic fins, 9–10 rays each; and the pectoral fins, 12–15 rays each. The tail is moderately forked. There are 15–28 gill rakers on each arch and 9–12 branchiostegal rays. Scales are typically smaller (usually 110–130 in the lateral line) than those of rainbow trout.

Parr have 9–10 oval parr marks centering on the lateral line that are covered with black speckles dorsally. The interspaces are wider than the parr marks. The fins are generally plain except for a dark leading edge on the dorsal fin and a few spots on the adipose fin.

Taxonomy Four subspecies of cutthroat trout live in California: coastal cutthroat trout (*O. c. clarki*), Lahontan cutthroat trout (*O. c. henshawi*), Paiute cutthroat trout (*O. c. seleneris*), and Colorado cutthroat trout (*O. c. pleuriticus*). The latter subspecies was introduced in 1931.

Cutthroat trout are an old (over 2 million years) lineage of salmonids that apparently evolved in the Columbia River basin and diverged into four distinct groups: coastal cutthroat, Columbia and Missouri river cutthroat, Great Basin cutthroat, and southern Rocky Mountain cutthroat (1, 2). Coastal cutthroat trout in turn diverged from interior forms perhaps a million years ago and are distinct from them in many ways (e.g., 68 versus 66 or 64 chromosomes). Although coastal cutthroat are regarded as just one subspecies throughout their vast range, their populations fall into regional groupings (ESUs) with common characteristics. Six ESUs are recognized in Washington, Oregon, and California, although all populations in California are part of the southern Oregon–California coastal ESU (SOCC-ESU), which has its upper geographic limits at Cape Blanco, Oregon (36).

The three interior forms are also further divided, and three to eight subspecies have been recognized for each (1, 2). The systematics of these subspecies are complicated and subject to change, but in California there are three generally recognized native subspecies: coastal cutthroat and two Great Basin forms, Lahontan cutthroat trout and Paiute cutthroat trout. The Paiute cutthroat is a recent derivative of the Lahontan cutthroat and is differentiated from it mainly by the near-absence of spots on the body (1). The San Gorgiono trout (described as *Salmo evermannii*) is an extinct population of Lahontan cutthroat that was apparently temporarily established in the upper reaches of the Santa Ana River, Los Angeles County, following a very early transplant from Lake Tahoe (3).

Cutthroat trout will hybridize with rainbow trout both naturally in coastal regions and through introductions of rainbow trout into interior basins. Distinct populations of hybrid forms sometimes result. Hatchery strains of hybrids, called "cutbows," have also been developed.

Names Cutthroat trout is a name given these fish by anglers in the 1880s, who were struck by the distinctive red slashes below the gill openings. The name persisted despite early opposition from ichthyologists (1). Coastal cutthroat trout are often called sea-run cutthroat trout. *Clarki* is for Captain William Clark, co-leader of the Lewis and Clark expedition to the Pacific coast; *henshawi* is after naturalist H. W. Henshaw, who provided the specimen upon which the subspecies is based. *Seleneris* refers to Selene, goddess of the moon, and is a reference to the distinctive body coloration of live fish. Snyder (21, p. 472), in his description of Paiute cutthroat (as a full species), stated: "The color is pale, the whole body suffused with yellow. . . . The entire body exhibits evanescent opaline reflections, and the skin is translucent, so much so that the dorsal cranial bones are outlined through the overlying tissue." *Pleuriticus* means side, a reference to the bright lateral band on Colorado cutthroat.

Distribution Coastal cutthroat trout live in coastal drainages from the Eel River (Humboldt County) north to Seward, Alaska. An isolated population may also exist on the Kamchatka Peninsula in Asia (2). The diverse interior forms are widely distributed in interior basins of western North America, occupying headwater streams of the Columbia, Missouri, Platte, Colorado, Rio Grande, and Saskatchewan river systems, as well as the Bonneville and Lahontan drainages of the Great Basin.

In California coastal cutthroat occur in coastal streams from the Oregon border south to tributaries of Salt Slough at the mouth of the Eel River and to Fox Creek, a tributary of the Van Duzen River, a fork of the Eel (4, 5). The streams and lagoons in which they occur are largely within the coastal rain forest, so most populations are within a coastal zone that is 8 km wide at the mouth of the Eel and 48 km wide at the Oregon border (5). Upstream from this zone is a population in Elliot Creek, a tributary to Applegate River in the Rogue River watershed of Oregon, 120 km from the ocean (5). In 1958 fish from Elliott Creek were introduced successfully into Twin Valley Creek, Siskiyou County, in the Klamath River watershed (31). Although the transplant involved just six fish, the population is apparently still extant.

Lahontan cutthroat are native to streams and lakes of the Lahontan basin in California, Oregon, and Nevada (6). Behnke (2) considers populations in the Oregon and eastern parts of the basin to be separate subspecies, still undescribed. In California they are native to streams and lakes on the east side of the Sierra Nevada (Carson, Walker, and Truckee Rivers, and perhaps Susan River as well). Today only scattered populations exist within their native range, including California. The only California populations that seem to represent authentic endemic fish are in Independence Lake (Placer County) and By-Day Creek (Mono County), although Heenan Lake may contain vestiges of a population of Lahontan cutthroat from the West Fork Carson River, planted there in the 1860s (32). However, Heenan Lake is used to rear Independence Lake cutthroat for stocking elsewhere (14). By 1999 ten other populations had been established in California within the native range, including populations upstream of native waters in formerly fishless areas (6, 32). All but one of these populations are small and isolated, so unlikely to be self-sustaining through long periods of time (32). In addition, small populations resulting from introductions into formerly fishless high-elevation streams exist in the Owens watershed (O'Harrel Creek) in the eastern Sierra and in the Yuba watershed (Macklin Creek, East Fork Creek), Stanislaus watershed (Disaster Creek), Mokelumne watershed (Marshall Canyon Creek, Milk Ranch Creek), and upper San Joaquin watershed (West Portuguese Creek, Cow Creek) in the western Sierra (6).

Paiute cutthroat are native only to Silver King Creek, Alpine County, below Llewellen Falls. Today they exist only where introduced: Silver King Creek above the falls and three tributaries (Fly Valley, Four Mile Canyon, and Bull Creeks), two tributaries of Silver King Creek below the falls (Coyote Valley Creek and Corral Valley Creek), and three creeks in other areas: Cottonwood and Cabin Creeks (Inyo County) and Stairway Creek (Madera County) (4, 32).

Colorado cutthroat trout are native to the upper Colorado River basin of Wyoming, Utah, and Colorado, in the southern Rocky Mountains. California populations of these brightly colored trout originated from Trappers Lake, Colorado. In 1931 they were successfully planted in the Williamson Lakes (Inyo County), a small chain of isolated, high-elevation lakes, but they were largely forgotten until their rediscovery in the 1970s (30).

Life History Because coastal cutthroat trout are ecologically quite distinct from interior forms, the life histories of coastal cutthroat and Lahontan-Paiute cutthroats will be discussed in two separate sections.

Coastal cutthroat trout. Coastal cutthroat trout are more strongly tied to fresh water than most anadromous fishes, especially in California. Most sea-run populations leave their streams only for the summer months and return to overwinter in fresh water, even as nonspawning fish (7). Many of these fish never leave estuaries or lagoons or, if they do go out to sea, remain close to the coast, often in low-salinity plumes of big rivers (7). Other populations do not go to sea at all, although some of these populations exist upstream of natural barriers to migration. In California most populations are weakly anadromous and migrate mainly between large and small streams or between rivers and estuaries (5). In the Smith River fish that are resident, potadromous (migrating within the river system), and anadromous live together in the same stream sections (8). Even fish in populations isolated in streams above barriers

to anadromous trout show considerable local movement and many wind up in downstream areas, mixing with anadromous fish (33). In anadromous forms some fish migrate to sea during their first year, but others spend up to 5 years in fresh water before migrating to coastal waters or estuaries. Cutthroat trout evidently remain in shoals during their saltwater residence (10).

Coastal cutthroat live mainly in small, low-gradient coastal streams and estuaries. Even in large river systems they tend to be most abundant in small tributaries and to move into larger waters mainly when they are large enough to prey on other salmonids. In Blue Creek, a tributary to the Klamath River, downstream movement of yearling and older cutthroat (12–20 cm FL) occurs mainly in April–June, coinciding with the outmigration of juvenile salmon, a source of food for migrants (34). The apparent preference of juvenile cutthroat for small streams and for shallow riffles within larger streams is probably the result of interactions with more aggressive coho salmon and steelhead juveniles, which keep small cutthroat from occupying pools or larger waters (7). Typical streams are cool (<18°C) and well shaded, with an abundance of instream cover (9). Preferred temperatures are 9–12°C, with spawning temperatures of 6–17°C (9). Coastal cutthroat trout generally avoid water with dissolved oxygen levels less than 5 ml/liter, and feeding and movement of adults are inhibited at turbidities greater than 35 ppm. Embryo survival can be reduced to less than 10 percent if sediment levels exceed 103 ppm, combined with dissolved oxygen levels lower than 6.9 mg/liter and water velocities in the redd of less than 55 cm/sec (9). Cutthroat fry are typically found in water with velocities of less than 0.30 m/sec, with the apparent optimum being less than 0.08 m/sec; they prefer shallower and slower water than do older life stages. Summer flows in natal streams average 0.12 m^3/sec (9). Adults that spend winter in streams inhabit pools with fallen logs or undercut banks, but boulders, depth, and turbulence provide alternative forms of cover (5).

Coastal cutthroat in streams feed opportunistically. Juveniles feed mostly on benthic and drift insects, microcrustaceans, and occasionally smaller fish, including other salmonids (7). Larger fish feed on insects, crustaceans, salmon eggs, and other fish, becoming more piscivorous as they increase in size. In fresh water adult cutthroat trout prey on small fishes, such as threespine stickleback, sculpins, and juvenile salmon and trout. They are the top predator in some streams, such as the Smith River, and therefore may significantly affect community structure. In the marine environment cutthroat trout feed on various crustaceans and fishes, including Pacific sand lance (*Ammodytes hexapterus*), salmonids, herring, and sculpins. Cutthroat trout returning to spawn from the ocean tend to feed on insects and other stream prey during their first spawning but in subsequent years may not feed if they have sufficient fat stored from marine feeding (7). Marine predators include Pacific hake (*Merluccius productus*), spiny dogfish (*Squalus acanthias*), harbor seals (*Phoca vitulina*), and adult salmon (8). Freshwater predators include the usual array of herons, mergansers, kingfishers, otters, snakes, piscivorous fishes, and humans.

Coastal cutthroat trout attain 7–20 cm FL in their first year of life and typically measure 25–30 cm after 2–3 years of stream life. Once they have migrated to the ocean, an estuary, or a large river and begin feeding heavily on fish, their growth rate is 5–10 cm/year (7, 9). Maximum length is around 50 cm FL. The angling record is a fish weighing 2.72 kg. In resident populations growth is much slower, and fish may reach only half the size of anadromous forms at a given age (11). Coastal cutthroat rarely live more than 7 years, but 10-year-old fish have been recorded (7).

Anadromous coastal cutthroat trout spawn first at 2–4 years of age and may return two to five times to overwinter and spawn. In northern California they begin to migrate up spawning streams in August–October following the first substantial rainfall. Ripe or nearly ripe females have been caught from September to April, indicating a prolonged spawning period. Sexually mature cutthroat trout seem capable of precise homing migrations to their natal streams. Females excavate redds in clean gravel with their tails. The completed redd measures approximately 350 mm in diameter by 100–120 mm in depth. After spawning is completed, the female will cover the redd with about 150–200 mm of gravel by displacing the substrate upstream of the redd. Each female will dig a number of redds sequentially. Spawning can take place during the day or night (9).

Stream sections with small or moderate-size gravel substrates are essential for spawning. The size of gravel used for spawning ranges from 0.2 to 10.2 cm in diameter, with intermediate sizes presumably being optimal. Finer material reduces the survival of embryos, and larger substrates can be excavated only with difficulty. Cutthroat trout usually choose the tails of pools in small streams for spawning, preferring headwater tributaries of larger streams. Spawning occurs at water velocities of 0.3–0.9 m/sec in northern California, but cutthroat trout have been observed to spawn in small streams in Oregon with flows as low as 0.01–0.03 m³/sec, where velocities over the redds were very low (9).

Fecundity increases with the size and age of females. Eggs of large females are also larger in size than those of smaller, first-spawning females. Fecundities are 250–4,400 eggs, with a mean of 1,100–1,700 eggs for females measuring between 200 and 400 mm TL (9). Embryos hatch following 6–7 weeks of incubation, depending on temperature. Alevins remain in the gravel for an additional 1–2 weeks until the yolk sac is absorbed. Thus in California fry emerge from March to June (12). Newly emerged fry move into shallow habitats on the edges of streams, where currents are slow, temperatures warm, and small invertebrates abundant; the best of these habitats are adjacent to areas with deciduous riparian vegetation, which provides cover, shade, and food (13).

Lahontan and Paiute cutthroat trout. Lahontan cutthroat trout were once the only trout (with the exception of Eagle Lake rainbow trout) found on the east side of the Sierra Nevada. They lived in a wide variety of cool waters, from large terminal desert lakes to small mountain lakes, from major rivers to small headwater creeks (15). This subspecies is particularly noted for its ability to thrive in highly alkaline (3,000–13,000 mg/liter total dissolved solids, pH 8.5–9.5) Pyramid and Walker Lakes, although high values (10,000 mg/liter and more) for total dissolved solids in Walker Lake in the 1990s severely reduced survival (29). Lahontan cutthroat are also noted for their ability to live in waters of Nevada streams, where temperatures may exceed 27°C for short periods and fluctuate 14–20°C daily (6, 39). They can survive prolonged exposure to temperatures of nearly 25°C, but growth ceases when 22–23°C is exceeded (28). In California most populations were historically found in coldwater streams, where they presumably used a wide variety of habitats as long as oxygen levels were high, temperatures rarely exceeded 23°C (and were typically less than 17°C), and cover and food were plentiful. Paiute cutthroat, for example, are typically most abundant in meadow sections of high-elevation streams, where there are plenty of undercut banks with overhanging vegetation and occasional deep pools for cover, as well as gravelly riffles with low amounts of sediment for spawning (16, 17). In lakes cutthroat trout require, for growth and high survival, temperatures less than 22°C, pH values of 6.5–8.5, and dissolved oxygen levels greater than 8 mg/liter (6). Lahontan cutthroat tend to stay close to the bottom but feed pelagically on small fish (18).

Stream-dwelling cutthroat trout may spend their entire lives in less than 20 m of stream (19), but fish in rivers presumably moved about as flows and prey availability changed. In lakes they seem to roam widely and then make extensive spawning migrations upstream. Trout in Pyramid Lake, for example, once ascended 160 km up the Truckee River into Lake Tahoe to spawn in tributaries to the lake, as well as in various tributaries to the Truckee itself (22).

Like other trout, stream-dwelling cutthroat trout feed mostly on drift, typically a mixture of terrestrial and aquatic insects. They are opportunistic, so whatever is most abundant in drift tends to be most abundant in their stomachs. In lakes small cutthroat trout feed on insects taken at the water's surface or on zooplankton, although if neither is abundant they will feed on bottom-dwelling insect larvae, crustaceans, and snails (20). Large trout in lakes (those measuring more than 30 cm FL) feed mainly on other fish, especially tui chubs (18).

Growth varies with water temperature and abundance of food organisms. Slow growth is seen in small mountain lakes, where Lahontan cutthroat reach 6–10 cm FL in 1 year, 18–22 cm FL in 2 years, 31–33 cm in 3 years, and 38–45 cm in 4 years (18). In Pyramid Lake, Nevada, where temperatures are fairly warm and forage fish are abundant, yearly average lengths are 22, 29, 36, 43, 50, 57, and 63 cm FL, respectively, with males and females being about the same size (18). Present-day Lahontan cutthroat seldom live longer than 9 years or reach more than 61 cm TL (2.2 kg), but larger fish were once common in Tahoe and Pyramid Lakes. The largest cutthroat trout known, from Pyramid Lake, measured more than 99 cm TL (18.6 kg) (22). In contrast Paiute cutthroat trout seldom exceed 25 cm FL, as might be expected of trout inhabiting a cold mountain stream. They typically reach about 9 cm FL in their first year, 13 cm in their second year, and 20 cm in their third year, although some populations have slightly more rapid growth rates (23). There are no records of Paiute cutthroat over 3 years old.

Maturity is achieved in their second to fourth year, and spawning takes place between April and early July. Lake-dwelling Lahontan cutthroat migrate up streams to spawn, seeking out gravel riffles. Spawners generally home to the same stream in which they were hatched. The stimulus for spawning in Lahontan cutthroat seems to be a combination of increasing daylight and increasing stream temperatures, resulting in spawning at 8–16°C. Spawning behavior is similar to that of rainbow trout. Lahontan cutthroat females produce 400–8,000 eggs (about 47 eggs per centimeter of fork length). In Heenan Lake average fecundity is 1,720 eggs (average length and weight 49 cm FL and 1.1 kg, respectively) (14). Each fish may spawn up to five times, but it is likely that most females spawn just once or twice (18). Paiute cutthroat have rather low fecundities: 325–350 eggs per 2- to 3-year-old female (23); they probably spawn just once.

Embryos hatch in 6–8 weeks, and fry emerge and begin feeding about 2 weeks after hatching. As in other trout, fry tend to occupy edge habitats in association with shallow water, low flows, and abundant food. Lahontan cutthroat juveniles tend to move into lakes in the first year.

Status IB, IC, or IIB. All cutthroat trout in California need special management if they are to thrive in the future.

Coastal cutthroat trout. IC. Coastal cutthroat in California (as of 1997) are present in 182 streams (many of them small tributaries to larger streams) that include 1,100 km of accessible habitat (compared with >9,650 km in Oregon) (5). They also occur in four coastal lagoons covering 1,875 ha (5). California drainages in which they live include the Smith River (30% of stream populations), the Rogue River (6%), the Klamath River (13%), Redwood Creek (8%), the Mad River (8%), Humboldt Bay (10%), the Eel River–Salt Slough (6%), other small coastal streams (14%), and coastal lagoon watersheds (5%) (5). There are more coastal cutthroat trout populations in northern watersheds because they are able to use streams farther inland.

The exact status of coastal cutthroat populations is hard to determine, because juveniles (those measuring <50 mm SL) are very difficult to distinguish from more abundant rainbow trout (steelhead) in the field. Migrating adult cutthroat may also sometimes be mistaken for steelhead at some localities. Although abundance estimates of coastal cutthroat are largely lacking, their numbers in most California streams are low (5, 24). Even in the Smith River drainage, where the largest California populations occur, cutthroat trout constitute a minor portion of the salmonids. Diving surveys in the north, south, and middle forks produced counts averaging 12 cutthroat trout (23–50 cm long) per kilometer (5). In smaller Smith River tributaries densities of juvenile cutthroat trout are typically 60 per kilometer, although where other salmonids are absent densities can exceed 300 per kilometer. Gerstung (5) estimated that the entire Smith River watershed supports, on average, 1,500 cutthroat measuring 20–50 cm TL and over 7,000 fish measuring <20 cm TL.

Overall California coastal cutthroat populations have stable or slightly decreasing trends, although it is likely the trends are within already depressed populations (5). Cutthroat trout populations have presumably declined considerably in historical times because they depend on small streams that have been damaged by logging and other human activities. A similar situation exists in Oregon, where anadromous populations are regarded as being in the greatest decline and resident headwater populations are regarded as being in reasonably good condition (25). Of the six ESUs recognized by NMFS on the West Coast, only one (the southwestern Washington–Columbia River ESU) is proposed for listing as Threatened (36). The SOCC-ESU was not thought by NMFS to be in danger of extinction, although most populations were thought to be depressed and concern was expressed about continuing threats to cutthroat habitats (36). In contrast, a 1993 analysis by the Wilderness Society suggests that coastal cutthroat trout could qualify as a threatened species throughout Washington, Oregon, and California. An earlier analysis indicated that coastal cutthroat populations in California faced a moderate risk of extinction (37).

The greatest cause of coastal cutthroat trout population declines in California and Oregon has been habitat alteration, particularly for developing embryos and fry in small streams. Probably the most significant cause of habitat loss is logging and its associated road building, which can result in increased temperatures, loss of cover, reduction in food supply, and increases in turbidity and siltation (5, 8, 25). For example, severe damage has been caused by tractor logging on steep and unstable slopes in the Klamath River drainage, where habitat recovery will take many decades in some places (38). Within the Smith River drainage, 44 percent of stream channels were rated as moderately to severely damaged by landslides, siltation, channel scouring, and removal of riparian trees (5). Other causes of declines include instream fisheries, pollution, interactions with hatchery fish (including steelhead), and poor ocean conditions. An additional problem is the frequently poor condition of estuaries, many of which have been substantially filled in as a result of accelerated erosion upstream and then further altered by dredging to provide boat access.

Lahontan cutthroat trout. IB. This subspecies was listed by USFWS as endangered in 1970, but the status was changed to threatened in 1975 to enable fisheries for hatchery-supported populations (e.g., Pyramid Lake). Lahontan cutthroat trout are now gone from most of the their native range, and a majority of existing populations in California are the result of reestablishment efforts. Populations now occupy less than 3 percent of their historical range in the Truckee, Walker, and Carson River basins (6). They are somewhat better off in Nevada, occupying about 54 streams, representing less than 15 percent of their historical habitat in the Quinn and Humboldt basins (6, 27). Most populations, especially in California, are isolated in small creeks. They are likely to be maintained only through continuous effort for two reasons. First, extinction risk from "natural" causes (e.g., floods, droughts) is much higher in small isolated streams with no natural source of fish to recolonize them (27). The probability of local extinction is greatly increased when streams have been degraded by grazing, logging, road building, mining, dams, diversions, and other human endeavors (6). Second, the major cause of decline, interactions with alien trout, is still a factor in most of their native streams and lakes. With few exceptions, populations decline and disappear following the introduction of rainbow, brown, and brook trout, often all three in the same stream or lake. Habitat degradation and introduced trout are synergistic in their effects, especially if the introduced trout are fall or winter spawners, as opposed to spring-spawning cutthroat trout. Not only are nonnative juvenile trout already occupying crucial habitat when cutthroat fry are still emerging (and subject to aggressive displacement and predation by the larger nonnatives), but they complete sensitive early life history stages before degraded streams become too warm or depleted of flows in late spring. In streams without alien trout, the downstream limits of cutthroat trout are determined by high stream temperatures, which in turn are tied to habitat degradation, especially from grazing (27).

A particularly tragic loss among Lahontan cutthroat populations was the extirpation of the populations of large, fast-growing fish that once occupied Pyramid Lake and

Lake Tahoe. Competition, predation, and diseases from introduced lake trout were presumably important factors in their complete elimination from Lake Tahoe. However, lake trout interactions probably just accelerated their extirpation as the result of (1) an excessive commercial fishery at the turn of the century, (2) logging-degraded spawning streams, and (3) construction in 1905 of Derby Dam on the Truckee River, which stopped migration from Pyramid Lake. The numbers of cutthroat trout of Pyramid Lake were also reduced by commercial fishing, but the main cause of their extinction was Derby Dam, which made the Truckee River inaccessible for spawning through reduction of flows and buildup of a large shallow delta in Pyramid Lake as lake levels dropped (18, 22). The last spawning run, of fish averaging 8–9 kg, took place in 1938, and wild trout became extinct a few years later. If diversions had continued at former levels, it is likely that Pyramid Lake would have dropped steadily lower until it became too alkaline for all fish life. By 1969 the lake had dropped nearly 30 m.

The loss of the Pyramid Lake fishery was felt keenly—enough so that in 1950, stocking of hatchery-reared Lahontan cutthroat trout began, using brood fish that originated from Heenan, Walker, and Summit Lakes. This effort was successful, and the Pyramid Lake fishery is still largely maintained by hatchery fish. However, listing of Lahontan cutthroat as Endangered meant that restoration of natural populations was required. Because restoration of flows to the Truckee River was initially unthinkable, in 1976 Marble Bluff Dam was built just above the large delta. This dam sends water down a 5.6-km fishway for passage around the delta and also has a fish elevator in the dam itself, so any fish that make it to the base can be lifted over the dam (18). This technological solution was at best modestly successful and worked only fortuitously (once) for the endangered cui-ui (*Chasmistes cujus*), which also required access to the Truckee River for spawning. Because of the failure of Marble Bluff Dam to recover cui-ui, Truckee River water has been appropriated for fish, and water stored in upstream reservoirs is now sent downstream in large amounts when cui-ui are spawning. Such flows also benefit cutthroat trout, mainly by helping to stabilize or even raise lake levels, because much of the trout spawning habitat is not in good condition.

For stream and small lake populations, numerous restoration efforts are under way, often involving eradication of nonnative trout above natural or human-made barriers. The Lahontan cutthroat trout recovery plan of USFWS has as its goal the delisting of the fish after sufficient numbers (to be determined) of self-sustaining populations are established (6). This is not likely to happen in the foreseeable future.

Paiute cutthroat trout. IB. The Paiute cutthroat trout was listed as Endangered by USFWS in 1969 and was reclassified to Threatened in 1975 (23). It would be extinct if a sheepherder had not moved Silver King Creek fish above Llewellyn Falls (a barrier to upstream movement of fish) in 1912. A population in the Corral Valley–Coyote Valley Creek drainage, tributary to lower Silver King Creek, may also have been established by transplants, although this is by no means certain. By the 1920s rainbow trout had invaded lower Silver King Creek and eliminated Paiute cutthroat through hybridization and competition. In 1949 and the 1950s rainbow trout and Lahontan cutthroat trout were planted above the falls, and the same process began, isolating pure Paiute cutthroat trout in a couple of tributaries. By that time (1946) Eldon Vestal, a CDFG biologist, had transplanted the fish to Cottonwood Creek (Mono County). In 1964 Silver King Creek above Llewellyn Falls was poisoned with rotenone to get rid of hybrid fish and was then restocked with fish from "pure" populations. Unfortunately, some hybrid fish remained in the subsequent population.

In 1964 apparent hybrid fish (i.e., fish with many spots) were found in Cottonwood Creek. In 1970 this stream was poisoned, but only after a large number of fish with fewer than five spots were removed by electrofishing. These fish were then reintroduced back into the creek (26). The resulting population looked like pure Paiute cutthroat trout, but biochemical studies indicated that the population still included hybrids. The pretreatment electrofishing had apparently selected for the "correct" phenotype even in hybrid fish! Thus by the late 1980s unhybridized Paiute cutthroat trout existed in only about 8 km of streams tributary to Silver King Creek, plus a few other transplanted populations.

In the late 1980s Coyote Valley Creek was poisoned to remove hybridized populations and then restocked. In 1991–1993 the mainstem of Silver King Creek and tributaries above the falls were retreated and restocked (35). By 1998, however, the population was low (32). Restoration efforts continue, but the situation for Paiute cutthroat trout remains precarious.

Colorado cutthroat trout. IIB. This introduced subspecies continues to maintain its populations in Williamson Lakes. In its native range it is in decline because of hybridization with introduced rainbow trout and other factors (1). In 1987 fish from Williamson Lakes were taken back to Colorado, and a population was established in a lake in Rocky Mountain National Park (1, 30).

References. 1. Behnke 1992. 2. Behnke 1997. 3. Benson and Behnke 1961. 4. Gerstung 1981. 5. Gerstung 1997. 6. USFWS 1994. 7. Trotter 1997. 8. Voight and Hayden 1997. 9. Pauley et al. 1989. 10. Giger 1972. 11. Michael 1983. 12. DeWitt 1954. 13. Moore and Gregory 1988. 14. Titus 1990. 15. Gerstung 1988. 16. Overton et al. 1994. 17. Kondolf 1994. 18. Sigler et al. 1983. 19. R. B. Miller 1957. 20. Calhoun 1944. 21. Snyder 1933a,b. 22. La Rivers 1962. 23. Ryan and Nicola 1976. 24. Moyle et al. 1995. 25. Hooton 1997. 26. Busack and Gall 1981. 27. Dunham et al. 1998, 1999. 28. Dickerson and Vinyard 1999a. 29. Dickerson and Vinyard 1999b. 30. Pister 1990. 31. M. Coots, pers. comm. 1996. 32. E. R. Gerstung, CDFG, pers. comm. 1999. 33. Harvey 1998. 34. Hayden and Gale 1999. 35. Flint et al. 1998. 36. Johnson et al. 1999. 37. Nehlson et al. 1991. 38. CDFG, unpubl. rpt. 39. Dunham et al. 1999.

Brown Trout, *Salmo trutta* Linnaeus

Identification Brown trout are the only trout in California with both red and black spots on the body. Black spots, which are large and variable in size, are present on the gill covers (many), tail (few and often indistinct), head, adipose fin, dorsal fin, and sides. Dark spots on the sides are typically surrounded by a pale halo. Red spots are present only on the lower sides. Adults are usually dark to olive brown on the back, shading to yellow brown on the sides, and white to yellow on the belly. The adipose fin is usually orange or reddish, and paired fins never have white on their leading edges. In large lakes brown trout can be silvery, with X-shaped clusters of small spots. Brown trout are slightly heavier bodied than other California trout, with thicker caudal peduncles. The mouth is large, the maxillary bone extending beyond the rear margin of the eye in fish over 14 cm TL. The jaw of spawning males is usually hooked. Well-developed teeth are present in both jaws, as well as on the head and shaft of the vomer, the palatines, and the tongue. Basibranchial teeth are absent. The dorsal fin has 12–14 rays; the anal fin, 10–12 major rays; the pelvic fins, 9–10 rays; and the pectoral fins, 13–14 rays. The tail is straight edged in large adults but may be slightly forked in young fish. The anal fin is rounded on males but falcate (the rear edge slightly indented) in females. There are 14–17 gill rakers on each arch and 9–11 branchiostegal rays. Scales are

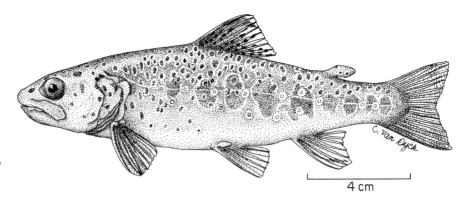

Figure 95. Brown trout, 15 cm SL, California.

4 cm

small, numbering 120–130 in the lateral line. Juveniles have 9–14 narrow parr marks, more or less centered on the lateral line, and are yellowish on the sides, with a few red spots.

Taxonomy There are many races and strains of brown trout in Europe, much like rainbow trout on the west coast of North America. Legend has it that two main strains of brown trout were introduced into North America: "German brown trout" from Germany and "Loch Leven trout" from Scotland. The original stocks have been so mixed in hatchery and planting programs that it is pointless to call these fish anything but just brown trout. Until the 1980s western trouts now placed in the genus *Oncorhynchus* were lumped with Atlantic salmon and brown trout in the genus *Salmo.* The split from *Salmo* recognizes that Atlantic salmon and trout have had a long, separate evolutionary history from Pacific salmon and trout (1).

Brown trout will occasionally hybridize with brook trout, and the sterile offspring are known as tiger trout because of the distinctive banding on their sides.

Names *Salmo trutta* is the name for brown trout originally assigned by Linnaeus, and it has remained remarkably stable. *Salmo* is Latin for salmon or leaper; *trutta* simply means trout. In California brown trout have been called at various times "German brown trout," "Loch Leven trout," and "Von Behr trout" (2).

Distribution Brown trout are native to Europe, North Africa, and western Asia. Because they are partially anadromous there, they are also found in the British Isles and Iceland. Brown trout were first introduced into North America in 1883. In 1893 embryos of brown trout were brought to California and successfully reared for planting in coastal streams (2). Since that time brown trout from various sources have been reared in California hatcheries and planted throughout the state. Brown trout are now present in a high percentage of suitable waters in the interior of the state, especially on both sides of the Sierra Nevada, although their distribution is spotty, often reflecting introduction histories. There are relatively few populations in North Coastal drainages (even though they were the site of the first introductions), although they are widespread in the upper Trinity River drainage. They have also become established in much of the United States and Canada, as well as parts of South America (at least five countries), the Falkland Islands, Africa (at least seven countries), Asia (India, Pakistan, Japan), Australia, New Zealand, and Papua New Guinea (3, 22).

Life History Brown trout are *the* trout of Europe and a favorite sport fish of serious anglers the world over. Therefore they are the most studied of trout and char next to rainbow trout, as indicated by summaries of their biology (4, 5, 6, 7, 8) and the hundreds of papers that are published annually.

Adult brown trout are largely bottom-oriented pool dwellers in streams and rivers, but younger, smaller trout are as likely to be found in riffles as in pools. The optimum habitat for brown trout seems to be medium to large, slightly alkaline, clear streams with both swift riffles and large, deep pools. They are found, however, in the complete range of trout waters, from spring-fed trickles to large lakes and reservoirs. Lake and reservoir populations typically spawn in streams, and young rear there for several years. Sea-run brown trout are rare in California but seem to be present in the Klamath and Trinity Rivers (9) and perhaps the Sacramento as well.

Temperature is an important factor limiting brown trout distribution (10). They can survive for short periods of time at temperatures up to 28–29°C (depending on acclimation temperature), but their preferred temperatures are 12–20°C. Optimal temperatures for growth seem to be around 17–18°C, although high growth rates are found in water of 12–18°C. If given a choice they will avoid streams in which temperatures do not exceed 13°C for extended periods of time (11). In Rush Creek (Mono County), a degraded stream undergoing restoration, temperature interacts with flow in affecting the growth and survival of brown trout (12). High summer flows reduce temperatures to a point at which brown trout growth is reduced, whereas low summer flows that increase temperature and temperature fluctuations also cause decreased growth. Stress caused by high temperatures at low flows could be partially offset by increased availability of food. Moderate flows, with associated moderate temperatures, result in predictably high growth rates.

Different life stages of brown trout select different combinations of depth, velocity, and cover in small Sierran streams (23). Fry (<50 mm TL) typically choose edge waters less than 30 cm deep, with low velocities. Juvenile or yearling trout select deeper water (50–75 cm) and higher velocities (0.1–0.4 m/sec), associated with large rocks, logs, or overhead cover. Adults are typically found in water 0.7–3.5 m deep, in deep cover, with variable (but often low) velocities. In bigger streams juvenile and adult brown trout are often found in deeper water. Otherwise Sierran results are fairly typical for brown trout.

In the Owens River (Inyo County) nonreproducing brown trout are rather sedentary, seldom moving more than a few meters from one spot, typically near or under dense cover. Trout measuring less than 25 cm TL establish feeding territories, and a dominance hierarchy is usually established as well. The largest, most aggressive fish defend the largest territories, which are usually located in the best positions in the stream for cover and food availability. Trout larger than 25 cm TL are more mobile and tend to remain under cover (e.g., undercut banks, logs) during the day and come out to pursue prey actively during the evening. Even these large fish, however, generally patrol rather restricted areas (13). This behavioral pattern appears to be fairly typical of brown trout in streams.

Brown trout diets in streams change with size and season. In general the smaller the trout, the greater the percentage of its diet made up of drift organisms, especially terrestrial insects. As trout grow larger, they spend more time selectively picking aquatic invertebrates from the bottom. Trout larger than 25 cm TL are active pursuers of large prey, such as other fish (including their own young), crayfish, and dragonfly larvae. Once they exceed 40 cm TL, their diet is almost exclusively fish (14). In the East Walker River (Mono County) the major prey of large brown trout are tui chubs, Lahontan redsides, and Tahoe suckers (14). There are, of course, many exceptions to this general description. In particular, trout of all sizes are prone to feeding on terrestrial insects during the late summer when the abundance of large aquatic insect larvae is low. They also feed on emerging aquatic insects when a large hatch is taking place. Most terrestrial insects are taken during the day, although feeding activity (mostly on aquatic organisms) is most intense at dawn and dusk. Active feeding, however, can be observed at nearly any time. In lakes small brown trout feed heavily on zooplankton, gradually switching to bottom-dwelling insect larvae (especially chironomid

midge larvae) and amphipods, and then (at sizes greater than 25–35 cm TL) to fish.

Juvenile brown trout show considerable variability in prey among individuals captured at the same time and location. Individual fish apparently specialize in certain types of prey for varying lengths of time, although some individuals will be persistent generalists (15). Fish living in pools are more likely to specialize than those living in riffles (15). There is considerable evidence that brown trout in streams engage in optimal foraging, balancing energy gain from efficient gathering of prey (specialists) with searching for alternate sources of food and being close to cover to avoid becoming prey themselves (16).

Growth in brown trout is as variable as the waters they inhabit. In California they reach anywhere from 3 to 8 cm TL (usually 5–7 cm) in their first year, 7–22 cm (usually 13–16 cm) in their second, 13–36 cm (usually 19–28 cm) in their third, and 23–45 cm (usually 35–41 cm) in their fourth (6). Brown trout can reach large sizes: the largest known was a sea-run individual from Scotland that measured 103 cm TL (18 kg); the largest recorded from California was a 12-kg fish from Upper Twin Lake (Mono County) caught in 1987. They can live as long as 38 years in alpine lakes in Norway (25), but the oldest one known from California was only 9 years old, from Castle Lake (Siskiyou County). Growth is usually faster in lakes than streams, but this generalization does not apply to high alpine situations, where growth is slow in both habitats. Growth is affected by temperature, alkalinity, total dissolved solids, turbidity, population density, and food availability. In streams growth (especially of small individuals) is affected by trout densities (27). Slower growth in turn helps to determine the upper limits of population size because smaller trout have lower survival rates (especially over winter) and smaller females have lower fecundities. When trout densities are low from natural causes, such as extreme high-flow events, increased growth rates allow rapid recovery of populations (27).

Brown trout usually mature in their second or third year, although a few may wait as long as 7 or 8 years. Spawning takes place in fall or winter, commonly in November and December in California. Most brown trout populations require streams with riffles that have pea- to walnut-size gravel for spawning. The most suitable locations are at the tails of pools, where water is deeper, current less turbulent, and cover close by (20, 23, 24). In a Wyoming stream depths (12–18 cm) and velocities (24–37 cm/sec) chosen by brown trout were consistent even with substantial changes in stream flow (17). Higher values for depths and velocities are likely in bigger streams or for bigger fish. In large lakes successful spawning will occasionally take place on gravel bars close to shore.

The reproductive cycle and spawning behavior of brown trout are described in the classic book by Frost and Brown (5). The initial stimulus for upstream movement to spawning grounds is often a rise in water level, although selection of the spawning site does not occur until water temperatures have dropped to 6–10°C. The redd site is selected by the female, and she soon starts a depression by turning on her side and digging with her tail ("cutting"). Gravel is moved downstream by suction created by the upward movement of the tail and by the stream current. The initial cutting attracts a male, who defends the female and redd from other males. The male does not help with redd construction but continually courts the female as she works. Courtship consists of swimming alongside the female and quivering. As the redd becomes deeper, courting becomes more intense. Finally the female sinks into the depression, with her anal fin resting on the bottom, and opens her mouth. The male immediately swims alongside her, quivering violently, mouth open, and releases his sperm as the female releases her eggs. The sperm is frequently visible for a few seconds as a white cloud on the bottom of the nest.

Following the spawning act, the female begins cutting again above the redd, simultaneously burying the newly fertilized eggs and digging a new redd. The spawning act must be repeated several times because each female normally lays only 100–250 eggs in each cut. Each female lays 200–21,000 eggs in all, the number depending in part on her size (about 30–40 eggs per centimeter of fork length). Egg numbers can also vary according to the habitat choices of individual females. Females that live in heavy cover grow more slowly, presumably because of lower food availability, and produce fewer, but larger, eggs. Females that live in more exposed situations produce more and smaller eggs (18).

The embryos hatch in 4–21 weeks (typically 7–8 weeks), depending on water temperature. Alevins emerge from the gravel and begin feeding 3–6 weeks later. The embryonic and alevin periods are critical for brown trout populations in California because high winter flows can scour the developing fish out of the gravel, resulting in small or absent year classes (21, 28). The fry live in quiet waters close to shore among large rocks or under overhanging plants. They are in shallower and slower water at night than during the day, when they seek the protection of deeper water and cover.

Status IID. Brown trout are abundant in more than 5,000 km of California streams and in numerous lakes (4). Their presence in the state is a mixed blessing. On the positive side they provide some of the finest angling for wild trout in California. Their bottom feeding and piscivorous tendencies,

coupled with their natural wariness, make them difficult for the inexperienced angler to catch. Brown trout essentially coevolved with stream trout angling and have the advantage of nearly 400 years of selection against being caught by anglers. Thus they can maintain substantial populations of large fish even in heavily fished streams. A number of streams in California (e.g., Owens River) are now being managed as wild brown trout streams. In some streams (e.g., Hot Creek, Mono County) the fishery is maintained by stocking juvenile brown trout (26).

On the other hand, brown trout often have a decidedly negative effect on other fishes, including other trout. In lakes and stream pools production of wild, catchable-size trout of all species can sometimes be increased considerably by removing large brown trout that subsist mostly on the other fish. Competition and predation from brown trout may be one factor that has contributed to the extinction of bull trout in the McCloud River. In competitive interactions for food, space, or spawning sites with other trout species, brown trout generally win, all things being equal (19). To restore golden trout and cutthroat trout to their native streams, brown trout eradication is necessary. Their predation on nonsalmonids in California streams can also be a problem. For example, brown trout predation is one of the factors limiting populations of the endangered Modoc sucker in Rush Creek (Modoc County) (20). In Martis Creek, a tributary to the Truckee River, when environmental conditions (especially the absence of scouring flows in winter) favor brown trout for a number of years, they greatly reduce the abundance of native minnows and suckers, nearly eliminating some species (21). Thus brown trout should not be introduced into any more waters in California, and they will have to be removed from a number of small streams to enable conservation of native fishes.

References 1. Stearley and Smith 1993. 2. Dill and Cordone 1997. 3. MacCrimmon et al., 1970. 4. Staley 1966. 5. Frost and Brown 1967. 6. Carlander 1969. 7. Scott and Crossman 1973. 8. Elliott 1994. 9. Fry 1973. 10. Armour 1994. 11. Vincent and Miller 1969. 12. Mesick 1995. 13. Jenkins 1969. 14. Leipzig and Deinstadt 1997. 15. Bridcut and Giller 1995. 16. Ringler 1979. 17. Grost et al. 1990. 18. Lobon-Cervia et al. 1997. 19. Sorenson et al. 1995. 20. Moyle and Marciochi 1975. 21. Strange et al. 1992. 22. Lever 1996. 23. G. E. Smith and Aceituno 1987. 24. Taylor 1993. 25. Svalastog 1991. 26. Deinstadt 1998. 27. Jenkins et al. 1999. 28. Spina 2001.

Bull Trout, *Salvelinus confluentus* (Suckley)

Identification The bull trout is a charr, which means it has fine scales (110 or more in the lateral series), light-colored

spots on the body, and paired and anal fins with white leading edges. The body color is usually olive green with tiny yellowish spots on the back and small conspicuous red spots on the sides. There are no black spots on the body, and fins

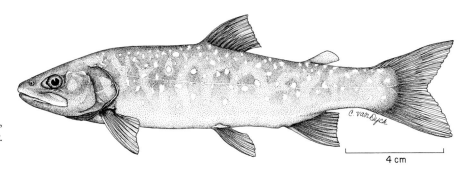

Figure 96. Bull trout, 16 cm SL, McCloud River, Shasta County. CAS 19889.

are free of any spotting except for a few yellow spots at the base of the tail. The head is exceptionally broad and long for a salmonid; it typically accounts for more than 25 percent of the body length and is markedly flat between the eyes. The eyes are placed closer to the top of the head than in most salmonids. The mouth is large and contains sharp teeth; the maxillary bone of the upper jaw extends beyond the eye. There is a fleshy nob at the tip of the lower jaw that fits into a notch on the top of the upper jaw (between the premaxillary bones). The adipose fin is the largest in North American salmonids, its length being 50–85 percent of the depth of the caudal peduncle (15). For McCloud River fish, the branchiostegal rays number 13–15 per side; the mandibular pores, 7–9 per side; and the gill rakers, 15–18 per arch, with visible teeth on the anterior margin of each (15).

Taxonomy Bull trout are part of the Arctic charr complex of *Salvelinus* species native to North America and Eurasia (1, 2, 3). The species within this group are variable and highly overlapping in their characteristics; species are often hard to distinguish where they co-occur. Within this complex bull trout are one of the most recognizable forms and have periodically been recognized as distinct by taxonomists since the 1850s. However, for nearly a hundred years, starting in the 1880s, the species was lumped with Dolly Varden (*S. malma*), a largely anadromous coastal species. In fact the name "Dolly Varden" was originally applied to a bull trout population (see the Names section of this account). In 1978 Cavender (1) provided convincing evidence that Dolly Var-

den and bull trout were indeed separate species, a conclusion that is supported by more recent analyses (3). Although two poor, very old specimens of charr from the McCloud River were identified as Dolly Varden (1), it is most likely that the poor state of the specimens led to misidentification of bull trout (3). The California population of bull trout is distinct morphologically from other populations, but probably not sufficiently so to label it a subspecies (1, 15).

Names Salvelinus was originally assigned by Linnaeus to arctic charr (*S. alpinus*), but the origin of the name was not explained. It was considered by Jordan (13) to be a "low Latin" diminutive of *Salmo,* meaning "little salmon." Another, perhaps more likely, hypothesis about its origin is that it is derived from an old German word for arctic charr present in an account to which Linnaeus had access (14). *Confluentus* translates as, roughly, "flowing together," presumably a reference to the larger streams bull trout inhabit. The rather ugly (but widely used) common name "bull trout" refers to the broad head. I would actually prefer to call it "bull charr" because the trout and charrs represent two distinct salmonid lineages. Charr is also the original English name for members of this genus, a word with Celtic origins meaning blood, referring to the bright red sides of arctic charr (14). However, I yield to the American Fisheries Society committee on names of fishes for the use of "bull trout."

Until the bull trout was separated from the Dolly Varden, it was also officially called Dolly Varden, a name that originated with the California population of bull trout. There are a number of variations on the theme of how the name Dolly Varden came to be (4). I prefer the one recounted to me in a letter (24 March 1974) from Mrs. Valerie Masson-Gomez:

> My grandmother's family operated a summer resort at Upper Soda Springs on the Sacramento River just north of the present town of Dunsmuir. She lived there all her life and related to us in her later years her story about the naming of the Dolly Varden trout. She said that some fishermen were standing on the lawn at Upper Soda Springs looking at a catch of the large trout from the McCloud River that were called "calico trout" because of their spotted, colorful marking. They were saying that the trout should have a better name. My grandmother, then a young girl of 15 or 16, had been reading Charles Dickens' *Barn-*

aby Rudge in which there appears a character named Dolly Varden; also the vogue in fashion for women at that time [mid-1870s] was called "Dolly Varden," a dress of sheer figured muslin worn over a bright-colored petticoat. My grandmother had just gotten a new dress in that style and the red-spotted trout reminded her of her printed dress. She suggested to the men looking down at the trout, "Why not call them Dolly Varden?" They thought it a very appropriate name, and the guests that summer returned to their homes (many in the San Francisco Bay area) calling the trout by this new name. David Starr Jordan, while at Stanford University, included an account of this naming of the Dolly Varden trout in one of his books. Jordan's prestige as the world's preeminent ichthyologist led to the widespread use of the name.

Distribution Bull trout in California were known from only about 100 km of the McCloud River, Shasta and Siskiyou Counties, from the mouth to Lower Falls (4). It is likely that they also occurred in similar spring-cold waters of the upper Sacramento and Pit Rivers, but solid records are lacking. Campbell noted in 1881 that they were scarce near the mouth of the river (17). This was the southernmost population of the species. Today the southernmost populations are found in the Jarbridge River, Nevada, and small streams in the upper Klamath Basin, Oregon (3). The northernmost populations appear to be in the headwaters of the Yukon River, British Columbia (3). The easternmost populations are found in Columbia River tributaries in Alberta and Montana (3). In between these points they are widely scattered in the Columbia River system, in the headwaters of coastal rivers of British Columbia, and in interior drainages of British Columbia and Alberta (Saskatchewan, Athabasca, and Peace Rivers). The presence of many disjunct populations in their present range indicates a wider distribution in the Pleistocene period, under wetter and cooler conditions.

The presence of bull trout in the McCloud River is not easy to explain, but it is tied to both the complex geologic history of the upper Klamath and Pit River basins and to the migratory nature of bull trout, combined with the need of bull trout for habitat conditions that exist mainly in spring-fed headwater streams. One scenario is that they originated in the Columbia River basin (3) and colonized the upper Klamath-Pit basin when it was connected to the ancestral Snake River. Bull trout would have to have first been isolated in the Klamath-Pit system when it became connected to the lower Klamath River and then become isolated further in the Pit basin when volcanic activity severed it from the Klamath and it became connected to the Sacramento River basin. Just as Sacramento River fishes were then able to invade the Pit River basin, bull trout were able to colonize the Sacramento system, and they persisted only in the coldest streams. The weakness of this scenario is the lack of records for bull trout in suitable habitat in the Pit River, including the Fall River. Another scenario is colonization of the river by anadromous bull trout moving up the Sacramento River during a cooler period of the Pleistocene (15).

Life History Little information is available on McCloud River bull trout (4, 5), so most of the information summarized here is from other areas (4, 6). In terms of basic life history, bull trout can be adfluvial (adults in lakes, spawning and rearing in streams), fluvial (all stages in streams, but adults migrate up tributaries for spawning), or resident (no separation of life history stages). A few populations may also be anadromous (3). Most resident populations occur in small streams, and it is possible that many, if not all, of these populations are remnants of populations that were once fluvial (e.g., populations in Klamath basin tributaries in Oregon) (7). In the McCloud River the population was apparently fluvial, with adults concentrating in pools in the lower reaches of the river, migrating upstream to spawn in higher-gradient reaches below Lower Falls (4).

Regardless of life history strategy, the defining characteristic of streams containing bull trout is exceptionally cold, clear water, often originating from springs. They are rarely found in streams that have maximum temperatures greater than 18°C, and optimum temperatures appear to be 12–14°C for adults and juveniles and 4–6°C for embryo incubation (4, 6). The McCloud River prior to the construction of McCloud Dam provided near-ideal temperatures for bull trout, with its major source (Big Springs) flowing in at 7.5°C year round and temperatures in the lower river rarely exceeding 13°C during the summer (4). The river also had other characteristics favorable to bull trout: good conditions for spawning and rearing in the reach below Lower Falls, deep pools in the lower river for adults, and abundant food in the form of juvenile chinook salmon. The McCloud River was once a major spawning stream for winter- and spring-run chinook salmon as well as steelhead and coho salmon. It also supported resident populations of riffle sculpin and Sacramento sucker.

Adult bull trout in rivers prefer to live on the bottom in deep pools; they are also associated with pools in smaller streams. Adfluvial populations thrive in large coldwater lakes and reservoirs (e.g., Flathead Lake and Hungry Horse Reservoir, Montana). In California bull trout were unable to maintain populations in either McCloud or Shasta Reservoir, the two to which they had access. Juvenile trout (to 20 cm TL) are strongly bottom oriented, hanging out near or under large rocks and large woody debris, in stream reaches with coarse, silt-free substrates. They seem to prefer pockets of slow water near faster-moving water that can deliver food (6). As they grow larger they move into pools (4). They seem to be most active at night.

Juvenile bull trout (<11 cm TL) feed heavily on aquatic insects (6). Fish gradually become more important in the diet as they grow larger. Bull trout more than 25 cm TL feed primarily on fish, including juvenile trout and salmon, sculpins, and their own young. Frogs, snakes, mice, and ducklings have also been found in their stomachs. Bull trout typically lie in wait underneath a log or ledge and then dash out to grab passing fish. Feeding is probably most intense in evening and early morning, but I have watched a bull trout that measured 20 cm TL in a Montana stream capture small cutthroat trout at midday. High bull trout densities are often associated with concentrations of small fish, often from migratory populations. Chinook salmon that once spawned in the McCloud River were presumably once a major source of food for local bull trout, both as loose eggs and as juveniles that reared in the river year round.

Bull trout grow slowly but have long life spans (up to 20 years), and so are capable of achieving large sizes. They typically reach 5–8 cm TL in their first year, 10–14 cm in their second, and 15–20 cm in their third. Growth is slowest thereafter in resident populations and fastest in adfluvial populations, members of which may reach 40–45 cm TL in 5–6 years. The largest bull trout on record, from Lake Pend Oreille, Idaho, measured 103 cm TL (14.5 kg) (8). Bull trout from the McCloud River were purported to reach over 7.3 kg (about 70 cm TL), and the California angling record is a fish from McCloud Reservoir that weighed about 5.1 kg. A fish that lived for 19 years in the Mt. Shasta hatchery weighed around 6 kg at the time of death; a second display fish at the hatchery reached a similar size (4). The last two bull trout caught from the McCloud River (in 1975) measured 37 cm SL and 42 cm SL and were 4–6 years old (9).

Bull trout from fluvial and adfluvial populations spawn for the first time in their fourth or fifth year, at lengths of 40 cm TL or more. Fish from resident populations spawn at smaller sizes (25–30 cm TL) and presumably younger ages. They usually migrate upstream to spawn in gravel riffles of clear, cold streams. Migrations of 150–250 km are not unusual in adfluvial populations (10). Movements toward spawning grounds can begin in July or August, but spawning does not begin until water temperatures have dropped below 9–10°C (6, 10) in late summer or fall, apparently in September and October in the McCloud River. Female spawners choose sites that have relatively low gradients, expanses of loose gravel, groundwater or spring inflow, and nearby cover, such as pools. Spawning behavior is similar to that of brook trout (6, 11), although males may spawn with multiple females (10). Small jack males are present among the spawners as well (10). Each female, depending on her size, lays 1,000–12,000 eggs; a mean of 5,482 eggs was found for 32 females averaging 65 cm FL in Montana (10).

Embryos are buried at a depth of 10–20 cm and hatch in 100–145 days (6). After hatching they remain in the gravel for another 65–90 days, absorbing their yolk sacs. They begin feeding while still in the interstices of the gravel and emerge at 23–28 mm TL to fill their air bladders, usually in April or May (6, 10). Young-of-year spend much of their first summer along stream edges or in backwaters, until they reach about 50 mm TL, when they move out into faster and deeper water. Juveniles from adfluvial and presumably fluvial populations will spend 2–3 years in their rearing streams before moving down into adult habitat.

Status IA. Bull trout are extinct in California, increasingly rare in much of their range in the United States, and declining in Canada. They have been proposed for listing as Threatened in the United States. The last known bull trout caught in California was captured by graduate student Jamie Sturgess in 1975, by hook and line. It was tagged and released. Attempts to reintroduce bull trout to the McCloud River from Oregon have failed, and additional attempts are unlikely unless the best source populations (in the Klamath River basin) recover their former abundance (4).

It is hard to be optimistic about the success of a reintroduction program in any case, because the conditions that caused the demise of the trout are still present. They were apparently in decline through most of the 20th century because the McCloud was famous for its "Dolly Varden" fishery in the late 19th century. By the 1930s they were regarded as common but not particularly abundant (5). By the 1950s they were scarce although present in low numbers (4). They became increasingly rare in the 1960s and were gone by the late 1970s. The factors that contributed to the extirpation of bull trout, in rough chronological order, are as follows.

Depletion of salmon. In the 19th century the McCloud River supported at least two runs of chinook salmon, a run of steelhead, and a small run of coho salmon. Juveniles of these fish as well as the annual influx of energy from salmon carcasses quite likely supported fairly large bull trout populations. The 19th-century Sacramento River fishery combined with sediments from hydraulic mining severely depleted salmon runs coming into the McCloud. The Baird Hatchery, established on the lower river in 1874 to take eggs from chinook salmon in order to help restore depleted runs, may, ironically, have contributed to the further decline of McCloud River salmon because the weir next to the hatchery blocked much of the run at times. In the early 20th century the runs recovered somewhat, but not to former levels (16). Then in 1942 Shasta Dam closed and blocked access for all salmon. Salmon were a major driving force in the McCloud River ecosystem, so their depletion and loss undoubtedly had a major impact on the piscivores in the river, including bull trout.

Introduction of brook trout. Brook trout were established in the McCloud River watershed by 1910 or so (4). They are present in small tributaries that juvenile bull trout may once have used for rearing (10). Brook trout will hybridize with bull trout, and this hybridization is a major cause of the decline of resident populations in Oregon and elsewhere (12). However, there is no evidence that hybridization took place in the McCloud River.

Introduction of brown trout. Brown trout probably entered the McCloud River in the 1920s, although they do not seem to have been especially abundant until after the creation of Shasta Reservoir in the 1940s. The reservoir allowed a substantial migratory population of large fish to develop. Large brown trout are ecologically similar to bull trout, hanging out in large pools and preying on other fish. They may have contributed to bull trout decline through a combination of competition and predation.

Shasta Dam and Reservoir. When Shasta Dam closed in 1942, it blocked access of major salmon runs, provided better habitat for migratory brown trout, and flooded about 26 km of the lower river, about a quarter of the bull trout's habitat. Although fluvial bull trout elsewhere have become adfluvial following the construction of reservoirs, this did not happen with Shasta Reservoir. Small numbers of bull trout appeared in the reservoir fishery, but runs from the reservoir never developed. Presumably the reservoir was just too warm for the growth and survival of bull trout (4).

McCloud Dam and Reservoir. McCloud Dam, completed in 1965 and blocking the river about 45 km upstream from Shasta Reservoir, was the final blow to bull trout. First, it flooded 8 km of prime habitat for bull trout. Second, it probably severed the connection between juvenile and adult habitats by blocking adult migrations to upstream areas. Third, it altered conditions downstream of the dam, reducing flows, reducing recruitment of spawning gravel, reducing the frequency of flushing flows, increasing turbidity in the fall, and, most important, raising water temperatures in the river by 5–10°C (4). Once the dam was in place, the long-lived bull trout hung on for 10–12 years before dying out completely.

CDFG has developed a plan for restoring bull trout, mainly by establishing resident populations in some tributaries upstream of McCloud Reservoir and in the lower river (4). These populations would be boosted periodically by hatchery fish if they could not sustain themselves. "Booster" stocking would be inevitable if a fishery became established again. The plan is now on hold although, realistically, this is probably the only way to have bull trout in the river again. Sadly, it reflects the fact that the two dams have radically altered the McCloud River as an ecosystem. As long as they remain, restoration of bull trout will be an uncertain activity at best.

References 1. Cavender 1978. 2. Cavender 1980. 3. Haas and McPhail 1991. 4. Rode 1990. 5. Wales 1939. 6. Pratt 1982. 7. Ziller 1992. 8. Scott and Crossman 1973. 9. Sturgess and Moyle 1978. 10. Fraley and Shepard 1989. 11. Needham and Vaughan 1952. 12. Markle 1982. 13. Jordan 1894b. 14. Karas 1997. 15. Cavender 1997. 16. Yoshiyama 1999. 17. Campbell 1882.

Brook Trout, *Salvelinus fontinalis* (Mitchell)

Identification Brook trout are distinguished from other trout by the combination of a dark, olive green back with lighter-colored wavy lines (vermiculations), red spots on the sides surrounded by blue halos, and white leading edges on the pectoral, pelvic, and anal fins. The tail is slightly forked to a nearly straight shape. The mouth is large and slightly oblique, with the maxillary bone extending past the posterior margin of the eye. Teeth are present in both jaws, on the head of the vomer, and on the tongue and palatine bones, but absent from the shaft of the vomer and basibranchial bones. There are 110–132 scales in the lateral line, 10–14 dorsal fin rays, 9–12 anal fin rays, 11–14 pectoral fin rays, and 8–10 pelvic fin rays. Spawning males are deep bodied with hooked lower jaws (kype); the females develop a protruding genital papilla. Both sexes may become brightly colored when spawning, with dusky to black bellies, red sides, and red lower fins. Young fish have 8–10 wide parr marks, some as wide as the eye, and usually a few red, yellow, or blue spots.

Taxonomy In their native range brook trout show considerable variability in color patterns, morphology, and life history (3), but the species has not been broken up into subspecies. Brook trout in California seem to have rather limited origins and are fairly uniform from place to place.

Brook trout occasionally hybridize with brown trout, producing offspring known as tiger trout—a name that seems to fit both the hybrid's striped color pattern and its voracious feeding habits. Such hybrids are sterile. In hatcheries brook trout have been crossed with both rainbow trout and lake trout. The brook trout–lake trout cross has pro-

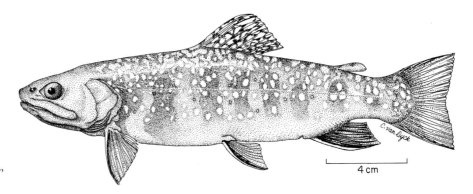

Figure 97. Brook trout, 19 cm SL, First Lake, Inyo County.

4 cm

duced splake, a fertile hybrid that has been stocked in a number of lakes in the eastern United States and Canada.

Names Brook trout are frequently called eastern brook trout in California and speckled trout in Canada. Brook charr is actually a better name, because most members of the genus *Salvelinus* are called charrs. There is a movement afoot, lead by Dr. Eugene Balon of Guelph University, to change the name officially to charr, but the conservatism of fisheries biologists and anglers with regard to fish names makes the likelihood of its use beyond a few scientific journals slim. The word *trout* has been applied to this species since at least 1815, when E. Mitchell described the species from a stream in New York. He placed it in the genus *Salmo* ("true" trouts), from which it was removed in 1878 by D. S. Jordan, who failed to call it a charr. It was then called eastern brook trout in California to distinguish it from native western brook trout, now called rainbow trout (14). Among anglers the name eastern brook trout still persists, so it is probably too much to expect them to call the fish brook charr. *Fontinalis* means living in or near springs. For *Salvelinus* and other names see the account of bull trout.

Distribution Brook trout are native to the northern half of the eastern United States and to eastern Canada, west to eastern Minnesota and Manitoba and northeastern Iowa. A few populations are native as far south as northern Georgia in Appalachian mountain streams. The species has been widely planted in suitable and unsuitable waters throughout the United States and Canada. The first introductions to California were 5,000 embryos brought in by the California Acclimatization Society in 1871 and raised in a hatchery in San Francisco (1). Additional shipments from the East followed. In 1872 the California Fish Commission purchased 6,000 brook trout and planted them in several places. By the 1890s they were being raised in large numbers and being distributed throughout the state (1). They are now established in mountain streams and lakes from the San Bernardino Mountains north to the Oregon border, but they are most abundant in the Sierras. They are also widely established in

mountainous regions of other western states and provinces, as well as South America (5 countries), South Africa, Zimbabwe, Japan, Australia, New Zealand, Europe (at least 13 countries), and a few Pacific islands (2, 14, 15, 18).

Life History Brook trout are fish of clear, cold lakes and streams. Despite their continuous and widespread planting throughout California, populations have become established mostly in small, spring-fed headwater streams and in isolated mountain lakes. These are the coldest of California's trout waters, so it is not surprising that brook trout are among the most cold tolerant of salmonids, feeding at temperatures as low as 1°C (7). They prefer temperatures of 14–19°C but can survive temperatures up to 26°C if acclimated to them. However, growth is poor at temperatures much above 19°C.

In streams brook trout show a wide variety of social behaviors (4). When flows are moderately fast and food is abundant, they defend feeding areas against other trout. Such territories are generally located behind rocks that break the current, permitting the trout to stay in back eddies without expending much energy. As currents become either very slow or very fast, they exhibit less aggressive behavior and engage in other foraging modes. In lakes brook trout tend to swim about as individuals, schooling mainly when alarmed. However, observations in Chiquito Lake, Madera County (elevation 1,700 m), during August 1973 showed that they will congregate in large numbers over springs, presumably attracted to the lower water temperatures (17).

Brook trout in streams feed mainly on terrestrial insects and aquatic insect larvae. Both types of food are taken primarily as drift, on or close to the surface of the water. Brook trout are not particularly selective in their feeding but concentrate on whatever organisms are most abundant. They also do some bottom feeding, indicated by the fact that 20 percent of their summer diet in Sagehen Creek is sculpins (5). Their diet in lakes is similar to that in streams, although zooplankton may also be important. Brook trout juveniles in Castle Lake (Siskiyou County) feed mainly on benthic insects in the littoral zone of the lake; larger trout are also benthic feeders but consume terrestrial insects and zooplankton as well (6). Large trout in lakes may also become piscivorous. Rainbow trout fry dropped into alpine lakes by airplane frequently wind up in brook trout stomachs. Feeding in both lakes and streams has definite daily and seasonal rhythms. Brook trout will feed any time there is sufficient light to see prey, but the most intensive feeding occurs in the evening, when insects are most active, and in early morning. In mountain lakes some feeding takes place under ice in winter, mostly on aquatic insect larvae, zooplankton, and molluscs, but the amount consumed is small compared with that in summer feeding (7). There is also frequently a period in midsummer when the pace of feeding slackens owing to high water temperatures. This is particularly noticeable in shallow lakes and small streams.

Growth in brook trout is highly dependent on length of growing season, water temperature, population density, and availability of food, although other factors—such as water chemistry, the presence of other trout species, heredity, and fishing pressure—also frequently affect growth. In California the fastest growth occurs in lakes and streams of moderate elevation that do not contain large populations either of brook trout or of other fishes. In such situations brook trout will reach 15 cm TL by the end of their first year, 18–20 cm in their second year, and 23–25 cm in their third year. Somewhat slower growth, however, is typical of most California populations, so they seldom exceed 30 cm TL (340 g). The largest brook trout from California, caught in 1932 in Silver Lake (Mono County), measured more than 60 cm TL and weighed 4.4 kg. On the opposite end of the size spectrum are brook trout from Bunny Lake (Mono County). Here poor growing conditions produced fish that measured only 24–28 cm TL, even though some lived as long as 24 years (8). In their long lives these trout reproduced only once (at age 15!) and showed distinct signs of senescence as they got old. The Bunny Lake trout are the oldest brook trout on record from anywhere. Brook trout that live longer than 4 or 5 years are rare.

Accompanying this short life span is a generally early age of maturity. Male brook trout may spawn at the end of their first summer of life at less than 10 cm TL; females may mature at the end of their second summer at 11–12 cm TL. It is more common, however, for males to mature in their second or third year at 12–15 cm TL and for females to mature in their third or fourth year at 14–20 cm TL.

Brook trout are fall spawners, but the specific time depends on water temperature. They usually spawn in California from mid-September to early January at 4–11°C. However, some reproductive activity was observed in Frying Pan Lake, a high-altitude lake in Madera County, in mid-August, when water temperatures were considerably higher (17).

Spawning sites are chosen by females, who seek out areas with the following characteristics (in approximate order of importance): depth greater than 40 cm, upwelling through the substrate, water temperatures colder than those of the surrounding water, pea- to walnut-size gravel, and nearby cover. The preferred site for redd construction is a gravel-bottomed spring in a stream or lake, close to an undercut bank or log. Such a site presumably ensures maximum egg survival. Upwelling through coarse gravel provides a constant flow of cold water around the embryos and slows development so that hatching does not occur before spring; it also prevents ice from infiltrating the gravel in shallow water (9, 10). The presence of nearby cover offers protection from predators for the brilliantly colored spawners. Frequently one or more of the ideal site characteristics may be missing from water where brook trout are established. They will then spawn in suboptimal areas and usually can still maintain populations. Thus brook trout have been observed spawning in gravel riffles, sandy-bottomed springs, and gravel-bottomed shallows of lakes, as well as over piles of boulders. Their adaptability to lake conditions in particular has permitted brook trout to maintain populations in mountain lakes that lack the accessible inlets or outlets most other salmonids require.

Once a female has chosen a spawning site, she begins to dig the redd by turning on her side and shoveling up gravel with rapid movements of her tail. Usually this behavior does not begin unless there are males in the vicinity. Males are attracted to the digging female, and one quickly becomes dominant and defends the redd site against all other males. Often redds are located in territories already defended by males. The female chases away other females, although the male will also perform this task on occasion. As the female digs the male courts constantly by swimming alongside her, nudging and quivering. When the redd is complete (its size depends on the length of the female), the female swims slowly to the bottom and the male quickly swims alongside her, quivering. Together they swim over the bottom of the redd, releasing eggs and sperm simultaneously, the milt visible as a white cloud. The female almost immediately begins to sweep gravel over the eggs with her tail. This new digging activity covers the embryos and serves to start a new redd just upstream from the old one. As only 14–60 eggs are laid at one time and because wild brook trout females contain anywhere from 50 to 2,700 eggs, each female has to repeatedly dig new redds. In California the average fecundity seems to be between 200 and 600 (11). Males also spawn repeatedly, usually with more than one female, and females frequently switch mates between spawns. Spawning activity can occur at any time of day or night but tends to peak in the early morning or at dusk.

Because spawning occurs in autumn, just before the long, hard winter, it is a risky business. The energy drain of spawning reduces survival chances, and as a consequence brook trout spawn just once in many high mountain lakes. Small mature males and large mature females seem to have a particularly hard time surviving (13). This characteristic presents strong local selection pressure on life history traits in brook trout, which may therefore show considerable variability.

Because embryos have to overwinter at low water temperatures, development time is long, usually 100–144 days at water temperatures of 2–5°C. For the first 3–4 weeks after hatching, alevins remain in the gravel. They gradually become more active as the yolk sac is absorbed and the water warms up. Fry in streams move into shallow edges, among emergent plants, or into backwaters of pools, where they feed on small crustaceans. In pools individual brook trout fry show a wide variety of feeding patterns, from benthic feeding to feeding in the water column, with individual fish showing distinct preferences for modes of feeding (12). In lakes they move into shallow water, concentrating in areas protected from wave action.

Status IID. Brook trout are present in more than 1,000 lakes and 2,200 km of stream in California (11). In most of these waters their populations are self-sustaining and support angling. Small numbers are still raised in California hatcheries despite low demand and abundant wild fish. This has not always been the case. In the 1890s and early 1900s large numbers were raised and planted, many in fishless waters of the high Sierras. Stocking was done by fisheries workers, foresters, and laymen enthusiastic about the beautiful colors, edibility, and angling qualities of brook trout—but unfortunately ignorant of their biology. Alpine lakes are also typically in headwaters and so provide a continuous source of trout to invade downstream areas, leaving few places as refuges for rare native trouts (19, 21). Their continuing impact on high mountain lakes that were originally fishless is a major problem. Brook trout introductions fundamentally change alpine lake ecosystems, including stimulating algae blooms through increased nutrient cycling (19). Repeated introductions of large numbers of juvenile brook or rainbow trout into lakes may also be the equivalent of fertilizing the lakes on a regular basis. More directly, brook trout (and other trout in these lakes) have eliminated populations of mountain yellow-legged frogs, Yosemite toads, other amphibians, and large invertebrates through predation (20). Mountain yellow-legged frogs, for example, spend 2 years as tadpoles and have to survive winters in deep lakes, making them easy prey for trout. The lakes they require are the same ones that are best for fish. Although other factors, such as airborne pollutants, contribute to amphibian declines, in the Sierra Nevada fish appear to be the single biggest cause (16, 18). If present trends continue, high mountain amphibians that depend on lakes will all be endangered.

Brook trout can also contribute to the elimination of native trout populations in streams. Where they are introduced into waters with native cutthroat trout, they typically displace and eventually eliminate native trout through competitive interactions.

In mountain lakes and streams they often do not even provide much of a fishery because of their inability to reach large sizes, through either stunting (intraspecific competition for limited food) or poor overwinter survival (owing to postspawning stress). Obviously, eliminating brook trout from hundreds of high mountain lakes and streams is not possible, but there are ways for humans and trout to share the mountains with the native aquatic fauna, especially frogs and toads. The best solution is to select some of the more remote watersheds in the Sierras and elsewhere and systematically eliminate fish from them, creating special fish-free watersheds as refuges (16).

References 1. Dill and Cordone 1997. 2. MacCrimmon and Campbell 1969. 3. Hutchings 1996. 4. Grant and Noakes 1988. 5. Dietsch 1959. 6. Wurtsbaugh et al. 1975. 7. Elliott and Jenkins 1972. 8. Reimers 1958, 1979. 9. Curry and Noakes 1995. 10. Curry et al. 1995. 11. McAfee 1966. 12. McLaughlin et al. 1994. 13. Hutchings 1994. 14. Karas 1997. 15. Lever 1996. 16. Knapp and Matthews 2000. 17. J. P. Bartholomew, CDFG, pers. comm. 1974. 18. Bradford et al. 1998. 19. Schindler et al. 2001. 20. Knapp et al. 2001. 21. Adams et al. 2001.

Lake Trout, *Salvelinus namaycush* (Walbaum)

Identification Lake trout can be readily recognized by their deeply forked tails with pointed lobes and by their color pattern of irregular white to yellow spots on a background of light green to gray that covers the entire body, including the head and fins. The spots may be obscured if the trout has assumed a silvery color overall. There is a pale white border on the leading edges of the paired and anal fins. Lake trout are heavy bodied. The head is broad and constitutes about 25 percent of the standard length. The mouth is large, the maxillae extending past the posterior margin of the eye. Well-developed teeth are present on the jaws, head of the vomer, palatines, tongue, and basibranchial bones. There are 16–26 gill rakers per arch and 10–14 branchiostegal rays on each side. The dorsal fin has 8–10 major rays; the anal

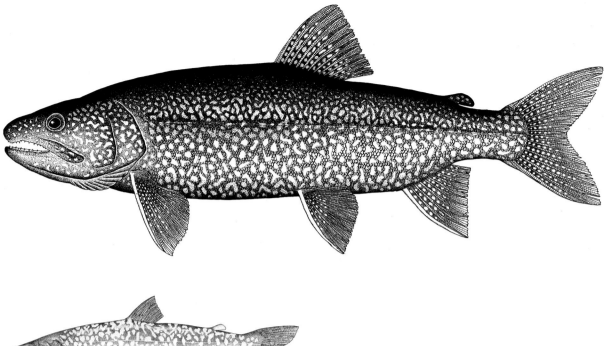

Figure 98. Lake trout. Top: Adult, ca. 60 cm SL, drawing by Paul Vecsei. Bottom: Juvenile, 23 cm SL, Lake Tahoe, Placer County.

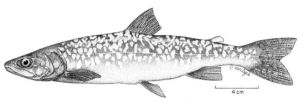

fin, 8–10 major rays; the pelvic fins, 8–11 rays; and the pectoral fins, 12–17 rays. The scales are small, 116–138 in the lateral line.

Parr have 7–12 irregular parr marks, which are equal in width to or narrower than the interspaces. Their fins are without color and their backs have small, irregular white spots.

Taxonomy Lake trout are distinct enough from other *Salvelinus* species that they have been placed in the past in their own genus, *Cristivomer*.

Names Lake trout in California are sometimes referred to as Mackinaw trout or just Mackinaws. *Namaycush* is a version of the name given this fish by Native Americans. For *Salvelinus,* see the account of bull trout. For the question of trout versus charr, see the account of brook trout.

Distribution Lake trout are native to most of the interior of Canada, coastal drainages of Alaska (except the Yukon River), and the Great Lakes and St. Lawrence drainages of the United States (1). Relict populations are present in some Montana lakes. They have been widely introduced into deep, cold lakes in the United States, New Zealand, Sweden, South America, and probably elsewhere. Lake trout from the Laurentian Great Lakes were first introduced into California (Lake Tahoe tributaries) in 1886 by the Nevada Fish Commission, but it is not known if any of those initial fish survived. However, in 1889 and subsequent years many more lake trout were propagated and planted (2). They are present today in California in Tahoe, Donner, Fallen Leaf, and Stony Ridge Lakes, all in the Tahoe basin. They have also become established in Sly Park Reservoir (El Dorado County), Caples Reservoir (Alpine County), and Gold Lake (Sierra County). Large numbers were planted in Oroville Reservoir, and a number were caught in the fishery, but apparently no reproduction has taken place.

Life History Because lake trout are the most sought-after game fish in Lake Tahoe, their life history has been extensively investigated in the lake (3, 4, 5, 6, 7).

Lake trout ordinarily inhabit deep, cold waters of lakes, although in more northern parts of their range they also live in shallow water and in rivers. In Lake Tahoe they are usually found deeper than 30 m and have been collected as deep as 430 m. In spring and fall, however, they may move into shallow water to feed. They are one of the least tolerant salmonids to high temperatures, preferring water less than 13°C and dying if it becomes much warmer than 23°C. Their salinity tolerance is also low for a salmonid: 11–13 ppt is usually the maximum they can withstand, although there are a few anadromous populations in eastern Canada. Lake trout dwell on or close to the bottom. They often concentrate around deep reefs, but they exhibit little social behavior outside the breeding season.

The diet of Lake Tahoe lake trout changes with the size of the fish as well as with the season (7). Prior to the introduction of opossum shrimp, *Mysis relicta*, into Lake Tahoe, trout measuring less than 13 cm FL fed mostly on zooplankton (91% by weight, primarily *Daphnia pulex*) but also on chironomid midge larvae and pupae. Zooplankton continued to be important (33%) to trout measuring 13–25 cm FL, but Paiute sculpins were the main item in their diet (56%). As the trout increased in size, zooplankton ceased to be of much importance in the diet, and they started preying on virtually every available fish species in the lake and, to a lesser extent, on crayfish. For trout greater than 50 cm FL, the favorite prey was Tahoe sucker (45%), followed by other trout (17%) and mountain whitefish (11%). Very few Lahontan redside, speckled dace, or kokanee salmon were taken by lake trout of any size, reflecting the usual restriction of trout to bottom habitats in deep water. They also took surprisingly small numbers of their own young. Feeding activity was most intense during the spring and fall months. After the opossum shrimp became abundant in the lake, *Daphnia* ceased being part of the diet of small trout because the shrimp had consumed them all (8). The shrimp became important in the diet instead, and they now typically constitute 30–50 percent of the diet of lake trout of all sizes, although their importance decreases once the trout exceed 50 cm TL (9). In large trout fish still predominate, especially Paiute sculpin and Tahoe sucker.

Growth in Lake Tahoe is slow even for lake trout, which is a slow-growing species in general (6). The trout, however, are long lived (up to 17 years in Lake Tahoe, up to 41 years elsewhere), so they can achieve large sizes (more than 1 m FL and 9.1 kg in Lake Tahoe and 1.25 m and 28.6 kg in Lake Athabasca, Saskatchewan). In Lake Tahoe average fork lengths for ages 1–10 are, respectively, 12, 18, 25, 32, 38, 43, 48, 53, 58, and 62 cm. This pattern of growth has remained relatively constant during the over 60 years for which records have been kept, even following the introduction of mysid shrimp. Most growth takes place in June through September. It does not cease in the winter but only slows down.

Lake trout in Lake Tahoe become mature for the first time in their fifth through eleventh years, but they spawn every year thereafter (5). They spawn from mid-September to mid-November in deep (but <37 m) water over bottoms covered with rubble and boulders. Lake trout are unique among North American charrs, trout, and salmon in that they do not build redds or defend breeding territories. Instead males arrive first in the breeding area and sweep rocks clean of silt and debris by fanning them with their fins or rubbing them with their bodies. Curiously, in Lake Tahoe they spawn on beds of macrophytes, at 40–60 m, the only population known to use such a substrate for spawning (10). Most spawning takes place at night. Each female spawns with one or more males simultaneously after a brief courtship ceremony. Fertilized eggs fall between crevices of the rocks and are left unattended by the adults. In Lake Tahoe each female lays an average of 3,400 eggs, with a range of 900–11,500, depending on body length.

Embryos hatch in 4–6 months, and alevins remain among the rocks for the first month or so. Little information is available on the ecology of lake trout for the next 1–2 years in Lake Tahoe, although it is generally assumed that they continue to live on the bottom in deep water and feed on benthic invertebrates and zooplankton.

Status IIC. Lake trout are a well established and popular game fish with self-sustaining populations in the Tahoe basin and a few reservoirs. Most are taken by trolling with bait and lures close to the bottom in deep water. The specialized nature of the fishery makes it a stable one, and there is no evidence of overharvesting. The obsession of management agencies with the lake trout fishery in Tahoe led to the introduction of opossum shrimp into the lake. The shrimp was assumed to be a detritus feeder that would become a major food source for juvenile trout and improve their growth and survival, on the assumption that small trout were food limited. Unfortunately, the shrimp preys on larger zooplankton (the same species important to juvenile lake trout), which thus largely disappeared from the lake. This miscalculation caused a major shift in the lake's ecosystem to the detriment of plankton-feeding fishes (8).

Even earlier, lake trout may have been one of the main reasons Lahontan cutthroat trout, which they (and rainbow trout) replaced ecologically, are now extinct in Lake Tahoe. Presumably the combination of competition, predation, and disease from lake trout made it impossible for cutthroat trout to recover from the ravages of the turn-of-the-century commercial fishery. Overall the suite of introduced species in Lake Tahoe—lake trout, rainbow trout, kokanee, crayfish, and opossum shrimp—have caused major changes in the way the ecosystem functions, probably to the detriment of native fishes. The lake trout's slow growth rate, late age of maturity, and vulnerability to trolling, however, do make it susceptible to overfishing—which might actually be a good idea if restoration of cutthroat trout is to be seriously considered for Lake Tahoe.

References 1. Scott and Crossman 1973. 2. Dill and Cordone 1997. 3. R. G. Miller 1951. 4. Cordone and Frantz 1966. 5. Hanson and Wickwire 1967. 6. Hanson and Cordone 1967. 7. Frantz and Cordone 1970. 8. Richards et al. 1991. 9. Frantz 1979–1981. 10. Beauchamp et al. 1992.

Silversides, Atherinopsidae

Silversides are 70 or so species of silvery, streamlined pelagic fishes abundant in tropical and temperate coastal environments of North and South America. A number of species, including three in the eastern United States, live in fresh water. Silversides are often called smelt because of their superficial resemblance to the true smelts (Osmeridae). However, they lack adipose fins and lateral lines and have *(1)* two widely separated dorsal fins, the first composed of spines; *(2)* a single spine on the anal fin; *(3)* pectoral fins at or above the midline of the body; *(4)* a silvery band on each side; and *(5)* a small terminal mouth. The three silversides native to California, all marine, are the California grunion (*Leures-thes tenuis*), jacksmelt (*Atherinopsis californiensis*), and topsmelt (*Atherinops affinis,* covered here). Topsmelt are often abundant in estuaries, and juveniles occasionally occur in tidal reaches of streams, although rarely in fresh water. The only completely freshwater silverside in California is the alien inland silverside, which can also tolerate a wide range of salinities.

The New World silversides, Atherinopsidae, were formerly lumped together with the Old World silversides in the family Atherinidae. The distinctiveness of the New World forms led to them being split off as a separate family (Dyer and Chernoff 1996).

Inland Silverside, *Menidia beryllina* (Cope)

Identification Inland silversides are elongate (standard length is 6–7 times body depth) and slender, with large eyes, oblique mouths, and pointed heads that are flattened on top. The two dorsal fins are widely separated; the first fin is small, with 4–5 weak spines, while the larger second fin has 1 spine and 8–9 rays. The sickle-shaped anal fin has 1 spine and 16–18 rays. The lateral line has 36–44 scales. Live fish are yellowish on the back with a silvery band on each side. The lower half of the fish often appears translucent with a greenish cast. The upper rows of scales are outlined with pigment spots, and there are two rows of pigment spots on the bottom of the caudal peduncle between the anal fin base and the caudal fin base. Individuals are rarely longer than 10 cm SL.

Taxonomy When first introduced into California, inland silverside were listed as Mississippi silverside, *M. audens,* which were regarded as a freshwater species distinct from *M. beryllina,* an estuarine species. The two forms merge imperceptibly in many areas, so there is no reason to separate them as species (1). However, the systematics of this widely distributed species is fascinatingly complicated. For example, hybrids between *M. beryllina* and *M. peninsulae* in coastal regions of Texas form all-female self-reproducing clones (2).

Names Menidia seems to be derived from an old Italian word for small silvery fish. The genus was first described by

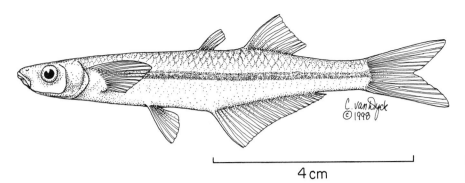

Figure 99. Inland silverside, 7 cm SL, Putah Creek, Yolo County.

Charles Lucien Jules Laurent Bonaparte, a nephew of Napoleon and author of *Fauna Italica. Beryllina* means emerald color. A commonly used, if unofficial, alternative name for inland silverside is tidewater silverside.

Distribution Because of their complicated taxonomic history and similarity in appearance to other *Menidia* species, the exact native range of inland silverside is not clear. They appear to live in estuaries and lower reaches of coastal streams along the Atlantic coast from Maine to Florida and along the Gulf Coast from Florida to Veracruz, Mexico. In the Mississippi River they occur in backwaters and reservoirs from southern Illinois and Missouri to the coast, including a wide distribution in tributaries in Texas and Oklahoma (3). Fish from Texoma Reservoir, Oklahoma, were introduced into Blue Lakes and Clear Lake, Lake County (the latter introduction unauthorized) in 1967 (4). By 1968 they were enormously abundant in Clear Lake and quickly (by 1972) spread down its outlet stream, Cache Creek, and to Putah Creek, which is connected to Cache Creek by irrigation ditches (5). In 1968 experimental introductions by CDFG were made in small, isolated lakes in Alameda and Santa Clara Counties, and fish were subsequently spread, mostly illegally, into a number of other reservoirs in these counties, including Anderson, Del Vall, and Lexington (5), and elsewhere (e.g., Coyote Reservoir, Mendocino County, in the Russian River drainage). By 1975 they were well established in the San Francisco Estuary. From there they spread south into the San Joaquin Valley and via the California Aqueduct into reservoirs in southern California, where they became established sometime before 1988 (6). They can thus be expected anywhere habitat conditions are suitable in the Central Valley, Russian River, and coastal southern California. Unauthorized introductions by bait fishermen may also put them into other watersheds.

Life History Silversides are most abundant in shallows of warmwater lakes, reservoirs, and estuaries, where they shoal in enormous numbers, concentrating in or near protected areas with sand or gravel bottoms. Despite their delicate ap-

pearance and tendency to die when handled, silversides survive a wide range of conditions. Large numbers were taken in August 1972 from a warm, stagnant, and extremely turbid pool in an intermittent section of Cache Creek, the outlet of Clear Lake. These fish must have been derived from fish that arrived in the pool in spring, after being washed downstream more than 87 km over a steep series of rapids. They occur in water of 8–34°C, but environments of 20–25°C are probably optimal for growth and survival. They can survive salinities over 33 ppt (14, 27) and are commonly found at salinities of 10–15 ppt. For larvae, optimal salinity for survival and growth appears to be around 15 ppt (25), although they do well in fresh water as well. They are most typical of areas that are at least seasonally freshwater. In the lower San Joaquin River they are abundant in highly disturbed areas with high levels of pollutants from agricultural areas (7).

Shoaling is the predominate social pattern of silversides. Fish of similar size group together and swim in huge schools during the day. In Clear Lake segregation of size groups is especially striking when a school of large fish passes over one of smaller fish without any mixing taking place. Wurtsbaugh and Li (8, pp. 568–569) observed a striking pattern of daily behavior in Clear Lake involving both inshore-offshore migrations and along-shore movements:

At night the fish were in the nearshore area in high-density, inactive non-schooling aggregations. . . . At dawn . . . , the fish disappeared from the nearshore zone; they returned 3–4 h later and were first seen in schools of several hundred, swimming parallel to the beach in the nearshore zone. As the numbers of fish increased, the actively moving schools became a band 5–6 km long and 1–30 m wide, with its closest edge 20 m from shore. . . . The numbers of fish in the stream increased until about noon. Extremely large numbers of fish were involved in the alongshore migration, and on two occasions we visually estimated . . . that between 400,000 and 800,000 fish per hour were passing a fixed point on the beach. Swimming speeds of the fish were high, ranging from 27 to 55 cm/sec. . . .

During the morning, the direction of movement was counterclockwise. . . . Streaming behavior of the band of fish was interrupted or ended when the fish reached cover. As the fish streamed into a major resting area, huge aggregations built up and we measured more than 500 fish per m². . . . Aggregations of this density were seen near cover such as inundated terrestrial vegetation or docks. In the afternoons, streaming was again observed along the beaches and was clockwise in 11 of 13 observations—the reverse of the morning migration. . . . Although some fish began streaming in the afternoon, aggregations of fish near cover persisted until dark.

In addition to these movements, during summer most silversides in Clear Lake move offshore at dawn, 1–2 hr before sunrise, and then return to inshore areas in midmorning. A similar movement, but of lesser magnitude, can occur in the afternoon (8). The fish spend the night inshore. Once out in open water, silversides may move downward to depths as great as 5 m to forage on zooplankton (8). The complex behavior of silversides observed in Clear Lake is apparently variable, because in some situations no diurnal movement patterns are known (9). The complex movements may also cease during winter.

Silversides move out into open waters to feed on zooplankton and presumably move into shallow water to reduce predation by piscivorous fishes. They are eaten by virtually every fish in Clear Lake large enough to capture one, including bluegill, largemouth bass, black and white crappie, white catfish, brown bullhead, and channel catfish. However, in the 1970s it was clear that predation was not limiting their populations because shoals of silversides contained many deformed individuals and many fish with infections of parasitic copepods. Many copepods embedded in fish had long streamers of green algae growing from them. Both deformed and algae-infected fish were easy to pick out in shoals and to capture with dipnets, a simulation of predation. The recovery of populations of western grebes and other piscivorous birds in Clear Lake may change silverside predator-prey dynamics.

The immense shoals of fish streaming into inshore waters to avoid predators deplete local zooplankton, forcing fish into offshore waters, where predation risk is higher, to feed. In Clear Lake silversides feed primarily on zooplankton (mainly copepods and cladocerans), selecting larger species (>1.0 mm) especially among cladocerans (10). Prey size increases with fish (mouth) size (11). They also consume planktonic instars of chironomid midges and chaoborid gnats, but only when the instars are abundant. As indicated by movement patterns, silversides in Clear Lake show a peak of feeding 1–3 hr after daybreak and another just before dusk; they feed at night, but the average size of prey increases as the level of light decreases (10). Thus, even though cladocerans dominate daytime diets in Clear Lake, amphipods and insect larvae dominate nighttime diets, although many individuals apparently are unable to find food

at night. A similar diet is found in fish living in brackish Suisun Marsh, Solano County. Silversides (N = 129) collected at various times and places in the marsh fed mainly on planktonic copepods, but at times chironomid midge larvae and pupae, cladocerans, and harpactacoid copepods dominated the diet. Individuals fed opportunistically on a wide variety of aquatic insects and crustaceans.

The life cycle of silversides is characterized by fast growth and short life. In their first year, despite a virtual cessation of growth in winter, they reach 8–10 cm TL (9). Most spawn and die in their first or second summer of life, although a few females may live through another year, reaching lengths of 11–12 cm SL. Females grow faster and larger than males. Silversides are fractional spawners, so spawning can take place on nearly a daily basis at temperatures of 15–30°C (12, 13). Under the right conditions females can produce 200 (50 mm SL fish) to 2,000 (100 mm SL fish) eggs per day for at least a 3-month period (13). Hubbs et al. (14) estimated that a single female could lay more than 15,000 eggs in a summer if spawning were continuous. However, this does not seem to be the case in California, because there is apparently high mortality in spawning fish. Spawning occurs April–September (15). There appear to be two peaks of reproductive activity in Clear Lake, one in May (from overwintering fish) and the other in August. At least three size classes of silversides are evident by the end of summer, suggesting that silversides are capable of maturing by the end of their first summer. In Putah Creek (Yolo County), and the Cosumnes River (Sacramento County), larval silversides are present from March through August, with a strong peak in April and May (24).

Spawning takes place during the day (25) over beds of aquatic plants, among emergent vegetation, or among organic debris, where groups of males apparently station themselves while waiting for passing schools of females. These males are presumably more vulnerable to local predatory fishes, so the male:female ratio drops considerably as spawning season progresses (14).

The spawning act was described by Fisher (16, p. 315):

On May 11, 1973, during mid-morning, silversides were observed spawning in Lexington Reservoir, over a gentle slope in water 1 to 24 inches deep. Water temperature was 68°F. The slope was covered with rooted aquatic plants and some inundated terrestrial plants. The vegetation formed a mat 1 to 2 inches thick. A school of 25 to 150 individuals would approach parallel to the shoreline led by one or more females with visibly swollen abdomens. When sampled, one school contained 2 large females and 39 smaller males. When a school turned onto the slope, males began to swim vigorously around the female, nipping and prodding at her abdomen. Occasionally during this frenzied activity, a female would suddenly break fee, closely accompanied by 3 to 5 males. She would dive, along with the males, into the rooted vegetation. There, both sexes began trembling violently. While lying on her side in close contact with the males, the female laid her eggs. Upon completion, she would rapidly swim away, still closely pursued by several males. The spawning group rejoined the larger school and left the shallow area. Examination of the vegetation showed each female deposited from 10 to 20 eggs. As each school passed, the females made a single spawning pass and were not observed to repeatedly broadcast eggs.

Newly fertilized eggs are adhesive and also have chorionic filaments that attach them to the substrate. The larvae hatch in 4–30 days, depending on the water temperature, and measure 3.5–4.2 mm TL (15). Larvae are planktonic for several weeks following hatching, before joining larger fish in shallow water.

Status IIE. Silversides became the most abundant fish in Clear Lake within 2 years after their introduction, a testament to their powers of reproduction. Their continuing spread to other waters in California has been so rapid that their impact is just beginning to be appreciated. They are now one of the most abundant fish in many areas, including the San Francisco Estuary. Here I discuss their impacts, real and speculative, on two ecosystems where they have been studied: Clear Lake and the San Francisco Estuary.

Clear Lake. Silversides' effects on the Clear Lake ecosystem can be summarized by examining their interactions with *(1)* native planktivores, *(2)* piscivores, *(3)* zooplankton, and *(4)* algae.

Planktivores. The ability of inland silversides to replace other littoral zone fishes was first demonstrated in Oklahoma, where they almost completely replaced ecologically similar brook silversides (*Labidesthes sicculus*) in at least two reservoirs (17). In Clear Lake native planktivores and other small fishes became much less abundant following the silversides explosion. The original inshore planktivores were young of native minnows, especially Clear Lake splittail, which, early residents noted, were enormously abundant in shallow water (and, ironically, were called "silversides"). The minnows were gradually replaced by young bluegill; Cook et al. (18) reported taking as many as 10,000 small bluegill in one seine haul. Despite the abundance of bluegill, about 20 percent of fish in their hauls were hitch and blackfish (18). Today these minnows make up less than 1 percent of such samples. Competition from large populations of silversides may have been the final blow to Clear Lake splittail, which became extinct a few years after the silversides introduction. The mechanism of replacement is not known, but it presumably relates to depletion of inshore zooplankton, crowding of juvenile fish into deeper water where they would be more vulnerable to predation, and direct predation on larvae of native fishes. Splittail were (and the hitch still are) stream spawners, with larvae that wash into inshore areas of the lake, so their larvae would be exceptionally vulnerable to predation by silversides.

Piscivores. One justification for introducing silversides into Clear Lake was that they would have a positive effect on game fish, especially largemouth bass and crappies, by providing abundant year-round forage for them (4). Indeed virtually every game fish in the lake, including large bluegill (not normally piscivorous), feeds on them. However, there is no evidence that the silversides have increased either the population sizes or the individual growth rates of game fish; they may have simply replaced forage species already present. In the 1940s the dominant fish in the diet of largemouth bass were still native minnows, but by the 1950s the diet was mainly bluegill (19). Silversides may actually have a negative impact on game fishes by decreasing the growth and survival of juveniles that also feed on zooplankton. Thus black and white crappie showed significantly slower growth as juveniles following the silverside introduction, suggesting that competition was affecting them. However, growth of adults increased after the invasion, so that by the end of their fifth year adult crappie were slightly larger than they had been prior to the invasion (20). Crappie populations nevertheless collapsed in the late 1970s and never recovered, suggesting that interactions with silversides (perhaps larval predation) have been a factor in their decline and failure to recover. Bass populations, in contrast, have flourished. One factor contributing to the success of bass may be that juvenile bass switch from feeding on zooplankton to feeding on silversides at a smaller size than they did when their prey were bluegill and native minnows (21).

Invertebrates. The main justification for introducing silversides into Clear Lake was control of Clear Lake gnat (*Chaoborus astictopus*) and chironomid midges (4). These tiny, nonbiting flies live in the lake as larvae, the gnat as a planktivore, the midges as a members of the benthos. Gnats were regarded as a particular nuisance because they emerged in huge numbers at times, swarming to any source of light at night. During the 1950s through early 1970s gnat control was tried with pesticides, first DDE and then parathion applied in massive quantities. The lake ecosystem is still recovering from the effects of persistent DDE, which wiped out many of the piscivorous birds. Silversides were introduced as a form of biological control; it was assumed they would eat gnat larvae in the plankton and midge larvae as they emerged. Since the introduction gnat and midge populations have been small and the nuisance problem has abated. Whether this is causation or just correlation is not known. It is reasonable to assume that a superabundant planktivore that moves offshore to feed on a daily basis would be able to consume large numbers of gnat and midge larvae and pupae. One unresolved apparent contradiction is that gnats feed mainly at night and stay close to the bottom during the day, to avoid predation by fish, whereas silversides are diurnal feeders. It is also curious that gnats were an occasional problem when planktivorous native minnows were abundant both inshore and offshore. Thus it is possible that the low numbers of gnats are related to some other factor—perhaps even tied to other ecosystem changes wrought by silversides.

San Francisco Estuary. The effects of silversides on other organisms in the San Francisco Estuary have been poorly studied, yet there is reason to be concerned because they are the most abundant fish in shallow-water areas in many areas. These areas are important for rearing of juvenile salmon, splittail, and other fishes. If resident populations of silversides deplete the abundance of inshore plankton and insects as they have in Clear Lake, then they may reduce the growth and survival rates of juveniles. For species that spawn in these areas, silversides may prey on embryos and larvae. One species likely to have been negatively affected by silversides is the endangered delta smelt. Bennett and Moyle (22) noted that *(1)* the crash of delta smelt populations coincided with invasion of silversides into the estuary, *(2)* silversides have been found experimentally to be voracious predators on fish larvae, and *(3)* delta smelt spawn in shallow water areas where silversides are abundant, especially during dry (low-outflow) years. Delta smelt are affected by many other factors that were presumably also tied to the collapse of their populations, but the abundance of silversides may inhibit recovery.

A major effort is under way to restore shallow-water habitat in the estuary. There is considerable concern that this effort may just increase habitat for silversides at the expense of other species. At the very least, this habitat will have to manipulated in ways that discourage silversides.

Overall the introduction of inland silversides is a classic case of the "Frankenstein effect" (23), in which a species introduced for a positive reason turns out to be a monster. Silversides were introduced to control a nonbiting nuisance gnat in Clear Lake, and may even have done so. They have not, however, proven useful for insect pest control in other situations (26). Their rapid spread to other waters may have had many unintended consequences, such as suppression of delta smelt populations, which are just beginning to be appreciated. The spread of silversides could have been anticipated, but the biologists making the introduction were so focused on a local problem that they failed to appreciate the bigger picture.

References 1. Chernoff et al. 1981. 2. Echelle and Echelle 1997. 3. Lee et al. 1980. 4. Cook and Moore 1970. 5. Moyle et al. 1974. 6. Dill and Cordone 1997. 7. Brown 2000. 8. Wurtsbaugh and Li 1985. 9. Mense 1967. 10. Elston and Bachen 1976. 11. Shoup and Hill 1997. 12. Cl. Hubbs and Bailey 1977. 13. Cl. Hubbs 1982. 14. Hubbs et al. 1971. 15. Wang 1986. 16. Fisher 1973. 17. Gomez and Lindsay 1972. 18. Cook et al. 1964. 19. McCammon et al. 1964. 20. Li et al. 1976. 21. Moyle and Holzhauser 1978. 22. Bennett and Moyle 1996. 23. Moyle et al. 1986. 24. Rockriver 1998. 25. Middaugh et al. 1986. 26. Kramer et al. 1987. 27. Baxter et al. 1999.

Topsmelt, *Atherinops affinis* (Ayres)

Identification Topsmelt are small and slender, with flattened backs and small oblique mouths. The first dorsal fin has 5–9 spines and is separated from the second dorsal fin (1 spine, 8–14 rays) by 5–8 scale rows. The anal fin has 1 weak spine and 19–25 rays; the pectoral fins have 13 rays. The lateral line is absent, but there are 63–65 scales in the lateral se-

ries. The jaw teeth are tiny, forked, and arranged in one row on each jaw. Adults have bright green backs and silvery bellies, separated by a wide silver to pale stripe. Juveniles are almost uniformly translucent white, although their midline stripe is still conspicuous and they have more than 3 scale rows on the back that are outlined with pigment spots. Pigment spots on the base of the caudal peduncle are in a scattered pattern, not in rows. Topsmelt reach 37 cm TL, but in low-salinity water it is unusual to find one over 10 cm TL.

Taxonomy Five subspecies have been described for topsmelt for different regions of the coast (1). Given the lack of real geographic isolation among most of the subspecies, the validity of these subspecies is doubtful, and they should be reexamined using modern biochemical and statistical techniques.

Names The word *smelt* has always been associated with this fish, which is unfortunate because it is only superficially similar to the true smelts. The common name otherwise is descriptive, because they are commonly seen in silvery schools

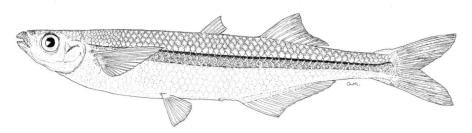

Figure 100. Topsmelt, 18 cm SL, Navarro River estuary, Mendocino County. Drawing by A. Marciochi.

near the surface. *Atherin-ops* literally means *atherina*-appearing, denoting the resemblance of topsmelt to European silversides, genus *Atherina*. *Atherina* is the Latinized version of the ancient Greek word for silversides, which was derived from their word for spike or arrow. *Affinis* means related, referring apparently to jacksmelt (*Atherinopsis californiensis*), with which topsmelt are often found.

Distribution Topsmelt are found in coastal waters, bays, and estuaries from Vancouver Island, British Columbia, to the Gulf of California, although they are uncommon north of central Oregon (2).

Life History Topsmelt are primarily saltwater fish, preferring shallow bays, sloughs, and estuaries. They are, however, one of the most common marine fishes found in lower reaches of coastal streams and upper estuaries, although seldom are they found in completely fresh water. Because estuaries are used mainly for spawning and as nursery areas, most topsmelt in fresh or brackish water are young-of-year or yearlings. In San Francisco Bay they are most abundant in shallows in March–September but move into deeper water or the ocean in winter (8). They occur at salinities of 0–34 ppt and temperatures of 5–29°C (8). In the Navarro River in August 1973, small shoals were observed in areas of slow current only at salinities greater than 9–10 ppt. In Mugu Lagoon, Ventura County, they were found to be abundant in September–November at salinities ranging from nearly fresh water to seawater, temperatures of 9–25°C, and dissolved oxygen levels of 2–18 mg/liter (3). They can live in hypersaline environments as well, at salinities up to 80 ppt (4).

Given their large eyes and surface-feeding body shape, it is surprising to find that topsmelt are, in general, bottom-grazing or algae-browsing omnivores. Topsmelt in the Navarro River estuary, measuring 49–56 mm SL, fed on diatoms and filamentous algae (50% by volume), detritus (29%), chironomid midge larvae (10%), and amphipods (10%). The type of organisms, coupled with the presence of detritus and sand grains in each stomach, indicates bottom feeding. In Anaheim Bay juvenile topsmelt (<81 mm SL) also consumed zooplankton, especially cladocerans, copepods, and various crab larvae (5). Adults, however, were more benthically oriented, feeding on a variety of benthic invertebrates.

The spawning period of topsmelt is long, March–October, although its timing and extent depend on location and temperature. Spawning occurs at 10–27°C and at salinities up to 72 ppt, but optimal conditions seem to be 13–27°C with salinities of around 30 ppt (2, 6). Females are fractional spawners and apparently spawn off and on all summer. Spawning occurs at night over submerged vegetation, and embryos are attached to plants by tangling chorionic filaments (7). Each ripe female contains 200–1,000 eggs, depending on her size (2), but actual fecundities may be higher because they are fractional spawners (see the account of inland silverside). After hatching, larvae aggregate in shallows, gradually moving into more open and slightly deeper water as they grow larger (7). Juveniles may move up higher into the estuary, even into fresh water, but as they mature they prefer more saline bays and estuaries.

Status IE. Topsmelt are basically a euryhaline marine species highly tolerant of extreme environmental conditions. They are typically one of the most abundant fish in bays and estuaries of California, even those highly altered by human activity, such as San Francisco Bay (8). Small numbers are caught by both commercial and sport fishermen.

References 1. Schultz 1933. 2. Emmett et al. 1991. 3. Saiki 1997. 4. Carpelin 1955. 5. Klingbeil et al. 1975. 6. Middaugh and Shenker 1988. 7. Wang 1986. 8. Baxter et al. 1999.

Killifishes, Fundulidae

The killifishes or topminnows are small, hardy fishes with only two species in California, one native (California killifish) and one alien (rainwater killifish). The family contains only about 48 species, which are widely distributed in freshwater, brackish, and shallow marine habitats in North and Central America. It is one of three families of similar fishes found in California that belong to the order Cyprinodontiformes: Fundulidae (killifishes), Cyprinodontidae (pupfishes), and Poeciliidae (livebearers). Members of these families are characterized by small size (<75 mm SL, usually <50 mm SL); rounded to square tails; abdominal placement of pelvic fins (if present), with pectoral fins located below the midline; location of lateral line canal and pores mainly on the head; 3–7 branchiostegal rays; a protrusible mouth, with the upper jaw bordered by the premaxillary bone; strong sexual dimorphism; cycloid scales; no spines in the fins; and numerous osteological characters (Nelson 1994). Members of the Fundulidae are technically distinguished from other families by the structure of the jaw (e.g., the tiny maxillary bone, which acts as a lever to protrude the premaxilla, is twisted rather than straight). They can usually be recognized, however, by their upturned mouth and large eyes set in heads that are flat on top, as is the back. The single dorsal fin is typically closer to the tail than to the head, and there is no lateral line. The two species of killifish in California are basically brackish water or estuarine species that can be found in fresh water on occasion. Their tolerance of a wide range of temperatures, salinities, and dissolved oxygen levels is typical of the family.

Two other killifish-like species have been found in California, both introduced from South America: the giant rivulus, *Rivulus harti,* and the Argentine pearlfish, *Cynolebias bellottii.* They are both members of subfamily Rivulinae (sometimes given as family Rivulidae) in the Aplocheilidae. They will not be discussed separately because the giant rivulus is probably not permanently established and the Argentine pearlfish has apparently died out. Giant rivulus, attractive brown and green fish from Colombia and Venezuela, were reported as established only in a small ditch flowing into the Salton Sea, Imperial County (St. Amant 1970). The ditch once drained a now-defunct tropical fish farm. They were thought to have died out, but were collected again in the early 1990s in the same locality (Swift et al. 1993). Argentine pearlfish, a bright blue species adapted for living in temporary ponds in the Rio de la Plata basin, Argentina, was established in the 1960s for a few years in some experimental ponds on the University of California, Riverside, campus (Bay 1966, E. F. Legner, pers. comm. 1974). Attempts were made to establish this species and other "annual" fishes for mosquito control in southern California ponds and in rice fields of the Central Valley, but fortunately the attempts failed (Dill and Cordone 1997).

California Killifish, *Fundulus parvipinnis* Girard

Identification California killifish are small (to 115 mm SL), thick-bodied fish with small pelvic fins, a nearly square caudal fin, and several rows of conical teeth in small oblique mouths. They have 3–37 scales in the lateral series, 11–13 rays in the anal fin, 12–15 rays in the dorsal fin, and 7–11 gill rakers. Their overall color is olive green on the back and sides and yellowish brown on the belly. Males have about 20 short, black bars on the side and a long anal fin. When breeding they turn dark brown on the back with bright yellow coloring on the belly, lower head, and paired fins. Females lack bars but have a faint lateral band. An oviducal pouch covers the first rays of the anal fin of females.

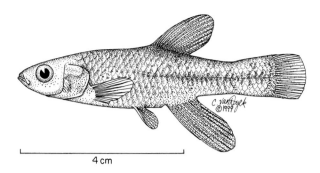

Figure 101. California killifish, male, 16 cm SL, Morro Bay, San Luis Obispo County.

Taxonomy Two species of *Fundulus* are recognized from the Pacific coast of North America: the Baja killifish (*F. limi*), found in freshwater lagoons in Baja California, Mexico, and the California killifish. These two species show evidence of long isolation from other *Fundulus* species, which are found throughout eastern North America (15). They may belong in a new genus. The various populations of California killifish along the coast show genetic signs of isolation from one another (16). The southernmost populations, found south of Punta Eugenia, Baja California, have been described as a subspecies (*F. p. brevis*) separate from the rest of the species (*F. p. parvipinnis*) (5), an approach validated by genetic evidence (16).

Names Killifish is an American word for cyprinodont fishes dating back to colonial days, but its origins are obscure. It is perhaps derived from kill-fish because some eastern *Fundulus* are poisonous (1). Another explanation is that killifish is a contraction of "killing fish," because the word *killing* was used to describe exceptionally effective (deadly) fishing bait (2). However, it is most likely that the name was given by Dutch colonists in New York to small fish that lived in "kills," small streams or canals (hence "killivisch") (13). Killifish are also known as topminnows. *Fundulus* is derived from *fundus*, Latin for bottom, and *ulus*, Latin for small, because the first species of the genus described (*Fundulus heteroclitus*) is a small fish with the habit of burrowing in bottom mud. *Parvi-pinnis* means small-finned, referring to the pelvic fins.

Distribution California killifish are found in shallow bays, estuaries, marshes, and lower reaches of streams from Morro Bay to central Baja California (Magdalena Bay). In southern California they have a more or less continuous distribution from Goleta Slough, Santa Barbara County, to the lower Tijuana River on the Mexican border (3). The population in Morro Bay is a disjunct one because killifish are absent from bays or streams between there and Goleta Slough. They have been recorded from the Salinas River (Monterey County) (4), but this record is apparently erroneous (3). They are occasionally used as bait, so individuals may be found in unexpected places.

Life History California killifish are most abundant in salt-water lagoons and estuaries. Because they can tolerate a wide range of salinities, populations occasionally become established in lower reaches of freshwater streams (3, 5), usually within 1–2 km of the mouth. They can live in water ranging from completely fresh to salinities up to 128 ppt (18). In Mugu Lagoon (Ventura County), a fall survey found them to be abundant at salinities of 0.5–33.8 ppt and temperatures of 11–25°C (6). Small fish are more resistant to sudden changes of salinity than are larger fish, but larger fish can withstand lower oxygen levels (7). The presence in fresh water of fish of all sizes, including ripe males and females, indicates that they can complete their entire life cycle in fresh water. In many lagoons the environmental tolerances of California killifish allow them to be extremely abundant, even where the water is polluted. For example, they can tolerate levels of sulfides that will kill most other fishes (9). In Ojo de Liebre Lagoon, Baja California, killifish show high production rates and are presumably a major source of energy transfer through the ecosystem (8).

California killifish spend most of their time in shallow water, often near beds of aquatic plants, but in bays and estuaries they move in and out of areas flooded by tides (10). In open water they form loose shoals, although they forage as individuals. When pursued by predators they seek cover in vegetation or in the burrows of tidal flat invertebrates (10, 16).

Like other members of the genus *Fundulus*, they feed on a wide variety of benthic and planktonic invertebrates. In

Mugu Lagoon the most abundant items in their diet were small snails, a variety of small crustaceans, siphons of clams, and fish eggs (11). In other bays fish of all sizes feed mainly on amphipods, copepods, tanaids, ostracods, beetles, and dipteran larvae, with the importance of each item varying with the season and place of feeding (10, 14, 17). Juveniles are more likely than adults to feed on terrestrial insects and zooplankton (14). Most feeding occurs at slack high tide, among flooded vegetation (10). Fish without access to flooded vegetation feed less and on fewer kinds of organisms (17).

The length-frequency distributions of California killifish indicate that they are largely an annual species, although some may live 2–3 years (8, 10). They seem to quickly (6–7 months) reach an adult length of 50–70 mm SL. In Anaheim Bay 1-year-old fish measured 44–79 mm SL and 2-year-olds measured 79–92 mm SL (10). Fish over 85 mm SL are rare.

Breeding takes place mostly from May through June but may continue through July and into September (7, 10). Breeding may occur as early as February in southern parts of their range (8). There is some evidence that breeding in killifish follows lunar cycles, with spring peaks at the new moon and consequent high tides (12). In Anaheim Bay spawning was observed in isolated permanent pools in grassy tidal flats (10). Killifish are fractional spawners, laying large eggs over a period of several weeks or months. Total fecundity ranges from 61 to 439 eggs, with the relationship $F_{ln} = 0.23SL_{mm} + 3.70$ (10). Embryos are attached to vegetation with sticky strands (10). Most fish apparently die after spawning.

Status IE. Despite a limited distribution that includes some of the most altered habitats in California, California killifish seem to be doing reasonably well as a species, presumably because of their tolerance of adverse conditions and their ability to recolonize areas by moving through salt water. However, the destruction of coastal wetlands within the range of the killifish is likely causing loss of genetically distinct populations, fragmenting populations, and reducing the genetic diversity of the species. This is a species that should be monitored, both for its own sake and as an indicator of the health of coastal wetlands (16).

References 1. White et al. 1965. 2. *Oxford English Dictionary* 1971. 3. Swift et al. 1993. 4. Kukowski 1972. 5. Miller 1939, 1943a. 6. Saiki 1997. 7. Keys 1931. 8. Perez-Espana et al. 1998. 9. Bagarinao and Vetter 1993. 10. Fritz 1975. 11. Onuf 1987. 12. Foster 1967. 13. Jenkins and Burkhead 1994. 14. Hartney and Tumyan 1998. 15. Bernardi 1997. 16. Bernardi and Talley 2000. 17. West and Zedler 2000. 18. Feldmeth and Waggoner 1972.

Rainwater Killifish, *Lucania parva* (Baird and Girard)

Identification Rainwater killifish are tiny (usually <41 mm SL), guppylike fishes with large eyes, chunky bodies, oblique mouths with a single row of tiny conical teeth, rounded dorsal and caudal fins, and small pelvic fins. Scales in the lateral series number 23–29 (usually 27); anal rays, 8–13 (usually 9–10); dorsal rays, 9–14 (usually 11); caudal rays, 12–18 (usually 14–16); pectoral fin rays, 10–15 (usually 13–14); pelvic rays, 4–7 (usually 6); and gill rakers, 4–12 (usually 6–9). Their basic color is olivaceous on the back, silvery blue-gray on the sides, and yellowish on the belly. Males have black edges on the anal and pectoral fins and blackened anterior dorsal rays. Females have a membranous oviducal pouch covering the anterior portion of the anal fin.

Names Lucania seems to be a nice-sounding name without any meaning, coined by Charles Girard; *parva* means small. Rainwater nicely describes the shallow-water ponds in which they are sometimes found. For killifish see the account of California killifish.

Distribution Rainwater killifish are native to coastal waters from Massachusetts and Florida, around the Gulf of Mexico, to lower Rio Panuco in Mexico. Their native range also includes the Rio Grande and Pecos River, Texas and New Mexico. In southern California they are established in California in Irvine Lake (Orange County); the lagoon at the University of California, Santa Barbara (near Goleta Point); Arroyo Seco Creek above Vail Reservoir (Riverside County); and perhaps upper Newport Bay. In northern California they are found in sloughs and streams flowing into Suisun Bay and San Francisco Bay and Lake Merritt in Oakland (1, 2). They are also present in Yaquina Bay, Oregon, and some springs in Utah. How they were introduced into these areas is a bit of a mystery, but Hubbs and Miller (1) present circumstantial evidence that populations in San Francisco and Yaquina Bays started from embryos attached to live oysters imported from

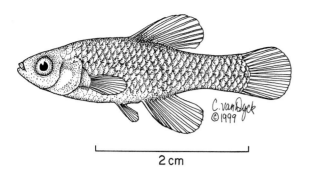

Figure 102. Rainwater killifish, male, 3 cm SL, Lake Merritt, Alameda County.

the East Coast for culture around 1958. The other fish apparently came in with shipments of game fishes from a federal hatchery along the Pecos River, New Mexico (1).

Life History Information on the life history of rainwater killifish in California is limited, so this account is based largely on information from elsewhere (1, 3, 4, 5, 9). They are primarily inhabitants of brackish water, but they can live at salinities from 0 to 80 ppt. In the San Francisco Bay region they are most abundant in tidal ditches (6) in the lower, channelized ends of inflowing streams (11), in ponds in diked areas, and in shallow salt evaporation ponds (7). In San Francisco Bay streams they are typically associated with unshaded, silt-bottomed pools, with scattered aquatic plants (11). In evaporation ponds the killifish live in water as salty as 70 ppt with low dissolved oxygen levels (2–3 mg/liter), conditions that exclude most other fishes. There is evidence of reproduction at salinities as high as 55 ppt (7). Typical associates of rainwater killifish are western mosquitofish and threespine stickleback.

In their native areas rainwater killifish will migrate into fresh water to breed and then move back into brackish water in areas dominated by emergent and submerged vegetation, although this has not been observed in California. Such migrations may be spectacular: as many as 270,000 fish have been observed moving downstream in a Virginia river in a 4-hr period (8). Regardless of salinity, killifish populations are typically associated with aquatic vegetation.

Rainwater killifish feed on whatever invertebrates are abundant in their habitat. In a Florida marsh they ate mostly mosquito larvae and copepods; the remainder of the diet consisted of miscellaneous crustaceans and aquatic insects (4). Diversity of diet increased with size; fish measuring less than 15 mm TL fed almost exclusively on mosquito larvae while fish measuring greater than 35 mm TL fed on a variety of crustaceans (4). Their diet in San Francisco Bay ditches and ponds is apparently similar (9). In evaporation ponds they feed on amphipods, brine shrimp, and copepods (7).

Growth is rapid and sexual maturity can be reached in 3–5 months at a minimum length of 25 mm TL. Females grow larger than males, reaching a maximum size of 62 mm TL (5). Time of breeding in northern California is May–July (9).

Breeding begins when males set up territories over or near beds of aquatic plants or algae (5). Spawning has been observed at temperatures of 17–25°C and salinities of 0–18 ppt (9). Spawning males acquire orange fins, assume a cross-hatched pattern on the sides, and display vigorously to other males holding nearby territories. When a female approaches, the male circles rapidly around her. If she is interested in spawning, she stops and he moves quickly beneath her, rubbing the top of his head against the underside of her head. In this position they swim toward the surface to just above a suitable substrate for egg attachment. The male then wraps himself around the female, placing his vent close to hers, and fertilizes eggs as they are released. The fertilized eggs are attached to vegetation by adhesive threads from the chorion. Each female spawns repeatedly, but fecundities are low (7–104 eggs) (9). In one wild population females contained a mean of 72.5 ova (10).

The developmental stages of rainwater killifish have been described in detail (9, 10). The embryos hatch in about 12 days at 24–25°C and the lightly pigmented larvae settle down to the bottom (9). They assume an active existence, feeding in vegetation, about a week after the yolk sac is absorbed (5). Juveniles either remain in the vegetation, forming loose aggregations, or move into water only a few millimeters deep, where they are relatively safe from fish predation (9).

Status IIC. Rainwater killifish are well established in the San Francisco Bay area and as scattered populations in southern California. Their gradual spread southward seems likely, although special efforts will have to be made to document this because they superficially resemble the ubiquitous mosquitofish. Their spread may be assisted by attempts to use them for mosquito control in salt marshes, although they are apparently no more (or less) effective than mosquitofish.

References 1. Hubbs and Miller 1965. 2. McCoid and St. Amant 1980. 3. Renfro 1960. 4. Harrington and Harrington 1961. 5. Foster 1967. 6. Balling et al. 1980. 7. Lonzarich and Smith 1997. 8. Beck and Massmann 1951. 9. Wang 1986. 10. Crawford and Balon 1994. 11. Leidy 1984.

Livebearers, Poeciliidae

The more than 200 species of livebearers are closely related to pupfishes and killifishes, differing from them most obviously in their method of reproduction. Instead of laying eggs, female poeciliids incubate them internally, giving birth to free-swimming young. Males have elongated rays in their anal fins for use as intromittent organs (gonopodia) to place packets of sperm in the vents of females. Livebearers are small, active fishes, seldom exceeding 10 cm TL, and are diverse in morphology and coloration. Their colorfulness, hardiness, and readiness to breed in captivity have made them one of the most popular groups of aquarium fishes. Intensive breeding in aquaria of guppies, mollies, and platys has produced many bizarre body shapes and color patterns and has given scientists insights into evolution, genetics, and behavior. Studies of these fish in the wild have also provided many evolutionary, behavioral, and ecological insights (Meffe and Snelson 1989).

Central America is the center of poeciliid abundance and diversity, but the family has spread to both North and South America. Although many species are capable of living in saline water, their distribution patterns are best explained by dispersal through fresh or inland waters. At least 12 species are native to the United States, including the mosquitofishes. Because of the high regard in which they are held as mosquito control agents, the two species of mosquitofish have become, through introductions, among the most widely distributed freshwater fishes in the world.

All livebearers found in California are introduced. Four species—sailfin molly, shortfin molly, porthole livebearer, and western mosquitofish—are common enough to warrant separate treatment here, although the shortfin molly and porthole livebearer are largely confined to drains in the Salton Sea area. Three platyfishes—variable platyfish (*Xiphophorus variatus*), southern platyfish (*X. maculatus*), and green swordtail (*X. helleri*)—have been collected from drains around the Salton Sea and other localities in southern California at various times (Dill and Cordone 1997). These are all common aquarium fishes, and there is no evidence of permanent establishment, although variable platyfish and green swordtails may persist in small numbers in some Salton Sea drains. Guppies (*P. reticulata*) frequently establish populations in sewage treatment ponds, but so far no populations have been recorded in natural waters in California. Ephemeral populations of aquarium fishes are either the result of releases by aquarists tired of their charges or escapees from tropical fish farms.

Western Mosquitofish,
Gambusia affinis (Baird and Girard)

Identification Mosquitofish are small (to 60 mm TL), stout-bodied fish with short, flattened heads, small oblique mouths, a stout caudal peduncle, and a rounded tail. The dorsal fin (6 rays) is placed behind the origin of the anal fin (9 rays, although branching often gives the impression of more). The anal fin of males is rodlike because it is used as an intromittent organ (gonopodium). The third ray on the anal fin lacks prominent toothlike projections. Pectoral fin rays number 12–14 and pelvic rays, 6. Scales are large, number 29–32 in the lateral series, and are frequently outlined with pigment on the back and sides. The intestine is short, with a single loop. Males are much smaller (to 35 mm TL) and less stout-bodied than females. Overall color is generally gray or olivaceous, lighter on the belly, with few conspicuous markings. When examined closely, mosquitofish

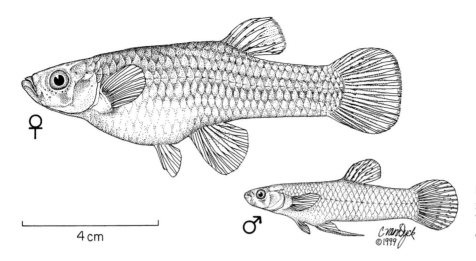

Figure 103. Western mosquitofish. Top: Female, 5 cm SL, pond,
Yolo County. Bottom: Male, 2.5
cm SL, pond, Yolo County.

4 cm

will often have a teardrop-like black streak below each eye
and rows of speckles on the caudal and dorsal fins.

Taxonomy The genus *Gambusia* contains about 45 species
(31), 20 of them confined to rivers and spring systems in the
southwestern United States and Mexico. Many are endangered or extinct. Curiously, there are also two species with
wide distributions, including populations introduced into
many other parts of the world: western mosquitofish and
eastern mosquitofish (*G. holbrooki*). Eastern mosquitofish
were formally described by Girard in 1859 (1) and were long
recognized as a subspecies of *G. affinis*. They are regarded
once again as a distinct species (2). Because it is possible
(but not confirmed) that eastern mosquitofish were also introduced into California (3), it is worth noting that they can
be distinguished from western mosquitofish by their 7 dorsal rays, 10 anal rays, and small serrations on the third ray
of the anal fin.

Names Gambusia was coined by F. Poey, a Cuban naturalist,
who is quoted by Jordan and Evermann (4, p. 679): "The
name owes its etymology to the provincial Cuban word
Gambusino which signifies nothing, with the idea of a joke or

farce. Thus one says 'to fish for *Gambusinos*' when one catches
nothing." *Affinis* means related, apparently to *G. holbrooki*, a
species described first by Louis Agassiz (but the description
was not published until after *G. affinis* was described).

Distribution Western mosquitofish are native to central
North America, in watersheds tributary to the Gulf of Mexico from southern Illinois to Alabama, Texas, and eastern
Mexico, although the western extent of their native range
is uncertain. Mosquitofish have been introduced for
mosquito control throughout the world; a majority of the
introduced populations outside North America may be
eastern mosquitofish (5). Western mosquitofish were introduced into California in 1922 from Texas and have since
been spread throughout California. Within 4 years, they had
been enthusiastically planted in 30 counties (3) with fish
provided by the California State Board of Health. The
weekly bulletin of the Board documents their spread and
the consequent reduction in the need to spray oil on the surface of ponds, the main method of mosquito control used
at the time. The bulletin notes that mosquitofish spread
rapidly on their own; on 29 September 1923 it commented
that a single spring introduction into Putah Creek (Yolo
County) led to their becoming abundant in several miles of
creek by the end of summer. It is safe to say that today western mosquitofish are found throughout California where
the water does not get too cold for extended periods of time.
Because they are still widely planted for mosquito control,
they are also found in many habitats unsuitable for self-
sustaining populations.

Life History Western mosquitofish survive in a wide range
of environmental conditions (6). They are found in brackish sloughs and salt marshes as well as in warm ponds, lakes,
and streams. They are particularly well adapted for life in
shallow, often stagnant, ponds and along shallow edges of

lakes and streams, where predatory fishes are largely absent and temperatures are high. They occur at temperatures of 0.5–42°C but persist mainly where temperatures remain within the range 10–35°C over the course of a year; optimal temperatures for growth and reproduction are 25–30°C (6). They are also capable of withstanding the extreme daily temperature fluctuations that characterize shallow-water habitats. They generally cannot withstand prolonged exposure to cold (less than 4°C), although the classic study by Krumholz (7) demonstrated that they can be acclimated to climates as severe as that of northern Illinois. They can survive in a pH range of 4.7–10.2 but are typically found in waters within the range of 7–9. Likewise, for salinity they can tolerate a range from 0 to 58 ppt, persisting mainly in areas with salinities under 25 ppt (6). Low oxygen also poses few problems for mosquitofish. Their small size, flat head, and oblique mouth permit them to use the few millimeters of water close to the surface into which oxygen diffuses from the air. Thus they can live in pools with dissolved oxygen levels as low as 0.2 mg/liter, levels that would be lethal to most fish (6). Although mosquitofish occur in some of the most polluted waters in California, even they have limits. For example, high levels of selenium found in agricultural drains in the San Joaquin Valley are related to low reproductive rates in mosquitofish, mainly through production of malformed young (8). In California streams mosquitofish are most abundant in disturbed portions of low-elevation streams, especially warm, turbid pools with beds of emergent aquatic plants. Large schools can often be observed swimming in shallows where water temperatures approach near-lethal limits (9). If submerged and emergent aquatic plants are present, mosquitofish tend to be found among them, but only if plant growth is not too heavy. When beds are dense, mosquitofish remain close to the edges, seldom penetrating very far inward.

Mosquitofish are omnivorous, opportunistic feeders that feed mainly during the day. They usually take their prey close to the water surface but will feed on the bottom or pick prey from plants. Mosquito larvae and pupae can form a substantial portion of their diet, but they do not feed selectively on them. Instead they feed on whatever organisms are most abundant, typically algae, zooplankton, terrestrial insects, and miscellaneous aquatic invertebrates (6, 10). They feed selectively on the largest prey available, with size of prey varying with size of fish (6). Under crowded conditions or during periods when animal food is scarce, mosquitofish may feed extensively on filamentous algae and diatoms (11). When animal food is abundant, growth and reproductive rates of mosquitofish are high (12, 13). Cannibalistic fish may show particularly high growth and reproductive rates (28).

Growth in mosquitofish also varies with sex, productivity of the water, temperature, and other environmental factors (6, 7, 14). Males almost stop growing once the formation of the gonopodium is complete. Thus males seldom exceed 31 mm TL, and most reach maximum size in one growing season. Females reach 50 mm in one growing season, although smaller sizes are more common. Maximum size is about 35 mm TL for males and 65 mm for females. Maximum growth is possible when productivity is high and warm temperatures prevail for extended periods of time. Most fish die in the same summer they reach maturity, so few fish live more than 15 months (7). Overwintering fish are generally juveniles or adults that achieved maturity late in the growing season. However, as the end of summer approaches, mosquitofish reduce their reproductive efforts in favor of fat storage to allow overwinter survival (27). In subtropical climates mosquitofish are capable of reproducing all year round (15), but in California reproduction ceases in winter.

Males become mature at 19–21 mm TL; females usually reach at least 24 mm TL before becoming pregnant. Under optimal conditions females can become pregnant at 6 weeks, but life history traits (such as time to maturity, size at maturity, and growth rates) vary with environmental conditions. The variation can have a genetic basis even in introduced populations established for a relatively short period of time (29). In warmer environments in California the gestation period is only 3–4 weeks, and 3–4 generations per year are possible. Two generations are most typical in the Central Valley. In addition, under the right conditions of temperature and food availability, large females give birth to at least four broods of young in a season (usually April through September). Females living in water of 25–30°C can produce a brood every 18–23 days (13). Thus the potential for rapid population growth in mosquitofish is very high. Each female contains 1–315 embryos, the number increasing with the size of the fish. Typical brood size is around 50. However, the number may decrease with age, toward the end of the reproductive season, and under crowded conditions (7). Cannibalism may help limit population size, as well as providing an excellent source of food for some individuals.

In rice fields in central California mosquitofish introduced for pest control show rapid recruitment as initial females give birth, followed by a period of low recruitment in which populations may actually decline (16). A second peak develops late in the season (September), apparently in response to "new" females beginning to reproduce and older females producing their second brood.

Courtship and copulation are constant activities among mosquitofish, although individual acts are brief and variable in pattern (17). Two types of courtship displays can be observed. In the infrequent but conspicuous frontal display, the male swims in front of the female, orienting his body at a 90° angle to hers. He partially folds his dorsal and anal fins, bends his body into an S, and quivers for a few seconds. Then he quickly swims around behind and below the female and attempts to shove his gonopodium into the female's genital opening. The lateral display is similar to the frontal display except that it is performed alongside the female, close to her head, and the male does not bend or quiver as much. Often no courtship display precedes attempts at copulation (18). With or without courtship display, most attempts at copulation are unsuccessful, usually because the female is not receptive. Receptive females tend to swim slowly or remain nearly stationary. In aquaria they are quickly surrounded by males, although one male is usually dominant over the rest (17). Large females are strongly preferred by males and may be inseminated by several males; large females also have a strong preference for large males (19). Females store sperm, and eggs may be fertilized from several copulations.

The early life history stages take place inside the female, although developing young are mainly dependent on the yolk sac for nutrition. The young are expelled by the female through her urogenital opening, usually in very shallow water or among aquatic vegetation. Because of the cannibalistic tendencies of females, the young must find cover as quickly as possible.

Status IIE. Western mosquitofish are popular mosquito control agents in California, so they continue to be planted in warm waters throughout the state, even in places where they are unlikely to survive over the long term. Because mosquitoes and similar insects are increasingly difficult to control using pesticides, biological control methods, including the use of mosquitofish, have increased in popularity. There are many areas where mosquitofish are effective mosquito control agents if used properly, such as rice fields, isolated ponds, agricultural drains, and urban sumps. There is a general lack of evidence for their control of mosquitoes in natural situations where other fish or predatory invertebrates are already present.

The success of mosquitofish as mosquito control agents can be attributed to a number of factors: a high reproductive rate that permits rapid population buildups from small initial plants of fish; a method of reproduction that frees them from special spawning substrates; the ability to live in extreme environmental conditions, which also favor mosquitoes; omnivorous feeding habits that allow them to live on other food, including algae, when mosquitoes are not abundant; a preference for habitats where predators are usually absent; an ability to develop resistance to pesticides, so that they can be used in areas where pesticides are also used (20); and their ease of culture.

Although these characteristics have made mosquitofish useful in a number of situations, they also make them a problem species. Mosquitofish have been accused of eliminating small fish species the world over through predation and competitive interactions (21), and a number of such cases in the southwestern United States and Australia have been well documented (22). In general mosquitofish have negative effects mainly on fishes of similar size in small or isolated habitats where they can become the dominant species (32). They can also have negative impacts on endemic invertebrates and amphibians. For example, in small streams in southern California mosquitofish can eliminate or reduce the abundance of eggs and larvae of California newts, *Taricha torosa* (30), and Pacific treefrogs, *Hyla regilla* (33). Other fishes and amphibians are eliminated through a combination of direct predation on small individuals and harassment of adults, which keeps them from breeding. In California it is quite likely that mosquitofish have contributed to the decline of isolated pupfish populations (see the account of Amargosa pupfish). However, destruction of habitat and introduction of predatory fishes have often initiated pupfish declines while simultaneously creating conditions that favor mosquitofish. There are also many situations in which mosquitofish coexist with similar small native fishes, although these areas often experience environmental fluctuations that keep mosquitofish from becoming abundant. In a canal near the Salton Sea mosquitofish, desert pupfish, and other species showed considerable segregation by habitat until flooding and construction eliminated habitat complexity (23).

A problem created by the popularity of mosquitofish is that serious consideration of native fishes and invertebrates for mosquito control has been neglected, even though they might be more effective locally than mosquitofish. For example, pupfish can be more effective mosquito predators in emergent vegetation (34). This may explain why mosquitoes became a problem in the Owens Valley after Owens pupfish were eliminated, despite the introduction of mosquitofish. The omnivorous nature of mosquitofish also creates problems because they can alter food webs in small bodies of water by reducing populations of invertebrate predators and grazers. In small experimental ponds introduction of mosquitofish resulted in large blooms of phytoplankton after zooplankton grazers had been eaten (24). An improper introduction of mosquitofish may actually result in an increase in mosquitoes, because they may reduce populations of invertebrate predators of mosquitoes before the fish build up populations large enough to control mosquitoes. Thus mosquito populations in rice fields stocked with only a small number of mosquitofish can be larger than populations in fields either with substantial numbers of invertebrate predators (mostly Notonectidae) or into which large numbers of mosquitofish had been stocked (25).

Ornamental ponds containing goldfish and koi are examples of ponds where mosquitofish are usually of little benefit. Small goldfish and koi will eliminate most mosquitoes from the pond, and mosquitofish may seriously limit the ability of the other fish to reproduce, through predation on their eggs and larvae. Overall, caution should be exercised when using mosquitofish for mosquito control, because the harm they can do may outweigh the good. It is especially important to be cautious when introducing them into natural environments, such as ponds. Amed et al. (26) present a rating system for choosing fish for biological control, and Swanson et al. (6) provide guidelines for use of mosquitofish specifically.

References 1. Girard 1859. 2. Wooten et al. 1988. 3. Dill and Cordone 1997. 4. Jordan and Evermann 1896. 5. Lever 1996. 6. Swanson et al. 1996. 7. Krumholz 1948. 8. Saiki and Ogle 1995. 9. Moyle and Nichols 1973. 10. Greenfield and Deckert 1973. 11. Rees 1958. 12. Wurtsbaugh and Cech 1983. 13. Vondracek et al. 1988a. 14. Goodyear et al. 1972. 15. Haynes and Cashner 1995. 16. Botsford et al. 1987. 17. Itzkowitz 1971. 18. Peden 1972. 19. McPeek 1992. 20. Rosato and Ferguson 1968. 21. Myers 1965. 22. Meffe and Snelson 1989. 23. Schoenherr 1979. 24. Hurlbert et al. 1972. 25. Hoy et al. 1972. 26. Ahmed et al. 1988. 27. Reznick and Braun 1987. 28. Meffe and Crump 1987. 29. Stearns 1983. 30. Gamradt and Kats 1996. 31. Rauschenberger 1989. 32. Minckley 1999. 33. Goodell and Kats 1999. 34. Danielsen 1968.

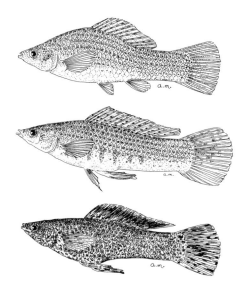

Figure 104. Sailfin molly. Top: Female, 5 cm SL, Salton Sea, Imperial County. Middle: Normal male, 6 cm SL, Salton Sea, Imperial County. Bottom: Male color variant, 5 cm SL, Salton Sea, Imperial County. Drawings by A. Marciochi.

Sailfin Molly, *Poecilia latipinna* (Lesueur)

Identification Sailfin mollies are small (to 15 cm but usually 5–8 cm as adults), chunky-bodied fish with thick caudal peduncles, rounded tails, and oblique mouths. Males are brightly colored with long (13–16 rays), sail-like dorsal fins and gonopodia modified from the first rays of the anal fin (9–10 rays). In females the dorsal fin is shorter and the anal fin is unmodified. The pectoral fin has 13 rays, the pelvic fins have 6 rays, and the lateral line has 23–28 scales. Color patterns are highly variable but distinctive, especially for males. Wild-type sailfin molly males are brown on the back, and that color gradually changes to iridescent blue or pink on the sides and white on the belly. There are 5–6 horizontal rows of dark spots on the sides, often interspersed, in males, with red, blue, or green spots. The dorsal fin is pale blue, with rows of black spots and a yellow border. Female coloration is similar to that of males, but paler. California populations were derived from black or checkered mollies of the aquarium trade, so some wild fish may be solid iridescent black with an orange border on the dorsal fin or heavily speckled on the sides with varying numbers of black "checkers," which merge into one another.

Taxonomy Sailfin mollies are often listed under the generic name *Mollienesia,* even though Rosen and Bailey (1) synonymized it with *Poecilia.* Within their native range they are variable in characters and form a species complex with other similar forms (19). Domestication, however, has produced much variety in color and body shape. Black mollies of the aquarium trade are usually sailfin mollies, although the large fin often does not develop under aquarium conditions.

Names Poecilia means many-colored; *latipinna* means broad fin. The common name molly is derived from the original generic name (*Mollienesia*) given in 1821 by C. A. Lesueur, after Count F. N. Mollien (1758–1850).

Distribution Sailfin mollies are native to lowland streams and estuaries of coastal North America from Cape Fear, North Carolina, to northeastern Mexico. They have been introduced into the Philippines, Singapore, Colombia, Australia, New Zealand, Guam, the Hawaiian Islands, and elsewhere, either for mosquito control or as released pets (2). In North America they have been successfully introduced into thermal springs and warm waters in Alberta, Colorado, Nevada, and Arizona. In California they first appeared in the wild in the 1950s as escapees from a tropical fish farm in the Salton Sea area and subsequently spread (or were spread) throughout the Salton Sea drainage and the lower Colorado River (3). They are also present in a number of drainages in coastal California, including the Tijuana River estuary, the Santa Ana River, and various sloughs and estuaries in Ventura, Los Angeles, and San Diego Counties (4, 5).

Life History Sailfin mollies are native to warm, brackish coastal swamps, so it is not surprising that they have done so well in salty irrigation water of the Imperial Valley and estuaries of southern California. Typical habitats in the Salton Sea basin are stagnant drains and shoreline pools with abundant algae and emergent vegetation (18). Because they tolerate salinities up to 87 ppt (6), they also live in the Salton Sea proper and are likely to persist for a long time after rising salinities kill off other species. They can withstand a fairly wide range of temperature, but 24°C or higher is necessary for breeding and growth, and this requirement limits their distribution in California. Males prefer temperatures of 25–29°C, whereas pregnant females prefer warmer temperatures (28–32°C), reflecting the need for young to live in shallow water (7). In Sweetwater Marsh (San Diego County) sailfin mollies were found to be most abundant in first-order drainage channels and shallow pans within the marsh, where temperatures were 22–33°C, salinities 23–46 ppt, depths 5–50 cm, and dissolved oxygen as low as 1.5 mg/liter (5).

Mollies collected in ditches near the Salton Sea in July 1973 were feeding on detritus (86% by volume) and algae (14%). Likewise, mollies in the lower Colorado River are almost exclusively herbivores and detritivores (8). Invertebrates are eaten only when superabundant, and even then they are consumed mainly by smaller mollies (9). Mollies grow rapidly and reach 8–10 cm TL in a year under optimal conditions of temperature and salinity. In the Salton Sea individuals measuring 12–15 cm TL are occasionally caught, although few exceed 8 cm TL. The largest fish are usually females, but males also exhibit indeterminate growth (10).

Female sailfin mollies become mature at 25–40 mm SL, whereas males may mature at 20–30 mm (11, 12). Time from hatching to maturity is typically 110–150 days, but it varies with temperature and salinity. Optimal conditions for maturation and growth seem to be offered by warm (24–32°C), salty (about 20 ppt) water (11). Mature males show wide variation in morphology and color, as well as size, depending on environmental conditions and the abundance of predators (13). It appears that in harsh environments (e.g., those prone to desiccation) or those with an abundance of predators, it pays males to be small and inconspicuous. Under more favorable conditions large, brightly colored males develop. Larger and brighter males are preferred by females, so there is strong selection pressure in favor of them (14). Larger fish of both sexes are selected by herons, a fact that contributes further to the complexity of population structure in mollies (17).

Courtship and copulation are quick. Females can store sperm, so several batches of eggs can be fertilized internally from one copulation. For this reason it is possible to start a population with one pregnant female. Ripe eggs are shed by the ovaries into the ovarian cavity, where they are fertilized and incubated. Although most nutrition for developing embryos comes from yolk, females will provide additional nutrients in utero when food is abundant and environmental stress is low (15). This practice results in bigger young being born. At birth the sex ratio of males to females is normally about 1:1 (16). However, it is not uncommon for wild populations to have sex ratios skewed in favor of females. For example, in Salton Sea ditches in July 1973, the male: female ratio was 1:13. The number of young produced by each female depends on size, water temperatures, and salinities. In Salton Sea ditches mollies normally produce 20–60 young, but one large (59 mm SL) female contained 141 embryos. The young are large (9–12 mm TL) and self-sufficient at birth.

Status IIC. Sailfin mollies have become widely established in drains of the Salton Sea and lower Colorado River and appear to be gradually spreading through coastal marshes of southern California. They are often the most abundant fish in drains flowing into the Salton Sea and may be limiting desert pupfish populations through aggressive interactions. Likewise, in desert springs their presence is generally associated with declines in pupfish populations. Because of their abundance in shallow-water marshes and drains, they are presumably important prey for predatory birds, although their role in these ecosystems has not been studied.

References 1. Rosen and Bailey 1963. 2. Lever 1996. 3. Dill and Cordone 1997. 4. Swift et al. 1993. 5. Williams et al. 1998. 6. Barlow 1958a. 7. Stauffer et al. 1985. 8. Minckley 1982. 9. Harrington and Harrington 1961. 10. Snelson 1982. 11. Trexler 1989. 12. Snelson 1984. 13. Snelson 1985. 14. Ptacek and Travis 1997. 15. Trexler 1985. 16. Snelson and Wetherington 1980. 17. Trexler et al. 1994. 18. S. Keeney, CDFG, pers. comm. 1999. 19. Ptacek and Breden 1998.

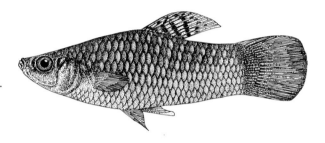

Figure 105. Shortfin molly, male, 50 mm SL, Veracruz, Mexico. Drawing by F. M. Watson, from Lee et al. (1980).

Shortfin Molly, *Poecilia mexicana* Steindacher

Identification Shortfin mollies are olivaceous with a short dorsal fin (8–11 rays) originating behind the origins of the pelvic fins. The caudal peduncle is thick (nearly as wide as the body) and supports a stubby, rounded tail that is often metallic blue in color with an orange fringe. The scales are large (25–30 in the lateral series). There are often several rows of orangish spots on the sides, and the belly is white to pale orange, as are the anal and pelvic fins. Large males are typically dark green with black spots on the dorsal fin. Shortfin mollies always measure less than 11 cm TL.

Taxonomy The shortfin molly is part of a complex of interrelated species originally described as *Poecilia sphenops*

(1, 9). Introduced populations in California presumably have a limited ancestry but fit *P. mexicana* descriptions well.

Names Shortfin mollies are also known as liberty mollies. *Mexicana* refers to their native range. Other names are as for sailfin molly.

Distribution Shortfin mollies are native to streams along the Atlantic coast of Central America from the Gulf Coast of Mexico to Guatemala and Nicaragua, but the exact range within this area is uncertain. They have been introduced into scattered oceanic islands (Fiji, Hawaii [Oahu], Samoa, Tahiti) for mosquito control (2). They are also established in a number of warmwater springs and spring-fed streams in the western United States, including Ash Meadows, Nevada, part of the Amargosa River drainage (3). Their distribution in California is uncertain because they are rarely distinguished from sailfin mollies in collections. They are established for certain in small numbers in drains feeding the Salton Sea and may also be present along the lower Colorado River (4, 5).

Life History Comparatively little information is available on their ecology and life history. Shortfin mollies live at temperatures of 15–36°C but prefer temperatures around 25–30°C (6, 7). In drains of the Salton Sea they have been found at salinities of up to 18 ppt, but they can presumably handle higher salinities as well. In one drain shortfin mollies seemed to select shallow (10–18 cm), flowing, well-oxygenated water (7).

Shortfin mollies are herbivores (8) and presumably have a life cycle much like that of sailfin mollies.

Status IIC. This species may be widespread, but it may also be highly localized in a few Salton Sea drains, where pollution and other changes could drive it to extinction. Like other nonnative poeciliids in the Salton Sea basin, it may have a negative effect on endangered desert pupfish, but this is not known for certain.

References 1. Rosen and Bailey 1963. 2. Lever 1996. 3. Sigler and Sigler 1987. 4. Lau and Boehm 1991. 5. S. Keeney, CDFG, pers. comm. 1998. 6. Deacon and Bradley 1972. 7. Schoenherr 1979. 8. Lee et al. 1980. 9. Ptacek and Breden 1998.

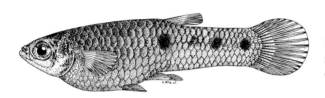

Figure 106. Porthole livebearer, male, 21 mm SL, drainage ditches near Mecca, Riverside County. Drawing by F. M. Watson, from Lee et al. (1980).

Porthole Livebearer, *Poeciliopsis gracilis* (Heckel)

Identification This small (4–7 cm TL), chunky fish is distinguished by the row of 3–8 large black spots running down each midline, mostly on the rear half of the body. Males have an extremely long gonopodium (modified rays on the anal fin) that extends nearly to the caudal fin when depressed. The tail is rounded and the dorsal fin short, with its base entirely behind the base of the equally small anal fin. The scales are large (about 29 in the lateral series), as are the eyes, dominating the small flat head. Live fish are pale brown and translucent.

Taxonomy The porthole livebearer appears to be a stable, well-defined taxon (1).

Names *Poeciliopsis* means "has the appearance of *Poecilia*," which in turn means many-colored. *Gracilis* means slender, presumably a reference to their body shape relative to other livebearers. The common name is a good description of the fish. In the aquarium literature they are often listed as porthole fish, *Poecilistes pleurospilus*.

Distribution Porthole livebearers are native to Atlantic and Pacific coastal watersheds from Rio Coatzacoalcos in southern Mexico to Rio Choluteca in central Honduras (1, 7). Their range in Mexico has expanded northward through introductions (7). They were first found in California in a drainage ditch near Mecca, on the Salton Sea, Riverside County (2). Presumably they became established as escapees from a tropical fish farm. They are now fairly common in drains on the north side of the Salton Sea (3, 4) and may spread to similar habitats throughout the region.

Life History Most information on porthole livebearers comes from observations of aquarium fish, which apparently do best in fresh water at 22–24°C (2). However, they have been collected in Salton Sea drains at salinities around 18 ppt and temperatures of 26–28°C (5), so it is likely they tolerate conditions even more severe. They also seem to have a preference for flowing water, which may explain why they are found largely in shallow (10–45 cm) parts of the drains (2, 5) and why they have not spread easily to other parts of the Salton Sea drainage. Dissolved oxygen levels and turbidity are both variable within muddy-bottomed drains. Porthole livebearers occur with a mixed assemblage of small fishes that include, depending on conditions, desert pupfish, sailfin and shortfin mollies, mosquitofish, swordtails, tilapia species, and longjaw mudsuckers.

They are omnivores that feed both on the bottom and in the water column. They rarely live more than 1 year or exceed 75 mm SL (7). Males become mature at around 22–25 mm SL and apparently will defend small territories while breeding (7). Females measure at least 36 mm SL before they mature (7). Each female carries 1–140 embryos, but average fecundity in one population was only 18 (7). They produce broods in 10–12 days when conditions are favorable (6). They are capable of reproducing all year round (7) but probably only reproduce in summer around the Salton Sea.

Status IIB. Porthole livebearers are locally common at the north end of the Salton Sea, in drains that enter along about 10 km of shoreline. They have the capacity to spread throughout similar degraded environments in the region but are probably too sensitive to low temperatures to live outside the Salton Sea basin in California. Although they co-occur in a few places with desert pupfish, their impact on this fish (if any) is likely to be small compared with that of the more abundant mollies, mosquitofish, and tilapias.

References 1. Rosen and Bailey 1963. 2. Mearns 1975. 3. Lau and Boehm 1991. 4. S. Keeney, CDFG, pers. comm. 1998. 5. Black 1980. 6. Breder and Rosen 1966. 7. Contreras-MacBeath and Espinoza 1996.

Pupfishes, Cyprinodontidae

Pupfishes—along with killifishes (Fundulidae), livebearers (Poeciliidae), and five other families—make up the order Cyprinodontiformes. This order is a fascinating collection of small fishes found the world over, often in extreme habitats that exclude other fish. Species live in hot springs, temporary ponds, fluctuating estuarine edges, highly saline inland lakes, and desert pools. The Cyprinodontidae epitomize this group, with over 100 species living in habitats as diverse and widespread as high mountain lakes in South America, islands in the Caribbean, desert springs in North America and North Africa, and saline marshes in North America and the Mediterranean region. California has four species of pupfish, all in southern deserts.

Pupfishes are small and aggressive, usually with strong sexual dimorphism. All have large eyes, deep bodies, and small terminal mouths with highly protrusible lips. Unlike minnows, with which they are frequently confused, they possess teeth in the jaws. All pupfishes lay eggs.

Much has been learned about fish evolution, ecology, behavior, and physiology through study of pupfishes living in deserts in the western United States and Mexico. Scientific and public attention has focused particularly on pupfishes of the Death Valley region, California and Nevada. In springs, streams, and swamps of this area, four species and eight subspecies have evolved, each in a different locality. Most remarkable is the Devils Hole pupfish (*Cyprinodon diabolis*), a tiny species that occupies the smallest known range of any vertebrate animal, 20 m^2 of submerged limestone shelf in a deep Nevada spring. For 20,000 or more years this species has maintained itself in a population fluctuating between 200 and 700 individuals (R. R. Miller 1961b).

As discussed in the species accounts, even isolated and harmless pupfishes are threatened by human activities. At least three Death Valley forms are extinct, and all others are threatened to one degree or another. Concern for Death Valley pupfishes resulted in the formation in 1969 of the Desert Fishes Council, made up of individuals from numerous public and private agencies. The council succeeded in coordinating efforts of agencies to preserve and study pupfishes, in publicizing their plight, and in supporting legal actions to protect them (Pister 1991). Their future, however, is still not secure.

Desert Pupfish,
Cyprinodon macularius Baird and Girard

Identification Desert pupfish are small (<75 mm TL), chunky fish with a single series of incisorlike tricuspid teeth in the jaws. The middle cusp of each tooth is spatulate. Scales are large and regular, numbering 25–26 in the lateral series. Circuli on the scales have spinelike projections, and interspaces are regular, without reticulations. Fin rays number 9–12 in the dorsal, 9–12 in the anal, 14–18 in each pectoral, 14–20 in the caudal, and 2–8 (usually 7) in each pelvic fin. The dorsal fin is equidistant between the base of the caudal fin and the snout. Males are larger and stouter than females and when breeding are bright blue with lemon yellow tails and caudal peduncles. The tail has a black terminal band. Females are tan to olive with a lateral series of 5–8 disrupted vertical bars. Males also possess these bars, but less prominently.

Taxonomy Prior to Robert R. Miller's definitive work (1, 2, 3) most *Cyprinodon* in the Southwest were lumped with *C. macularius.* Today desert pupfish are just the species of the lower Colorado River and Salton Sea (4, 32). Quitobaquito pupfish, *C. eremus,* of Quitobaquito Spring, Pima County,

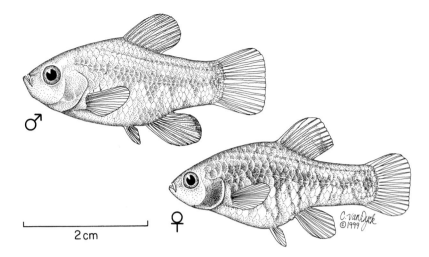

Figure 107. Desert pupfish. Top: Female, 3.5 cm SL, Salton Sea, Imperial County. Bottom: Male, 3.2 cm SL, Salton Sea, Imperial County.

Arizona, and Rio Sonoyta, Sonora, Mexico, was considered a subspecies of desert pupfish until recently, when it was elevated to species status (32). The closest relative of the desert pupfish in California is the Owens pupfish (4). The Salton Sea population was described as a subspecies of desert pupfish, but there is little justification for this designation (27, 28). Small genetic differences between Salton Sea and Colorado River delta populations indicate that they should be treated as separate units for conservation purposes (32).

Names *Cyprino-don* means carp with teeth; *macularius* translates as spotted. The name pupfish was coined by Carl L. Hubbs, one of the first people to take an interest in them, after he observed their "playful" behavior, which is actually the aggressive behavior of males. The name pupfish was fortuitous, however, because it has undoubtedly made promoting their conservation easier.

Distribution Desert pupfish were originally found in California, Arizona, and northern Mexico: along the lower Colorado River and in its delta, in San Filipe Creek in the Salton Sea basin, and throughout the Gila River watershed in Arizona (5, 6). Today they are still present in much of their historical native range in California and Mexico, but only as

small isolated populations around the Salton Sea and in the Colorado River (32). Their presence in 11 localities in Mexico in 1987 was probably partially the result of unusually wet conditions allowing them to expand their populations (7). In 1997–1998 eight populations were found in the same region (32). They became established along the edges of the Salton Sea and in its main tributary, San Felipe Creek, after flooding of the basin in 1905–1907. In 1991 and 1998 they remained widely distributed in irrigation drains and shoreline pools (8), but their long-term persistence is problematic. They have been introduced into three sanctuaries in Anza-Borrego State Park (San Diego County) and into several other sanctuaries elsewhere (9, 31).

Life History Few fish can live in the extreme range of environmental conditions tolerated by desert pupfish (10): salinities ranging from nearly twice that of seawater (68 ppt) to that of fresh water (11), temperatures from nearly 45°C in summer to 7°C in winter; and oxygen levels from saturation down to 0.1–0.4 mg/liter (12). They also survive 10- to 15-ppt changes in salinity as well as daily temperature fluctuations of up to 26°C (13, 14, 15). In the Salton Sea they are common in polluted and fluctuating conditions in the lower ends of irrigation drains. They prefer quiet water, and flash floods often reduce populations (16). By and large they tend to do poorly when forced to live with predaceous or competing fishes.

Much of our knowledge of the behavior and life history of desert pupfish is based on studies of Salton Sea populations when they were abundant (17, 18, 19). They typically swim in loose shoals of similar size and age from which small groups break off to forage. While breeding, males are territorial, and shoals then consist entirely of either adult females or juveniles. Smaller fish tend to be found in shallower water than larger fish. In the Salton Sea pupfish of all sizes once moved in and out of the shallows during the day, apparently to avoid temperatures higher than 36°C (18).

Thus they moved in to forage in the morning and then moved back to deeper water (about 40 cm) as the sun warmed the shallows. They remained there, relatively inactive, until evening. As the water cooled they again moved inshore to forage until it become dark, when foraging ceased. As dawn approached they moved back into the warmer deep water, remaining there until the shallows warmed up. Desert pupfish may bury themselves in loose debris on the bottom and become dormant during winter (20). They may also avoid excessively high temperatures by burrowing.

Desert pupfish forage on small invertebrates and algae picked off the substrate. In the Salton Sea this meant ostracods, copepods, and occasionally insects and pile worms. Elsewhere aquatic crustaceans, aquatic insect larvae, and snails are important (21). Because they are rather unselective in feeding, their long gut usually contains algae and detritus as well as invertebrates; occasionally they eat their own eggs and young (22, 29).

Growth is rapid and varies with age, temperature, and salinity, although fish of all ages grow at temperatures of 15–35°C and salinities from 0 to over 40 ppt (29). For juveniles the most rapid growth (in the laboratory) occurs at 30°C and 35 ppt salinity; for older fish optimum temperatures for growth are 22–26°C at salinities of about 15 ppt (13). In the Salton Sea desert pupfish measure 4–5 mm TL at hatching and double in size in less than 8 weeks. At 24 weeks they measure 15–28 mm TL. Maximum length at the end of the first growing season is 45–50 mm TL (13). Because desert pupfish can mature at 15 mm SL, it is possible for them to complete an entire life cycle in one summer. Most, however, do not breed until they have reached 30–50 mm SL.

Desert pupfish have a lek system of mating, in which males gather on a patch of silt-free bottom and try to lure females to spawn (23). The largest establish and defend territories, while smaller males hang out along the edges, dashing in to add their sperm when a territorial male spawns. Spawning takes place from April to October, whenever temperatures exceed 20°C. The first sign of reproduction is a few brightly colored males busily patrolling territories. Territories are usually located together in water less than 1 m deep and center on small submerged objects on the bottom (18). The depth of territories depends somewhat on water temperature, so they are often found in deeper water in summer than in spring (20). The size of the territory depends on the size of the male, number of males, and water temperature. Normally each male defends 1–2 m².

Basic spawning behavior, as described by George Barlow (18), is as follows: A female, when ready to spawn, is attracted to a territorial male and leaves her shoal. She approaches the male, who moves toward her. The female then tilts head-first toward the bottom and nips it, usually taking a small piece of substrate into her mouth. Renewing her horizontal position, she spits out the piece and halts close to the bottom. The male swims up to the female and lies parallel to her. The two fish then bend together into an S, and the male cups his anal fin around the caudal peduncle of the female. The female trembles and lays a single egg, which is fertilized by the male. Up to 4 eggs may be deposited on the bottom in this manner in quick succession, each spawning act taking a few seconds. Depending on size, a female lays 50–800 eggs or more during a season (24). Territorial males show strong preference for the largest (over 45 mm TL) females (23), apparently because large females produce more eggs per spawn. One cost is that larger females are more likely to eat eggs from previous spawns of other females!

Embryos hatch in 10 days at 20°C, and larvae start feeding on small invertebrates within a day after hatching (24). Larvae are frequently found in shallow water where environmental conditions are severe, but they survive higher salinities (to 90 ppt) than do adults, and sudden salinity changes up to 35 ppt (24).

Status IB. Desert pupfish were listed as Endangered by California in 1980 and by the federal government in 1986. Colorado desert pupfish are gone from most of their original habitats in the United States, which included marshes and backwaters of the lower Colorado and Gila River drainages. They persist in only a few refuges in which they have been planted, in scattered locations in Mexico, and in drains, shoreline pools, and tributaries to the Salton Sea. Although these fish have enormous physiological capacity to survive environmental stress, the environments in which they live in Mexico and in the Salton Sea basin are continually threatened with agricultural pollution, dewatering, or temperatures and salinities too severe even for pupfish. Reports of pupfish abundance vary widely from year to year and survey to survey. This is partly because pupfish populations can respond rapidly to favorable conditions through reproduction and emigration of surplus fish from "safe" areas (25). After a wet year they may be temporarily widespread and abundant, but during dry years many of the habitats may disappear and the fish become confined to a few small areas. It is ironic that many desert pupfish populations, but especially those in California, depend on agricultural drain water, unreliable in both quantity and quality, for maintaining their habitats. As long as most remaining wild populations of pupfish depend on irrigated agriculture, the desert pupfish will be an endangered species.

An additional continuous threat to desert pupfish are alien fishes that dominate the waters of the region. In deeper waters piscivores such as largemouth bass and various catfishes are a continuous threat, although native predators, such as Colorado pikeminnow, presumably kept pupfish out of deeper waters prior to introductions. In shallows, especially those in marshy areas or saline pools, pupfish must now interact with many species, including tilapia species and various poeciliids (mainly mosquitofish, sailfin mollies, and porthole livebearers). Predation on pupfish embryos and larvae by poeciliids may be a major limiting factor. Male pupfish defending breeding territories have to chase away mollies and juvenile tilapia, leaving embryos more vulnerable to predation and using energy needed for courtship and spawning (29). Although pupfish are most abundant in areas where they are by themselves, they nonetheless can coexist in shallow-water drains containing alien species despite competition and predation. Coexistence is presumably related to the fluctuating, often severe nature of the environments, where pupfish have an advantage, and perhaps also to the existence of inadvertent refuges where pupfish can breed in peace and produce large numbers of emigrants. The refuges are most likely to be isolated, shallow waters from which extremes of temperature and salinity exclude most other fish (7, 15, 26). The abundance of pupfish in drains and pools along the Salton Sea in 1991 (8), for example, may have resulted from die-offs of normally superabundant tilapia due to cold weather, leaving more "space" for pupfish. By 1998 pupfish were again scarce (30).

In fact the long-term prognosis for wild populations in the Salton Sea drainage is not good. Schoenherr (29) noted that in the late 1960s pupfish disappeared from shoreline pools, replaced by sailfin mollies, but they were common in drains feeding the sea. After the introduction of redbelly tilapia in the 1970s pupfish became largely confined to habitat less than 10 cm deep. Even this habitat is shared with juvenile tilapia and other species. As indicated previously, changing conditions may create a fluctuating coexistence. However, further introductions may eliminate pupfish. For example, they disappeared from Salt Creek, a former stronghold, when largemouth bass were introduced (30). Their habitats in drains must be regarded as temporary because they are subject to the maintenance activities of irrigation districts, uncertainty of freshwater input, and high abundances of tilapia and other alien fishes (30). Nevertheless, management of some drains to favor pupfish may be important for their long-term persistence in the Salton Sea basin (30).

Fortunately, desert pupfish are easily bred and maintained in aquaria (24) and are easily established in natural and artificial refuges. There is always the danger, of course, that planting pupfish in isolated natural waters will result in extinction of endemic invertebrates, especially tiny snails. While refuges outside the native range may be necessary, they should be a minor part of the mix of active management programs needed to keep this species in existence. Some refuge populations already show significant genetic differences from wild populations, suggesting that refuges cannot be relied on for long-term survival of the species (31). In the long run natural habitats such as San Filipe Creek should be managed for pupfish, even though eradicating alien fishes will be necessary and difficult. Any program to "restore" the Salton Sea should include experimental management of drains, shoreline pools, and other habitats to favor pupfish. Despite their amazing tolerance for adverse conditions, desert pupfish require flexible and intensive management for the indefinite future.

References 1. Miller 1943b. 2. Miller 1943c. 3. Miller 1948. 4. Echelle and Dowling 1992. 5. McMahon and Miller 1985. 6. B. Turner 1983. 7. Hendrickson and Romero 1989. 8. Lau and Boehm 1991. 9. Bolster 1990. 10. Cowles 1934. 11. Barlow 1958b. 12. Lowe et al. 1967. 13. Kinne 1960. 14. Lowe and Heath 1969. 15. Schoenherr 1979. 16. Schoenherr 1992. 17. Barlow 1958a. 18. Barlow 1961b. 19. Walker et al. 1961. 20. Cox 1966. 21. Walters and Legner 1980. 22. Cox 1972. 23. Loiselle 1982. 24. Crear and Haydock 1970. 25. McMahon and Tash 1988. 26. Zengel and Glenn 1996. 27. Miller and Fuiman 1987. 28. Minckley et al. 1991b. 29. Schoenherr 1988. 30. S. Keeney, CDFG, pers. comm. 1999. 31. Dunham and Minckley 1998. 32. Echelle et al. 2000.

Owens Pupfish, *Cyprinodon radiosus* Miller

Identification Owens pupfish resemble desert pupfish, but they lack spinelike projections on the scale circuli and have reticulated spaces between circuli. The scales are large, numbering 26–27 in the lateral series. The middle cusps of the jaw teeth are more truncate than spatulate. Dorsal rays number 10–12 (usually 11); anal rays, 9–12 (usually 10); pectoral fin rays, 13–17 (usually 14–15); pelvic rays, 6–8 (usually 7); and caudal rays 16–19. The dorsal fin has a thickened first ray and is equidistant between the base of the caudal fin and the tip of the snout. Gill rakers number 15–20 (usually 16–19). The head is more slender and the caudal peduncle longer than in Death Valley pupfishes. Males are larger and deeper bodied than females, but the differences are not as striking as in many other *Cyprinodon* species. Breeding males are bright blue, with purplish lateral bars (1). Unlike those of other pupfishes, these bars do not narrow ventrally. Females are similar to other pupfish females: olive brown with a purplish sheen, with lateral blotches and bars.

Names Radiosus "refers to the high number of dorsal, anal, and pelvic rays" (2, p. 98). The common name is a reference to the region in which it is endemic, named by John Charles Frémont in 1845 for Richard Owens, a member of one of his expeditions. Other names are as for desert pupfish.

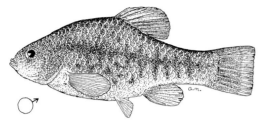

Figure 108. Owens pupfish. Left: Male, 41 mm SL, Owens Valley, Inyo County. Right: Female, 35 mm SL, Owens Valley, Inyo County. Drawing by A. Marciochi.

Taxonomy The Owens pupfish was described as a species distinct from desert pupfish by R. R. Miller in 1948 (2). Genetic analysis demonstrates that it is distinct but closely related to the desert pupfish and more distantly related to the Amargosa pupfish (3). The Owens pupfish may have been isolated for as long as 2 million years (4). At present all are descended from about 400 fish rescued from the last "natural" population in Fish Slough (5).

Distribution Owens pupfish were originally found in the Owens River from Fish Slough and its springs (Mono County) near Bishop downstream to Owens Lake (Inyo County), as well as in springs around the lake (6). Their historical presence in Owens Lake is not known. Now they exist only in special refuges. The first refuges were established in Warm Springs near Big Pine (1) and in two areas of Fish Slough. Unfortunately one of these areas, the Owens Valley Native Fishes Sanctuary, no longer contains pupfish because of the introduction of largemouth bass in 1990. Other refuges have subsequently been established (7, 8).

Life History Owens pupfish once occupied a wide variety of shallow-water habitats: spring pools, sloughs, irrigation ditches, swamps, and flooded pastures along the Owens River (9). The water was clear and warm, often with heavy stands of emergent bulrushes in shallows and dense mats of *Chara* on the bottom of pools. The fish tended to live among stems of emergent plants and breed in open areas. Water temperatures probably ranged annually from about 10 to 25°C. Curiously, despite the relatively benign environment, Owens pupfish retained their inherited ability to tolerate

wide temperature ranges (10). The sanctuaries to which they are now confined have large, clear pools with extensive shallows as well as holes up to 2 m deep. Like other pupfishes they are most abundant where other fishes, especially predators, are absent. They were originally associated with Owens tui chub, Owens sucker, and Owens speckled dace, but these fishes, as adults, were found in deeper water than pupfish.

Adult Owens pupfish forage on the bottom or on aquatic plants in small shoals, often of 30–50 fish (8). They feed mostly on aquatic organisms such as chironomid midge, mayfly, beetle, and dragonfly larvae as well as snails and small crustaceans (8, 9). Juveniles (<20 mm TL) also feed in small groups and stay close to the substrate. The presence of Owens pupfish in shallow water among emergent plants, coupled with the comparative absence of mosquitoes in Owens Valley, led one early observer to conclude that pupfish kept mosquitoes under control (9). As pupfish became rare, mosquitoes became a problem (2). At night Owens pupfish are inactive, lying on or in the substrate (8). They are also inactive during the low-temperature periods of winter, moving about only on unusually warm days when temperatures exceed 15°C (8).

Owens pupfish males can grow to about 65 mm TL, and females, to about 61 mm. They are similar to other pupfishes in their ability to reach 35–50 mm TL during the first growing season and breed before they are a year old. Few if any are likely to live more than 2 years. Unlike other pupfish, Owens pupfish grow at low temperatures during winter months.

The breeding season varies with location (8). In Warm Springs, where temperatures do not drop below 21°C, females start producing eggs in mid-January; spawning commences by early February and finishes by the end of June. In habitats with strong seasonal fluctuations in temperature (7–26°C), spawning may not begin until mid-March and may continue through August. Eggs are produced by females at temperatures of 13.5–27°C (8). Successful spawning can also occur in environments that are almost always below 20°C, unusual for a pupfish.

Spawning behavior is lek-like, similar to that of desert pupfish (8, 11). Males set up temporary spawning territories in open areas close to shore, with territorial males in clusters. The spawning substrate is highly variable and includes silty bottoms, submerged plant stems, clumps of algae, flat rocks, and crevices (8) at depths from 2 cm to 2 m (12). Females stay in vegetation, emerging only to spawn, while males vigorously defend territories. For the most part females initiate spawning by hovering above a territory, entering it, and then resting on the substrate or nipping at the bottom. The male then courts the female. The female may interrupt courtship or not release an egg (which happens 60–70% of the time), but a single female may spawn up to 200 times a day (12). At each successful spawn only 1 or 2 eggs are typically laid and fertilized.

Males that initiate spawning are mainly nonterritorial ones hanging out around the edge of a lek to intercept females on their way to spawn with territorial males. Such spawning can succeed unless interrupted by a territorial male, who may then spawn with the female. Each female spawns with multiple males over several months, but the criteria used by females to choose mates are not clear and vary from place to place (8). Males seem to choose females largely (but not entirely) on the basis of availability and not by size, perhaps reflecting the small number of eggs laid each time and the lack of a relationship between female size and egg size (12).

Embryos hatch in 4–10 days at 24–27°C (12), and larvae and juveniles stay close to the substrate.

Status IB. Owens pupfish were listed as Endangered by the federal government in 1967 and by California in 1971. They were once thought to be extinct, and their rediscovery and preservation make for a dramatic story (1, 5). When Owens pupfish were originally described by Miller (2), much of their habitat had already disappeared owing to removal of Owens River water by the city of Los Angeles. The presence of largemouth bass, carp, and mosquitofish in what habitat remained made it unlikely that any pupfish survived. Thus it came as a surprise when, in 1963, R. K. Liu, then a graduate student at the University of California, Los Angeles, discovered that two biologists from CDFG had collected Owens pupfish in 1956 but had not told anyone because they did not realize the pupfish was thought extinct. In 1964 C. L. Hubbs, R. R. Miller, and E. P. Pister located a slough containing a few hundred fish. Realizing that the position of the pupfish was precarious, in 1967 they carefully examined

the slough and laid plans for construction of a refuge, complete with barriers to keep out alien fish.

The sanctuary pools were ready for stocking in June 1970, none too soon. In August 1969 natural events nearly dried up the marshy pool containing the last pupfish. Fortunately R. E. Brown, also a UCLA graduate student, saw that the pool was nearly dry and reported the impending disaster to Pister of CDFG. Pister and his crew took immediate action, rescuing about 800 fish. These fish were placed in live cages in a remaining part of the slough, prior to being moved the next day to a new location. Unfortunately, poor water quality started killing them, and Pister, back to check on the fish, rescued the survivors. At one point he was carrying the entire species in two buckets, praying that he would not stumble and fall in the marshy terrain. Temporary sanctuaries were then hastily built. Construction of both temporary and permanent sanctuaries, another miracle engineered by Pister and his colleagues, required the bureaucratic cooperation of numerous public and private agencies, including CDFG, the California Division of Forestry, the Los Angeles Department of Water and Power, the Bureau of Land Management, The Nature Conservancy, and the John Muir Institute (1).

Populations of the Owens pupfish are now regarded as stable in two refuges, but they remain threatened because their small sanctuaries must be actively managed to keep habitat favorable for pupfish and to keep out alien species. A major problem is inconsiderate or ignorant anglers, who illegally plant largemouth bass in the sanctuaries; the bass can only be removed by poisoning all fish and starting again. Thus by 1998 Owens pupfish were gone from BLM Spring and Owens Valley Native Fishes Sanctuary as a result of largemouth bass predation. Another major problem is the continual encroachment of cattails, tules, and other emergent plants, which can completely choke out the open water needed by fish (13). Such plants must be periodically controlled either mechanically or with herbicides.

References 1. Miller and Pister 1971. 2. Miller 1948. 3. Echelle and Dowling 1992. 4. Miller 1981. 5. Pister 1993. 6. USFWS 1984b. 7. Minckley et al. 1991b. 8. Mire 1993. 9. Kennedy 1916. 10. Brown and Feldmeth 1971. 11. Liu 1969. 12. Mire and Millett 1994. 13. S. Parmenter, CDFG, pers. comm. 2001.

Amargosa Pupfish,
Cyprinodon nevadensis Eigenmann and Eigenmann

Identification Amargosa pupfish are small fish that rarely exceed 50 mm TL. The body is deep, especially in reproductive males. The head is blunt and slopes steeply in front to a small, terminal, oblique mouth. There is one row of tricuspid teeth on each jaw, with the central cusps truncated or pointed. Characters of Amargosa pupfish are variable, but it can be distinguished from other pupfish by *(1)* the presence of large scales with circuli lacking spinelike projections and having reticulated interspaces; *(2)* having 23–28 scales (usu-

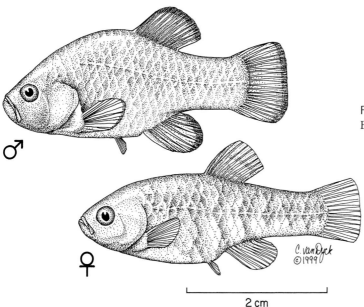

Figure 109. Amargosa pupfish. Top: Male, 3.5 cm SL. Bottom: Female, 3.5 cm SL. Source unknown.

2 cm

ally 25–26) along the lateral line and 15–24 scales (usually 16–18) anterior to the dorsal fin; *(3)* small or absent pelvic fins; and *(4)* the presence of 8–11 anal fin rays (usually 10), 11–18 pectoral fin rays (usually 15–17), 0–9 pelvic fin rays (usually 6), and 14–22 caudal fin rays (usually 16–19). Gill rakers range from 14 to 22 (usually 15–17) and preopercular pores, from 7 to 17 (usually 12–14). Reproductive males are bright blue on the sides and caudal peduncle with a black band at the posterior edge of the caudal fin. Reproductive females are drab olive brown and develop 6–10 lateral vertical bars, which may be faint. A faint ocellus is typically present on the posterior base of the dorsal fin of females.

Names "The name *nevadensis* indicates a desire on the part of the describers to honor the region east of California, but the reasons for such a desire apparently are lost in obscurity, for the Eigenmanns were well aware that the type locality was in California" (2, p. 503). Other names are as for desert pupfish.

Taxonomy Cyprinodon nevadensis was first described from Saratoga Springs, California, by Eigenmann and Eigenmann in 1889 (3). Following the initial description the species was lumped with desert pupfish (*C. macularius*) until Miller (4) resurrected it after additional analysis. Its closest relatives are the Salt Creek pupfish and Devils Hole pupfish (Nevada); the Owens pupfish appears more distantly related despite its geographic proximity (5). R. R. Miller (1) recognized and described six subspecies, four in California and two in Nevada, as discussed under Distribution.

Distribution Amargosa pupfish are confined to the Amargosa River basin in southern California and Nevada. The species is divided into isolated populations, six of which are recognized as subspecies:

1. Saratoga Springs pupfish, *C. n. nevadensis,* occur only in Saratoga Springs in the southeastern corner of Death Valley, San Bernardino County.

2. Tecopa pupfish, *C. n. calidae,* now extinct, formerly occurred in north and south Tecopa Hot Springs, Inyo County.

3. "True" Amargosa pupfish, *C. n. amargosae,* are most abundant in permanent waters of the Amargosa River, in Amargosa Canyon, and in ditches draining Tecopa Hot Springs and Tecopa Bore (an outflow of an artesian well), San Bernardino County. Another population occurs in the lower Amargosa River around Sperry, in Death Valley National Park. In 1940 R. R. Miller transplanted about 350 fish to River Springs in Adobe Valley (Mono County), where they still exist (23).

4. Shoshone pupfish, *C. n. shoshone,* were confined to Shoshone Springs (Inyo County) but now exist mainly in artificial refuges.

5. Ash Meadows pupfish, *C. n. mionectes,* remain in a number of springs in lower Ash Meadows, Nye County, Nevada.

6. Warm Springs pupfish, *C. n. pectoralis,* live in School and Scruggs Springs, upper Ash Meadows, Nye County, Nevada.

The subspecies can be further broken down into recognizable populations, each inhabiting a different spring or stretch of river.

Life History Amargosa pupfish are adapted to a wide variety of aquatic environments. The populations in Ash Meadows inhabit freshwater springs with temperatures that range collectively from 21 to 33°C, although each spring varies only 2–7°C annually. In contrast, the highly saline Amargosa River varies from close to freezing in winter to nearly 40°C in summer, with daily fluctuations as great as 15–20°C. Most spring habitats are rarely more than 2 m deep; have soft, flocculent bottoms; and support heavy growths of algae and associated invertebrates, with emergent cattails and rushes along the edges.

The lower Amargosa River contains a variety of habitats in which pupfish thrive. The upper reach is divided into two distinct areas, one near Tecopa, with broad marshes fed by hot springs, and another immediately downstream, where the river flows through a narrow, steep-sided canyon. There it is less than 2 m wide and up to 2.5 m deep. Flows are swift in runs between pools, and the substrate consists mostly of gravel and sand, with some boulders and rubble (1, 6). The preferred depth range of pupfish in this reach is 10–35 cm (6). The water is clear and saline, with the pH ranging from 8.2 to 8.7. Total dissolved solids are fairly high and variable at 1,390–3,890 ppm, and the dissolved oxygen level is 7.3–11.6 mg/liter. Shoreline vegetation is abundant. Pools are numerous, both in the river and on the flood plain, the largest being about 8 × 5 m. The substrate in pools is mostly mud and clay.

Downstream the river flows into Death Valley at an elevation of about 33 m (1). The river bottom consists of fine silt, clay, mud, and sand; there is no instream vegetation except cyanobacteria and algae. Salinities are 3–18 ppt, depending on location and season (7). The current is moderate to swift between pools that are 0.75–1.25 m deep. Depths selected by adults seem to depend on local habitat conditions. In one reach they selected depths of less than 16 cm, while in another they selected depths of 25–50 cm (7). Juvenile pupfish, however, consistently select depths of less than 5 cm, staying close to shore. All fish occupy areas where water column velocities are less than 2 cm/sec, substrates are fine, and there is little aquatic vegetation (7). Water temperature varies seasonally from 10 to 38°C and during severe winters may approach freezing. Large daily variation in water temperature is also present. Younger fish tolerate higher water temperatures than adults (8) and are commonly found in warmer places (1), which serve as refuges from predation or competition for food.

One of the more interesting habitats of Amargosa pupfish is Tecopa Bore, an outflow of an artesian well colonized by pupfish from a marsh connected to the Amargosa River. The water temperature at the head is 47.5°C, but by the time the water flows about 1 km it may be close to freezing, depending on the season. The maximum temperature pupfish can withstand is about 42°C. In the stream they tend to concentrate in the reach that is 42°C, because the cyanobacteria upon which they feed are most abundant there (9). Wind frequently blows cooler water upstream, and the pupfish quickly take advantage of the temporary availability of ungrazed pasture. When the wind dies the fish move back downstream. Pupfish are occasionally caught in water warmer than 42°C, dying unless they are quickly washed into cooler water.

A habitat in marked contrast to Tecopa Bore is Saratoga Springs, circular in shape, approximately 10 m in diameter, and 1–2 m deep (1). Water in the main spring pool is clear with some algae and detritus on the soft bottom. Water temperature is constant at 28–29°C. The spring overflows into a larger pond that in turn drains into several shallow interconnected "lakes" 4–6 ha in area. The lakes have grassy bottoms of mud and sand. Water temperatures fluctuate daily with ambient air temperature and vary from 4 to 49°C on a seasonal basis, although fish avoid areas in which temperatures regularly exceed 38°C. Temperatures of 20–30°C are selected when possible (7). Fish remain along shore, in water less than 40–50 cm deep, and move into the marshy meadows when disturbed. They seem to select areas with aquatic vegetation, in contrast to their counterparts in the Amargosa River (7). Juvenile fish abound in the lakes but are absent from the spring, suggesting that spawning occurs only in the lakes. In autumn, as temperatures cool in marshy environments, about a third of the fish bury themselves in the substrate (7).

Despite their ability to survive in a wide variety of conditions, Amargosa pupfish grow and reproduce under a much more limited set. Reproduction is greatly diminished at pH values below 7 (10) and temperatures below 25°C or above 31°C (11, 24). Optimal temperatures for reproduction are 28–29°C, and those for growth, 22°C, with growth ceasing below 17°C and above 31°C (11, 24). Under fluctuating conditions Amargosa pupfish reproduction apparently does not suffer if the range is 4–6°C during the day and includes optimal temperatures (12).

The main food of Amargosa pupfish seems to be cyanobacteria and algae. They have long, convoluted intestines characteristic of aquatic herbivores and teeth adapted for nipping. However, they also feed seasonally on small invertebrates, mostly chironomid larvae, ostracods, and copepods (13, 14), and they can be effective predators on mosquito larvae in heavy vegetation (15). They forage continuously from sunrise to sunset, becoming inactive at night.

Growth is rapid, especially in warm, constant springs where they can reach 25–30 mm TL and sexual maturity in a few months (1). Growth rates, however, are slower in fluctuating environments (16). Maximum length is about 65 mm TL, but fish over 50 mm are rare. It is unlikely that many fish live longer than a year. Their short generation time has allowed some populations of Amargosa pupfish to maintain themselves with extremely small numbers of fish. Mexican Spring in Ash Meadows, which contains about 80 gallons of water, is estimated to support 20–40 fish, half of which are adults at any one time (9).

The spawning behavior of Amargosa pupfish varies with habitat. In springs their behavior is similar to that of Owens pupfish, with males defending territories in leks. In the Amargosa river pupfish are group spawners (17). Males do not establish and defend territories. Instead a male usually directs a receptive female to the periphery of the group, where spawning occurs, although it may take place in the center of the group as well.

Status IB. Most populations of Amargosa pupfish are at or near historical levels, yet the species as a whole must still be regarded as threatened, as should any species that depends on water in a desert region. Of the six subspecies, one is already extinct, two are listed as Endangered, and the remaining three are Species of Special Concern. In this section the general reasons why the species is at risk of extinction are discussed, followed by problems specific to each subspecies.

The major threat to Amargosa pupfish is potential dewatering of their unique spring and stream habitats. The aquifer that feeds this system is apparently fed by a large, ancient groundwater source that extends into western Utah and central Nevada. The Las Vegas Valley Water District proposes to mine this water in large quantities to supply the ever-growing human population of Las Vegas, southern Nevada. The amounts of groundwater to be pumped could be immense. As John McPhee writes in *The New Yorker* (26 April 1993):

> Around Las Vegas, well casings stand in the air like contemporary sculpture, and so much water has been mined from below that the surface of the earth has subsided six feet. While new wells are no longer permissible, Las Vegas desperately needs water for its lakes. They are not glacial lakes. If you want a lake in Las Vegas, you dig a hole and

pour water into it. In one new subdivision are eight lakes. Las Vegas has twenty-two golf courses, at sixteen hundred gallons a divot. Green lawn runs down the median of the Strip. Here is the Wet 'N' Wild park, there the new M-G-M water rides. Outside the Mirage, a stratovolcano is in a state of perpetual eruption. It erupts water.

If the proposed large-scale removal of groundwater happens, it is quite possible that springs will stop flowing and the Amargosa River will have its base flow reduced or eliminated. Already diversions of springs and outflows on private land in the Tecopa area have undoubtedly reduced local flows in the river and local pupfish populations as well. A comparable situation existed for Devils Hole in Nevada, which experienced lowered water levels as a result of water pumped from its aquifer for irrigation. It took an order from the U.S. Supreme Court to protect Devils Hole and the Devils Hole pupfish by stopping the pumping (18). After rising and then stabilizing for a number of years, the water level in Devils Hole is now dropping again, most likely as a result of groundwater pumping (22).

One source of this pumping may be the city of Pahrump, a bedroom community for Las Vegas located south of Ash Meadows. Pahrump is the fastest-growing city in Nevada and depends entirely on groundwater, probably from the same aquifer that feeds springs and streams of the region. In recent years groundwater levels have dropped over 13 m, drying up Manse Spring, the evolutionary home of the Pahrump poolfish, *Empetrichthys latos latos* (23). With an increasing human population in Pahrump, Tecopa, and the upper Amargosa Valley, demand for water and, ironically, protection from floods is increasing. In what may be a futile effort in the long run, USFWS has established Ash Meadows National Wildlife Refuge to protect the endemic flora and fauna of the springs and river.

An additional threat is the introduction of potential competitors and predators, because the Amargosa River and most springs are accessible to the public. Mosquitofish are associated with declines of other pupfish species and are often abundant in Amargosa Canyon, yet Amargosa pupfish appear able to coexist with them (6). Flash floods periodically reduce mosquitofish populations, to the advantage of pupfish. The possibility exists, however, for additional introductions of alien fishes into the Amargosa River. A catfish farm located in Shoshone will require careful management to prevent escape of unwanted species into the river. Many springs already contain alien species (Saratoga Springs is a fortunate exception), and control measures are often needed. For springs there is the continuous threat of perverse or ignorant individuals introducing additional fishes or invertebrates, which compete with or prey on pupfish, or bring diseases to which pupfish are not resistant.

Saratoga Springs pupfish. IB. Although listed as a Species of Special Concern by CDFG, it is fairly secure because Saratoga Springs is located in Death Valley National Park. It is at risk from distant water removal and introductions of nonnative fishes.

Tecopa pupfish. IA. This form disappeared when Tecopa Hot Springs were converted into bathhouses (19).

Amargosa pupfish. IB. The most widespread of all the subspecies, their population was considered "stable" in 1998 (23). They are threatened by water removal and introduced species. A single transplanted population provides some security. It is listed as a Species of Special Concern by CDFG.

Shoshone pupfish. IB. The Shoshone pupfish is clearly endangered, but it is listed only as a Species of Special Concern by CDFG. The original habitat of this form, Shoshone Spring and its outfall, was converted into a water supply for the town of Shoshone, including a swimming pool and catfish farm. The pupfish were thought extinct until 1986, when a population was discovered living precariously in the cement-lined outflow ditch (20). In 1988 the ditch was dominated by mosquitofish. About 20 Shoshone pupfish were rescued by J. E. Williams and a class of graduate students from the University of California, Davis (21), and raised in captivity until a refuge was built at Shoshone Spring. These and some fish in captivity at the University of Nevada, Las Vegas, were introduced in 1990 into the pond, where they still survive in modest numbers. There also remains a population in the outflow ditch that is constantly threatened by chlorinated water from the swimming pool, competition from mosquitofish, and alteration of habitat by invasion of tamarisk. Unfortunately, genetic evidence indicates that the rescued fish may be the same as Amargosa pupfish, in which case the Shoshone pupfish is extinct (23).

Ash Meadows pupfish. IB. This Nevada form, which is federally listed as Endangered, is present in springs scattered throughout Ash Meadows. The springs were drying up in the 1970s as a result of attempts to farm and develop Ash Meadows. They were saved by complex legal action that included saving habitat for the endangered Devils Hole pupfish through a Supreme Court decision (18). Pumping from the aquifer and alien species continue to be threats.

Warm Springs pupfish. IB. Also federally listed as Endangered, this Nevada pupfish was one of the first pupfish to be protected. In the late 1960s BLM constructed cement ponds and made other changes to School Spring in order to keep the pupfish population extant when flow was reduced. The Warm Springs pupfish continues to be threatened by the same factors affecting all aquatic organisms that live in the Ash Meadows area.

References 1. Miller 1948. 2. La Rivers 1962. 3. Eigenmann and Eigenmann 1889. 4. Miller 1943c. 5. Echelle and Dowling 1992. 6. Williams et al. 1982. 7. Sada et al. 1997. 8. Shrode 1975. 9. J. H. Brown 1971. 10. Lee and Gerking 1980. 11. Gerking and Lee 1983. 12. Shrode and Gerking 1977. 13. Naiman 1975. 14. Naiman 1976. 15. Danielsen 1968. 16. Miller 1961b. 17. Kodric-Brown 1981. 18. Deacon and Williams 1991. 19. Minckley et al. 1991b. 20. Taylor et al. 1988. 21. Castleberry et al. 1990. 22. L. L. Lehman and R. G. Atkins, unpubl. rpt. 1991. 23. E. P. Pister, Desert Fishes Council, pers. comm. 1999. 24. Gerking et al. 1979.

Salt Creek Pupfish, *Cyprinodon salinus* Miller

Identification Salt Creek pupfish are the most slender of the Death Valley pupfishes, distinguished by small scales (28–29 in the lateral series) that have reticulated interspaces between the circuli and by tricuspid teeth with prominent median ridges. The dorsal fin, with 8–11 rays (usually 9–10), is closer to the base of the caudal fin than to the snout. Anal rays number 9–11 (usually 10); pectoral fin rays, 14–17 (usually 15–16); and caudal rays, 15–19 (usually 16–17). The pelvic fins are small (6 rays) or, occasionally, absent. Gill rakers number 18–22 (usually 19–21) and are shorter and more compressed than those of other pupfishes. Scales are absent from most of the preorbital region of the head. Breeding males become deep blue on the sides and iridescent purple on the back. The caudal fin has a conspicuous black terminal band. The sides of spawning males have 5–8 broad vertical bands, which may be either continuous or interrupted. Females have less conspicuous coloration: brownish with a silvery sheen (1). They do, however, have 4–8 vertical lateral bars that are less intense than the barring pattern of males, except during spawning. Males are deeper bodied than females, with a noticeable arch to their anterior profile.

Taxonomy Salt Creek pupfish were described by R. R. Miller (1) in 1943 from Salt Creek, Death Valley. Characteristics such as reduced or absent pelvic fins, posterior position of the dorsal fin, short head, small eyes, low mean fin ray

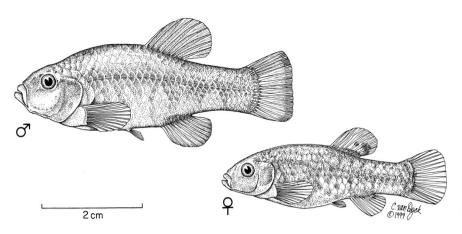

Figure 110. Salt Creek pupfish. Left: Male, 28 mm SL, Salt Creek, Inyo County. Right: Female, 25 mm SL, Salt Creek, Inyo County.

2 cm

counts, and inconspicuous humeral process indicate a close relationship to *C. nevadensis* (1). Mitochondrial DNA analysis confirmed this but suggested that the two species have been isolated for a long time (2). Miller (3) assumed that desiccation of Lake Manly, which occupied the floor of Death Valley during the late Pleistocene, was the force isolating the two populations, resulting in speciation. However, the two species had diverged even prior to the formation of Lake Manly. *Cyprinodon salinus* has subsequently been divided into subspecies: *C. s. salinus* from Salt Creek and *C. s. milleri* from Cottonball Marsh; the latter was first described as a full species (4).

Names *Salinus* means salt. Other names are as for desert pupfish.

Distribution Salt Creek pupfish are restricted to Salt Creek (Inyo County) in the northern part of Death Valley, Death Valley National Park, about 49 m below sea level. The amount of habitat varies seasonally, but normally shrinks to about 1.5 km of channel below McLean Springs. They were introduced into Soda Lake (San Bernardino County) and River Springs (Mono County) (5), but neither population has persisted (12). The Cottonball Marsh pupfish lives in 259 ha of marsh on the floor of Death Valley, adjacent to the sink for Salt Creek, about 80 m below sea level. Only about 10 percent of the marsh contains pupfish habitat (6).

Life History Salt Creek and Cottonball Marsh are among the most severe habitats inhabited by any fish. Both are on the floor of Death Valley, where summer air temperatures often exceed 50°C. Upper Salt Creek begins with seepages and exists as surface water, uninhabited by pupfish, only in winter and spring, when it meanders muddily across Mesquite Flat for 1–2 km before dropping into a narrow, shallow canyon. Thanks to inflow from McLean Springs, the reach through the canyon is permanent and thus supports pupfish year round for about 1.5 km. When rains increase

flows, the fish may occupy as much as 5 km of stream. Occasional flash floods cause high mortalities, washing fish into reaches that dry up (7). The stream channel on the canyon floor is carved 3–7 m deep into alkaline mud and consists of a series of interconnected pools, some as large as 250 m² and 2 m deep. The pools are edged with saltgrass, pickleweed, and wiregrass. The plants may completely roof over interconnecting channels and hang over pool edges, providing shelter for pupfish. The pools contain beds of aquatic plants that are favored by pupfish (1). Below the canyon the stream is quite shallow and exposed, but fish will inhabit as much of this area as fluctuating flow permits.

Temperatures in the creek range from near freezing in winter to nearly 40°C in summer, although they seldom exceed 28°C in the deep water of pools. In the laboratory Salt Creek pupfish can tolerate temperatures to 38°C and survive short-term exposures to 43°C (8). Salinities in the canyon reach vary, but in summer they approach that of seawater. However, the pupfish can live at salinities about twice that of seawater (4) as well as in fresh water (9). Boron levels (39 ppm) and total dissolved solids (23,600 ppm) are exceptionally high for any inland water containing fish (1).

Cottonball Marsh is an extraordinary habitat, even for pupfish. They live in shallow (about 75 cm) pools and even shallower (about 15 cm) streams, with rims encrusted with gypsum and salt (mostly sodium sulfate). The pools become smaller and shallower as summer progresses, exposing fish to extremes of temperature and salinity. The soil is too saline to support even the hardiest plants, such as pickleweed and saltgrass. In the water, however, there are mats of algae and, in less saline pools, stands of emergent rushes. The algal mats support amphipods, ostracods, and small snails, which the pupfish consume along with the algae. Salinities of the pools are 14–160 ppt (about 4.6 times that of seawater), depending on time of year and closeness to the seeps that are the water source for the marsh. Pupfish are rarely found in water more saline than 70 ppt (6). Temperatures vary from close to freezing in winter to nearly 40°C in summer, which approaches the maximum temperature pupfish can withstand (8). Daily fluctuations in temperature can also be extreme, paralleling those of the air. In shallows daily temperatures can fluctuate by as much as 15°C, although in deeper areas (33 cm maximum) fluctuations may be only 2–3°C (10). During summer places inhabited by pupfish rarely exceed 34°C because of the cooler temperatures of inflowing springs and the high evaporation rates (10). In fact Sada and Deacon (6) found that habitats selected by pupfish in Cottonball Marsh were less severe than those in which pupfish often lived in Salt Creek.

One of the most remarkable aspects of the biology of Salt Creek pupfish is their population fluctuation. When flows are high the population grows rapidly, spreading beyond the limits of permanent flow. Miller (1) estimated peak populations to number in the millions. This estimate may be high, but densities as high as 527 fish per square meter have been measured (6). The rapid population buildup suggests that the fish may go through several generations in a year. As water temperatures rise and the stream shrinks in summer, fish die by the thousands. Ravens and herons take advantage of this accessible food supply, as did the Panamint People, who caught pupfish in baskets and baked them (1).

The diet of Salt Creek pupfish is presumably mostly cyanobacteria and algae, although they no doubt also feed on the endemic snails and crustaceans that share their habitat.

Salt Creek pupfish become mature at 30–40 mm TL. They occasionally grow as large as 63 mm but rarely exceed 50 mm (6). Breeding behavior is similar to that of desert pupfish (11).

Status IB. Salt Creek pupfish should be considered as a threatened species, but they are listed only as a Species of Special Concern by CDFG. Salt Creek and Cottonball Marsh are located entirely within Death Valley National Park, so the pupfish continue their dramatic population cycles interrupted only by tourists and biologists. Present management by the Park is sufficient to maintain pupfish populations, as well as the unique Salt Creek ecosystem. Despite the protected and relatively pristine nature of their environments, the two subspecies cannot be regarded as secure, given their limited habitats. Unauthorized introductions of exotic species are always a possibility, as are catastrophic events, natural and unnatural. The springs that feed both habitats may be connected to the aquifer that provides water to Furnace Creek, the town center of Death Valley's tourism, so excessive pumping could harm the fish. It would seem advisable to maintain captive or refuge populations of both subspecies in case unexpected events threaten wild populations.

References 1. Miller 1943b. 2. Echelle and Dowling 1992. 3. Miller 1981. 4. LaBounty and Deacon 1972. 5. Miller 1968. 6. Sada and Deacon 1995. 7. Williams and Bolster 1989. 8. Brown and Feldmeth 1971. 9. Baugh 1981. 10. Naiman et al. 1973. 11. Baugh 1982. 12. R. R. Miller, University of Michigan, pers. comm. 1974.

Sticklebacks, Gasterosteidae

Sticklebacks are a small, cohesive family of fishes, abundant in Europe, northern Asia, and North America. They are famous for their pugnacity and for their stereotyped breeding behavior, which they perform readily in aquaria. As a result their behavior has probably been studied in greater detail than that of any group of freshwater fishes. Sticklebacks are also well known to students of evolution because of their proclivity to rapidly develop distinctive behavioral and morphological characteristics in isolation. If a strict biological definition of species was adhered to in sticklebacks, there would arguably be hundreds of species rather than the seven currently recognized.

The bodies of sticklebacks are distinctive: they are spindle-shaped with bony plates on the sides (usually) and pointed heads. They have narrow caudal peduncles and are protected with sharp spines on the back and pelvic fins. Some species have both resident freshwater and anadromous populations, whereas others are exclusively marine or freshwater. They typically inhabit quiet water among beds of aquatic plants and feed on small invertebrates. Two of the five species currently recognized for North America occur in California: one native, one alien.

Threespine Stickleback,
Gasterosteus aculeatus Linnaeus

Identification Threespine sticklebacks are small (typically 3–5 cm TL), laterally compressed fish with 3 sharp spines in front of a soft dorsal fin. Their eyes are large, their mouths terminal (but slanting upward), and their caudal peduncles narrow, and their pelvic fins are each reduced to a single stout spine and small ray. They have 1–35 bony plates on the sides but lack scales. A few populations lack plates altogether. Gill rakers number 17–26; dorsal rays, 10–24; anal rays, 6–10; and pectoral fin rays, 9–11. Adults in fresh water are usually olive to dark green on the back and sides and have white to golden bellies. The fins are generally colorless. Breeding males usually have bright red bellies and undersides of the head, blue sides, and iridescent blue or green eyes, but coloration is variable and often muted in inland populations. Breeding females are pale green-brown dorsally and silvery ventrally. Individuals from anadromous populations tend to be larger and more silvery than nonanadromous populations.

Taxonomy The study of threespine sticklebacks gives headaches to taxonomists who are mainly interested in pigeonholing distinct forms, each with a nice name. To a student of evolution, however, their study can provide fascinating insights into speciation. The confusing taxonomy of threespine sticklebacks is the result of their wide distribution,

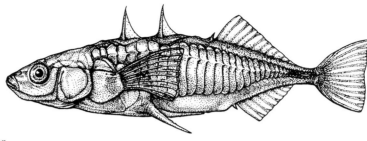

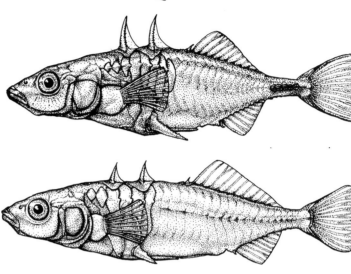

Figure 111. Threespine sticklebacks from California. From top: Fully plated form, partially plated form, low-plate-count form, unarmored form. Courtesy M. A. Bell, SUNY, Stony Brook.

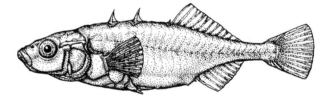

anadromous nature, and ability to repeatedly establish nonanadromous freshwater populations. Michael Bell, who has studied California populations in great detail, characterizes the threespine stickleback as a species complex ("superspecies") made up of a large number of closely related but morphologically distinct species ("semispecies"), and subspecies (1, 2, 6).

In the coastal areas of North America (Atlantic and Pacific), Asia, and Europe two basic forms generally exist: *(1)* a robust anadromous form with a row of plates extending along each side to the caudal peduncle, a keel on the peduncle, and strong, sharp dorsal and pectoral spines; and *(2)* a smaller nonmigratory freshwater form with bony plates on only the anterior portion of the body, no keel, and smaller spines. Hagen (3), in a study of evolutionary mechanisms, found that the two forms in a British Columbia stream behave as separate species, an observation consistent with populations in many California streams (4). Because each nonmigratory population is independently derived from anadromous sticklebacks, each stream with a nonmigratory population arguably has its own endemic species or subspecies of stickleback. No one, however, appears willing to name the hundreds of forms, for good practical reasons. On the other hand, the presence of so many distinctive forms is a wonderful example of parallel evolution by natural selection (5).

Even the two-form picture just presented is a simplification of reality. In California there are often three distinct

morphs in each coastal drainage, usually described on the basis of their easy-to-count lateral plates: *(1)* a low-plate-count form with only 1–5 plates on the anterior part of the body, *(2)* a partially plated form that has anterior plates and plates on the caudal peduncle with a wide area in between with no plates, and *(3)* a fully plated form with plates along the entire body and caudal peduncle (6, 7). The morphs may be mixed together or found in different habitats (8). There is evidence that plate numbers and size are related at least partly to predation intensity acting on traits that have a strong genetic basis (9). The anadromous forms have to dodge many kinds of predators, and they improve their survival rate by being completely armored, having larger spines, and being bigger in general. In southern California streams higher plate numbers are found in areas where sticklebacks co-occur with predatory trout (10) or face intense predation from garter snakes (11). Presumably the completely plateless forms living in some upstream areas experience relatively low levels of predation and find it energetically advantageous not to "grow" plates. In many situations, however, the reasons for varying plate numbers are not obvious. There are low-plate-count populations that breed in salt water (2) and high-plate-count populations that occur far from the ocean in the Central Valley (13). The population in the San Joaquin River near Friant has plate counts with a bimodal distribution (6–11 and 28–34), which is genetically based (14). In the Pajaro River high-plate-count morphs occur mainly in cool, relatively fast-flowing habitats while low-plate-count morphs occur in warm, slow-water habitats (40).

Despite the complexity of the interrelationships among stickleback populations, there is some utility in giving California sticklebacks representing distinct evolutionary phenomena subspecies names, as in Miller and Hubbs (13). The following forms are widely recognized:

Anadromous threespine stickleback, *G. a. aculeatus.* This is the common form, breeding in lower reaches of streams, that migrates between the streams and estuaries and bays. For the most part they are robust, colorful, and completely plated. Arguably, anadromous sticklebacks along the Pacific coast form one continuous population, although no doubt, as in salmon, local "stocks" (ESUs) exist as well.

Resident threespine stickleback, *G. a. microcephalus.* This name covers isolated populations found in many coastal streams that show genetic and morphological evidence of divergence from anadromous forms. The forms in different streams are similar to one another because of similar selection pressures. This name represents a polyphyletic group of fish, not a classic interbreeding subspecies.

Unarmored threespine stickleback, *G. a. williamsoni.* Unarmored threespine sticklebacks are small forms with no or at most 1 or 2 plates on their sides, characteristic of some southern California streams (44). They were apparently once widely distributed, especially in the Los Angeles basin, but are now rare. From a genetic perspective only the population in the upper Santa Clara River system seems to be distinctive enough to merit subspecies status (15). However, populations in San Antonio Creek (Santa Barbara County) and Sweetwater River (San Diego County) are morphologically similar and can be included under this name (16). Swift et al. (17) recommend separate status for the San Antonio Creek population.

Shay Creek stickleback, *G. a.* subsp. This stickleback, occurring in an isolated basin, is genetically the most distinctive of all southern California sticklebacks (18), although morphologically it is similar to unarmored threespine sticklebacks. Recognition of the Shay Creek stickleback as a "real" subspecies or even a species is also warranted by its unusual habitat requirements, the distinctive breeding coloration of males (especially the large amount of red on the body), and its unusual nest-building behavior (19, 20).

Other forms. Not surprisingly, distinctive sticklebacks exist in many places. A form with black, rather than red, male breeding colors is found in Holcomb Creek (San Bernardino County), a tributary to the Mojave River (21). Genetic evidence indicates that this form probably originated as an introduction from Sespe Creek, where a trout hatchery exists (17), but its unusual coloration says something about the ability of sticklebacks to develop distinctive forms in isolation. Another distinctive form exists in Bodega Bay, Sonoma County, and probably elsewhere along the coast. This form is exceptionally large as adults (70–90 mm TL) and breeds in salt water (41). No doubt other unusual forms await discovery and study in California, such as the isolated population of small fish found in spring-fed Cold Creek in the Clear Lake drainage, Lake County.

Names Gaster-osteus means belly-bone; *aculeatus* means spined.

Distribution Threespine sticklebacks are found in coastal streams, estuaries, and bays from Mediterranean Europe, north to Russia, and across to Japan and Korea. North American populations are found on the East Coast south to Chesapeake Bay and on the West Coast south to Baja California. They are absent from arctic and interior North America. In California anadromous threespine sticklebacks are found in streams from the Oregon border south to Monterey Bay (7, 13), and fully plated nonmigratory forms are found as far south as San Luis Obispo Creek (7, 13, 17). They have been present in the state for at least 16 million years (39). During wetter and cooler periods in the Pleistocene they presumably extended as far south as northern Baja California, giving rise to instream populations. Resident threespine sticklebacks were historically found in coastal streams along the entire coast upstream to major fish barriers such as falls, although their distribution was discontinuous south of Point Conception (Santa Barbara County) (17). In the Central Valley populations are scattered from the lower Kings River (below Pine Flat Dam) and the San Joaquin River (below Friant Dam and in a small stream above Kerckoff Reservoir) (22) to roughly Redding in the Sacramento River drainage.

Resident sticklebacks were widely introduced into southern California and on the east side of the Sierra Nevada as contaminants in shipments of trout. Presumably the sticklebacks either were in the water supply of hatcheries or were scooped up when trout were rescued from drying streams (such as the Santa Inez River) and moved to new locations. As a result sticklebacks are now established in the Mojave and Owens Rivers and in the Mono Lake basin (June Lake, Rush Creek). Introduced populations are also present in some high-elevation reservoirs, such as Big Bear Reservoir (San Bernardino County) (17).

Unarmored threespine sticklebacks were once abundant

in streams in the Los Angeles basin, but they now occur naturally only in the upper Santa Clara River, San Antonio Creek, and Whitewater River (16, 17). A few populations have been established through introductions into other small streams in the area, but they tend not to persist. Recent transplants include those in San Filipe Creek in the Salton Sea drainage (12).

The Shay Creek stickleback is endemic to Shay Creek (San Bernardino County) and, during wet years, Baldwin Lake, a playa lake that is often dry, located just east of Big Bear Reservoir. The basin is closed, although with enough water it could overflow into the Mojave River. The area is situated at 2,000 m elevation, 1,200 m higher than other native stickleback populations. The Shay Creek stickleback currently survives in three ponds, which are covered with ice for long periods during the winter (19). One pond is an artificially maintained pool in Shay Creek; the other two ponds support introduced populations. One of the introduced populations is in Sugarloaf Meadows in the upper Santa Ana drainage, and the second is in a privately owned pond 1 mile east of Baldwin Lake (19). The remaining native habitat represents less than 1 percent of the range of the species when the sticklebacks were first discovered (19). Their historical distribution possibly included much of the Santa Ana River drainage.

Life History Few fish have been studied as much as threespine stickleback, thanks to their wide distribution in western countries and ease of maintenance in the laboratory (23). Threespine sticklebacks are quiet-water fish, living in shallow, weedy pools and backwaters or among emergent plants at stream edges over bottoms of gravel, sand, and mud. The water has to be clear enough so that aquatic plants, required for building nests, will grow. As long as water quality is fairly high, sticklebacks are capable of living in shallow, urbanized streams that support few other fish, at least in northern California. The species is consequently one of the commonest fish found in San Francisco Bay–area streams (24). Anadromous adults are largely pelagic, staying close to shore, although they have been collected up to 800 km offshore (25). By and large they require cool water (<23–24°C) for long-term survival. They will seek areas with optimum temperatures for various activities such as feeding and digestion (29). It is also unusual to find them in turbid water because they are visual feeders. Sticklebacks have broad salinity tolerances; even individuals from freshwater populations can be readily reared in seawater (26).

Threespine sticklebacks are capable of completing their entire life cycle in either fresh or salt water, or migrating between the two environments (2). The life history characteristics of populations adapted to living in different areas are genetically based, indicating that selection pressures to maintain them are strong (26, 27). R. J. Snyder found that migratory versus nonmigratory sticklebacks from the Navarro River differed in such characters as age at first reproduction (195 days versus 220 days), initial clutch size of females (63 versus 57 eggs), egg size (1.70 mm versus 1.76 mm), and growth rates (26, 27, 28, 29). The tendency to migrate is also an inherited trait.

Unless they are breeding, sticklebacks shoal, albeit loosely. Shoaling seems to help fish find concentrations of food. When one starts to feed, the others come to investigate. Freshwater populations feed primarily on organisms living on the bottom or in aquatic plants (2, 30, 43). Anadromous populations, perhaps as a reflection of their pelagic existence while in salt water, feed more on free-swimming crustaceans, although bottom organisms are also taken. While feeding, sticklebacks tend to move about in a jerky fashion, stopping frequently to investigate potential prey organisms. They hover at an angle to the prey, using the opposing forces of the pectoral fins and tail, and then quickly pick it off the substrate. Individual sticklebacks (and populations) specialize in feeding on a rather limited number of organisms (e.g., chironomid midge larvae, ostracods) and are rather slow to learn to exploit new sources of food (31). In San Pablo Creek (Contra Costa County) sticklebacks fed mainly on aquatic insects (41% by volume), crustaceans (28%), and earthworms (10%), although the relative proportions varied seasonally, with insects most abundant in summer and earthworms in winter (32). Eggs of other sticklebacks are also common in the diets, especially of males, during the breeding season, and raiding by males on nests of rival males is a habit that has been the subject of considerable study.

The small size, slow movements, and shallow-water habits of sticklebacks seem to make them ideal prey for avian and piscine predators. Their spines, however, evolved to make them a less ideal prey, because fish with rigidly erect dorsal and pelvic spines are very hard for predators to swallow (33). Similarly, the bony plates appear to function as armor. Freshwater populations that occur where predators are common tend to have more plates than populations where there are few potential predators (34). Nevertheless, sticklebacks are frequently important prey of salmonids and birds. One bit of evidence for this is that sticklebacks are often infested with intermediate stages of bird tapeworms. These larval tapeworms can occupy most of the body cavity, causing the fish to swim sluggishly near the surface or to turn white (35). This in turn makes them more vulnerable to kingfishers and herons. Digestion releases the larval tapeworm into the gut of the bird, where it grows to adulthood.

Predation is also a strong determinant of stickleback distribution within habitats. In waters with piscivorous fish they are typically found in beds of aquatic plants or in other dense cover, such as among branches of fallen trees. In the Pajaro River sticklebacks are found either in slow, shallow pools where predatory fish are absent and algal mats are abundant or in deep pools with cover (36). In the Eel River invasion of predatory pikeminnows caused sticklebacks to shift from using a wide variety of habitats in pools (mean depth used, 95 cm) to concentrating in shallow edge water (mean depth, 46 cm), among cover. In flowing water mean depth shifted from 55 cm to 37 cm (37). In some areas pikeminnow predation largely eliminated sticklebacks, which showed relatively poor avoidance response. This may explain in part the scattered distribution of sticklebacks in many California river systems, including those of the Central Valley. For example, in San Francisco Bay streams they are largely absent from areas containing introduced predatory fish (24).

Most sticklebacks complete their life cycle in 1 year. Usually a majority in one area are of uniform size, although the presence of occasional large individuals indicates that a few live for 2 or possibly 3 years. Freshwater sticklebacks in California seldom exceed 50 mm TL; anadromous sticklebacks commonly reach 80 mm TL. Females are usually larger than males.

As daylight increases in spring and the water warms (April through July), threespine sticklebacks move into breeding areas (42). For anadromous forms this means moving first into shallow regions of estuaries and then into fresh water. They generally spawn earlier than freshwater populations. The males soon begin to assume their breeding colors and move away from their schools to set up territories among beds of aquatic plants. Once a territory is established, nest construction begins. The male excavates a shallow pit by taking up mouthfuls of sand and dropping them several centimeters away from the nest site. Next he gathers strands of algae and pieces of aquatic plants and deposits them in the pit. The material is pasted together with a sticky kidney secretion. When the sticky pile is large enough, the male wriggles through it until a tunnel is created.

Meanwhile schools of females, glistening and egg-swollen, cruise back and forth in the vicinity of male territories. Once his nest is finished a male approaches passing females, performing a zigzag courtship dance described by Tinbergen (38). If a female is ready to spawn she will respond to the dance by following the male to the nest. She enters the nest and after a few nudges begins laying eggs. When finished laying, she leaves the nest and the male enters and fertilizes the eggs. He then chases the female away and repairs the nest. Usually several females are courted in this fashion for each nest. Each female lays 50–300 eggs, in several spawns.

Once the nest has its complement of embryos, the male begins incubation behavior, which consists of maintaining a headstand, at an angle, in front of the nest while using his pectoral fins to fan water over the developing embryos. The movement of the pectoral fins is counterbalanced by sweeps of the tail. When not fanning the embryos, the male defends the nest vigorously from other sticklebacks and potential predators. Male Shay Creek sticklebacks at the beginning of incubation cover the nest with small pebbles until only the entrance is visible, a highly unusual behavior (19).

Embryos hatch in 6–8 days at 18–20°C, but fry remain in the nest for the first couple of days. Once they begin swimming about, the male guards the shoal carefully, grabbing wanderers in his mouth and spitting them back into the main group. As the fry become more active and the male's task of guarding them becomes more difficult, he gradually ceases his caregiving. The male may then begin the spawning cycle again or go off and join a shoal of other fish that have finished reproducing. The young fish likewise join shoals of similar-size fish.

Status IE. As a superspecies, threespine sticklebacks are still common and widely distributed in California, in both freshwater and marine environments. They tend to disappear, however, from streams that are heavily altered or polluted or that have suffered introductions of predatory fishes.

Anadromous threespine stickleback. IE. This form is still present and common throughout its native range.

Resident threespine stickleback. IA–E. The myriad forms in this category vary in status. In coastal streams north of Point Conception most populations are doing reasonably well. However, in southern California they have disappeared from many streams or stream reaches. If the independently evolved forms were recognized as discrete taxa (e.g., ESUs as in salmonids), many would be listed as threatened or endangered species.

Unarmored threespine stickleback. IB. Unarmored threespine sticklebacks were listed as Endangered by the federal government in 1970 and by the state in 1971. Thanks to their status, remaining populations have managed to persist, although long-term survival in such a heavily populated region is always problematic. In addition to the three remaining natural populations, two transplanted populations exist in Honda Creek (Santa Barbara County) and San Felipe Creek (San Diego County). The causes of disappearance of unarmored threespine sticklebacks from their native haunts in southern California are multiple, from dewatering of streams, to pollution, to habitat alteration, to introductions of exotic predators, but are all tied to increased urbanization of the Los Angeles region.

Shay Creek stickleback. IB. Shay Creek sticklebacks have a very limited distribution, mainly on private land. The population undergoes major fluctuations as its pond and stream habitats wax and wane. Catastrophic mortality was observed in 1985 and 1986 as their habitat dried up. Most fish in Shay Creek died of desiccation, and hundreds were observed dying in Baldwin Lake, evidently of temperature- and salinity-related stress. Low rainfall, coupled with water removals for human use, caused most of Shay Creek to dry up during 1985. Two deep pools with fish persisted, one of which received water pumped in by the local water company. Drought conditions caused Baldwin Lake to evaporate and most of Shay Creek to remain dry. In 1989 the pool that did not receive pumped water was reduced to a low level, and its sticklebacks died. In October 1988 about 400 sticklebacks from Shay Creek were introduced into a pond in Sugarloaf Meadow, about 11 km southeast of Baldwin Lake in the San Bernardino Mountains (17, 19). A second, accidental, introduction occurred when aquatic plants were moved from Shay Creek to a private pond east of Baldwin Lake (19). Mark-and-release estimates of population size indicate that there were at least 5,000–6,000 sticklebacks in the Shay Creek pool and at least 6,000–7,000 in Sugarloaf Meadow pond in the early 1990s (19). The most immediate threat to the Shay Creek population is drying up of the remaining natural pool. The Shay Creek pool remained stable for 5 years, but it decreased in size in 1991–1992. Survival of this unusual form in the long run will thus depend on restoration of flows to Shay Creek and protection of ripar-

ian and instream habitats from livestock and other factors. Regular monitoring and protection of pond populations is also important.

References 1. Bell 1976. 2. Bell 1979. 3. Hagen 1967. 4. Snyder 1991a. 5. Schluter and Nagel 1995. 6. Bell 1981. 7. Baumgartner and Bell 1984. 8. Bell 1982a. 9. Baumgartner 1995. 10. Bell and Richkind 1981. 11. Bell and Haglund 1978. 12. C. C. Swift, pers. comm. 1999. 13. Miller and Hubbs 1969. 14. Avise 1976. 15. Buth et al. 1984. 16. White 1993. 17. Swift et al. 1993. 18. Haglund and Buth 1988. 19. Malcolm 1992. 20. Moyle et al. 1995. 21. Bell 1982b. 22. Brown and Moyle 1993. 23. Wooten 1976, 1984. 24. Leidy 1984. 25. McPhail and Lindsey 1970. 26. Snyder and Dingle 1989, 1990. 27. Snyder 1991b. 28. Snyder 1990. 29. Magee et al. 1999. 30. Hynes 1950. 31. Beukema 1963. 32. Snyder 1984. 33. Hoogland et al. 1956. 34. Hagen and Gilbertson 1972. 35. LaBue and Bell 1993. 36. Smith 1982. 37. Brown and Moyle 1991. 38. Tinbergen 1953. 39. Bell 1994. 40. J. J. Smith, San Jose State University, pers. comm. 1999. 41. H. V. S. Peeke, pers. comm. 1998. 42. Baggerman 1957. 43. Markley 1940. 44. Ross 1973.

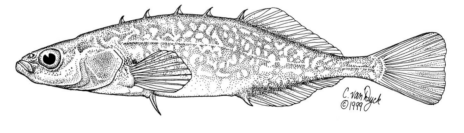

Figure 112. Brook stickleback, 50 mm SL, Scott River, Siskiyou County.

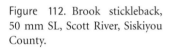

Brook Stickleback, *Culaea inconstans* (Kirtland)

Identification Brook stickleback are small (adults, 40–70 mm TL) and laterally compressed, with 5 short spines preceding the dorsal fin (9–11 rays); a short, narrow caudal peduncle; and a small pointed head (1). The mouth is small and oblique, with the lower jaw projecting slightly beyond the upper jaw. The 4–6 dorsal spines (usually 5) are clearly separated from the soft dorsal fin, and each has a tiny membrane on its rear, connecting it to the back. The anal fin has 1 spine and 9–11 rays, with the rayed part symmetrically placed below the dorsal rays, just before the caudal peduncle. Each pelvic fin is small and has a single spine next to a stout ray. The pectoral fins have 9–11 rays. The tail is round to truncated. Each side has a row of 30–36 tiny bony plates running along the lateral line. Live fish are typically olive green on the back and sides, lighter on the belly, with varying degrees of mottling. Breeding males are dark green to black, sometimes with red on the fins.

Taxonomy Like the threespine stickleback the brook stickleback is probably a species complex, but it is much less studied (2). The source of the fish in California is not known.

Names The generic name of brook stickleback was originally *Eucalia*, meaning "true nest," referring to its breeding behavior. However, it was later discovered that *Eucalia* had already been assigned to a butterfly, so the name was shortened to *Culaea,* which really has no meaning but sounds similar. *Inconstans* means changing or variable, a good name for any "species" of stickleback.

Distribution Brook sticklebacks are native to fresh waters of much of Canada east of the Rocky Mountains, including the Great Lakes and the upper Mississippi-Missouri River drainage. They occur north to Hudson Bay, west to Montana, east through much of New York, and south to southern Ohio and Nebraska (3). There is also an isolated population in New Mexico. Introduced populations are known from Alabama, Kentucky, Tennessee, Colorado, and Utah (4). They were first noticed in California by CDFG biologists Dennis Maria and Shaun Smith in the Scott River drainage (Siskiyou County) in 1991 (5). By that time they were already present in the river and its tributaries in Scott Valley, from French Creek down to Shackleford Creek, encompassing 60–70 km of stream. Subsequently a population was discovered in Lily Pad Lake (elevation, 1,811 m), in the headwaters of Rail Creek, which flows into the South Fork Scott River about 40 km above the mouth of French Creek (9). Whether or not Lily Pad Lake is the original introduction site or the site of a subsequent introduction is not known. In any case brook sticklebacks are likely to appear downstream in the main Klamath River at some point, although there seems to be little suitable habitat until the lower 30 km or so, where threespine sticklebacks occur.

Why or how brook sticklebacks were introduced is not known. Either they were contaminants in a bucket of bait minnows carried in from the Midwest (6) or someone released pet fish brought in from their native range.

Life History Brook sticklebacks have not been studied in California, although their biology in their native range is well documented (1, 7). This account will consequently be brief.

In their native range sticklebacks are associated with cool (8–25°C in summer), usually clear waters with extensive beds of aquatic plants. These habitats range from spring ponds to lakes to slow-moving streams to river edges. Their habitats are typically shallow (<1.5 m) with fine substrates. Not surprisingly, their habitats in the Scott Valley are similar: shallow, weedy areas in low-gradient creeks, backwaters of the Scott River, irrigation ditches, and clear lakes. The fish are typically never far from vegetation while foraging and plunge deeply into it when threatened. To escape predators they will dive into loose organic matter on the bottom and completely cover themselves.

Brook sticklebacks are active, diurnal predators on small invertebrates. They pick insect larvae, especially chironomid midge larvae, from plants and other substrates or capture swimming crustaceans associated with beds of aquatic plants (1).

Brook sticklebacks can live up to 3 years, but most adults are 1–2 years old, breeding in their second summer of life. A collection of 124 fish from Shackleford Creek on 5 December 1991 measured 42–71 mm TL, with a mean of 61 mm; most fish measured 54–68 mm (10). This suggests that the smallest fish were young-of-year, while the largest ones were 2 years old (going into their third summer). In their native range fish over 65 mm TL are rare; the maximum size recorded is 87 mm (1).

In their native range spawning takes place from April through June, at 15–19°C. Breeding behavior is similar to that of the threespine stickleback, with males building and guarding nests and then enticing females to spawn with a courtship dance (1, 7). The nest is globular, constructed of pieces of debris, and is attached to the stem of an aquatic plant a few centimeters off the bottom. Each female produces 100–200 eggs, which are spawned in small batches. Males guard and aerate developing embryos for the 7–11 days (at 16–18°C) required for development to larvae. The larvae are guarded until they become too active and escape from the nest.

Status IIC. The brook stickleback should spread downstream to the lower Klamath River, where it will most likely occupy habitats now occupied by the threespine stickleback. It is possible, however, that threespine sticklebacks and associated predators may prevent their establishment. Brook stickleback in California may present an interesting opportunity for evolutionary studies because in their native range they show considerable variation in spine strength and behavior, related to the kinds of predators present in their environment (8).

References 1. Becker 1983. 2. Wooten 1976. 3. Lee et al. 1980. 4. Modde and Haines 1996. 5. Dill and Cordone 1997. 6. Ludwig and Leitch 1996. 7. Winn 1960. 8. Reist 1983. 9. J. Schlosser, Humboldt State University, pers. comm. 1998. 10. D. Maria, CDFG, pers. comm. 1991.

Sculpins, Cottidae

Sculpins are small bottom-dwelling fishes with large flattened heads, fanlike pectoral fins; and smooth, scaleless, but occasionally prickly bodies. These features, combined with the absence of an air bladder, enable sculpins to stay on the bottom even in fast-flowing streams. A further adaptation is dark, mottled coloration that blends with their rocky habitat, concealing them from predators and prey alike. The large mouth with numerous small teeth and the short gut with a muscular stomach reflect the sculpin's predatory habits. Sculpins are also characterized by long, narrow pelvic fins located between the pectoral fins and by two dorsal fins. The first dorsal fin is short and composed of soft spines, while the second is long and composed only of rays.

Freshwater sculpins, genus *Cottus*, are a small branch of a large family of marine fishes. Their ancestry is reflected by the number of species that enter salt water, especially as larvae. As a result of its salinity tolerance the genus has been able to colonize coastal and inland streams throughout North America and Eurasia. Curiously, although the family as a whole can be classified as saltwater dispersants (Moyle and Cech 2000), many species show distribution patterns reflecting only dispersal through streams. In addition to freshwater sculpins the lower reaches of coastal streams and estuaries may support true marine sculpins. Of two such species recorded for California, the sharpnose sculpin (*Clinocottus acuticeps*) has been found only once in fresh (?) water; the staghorn sculpin regularly lives as a juvenile in fresh or brackish water and is therefore treated as a freshwater fish.

Sculpins are most abundant in coldwater streams, and their presence is usually an indicator of high water quality. Because they share these streams with salmon and trout and can occasionally be found feeding on small salmon or salmon eggs, they have been accused of limiting salmonid populations. There is in fact no solid evidence that they have much effect on salmonids in streams (Moyle 1977). Eggs found in sculpin stomachs are usually loose eggs that salmon or trout failed to bury or embryos that have been dug up by spawning fish, although sculpins will also hold in redds and capture newly laid eggs (Foote and Brown 1998). Salmon juveniles are most often eaten when they are trapped with sculpins in fry traps. In some lakes and streams, sculpins are important forage for game fishes. Under natural conditions sculpins rarely prey on other fish, and when they do it is usually other sculpins.

Perhaps no group of fishes occurring in the fresh waters of California has given biologists (including the author) more identification headaches than sculpins. In addition to being highly variable in structure and color, most species are widely distributed. Similar species frequently occur together, making careful examination of each specimen necessary. Only in recent years has enough attention been paid to variation within species that most of them can be characterized with a reasonable degree of certainty. However, much work still needs to be done on sculpin systematics.

There are eight species of *Cottus* in California streams, and it is common to find two or three species living together. Two additional species, Klamath Lake sculpin (*C. princeps*) and slender sculpin (*C. tenuis*), have not been recorded from California but can be expected in Copco and Iron Gate Reservoirs on the Klamath River.

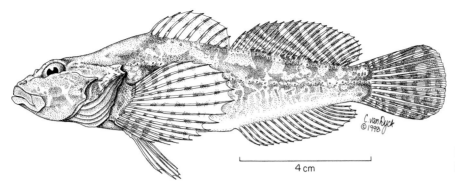

Figure 113. Prickly sculpin, 10.5 cm SL, Eel River, Mendocino County.

Prickly Sculpin, *Cottus asper* Richardson

Identification Prickly sculpins are distinguished by their long anal fin (16–19 rays), which is about three times longer than the caudal peduncle. The dorsal fin is also long: the first dorsal fin has 7–10 soft spines, and the second dorsal fin has 19–23 rays. The two dorsal fins are joined at the base. The palatine teeth are well developed and are usually visible without magnification or dissection. There is a single pore on the chin (occasionally two). The pelvic fins have 1 spine and 4 rays (the spine is fused with the first ray, so the count is usually 4 "elements"); the pectoral fins, 15–18 rays. The gill rakers number 5–6; the branchiostegal rays, 6 on each side; and the preopercular spines, 2–3 (usually only the uppermost one is conspicuous, the lower ones being under the skin). The lateral line is complete with 28–43 pores. The extent of visible prickling on the body ranges from nearly complete coverage to only a very small area behind the pectoral fins. Younger individuals and those in inland populations usually have more extensive prickling than older individuals in coastal populations. Maximum size is about 20 cm SL in California. The caudal peduncle is relatively narrow and rounded. Coloring is highly variable, but the back and sides are usually mottled reddish brown to dark brown. It is usually possible to recognize 4–5 dark saddles of various sizes. The belly is white to yellow. The fins are generally barred, and the first dorsal fin often has a dark oval spot on

the posterior end. Both sexes develop an orange edge on the first dorsal fin while breeding, and the body color of males turns very dark. When not breeding, males can be distinguished from females by their long, V-shaped genital papilla.

Taxonomy Prickly sculpins are highly variable, with a wide distribution and poorly understood systematics (1). Although their pelagic larvae ensure wide dispersal and some mixing of populations up and down the Pacific coast, it is likely that a genetic analysis would reveal distinct subgroups. In California there seem to be three distinct forms: a coastal form, a Central Valley form, and a Clear Lake form, although none has been formally described. The coastal form is usually lightly prickled and found in coastal streams and estuaries; most freshwater populations are amphidromous (larval stages in salt water). The Central Valley form is usually fully prickled and apparently confined to fresh water and the San Francisco Estuary. The Clear Lake form has limited prickling and is confined to Clear Lake proper; Hopkirk (2) did not find any distinctive morphological features, but its isolation and life history would seem to imply some differentiation from Central Valley fish. Although this has not been confirmed, Central Valley prickly sculpin may hybridize with riffle sculpin.

Names *Cottus* is apparently the Latin name for European sculpins; it was derived from the Greek word for head—a good name for a fish that seems to consist mostly of head. Both the trivial name *asper,* meaning rough, and the official common name reflect the fish upon which Richardson based his original description. He was unaware of the smooth forms typical of the South Coast. Sculpins are frequently referred to as bullheads, miller's thumbs, or muddlers. The name sculpin seems to be a corruption of the ancient Greek name for various marine cottids (*Scorpaena*) (3).

Distribution Coastal prickly sculpins are found in coastal streams and estuaries from the Kenai Peninsula, Alaska, down to the Ventura River, southern California. However, sculpins found in streams south of Point Conception are apparently the result of colonization by larvae drifting south from more northern populations (4). There is no evidence of reproduction in southern California streams, although the Central Valley form may now be invading from upstream, from reservoirs connected to the California Aqueduct. Coastal forms are found mostly within 50 km of an estuary, although in the Eel River there appears to be a resident population at least 80 km upstream (5). Large adults, usually solitary, are occasionally found 120 km or more upstream (5). Streams without estuaries or lagoons typically lack prickly sculpin populations. The Central Valley form is found in the Sacramento Valley at low elevations in most streams roughly up to Keswick Dam on the Sacramento River and in the San Joaquin Valley south to the Kings River (6). They are common in reservoirs and have spread via the California Aqueduct into reservoirs in southern California and to streams below them, where they may eventually come into contact with the coastal form (4). The Clear Lake form is confined to Clear Lake, although it presumably gets washed into Cache Creek as well.

Life History Few fishes live in as many environments as prickly sculpin. They live in fresh water to brackish water to seawater; in streams ranging from small, cold, and clear to large, warm, and turbid; and in lakes and reservoirs ranging from small to large, eutrophic to mesotrophic. In one small area near Friant, Fresno County, prickly sculpins are abundant in a cool trout stream (San Joaquin River); a large, warmwater reservoir (Millerton Reservoir); and a small, shallow lake with bottom temperatures that exceed 26°C in the summer (Lost Lake). In lowland rivers they may experience summer temperatures of 28–30°C (7, 8). In the Central Valley they are typically found in moderate-size, clear, low-elevation streams with bottoms of rubble, sand, and scattered logs and boulders. Their absence from warm, polluted habitats on the San Joaquin Valley floor indicates that water quality limits distribution, because they are found in streams above the valley floor and in the estuary below it (9). In Clear Lake they are most abundant inshore around beds of tules on gravelly bottoms (10). They are also usually the most abundant sculpin in coastal streams and estuaries, although they often share this honor with coastrange and staghorn sculpins. In inland waters they may co-occur with riffle sculpins. In streams they use a wide variety of habitats, with the presence of cover (rocks, logs, overhanging vegetation) probably the most important habitat characteristic (5).

As their body shape and cryptic coloration indicate, prickly sculpins spend most of their time quietly lying on the bottom. During the day they hide underneath or in submerged objects such as rocks, logs, and pieces of trash. Before the days of recycling, it was common to find them inhabiting disposable beverage cans in streams (11). At night they come out to forage actively. Prickly sculpins are usually not gregarious, but neither do they appear to be territorial outside the breeding season, a behavior pattern that might be expected of sedentary bottom fish. Apparent shoaling behavior, however, was observed among prickly sculpins moving along the shore of a British Columbia lake (12).

The basic life history pattern of prickly sculpins is for adults to move to a suitable spawning area (a place in flowing water with loose rocks under which males can locate nests), spawn, and guard embryos until they hatch. After hatching larvae swim up into the water column. In streams they may be washed into an estuary or large pool for rearing, whereas in a lake or estuary they simply become pelagic. Larvae eventually transform into juveniles, which settle onto the bottom and move into areas with plenty of food and cover, in some cases making extensive upstream migrations. These migrations are easily blocked by low human-made barriers (13). Downstream migration of adults and upstream migration of young-of-year sculpins is typical of coastal populations. Shapovalov and Taft (13) noted a pronounced downstream movement of adult prickly sculpins in Waddell Creek (Santa Cruz County) during winter, especially in January and February. The function of the downstream movement was presumably to find spawning sites close to the estuary, reducing the distance larvae would have to be flushed to reach the estuary, where they rear. In the Eel River the larvae of upstream populations settle in large pools or at stream edges and then rear in riffles (5), as do inland Central Valley populations. In Clear Lake and many reservoirs the larvae are planktonic and settle out on the bottom in a more or less haphazard fashion. Juvenile fish tend to move into inshore areas to live, but as they become larger they shift into offshore areas, moving inshore at night to feed (14).

Prickly sculpins typically do not have strong competitive interactions with other species of fish, although they will prey on small fishes (including other sculpins and tidewater gobies) if given the opportunity. In the Eel River prickly and coastrange sculpins are nearly identical in their distribution and ecology, presumably as a result of the relatively low densities of both species, a situation noted elsewhere in their range as well (5). However, in the Smith River (Del Norte County) prickly sculpins largely occupy deep (2–14 m), rocky-bottomed pools while coastrange sculpins occupy riffles less than 1 m deep (15). They are regularly eaten by predatory fish, but are less common in diets than their abundance would suggest. However, the strong association of sculpins with cover suggests that predation does have a major effect on their behavior (16).

As might be expected, prickly sculpins feed mostly on large benthic invertebrates, particularly blackfly, midge, mayfly, stonefly, and caddisfly larvae. Other aquatic insects, molluscs, isopods, amphipods, and small fish and frogs are also eaten. In the Eel River small (<45 mm SL) sculpins feed mainly on baetid mayfly larvae and hydropsychid caddisfly larvae, while larger fish feed on larger invertebrate prey, especially the mayfly *Isonyschia* (5). Prior to the introduction of inland silverside into Clear Lake (Lake County), 74 percent of the summer diet of sculpins was chironomid midge larvae and pupae (16). After the introduction, chironomids made up less than half the diet, and amphipods became the most abundant food item (14). However, amphipods are typically eaten in inshore areas and chironomid larvae are eaten in offshore areas, so dietary differences may reflect collection differences. Clear Lake sculpins feed around the clock, with peaks at sunrise and sunset. Those collected on sandy substrates have much less varied diets than those collected from rocky substrates, reflecting the opportunistic nature of their feeding. In Clear Lake the diet seems to vary little with the size of the fish, although pelagic larvae feed on planktonic copepods and cladocerans (14). Likewise, diet varies little with the size of fish in the San Francisco Estuary, with major prey being gammarid amphipods, mysid shrimp, and postlarval fish (17). When mysid shrimp populations collapsed in the estuary, sculpins switched to feeding more heavily on gammarid amphipods and *Corophium* (also amphipods) (24).

Growth in prickly sculpins is subject to much variation. In the San Joaquin River they reach 51–71 mm SL in their second year, 61–85 mm in their third, 64–90 mm in their fourth, and 75–90 mm in their fifth (18). The oldest fish found was 7 years old (105 mm SL), although much larger (up to 200 cm SL) and presumably older fish have been collected elsewhere. In Clear Lake fish collected in July had the following age–mean length relationship: young-of-year, 26 mm SL; 1+, 34 mm; 2+, 44 mm; 3+, 48 mm; and 4+, 55 mm (19). In addition, a single sculpin measuring 95 mm SL was aged at 5+. Variability in length at a given age was high: 3-year-old sculpins ranged from 30 to 90 mm.

They become mature at 40–70 mm SL during their second, third, or fourth year (20). Spawning occurs from late February through June, although most spawning in California probably takes place in March and April. In Clear Lake spawning may continue through May. Spawning in streams usually requires temperatures of 8–13°C (1). In Putah Creek and the Mokelumne River larval prickly sculpins start appearing in mid-February and continue appearing through mid-June (23, 26). In the Delta and Suisun Marsh larvae can be found from February through May, peaking in March in response to increased outflow and cool (<15°C) temperatures (21, 28).

Prior to spawning prickly sculpins move into freshwater or intertidal areas that contain large flat rocks and moderate current. Males are ready for spawning before females and select nest sites underneath rocks (or in beer cans, auto bodies, and other trash), while females congregate upstream from the spawning area (1). Each male then prepares a nest by digging a small hollow underneath the rock and cleaning off the ceiling of the nest. When a female is ready to spawn she moves into the spawning area and is courted by a male, who lures her into his nest. Further courtship and spawning take place within the nest, mostly at night (1). During spawning eggs are attached to the ceiling of the nest in a cluster. The male then chases the female from the nest and guards the embryos until they hatch. Movements of the male help keep water circulating over the embryos, ensuring development. Mechanical agitation of fully developed embryos by the male seems to be necessary for hatching (27). Males frequently spawn with more than one female, so as many as 25,000–30,000 embryos have been found in one nest (1), although the number is usually much smaller. Individual females produce anywhere from 280 to 11,000 eggs, the number depending on both size and age (20). Prickly sculpins are apparently the most fecund species in the genus *Cottus*, presumably because they are also the largest species.

The fry when hatched measure 5–7 mm TL. They start swimming fairly soon after hatching. As a result they are swept downstream to large pools, lakes, and estuaries, where they assume a planktonic existence for 3–5 weeks. Soon after settling down to the bottom, at lengths of 15–20 mm, they start a general upstream movement in streams (22) or a movement into shallow water in lakes and pools.

Status IE. Prickly sculpins are abundant where found and have managed to adapt to altered environments. They are one of four native fish species that have managed to remain abundant in Clear Lake (14). In Millerton Reservoir they are the only native fish still abundant and are important forage for largemouth bass. They are especially abundant in the San Francisco Estuary and have spread from there into southern California reservoirs and streams via the California Aqueduct. Nevertheless, it is likely that many populations have been eliminated or reduced by the construction of barriers on streams. Even low barriers, such as those constructed for flow gauges in small streams, can block upstream movement of small fish, preventing completion of migratory life cycles (25). Thus, when Shapovalov and Taft (13) constructed a weir across a stream to count salmonids, they noted a marked decline in sculpins moving downstream over the weir over a 5-year study period.

The systematics of the species should be examined using biochemical techniques in order to determine the degree of genetic isolation of various populations, such as the one in Clear Lake. This would help in developing conservation strategies.

References 1. Kresja 1965, 1970. 2. Hopkirk 1973. 3. *Oxford English Dictionary* 1971. 4. Swift et al. 1993. 5. Brown et al. 1995. 6. Brown and Moyle 1993. 7. Bond 1963. 8. Smith 1982. 9. Brown

2000. 10. Week 1982. 11. Kottcamp and Moyle 1972. 12. North-cote and Hartman 1959. 13. Shapovalov and Taft 1954. 14. Broadway and Moyle 1978. 15. White and Harvey 1999. 16. Cook 1964. 17. Herbold 1987. 18. Kottcamp 1973. 19. L. Decker and M. LeClaire, University of California, Davis, unpubl. rpt.

1978. 20. Patten 1971. 21. Wang 1986. 22. McLarney 1968. 23. Rockriver 1998. 24. Feyrer 1999. 25. C. C. Swift, pers. comm. 1999. 26. Marchetti 1999. 27. R. J. Kresja, California Polytechnic University, San Luis Obispo, pers. comm. 1974. 28. Meng and Matern 2001.

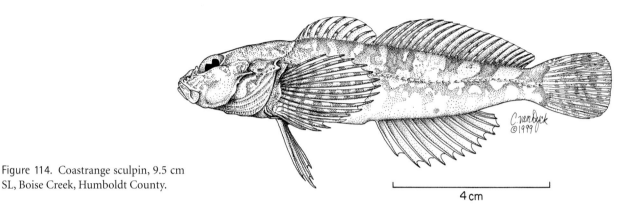

Figure 114. Coastrange sculpin, 9.5 cm SL, Boise Creek, Humboldt County.

Coastrange Sculpin, *Cottus aleuticus* Gilbert

Identification Coastrange sculpins can usually be recognized by pelvic fins that reach the vent when depressed (for fish measuring less than 100 mm TL), the large size of the posterior pair of nostrils, the lack of palatine teeth, dorsal fins that are distinctly separate, a complete lateral line (34–44 scales), and a single chin pore. The pelvic fins have 1 spine and 4 rays (4 "elements"); the first dorsal fin, 8–10 weak spines; the second dorsal fin, 17–20 rays; and the pectoral fins, 13–16 rays. The gill rakers number 5–7; the branchiostegal rays number 6 on each side. There is only one sharp preopercular spine. Prickling is confined to a small area behind the pectoral fins. Body coloration is typical of sculpins: dark mottling on the back and white on the belly. Often there are 2–3 more or less distinct vertical bands below the second dorsal fin and one on the caudal peduncle. The posterior edge of the first dorsal fin lacks a dark spot. Most fins are barred except on juvenile fish. When breeding,

males have a long genital papilla, an orange band on the edge of the first dorsal fin, and an orange spot on the top of the caudal peduncle. Many individuals, especially small ones, frequently lack one or more of the diagnostic features.

Taxonomy This widely distributed species presumably has local and regional forms, but its systematics has never been seriously examined (1, 2). Most records of riffle sculpins from coastal streams are probably misidentified coastrange sculpins (3).

Names Coastrange aptly describes the wide distribution of the species; *aleuticus* refers to the Aleutian Islands, from which the type specimens were collected. They are sometimes called Aleutian sculpins. Other names are as for prickly sculpin.

Distribution Coastrange sculpins are found in coastal streams from the Aleutian Islands and Bristol Bay, Alaska, south to Oso Flaco Creek, Santa Barbara County (1, 3). Although they are often locally abundant, their distribution in coastal streams south of Mendocino County is sporadic. There are a few records from streams tributary to San Francisco Bay, but it is likely that they are no longer present (4). In the Eel River they have been found as much as 120 km upstream from the ocean (5), but in most streams they occur less than 20 km above the stream mouth, presumably because they use coastal lagoons for larval rearing.

Life History Coastrange sculpins are typically found in swift gravel to rubble riffles (6) in the lower reaches of coastal streams that also have some sort of lagoon or estuary at the mouth. The presence of cover, especially large rocks and logs, is crucial for protection from predation. Temperatures are usually less than 20–22°C in these streams. In the Eel River coastrange sculpins are most abundant in larger tributary streams associated with coarse substrates, riffles, and slightly turbid water (5). Similarly, in the Smith River they are most common where water column velocities are greater than 5 cm/sec and depths are less than 1 m (7). However, there is a population in Lake Washington, Washington, that completes its entire life cycle in the lake (14).

Coastrange sculpins commonly co-occur with prickly sculpins, threespine sticklebacks, and anadromous salmonids. In the Eel River they do not segregate from the similar prickly sculpin, although in the Smith River they occupy shallow riffles while prickly sculpins occupy deep pools (7).

They are most active at night and, except during the breeding season, exhibit little social behavior. Large daytime aggregations, however, have been observed in Alaskan lakes, often in association with shoreline spawning of salmon (8, 16). Like prickly sculpins, California coastrange sculpins in small streams migrate downstream in January, February, and March (9), presumably so that spawning will take place fairly close to the estuary where the larvae live. The importance of this amphidromous life style is indicated by the tendency of adult fish to be increasingly large in upstream areas, while juvenile sculpins predominate in downstream areas (5). However, the presence of large sculpins more than 80 km from the estuary in the Eel River makes it likely that at least some spawning takes place in upstream areas, because a 160-km round trip migration within a few months seems unlikely (5). It is possible that spawning in upstream areas is timed in response to high flows so that larvae can be washed into the estuary.

Feeding is primarily on aquatic insect larvae and other bottom-dwelling invertebrates, especially clams and snails. In the Eel River juvenile sculpins eat primarily riffle-dwelling hydropsychid caddisfly larvae, chironomid midge larvae, and baetid mayfly larvae. Adult sculpins eat large hydropsychids and large mayfly nymphs, although fish and small frogs are taken on occasion (5). When salmon are abundant, coastrange sculpins also consume loose eggs left from spawning (10).

In Alaska and Oregon coastrange sculpins measure around 25 mm TL at the end of their first summer, 35–40 mm at the end of their second, 45–50 mm at the end of their third, and 60–70 mm at the end of their fourth; they increase in length 3–5 mm/year thereafter (10, 11). Length-frequency distributions suggest that growth rates in California populations are considerably faster, with juveniles reaching 35–45 mm TL at the end of their first summer. They apparently can live up to 8 years and reach 170 mm, although the largest fish I have encountered in California have been about 145 mm (135 mm SL). Ricker (12) described a dwarf population of coastrange sculpins living in the deep waters of a British Columbia lake in which 4-year-old fish measure only 49 mm TL.

Coastrange sculpins mature during their second or third year and usually spawn in early spring. Some females have two separate spawning periods (13). In California spawning presumably takes place from January through March, because Shapovalov and Taft (9) noted large numbers of young-of-year sculpins moving upstream during the period March–May. The usual spawning site is the underside of a flat rock in swift water, to which clusters of orange embryos are attached. The number of eggs laid by a single female varies with size and age. Females measuring 5–10 cm TL produce 100–1,764 eggs each (13).

Immediately after hatching, coastrange sculpin larvae are carried into estuaries, lakes, or large riverine pools, where they are planktonic for 3–5 weeks (15). Once they assume a bottom-dwelling existence they gradually move upstream seeking riffle habitats.

Status IE. Coastrange sculpins are still widespread, even though pollution, reduced stream flows, and filling of estuarine lagoons have eliminated them from some streams, especially in the southern portions of their range.

References 1. McPhail and Lindsey 1970. 2. Robins and Miller 1957. 3. Swift et al. 1993. 4. Leidy 1984. 5. Brown et al. 1995. 6. Bond et al. 1988. 7. White and Harvey 1999. 8. Greenbank 1957. 9. Shapovalov and Taft 1954. 10. Wydoski and Whitney 1979. 11. Bond 1963. 12. Ricker 1960. 13. Patten 1971. 14. Ikusemiju 1975. 15. Heard 1965. 16. Foote and Brown 1998.

Riffle Sculpin, *Cottus gulosus* (Girard)

Identification Riffle sculpins are highly variable but are defined by the following combination of characteristics: four pelvic "elements" (1 spine and 3–4 rays); 7–8 soft spines on the first dorsal fin; 16–19 rays in the second dorsal fin; 15–16 rays in each pectoral fin, some of which may be branched; 12–16 rays (usually 13–15) in the anal fin; palatine teeth that are usually present; prickles that are present only behind the pectoral fin (axillary patch); 2–3 preopercular spines; a lateral line that is complete or incomplete with 22–36 pores; and dorsal fins that are usually joined.

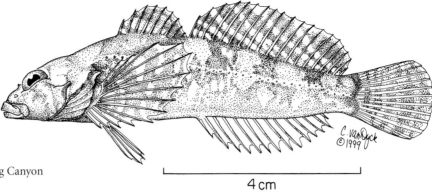

Figure 115. Riffle sculpin, 7 cm SL, Big Canyon Creek, Lake County.

The mouth is large, so the maxillary may reach as far as the rear edge of the eye. The pelvic fins usually do not reach the vent when depressed. There is usually one median chin pore. They have the typical sculpin mottled body color, with a large black blotch on the rear of the first dorsal fin. Spawning males are dark, often with an orangish edge to the first dorsal fin.

Taxonomy Riffle sculpin were originally described by Charles Girard in 1854, from San Mateo Creek, San Mateo County, as *Cottopsis gulosus.* The identity of local populations has been in a state of confusion ever since. The systematics of this sculpin badly needs to be examined using modern biochemical techniques. The biggest problem is that, as currently constituted, the species consists of two disjunct groups. One group is found in central California, while the other occupies coastal streams in northern Oregon and Washington, with a wide gap separating the two. It is highly likely that the two groups are distinct species. Because riffle sculpin have demersal, rather than planktonic, larvae, they are slow dispersers and colonize areas only by moving through fresh water. In California, at least, it is exclusively a freshwater fish. It is difficult to explain how one species with such limited dispersal characteristics could have colonized two such widely separated regions.

The confusion is amplified by the assertion that riffle sculpin are found in "coastal streams" in California. I am convinced that most of these records are misidentifications of either prickly or coastrange sculpins. The only coastal streams in California in which I have confirmed the presence of riffle sculpin (e.g., Pajaro River) have historical connections to the Sacramento–San Joaquin drainage. In addition there are a number of populations (e.g., upper Uvas Creek, San Leandro Creek, lower Putah Creek, Kings River) that seem to have fish intermediate in their characteristics between riffle and prickly sculpins; they may be either hybrids or one of the two species with "extreme" characteristics (11, 12). Pit sculpin are apparently a recent derivative of riffle sculpins that managed to colonize the Pit River drainage along with other Sacramento River fishes.

Names *Gulosus* means big-mouthed (literally, full of gullet) or, alternatively, gluttonous. Other names are as for prickly sculpin.

Distribution Riffle sculpins, as constituted today, have a rather unlikely disjunct distributional pattern (see the Taxonomy section of this account). One group of populations occurs in coastal streams from the Coquille River, Oregon, north to Puget Sound, Washington (1). The second group occurs in the Sacramento–San Joaquin drainage. This second group is almost certainly a separate species endemic to California. On the San Joaquin side they are present south to the Kaweah River, whereas on the Sacramento side they are present north into the upper Sacramento and McCloud Rivers. In the San Francisco Bay region they are found in a scattering of stream systems, including Coyote Creek, the Guadalupe River, the Napa River, Sonoma Creek, Corte Madera Creek, and Green Valley Creek (2, 11). They are absent today from San Mateo Creek, the type locality, and from Alameda Creek (11). They are found in coastal streams that have had historical connections to the Central Valley drainage, including the Pajaro and Salinas Rivers, Salmon and Redwood Creeks (Marin County), and possibly the Russian and Navarro Rivers. Recent (1999–2001) sampling of the Navarro River failed to locate any. They are absent from the Trinity, Klamath, and Rogue Rivers. The absence of riffle sculpins from many tributary streams in which they might be expected within their known range is a good indication of the difficulties this species has in dispersing from one watershed to the next.

Life History Riffle sculpins are well named, for they are most common in California in permanent, cold, headwater streams where riffles and rocky substrates predominate. In Deer Creek (Tehama County) they select areas where water flows swiftly (mean water column velocity, 42–44 cm/sec) and is fairly shallow (mean depth, 38–39 cm). Even in swift water they spend much of their time in areas sheltered from strong currents, under rocks or logs (mean velocity of water, 8–9 cm/sec) (3). They are most abundant in water that does not exceed 25–26°C for extended periods of time; temperatures over 30°C are usually lethal (5, 6). In warmer reaches they tend to be replaced by prickly sculpins. In small, well-shaded, cool streams riffle sculpins live in pools as long as the pools contain undercut banks, rubble, or other complex cover (4). When gold dredging disturbs a riffle and buries the loose rocks under which sculpins hide, riffle sculpin numbers are reduced (7). A key aspect of their habitat is oxygen levels at or near saturation (6), a requirement that also restricts them to flowing water.

Riffle sculpins almost invariably occur with rainbow trout, but interactions are few because microhabitats and diets are different. Sculpins will eat an occasional small trout, and adult trout, an occasional sculpin. More importantly, sculpins are a good indicator of water and habitat quality; their presence usually indicates first-class salmonid habitat. They are strongly aggressive toward other benthic fishes, however, and will displace speckled dace from riffles (5). Dace that do not leave get eaten.

The scattered distribution of riffle sculpins reflects their narrow habitat requirements and their poor dispersal abilities. Following a severe drought, it took over 18 months for them to recolonize a riffle that went dry only 500 m downstream from a large permanent population (4). The fact that they have benthic larvae that do not move far after hatching also reduces their ability to disperse.

Riffle sculpins are opportunistic feeders on benthic invertebrates. In riffles of Deer Creek and Big Chico Creek (Tehama County) they eat mainly hydropsychid caddisflies and various mayfly larvae (5, 10). Their diet in Putah Creek (Solano County) is similar, although they also consume *Corophium* amphipods, unusual in fresh water. Large sculpins occasionally eat smaller ones. Although they forage most actively at night, their stomachs contain food at most times of the day. Presumably they chase down prey at night but ambush it during the day.

Age and growth of northern (Washington, Oregon) riffle sculpins are similar to those of other sculpins (1, 8). Most growth occurs in the spring and summer. During their first summer they grow about 6 mm per month, reaching 25–45 mm SL by the end of the growing season; 2-year-old fish average 40–50 mm SL, and 3-year-old fish, 50–60 mm (8). California populations grow faster and larger than the northern ones (9, 10). In most streams adults, presumably 2 years old, measure 60–80 mm SL. Presumed 3-year-olds measure 75–100 mm. Larger fish are rare. In the North Fork Feather River, in an area in which fish populations have been periodically suppressed by fish poisons, many sculpins measure 100–160 mm TL (9). The maximum age for the species is not known, but it is likely that riffle sculpins seldom live longer than 4 years.

Maturity sets in at the end of the second year and culminates in spawning in late February, March, and April. Riffle sculpins spawn either on the underside of rocks in swift riffles or inside cavities in submerged logs (1, 8). Embryo counts in nests range from 462 to more than 1,000; embryos may be in different stages of development, indicating that more than one female spawns in each nest. Males stay in the nest guarding embryos and fry, often becoming emaciated; this behavior suggests that they do not feed while guarding (8).

The embryos hatch in 11 (at 15°C) to 24 (at 10°C) days. After absorbing the yolk sac, at about 6 mm TL, fry assume a benthic existence (8).

Status ID. Riffle sculpins are still locally abundant and widely distributed. Yet their populations are increasingly isolated from one another and subject to local extinction. They seem to be exceptionally vulnerable to habitat change, especially changes that reduce flows or increase temperatures. As a result they are found mainly in relatively undisturbed streams, especially headwaters (2), or in areas immediately below dams where there are permanent flows of cold water. Despite their slowness at dispersing, populations decimated by drought (4) and toxic substances (9, 10) show considerable ability to recover, although recovery is never rapid. In the upper Sacramento River riffle sculpin populations largely recovered by 1998, after being nearly wiped out in 1991 by a spill of a highly toxic fungicide at Cantara, near Dunsmuir (Shasta County).

References 1. Bond 1973. 2. Leidy 1984. 3. Moyle and Baltz 1985. 4. Smith 1982. 5. Baltz et al. 1982. 6. Cech et al. 1990. 7. Harvey 1986. 8. Millikan 1968. 9. Moyle et al. 1983. 10. Travanti 1990. 11. R. A. Leidy, U.S. Environmental Protection Agency, pers. comm. 1999. 12. J. J. Smith, San Jose State University, pers. comm. 1999.

Plate 1   Pacific lamprey

Plate 2   Western brook lamprey

Plate 3   Kern Brook lamprey (ammocoetes)

Plate 4   White sturgeon

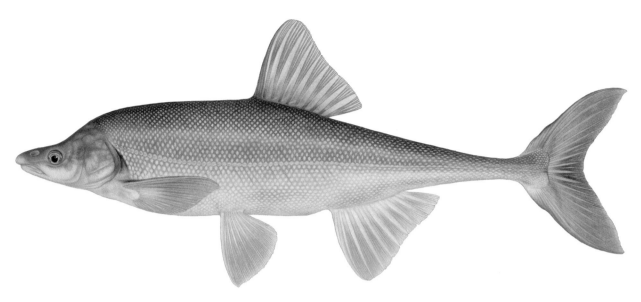

Plate 5   Bonytail

Plate 6   Colorado pikeminnow

Plate 7   Sacramento pikeminnow

Plate 8   Hardhead

Plate 9   California roach

Plate 10   Sacramento blackfish

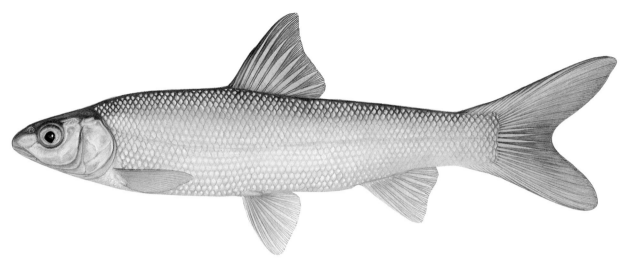

Plate 11 Sacramento splittail, Suisun Marsh

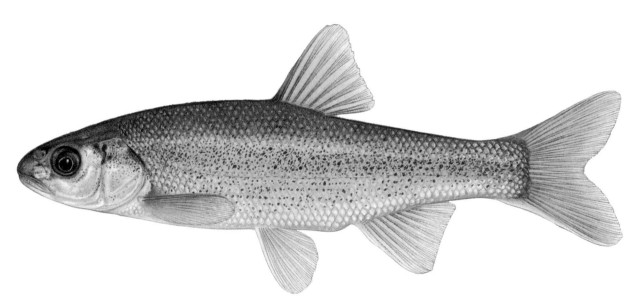

Plate 12   Lahontan redside

Plate 13   Mountain sucker

Plate 14   Razorback sucker

Plate 15   Sacramento sucker

Plate 16   Klamath largescale sucker

Plate 17   Lost River sucker

Plate 18   Shortnose sucker

Plate 19 Coastal rainbow trout (steelhead)

Plate 20   Coastal rainbow trout (resident)

Plate 21   Eagle Lake rainbow trout

Plate 22   Kern River rainbow trout

Plate 23   California golden trout

Plate 24   Little Kern golden trout

Plate 25   McCloud redband trout

Plate 26   Lahontan cutthroat trout

Plate 27   Paiute cutthroat trout

Plate 28   Coastal cutthroat trout

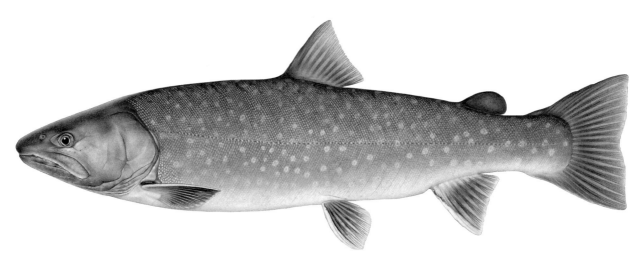

Plate 29   Bull trout

Plate 30   Coho salmon (ocean colors)

Plate 31   Chinook salmon (ocean colors)

Plate 32   Desert pupfish

Plate 33   Prickly sculpin

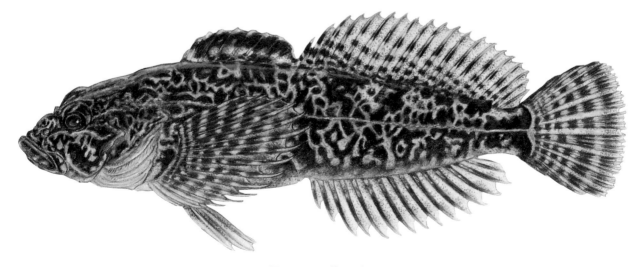

Plate 34   Riffle sculpin

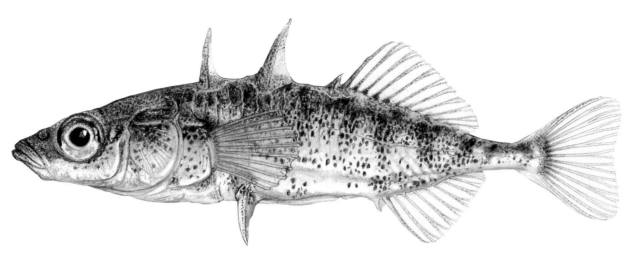

Plate 35   Threespine stickleback

Plate 36   Sacramento perch

Plate 37   Tule perch

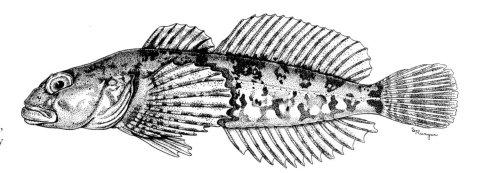

Figure 116. Pit sculpin, holotype, 7 cm SL, Pit River. From Bailey and Bond (1963).

Pit Sculpin, *Cottus pitensis* Bailey and Bond

Identification Pit sculpins closely resemble riffle sculpins but differ in the following ways: the palatine teeth are absent; there are usually only 2 preopercular spines (but occasionally 3); the lateral line in fish measuring 5 cm TL or larger is usually complete, with 31–39 pores (usually 33–37); the pectoral fin rays number 12–16 (usually 13–15) and are mostly unbranched; and the dorsal spines number 8–9 (occasionally 7 or 10). The anal fin rays number 12–15, and the median chin pore is usually absent. There are patches of prickles just behind and slightly above the bases of the pectoral fins. Color patterns are variable, but there are usually 5–6 faint dark "saddles" on the back, 2 beneath the first dorsal and 3–4 beneath the second dorsal. A dark band often encircles the end of the caudal peduncle. The pectoral, caudal, and second dorsal fins are banded, and the first dorsal fin usually has a large blotch on the posterior end. The belly is light colored.

Taxonomy This recently described species is closely related to the riffle sculpin, from which it is presumably derived through isolation in the upper Pit River drainage (1). However, sculpins from regions where the two might be in contact (e.g., Squaw Valley Creek, Shasta County) have not been closely examined.

Names Pitensis is after the Pit River. Other names are as for prickly sculpin.

Distribution Pit sculpins are widely distributed throughout the Pit River watershed, Modoc and Shasta Counties, including its north and south forks as well as the mainstem and accessible tributaries down to Squaw Valley Creek (Shasta County) (2). They are also present in tributaries to Goose Lake in both California and Oregon, including Lassen and Willow Creeks (Modoc County, California) and Drews, Cottonwood, and Thomas Creeks (Lake County, Oregon).

Life History Like riffle sculpins, Pit sculpins are most abundant in fast-flowing rocky riffles in smaller, well-shaded streams (2, 3). However, they also occupy a wide variety of other habitats, from large boulder-strewn rivers to spring-fed creeks, as long as oxygen levels are near saturation and temperatures are less than 25°C most of the time (2, 8). Preferred temperatures are 10–16°C, presumably reflecting optimal temperatures for growth and reproduction (4). They cannot survive temperatures greater than 27.5°C (4). Pit sculpins are typically found in runs and riffles 20–50 cm deep. Their strong preference for riffles is shown by their ability to remain abundant in channelized sections of a small coldwater stream (5) and by their occurrence in water with flows as swift as 110 cm/sec (3). Pit sculpins commonly co-occur with rainbow trout, speckled dace, and Sacramento sucker, forming a distinct assemblage in second- and third-order streams in the Pit River system (2). In some spring-fed streams they also co-occur with marbled and rough sculpins, but usually Pit sculpins are absent or rare where the other species are common and vice versa (6). The reasons for this distribution are uncertain, although the tendency of Pit sculpins to be more active than the other species may make them more vulnerable to predation in open, sandy areas where the other species are abundant. In addition their aggressiveness may keep the other species out of rocky riffle areas (6).

Like other sculpins they feed primarily on aquatic insect larvae and are selective in their feeding. In Ash Creek (Lassen County) they select for the rounded baetid mayfly larvae, large stonefly nymphs, and web-spinning moth larvae (Pyralidae) but against flattened heptageneid mayfly larvae (7). The sculpins also select against caddisfly larvae with cases, snails, and other hard-shelled invertebrates. Two species of web-spinning, caseless caddisfly larvae, among the most abundant larvae in riffles, are eaten but not preferred. They feed throughout the day and night, but most intensely in early morning.

Growth is similar to that of other sculpins living in small, cold streams. They typically grow to around 40 mm SL in their first year, 56 mm in their second, 70 mm in their third, 84 mm in their fourth, and 109 mm in their fifth (3). Thus in June 1973 Pit sculpins starting their third summer of life (II+ age class) measured 57–72 mm SL (mean, 62 mm); III+ fish measured 59–89 mm (mean, 80 mm), and IV+ fish measured 75–104 mm (mean, 82 mm). The largest Pit sculpin recorded was 127 mm SL. The various age classes can often be recognized in length-frequency diagrams, with steadily decreasing abundance with age; about half of each age class dies each year (3). Both males and females become mature in their second year.

Spawning occurs in February through early May, al-though most apparently takes place in late February and March. Fecundities are relatively low: each female produces 61–320 eggs, depending on size ($F = 6.4SL_{mm} − 253$) (3). When ready to spawn, males become dark in color with an orange fringe on their dorsal fins, and they begin to guard nesting sites, usually a small "cave" under a rock or submerged log. Each male entices several females to lay their sticky eggs on top of the nesting cavity, where they are fertilized. Larval sculpins are apparently demersal and stay close to the nest.

Status IE. Pit sculpins are still common throughout the Pit River drainage and have adjusted to degraded habitats in some cases. In Oregon they are regarded as a sensitive species because of their limited distribution in the state.

References 1. Bailey and Bond 1963. 2. Moyle and Daniels 1982. 3. Daniels 1987. 4. Brown 1989. 5. Moyle 1976. 6. Brown 1991. 7. Li and Moyle 1976. 8. Brown 1988.

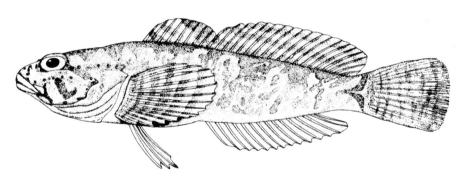

Figure 117. Reticulate sculpin, 5 cm SL, Oregon. From Bond (1961).

Reticulate Sculpin, *Cottus perplexus* Gilbert and Evermann

Identification Reticulate sculpins resemble riffle and marbled sculpins. They are identified by the following complex of characters: palatine teeth absent; dorsal fins usually broadly joined, with 7–8 spines in the first dorsal fin and 18–20 rays in the second; pectoral fin rays 13–16 (but usually 14–15 and unbranched); anal fin rays 13–16; preopercular spines 1–4 (but usually only 2 visible); body prickling highly variable but axillary patch always present; and maxilla extending to just below the anterior portion of the eye. The mouth is narrower than the body width behind the pectoral fins. The body is usually patterned with faint vermiculations and small patches of dark pigment; the pectoral fins frequently have a checkerboard pattern like that of marbled sculpins. The first dorsal fin has a dark blotch on the posterior portion.

Taxonomy Reticulate sculpins, first described in 1894 (1), were lumped taxonomically with riffle sculpins, but Robins and Miller (2) resurrected the taxon. Its status and relationships are in need of careful examination.

Names *Perplexus* translates as perplexing, reflecting the difficulty in defining the species, although Gilbert and Evermann may have had the reticulated appearance in mind when assigning the name. Other names are as for prickly sculpin.

Distribution Reticulate sculpins occur in coastal streams from the lower Columbia River, Washington, down to the Rogue River, Oregon. In California they are found only in a few creeks that drain north into the Rogue River, where the species is the dominant sculpin. I have examined sculpins assigned to this species only from Elliot Creek, a tributary to the Applegate River (a branch of the Rogue River) (3). Their distribution seems to be limited in part by the presence of other sculpins, especially riffle sculpins, sensu lato, with which they are rarely sympatric (4).

Life History Reticulate sculpins are one of the most common sculpins in coastal Oregon. They live primarily in slower portions of small coastal and headwater streams but occupy various stream habitats (4, 5). In general they inhabit small streams with summer temperatures of less than 20°C, in areas with sandy and gravelly bottoms and moderate current velocities (5). In the absence of other sculpin species, they seek out rubble or gravel bottoms in riffles, preferably near dense cover, such as overhanging vegetation (6). They tend to live in pools or along stream edges on sandy or silty bottoms when other sculpins are present. They may be adapted in part for living in streams where other sculpins cannot live, especially streams with high or fluctuating temperatures. They can withstand temperatures up to 30°C and salinities up to 18 ppt (4).

Reticulate sculpins feed mostly on aquatic insect larvae, especially those of mayflies, stoneflies, chironomid midges, beetles, and caddisflies. Fish are only rarely taken. Under experimental conditions (uniform gravel size in small tanks) reticulate sculpins have been shown to prey on trout eggs and fry, even those buried in gravel (7). However, the applicability of these results to natural conditions is questionable.

Growth is slower in reticulate sculpins than in most species. Bond (4) found that yearling (age I) fish averaged only 27 mm SL; age II, 42 mm SL; age III, 56 mm SL; and age IV, 64 mm SL. Maximum size is about 110 mm SL. Despite their small size they mature early; females may mature at 30–39 mm SL (4). Fecundity ranges from 35 to 315 eggs per female, depending on the fish's size.

Spawning occurs from March through May when stream temperatures exceed 6–7°C. The eggs are laid on the underside of rocks 10–45 cm in diameter, although various types of trash can also be used as nests. Males, however, show a strong preference for moderately embedded cobbles in both pools and riffles (8). Usually more than one female contributes eggs to each nest. When other species of sculpin are absent or rare, reticulate sculpins spawn in riffles. They spawn in slower-flowing areas when other species are abundant. The males guard the nest from egg predators, probably until the fry are past the yolk sac stage and can fend for themselves. The fry assume a benthic existence immediately after leaving the nest and tend to stay in quiet water at stream edges (4).

Status ID. Reticulate sculpins are uncommon in California but abundant in Oregon and Washington. Their retention as part of the California fauna requires that the few creeks in which they live be kept in good condition.

References 1. Gilbert and Evermann 1894. 2. Robins and Miller 1957. 3. Bond 1973. 4. Bond 1963. 5. Bond et al. 1988. 6. Finger 1983. 7. Phillips and Claire 1966. 8. Bateman and Li 2001.

Marbled Sculpin, *Cottus klamathensis* Gilbert

Identification Marbled sculpins can be distinguished by the following suite of characteristics: 7–8 dorsal fin spines, broadly joined dorsal fins, incomplete lateral line with 15–28 pores, smooth skin (except for a small patch of prickles in some populations), 2 chin pores, absent palatine teeth, and only 1 well-developed preopercular spine (although one or two inconspicuous protruberances may be present below it) (1). In addition they have a wide head, with widely separated eyes, a blunt snout, a maxillary bone that does not reach the posterior edge of the eye, and usually no conspicuous dark patch on the rear portion of the first dorsal, although a band may run across most of the fin. There are 7–8 spines in the first dorsal fin, 18–22 rays in the second dorsal fin, 13–15 anal fin rays, 14–16 pectoral fin rays, and 11–12 principal rays in the caudal fin. The pelvic fins have four "elements" and may or may not reach the vent when depressed. Prickling may be well developed on young fish but is confined to a small region behind the pectoral fins in adults or is absent. The pectoral fins often appear checkered, with alternating dark and light spots on the rays. Fish

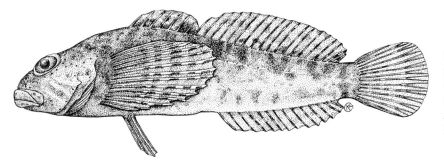

Figure 118. Marbled sculpin, *macrops* subspecies, 6 cm, Fall River, Shasta County. From Daniels and Moyle (1984); reprinted by permission of the American Society of Ichthyologists and Herpetologists.

from the Klamath drainage system have a strikingly marbled appearance and barred fins. Fish from the Pit River are usually darker and less strikingly marked.

Taxonomy Marbled sculpins were originally described in 1898 from the Klamath River. In 1908 Rutter described *C. macrops* from the Fall River (2), but Robins and Miller (3) concluded that they belonged in *C. klamathensis*. Reappraisal of the taxonomy of the marbled sculpin (1) indicated that there are three subspecies: *C. k. klamathensis* from the Klamath River basin above Klamath Falls, *C. k. polyporus* from the lower Klamath River, and *C. k. macrops* from the Pit River drainage. The *macrops* form is fairly distinctive in morphology, ecology, and behavior, and so may deserve separate species status.

Names The common name reflects the color pattern of Klamath River populations. *Klamathensis* is after the Klamath River, *macrops* is after the large eyes of the Pit River subspecies, and *polyporus* refers to the large number of lateral line pores (22–28) possessed by the lower Klamath fish in contrast to other subspecies (15–22). In this account the three subspecies are called upper Klamath marbled sculpin (*C. k. klamathensis*), lower Klamath marbled sculpin (*C. k. polyporus*), and bigeye marbled sculpin (*C. k. macrops*).

Distribution The upper Klamath marbled sculpin is found throughout the upper Klamath River basin, Oregon, down to Klamath Falls. In the Lost River drainage of northeastern California they are found in only a few small headwaters (Willow, Rock, and Fletcher Creeks) (4). Lower Klamath marbled sculpin are common in the Klamath River and tributaries from Iron Gate Dam roughly to its confluence with the Trinity River. They are rare or absent in the lowermost river and perhaps from the Trinity as well. Bigeye marbled sculpin are confined to the Fall River (including Spring Creek and Tule River), lower Hat Creek, lower Burney Creek, and portions of the main channel of the lower Pit River from Britton Reservoir down to Tunnel Reservoir, all in Shasta County (5, 6).

Life History Bigeye marbled sculpins are found mainly in low-gradient spring-fed streams and rivers, where the water

is cold (<20°C in summer) and there is enough fine substrate to support beds of aquatic plants. Sites where they are found are typically runs or pools that have mean depths of 57–72 cm and mean water column velocities of around 23 cm/sec (5, 6). They select deeper parts of the environment, close to cover, especially aquatic plants (7). Bigeye marbled sculpins are capable of living in standard rubble-dominated riffles, but most such habitats in the Pit River drainage are too warm in summer; the sculpins prefer temperatures of 11–15°C and become stressed when temperatures are higher, especially if there are wide temperature fluctuations as well (8). Extended exposure to temperatures of 25–27°C is lethal. The presence of the more eurythermal Pit sculpin in most riffles may also inhibit their use by bigeye marbled sculpins.

In the upper and lower Klamath River habitat requirements of marbled sculpins are less well documented, but they seem to be the standard riffle-dwelling sculpin there, living in shallower, warmer, and more fluctuating conditions than do sculpins of the Pit River. They are also present in upper Klamath Lake, living on bottoms of mud and sand. The upper Klamath marbled sculpin are most abundant in streams with summer temperatures of 15–20°C, coarse substrates, widths greater than 10 m, and moderate to slow current velocities (9).

Bigeye marbled sculpins can live about 5 years, attaining 35 percent of their maximum length during their first year (6). The growing season begins in spring and lasts until early autumn. They average about 39 mm SL at the end of their first growing season, 55 mm at the end of their second, 62 mm at the end of their third, 70 mm at the end of their fourth, and 79 mm at the end of their fifth. The largest recorded specimen measured 111 mm SL, but fish over 80 mm are unusual.

They attain sexual maturity after 2 years. Males and females begin to mature during winter, and spawning occurs from late February to March. Fecundity is low; females produce only 139–650 large ova per fish. Adhesive eggs are deposited in nests under flat rocks in globular clusters. Nests, which are guarded by large males, typically contain spawn from multiple females. Daniels (6) found that the number of embryos per nest was 826–2,200. The low fecundity, late reproductive maturation, and relatively long life span reflect bigeye marbled sculpin's adaptation to relatively cold, constant, spring-fed streams (6). The larvae measure 6–8 mm TL after hatching and are benthic, so juvenile fish presumably rear close to the nesting site.

Status ID. The bigeye marbled sculpin is the least abundant of the three sculpins endemic to the Pit River drainage, but it still remains fairly common in much of its limited range. It co-occurs with the rough sculpin, listed by the state as a threatened species, and with the Shasta crayfish (*Pascifasticus fortis*), listed by the federal government as an endangered species. As long as habitats for rough sculpin and Shasta crayfish are protected, co-occurring marbled sculpin will be as well. However, it is apparently less abundant in the Fall River than when Cloudsley Rutter (2) first collected it there in 1898. This may indicate subtle long-term changes in its native habitats, such as a change in water quality caused by agricultural effluent and watershed degradation (Fall River) or a change in flow created by the variable operation of hydroelectric projects (Pit River). Predation by trout may also affect their distribution and abundance (7).

For the most part native rainbow trout have dominated the streams, and more piscivorous alien brown trout have not become common. Changes in water quality or fisheries management that favor brown trout might have negative effects on marbled sculpin through increased predation.

Upper Klamath marbled sculpin are still widespread in Oregon, but many of their habitats are degraded. In California they probably qualify for state listing as a threatened species because they are confined to only a few isolated sections of small tributaries to the Lost River (4). The streams in which they occur are heavily degraded through diversion and impoundment of water and through grazing of riparian areas by livestock, raising temperatures, reducing cover, and increasing siltation of riffles.

The lower Klamath marbled sculpin is presumably still abundant throughout its native range, although thorough surveys are lacking.

References 1. Daniels and Moyle 1984. 2. Rutter 1908. 3. Robins and Miller 1957. 4. Buettner and Scoppetone 1991. 5. Moyle and Daniels 1982. 6. Daniels 1987. 7. Brown 1991. 8. Brown 1989. 9. Bond et al. 1988.

Paiute Sculpin,
Cottus beldingi Eigenmann and Eigenmann

Identification Paiute sculpins can be distinguished by their possession of the following combination of characteristics: anal fin rays 11–14; palatine teeth absent; prickling completely absent; pectoral fin rays all unbranched; dorsal fins separated; upper preopercular spine usually long and slender, lower spine inconspicuous; more than 23–35 pores on an incomplete lateral line; and median chin pores usually 2 (rarely 1 or none). Pelvic "elements" number 4; first dorsal spines, 6–8; second dorsal fin rays, 13–16; and pectoral fin rays, 14–15. The maxilla is long, often reaching past the middle of the eye. The pelvic fins occasionally will reach the vent, and the caudal fin is rounded. Coloration is highly

variable, although 4–5 vertical bands are frequently discernible on the sides. The fins are barred or mottled.

Males can be distinguished from females by their long anal papillae (2–3 mm in mature fish) and their larger mouths. In males the width of the mouth is greater than the distance from the pelvic fins to the anus; in females it is less (1).

Taxonomy This species is badly in need of taxonomic reappraisal using both biochemical and conventional techniques. It is likely that the forms in the Columbia, Bonneville, and Lahontan basins are distinct taxa. The forms on the east side of the Sierras in particular have long been isolated from other sculpins.

Names *Beldingi* is after C. Belding, who collected the specimens used by Carl and Rosa Eigenmann to describe the species. The common name is after the Paiute people, who live in the Lahontan basin, where the sculpins are abundant. Snyder (2) gave them the descriptive name "desert rifflefish." For other names see the account of prickly sculpin.

Distribution Paiute sculpins are the only sculpins in the Lahontan drainage of California and Nevada, including the Susan, Truckee, Walker, Carson, Quinn, and Humboldt River watersheds. They are widely distributed in the Snake River system in Oregon, Washington, Idaho, and Wyoming and in the lower Columbia drainage (below the confluence

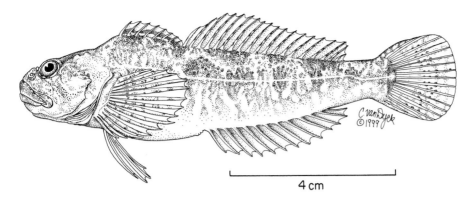

Figure 119. Paiute sculpin, 9 cm SL, Desert Creek, Mono County.

of the Snake River) and its southern tributaries. Isolated populations are found in the Bear River of the Bonneville system in Utah and Idaho. Records of this species in the upper Colorado River (3) are probably *C. bairdi*.

Life History The life history of Paiute sculpin is well documented thanks to studies in Lake Tahoe (1, 4, 5), Sagehen Creek (6, 7, 8, 9, 10), and elsewhere (11, 12, 13). In all situations Paiute sculpins prefer bottoms of rubble and gravel, although it is not unusual to find them living on other substrates. In Lake Tahoe they are abundant in or near aquatic macrophyte beds in deep water. They have been collected down to 210 m in Lake Tahoe, but the largest numbers live in water less than 60 m deep (5). In streams they are largely absent from high-gradient headwaters, as well as from warm, low-gradient stretches. Their typical habitat is shallow (<50 cm), rocky riffles in clear, cold mountain streams, where they are almost always associated with trout. They can tolerate fairly warm temperatures as long the water is flowing rapidly; I have found them in streams with daytime temperatures regularly exceeding 20–25°C. However, they are found mostly in streams with maximum temperatures of less than 20°C (13).

Paiute sculpins tend to be sedentary, even in Lake Tahoe (5). Daylight hours are spent hidden among rocks and aquatic plants, from which they can ambush passing prey, whereas during the night they are often out actively foraging. In streams most of their feeding may take place at night (11). Not surprisingly, Paiute sculpins selectively consume active prey (14). When a prey organism approaches, the fish lunge out suddenly and engulf it. The numerous small teeth on the roof of the mouth prevent prey from wriggling out.

The prey of Paiute sculpin are invertebrates that live on or close to the bottom. In most creeks their diet is primarily aquatic insect larvae associated with riffles, especially mayflies, stoneflies, and caddisflies. The remainder of the diet consists of miscellaneous organisms, such as snails, water mites, aquatic beetles, and algae, as well as detritus accidentally ingested (4, 12). In meadow streams their diet is mainly dragonfly larvae and similar invertebrates charac-

teristic of slow-moving water (17). Most of the diet of sculpins in shallower waters of Lake Tahoe is also bottom organisms, especially chironomid midge larvae. The remainder is planktonic crustaceans and algae (1). In sculpins taken from deep water (>30 m) detritus and filamentous algae make up the bulk of the stomach contents, and this finding presumably reflects their capture of prey living on soft bottoms. Snails, amphipods (scuds), aquatic insect larvae, and other sculpins are the most important animal foods. Snails are most abundant in sculpins taken at 30–60 m; oligochaetes and deep-water amphipods (*Stygobromus*) are most abundant in fish taken at 60–90 m (4). Paiute sculpins occur in trout spawning areas, but the presence of trout eggs in their diet is unusual.

The food of Paiute sculpins varies with the seasonal availability of prey. They feed year round, but feeding activity diminishes in fall and winter. In Sagehen Creek chironomid midge and caddisfly larvae are most important in fall; mayfly larvae gradually assume importance in winter. As spring approaches sculpins concentrate more and more on heptageniid mayfly larvae (6). In the inshore waters of Lake Tahoe aquatic insect larvae are consumed in fairly consistent amounts between May and September, but planktonic forms vary in abundance, presumably reflecting seasonal succession in the lake (1). In deep-water populations snails are eaten mostly in summer and fall; amphipods are eaten in spring and summer. Interestingly enough, other sculpins—which make up 8–13 percent of the diet during summer, fall, and winter—are absent from the diet in spring (4), indicating that reproductive activity may inhibit cannibalism.

Diet varies with size of fish. Stream sculpins measuring less than 59 mm TL feed mostly on chironomid larvae; those greater than 80 cm feed mostly on large aquatic insect larvae, such as those of caddisflies and mayflies (6). In Lake Tahoe fish measuring greater than 41 mm TL feed on a greater variety of invertebrates than do smaller fish, which tend to concentrate on ostracods, chironomid larvae, and amphipods. Cannibalism is observed only in sculpins measuring greater than 82 mm TL (4).

Growth rates, as determined by otoliths, are similar for

both Lake Tahoe and Sagehen Creek populations. By the end of their first summer they measure 26–36 mm TL; at the end of the second summer, they average 54 mm; third summer, 68 mm; fourth summer, 83–84 mm; fifth summer, 93–97 mm (4). Most growth takes place from May to October, although it can continue during winter months. The largest sculpin on record from Lake Tahoe measured 127 mm TL; the largest from Sagehen Creek was 110 mm TL. Such large Paiute sculpins are rare and are usually males.

Maturity generally sets in in the second or third year, and spawning occurs mostly in May and June. Jones (8) noted that most spawning in Sagehen Creek took place during a 1-week period in early June, but that the exact time and length of the spawning period depend on when and how rapidly the water warms up. In Lake Tahoe the spawning season is long. Ripe females have been collected from early May to late August (1). Embryos, however, have been found only in shallow water from May to early July, indicating that peak spawning takes place during that period. As in other sculpins, the embryos are glued in clusters to the underside of rocks, and the nest is tended by a male. Spawning males defend nests from other males, partly to reduce embryo predation. Although there seems to be little actual preparation of the spawning site, sites are selected carefully. Most nests are located under rocks over gravel bottoms; cavities over bedrock or mud bottoms are avoided (1). Most spawning sites in Lake Tahoe seem to be located in wave-swept littoral areas or just off the mouths of streams, yet some spawning must take place in deeper water because no dramatic inshore movements of sculpins have been observed. In streams most spawning sites are in riffles.

The number of embryos in each nest is around 100–200. This range is similar to the fecundity range of females, indicating that multiple spawns in one nest are uncommon, unless each female lays only a small number of eggs at one time. The mean number of eggs from Lake Tahoe sculpins is 123, with a range of 11–387 (4). The fecundity of Sagehen Creek fish is similar. The number of eggs increases with age and size of fish (8, 15). The mean number of eggs in 2-year-old fish is 73, and in 3-year-old fish, 130 (15).

After fry hatch, at about 10 mm TL, they drop down among the gravel on the floor of the nest. There they remain for 1–2 weeks until they absorb the yolk sac. In Sagehen Creek, although some postlarval sculpins stay in the vicinity of the nest, many swim up off the bottom, are caught by the current, and are washed downstream. Most larval drift occurs at night (7). There are two peaks in numbers of drifting sculpins, one immediately following yolk absorption and another, of slightly larger individuals, about 2 weeks later. The initial urge of young sculpins to forsake their sedentary bottom-living habits for a few hours is presumably a mechanism that ensures wide dispersal of young sculpins and

keeps populations at reasonable levels in spawning areas. The later peak in drifting may be a result of interactions among individuals, the losers of aggressive encounters over food and space being forced to drift downstream (7).

Paiute sculpins show tremendous variation in abundance through time; density can vary 20–50 times over a period of years (10). Most likely this variation is caused by a reduction in numbers when high winter flows scour the bottom, turning over rocks under which sculpins are seeking refuge. Dead sculpins have been found among rocks moving down the channel of Sagehen Creek in winter (16). In the absence of severe winter flooding, not only do sculpin populations become dense (20–30 fish per square meter), but individuals will be found in areas less suitable for sculpins, where substrates are less coarse and temperatures are stressful.

Status IE. Paiute sculpins are still one of the most abundant fish in streams on the northeast side of the Sierras. Because of its large numbers and carnivorous reputation, the Paiute sculpin's relationships with trout have been intensively studied. The results of those studies indicate that sculpins have a mostly positive effect on trout populations. No evidence of egg predation has been found in Lake Tahoe, and trout eggs are found only rarely in the stomachs of stream sculpins. Competition in streams between sculpins and trout is minimal because trout feed mainly on drift organisms, whereas the sculpins concentrate on bottom forms. Thus in Sagehen Creek the diet of sculpins overlaps that of trout by only 20 percent (6). In Lake Tahoe the overlap is even less, and sculpins form a large part of the diet of both lake trout and rainbow trout. Similarly, in Sagehen Creek 20 percent of the diet of brook trout is sculpins. Because brook trout can also be cannibalistic, sculpins may serve as a "buffer" prey, their availability preventing extensive cannibalism by trout (6).

Despite their abundance and wide distribution, Paiute sculpins are probably less abundant in many areas than they were formerly. They are fairly sensitive indicators of high-quality water and stream habitats. Their disappearance from, or low abundance in, a stream reach may be indicative of land use practices—such as water diversions, grazing, logging, or urbanization—that degrade stream environments. Thus they are a useful organism to incorporate into schemes to monitor the health of eastern Sierran streams.

References 1. R. G. Miller 1951. 2. Snyder 1918. 3. Sigler and Miller 1963. 4. Ebert and Summerfelt 1969. 5. Baker and Cordone 1969. 6. Dietsch 1959. 7. Sheldon 1968. 8. Jones 1972. 9. Gard and Flittner 1974. 10. Erman 1986. 11. Johnson 1985. 12. Moyle and Vondracek 1985. 13. Bond et al. 1988. 14. Kratz and Vinyard 1981. 15. Patten 1971. 16. Erman et al. 1988. 17. Moyle et al. 1991.

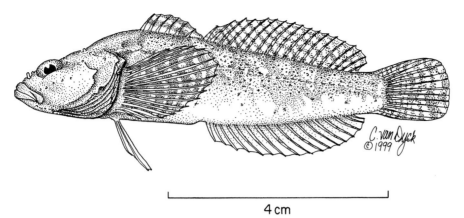

4 cm

Figure 120. Rough sculpin, 7 cm SL, Hat Creek, Shasta County.

Rough Sculpin, *Cottus asperrimus* Rutter

Identification Rough sculpins are relatively small, slender sculpins that consistently have 1 spine and 3 rays in the pelvic fins (3 "elements"). Palatine teeth are absent, and the lateral line does not extend past the end of the second dorsal fin. Their sides are covered with prickles, making them rough to the touch. There are 5–7 spines in the first dorsal fin, 17–19 rays in the second dorsal fin, 13–17 anal fin rays, 14–16 rays (many of them branched) in each pectoral fin, and 19–29 lateral line pores. The upper preopercular spine is well developed, the lower a blunt nob. They are light brown to purplish brown on the back, dusky on the sides with 4–5 irregular blotches, and speckled on the belly. The dorsal fins are brown to reddish, with light streaks.

Taxonomy Cloudsley Rutter (1) described this distinctive fish from the Fall River. It is clearly most closely related to the slender sculpin, *C. tenuis*, of Upper Klamath Lake, a species that is similar both morphologically and ecologically (2). Small differences in meristic characters and differences in reproductive timing suggest that Fall River sculpins have been isolated for some time from those in Hat Creek and elsewhere.

Names *Asperrimus* means very rough, referring to the prickling on the sides. Other names are as for prickly sculpin.

Distribution Rough sculpins are largely restricted to spring-fed tributaries of the Pit River in northeastern Shasta County (3, 4, 5). They are most abundant throughout Fall River and its major tributary, Tule River (but are absent from a nonspring tributary, Bear Creek). They also occur in Sucker Springs Creek and lower Hat Creek (and associated springs) up to and including the major spring system at Crystal Lake. Prior to construction of power-generating facilities in this region in the 1920s, they were presumably also abundant in interconnecting reaches of the Pit River. They are rare or absent there today, except in the reach immediately below the confluence with Hat Creek. However, these same facilities have apparently created habitat for rough sculpin in run-of-river reservoirs on the Pit River; they have been collected from Britton Reservoir and from Tunnel Reservoir, 22 km farther downstream.

Life History The major habitat of rough sculpins reflects their distribution: large, spring-fed streams where water is cool (usually around 15°C), deep (often >1 m, although they are most abundant at 50–75 cm), rapidly flowing, and remarkably clear (3, 6). In these streams they are associated with gravel or sand bottoms and beds of aquatic plants. When given an opportunity (in part by the absence of Pit sculpin), they live in other habitats as well, such as small riffle-dominated streams; Big Lake (drained by the Tule River), which is turbid and has summer surface temperatures up to 30°C; and run-of-river reservoirs (7). In the laboratory rough sculpins prove to be stenothermal, preferring temperatures between 13.3 and 14.4°C (8). They become increasingly stressed as temperatures become higher than 15°C and die as temperatures exceed 25–27°C for extended periods of time. The commonest associate of rough sculpin is marbled sculpin, although other native fishes (especially Sacramento sucker, rainbow trout, tui chub, and Pit-Klamath brook lamprey) are frequently found with it as well.

Rough sculpins are relatively inactive during the day, remaining quietly on the bottom under or near cover, moving mainly to ambush passing prey. In sandy areas they burrow to escape predators (6). They feed around the clock but are most active at dawn and dusk (3). However, prey taken at night tends to be larger and more active (e.g., amphipods and isopods) than prey taken during the day. Dominant prey year round are chironomid midge and baetid mayfly larvae, although other prey will be consumed opportunistically. Larger fish capture larger and more varied prey than small fish. Curiously, rough sculpins do not eat two common invertebrates found in their habitats: stonefly larvae and snails.

Compared with other sculpins, rough sculpins are slow growing and small. They average 34 mm SL at the end of their first year, 44 mm at the end of their second year, 53 mm at the end of their third year, 60 mm at the end of their fourth year, and 66 mm at the end of their fifth year (5). Both sexes grow at about the same rate, but the oldest and largest (to 81 mm SL) fish are males. Growth rates decrease with age (5). Most sculpins become mature in their second year, at sizes greater than 35 mm SL. Females produce 140–580 large (1–2 mm diameter) eggs, with larger females producing more eggs ($F = 15.3SL_{mm} - 570$) (5).

Spawning occurs between September and January in Fall River, but fish in Hat Creek apparently spawn from mid-February to early May (3, 7). Nest sites are in a wide variety of habitats, from riffles to pools, often over active springs, with bottom substrates ranging from sand to cobble (5). Like other sculpins, males choose a nest site under a rock or submerged log, entice one or more females to spawn with them on the ceiling of the "cave" to which the fertilized eggs are attached. Nests typically contain multiple clutches of developing embryos, 800–3,000 in all (mean, 1,720; $N = 7$). Males defend nests and embryos through hatching, a period of 2–3 weeks. Although most females produce just one brood of eggs per year, there is some evidence that a few may spawn twice at widely spaced intervals (5). Larvae are benthic and rear close to the nest site.

Status IC. The rough sculpin was one of the first species of fish to be officially protected by the California Fish and Game Commission, in 1973, an action that preceded the official state endangered species act. Its protected status was relatively noncontroversial because of its apparent rarity and occurrence in waters with a high degree of natural protection. In the past the biggest threats to rough sculpin were attempts to manage its streams for trout fishing. Predatory brown trout were introduced, and Hat Creek was poisoned in the 1960s to get rid of nongame fish, including sculpins. Today management of Fall River and Hat Creek to support wild trout fisheries seems to favor the sculpins. There is also a growing appreciation of the unusual nature of spring-fed river ecosystems in the region, which support endemic invertebrates as well as endemic rough and bigeye marbled sculpin. The listing of Shasta crayfish (*Pascifasticus fortis*) as a federally endangered species (largely because of its displacement by alien crayfishes) should provide protection for rough sculpin, because the two species have very similar habitat requirements.

The extraordinary development of the Fall and Pit Rivers for power production has been a mixed blessing to the rough sculpin. Sections of the rivers have had much of their water removed and sent through penstocks (giant pipes) to powerhouses. Rough sculpins are absent today from dewatered sections that presumably once supported them. On the other hand, a stairstep series of reservoirs, created to shunt water into penstocks, have apparently been colonized by rough sculpins, extending their range downstream by about 22 km.

More recent and more troublesome threats to Fall River sculpins are related to changes in land use. Large amounts of sediment have washed into the Fall River from Bear Creek, from logging, grazing, and fires in the watershed, although this problem is being dealt with through cooperative efforts of private and public landowners. Warm, polluted water enters the river from wild rice paddies. In short, although rough sculpins seem to have a reasonably secure future in their limited range, they also need continued protection. Populations in Hat Creek and Fall River should be resurveyed every 5–10 years to determine their status.

References 1. Rutter 1908. 2. Robins and Miller 1957. 3. Daniels and Moyle 1978. 4. Moyle and Daniels 1982. 5. Daniels 1987. 6. Brown 1991. 7. Daniels and Courtois 1982. 8. Brown 1989.

Pacific Staghorn Sculpin, *Leptocottus armatus* Girard

Identification Pacific staghorn sculpins are readily recognized by the conspicuous, antlerlike projections on their gill covers, which bear 3–4 small sharp spines. The head is large and flat with small eyes. When captured they flatten their heads and erect the "stag horns." The maxillae reach past the eyes. The teeth, including the palatines, are well developed. The pelvic fins have 4 rays; the first dorsal fin, 7 spines; the second dorsal fin, 17 rays; and the anal fin, 17 rays. The skin is smooth (no scales or prickling), the lateral line is complete, and the caudal peduncle is narrow. Staghorn sculpins

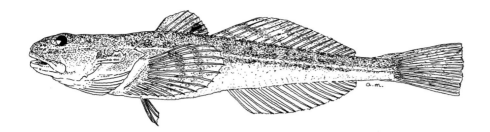

Figure 121. Staghorn sculpin, 9 cm SL, Navarro River estuary, Mendocino County. Drawing by A. Marciochi.

are grayish olive on the back, creamy yellow on the sides, and abruptly white on the belly. The first dorsal fin usually has a dark spot on the posterior half, and other fins are barred.

Taxonomy Staghorn sculpins show considerable variation in meristic characters, following a north-south cline with a distinct shift in populations from Monterey Bay south (11, 12). Consequently, C. L. Hubbs considered populations from Santa Cruz north to be in the subspecies *L. a. armatus* and southern populations to be in *L. a. australis*.

Names Lepto-cottus means slender sculpin, presumably after the narrow caudal peduncle; *armatus* means armed. Anglers often call them bullheads.

Distribution Staghorn sculpins are found in shallow coastal waters, especially bays and inlets, from the southern Bering Sea, Alaska, to San Quentin Bay, Baja California.

Life History Although staghorn sculpins are primarily estuarine and marine, their frequent occurrence in the lower reaches of coastal streams and their abundance in the upper San Francisco Estuary justifies their inclusion among freshwater fishes. They are often found in coastal lagoons that have become fresh when sandbars close them off from the sea during late summer. Even in streams, however, they are seldom more than a kilometer or two removed from salt water. In the San Francisco Estuary they are found at all salinities (0–34 ppt) but become progressively less abundant from San Francisco Bay to San Pablo Bay to Suisun Bay to

the Delta (10). Staghorn sculpins are truly euryhaline and often move freely between waters of varying salinities. In fresh water almost all are juveniles or newly mature adults, measuring mostly 2–14 cm TL but occasionally as large as 22 cm (1). Most are less than 2 years old. The probable reason for the predominance of juveniles is their greater tolerance of low salinities compared with adults or larvae (1). They are also tolerant of high salinities, rearing in salt ponds and marshes at salinities of over 67 ppt and at temperatures of over 25°C (2).

Most staghorn sculpins complete their entire life cycle in salt water, usually estuaries, and a relatively small fraction of the population moves into fresh water. Juveniles are most numerous in coastal streams and upper estuaries in spring (1). Once in fresh or brackish water, they gradually move upstream, and more (usually younger) juveniles move in from salt water to take the place of the upstream movers. The net effect of these movements is that the largest fish are found farthest upstream. At the upstream end staghorn sculpins tend to be associated with other euryhaline freshwater fishes (such as threespine stickleback, splittail, and prickly sculpin) and anadromous species (such as steelhead rainbow trout and Pacific lamprey). In lower reaches their most common associates are other euryhaline marine species, especially starry flounder, topsmelt, and shiner perch.

Milton Love (3, p. 218) notes that in estuaries and tidal marshes staghorn sculpins "follow the tides, coming into mud flats at high tide to feed. Not having accurate time pieces, the little bozos sometimes find themselves stranded on the flats. When this unfortunate event occurs, the sculpins bury themselves in the mud and await the next tide. Fortunately, staghorn sculpins can breathe air. Unfortunately, blue herons, which eat large quantities of staghorns, can search out buried ones, apparently with ease."

The dominant food of sculpins living in fresh water is bottom-dwelling euryhaline amphipods (*Corophium* spp.); invertebrates such as mysid shrimp, gammarid amphipods, nereid worms (*Neanthus* sp.), and aquatic insect larvae are of secondary importance (1, 4, 9). Small fish may be eaten on occasion as well (9). Aquatic insects are most important in the diet during times of high inflows to lagoons and estuaries. In Anaheim Bay and other saltwater environments young-of-year feed on a wide variety of prey items, but amphipods, decapods, and small fish (mainly small gobies) dominate the diet (1, 5). Adults feed more on crabs, shrimp, and fish. Staghorn sculpins feed at all times of day, but feeding is most intense at night (5).

Jones (1) found that there were three age classes of staghorn sculpin in Walker Creek, Marin County: young-of-year, yearlings, and 2-year-olds. Most young-of-year entered the stream at less than 40 mm TL. Few remained longer than a year; the largest fish encountered were 22 cm and 2 years old. In salt water they seldom live longer than 3 years or exceed lengths of 31 cm TL. In Anaheim Bay virtually all fish were young-of-year, growing from 15–20 mm SL to 120–150 mm SL in a year, roughly 9 mm per month (5). Growth is considerably slower in more northern populations. In Friday Harbor, Washington, sculpins the same size as yearlings in Anaheim Bay would be 4–5 years old, and a large adult (20–25 cm TL) is likely to be 10 years old (6). However, sculpins in Grays Harbor, Washington, had growth rates comparable to those of the Walker Creek fish (7). They are reputed to reach 46 cm TL, but any fish over 22 cm TL would be regarded as unusual in California (3).

Staghorn sculpins mature when they measure 12–15 cm SL, typically at 1 year of age (1, 10), so mature sculpins occur in fresh water (1). It is nevertheless unlikely any spawning takes place in streams. In the San Francisco Estuary most adults in low-salinity areas seem to move down into San Francisco Bay as they mature and then either die, move out of the Bay, or cease growing after spawning for the first time (10). They will spawn in more upstream areas during dry years (10). Spawning takes place from October to April, but it occurs mostly in January and February in California, in areas of stable salinity. Development of embryos requires salinities of 10–34 ppt but is optimal at 18–26 ppt (1). Depending on their size, females lay 2,000–11,000 eggs (1). Mean fecundity for 11 females (90–158 m SL) was 3,200 eggs (5).

Spawning behavior has not been described, but eggs are laid in clusters and are adhesive and demersal, which implies that developing embryos are protected by males. Development of the embryos and larvae has been well described, however (1, 8). Larvae measure 4–5 mm TL after hatching and are planktonic, swimming near the surface. They settle out at 10–15 mm TL, usually on sandy or soft-bottomed areas. The juveniles then (February–May) disperse widely, including into fresh water.

Status IE. Staghorn sculpins are one of the most common species in California's bays and estuaries and are frequently found in the lower reaches of streams along the entire coast. In the tidal reaches of streams in San Francisco Bay they are one of the most common fishes, often living in habitats degraded by channelization and pollution (13). Although often caught by anglers, they are seldom eaten. They are an important component of the food web of the San Francisco Estuary (10).

References 1. Jones 1962. 2. Morris 1960. 3. Love 1996. 4. Porter 1964. 5. Tasto 1975. 6. Wydoski and Whitney 1979. 7. Armstrong et al. 1995. 8. Wang 1986. 9. Martin 1995. 10. Baxter et al. 1999. 11. Hubbs 1921. 12. Morris 1977. 13. R. Leidy, USEPA, pers. comm. 1999.

Striped Basses, Moronidae

Striped basses are a small family of piscivorous fishes within the giant order Perciformes, an order regarded by some as the pinnacle of teleost evolution. The Moronidae seem to represent a rather generalized condition in the order, with their moderately deep bodies, large mouths, 2 dorsal fins (8–10 spines, 10–13 rays), and anal fin with 3 spines and 9–12 rays (Nelson 1994). They were once considered to be members of the Serranidae, a family that had become a taxonomic garbage can for the more generalized perciform fishes (Gosline 1966), and later part of the Percichthyidae, a slightly smaller garbage can. Today there are just six species in the family: four in the genus *Morone* in North America and two in the genus *Dicentrarchus* in Europe and North Africa. All species are tolerant of a wide range of salinities, although some live mostly in salt water and others largely in fresh water, whereas others move freely between the two environments. All move about in voracious schools, pursuing schools of small pelagic fishes. All are favored sport and commercial fishes, famous for their culinary qualities.

Two species of *Morone* have been introduced into California, the striped bass and the white bass. They can be distinguished from other spiny-rayed fishes in the state by the presence of a small gill (pseudobranch) on the underside of each gill cover, the separation of the spiny- and soft-rayed portions of the dorsal fin, a complete lateral line, 1–2 spines on the operculum, and narrow, horizontal black stripes on the sides.

Striped Bass, *Morone saxatilis* (Walbaum)

Identification The streamlined, silvery-white body—with its 6–9 black horizontal stripes and sharply separated spiny- and soft-rayed portions of the dorsal fin—makes adult striped bass instantly recognizable. The body is deepest below the gap between the 2 dorsal fins. The tail is pointed at the tips and slightly forked. There are 9–10 spines in the leading half of the dorsal fin, and 1–2 spines and 11–12 rays in the following half. The anal fin has 3 spines and 9–11 rays; the pectoral fin, 13–17 rays; and the lateral line, 53–65 scales. There are 2 small but distinct spines on the operculum. The mouth is terminal and large, but the maxilla does not reach past the hind margin of the eye. The tongue has 2 distinct patches of teeth on its surface. The eye is moderate in size, less than one-fourth of the head length.

Taxonomy The striped bass is a well-defined species, with genetically defined races in various parts of its native range. Striped bass on the West Coast had a limited geographic origin and so are presumably fairly uniform genetically, although it would be interesting to see how much, if any, they have diverged from the source populations.

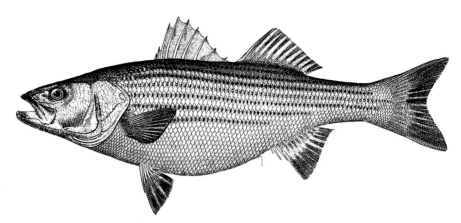

Figure 122. Striped bass, 41 cm SL, Washington, D.C. USNM 25219. Drawing by H. L. Todd.

Names The generic name for striped bass ("stripers") has changed back and forth over the years between *Morone* and *Roccus*. It has remained *Morone* since the American Fisheries Society recognized the name. Unfortunately, Samuel Latham Mitchill (1764–1831), who first used *Morone* in his 1814 book *Fishes of New York,* neglected to explain how he arrived at the word. *Saxatilis* means "living among rocks," apparently a reference to the rather inappropriate common name rockfish, widely used on the East Coast. Similarly, *Roccus* is a Latinization of the word rock.

Distribution Striped bass are native to streams and bays of the Atlantic coast, from the St. Lawrence River in the north to the St. Johns River, Florida, in the south, and into streams, bays, and estuaries connecting to the Gulf of Mexico from Florida to Louisiana. They also range widely in the ocean along the Atlantic and Gulf coasts of North America. They are commonly introduced into reservoirs within or adjacent to their native range. They were first introduced to the Pacific coast in 1879 when about 135 fish from the Navasink River, New Jersey, were planted in the San Francisco Estuary. In 1882 an additional plant of about 300 fish from the Shrewsbury River, New Jersey, was made (1). So successful was the introduction that by 1888 a commercial fishery for bass had started up; the catch reached more than 1.2 million pounds by 1899 (2). They have since been found in salt water from 25 miles south of the Mexican border to southern British Columbia. They are most abundant and widely distributed in the ocean and estuaries during years of the El Niño phenomenon, when ocean temperatures are warmer.

The main breeding population, however, is still in the San Francisco Estuary, although a smaller population is present in Coos Bay, Oregon. In the Sacramento Valley they regularly penetrate upstream as far as barrier dams, such as Folsom Dam on the American River, Daguerre Point Dam on the Yuba River, or Red Bluff Diversion Dam on the Sacramento River. An apparently self-reproducing landlocked population now lives in Millerton Reservoir, Fresno-Madera Counties. Another population in San Luis Reservoir, Merced County, is continually replenished with small bass pumped in through the California Aqueduct from the Delta. In fact every reservoir in southern California fed by the aqueduct supports striped bass (and a fishery for them), as does the aqueduct itself. Reproducing populations are also established in the lower Colorado River, the result of transplants from the San Francisco Estuary from 1959 through 1964 (1). As a result striped bass are found down the river into Mexico and in the various canals that take water from the river in California. Striped bass are also raised in hatcheries and planted in various California reservoirs, mainly on rivers flowing into the Central Valley.

Life History Striped bass are one of the most thoroughly studied fish in the eastern United States and in California, so this short account is mostly a summary of more detailed reviews (2, 3, 4, 5, 6).

Striped bass move regularly between salt and fresh water, and they usually spend much of their life cycle in estuaries. It is not surprising, therefore, to find that they are remarkably tolerant of a wide range of environmental conditions. Adults and juveniles can survive temperatures as high as 34°C for short periods of time, although they are under stress once temperatures exceed 25°C, and temperatures over 30°C are usually lethal. Adults are capable of withstanding abrupt temperature changes (up to 27°C) that are simultaneous with shifts from seawater to fresh water. Younger fish are less tolerant of such changes. They can also withstand low oxygen levels (3–5 mg/liter) for short periods, as well as high turbidity, although extreme conditions inhibit reproduction. Besides these rather broad water quality requirements, striped bass have three basic requirements for successful completion of their life cycle: *(1)* a large cool river for spawning, with sufficient flow to keep embryos and larvae suspended off the bottom until they reach the estuary and become free-swimming; *(2)* a large body of water (e.g., San Francisco Bay, the Pacific Ocean) with large populations of small fishes for forage; and *(3)* a productive estuary where larval and juvenile striped bass can take advantage of large invertebrate populations. In California only the San Francisco Estuary has satisfied all these conditions, although small landlocked populations maintain themselves in Millerton Reservoir (they use the upper San Joaquin River for spawning) and in the lower Colorado River.

Atlantic populations of striped bass are focused on large bays, but large fish, mainly females, move into the ocean in summer and make extensive migrations along the coast, foraging on abundant shoaling fishes. In contrast, California striped bass usually spend most of their lives in San Pablo and San Francisco Bays. During El Niño years, when ocean temperatures are warmer and anchovies and other prey are abundant close to shore, large bass move from the estuary into the ocean, often traveling long distances both north and south. There is a general movement of adult bass out of bays into fresh water in fall. Many spend winter in the Delta and move back into salt water in spring following the upstream spawning migration.

Striped bass are gregarious pelagic predators. This life style is reflected in their streamlined body shape, silvery coloration, and feeding habits (7). Larval and juvenile striped bass are primarily invertebrate feeders. Larval and postlarval bass feed mainly on copepods, historically principally *Eurytemora affinis* but after the mid-1980s various alien species. Young-of-year (<10 cm FL) rely mostly on opossum shrimp, *Neomysis mercedis* and, increasingly, *Acanthomysis* spp., although amphipods, copepods, and small threadfin shad may be important foods on occasion. The diet of larger juveniles (10–35 cm FL) is similar to that of young-of-year, but fish are increasingly important as bass increase in size (42). Subadult bass (age 2+, 26–47 cm FL) are primarily piscivorous; invertebrates can be important in winter and spring when small fishes are hard to find. In the Delta adults feed mostly on threadfin shad and smaller striped bass, whereas in San Pablo Bay and the Pacific Ocean they take a wide variety of pelagic fishes (e.g., anchovies and herring) as well as bay shrimp (*Crangon* spp.).

Despite the seemingly limited nature of the striped bass diet, they are rather opportunistic feeders, and almost any fish or invertebrate found with them sooner or later appears in their diet, depending on time and place. Thus in the Sacramento River adult striped bass feed largely on juvenile salmon (8), whereas in the American River they feed largely on crayfish and various native fishes (9). In Suisun Marsh large bass frequently feed heavily on threespine sticklebacks coming out of marsh drains. Adult bass often hang out near screened diversions, feeding on small fish, especially salmon, that concentrate near them. They are a major source of mortality of juvenile salmon and other fish entrained by the State Water Project pumps of the South Delta. They prey both on fish entering the fish rescue facility (in Clifton Court Forebay) and on fish that are trucked back to the Delta after being salvaged.

Growth is most rapid during the first 4 years and is also highly variable, depending on food supplies. In the estuary they typically reach 9–11 cm FL in the first year, 23–30 cm in the second year, 28–43 cm in the third year, and 44–54 cm in the fourth year, with growth increments of 5–10 cm/year thereafter. Fish over 10 years old and 85 cm are uncommon, but in the 1920s and 1930s 16- to 20-year-old bass were recorded that measured nearly 110–120 cm FL (10, 11). Growth in Millerton Reservoir is somewhat faster for the first few years, so by the end of the fourth year Millerton bass typically measure 55–56 cm FL. Striped bass will reputedly reach about 125 cm FL (41 kg) in California, and bass measuring 180 cm FL (56 kg) have been recorded from the Atlantic coast. The angling record for the state is a 30.6-kg bass taken from O'Neill Forebay (Merced County) in 1992. Large striped bass are difficult to age using either scales or otoliths, but the maximum age seems to be in excess of 30 years. The oldest (over 10 years) and largest bass are invariably females.

The age of maturity for females is 4–6 years. A few males may mature at the end of their first year, but most of them wait until they are 2–3 years old. As a result males typically measure 25 cm FL when they spawn for the first time, and females measure about 45 cm FL. Female bass are very prolific, and fecundity increases dramatically with size. Thus females in the estuary spawning for the first time at age 4 contain on the average 243,000 eggs, whereas females age 8 and older average 1.4 million eggs (12). Females are capable of spawning every year if conditions are right. The maximum fecundity seems to be around 5 million eggs. Large females not only produce more eggs than small females, they also produce larger eggs, with more yolk and oil, suggesting that their larvae should have higher survival rates (13).

Spawning may begin in April when bass, usually males first, start to move into suitable areas. In the eastern United States, there is some evidence that striped bass home to ancestral spawning grounds (36). Spawning peaks in May and early June. The exact time and location of spawning depend on the interaction of three factors: temperature, flow, and salinity (14). No spawning will occur until temperatures reach at least 14°C. Optimum temperatures appear to be 15–20°C, and spawning will cease above 21°C. In the Sacramento River most spawning occurs anywhere within a roughly 70-km reach starting above Colusa (about river km 195) and ending below the mouth of the Feather River (about river km 125). When flows are high water takes longer to warm up, so spawning takes place farther upstream than usual, because bass migrate upstream while waiting for temperatures to rise. It also takes place later in the year. Bass seeking a place to spawn may also be attracted to large outflows of agricultural return water from Colusa

Drain, which is significantly warmer than Sacramento River water and often laden with toxic materials (15). During wet years spawning may take place in the Sacramento River portion of the Delta. In the San Joaquin River successful spawning upstream of the Delta occurs mainly during years of high flow, when the large volume of runoff dilutes salty irrigation waste water that normally makes up much of the river's flow. In years of lower flow spawning occurs in the Delta itself. Because of interactions among these factors there are two main spawning areas in the Delta: the Sacramento River from Isleton to Butte City and the San Joaquin River and its sloughs from Venice Island down to Antioch. Most spawning, however, takes place in the Sacramento River.

Striped bass are mass spawners. In the Sacramento River thousands of large bass aggregate close to banks just off the main current (16). Groups of 5–30 fish, predominately males surrounding one or two females, break away from the main group and swim out into the main river, close to the surface. During the spawning act the group mills about. Individuals frequently turn on their sides, accompanying this action with vigorous splashing at the surface. Although spawning can occur at any time of the day, peak activity is usually during the late afternoon or early evening.

The newly fertilized eggs are slightly heavier than fresh water, so they slowly sink. If embryos remain on the bottom for any length of time they will not survive, but even a slight current will keep them suspended. They hatch in about 48 hr at 19°C. Larvae depend on their yolk sacs for nourishment for the next 7–8 days (17). As they become more capable of swimming, they begin feeding on small zooplankters. During this early period the embryos and larvae in the Sacramento River are carried into the Delta and Suisun Bay. In the San Joaquin River outflow is balanced by tidal currents, so that embryos and larvae stay suspended in the same general area in which spawning took place. Essentially, larval bass from both rivers are most abundant where salt and fresh water meet. Thus when they begin to feed they are concentrated in the most productive portions of the entire estuary. Larval growth and survival rates are also highest in areas of brackish water, presumably because of reduced energy costs for osmoregulation. Larvae often make vertical migrations to take advantage of the opposite directions in which riverine and tidal currents flow, to maintain themselves in food-rich areas. Larval striped bass swim rapidly compared with other fish larvae (3–4 body lengths per second) and need fairly dense concentrations of zooplankton to satisfy their high metabolic rates (38). Even as larvae they are voracious predators!

Survival of bass through the first year of life appears to depend in part on adequate river flows carrying them to the best places for rearing (12, 23), usually in Suisun Bay, although even this relationship has not been strong in recent years (39). Curiously, there is little connection between sur-

vival through the first year of life and number of adults in a year class, except when bass numbers are low, unlike striped bass populations on the Atlantic coast (25). This observation indicates that there is a bottleneck in survival of juvenile bass between the end of their first summer and the end of their second year (age 3) (39). The most likely causes of high mortality at this stage are predation and shortages of food.

Status IID. Striped bass are one of the most abundant fish in the San Francisco Estuary and are widespread along the Pacific coast. They are the most important sport fish in the estuary (37). Nevertheless they are much less abundant than they were during the first 75 or so years following their explosive invasion of the region. Their decline has been a matter of great concern because of their value as sport fish, and for several decades decisions regarding management of the estuary were often made as if they were the only fish that mattered. Their importance in management was partly due to the fact that there was more information available on their biology than on that of any other fish in the estuary. Management agencies made a basic assumption that an estuary managed for striped bass would favor other desirable species as well—an assumption that has proven to be, at best, only partially true. This is not surprising considering that the life history of the bass is quite different from that of any native species. Because of the importance of the striped bass and its fishery in the politics of California water, I discuss here four subjects: *(1)* the history of striped bass and its management, *(2)* why striped bass populations have declined, *(3)* their impact on native fishes, and *(4)* the future of striped bass in California.

History of striped bass and its management. The history of striped bass in California is one of the great success stories of fish introductions: a small introduction resulted in a major fishery that has persisted for over 125 years. In 1880, one year after the introduction, the first striped bass was caught (and eaten) from San Francisco Bay; within 10 years a commercial fishery had developed (1). The reasons for its spectacular success are not really known, but the availability of the Sacramento River as a near-optimal spawning stream was clearly a key factor. The suitable spawning areas were sufficiently restricted that even the small number of bass initially present had a high likelihood of finding one another once they started seeking a place to spawn (18). Equally important, striped bass, because of their semibuoyant embryos and pelagic larvae, could spawn successfully in the river despite its huge sediment loads from hydraulic mining in the Sierras. Most native species deposited eggs on the bottom, where they were easily smothered. The bass also came into an environment with abundant prey at all levels, from zooplankton to shrimp to fish (including small salmon), but without a resident schooling pelagic predator. Its main fish-eating rivals were sluggish Sacramento perch and thicktail chub, slow-growing pikeminnow, and cold water–requiring steelhead, none of which had the predation mechanisms or metabolic requirements that make striped bass both voracious and fast growing. The bass also had the advantage of being able to prey heavily on its own young and to switch readily to other alien species as they became established.

The commercial fishery for striped bass in the estuary lasted until 1935, when it was banned in favor of the growing sport fishery, after being subjected to increasingly severe regulation (1). By that time the sport fishery catch already exceeded the commercial catch, and it was thought that the best way to avoid overexploitation was to shut down the commercial fishery. This strategy seemed to work, and the sport fishery continued to sustain itself without any serious management except fishing regulations. The fishery even spread to the ocean during El Niño years and to estuaries on other parts of the coast, including Coos Bay, Oregon, where another breeding population became established. In an effort to expand the fishery further, numerous transplants (mainly in 1899–1933) were made into both marine and inland waters, with little long-term success (1). Successful transplants were made into Millerton Reservoir and the Colorado River (both highly altered environments) in the 1950s.

Despite these efforts, there were signs that the bass population was declining. The number of striped bass per angler, although not a particularly reliable statistic, declined fairly steadily, from 20 fish in the 1930s to 10 fish in the 1940s to fewer than 10 fish in the 1950s, at which point more restrictive angling regulations were instituted (2). In 1959 a major study of striped bass and other fishes in the estuary began in connection with the development of large state and federal water projects. In the early 1960s the bass population was estimated to be 2–3 million adult (legal size, >45.7 cm TL) fish (18). From 1969 to 1976 it hit a plateau of an estimated 1.5–1.9 million adult fish (18, 37).

The 1960s and early 1970s encompassed the period in which the State Water Project joined the federal Central Valley Project to further dramatically change the hydraulics of the estuary and its inflowing rivers. Therefore it was logical to assume that the projects were responsible for the decline, both by altering estuarine hydrodynamics and by sucking up embryos and larvae into the pumps (which were screened to prevent the entrainment of larger fish). These presumed effects were the reason for Decision 1379 of the State Water Resources Control Board in 1970, which directed that further studies be made to figure out what to do about the decline. The water agencies agreed to fund fish agencies to carry out the studies. In the interim D1379 also set minimal requirements to reduce pumping during times when larval striped bass were present, in addition to weak temperature and salinity standards to protect striped bass spawning in the San Joaquin River as well as opossum shrimp, an important food for striped bass (19). In 1978 Decision 1485 superseded D1379 as the main directive managing the estuarine ecosystem; it relied on abundance of 38-mm striped bass as the universal standard, on the weak assumption that there was a strong relationship between postlarval abundance and adult abundance. The goal was to bring striped bass numbers back to preproject levels by placing various operational constraints on state and federal pumping facilities. It didn't work. The number of adult striped bass dropped to less than a million by 1977 and continued to decline thereafter. By 1994 the number of legal-size bass was less than 580,000 (37). Simultaneously the 38-mm striped bass index dropped to record low levels and never recovered to numbers approaching historical levels.

The 38-mm striped bass index was used as the standard because for the first 15 years of study (1959–1975) it had a high correlation with Delta outflow, which in turn was partly related to the amount of water diverted. This relationship seemed to disappear after 1975, although it later became apparent that the correlation still existed, but at much lower numbers for the index. However, after 1989 the relationship became nonexistent, as index numbers continued to tumble despite wet years (40, 41).

Despite the continued indications of low production of juveniles, the numbers of legal-size striped bass have increased since the 1994 low. By 1998 they were estimated to number over 1.3 million, approaching the levels in the 1970s (41). The reasons for the unexpected increase are not known, but it appears to be related to increased survival of juvenile bass (bigger than those sampled to create the 38-mm index). There is no indication that the hatchery program (discussed later in this section) has contributed to the increase.

There were three main responses to the decline of striped bass:

1. Research into alternative or additional causes of decline expanded. This is the subject of the next section of this account.

2. Proposals were made to keep tinkering with water project operations to improve conditions for bass. Thus CDFG supported construction of the giant Peripheral Canal to carry Sacramento water around the Delta to the pumps, because they thought it would improve Delta hydraulics in ways that would favor striped bass. A proposition to build the canal was defeated by voters in 1982. Attempts to improve water project operations to favor bass were largely discarded following federal listing of winter-run chinook salmon as threatened in 1989 and delta smelt as threatened in 1993.

These steps forcibly brought to everyone's attention the fact that managing the estuary for striped bass—and even doing so badly—was not necessarily good for other species, especially native species. Striped bass therefore became a lower priority for beneficial project manipulations.

3. Bass were reared in hatcheries to supplement wild populations.

Hatchery rearing of striped bass in California was tried unsuccessfully in 1907–1910 and then not again until 1981 (1). For the next 10 years private aquaculturists were contracted to produce 11 million fingerlings and yearlings, which accounted for up to 30 percent of the bass population in some years (1989, 1990) (37, 41). The cost of each bass of hatchery origin caught by an angler was estimated by Dill and Cordone (1) to be $106 for fish planted as yearlings, $237 for those planted as advanced fingerlings, and $1,071 for those planted as fingerlings! More important, there was no indication that these fish affected the continuing decline of the bass population. If anything, the hatchery fish may have enhanced the decline by preying on their smaller wild cousins. Hatchery fish at a given age (for the first year or so at least) are larger than wild fish, so are no doubt preying on them, just as wild striped bass prey on their own young. In any case it is a rather peculiar strategy to enhance the populations of a top predator when the prey base is also in decline, which has generally been the case; the striped bass is only one of many species in decline in the estuary (20). Biologically the main justification for a rearing program for striped bass is that the population appears to be recruitment limited, as indicated in the next section.

In 1992 the hatchery program came to an abrupt halt when it was canceled by CDFG Director Boyd Gibbons, to his credit. His rationale was that, if even one bass of hatchery origin ate just one winter-run chinook salmon, CDFG would be in violation of the state and federal endangered species acts; therefore rearing bass was both legally and morally untenable. After Gibbons' sudden departure from the agency, planting of striped bass continued, beginning in 1993, using bass that were salvaged from the state fish trap in front of the SWP pumps and reared to larger size in net pens (1, 37). These pen-reared bass now account for about 2 percent of the adult population (41). It is not known, however, if this program actually increases the striped bass catch or if the hatchery fish are just replacing wild fish that have been eaten by larger hatchery fish. Despite its cost, doubtful effectiveness, and potential negative effects on endangered species, this program, funded by striped bass anglers, expanded to over a million fish per year (21, 37). The goal, according to a CDFG press release, is to "stabilize and restore the estuary's striped bass fishery." In recognition of the fact that striped bass do, on occasion, eat endangered species,

such as chinook salmon and splittail, a permit to rear the bass is required under Section 10 of the federal Endangered Species Act. The permit allows CDFG to continue to rear striped bass in exchange for reducing losses of fish by screening diversions in Suisun Marsh and the Sacramento River, and to monitor striped bass predation (41).

Why striped bass have declined. The decline of striped bass is clearly related to a number of factors acting simultaneously on different life history stages (18, 20). The causes are a mixture of those general to the fishes of the estuary and those specific to striped bass. The basic categories of interrelated causes include *(1)* climatic factors, *(2)* south Delta pumps, *(3)* other diversions, *(4)* pollutants, *(5)* reduced estuarine productivity, *(6)* invasions by alien species, and *(7)* exploitation.

Climatic factors. Survival of striped bass through the first year of life once had a strong relationship to outflow of fresh water (22, 23) and may still have a weak one (38). When outflows are high (wet years) survival is relatively high; when they are low (drought years) survival is relatively low. Thus natural fluctuations in climate can have dramatic effects on the striped bass population, especially when human demands for water accentuate natural fluctuations in outflows. Striped bass became established at a time when natural conditions were favorable, a succession of wet years in the late 19th century. Drought in the 1930s, heavy fishing, decline of salmon as a source of food, and diking and draining of the estuary presumably all contributed to their decline from the initial high numbers. Since about 1980 the climate in the region has been extremely variable, with long periods of drought and some exceptionally high outflow events as well. These events, probably related to human-induced global climate change, have decreased environmental predictability for fish, perhaps reducing survival. They have also occurred during a period of extreme human perturbation of the estuary and its inflowing rivers, increasing the likelihood that additional factors—such as toxic materials, decreased food abundance, or increased cannibalism (from hatchery fish)—will have negative effects on striped bass populations.

Since the mid-1970s changing ocean conditions have had a major effect on striped bass in the estuary (24). The frequency of ENSO events has increased, coinciding with a longer natural pattern of warming (the Pacific Decadal Oscillation). During these ENSO periods the ocean off the Golden Gate is 1–3°C warmer during much of the year, and upwelling decreases. With decreased upwelling, anchovies, herring, and other plankton-feeding fishes move closer to shore. When this happens large striped bass, mostly females, move out of the estuary, just as they do in their native habitats, to take advantage of warmer temperatures and abundant prey. When the ocean is colder and abundant prey are farther from shore, the bass apparently choose not to go out to sea. Once they leave the estuary, many (perhaps most) of the large females do not return, either because they are caught by shore and party boat anglers or because they have wandered into other estuaries. Although this scenario is based on correlations among diverse variables, a number of basic observations enhance its credibility:

1. It fits the behavior of bass in their native range.

2. The sudden downward shift in the 38-mm index in the 1970s coincides with the onset of changed ocean conditions.

3. Past colonization events for bass in other estuaries coincide with El Niño years, as does increased ocean catch.

4. The number of large bass has declined more dramatically than would be predicted.

5. The decline of large bass seems to be related to "natural" mortality rather than to catch in the estuary.

6. The recent striped bass decline is related at least in part to low egg supply or survival, which would fit with the disappearance of large females.

South Delta pumps. The large pumps of the CVP and SWP in the south Delta at times pump more fresh water than flows into San Francisco Bay. Even at less dramatic levels of diversion, pumping can significantly reduce Delta outflow, with numerous potential effects on bass, such as reduction in nursery areas, reduced productivity (less food), less dilution of pollutants, decreased turbidity (resulting in higher predation losses), and increased danger from entrainment (18). During the 1970s and 1980s pumping increased steadily, entraining millions of larvae and small juveniles. Large numbers of bass of all sizes have also been captured at the fish screens by the pumps and trucked back to the Delta. It is likely that many entrained fish do not survive the experience because of predation by larger bass, either before they enter the facility or immediately after they are dumped back into the Delta. Observations like these, combined with high correlations between striped bass young-of-year abundance and Delta outflows, led to the conclusion that the pumps have been the single biggest cause of striped bass declines, especially when combined with reduced flows from water being retained upstream by dams (12). The basic scenario is that increased entrainment leads to decreased recruitment into the adult population; decreased adult abundance leads to fewer eggs being produced, and this causes still lower larval and juvenile abundance. In short the pumps generate a downward spiral of an ever-decreasing population. However, the relationship between the abundance of young-of-year bass and adult pop-

ulations is weak at best (25). In addition, the failure of the outflow–bass abundance relationship in recent years suggests that other factors are now more important in regulating bass numbers than entrainment, although it is presumably still a contributing factor (20, 25).

Other diversions. To survive, larval and juvenile striped bass not only have to avoid the big pumps in the South Delta, they also have to avoid hundreds of small diversions, mostly unscreened, along the rivers and in the Delta. Fortunately, the diversions pump intermittently and probably do not take many small bass during most years, especially early in the season when agricultural water demand is low and flows are high. Larger diversions—such as the North Bay Aqueduct that sucks water out of Cache Slough (which connects to the Sacramento River) or the cooling water intakes of large power plants on the south side of the estuary—may be more of a problem. It is likely that these sources together kill large numbers of small fish, especially in low-flow years, but it is generally assumed that their impact on bass populations is small compared with other factors. One factor working in favor of larval bass is that the spawning behavior of adults results in most of them being swept quickly to the brackish waters of Suisun Bay, below diversions.

Pollutants. Pollutants have affected striped bass survival ever since they were first introduced. The bass may in fact have *benefited* from the big pollutant of the 19th century, sediment from the gold fields, because their embryos are semibuoyant and so would not have been smothered, as would the benthic embryos of potential predators and competitors. Yet as the 20th century progressed, their waters became increasingly polluted with organic wastes from sewage and agriculture. Initially these wastes may have increased productivity and bass food supplies or replaced nutrients lost through diking and draining of marshes and Delta islands. However, by the 1950s and 1960s the estuary was becoming an organic soup, increasingly hostile to fish life, in particular because more water was being stored behind dams, decreasing dilution. In addition to organic wastes there were a myriad of industrial wastes, including heavy metals and other toxicants. This heavy pollution may have contributed to the decline of bass during this period. Following the passage of the federal Clean Water Act in 1972, followed by the state Porter-Cologne Act, better sewage treatment plants were built in all surrounding cities, dumping of industrial waste was curbed, and water quality in the estuary improved markedly. Unfortunately, as organic and industrial pollution decreased, the input of toxic chemicals, mainly from agriculture, increased. The use of pesticides in California skyrocketed after World War II, initially with organochlorines such as DDT and then with a cocktail of herbicides and insecticides. In addition thousands of acres of desert land were put under irrigation in the San Joaquin

Valley, resulting in drainage water laden with salts and heavy metals toxic to striped bass and other fish (35). Heavy metals and pesticides can have both direct and indirect effects on the fish.

Direct effects occur when the toxic material kills fish outright. There are surprisingly few records of major kills of juvenile and adult bass, despite their frequent exposure to pesticides. Kills of larval bass are probably more frequent because of their susceptibility and occurrence in pesticide-laden water. Thus Bailey et al. (15) present evidence that Sacramento River water was frequently toxic to bass larvae, because of pesticides draining from rice fields, and that increased pesticide use during the period of recent bass decline seems tied to the decline. The effects were likely to be most pronounced during periods of low flow, when dilution is less (20).

Indirect effects are often the result of accumulation of toxic materials by the fish, stored in fatty tissue. Striped bass, at the top of the food chain and long-lived, are especially prone to the bioaccumulation of toxic substances in their tissues. One consequence of this problem is that regular consumption of striped bass by humans is not advised by health authorities (26). High concentrations of toxic materials in fish can impair reproductive function, decrease embryo and larval survival (through toxins passed into the egg), or even be lethal during times of stress. For example, high levels of toxic materials in the liver of bass have been tied to annual summer die-offs of large fish (27). Likewise, Bennett et al. (28) found that about a third of bass larvae in the Sacramento River showed signs of liver damage, presumably from rice herbicides, that would ultimately be lethal to the fish. When pesticide use changed and fewer larvae with damaged livers were present, overall larval survival nevertheless did not improve. Overall, pollutants presumably have a continuous, if erratic, impact on striped bass populations, but their actual effects are difficult to separate from those of other stressors.

Reduced estuarine productivity. Estuarine productivity may be much less than it was historically, because the estuary has been closed off from much of its presumed historical sources of nutrients: marsh and riparian systems, and, later, sewage. Thus the long-term decline of bass from 19th-century levels may reflect a fundamental change in energy and nutrient flow through the ecosystem. Kimmerer et al. (25) have noted that for the most part the number of young-of-year bass at the end of their first summer has no relationship to the number of bass entering the fishery at age 3. The numbers of older bass have in fact been consistently low, indicating that most bass die before they reach their third year. This period of mortality is also the period during which they are highly dependent on zooplankton and mysid shrimp for food. Reduced availability of food may result in starvation or increased susceptibility to predation. It

should also be reflected in increased rates of larval starvation (because of decreased zooplankton abundance), decreased fullness of juvenile stomachs, and decreased growth rates of fish at all ages. None of these outcomes has yet been observed for striped bass (28, 29), but they have been for other species (such as delta smelt), implying a general decline in the ability of the estuary to support fish. Thus reduction in productivity may have reduced the carrying capacity of the estuary for striped bass (25).

It is possible that striped bass populations are unusually responsive to reduced productivity because adults depend heavily on juvenile bass as food. During times of shortages of alternate prey, juvenile bass may become increasingly important in adult diets (including the diets of hatchery-reared bass), resulting in decreased survival rates of juveniles.

Invasions by alien species. The San Francisco Estuary has been labeled the "most invaded estuary in the world" because of the hundreds of species of alien invertebrates, plants, and fish that have become established in the past 150 years (30). One of the first major invaders, of course, was the striped bass, and it probably caused major changes to the estuarine ecosystem in its role as the most abundant piscivore. As a firmly established species, it is now vulnerable to the effects of new invaders. Indeed the rate of invasion, mainly by species carried in the ballast water of ships, has increased during the period of sharp striped bass decline. Some of the invaders have significantly affected the food supply of larval and juvenile bass. The most dramatic of these has been the overbite clam, *Potamocorbula amurensis,* which arrived in the 1980s. Its extraordinary numbers in Suisun and San Pablo Bays have severely reduced phytoplankton and zooplankton densities, decreasing the amount of food available to larval and juvenile striped bass. At the same time there have been major changes in the types of zooplankton eaten by bass. The species of copepod once dominant in the diets of small bass, *Eurytemora affinis,* has been largely replaced, at least seasonally, by alien copepods, which may be energetically less desirable (31). Likewise, opossum shrimp, *Neomysis mercedis,* may have been largely replaced in the diets of small bass by smaller *Acanthomysis* spp. (32), and the two species have a combined abundance much lower than historical levels of *N. mercedis* alone. Curiously, despite these changes evidence for food limitation in striped bass is only indirect, although it might be reasonably expected (18, 25). This story is subject to change, however, following the establishment of whatever major invader will arrive next, or after further study.

Exploitation. Harvest of striped bass, both legal and illegal, has likely been a contributing factor to the decline since at least the 1930s, mainly because harvest focuses on the largest fish, which are females. Removal of even a few exceptionally large and highly fecund females from the population has the potential to reduce recruitment in future years, especially when populations are declining and conditions for survival of young are poor (18, 20, 24). The main response to this problem has been to regulate the bass fishery to reduce legal catch in the estuary and to find ways to reduce poaching (which in fact has not been demonstrated to be a major problem). There is no sign that regulations and enforcement have increased bass populations, although they may have slowed the decline. Overall the annual harvest of striped bass has declined since the 1970s (12, 37).

Multiple and changing causes. Striped bass did not evolve to live in the unusual conditions of the San Francisco Estuary, so their initial establishment and extraordinary abundance must be regarded as an unusual event. Even if estuarine conditions had remained constant, striped bass populations would probably have declined as the estuarine ecosystem adjusted to their presence, especially through reduction of prey species, increase in predation on larval and juvenile bass, and increased exploitation. As it happens, the estuary has gone through a complete transformation, from a system dominated by natural processes and native species to one dominated by human-influenced processes and alien species. The rate of change accelerated in the latter half of the 20th century, when striped bass declines became most pronounced. Perhaps we should be surprised that the striped bass has done as well as it has, considering all the changes.

It is likely that all the factors listed here have contributed to the long-term decline of striped bass, with different factors having different importance at different times or acting in concert with one another. It is likely that the decline since the 1970s has resulted from multiple factors affecting recruitment, but especially those related to flow, juvenile survival, and egg production by large females (20, 25). However, even if these many layers of problems were all resolved, there is no guarantee that striped bass populations would bounce back. Conditions in the estuary and in the ocean may have changed in an irreversible fashion in ways that are less favorable to striped bass survival.

Impact on native fishes. The major impact striped bass had on native species, especially salmon, presumably took place after their initial establishment as voracious predators capable of eating their way through large populations of juvenile salmon and other species. They may have had major responsibility for the extinction of thicktail chub and Sacramento perch, through predation and competition, but we have no way of knowing for sure. Although chinook salmon declined in the Central Valley as bass increased, there was also a virtually unregulated fishery for salmon at the same time, and hydraulic mining was devastating to many salmon spawning and rearing habitats. It is likely that striped bass continue to be an important predator on small salmon and

that the decline of striped bass may have assisted recent increases in some salmon populations. On the other hand, large populations of other native fishes, such as delta smelt, longfin smelt, and splittail, thrived when bass were abundant, suggesting that they are capable of coexistence.

What we do not know is whether these species, now mostly depleted, can recover their populations in the presence of a large population of striped bass. For example, it has been estimated that 63–99 percent of juvenile salmon that are drawn into Clifton Court Forebay, just before hitting the screens of the SWP pumps, are consumed by striped bass, exacerbating the impact of the diversion (34). Problems like this provide a good argument for not artificially enhancing bass populations or for not managing the estuary in ways that favor bass over other species. A large population of bass, for example, could devastate a small population of salmon. It is worth noting that striped bass mostly spawn later than native fishes, so actions to benefit them (increasing outflows, decreasing pumping) are not likely to have much benefit for reproduction of native fishes.

The future of striped bass in California. The striped bass is primarily an Atlantic coast fish. There it is adapted to life in dozens of estuaries, chasing schools of small fish along the entire Atlantic coast while increasing its legendary status among anglers (33). Its fishery on the West Coast can never be anything but a pale imitation of that on the Atlantic shore. Although striped bass are not going to disappear from California, it is clear that the fishery will never again approach the extent of its halcyon days. The striped bass is a very resilient species, and it is now a permanent part of the California fish fauna and of the San Francisco Estuary ecosystem. The best thing that can be done for striped bass is to restore the estuary to a condition that allows it to support more fish of all kinds, but especially native species. The best thing to do with hatchery-reared striped bass is to plant them in reservoirs. Striped bass are clearly a good sport fish in reservoirs because they thrive on large populations of threadfin shad and other species and usually die out if they prove to be undesirable for any reason.

Because striped bass do maintain populations in some reservoirs, even reservoir planting should be undertaken with caution because of potential negative effects on existing populations or increased predation on native fishes. For example, they seem to be a permanent part of the fish fauna of reservoirs of the Colorado River, long after planting has stopped. In the river they have become part of the pantheon of predators that consume (or potentially consume) native minnows and suckers.

References 1. Dill and Cordone 1997. 2. Skinner 1962. 3. Raney 1952. 4. Turner and Kelley 1966. 5. Stevens et al. 1987. 6. Hassler 1988. 7. Stevens 1966b. 8. Tucker et al. 1998. 9. DeHaven 1977.

10. Scofield 1931. 11. Clark 1938. 12. Stevens et al. 1985. 13. Zastrow et al. 1989. 14. Turner 1972. 15. Bailey et al. 1994. 16. L. Miller and McKechnie 1968. 17. Wang 1986. 18. Herbold et al. 1992. 19. CDFG 1987a. 20. Bennett and Moyle 1996. 21. Retallack 1998. 22. Stevens 1977. 23. Jassby et al. 1995. 24. Bennett and Howard 1997, 1999. 25. Kimmerer et al. 2000. 26. Setzler-Hamilton et al. 1988. 27. Young et al. 1994. 28. Bennett et al. 1995. 29. Gartz 1998. 30. Cohen and Carlton 1998. 31. Meng and Orsi 1991. 32. Feyrer 1999. 33. Waldman 1998. 34. Gingras 1997. 35. Saiki et al. 1992. 36. Hocutt et al. 1990. 37. Kohlhorst 1999. 38. Meng 1993a,b. 39. Gartz 2000. 40. IEP newsletter, 2000. 41. D. Kohlhorst, CDFG, pers. comm. 2001. 42. Thomas 1967.

Figure 123. White bass, 29 cm SL, Nacimiento Reservoir, San Luis Obispo County. Fish print by Christopher M. Dewees.

The operculum has a single sharp spine and the margin of the preopercular bone (behind the eye) is distinctly sawtoothed. The first dorsal fin has 9 spines and the second, 1 spine and 13–15 rays. The anal fin has 3 spines, distinctly graduated in size, and 12–13 rays. The pelvic fins have 1 spine and 5 rays each and are located only slightly behind the pectoral fins (15–17 rays). The lateral line is complete and has 52–60 scales. Males have a single urogenital opening behind the anus; females have two.

Taxonomy White bass are fairly closely related to striped bass, with which they can hybridize. The hybrids are fertile and capable of reproducing in the wild (1). They have been produced artificially and are cultured (2). Restaurants listing striped bass on the menu are often in fact serving cultured hybrids.

Names White bass have, in the past, been placed in the genera *Lepibema* and *Roccus*. *Chrysops* means gold eye, although the eye is not conspicuously golden. Other names are as for striped bass. All white bass–striped bass hybrids are generally given the commercial name "sunshine bass," although the American Fisheries Society (3) has recommended that this name be reserved for crosses of male striped bass with female white bass, and that the name "palmetto bass" be used for the opposite cross.

White Bass, *Morone chrysops* (Rafinesque)

Identification White bass are deep-bodied, silvery-white fish with 4–7 dark stripes on the sides, usually interrupted, and a distinctly forked tail. Their body is laterally compressed, with the back rising up steeply behind the head; it is deepest at the point where the two dorsal fins separate. The head is small and the slightly oblique mouth is large, the maxillae extending to or slightly beyond the middle of the eye and the lower jaw projecting slightly beyond the upper jaw. The upper and lower jaws are lined with rows of tiny sharp teeth. There is a single patch of teeth on the tongue.

Distribution White bass are native to the Great Lakes region, the Mississippi River system, and the southern United States (including parts of Texas but exclusive of most Atlantic coastal drainages). They have been widely introduced into warmwater reservoirs in the United States, Canada, and, apparently, Mexico, including Lahontan Reservoir and other waters in Nevada. An introduction of 160 juvenile fish from Nebraska was made by CDFG into Nacimiento Reservoir (San Luis Obispo County) in 1965, followed by an introduction of 64 adults in 1966 from Oklahoma. By 1970 the species was well established in the reservoir and in the Salinas River above and below it (4). White bass were planted in the lower Colorado River in 1968 and 1969 but failed to become established. In 1977 they unexpectedly appeared in Kaweah Reservoir (Tulare County), where they became abundant and spread throughout the Tulare Lake basin on the floor of the San Joaquin Valley during a period of flooding (4).

By this time (1982–1983) CDFG had become concerned that their spread from the Tulare Lake basin into the rest of the Central Valley would exacerbate the decline of striped bass, chinook salmon, and other species. Farmers whose land was flooded by newly reemerged Lake Tulare decided to pump it dry again, sending the water into the San Joaquin River—and with it millions of white bass. In the face of the ensuing controversy CDFG managed to contain the bass with applications of piscicides to drainage canals and installation of filters on the outgoing water (5). CDFG also realized that white bass had to be eradicated from Kaweah Reservoir, the source population. They first proposed introducing striped bass to somehow eat the white bass out of existence, but after anglers protested this action (the white bass fishery had become very popular) they introduced hybrid "sunshine" bass instead—with, not surprisingly, no effect (4). In 1987 CDFG spent $7.5 million to apply rotenone to Kaweah Reservoir and all waters downstream from it. The complex operation was successful, but irresponsible anglers had by that time introduced white bass into Pine Flat Reservoir (Fresno County), where a small population is now established. This population is a time bomb of sorts: if it explodes and spreads downstream, white bass will become established in the Delta and other parts of the watershed. Pine Flat Reservoir is too big to treat with piscicides.

Life History White bass inhabit open waters of large lakes and reservoirs and slow-moving rivers. Warm, slightly alkaline lakes and reservoirs seem to provide the best conditions for growth and survival, but members of the species live in a wide variety of lakes and rivers (6) and estuaries along the Gulf of Mexico. They can survive and grow at salinities of 20 ppt but generally do better at lower salinities (7). Optimal temperatures for growth are around 28–30°C (9, 10), but individuals also live in water approaching 34°C for extended periods of time (8).

Most of the time white bass remain in surface waters (<6 m), roaming in schools. They tend to move offshore during the day and inshore at dusk, following the shoreline and foraging for food (10). At night they are quiescent, usually in deep water or near submerged objects. They become active again at dawn. They are capable of moving long distances within short periods both upstream and downstream and quickly colonize new areas. Tagged fish have moved as far as 211 km in 131 days (8). The classic studies of A. Hasler in Wisconsin have demonstrated that white bass can orient themselves with various celestial cues and home to spawning grounds or desirable areas (11).

White bass are voracious, visual piscivores, and their presence in a lake or reservoir is usually noted by disturbances made by bass driving schools of threadfin shad and other small fish to the surface. The surface of the water ripples as panicked prey jump out of the water (sometimes to be picked off by gulls and other birds); the pursuers may come partially out of the water as well. Although most populations depend on small pelagic fish, some rely almost entirely on zooplankton (12). Aquatic insects and crayfish may also be important on occasion. In Nacimiento Reservoir, adult bass feed mostly on threadfin shad (13), but they will feed on any fish available, including sunfish, crappie, native cyprinids, and their own young. Young-of-year are primarily pelagic zooplankton feeders, usually changing to the adult diet in their second year (6, 8, 12). However, even small fish will consume other fish. Larval white bass (7–12 mm SL) have an unusually large mouth and decurving teeth in the jaw, which allow them to capture and consume larvae of other fish as well as zooplankton (14).

Growth is extremely rapid, especially in the southern part of their range, but varies considerably from lake to lake. Thus at the end of their first year they can measure 9–31 cm TL; at the end of their second year, 17–39 cm; at the end of their third year, 26–43 cm; and at the end of their fourth year, 28–46 cm (6, 15). The growth exhibited by bass in Nacimiento Reservoir, however, typifies that of most white bass populations: at the end of their first year (I), they measure 22–25 cm FL; at the end of their second year, 30–33 cm; and at the end of their third year, 33–36 cm (13). White bass seldom weigh more than 1.5 kg (45 cm TL), but the largest on record weighed 2.4 kg. This fish was caught in 1972 in Ferguson Reservoir on the Colorado River, a survivor of plants made in 1968 or 1969 (none of which reproduced). White bass live 9 years in the northern parts of their range but seldom live more than 6 years in western reservoirs. Age and growth of both sexes seem to be about the same.

Spawning normally takes place for the first time in spring of the second year. Females spawn annually, with fecundity increasing with size. Fecundity estimates range from 61,700 eggs to nearly 1 million, but there seems to be enormous variability among populations (16). As the water warms up large schools of ripe fish congregate at the mouths of inlet streams or near suitable spawning areas in the lake (usually steep, rock- or gravel-covered bottoms with considerable wave action). Large streams seem to be preferred, however, and white bass will migrate up to 200 km upstream to spawn (6). Normally they move just a short distance from the lake to a gravelly or rocky area where the water is 1–3 m deep and begin spawning when temperatures reach 13–17°C. The Nacimiento Reservoir population apparently spawns in the Nacimiento River (13). Spawning can occur at any time of day or night. Spawning behavior is a mass affair, similar to that of striped bass, with spawning groups rising to the surface and releasing eggs and sperm. Eggs are fertilized as they sink to the bottom, where they stick to the substrate (17). Spawning lasts anywhere from 3–4 days to 3–4 weeks, with the largest fish spawning first. It ceases when temperatures exceed 26°C (8).

The embryos hatch in 40–46 hours at 16-21°C, optimal temperatures for spawning (17). Larvae initially stay in shallow water near spawning areas but eventually become planktonic. At this stage they are vulnerable to predation by threadfin shad, so large populations of shad aggregating near mouths of spawning streams may limit bass populations.

Status IIC. White bass were introduced into Nacimiento Reservoir by CDFG as a naïve "experiment" to see if white bass would do well in California reservoirs. The introduction was regarded as an experiment because CDFG thought the bass could be contained in Nacimiento Reservoir. From the outset CDFG biologists recognized that the bass could be a threat to salmon and striped bass in the Central Valley, but official enthusiasm for them allowed the introduction to go forward (4). The potential for anglers to move the fish to new locations—including a warning in the first edition of this book—was ignored.

The popularity of white bass in the West at their time of introduction seems to stem from the spectacular fishery in Texoma Reservoir, Oklahoma, that developed following its impoundment in 1944. One of the main justifications for the subsequent spread of white bass was that they would consume and control large threadfin shad populations in many Western reservoirs (also the result of optimistic planting programs). The shad were regarded as a problem because largemouth bass and other fishes with planktonic larvae sometimes did worse rather than better in the presence of the planktivore introduced as forage for them. There is little evidence that white bass have much effect on super-abundant threadfin shad populations, and there is some evidence that, at least in Nacimiento Reservoir, threadfin shad may control white bass populations rather than the reverse (13).

The fishery for white bass became very popular in Nacimiento Reservoir, so it was not surprising that anglers planted fish caught in Nacimiento into Kaweah Reservoir (4). By the time this illegal movement of fish was discovered, a sea change in official attitudes had taken place, and CDFG and other agencies recognized that the threat of white bass to Bay-Delta fish and fisheries was very real (5). The resulting massive poisoning operation was successful. However, despite new laws passed banning the possession of live white bass, irresponsible anglers introduced them into Pine Flat Reservoir. The population so far is small but is unlikely to stay that way.

The introduction of white bass into California was a big mistake and one that is not reversible. Chemical treatment of both Pine Flat and Nacimiento Reservoirs and chunks of their watersheds would be required to eliminate them from the state. Not only is the application of piscicides into water supplies increasingly less acceptable to the public, despite the minuscule threats they present to human health, but the two reservoirs are also very large. Their size makes treatment expensive, difficult, and prone to failure. In addition, negative effects of the poisoning of native fishes would have to be balanced against the potential negative effects of a continuing large population of white bass. At the very least, the lessons learned from the white bass introduction and spread—in combination with the painful lessons learned from the massive efforts in the 1990s to eradicate northern pike from two reservoirs—must be taught to the public, especially to anglers.

What can we expect if white bass invade the San Francisco Estuary? The most optimistic scenario is that they would not become very abundant because of predation from striped bass and unsuitable environmental conditions, especially water clarity. A more likely scenario is that they would find the estuary, especially the Delta, a very suitable place to live and become abundant. In this case it is likely they would come into conflict with striped bass through competition, predation, and hybridization. From the perspective of native fishes, especially salmon, splittail, and delta smelt, the best that could be hoped for in this case would be that white bass would just replace striped bass without any new impacts. It is equally likely that the ecology and foraging behavior of white bass are sufficiently different from those of striped bass that white bass will add additional unwelcome predation pressure on native fishes, including their larvae. This outcome will make recovery of threatened or endangered species more difficult and increase the likelihood of additional listings.

References 1. Gleason 1982a. 2. Tate and Helfrich 1998. 3. American Fisheries Society 1991. 4. Dill and Cordone 1997. 5. CDFG 1987b. 6. Chadwick et al. 1966. 7. Heyward et al. 1995. 8. Becker 1983. 9. Houston 1982. 10. Sigler and Miller 1963. 11. Hasler et al. 1969. 12. Priegel 1970. 13. C. von Geldern, pers. comm. 1974. 14. Clark and Pearson 1978. 15. Priegel 1971. 16. Baglin and Hill 1977. 17. Riggs 1955.

Sunfishes, Centrarchidae

The Centrarchidae is a small family (30 species) containing some of the most abundant, ecologically important fishes in warmwater ponds, lakes, and streams in North America: sunfishes (*Lepomis* spp.), "black" basses (*Micropterus* spp.), and crappies (*Pomoxis* spp.). In California they are often the most abundant fishes in reservoirs, sloughs, and low-elevation streams, where they support fisheries for everyone from professional bass anglers to children. They are all carnivorous and build nests to protect their embryos and young from predators. Structurally, centrarchids are characterized by united soft- and spiny-rayed dorsal fins, terminal mouths with small teeth in bands and protractile premaxillary bones, small membrane-covered pseudobranchs, strong pharyngeal teeth, ctenoid scales, and short intestines with pyloric cecae.

The family evolved in North America but now enjoys a worldwide distribution, thanks to enthusiastic stocking of various bass and sunfish species. Although the fossil record indicates they once occupied waters over much of the United States, mountain building and increasing aridity of interior drainage basins seem to have eliminated them from most of North America west of the Rocky Mountains, probably during the Miocene period (R. R. Miller 1959). One species that managed to survive in the West is the Sacramento perch of the Sacramento–San Joaquin drainage. As a result of its isolation and lack of competition from other related species, it has retained many ancestral structural and behavioral features. It is not surprising, therefore, that Sacramento perch have virtually disappeared from their native habitats following the introduction of 11 species of centrarchids from eastern United States. The predatory habits of basses and sunfishes have also contributed to the decline of many other native fishes, especially native minnows in lowland habitats and pupfishes in desert springs. Thus refuges for native fishes often have to be managed in ways that exclude centrarchids or minimize conditions that favor them.

Sacramento Perch, *Archoplites interruptus* (Girard)

Identification Sacramento perch are deep-bodied (depth is up to 2.5 times standard length) and laterally compressed, with long dorsal (12–14 spines, 10–11 rays) and anal (6–8 spines, 10–11 rays) fins. The mouth is large and oblique, with the maxilla extending just below the middle of the eye. Numerous small teeth are present on the jaws, tongue, and roof of the mouth. The 25–30 gill rakers are long. The scales are fairly large, numbering 38–48 along the lateral line. The spiny portion of the dorsal fin is continuous with the soft-rayed portion. Pectoral fin rays number 13–15 while vertebrae number 31–32, intermediate between the counts for bass and sunfish (1). Live fish are brown on the sides and top, with a metallic green to purplish sheen and 6–7 irregular vertical bars on the sides. Their bellies are white. The opercula have black spots. Breeding males become darker, especially on the opercula, which turn purple. Males also develop a distinct silvery spotting that shows through the darker sides, but in females the color is more uniform.

Taxonomy The Sacramento perch is the only member of the family Centrarchidae that occurs naturally west of the Rocky Mountains; it is believed to have been isolated from other centrarchids since the Miocene period (2). It was first described by Charles Girard in 1854 as *Centrarchus interruptus* (3) from the lower Sacramento River. Gill (4) assigned it to the monotypic genus *Archoplites*, recognizing

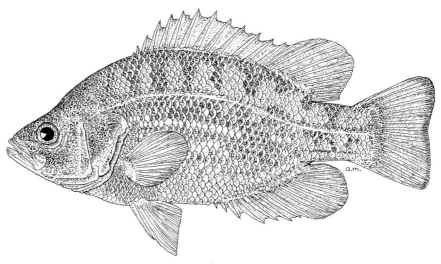

Figure 124. Sacramento perch, adult, 12 cm SL, Yolo Bypass, Yolo County. Drawing by A. Marciochi.

that it was distinct from other members of the family. However, recent phylogenetic analyses indicate that it is fairly closely related to flyer (*Centrarchus macropterus*) and crappies (*Pomoxis* spp.).

Meristic variation in Sacramento perch among populations from various areas is low, although there are some differences in color patterns (5). The Clear Lake population probably is genetically distinct, given its long isolation from other populations. Most extant populations are derived from Sacramento River fish, probably collected from Brickyard Pond (now Greenhaven Lake) in Sacramento, from which they are now gone. A likely exception is the population in Calavaras Reservoir on Alameda Creek (Alameda County), which is most likely derived from the original resident population (6).

Names *Archoplites* is derived from Greek words for anus and armature, referring to the conspicuous spiny anal fin; *interruptus* refers to the irregular bars on the sides.

Distribution Historically, Sacramento perch were found throughout the Central Valley, the Pajaro and Salinas Rivers,

and Clear Lake (Lake County) (7, 8) at elevations below 100 m. The only populations today that represent continuous habitation within their native range are those in Clear Lake and Alameda Creek. The Alameda Creek population apparently persists in gravel pit ponds adjacent to the creek and in Calavaras Reservoir. Within their native range they exist primarily in farm ponds, reservoirs, and recreational lakes into which they have been introduced, often upstream of their native habitats (Table 12). Outside their native range populations have become established in California reservoirs and associated streams in *(1)* the upper Klamath basin, including the Lost River and the mainstem Klamath River; *(2)* the Cedar Creek watershed in the south fork of the Pit River watershed; *(3)* the Walker River watershed (Lahontan Basin); *(4)* the Mono Lake watershed; and *(5)* the Owens River watershed. They were once established in the Russian River, probably from introductions (5), but the only population that may still exist is in Sonoma Reservoir. They were introduced into Nevada around 1877 and are still present in Pyramid, Walker, and Washoe Lakes, as well as in other localities in the Truckee, Carson, Walker, and Humboldt River drainages (9). They were widely planted in the western United States in the 1960s in alkaline lakes in Utah, Colorado, Nebraska, Texas, New Mexico, North Dakota, and South Dakota (10). From these introductions apparently only the population in Garrison Reservoir in Utah still persists (27). In the upper Klamath basin Sacramento perch were introduced by CDFG into Clear Lake Reservoir (Modoc County) in the 1960s, and they have since spread throughout the Lost River, in Oregon and California, to Tule Lake, and the Klamath River down to Iron Gate Reservoir, including Sheepy, Indian Tom, and Lower Klamath Lakes in California (Siskiyou County) (11).

Life History Sacramento perch were once, along with Sacramento pikeminnow, the dominant piscivorous fish in waters

Table 12

Major Localities Containing Sacramento Perch in California in the 1990s

Location	County	Watershed (subprovince)
Clear Lake[a]	Lake	Clear Lake
Calaveras Reservoir[a]	Alameda/Contra Costa	Central Valley
Gravel pit ponds, Alameda Creek near Niles[a]	Alameda	Central Valley
Lake Anza	Contra Costa	Central Valley
Jewel Lake	Contra Costa	Central Valley
Lagoon Valley Reservoir	Solano	Central Valley
Hume Lake	Fresno	Central Valley
Sequoia Lake	Fresno	Central Valley
San Luis Reservoir	Merced	Central Valley
Middle Lake[b]	San Francisco	Central Valley
Almanor Reservoir	Plumas	Central Valley
Butt Valley Reservoir	Plumas	Central Valley
Abbott's Lagoon	Marin	North Coast
Sonoma Reservoir[b]	Sonoma	Russian River
West Valley Reservoir	Modoc	Pit River
Moon (Tule) Reservoir	Lassen	Pit River
Honey Lake[b]	Lassen	Lahontan
Clear Lake Reservoir	Modoc	Upper Klamath River
Lost River, including Clear Lake Reservoir and Tule Lake	Modoc	Upper Klamath River
Iron Gate and Copco Reservoirs	Siskiyou	Upper Klamath River
Lower Klamath, Sheepy, and Indian Tom Lakes	Siskiyou	Upper Klamath River
Bridgeport Reservoir	Mono	Lahontan
East Walker River	Mono	Lahontan
West Walker River	Mono	Lahontan
Topaz Lake	Mono	Lahontan
Gull, June, Silver, and Grant Lakes	Mono	Mono Lake
Crowley Reservoir	Mono	Owens River
Lower Owens River, including Pleasants Valley Reservoir	Inyo	Owens River

Note: This record is by no means comprehensive in that it does not take into account small farm ponds and other temporary introduction sites, including those listed by Aceituno and Nicola (8).
[a]Native populations.
[b]Status uncertain.

of the Central Valley floor. Early observers were impressed with their abundance and potential as food fish (12), and they were one of the most common fishes caught by native peoples (13, 14, 15). They formerly inhabited sloughs, slow-moving rivers, and lakes, but are now mostly found in reservoirs and farm ponds. They are often associated with beds of rooted, submerged, and emergent vegetation and submerged objects. In moderately clear water, beds of aquatic plants seem to be essential for young-of-year, which inhabit shallow areas close to or in them. However, Sacramento perch can achieve high numbers in shallow, highly turbid reservoirs with no aquatic plants (e.g., Moon Reservoir, Modoc County). In large lakes they occur mainly in inshore areas, usually close to the bottom. In Pyramid Lake, Nevada, they are found mainly in water less than 15 m deep (16).

Because the waters they originally inhabited fluctuated tremendously with floods and droughts, Sacramento perch are adapted to withstand low water clarity, high temperatures, and high salinities and alkalinities. For example, they survive and reproduce in chloride-sulfate waters with salinities up to 17 ppt and in sodium-potassium carbonate concentrations of over 800 ppm (10). These waters exclude most other fish species. Most populations today are established in warm (summer temperatures, 18–28°C), turbid, moderately alkaline reservoirs or farm ponds. In the laboratory they readily acclimate to temperatures up to 30°C and prefer those in the range 25–28°C (17).

The key aspect of Sacramento perch habitat today, however, is the absence of other centrarchids, especially black crappie and bluegill. Nonnative fishes are excluded either by high alkalinities or by lack of introductions. The one exception to this "rule" seems to be Clear Lake, where a small population of Sacramento perch persists despite the presence of six other centrarchid species. Unlike introduced sunfishes, Sacramento perch, except when breeding, show little intraspecific aggressive behavior in aquaria or small ponds. Adults also do not shoal strongly, although they congregate in favorable localities, especially for breeding. They are sluggish in their movements and spend most of their time on or close to the bottom near submerged objects, moving little except their opercula and paired fins. When a prey organism is sighted, they stalk it slowly until they are close enough to seize it with a sudden rush. Prey are seized by "inhaling" with a sudden expansion of the buccal cavity and then clamping down with the numerous small teeth in the mouth. They have a fairly difficult time capturing prey that has to be actively pursued (18).

The prey eaten depend on size of fish, availability, and time of year (19, 20, 21). Young-of-year feed mostly on small crustaceans (amphipods, cladocerans, ostracods, and copepods) that are usually associated with the bottom or with aquatic plants. In Clear Lake fish measuring less than 40 mm SL feed mainly on copepods, but cladocerans become more important as fish increase in size (22). As they grow larger, aquatic insect larvae and pupae, especially those of chironomid midges, become increasingly important. In large lakes (such as Pyramid Lake, Nevada) fish larger than 90 mm TL feed primarily on other fish, especially cyprinids. In small lakes and ponds chironomid midges and other aquatic insects continue to be important in the diet of large perch; small crustaceans and fish are of secondary importance. Adult perch occasionally feed heavily on their own young-of-year (19, 20). In general their diet is most varied in summer, when planktonic and surface organisms are eaten along with the usual bottom-dwelling invertebrates. In winter and fall they concentrate on insect larvae, mainly chironomid midge larvae, which they pick from the bottom or aquatic plants. However, Sacramento perch are highly opportunistic and occasionally glut themselves on abundant organisms, such as waterboatmen (Corixidae) or aquatic beetles. Feeding takes place at any time of day or night (19), but there seem to be peaks of activity at dusk and dawn.

Growth rates are variable and affected by both biotic and abiotic factors (10, 19, 20, 21, 23, 24). At the end of their first year (age I), fish typically measure 6–13 cm FL, while age II fish are 12–19 cm; age III fish, 17–25 cm; age IV fish, 20–28 cm; age V fish, 21–32 cm; and age VI fish, 28–36 cm (21). Nine-year-old fish from Pyramid Lake measure 38–41 cm FL. The maximum length recorded is 61 cm TL (25), and the highest weight was that of a 3.6-kg perch from Walker Lake, Nevada (9). The California angling record, however, is only a 1.64-kg fish, from Crowley Reservoir, although a fish measuring 43 cm TL and weighing 1.95 kg holds the angling record in Utah.

As in most fish, growth in older perch is mostly in weight rather than in length. Thus a perch measuring 10 cm TL from Pyramid Lake weighed about 15 g; a 20-cm perch, 150 g; a 30-cm perch, 550 g; and a 40-cm perch, 1,200 g (23). Females grow faster and have lower mortality rates than males, so large perch tend to be females. Overcrowding, diet, and gender will affect growth rates. Stunted populations occur where water temperatures are cool and large prey is not abundant. Thus in Lake Anza perch showed extremely slow growth after the second year, and 6-year-old fish measure only 15 cm FL (23). A similar situation seems to exist in Clear Lake, where four 6- to 9-year-old fish measured 16–19 cm SL (22), perhaps because of competitive interactions with introduced centrarchids (26). In Greenhaven Lake growth rates decreased as the population declined to extinction, a decline associated with construction of houses on the banks and establishment of other centrarchids in the water (24).

Sacramento perch breed for the first time during their second or third year. Fecundity of females is higher than that in most centrarchids, but varies with the size of the fish. The number of eggs in 16 females (120–157 mm TL) from the stunted population in Lake Anza ranged from 8,370 to 16,210 (mean, 11,438); 16 females (196–337 mm TL) from Pyramid Lake contained 9,666–124,720 eggs (23).

Spawning occurs in California from late March through early August, although late May and early June are generally peak times, when water temperatures are 18–29°C (10, 27). In Lagoon Valley Reservoir (Solano County) larval perch were collected from early April through July, indicating a long spawning period.

For spawning, perch congregate in shallow areas (20–50 cm deep) with heavy growths of aquatic macrophytes or filamentous algae nearby. Rock piles and submerged roots or sticks may also attract fish ready to spawn. Before spawning begins males start defending small territories over substrates ranging from clay and mud to rocks (21, 23, 28). The territories are approximately 40 cm in diameter and are aligned along the shore, rather than in colonies (21). Although some observers report no nest preparation (23, 28), male perch will create shallow depressions for spawning, mainly by "digging" with their caudal fins over a period of several days (29). The depth of nests ranges from 20 to 75 cm (23, 28). Nest areas are defended vigorously from other males by chasing, nipping, and flaring the opercular flaps. Fish of other species are also chased away from breeding areas. While patrolling their territories males frequently engage in a rapid quivering movement of their tails (23). When a female is ready to spawn, she becomes restless and approaches territorial males, who initially may chase her away. Usually a ready female is courted by a stiffly swimming male, who nips near her vent. During the spawning act either both sexes release sex products simultaneously while spawning side by side (sometimes at an angle to the bottom) or the female releases eggs first, to be immediately fertilized by the closely following male. While spawning a male flares its opercula and fins, opens its mouth, and quivers rapidly; occasionally both fish may engage in such behavior (29). After spawning the female leaves the territory and may quickly spawn with another male. The male continues to defend the territory for several days against other perch and potential egg predators, including other centrarchids and catfish, until larval fish are able to swim well enough to leave the nest (23).

The larvae are initially planktonic, presumably for 1–2 weeks, before settling into aquatic vegetation or shallow water. Young-of-year fish form shoals in inshore areas, often near overhanging trees or in clearings in aquatic plant beds.

Status IC. Sacramento perch are a CDFG Species of Special Concern and would undoubtedly be listed as an endangered species in California if there were not so many introduced populations (30). Only two native populations seem to be maintaining themselves, if tenuously: those in Clear Lake and in the Alameda Creek drainage. Of the introduced populations, the ones in the upper Klamath watershed; in Pyramid Lake, Nevada; in the lower Walker River; and in the Owens River are probably reasonably secure because of their abundance and fairly broad distribution within these waters. However, most reservoir and pond populations will not persist indefinitely because of changing conditions. Thus large populations in Moon and West Valley Reservoirs on Cedar Creek (Lassen and Modoc Counties) disappeared during an extended drought when reservoirs dried up; the perch had to be reestablished through planting (34). Isolated populations established in other states have gradually disappeared as anglers and agencies lost interest in them.

Because Sacramento perch are tolerant of a wide range of conditions, they would still be abundant throughout their native range in the absence of introduced centrarchids, especially crappie (*Pomoxis* spp.) and sunfishes (*Lepomis* spp.). The alien species compete successfully for food and space (26) and may prey on embryos and larvae as well. Decline of Sacramento perch in their native range was gradual, but was noticed even in the 19th century. Between 1888 and 1899, 40,000–432,000 lb were sold annually in San Francisco (31). However, Rutter (32) found that they were already rare outside the Delta in his 1898–1899 survey of Central Valley fishes. They were largely gone from the Delta by the time of the major fish surveys of the 1950s and 1960s.

In Clear Lake (Lake County) a 1930 fish survey found them to be abundant. By the late 1940s their numbers had been greatly reduced, but they were still common enough

for spawning to be observed (28). In the 1960s an exhaustive fish sampling program turned up only nine adult Sacramento perch and no juveniles (35). More recent surveys demonstrate a small but persistent population, centered around sloughs in Clear Lake State Park (22). Their numbers in the lake seem to have increased somewhat in recent years, associated with a decline in crappie numbers (33). Clear Lake Sacramento perch were transplanted to Sonoma Reservoir (Sonoma County) by CDFG to provide a "reserve" stock, and there are plans to stock an additional small pond (33).

Three hypotheses have been advanced to explain the decline of Sacramento perch: habitat destruction, embryo predation, and interspecific competition. Habitat destruction, especially draining of lakes and sloughs and reduction of aquatic plant beds, was the hypothesis favored by Rutter (32). However, perch have declined in areas where suitable habitat still exists (e.g., Clear Lake, sloughs of the Delta), so it is unlikely that this is the only reason, although it may have been a contributing factor.

Embryo predation, especially by catfish and carp, was first advanced as a cause of the decline by Jordan and Evermann (25) and was supported by the observations of Murphy (28) that the perch did not defend spawning sites. Later observations that they do defend the sites against potential embryo predators tend to make embryo predation less likely as a primary cause of decline. However, no Sacramento perch, no matter how aggressive, is able to defend its spawning area against a determined school of egg-eating bluegill or large carp. In addition, high abundance of small nonnative centrarchids and other fishes may result in heavy predation on larvae after they leave the nests.

Interspecific competition for food and space may be the single most important cause of the decline because, almost invariably, local declines of Sacramento perch populations have been associated with increases in numbers of introduced centrarchids, especially black crappie (8). In Clear Lake increases in Sacramento perch numbers in recent years seem to be associated with a decline in black and white crappie, which have diets most similar to those of perch. In aquaria and small ponds bluegill and green sunfish dominate Sacramento perch, chasing them away from favored places. Such behavior in the wild could force young perch out of shallow weedy areas and into more exposed waters, where they would be more vulnerable to predation by largemouth bass and other piscivores. They would also have less food available to them. In situations in which perch and bluegill compete for food, bluegill depress perch growth (and presumably survival) rates (26). The importance of interspecific competition is also reflected by the fact that Sacramento perch today are successful mostly in relatively simple fish communities where they can occupy the position of top littoral carnivore. Overall the decline of Sacramento perch populations is probably due to all three factors working together, because habitat alteration and fish introductions have occurred simultaneously throughout the Central Valley. Thus consistent defeats in interspecific encounters, especially of young fish, may have accelerated a decline started by other factors.

An additional concern with Sacramento perch is the limited genetic ancestry of the populations, which may limit their long-term survival potential. The Clear Lake population is presumably distinctive, given its long isolation and the distinctiveness of other fish in the lake (5), yet there is no assured self-sustaining population outside the basin. The only other native population appears to be in the Alameda Creek drainage, which may also be different from other populations because of its recent isolation. The majority of (if not all) introduced fish apparently originated from Greenhaven Lake (= Brickyard Pond) in Sacramento, which was a convenient source of fish for at least 100 years. That population, however, is now extirpated (24). Management of perch should therefore have as one of its goals maintaining remaining genetic diversity.

Given the fragility of most Sacramento perch populations, continued efforts should be made to propagate them and make them available for use in farm ponds and reservoirs in the Central Valley as a native sport fish. Special effort should also be made to establish additional populations of Clear Lake Sacramento perch. At least once every 10 years a review of the distribution and status of this unique endemic centrarchid should be conducted to determine if additional measures for its protection are needed.

References 1. Mabee 1993. 2. R. Miller 1959. 3. Girard 1854. 4. Gill 1861. 5. Hopkirk 1973. 6. Gobalet 1990. 7. Evermann and Clark 1931. 8. Aceituno and Nicola 1976. 9. La Rivers 1962. 10. McCarraher and Gregory 1970. 11. Buettner and Scoppetone 1991. 12. Lockington 1878. 13. Schulz and Simons 1973. 14. Broughton 1994. 15. Gobalet and Jones 1995. 16. Vigg 1980. 17. Knight 1985. 18. Vinyard 1982. 19. Moyle et al. 1974. 20. Imler et al. 1975. 21. Aceituno and Vanicek 1976. 22. Fong and Takagi 1979. 23. Mathews 1962, 1965. 24. Vanicek 1980. 25. Jordan and Evermann 1896. 26. Marchetti 1999. 27. P. Crain, University of California, Davis, unpubl. data 1998. 28. Murphy 1948a. 29. M. Dege, University of California, Davis, pers. comm. 1997. 30. Moyle et al. 1995. 31. Skinner 1962. 32. Rutter 1908. 33. R. Macedo, CDFG, pers. comm. 1998. 34. P. Chappell, CDFG, pers. comm. 1998. 35. Cook et al. 1966.

Bluegill, *Lepomis macrochirus* Rafinesque

Identification Bluegill are easily distinguished from other California sunfishes with deep, compressed bodies by the presence of flexible blue or black flaps on the rear of the opercula; long, slender gill rakers on the first gill arch; long, pointed pectoral fins (13–14 rays) that are contained about 3 times within the standard length; a black spot on the rear of the dorsal fin; and narrow, vertical black bars on the sides. The anal fin has 3 spines and 11–12 rays; the dorsal fin, 10 spines and 10–12 rays; and the pelvic fins, 1 spine and 5 rays each. There are fewer than 50 scales on the lateral line, typically 38–48. Nonbreeding fish usually have an iridescent purple sheen. Breeding territorial males become very dark olive to bronze on their back and sides and have orange breasts; their pelvic and anal fins turn an iridescent black, and a large dark spot develops on the soft-rayed portion of the dorsal fin.

Taxonomy Two subspecies have been recognized: northern bluegill, *L. m. macrochirus*, and southern bluegill, *L. m. purpurescens* (1). Northern bluegill are native to most of eastern North America and have been widely introduced in California and elsewhere. Southern bluegill are native to peninsular Florida and southern Georgia and have been introduced into Perris Reservoir (Riverside County) and a few other reservoirs in the state (2). In California bluegill commonly hybridize with green sunfish and may also hybridize with redear sunfish and pumpkinseed. The hybrids are sterile males.

Names Gills of bluegill are pink, as they are for most fish, so the common name is presumably derived from the sometimes blue flap on the operculum. California anglers often refer to bluegill (and other sunfishes) as "perch" or "bream." *Lepomis* means scaled cheek, because the scales present on the operculum were once considered to be a significant distinguishing feature. *Macrochirus* translates as large hand, referring to the long pectoral fins.

Distribution Bluegill were originally distributed throughout much of eastern and southern North America, north to Ontario and the Great Lakes region, west through the Mississippi drainage system, and south into Florida and northeastern Mexico. They were introduced into California in 1908 and became widely distributed in the next 10–20 years (2). They are now established throughout the state, including most reservoirs, and are probably the most widely distributed warmwater fish. They are also abundant in all other Western states and provinces and now have a worldwide distribution: Japan, Korea, the Philippines, Morocco, South Africa, Swaziland, Zimbabwe, Panama, Venezuela, Puerto Rico, the Hawaiian islands, and Mauritius (3).

Life History The ability of bluegill to survive and reproduce under many environmental conditions has made them one of the most abundant freshwater fishes in California. They do best in warm, shallow lakes, reservoirs, ponds, streams, and sloughs at low elevations, but occasional populations of slow-growing bluegills become established in colder lakes, such as Shaver Lake (Fresno County; elevation, 1,670 m). Temperature tolerances are very broad. They survive winter temperatures of 2–5°C and, when acclimated, summer temperatures as high as 40–41°C, at least for short periods of time (4). However, given a choice, bluegill select temperatures of 27–32°C, which seem to be physiologically optimal for growth (4). They prefer fresh water (<1–2 ppt) but occur in the San Francisco Estuary at salinities up to 5 ppt. Elsewhere they have been shown to have increased metabolic rates (indicating osmotic stress) at 8 ppt, with 12 ppt being lethal (25). They can survive in waters of surprisingly low oxygen content (<1 mg/liter), especially at low temperatures, but maximum growth and reproduction occur in fairly clear waters with moderate levels of dissolved oxygen (4–8 mg/liter). Bluegill are often associated with rooted aquatic plants, in which they hide and feed, and with bottoms of silt, sand, or gravel. They seldom live much deeper than 5 m.

Despite the association of bluegill with ponds, lakes, and sloughs, they are a common stream fish in areas where summer temperatures are warm and there are deep pools with beds of aquatic plants or other deep cover (26, 27). In lower Putah Creek (Yolo-Solano County) they persist through periods of high winter and spring flows by moving into temporary backwaters or areas of flooded vegetation: any place where there is refuge from high current velocities. Longtime persistence, however, depends on appropriate conditions being present in summer months (5). In California they are commonly associated with a complex, if variable, assemblage of other nonnative species: largemouth bass, green sunfish, redear sunfish, various catfish, golden and red shiners, common carp, inland silverside, and western mosquitofish (26, 27).

Individual bluegill spend most of their lives in a rather restricted area, even in large bodies of water. This behavior presumably gives each fish familiarity with an area within which it needs to find food and avoid predators, such as largemouth bass. It also reflects their long breeding season, during which colonies of breeding fish have fairly fixed locations. Perhaps one outcome of this behavior is that bluegill act as cleaners to other fishes, picking off parasites, loose scales, and other tissue (28). In Putah Creek I have observed juvenile (about 10 cm TL) smallmouth bass, swimming in a rigid manner, approach similar-size bluegill nesting at the edge of a colony. The bluegill proceeded to nibble at the side of each bass for a few seconds as they swam along the edge of the nest.

Bluegill are highly opportunistic feeders, feeding on whatever animal food is most abundant. Their mouth is relatively small but lined with small teeth, and the upper lip is protrusible. Bluegill are thus capable of ingesting many types of organisms. The larvae of aquatic insects—such as midges, mayflies, caddisflies, and dragonflies—seem to be preferred, but they also eat planktonic crustaceans, flying insects, and snails. Small fish, fish eggs, and crayfish may be eaten when available. In Pine Flat Reservoir (Fresno County) bluegill (10–26 cm FL) fed largely on fish eggs, midge larvae, and cladocerans from March through June, switched to flying insects from July through October, and went back to midge larvae and cladocerans from November through February (6). For larger fish threadfin shad also formed an important part of the diet in winter. In the Delta benthic organisms—such as amphipods (*Corophium*), isopods (*Exosphaeroma*), and chironomid larvae and pupae—dominate the summer diet of bluegill (7). In Clear Lake (Lake County) they fed mostly on zooplankton (cladocerans and copepods) until they reach about 45–50 mm SL, when they switched to benthic insects and, as size increases further, an occasional small inland silverside. When animal food becomes scarce the adult fish feed on algae and other aquatic plants, although they will become stunted on such a diet (8). Feeding is a nearly continuous activity in summer, reaching peak intensity in midafternoon and again at dusk (9).

Bluegill will feed on the bottom, in midwater, in aquatic vegetation, and on the surface. Their deep body and flexible fins are adapted for hovering at all levels and then darting forward to suck up a food item (10). The body is kept from rolling sideways by undulations of the nonspiny portions of the dorsal and anal fins as well as by movements of the upper lobe of the tail. The long pectoral fins, assisted by the pelvic fins, also help stabilize the fish, but their primary function is maneuvering. For this purpose they can be moved independently of each other with a wristlike action. This flexibility in feeding strategies makes them superb at optimal foraging. Although they usually gain the most energy from feeding on insects in aquatic vegetation, they readily switch to feeding on zooplankton when larger species (e.g., *Daphnia*) are very abundant and easy to capture (11). A shift in foraging tactics becomes more important as bluegill grow larger because growth (and ultimately reproduction) becomes increasingly limited by the availability of prey of appropriate size. Growth of small bluegill, in contrast, is limited more by competition with other bluegill (12).

Yearly growth of bluegill in California lakes and reservoirs is slower than that of bluegill in the southern United States, where the growing season is similar. California growth rates are similar to those in their native Midwest, but California bluegill seldom reach the sizes commonly achieved by bluegill from other areas (13). At the end of their first year they measure 4–6 cm FL, and they grow 2–5 cm in each subsequent year. Thus a typical 15-cm fish will be 4–5 years old and weigh about 90 g. A large (23 cm) individual is likely to be 8–9 years old and weigh more than 300 g, although few fish live longer than 6 years. Thus bluegill in the Delta average 45 mm FL at the end of year 1, 98 mm at year 2, 136 mm at year 3, 158 mm at year 4, 175 mm at year 5, and 189 mm at year 6 (14). The angling record for bluegill in California is a fish weighing 1.6 kg, from Otay Reservoir. Exceptionally cold or turbid water is likely to produce fish that rarely exceed 10 cm FL.

Spawning begins in spring when temperatures reach 18–21°C and may continue into September. Males are of three different types, each type genetically determined: parental males, satellite males, and sneaker males (15, 16, 17, 18). *Parental males* are typically the largest individuals, measuring 15–20 cm FL (or more) and 5 or more years old. They construct nests in shallow water by excavating, with vigorous fanning movements of their fins, depressions that are 20–30 cm in diameter and 5–15 cm deep. Nests are constructed on bottoms of gravel, sand, or mud that contain pieces of debris, such as twigs or dead leaves. Although parental males build their nests in colonies, each male defends his own nest and the area around it from all other males and from potential egg predators, such as minnows and catfish. The females swim about in schools in the general area of the nesting colony. When one is ready to spawn she approaches the nesting area and is approached and courted by a male, usually the largest one in the immediate vicinity. The male attracts the female to the nest, and the two spawn side by side. At each spawning the female dips toward the nest and releases about a dozen eggs, which are fertilized by the male. The fertilized eggs adhere to the debris on the bottom of the nest. Courtship movements are accompanied by distinctive grunting sounds (19). Each male courts and spawns with many females in succession, so a single nest can contain thousands of embryos (16). Each female also spawns with multiple parental males within the colony. All spawning within a single colony occurs rapidly, often on a single day, ensuring that the young will develop and emerge at about the same time. Parental males continue to defend their nests while embryos develop and then guard the school of young for several days after they hatch.

Satellite males mimic females, so they can deceive parental males and enter nests while spawning is taking place, fertilizing some eggs. They mimic females in part by being the same size (about 10–15 cm FL) and having the same coloration and behavior. They hover above the colony of parental males and slowly descend into a nest in which active courtship is occurring. When a female dips for the release of eggs, the satellite male presses against her on the side opposite the parental male and releases sperm. *Sneaker males* are small, inconspicuous fish. They hang out on the edge of the colony and dash in to spawn with the parental male and female. Parental males spend a good deal of time chasing satellite and sneaker males away from their territories, but females will readily spawn with satellite males when a parental male is chasing away another fish.

While sneaker and satellite males typically constitute less than 30 percent of the total male population (16), they definitely represent another way to be a successful male, in an evolutionary sense. These alternate male strategies evolved because otherwise only a few dominant parental males would do all the spawning. The advantages of being one of the two "cuckolder" male types include spawning at younger ages (2–3 years) than parental males and relying on the parental males to take care of their offspring. They do not live as long, however, in part because of the harassment they receive from parental males. Sneaker and satellite males often are fairly ragged looking by the end of the spawning season.

The complex mating system of bluegill (and some other sunfish) is permitted in good part by their habit of nesting in busy colonies, so that parental males are constantly being distracted by other parental males and schools of females. Colonial nesting, however, evolved largely as an antipredation measure (18, 20). Some of the best evidence for this is that territories in the center of the colony are typically held by the largest and most aggressive males, while smaller males hold territories on the periphery. The outside territories are under continual assault by embryo predators (such as snails and catfish), and survival of young is much lower than survival in interior nests. Nevertheless even outermost nests benefit from joint defense by many males as well as from the sheer number of nests, which swamp predators. The synchronization of breeding that takes place in colonies may also help reduce predation on young because of the large numbers emerging at one time. In addition embryos in colonial nests show a lower incidence of death from fungal disease than those in solitary nests, because each male spends more time fanning and otherwise caring for the embryos (21).

Bluegill are very prolific. Single females lay 2,000–50,000 eggs, depending on size. As many as 62,000 young bluegill have been reported hatching from one nest, although more typically 2,000–18,000 young are produced per nest (13).

At water temperatures of about 20°C embryos hatch in 2–3 days, and fry soon start to swim about. Males guard the embryos and fry for about a week before starting another breeding cycle. Gradually fry move away from the nests and into aquatic plant beds. In lakes they will enter the plankton for a short period of time and then settle into patches of plants. In streams they may enter the water column and be washed into backwaters (5, 22). Mortality due to predation by fish and invertebrates is very high at this time. In the vegetation of lakes they grow 10–12 mm TL and then move out into surface waters, where they remain for 6–7 weeks, feeding on planktonic crustaceans (23). By the time they return for good to the aquatic plant beds near shore they have grown to 21–25 mm TL.

Status IID. Bluegill are the most widespread and abundant game fish in California because they are always one of the first fish to be moved to new ponds and reservoirs. Fishing has little effect on their populations because of their high reproductive rates. In some waters severe intraspecific competition limits individual growth. The result is large populations of stunted fish, which may in turn limit populations of other fishes (including largemouth bass) both by eating embryos and young and by eating food that other young fish need to survive.

It is possible that the characteristic small size of bluegills in California is the result of their limited genetic background. Apparently most bluegill in the state are descended from fish collected in one or two Midwestern localities. In 1975 CDFG introduced southern bluegills from Florida, on the untested assumption that they would grow faster and become larger than northern bluegills. Their spread was halted, however, based on anecdotal evidence that they preyed heavily on embryos of bass (2).

The abundance, ubiquity, aggressiveness, and broad feeding habits of bluegill in lakes and lowland streams of California make it likely that they are one of the alien fishes that limit native fish populations, especially through predation on larvae, or through indirect effects that make natives more vulnerable to larger predators. Laboratory experiments indicate that they may have been a major factor in the demise of Sacramento perch (24).

References 1. Felley 1980. 2. Dill and Cordone 1997. 3. Lever 1996. 4. Houston 1982. 5. Marchetti 1998. 6. Goodson 1965. 7. Turner 1966b. 8. Kitchell and Windell 1970. 9. Keast and Welsh 1968. 10. Keast and Webb 1966. 11. Mittelbach 1983. 12. Osenberg et al. 1988. 13. Emig 1966. 14. Turner 1989. 15. Gross 1979. 16. Gross 1991. 17. Dominey 1980. 18. Dominey 1981. 19. Gerald 1971. 20. Gross and MacMillan 1991. 21. Cote and Gross 1993. 22. Rockriver 1998. 23. Werner 1967, 1969. 24. Marchetti 1999. 25. Peterson et al. 1987. 26. Moyle and Nichols 1973. 27. L. Brown 2000. 28. Powell 1984.

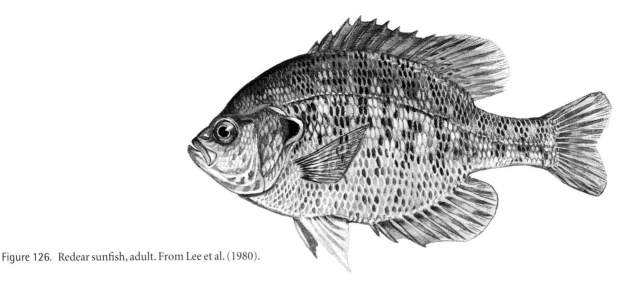

Figure 126. Redear sunfish, adult. From Lee et al. (1980).

Redear Sunfish, *Lepomis microlophus* (Günther)

Identification Redear sunfish are deep bodied and have small oblique mouths (barely reaching the front margin of the eye), long pointed pectoral fins, short stiff opercular flaps, and stubby gill rakers (2–3 times longer than wide). They differ from pumpkinseed and bluegill by the absence of any conspicuous patterns on either the sides or the fins, except for an orange-red edge ahead of the dark blotch on the opercular flap. The dorsal fin has 10 spines and 11–12 rays; the anal fin, 3 spines and 10–11 rays; the pelvic fins, 1 spine and 5 rays; and the pectoral fins, 13 rays. There are 34–43 lateral line scales. Adults are light olive on the back and a pale mottled brown to silvery on the sides, usually with some speckling, and often fairly bright yellow on the lower sides and belly. Young-of-year may have 7–8 faint vertical bars on the sides. The dorsal fin is dusky, without a dark spot on the rear portion.

Names Shellcracker is a widely used, unofficial common name for redear sunfish in much of their native southern United States. *Microlophus* means small crest, presumably because of the short opercular flap. Other names are as for bluegill.

Distribution Redear sunfish are native to the southeastern United States, including peninsular Florida, and to the Rio Grande and lower Mississippi drainage systems. They have been widely planted throughout the warmer regions of the United States, so their exact native distribution is poorly understood. Limited introductions have been made into other countries, including Morocco, South Africa, Panama, and Puerto Rico (11). They were introduced into the lower Colorado River in 1948 or 1949 by the Arizona Department of Fish and Game and were first collected in California in 1951 (1). They have since been introduced into southern California and the Central Valley, and into the Russian River drainage. As far as I know they are not present in reservoirs in any Great Basin watersheds or in any North Coast or Klamath Basin reservoirs. However, given the proclivity of sportsmen (and biologists) for moving fish around, they can be expected anywhere in the state, especially in farm ponds. They have been found, for example, in a farm pond in the upper Pit River watershed (2).

Life History The preferred habitat of redear sunfish is the deeper waters of warm, quiet ponds, lakes, and river backwaters and sloughs with substantial beds of aquatic vegetation. Reproduction and growth are inhibited in water that is too turbid, probably because low light penetration limits aquatic plant growth and forces them into shallow water, where they have to compete for food and space with other species, especially bluegill. Adults are usually most abundant in water more than 2 m deep. They are rarely found in brackish parts of the San Francisco Estuary, but in their native range they live in marshes with seasonal salinities of up to 5–12 ppt (3). They can adjust rapidly to changing salinities and tolerate salinities up to 20 ppt, making them one of the most euryhaline sunfishes (12).

Redear sunfish feed mostly by picking hard-shelled invertebrates, especially snails and clams, from the bottom and aquatic plants with their protrusible lips. The prey is then crushed by their molarlike pharyngeal teeth; the soft parts are swallowed and the hard parts are ejected. The short gill rakers make ejection of hard parts easier, especially pieces of snail shell, although most of the material is literally spit out. Short gill rakers also permit easy ejection of sand, mud, and bottom debris taken up while grabbing snails or burrowing mayflies (*Hexagenia*). Although snails form a major portion of their diet, especially in winter, when given a choice they select bottom-dwelling insect larvae (e.g., dragonfly, midge, and mayfly larvae) and amphipods in preference to snails (4). Such organisms often form the bulk of their diet in summer, especially in farm ponds and weedy lakes (13). It is possible that snails are important mostly where redear coexist with other species of sunfish that are better adapted for feeding on insects, especially when food is in short supply during the winter months. There are also distinct differences in the diets of juvenile and adult fish. In Rancho Seco Reservoir adults (>130 mm FL) were found to feed primarily on small clams and secondarily on chironomid midge larvae, while smaller fish fed primarily on chironomids and secondarily on other insect larvae and crustaceans (5). Diets are similar in the lower Colorado River (6).

Growth rates in central California lakes and reservoirs are usually slightly faster than those of bluegill in the same bodies of water and comparable to those of redear sunfish in their native southeastern United States. However, rates can be highly variable. An exceptionally slow-growing population occurs in Lost Lake (Fresno County), a small, turbid, gravel pit pond with only small amounts of aquatic vegetation. Lost Lake redear sunfish average 48 mm TL at the end of their first year, 92 mm in their second year, 135 mm in their third year, 163 mm in their fourth year, 189 mm in their fifth year, and 215 mm in their sixth year. In contrast Berryessa Reservoir (Napa County) has a fairly fast-growing population: 69 mm TL in the first year, 128 mm in the second year, 140 mm in the third year, and 170 mm in the fourth year. Growth slows down considerably after the fourth year, so redear sunfish in Berryessa exceeding 200 mm TL are uncommon (7). Growth rates in Rancho Seco and Folsom Reservoirs do not slow down as much; 3-year-old fish average 178–190 mm FL and 4-year-old fish, 216–224 mm (5). In Rancho Seco 5- and 6-year-old fish average 240 mm and 254 mm, respectively. The angling record for the United States is a 2.4-kg fish from Folsom South Canal (Sacramento County), a fish that had grown large feeding on *Corbicula* clams. The maximum age recorded is 7 years.

In their native range redear sunfish usually become mature at lengths of 13–18 cm TL when they are 1–2 years old (4). If the length-maturity relationship holds true for California populations, then they probably do not spawn until their third or fourth year. This is presumably one of the reasons why redear sunfish seldom have stunted populations in California, in contrast to bluegill.

Spawning takes place throughout summer, starting as soon as water temperatures reach 21–24°C (5, 8). This can be as early as mid-April in lowland areas. In Rancho Seco Reservoir two peaks of spawning were noted, one in mid-April through early May and one in July (5). The males construct nests in colonies, and each nest is defended vigorously by its owner from other males. The nests are depressions 25–61 cm in diameter (typically 35–45 cm) and 5–10 cm deep, constructed in bottoms of sand, gravel, or mud, usually at a wide range of depths. In Rancho Seco one cluster of nests occurred at 0.4–1.5 m, while another occurred at 4–6 m. Nest building and spawning behaviors are apparently similar to those of other sunfishes, especially pumpkinseed (4). Males make species-specific popping sounds during courtship (9). Females have fecundities of 9,000–80,000 eggs, depending on size, and spawn repeatedly.

Larvae presumably are planktonic at first, before settling into beds of aquatic plants, often along with bluegill of similar size. Juveniles typically stay close to or in aquatic plant beds, often in small shoals (10).

Status IID. Redear sunfish are not nearly as common in California as the ubiquitous bluegill and green sunfish, in part because they require somewhat warmer waters. They also are more restricted to ponds and reservoirs with beds of aquatic plants (which require fairly clear water to grow). Nevertheless they are favored for planting in reservoirs because they grow faster than bluegill and rarely stunt, sometimes producing "dinner plate"–size individuals. They also seem to be harder to catch than bluegill, their most common associate, apparently because they live in deeper water and feed on or close to the bottom. In many waters where they maintain substantial populations of large fish, fishermen do not even realize they exist and settle for catches of small bluegill. Increasing the angler harvest of redear sunfish is mostly a matter of letting anglers know where redear sunfish are and how to fish for them. In the southern United States they are much sought after as game fish.

Redear sunfish were introduced secondarily, well after other alien fishes had become established, so they have not been implicated in the decline of native fishes. In fact many of the molluscs on which they feed are introduced species as well.

References 1. Dill and Cordone 1997. 2. Moyle and Daniels 1982. 3. Jenkins and Burkhead 1994. 4. Wilbur 1969. 5. Christophel 1988. 6. Minckley 1982. 7. Skillman 1969. 8. Emig 1966. 9. Gerald 1971. 10. Wang 1986. 11. Lever 1996. 12. Peterson 1988. 13. Davis 1973.

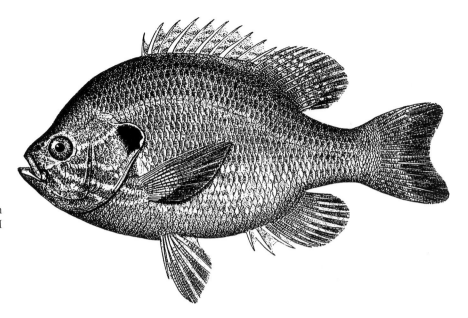

Figure 127. Pumpkinseed, 14 cm SL, Root River, Wisconsin. USNM 4163. Drawing by H. L. Todd.

Pumpkinseed, *Lepomis gibbosus* (Linnaeus)

Identification Pumpkinseed are deep bodied and have small oblique mouths, long pointed pectoral fins, short stiff opercular flaps, stubby gill rakers, and heavy molariform pharyngeal teeth, much like those of redear sunfish. They also have small spots on the soft portion of the dorsal fin, and adults have conspicuous orange and blue streaks on the dark operculum. The opercular flap is black with a red tip. The dorsal fin has 10 spines and 10–12 rays; the anal fin, 3–4 spines and 9–11 rays; the pelvic fins, 1 spine and 5 rays; and the pectoral fins, 12–13 rays. There are 36–44 scales in the lateral line. The background color of the body is gray-green to greenish brown with superimposed spots of orange, yellow, blue, and green and with 7–10 faint blue-green vertical bands. The cheek has a series of wavy blue lines running across it. The throat and belly are yellow to orange. Juveniles are a more uniform gray-green, with fairly conspicuous vertical bars.

Taxonomy Pumpkinseed apparently have relatively low morphological variability throughout their range, and no subspecies have been described. They will hybridize with most other *Lepomis* species but especially with bluegill and green sunfish. Hybrids are fast-growing, sterile males.

Names In their native East and Midwest pumpkinseeds are frequently called common sunfish or simply *the* sunfish. Pumpkinseed aptly describes their body shape; *gibbosus* means humped or rounded or, according to Jordan and Evermann (1, p. 1009), "formed like the full moon." Other names are as for bluegill.

Distribution Pumpkinseed are native to eastern North America from southeastern Canada, through the northern half of the Mississippi River system including the Great Lakes region, down the Atlantic coast to northern Georgia, and west into Missouri and South Dakota (2). They have been introduced throughout Canada and the United States, including Washington and Oregon, into coolwater lakes and rivers. They have also been introduced into much of Europe (where they are often regarded as pests) and into Morocco, Japan, Chile, and Guatemala (3). The exact date of their introduction into California is not known, but Dill and Cordone (4) suspect that they were introduced into the Susan River (Lassen County) in 1908 with a mixed shipment of sunfish. They also record an introduction into a private artesian pond near Mecca (Riverside County) in 1918, which was used as a source for unsuccessful plants all over southern California. The first record for the Klamath basin was 1942. The present distribution seems to be as follows:

1. Klamath Basin. They are widely distributed if uncommon in reservoirs of the Lost River watershed and in lakes and reservoirs of the upper Klamath

basin in Oregon and California, including Copco and Iron Gate Reservoirs (5, 6).

2. Susan River. Apparently a small population still exists in sloughs of the lower Susan River, above Honey Lake, although there is no recent confirmation of this. A population must exist in Mountain Meadows Reservoir (Lassen County), on a small tributary to the Susan River, because the state record pumpkinseed was caught there in 1996.

3. Sacramento–San Joaquin drainage. Populations have been recorded in the following reservoirs in the 1990s (26): San Justo (San Benito County), Lexington (Santa Clara County), and Elizabeth (Alameda County). Pumpkinseed are also present in San Ramon and Walnut Creeks (Contra Costa County), lower Coyote Creek (Santa Clara County), and Davis Reservoir (Plumas County) (4, 7). They are also found in small numbers in the south Delta and the lower San Joaquin River (26), where there are occasional records going back to at least 1986. They may be present elsewhere, but reports have not been confirmed.

4. Southern California. There are apparently no established populations in southern California, although attempts to establish them have been frequent (8).

Despite the lack of good records, pumpkinseed can be expected in cool, quiet waters anywhere in the state as the result of unauthorized introductions. Because of their attractiveness, they are often kept as aquarium fish and presumably released by owners tired of their aggressive charges. Some records of pumpkinseed may be of bluegill–green sunfish hybrids, which are brightly colored.

Life History Pumpkinseeds are ecologically similar to redear sunfish, preferring clear to slightly turbid lakes, sloughs, or sluggish streams with beds of aquatic plants that support large populations of snails. However, they seem to be adapted for living in much cooler waters than redears, especially lakes with large seasonal fluctuations in temperature. Curiously, in laboratory studies they show temperature preferences for warm water (24–32°C), and they can withstand temperatures up to 38°C (9, 10). At higher temperatures they live in water with oxygen levels as low as 4 mg/liter, but at low temperatures (3–4°C) they can withstand oxygen levels less than 1 mg/liter (10). Pumpkinseeds show a surprising tolerance of high salinities, considering their native distribution: up to 17 ppt (25). Juveniles will occur in loose aggregations around beds of aquatic plants, but adults tend to be more solitary, usually found as individuals or as small groups around submerged trees or other cover, close to aquatic plant beds.

Like redear sunfish, pumpkinseeds feed by picking hard-shelled invertebrates from the bottom or from aquatic plants. Snails are usually the most important item in their diet, but aquatic insects are apparently preferred. This preference is reflected in their pharyngeal jaw apparatus, which is used for crushing snails but not other prey (11). In lakes where snails form most of the diet, the pharyngeal teeth and jaws are much more robust than in locations where the fish feed mainly on insects (12, 13). In summer, when aquatic insect larvae are abundant, insects will frequently be the most common prey despite large snail populations (14). Aquatic insects also predominate in the diet when pumpkinseeds occur in the absence of other species of sunfish (15, 16). Pumpkinseeds of all sizes eat the same kinds of food, except that larvae feed on zooplankton. They will feed at almost any time of day, but peak feeding activity is generally at dawn or dusk. Feeding ceases when water temperatures drop below 6.5°C (17).

Pumpkinseeds grow rather slowly for sunfish, although this may be mostly an effect of the cooler waters they inhabit. Their growth in Honey Lake (Lassen County) approximates the average growth they achieve in their native Midwest: 1 year, 25 mm FL; 2 years, 66 mm; 3 years, 112 mm; and 4 years, 132 mm (16). Although they can live as long as 12 years, they seldom exceed 30 cm FL (10). The state angling record, from Lassen County, was a fish weighing 450 g. Growth in the first year of life may be enhanced by the presence of competing bluegills. Apparently, when bluegills dominate vegetation, juvenile pumpkinseeds spend more time in open habitats, where food is abundant but predation risk is higher. Survivors have increased growth rates until they reach a size at which they can feed safely on snails, with less competition (18).

Maturity sets in during the second or third year and does not seem to be strongly dependent on size, because populations of stunted pumpkinseeds under 10 cm FL are not unusual. Spawning takes place in April–June in California, as soon as water temperatures approach 13–17°C, and probably continues intermittently through August. The largest and oldest males spawn earliest in the year and dominate nest sites most favorable for survival of the embryos (19). The most favorable sites are in shallow water (<1 m) on bottoms of sand, gravel, or woody debris. The nests are built in loose colonies but defended individually. It is likely that pumpkinseeds, like bluegill, have males with alternate breeding strategies (20). Each parental male may spawn with several females and guards the embryos until they hatch, usually 3–5 days. The number of eggs laid by each female varies from 600 to 7,000, fecundity increasing with age and size (10, 21).

In their native range active pumpkinseed nests often attract golden shiners for spawning (22). Embryos of the shiner larvae have significantly higher survival when guarded by male sunfish, and sunfish embryos benefit from reduced predation, because of the "extra" embryos on the nest periphery. However, sunfish nests with many shiner embryos may also have higher mortality rates owing to fungus infections.

As soon as newly hatched young are able to swim, they leave the nest and venture into open waters, where they drift among and feed on zooplankton for several weeks (23). The presence of multiple spawning in pumpkinseeds, over the summer, is hypothesized to be a mechanism that allows larval sunfish to take advantage of zooplankton available in midsummer, reducing intraspecific competition (24).

Status IIC. Pumpkinseeds are well established only in the Klamath and Susan River drainages but seem to be spreading in the Sacramento–San Joaquin drainage. Because they are the most beautifully colored of all the sunfishes now in California, their spread to other waters by unthinking aquarists and anglers is likely. Such movement should be discouraged, although it is doubtful they can become widely established in warm waters containing a diversity of other sunfish. The ecologically similar redear sunfish grows faster in central and southern California and does not produce stunted populations. In Europe the widespread pumpkinseed is regarded as a pest because it is associated with declines of native fishes and is mostly too small to produce much of a fishery (3). Pumpkinseeds have the potential to play a similar role in middle- to high-elevation reservoirs and lakes in central California.

References 1. Jordan and Evermann 1896. 2. Lee et al. 1980. 3. Lever 1996. 4. Dill and Cordone 1997. 5. Buettner and Scoppetone 1991. 6. Simon and Markle 1997. 7. Leidy 1983, 1984 and pers. comm. 1999. 8. Swift et al. 1993. 9. Houston 1982. 10. Becker 1983. 11. Lauder 1983. 12. Wainwright et al. 1991. 13. Mittelbach et al. 1992. 14. Seaburg and Moyle 1964. 15. Keast 1966. 16. Kimsey and Bell 1956. 17. Keast 1968b. 18. Arendt and Wilson 1997. 19. Danylchuk and Fox 1996. 20. Gross 1982. 21. Hubbell 1966. 22. Shao 1997a,b. 23. Faber 1967. 24. Fox and Crivelli 1998. 25. Peterson 1988. 26. D. P. Lee, CDFG, pers. comm. 1999.

Green Sunfish, *Lepomis cyanellus* Rafinesque

Identification The bodies of green sunfish are stouter than but not as deep as those of other California sunfishes. Their pectoral fins (13–15 rays) are rounded and when bent forward will barely reach the eye. Their terminal mouths are large for a sunfish, the maxillae extending past the front margin of the eye. Their opercular flaps are short and stiff and their gill rakers long and slender. Their dorsal fins have 9–11 spines and 10–12 rays; anal fins, 3 spines and 8–10 rays; pelvic fins, 1 spine and 5 rays; and lateral line, 46–50 scales. There are no teeth on the tongue except in a few large individuals. Adults are dark olive on the back, becoming lighter on the sides with iridescent green flecks. The sides often have colored stripes. The breast and belly are yellow-orange, and there are usually green iridescent streaks on the cheeks. Both dorsal and anal fins have a large dark blotch on the rear of the soft-rayed portion. Young show fine, closely spaced chains of iridescent green that give the impression of a very fine blue-green grid on the sides. The backs, sides, and fins of breeding males turn very dark, and the fins usually have bright yellow margins.

Taxonomy Despite a wide native distribution, green sunfish have not been divided into subspecies. The species hybridizes with all other species of *Lepomis,* presumably as a result of male green sunfish sneak-spawning in nests of other species. The most common hybrid in California is with bluegill; these hybrids when small combine the vertical bars of bluegill with the horizontal stripes of green sunfish to form a colorful checkerboard pattern.

Names Cyanellus means green. Other names are as for bluegill.

Distribution Green sunfish were originally native to most of the Mississippi drainage system including the Great Lakes, but their original distribution has been obscured by widespread introductions. They were absent from Atlantic drainages (1). They are now present in virtually every state in the United States. They have also become established through introductions in Germany, Korea, the Philippines, Morocco, South Africa, Swaziland, Brazil, and Mauritius (2). They were introduced into California in 1891, first into San Diego County, and have since been spread to streams, ponds, and reservoirs throughout the state by well-intentioned fishermen and biologists, who often thought they were

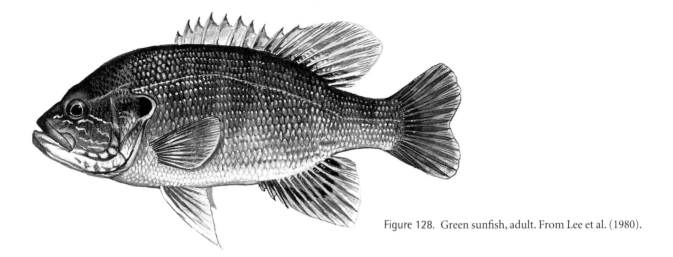

Figure 128. Green sunfish, adult. From Lee et al. (1980).

planting bluegill (3). In the upper Klamath River basin, they seem to be present only in the Lost River watershed (4). This means they are probably a fairly recent (1980s) introduction and are likely to spread to the rest of the region. In North Coast watersheds they have been recorded (so far) only from the Russian, Eel, and lower Klamath Rivers.

Life History Green sunfish are inhabitants of small, warm streams (especially those that become intermittent in summer), ponds, and lake edges. They are generally rare in habitats that contain more than three or four other species of fish. Thus in lakes and reservoirs they are usually only locally abundant in shallow, weedy areas that exclude larger or less tolerant species. In Clear Lake (Lake County), for example, they are found almost exclusively among beds of tules (5). In rivers they are often found in riprap and old car bodies. In central California they are abundant mostly in intermittent streams that have warm, turbid, muddy-bottomed pools containing beds of aquatic plants and populations of other introduced fishes, such as largemouth bass and mosquitofish (6, 7). They will often be the sole inhabitants of such streams, especially those that have been heavily disrupted or polluted by human activity. In less-disturbed streams at higher elevations it is not unusual to find a few large green sunfish in sections of stream dominated by native fishes, ready to take over at the first disturbance. In such streams they are typically associated with deep, bedrock-lined pools, where they hide in the crevices or under overhangs (8). In much of their native range green sunfish are a pioneer species, capable of surviving where other species cannot. They can survive high temperatures (>38°C), low oxygen levels (<1 mg/liter), and alkalinities up to 2,000 mg/liter (9, 10), although they prefer more moderate conditions (e.g., temperatures of 26–30°C seem to be optimal). They have rather low salinity tolerance, however, and avoid water with salinities higher than 1–2 ppt (27). They are extremely good dispersers, often the first fish to re-

colonize stream reaches dried up by drought. Thus they are capable of taking over small foothill streams that normally would contain only California roach and persisting in remnant pools through periods of extreme drought.

Green sunfish are very aggressive, although young-of-year frequently shoal. Older fish tend to be territorial for feeding. In a small aquarium, one fish, usually the largest, quickly assumes dominance and keeps other sunfish in a restricted area while it defends the rest as its territory. Green sunfish will also chase other species of fish from their territories. Such territoriality may not be as pronounced in the wild, but, once a large green sunfish has been located in a stream, it can generally be found in the same area for long periods of time. This aggressiveness may be one reason they are so quick to colonize new waters; small fish are presumably always seeking feeding areas not dominated by large fish. Despite their aggressiveness, green sunfish are frequent prey of largemouth bass and other piscivores. Given a choice, bass will select green sunfish over bluegill because their body shape makes them easier to handle and ingest (11). This may explain why green sunfish are relatively uncommon in many habitats, even though they are competitively dominant over bluegill (12).

As their large mouths and aggressive natures indicate, green sunfish are opportunistic predators on invertebrates and small fish, feeding on a wider spectrum of prey than other sunfishes. Young-of-year feed on zooplankton and small benthic invertebrates, such as chironomid midge larvae and mayfly larvae, and readily consume larvae of other fish (13). In Clear Lake they eat mostly invertebrates associated with emergent tules, such baetid mayfly larvae. As they increase in size they depend more on large aquatic insects, such as dragonfly larvae, terrestrial insects, crayfish, and fish, including their own young (14, 15, 16).

Compared with other sunfishes, green sunfish grow slowly and seldom reach sizes greater than 15 cm SL. They usually reach 3–5 cm SL in their first year, 5–10 cm in their second year, and 8–13 cm in their third year. Such growth seems to be typical for populations in California reservoirs (17). They can grow to more than 30 cm SL, achieve weights of nearly 1 kg, and live as long as 10 years. Yet green sunfish that even approach such longevity and size are extremely unusual, especially in California, because they are so aggressive, and large fish are usually caught by anglers. Some of the biggest fish I have seen have come from ponds in golf courses where fishing is forbidden. They are so prolific that large populations of stunted fish often develop. Thus it is not unusual to find 4- to 5-year-old fish that measure only 8–10 cm SL. For example, Eleanor Reservoir, Yosemite National Park (elevation, 1,420 m; summer surface temperature, 23–25°C) is dominated by remarkably small green sunfish that average 28 mm SL at the end of their first year, 47 mm at the end of their second year, 57 mm at the end of their third year, and 67 mm at the end of their fourth year (18). The largest sunfish observed measured 110 mm SL and was 6 or 7 years old. These fish have suppressed a wild rainbow trout population that once existed in the lake, presumably through competition with and predation on the small trout. The few trout that manage to grow large enough to prey on green sunfish, however, grow rapidly (18).

One of the reasons green sunfish often form stunted populations is that they can reproduce at 5–7 cm SL and usually mature at the beginning of their third year. They also are capable of spawning in disturbed waters (e.g., those with low dissolved oxygen levels) that exclude most other fishes (19). Spawning activity is most intense in May and June but often continues into July and August. Although green sunfish have been observed spawning at temperatures of 15–28°C (13), spawning in California usually does not begin until water temperatures reach 19°C. The first noticeable activity is the congregation of male green sunfish in shallow water; 1–2 days later the males start to dig nests, singly or in colonies, preferably on fine gravel bottoms near overhanging bushes or other cover. Nests are built in water 4–50 cm deep and are typically 15–38 cm in diameter (13).

Each male defends his nest against other males and, to a lesser extent, females. Females hover about the colony in small schools and are quickly courted and spawned with, sometimes two at a time (20). Courtship is accompanied by gruntlike sounds (21). During spawning females turn on their sides, vibrating and releasing eggs while the males remain alongside in an upright position, simultaneously releasing sperm (20). Each male may spawn several times with different females, and each female spawns with multiple males. Each female carries 2,000–10,000 eggs, depending on her size (19). Presumably their mating system is fairly complex, like that of bluegill (22).

Fertilized eggs adhere to the nest substrate, and males guard the embryos for 5–7 days, enough time for the young to hatch and become free-swimming. The larvae are planktonic and feed on small zooplankters for a few days before settling down on or near vegetation. Juvenile green sunfish suffer heavy predation losses at this stage and apparently have evolved a chemical alarm substance as a result (28).

Status IID or IIE. Green sunfish are found throughout California and are no doubt still spreading to new areas. Their introduction into California can only be regarded as unfortunate. They provide little in the way of sport or food and a great deal in the way of competition for (and predation on) native nongame fishes and other game fishes. In ponds and lakes they form large stunted populations that seriously affect the population size and growth of more desirable game fishes (23), as they do in Eleanor Reservoir (18). Whenever they invade a small stream or pool of a larger stream, including streams outside California (25), small native fishes such as minnows and sticklebacks tend to disappear (8). Because they are a common backwater species in the lower Colorado River, they are a significant part of the exotic predator complex that prevents reestablishment of native fishes. They have probably been responsible for elimination of California roach in a number of small streams in the foothills of central California (24). An example of their impact can be seen in Dye Creek (Tehama County), a stream whose upper reaches were dominated by California roach. The south fork of Dye Creek was invaded by green sunfish, and roach are now largely absent, except upstream of the farthest penetration of sunfish. In contrast they have recently invaded the cooler north fork of the creek, which contains a thriving population of native fish, but have so far not become the dominant species (26). There is consequently a need to prevent further spread of green sunfish to additional streams and to develop techniques for eradicating them from places where they do the most damage.

References 1. Lee et al. 1980. 2. Lever 1996. 3. Dill and Cordone 1997. 4. Buettner and Scoppetone 1991. 5. Week 1982. 6. Moyle and Nichols 1973. 7. Brown and Moyle 1993. 8. Smith 1982. 9. McCarraher 1972. 10. Smale and Rabeni 1995. 11. Savitz and Janssen 1982. 12. Werner and Hall 1977. 13. Becker 1983. 14. Applegate et al. 1967. 15. Moyle 1969. 16. Minckley 1982. 17. E. Miller 1970. 18. Moyle and Baltz 1985. 19. Wang 1986. 20. Hunter 1963. 21. Gerald 1971. 22. Gross 1982. 23. McKechnie and Tharratt 1966. 24. Moyle and Nichols 1974. 25. Lohr and Fausch 1996. 26. P. K. Crain and P. B. Moyle, unpubl. rpt. 1997. 27. Peterson 1988. 28. G. Brown and Brennan 2000.

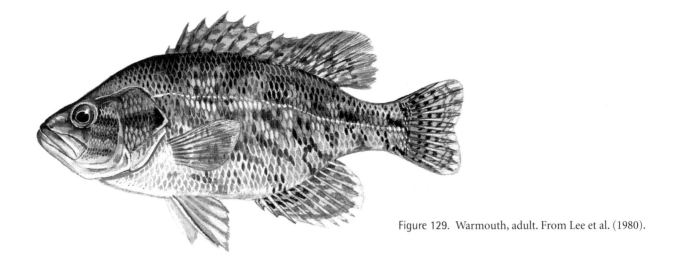

Figure 129. Warmouth, adult. From Lee et al. (1980).

Warmouth, *Lepomis gulosus* (Cuvier)

Identification Warmouth look like stout green sunfish, except that they are brown and have teeth on their tongues. Their terminal mouths are large, maxillae extending past the front margin of their eyes. Their opercular flaps are short and stiff, and their gill rakers are long, straight, and slender. The pectoral fins are short and rounded (12–14 rays) and when bent forward reach the edge of the eye. The dorsal fin is spotted with 10–11 spines and 9–11 rays, while the anal fin has 3 spines and 9–10 rays. The pelvic fins have 1 spine and 5 rays. There are 38–45 scales in the lateral line. Their overall color is brown, with an iridescent green to purple tinge to their scales, and they have dusky bellies. Three to five distinct lines radiate out from each eye. Faint vertical bars may sometimes be present on their sides. Four to six dark brown bars radiate across their cheeks from their eyes and mouths. Breeding males have bright red eyes and yellow bellies.

Taxonomy Warmouth were originally placed by themselves in the genus *Chaenobryttus*. Although phylogenetic analyses typically separate them from other members of *Lepomis*, the differences are usually not regarded as sufficient to justify a separate genus (1, 2, 3). Use of *Chaenobryttus* depends on whether the taxonomist wishes to emphasize differences or similarities in a classification scheme. Warmouth hybridize in the wild with other sunfish. It is not unusual to find bluegill-warmouth hybrids in sloughs along the Sacramento River.

Names The origin of the name warmouth is obscure, but it quite likely is based on a fancied resemblance of markings on the head to the war paint of Native Americans. Warmouth are frequently called warmouth bass and, occasionally, goggle-eye. *Gulosus* means large mouth or throat, interpreted to mean gluttonous. Other names are as for bluegill.

Distribution Warmouth are native to the Mississippi River drainage from northern Iowa on south, as well as to the Rio Grande River drainage, Gulf Coast drainages, Florida, and much of the Atlantic seaboard (3, 4); some of the populations at the edge of this distribution probably represent introductions. Further range expansion was made by introductions into Washington (where they have limited distribution), Oregon (where they are present in the Willamette River), and California (5, 6). They have also been introduced into Mexico and Puerto Rico (7). The exact date of their introduction into California is not known. It is possible that they were first planted in southern California and the Feather River in the Sacramento Valley in 1891, although they are not mentioned as part of the Central Valley fauna until the 1930s (8). They are now present, if uncommon, in waters of the Delta and Central Valley floor and in a few reservoirs at higher elevations (e.g., Bass Lake [reservoir], Madera County; Amador Reservoir, Amador County; McClure Reservoir, Merced County). They mysteriously appeared in the lower Colorado River in 1961 (9) and are now fairly common there. There are no recent records from southern California reservoirs, but they can be expected in those fed by the California Aqueduct.

Life History Most warmouth in California are found where there is abundant vegetation and other cover in warm, turbid, muddy-bottomed sloughs and backwaters of the Sacramento and Colorado Rivers. They also do well in reservoirs such as Bass Lake, a cool, fluctuating reservoir that supports substantial salmonid populations. They are uncommon in tidal portions of the estuary. These observations support scanty data on their environmental tolerances that suggest that their optimal summer temperatures are around 22–28°C; that they can withstand oxygen levels under 4 mg/liter in warm water; that they can live in fairly turbid water; and that they avoid salinities higher than 1–4 ppt, although they can survive salinities as high as 17 ppt (3, 10). Most of what is known about warmouth life history comes from a study in Illinois (11).

Warmouth are opportunistic predators that tend to hide quietly in ambush. Fish measuring less than 5 cm TL feed mostly on small crustaceans but start taking insect larvae and snails as they increase in size. By the time they reach 10–13 cm TL, they are feeding mostly on aquatic insects. Larger fish take larger organisms, and fish and crayfish are usually important to fish larger than 13 cm TL. In the Sacramento–San Joaquin Delta, warmouth of all sizes eat opossum shrimp (*Neomysis*), amphipods (*Corophium*), and aquatic insects, although larger warmouth also eat crayfish and fish (12). Peaks of feeding seem to be early morning and dusk.

Warmouth are fairly long-lived, but they grow so slowly that an individual more than 28 cm TL and 450 g would be a giant of its species. In their native range they typically reach 3–9 cm TL in their first year, 6–14 cm in their second year, 9–17 cm in their third year, 11–20 cm in their fourth year, and 13–21 cm in their fifth year. They often live 6–8 years. In stunted populations fish measuring 10 cm TL are 4–6 years old. Fish in newly established populations, on the other hand, may show fast growth for the first year or two, reaching 10–12 cm TL in their first year. Warmouth in the Colorado River appear to belong in this latter category (9). The angling record for California, a fish from the American River caught in 1982, weighed only 340 g.

Warmouth mature in their second or third summer at 7–10 cm TL. Spawning takes place in late spring and early summer when temperatures reach about 21°C. Warmouth are usually nongregarious when breeding. Males build nests near dense cover, at depths of 0.5–1.5 m, and their spawning and parental behavior is similar to that of green sunfish. Females produce 4,500–63,000 eggs, depending on size.

Status IIC. Warmouth are well established but uncommon in California, especially compared with other sunfishes. For example, in 1993 the fish traps at the giant state and federal pumping plants in the South Delta rescued only 200 warmouth, compared with 250,000 individuals of other centrarchid species (mostly bluegill and largemouth bass). Warmouth add little to the warmwater sport fishery of California, and the reason for their introduction is not known (8). Most likely the introduction was made either because the individuals brought in were thought to be the larger rock bass (*Ambloplites rupestris*) or because they were mixed with shipments of other sunfish. Their ecological role in sloughs and reservoirs is poorly understood, especially their interactions with other fish species.

References 1. Mabee 1993. 2. Etnier and Starnes 1993. 3. Jenkins and Burkhead 1994. 4. Lee et al. 1980. 5. Bond 1973. 6. Wydoski and Whitney 1979. 7. Lever 1996. 8. Dill and Cordone 1997. 9. Lanse 1965. 10. Becker 1983. 11. Larimore 1957. 12. Turner 1966b.

White Crappie, *Pomoxis annularis* Rafinesque

Identification White crappie have deep, laterally compressed bodies and small heads with a depression in the profile above the eyes fairly close to the pointed snout. Their mouths are large and oblique, so the lower jaws appear to project. The eyes are large. Their large, rounded dorsal and anal fins are nearly equal in size, giving a symmetrical appearance. The length of the dorsal fin base is less than the distance from the origin of the dorsal fin to the eye. There are 5–6 spines (occasionally 7) and 13–15 rays in their dorsal fins; 6–7 spines and 16–18 rays in their anal fins; 1 spine and 5 rays in their pelvic fins; and 13–14 rays in their rounded pectoral fins. The lateral line is arched, with 34–46 scales. Adults are iridescent olive green on their backs and silvery white on their sides, usually with 10 or fewer indistinct, dark vertical bars. Dorsal, anal, and caudal fins are checkered with dark spots. Breeding males become very dark, the head and breast turning nearly completely black.

Taxonomy See the account of black crappie.

Names The name crappie (politely pronounced "croppie") is, according to Jenkins and Burkhead (1, p. 712), "apparently . . . derived from the French Canadian word Crapet, the etymology of which is unclear. . . . Crapet may have been

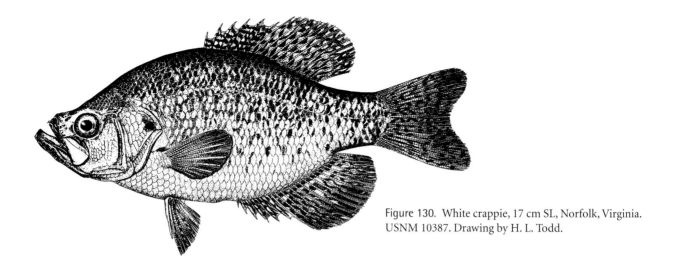

Figure 130. White crappie, 17 cm SL, Norfolk, Virginia. USNM 10387. Drawing by H. L. Todd.

derived from crapoud, which means toad, or it may have been applied to fish with a large head and big mouth." White crappie are sometimes called calico bass or strawberry bass, names that are more frequently applied to black crappie. *Pom-oxis* means cover-sharp, referring to the fact that the operculum ends in a blunt (not sharp!) point rather than in a distinct flap, as is characteristic of sunfish. *Annularis* means having rings, presumably a reference to the vague banding that seems to encircle the body.

Distribution White crappie were originally distributed throughout the Mississippi River basin north into Minnesota, east through the Great Lakes basin, and west and south to the Rio Grande River and Gulf Coast drainages of northern Mexico. They have been introduced successfully into reservoirs and lakes throughout the United States and northern Mexico. The exact date of their introduction into California is uncertain because of confusion between the two crappie species. Dill and Cordone (2) indicate that 1908 is the most likely year, with the introduction of white crappie from Illinois into reservoirs in southern California. They were not planted north of the Tehachapi Mountains until 1951 (3), about the same time they invaded the Lost River in California from Oregon introductions (2). They

may have been established in the Colorado River as early as 1920 (2). They are now well established, mainly in reservoirs, in all major basins of California, including Susan River, Russian River, and Clear Lake.

Life History White crappie are most abundant in warm, turbid lakes, reservoirs, and river backwaters. They seem to have slightly greater tolerances than black crappie for high turbidity, alkaline water, current, high temperatures, and lack of aquatic vegetation and cover (3, 4). They apparently have less tolerance for low dissolved oxygen levels, however, and may be replaced by black crappie in reservoirs that have low (2–4 mg/liter) dissolved oxygen levels during dry years (5). Optimal temperatures seem to be 27–29°C, and temperatures higher than 31°C are avoided (6). White crappie are rare in estuaries, but in Suisun Marsh they have been collected at salinities as high as 10 ppt. They can occur in streams, especially downstream from reservoirs, but have a hard time persisting through high-flow periods. In Putah Creek (Yolo-Solano County) they became very abundant in large pools during an extended period of drought but became rare after a series of high-flow events.

White crappie are shoaling fish, and their aggregations are often rather localized. Individuals, however, may move considerable distances within a body of water (7). During the day they tend to congregate around submerged logs or boulders in quiet water 2–4 m deep. They move into open water to feed during evening and early morning. Often they move closer to the surface as well (8). At low temperatures they are rather inactive, remaining close to the bottom in deep water (7).

The feeding mechanisms of white crappie are unusual in that they have long, fine gill rakers, suitable for retaining small zooplankters, combined with large, protrusible mouths that are suitable for ingesting large prey, including fish. Their deep bodies are not designed for extended pursuit of prey but rather for hanging in the water column, where their pale color, streaked sides, and flat shape help to make them less visible to prey, especially at low light levels. Their basic feeding strategy is to swim a short distance, halt, scan for prey, and then capture whatever is close by (9). Even their larvae use this strategy (10).

As the result of this combination of morphology and feeding strategy, the diet of white crappie is typically a mixture of planktonic crustaceans and small fish (3, 11). However, they are also opportunistic and will eat aquatic insects when they are readily available. Zooplankton are the main food of crappie measuring less than 140 mm FL. In Clear Lake (Lake County) copepods are the main prey of fish measuring less than 25 mm SL, and cladocerans (mainly *Daphnia*) become more important as fish grow larger. Fish and large invertebrates usually predominate in the diet of individuals larger than 140 mm FL. In California reservoirs threadfin shad are especially important prey. In Clear Lake inland silversides are important in their diet, and crappie move inshore at dusk to feed on them. Because of the abundance of small silversides, crappie switch to feeding on them at a small size (about 60 mm SL). Young-of-year feed mostly during the day with a peak of feeding in midafternoon (12).

Growth in California reservoirs is generally somewhat slower than growth where they are native. They reach 5–10 cm FL in their first year, 11–18 cm in their second year, 17–21 cm in their third year, and 20–27 cm in their fourth year (3). The introduction of threadfin shad into Isabella Reservoir (Kern County) increased the growth rates of white crappie, especially those in their second and third years (13). In Clear Lake, prior to the establishment of inland silverside, white crappie had the following standard lengths at the end of each year: 70, 145, 178, 189, 213, and 193 mm. After establishment of silversides the lengths were 56, 103, 138, 184, 225, and 226 mm (14). Apparently the depletion of zooplankton by silversides significantly reduced growth of juvenile crappie, but after their third year predation on silversides by crappie resulted in increased growth rates and larger sizes in adult fish. In California white crappie seldom live longer than 7–8 years or grow larger than 35 cm FL (0.8 kg) (3). The angling record for California, from Clear Lake, is for a 2.04-kg fish.

White crappie become mature in their second or third spring at 10–20 cm TL. Spawning usually begins in April or May at 17–20°C. The males construct nests in colonies underneath or close to overhanging bushes or banks in water less than 1 m deep (4, 15, 16). Nests are occasionally built in water as deep as 6–7 m. Nests usually consist of shallow depressions in hard clay bottoms (rarely in sand or gravel) near or in beds of aquatic plants, algae, or submerged plant debris (4, 15). The embryos adhere to plant material in the nest. Spawning behavior is similar to that of sunfishes, including, apparently, alternative male strategies (4). Nests are defended vigorously, and human swimmers will occasionally get bitten by an aggressive male. Fecundity is highly variable, and the number of eggs (970–326,000) is only partially related to size (6).

After leaving the nest the larvae are planktonic. Small juveniles also spend much of their time in the water column, feeding on zooplankton, but often aggregate in protected areas near shore during day (16).

Status IID. White crappie are a highly favored game fish and do fairly well in warmwater reservoirs of California. They have been stocked in virtually every one as a consequence. Their populations can show wide fluctuations. White crappie were introduced into Clear Lake in 1957 and quickly became very abundant. Their populations collapsed to low levels in the late 1970s and have not recovered. The reasons for the collapse, which also happened to black crappie, are not known. The effects of white crappie on native fishes are also not known, but they are likely to have been minimal because white crappie are primarily inhabitants of reservoirs and other highly disturbed habitats.

References 1. Jenkins and Burkhead 1994. 2. Dill and Cordone 1997. 3. Goodson 1966. 4. Hansen 1951. 5. McDonaugh and Buchanan 1991. 6. Becker 1983. 7. Grinstead 1969. 8. Markham et al. 1991. 9. O'Brien et al. 1986. 10. Browman and O'Brien 1992. 11. Mathur 1972. 12. Mathur and Robbins 1971. 13. Bartholomew 1966. 14. Li et al. 1976. 15. Hansen 1965. 16. Siefert 1968.

Black Crappie, *Pomoxis nigromaculatus* (Lesueur)

Identification The body shape of black crappie is similar to that of white crappie except that it is slightly heavier bodied. Black crappie also have a dorsal fin placed fairly far back on the body with a rounded end that is symmetrical with the end of the anal fin and a sloping head with a dip above the eye. They can be distinguished from white crappie by their longer dorsal fin (7–8 spines, 15–16 rays), the base of which is about the same length as the distance from the fin origin to the middle of the eye. They have 6 spines and 17–19 rays in their anal fins, 1 spine and 5–6 rays in their pelvic fins, 14–15 rays in their pectoral fins, and 38–44 scales in their arched lateral lines. Their body (side) coloring is whitish-silvery with heavy black spotting that is not arranged in vertical bands. The spots on the dorsal, anal, and caudal fins can be arranged in loose bands, but the pattern is often indistinct. The back is dark, the belly white.

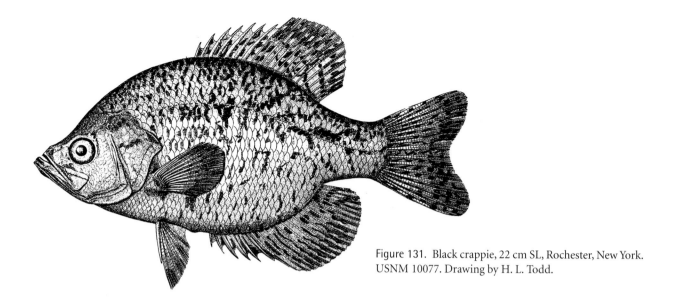

Figure 131. Black crappie, 22 cm SL, Rochester, New York. USNM 10077. Drawing by H. L. Todd.

Breeding fish turn nearly solid black on the anterior halves of their bodies.

Taxonomy The two species of crappie show comparatively little variation over their natural range and have not been broken into subspecies as a consequence. "Florida strain" black crappie were introduced into Clear Lake (Lake County) and a few other places in the late 1980s, on the assumption that they would do better in California than fish of more northern origin (1), but there is no evidence that this assumption was borne out. Although capable of hybridization, the two species rarely do so.

Names *Nigro-maculatus* means black-spotted. Other names are as for white crappie.

Distribution The native distribution of black crappie was apparently similar to that of white crappie except that they occurred considerably farther north in the Great Lakes region of Canada (2) and along the south Atlantic coast. Thus they were originally distributed throughout the Mississippi River basin from Quebec, Ontario, and Manitoba southward, throughout the Great Lakes basin, south to the Rio Grande River and Gulf Coast drainages into Texas and then in Gulf and Atlantic coast drainages north to Virginia, including Florida (3). They have been introduced successfully into reservoirs and lakes throughout the United States, southern Canada, and northern Mexico, as well as into Guatemala and Panama (4). The exact date of introduction into California is uncertain because of confusion between the two crappie species. Dill and Cordone (1) indicate that 1908 is the most likely year, with introduction of black crappie from Illinois into reservoirs in southern California. They were transplanted to the Central Valley in either 1916 or 1919 (or both) and quickly became abundant (1). They were established in the Colorado River by the 1940s (1). They are now well established, mainly in reservoirs, in all major basins of California, except the upper Klamath basin and a few Great Basin watersheds. They can be expected anywhere in the state where there is warm, quiet water.

Life History Black crappie are most successful in large, warmwater lakes and reservoirs. Optimal summer temperatures appear to be around 27–29°C (5); temperatures greater than 31°C are stressful, and those above 37–38°C are usually lethal (6). However, their distribution both within California and in their native range suggests a higher tolerance for lower and higher temperatures, as well as for other factors, than in white crappie. They are more abundant in the tidal sloughs of the San Francisco estuary than white crappie, although their tolerances for low dissolved oxygen levels (1–2 ppt for short periods of time) and salinity (up to 10 ppt) seem to be about the same.

Black crappie are usually found in highly localized shoals that hang around large submerged objects during the day but move offshore (or inshore if prey are abundant) in the evening and early morning. However, movement patterns are variable. One study found that movements increased as barometric pressure increased (7).

The feeding mechanisms of black crappie are almost identical to those of white crappie, so it is not surprising that their diets are similar. Black crappie may be somewhat less piscivorous. They are primarily midwater feeders: zooplankton and small dipteran larvae, especially chironomid midges, predominate in the diet of small fish (10–12 cm SL), whereas fish and aquatic insects predominate in the diet of larger fish (8, 9). However, it is not uncommon to find large amounts of planktonic crustaceans in the stomachs of fish up to 16 cm SL. In sloughs of the Sacramento–San Joaquin Delta in the 1960s opossum shrimp (*Neomysis*), amphipods (*Corophium*), and planktonic crustaceans were the main foods of black crappie measuring less than 10 cm FL, whereas fish (mostly threadfin shad and juvenile striped bass) were the main foods of adults (10). In Clear Lake crappie larger than 50 mm SL fed mostly on chironomid larvae. The importance of fish seems related to the frequency of infection of their main prey, inland silversides, by a parasitic copepod which makes affected individuals more visible to predators (11). In Britton Reservoir (Shasta County) adults feed on a variety of zooplankton, insects (but especially *Hexagenia* mayfly larvae), crayfish, and tule perch (19, 20). Because black crappie will feed at temperatures as low as 6–7°C (12), California populations feed throughout the year. Black crappie will forage at virtually any time of day or night, but tend to peak around noon, midnight, and early morning (9).

Growth in California is, on average, somewhat slower than growth in populations in the eastern United States, but some populations (e.g., that in Clear Lake) have excellent growth rates. In California black crappie measure 4–8 cm FL at end of their first year, 12–21 cm at end of their second year, 15–28 cm at end of their third year, and 17–33 cm at end of their fourth year (13). In Clear Lake, as is the case for white crappie, depletion of zooplankton by silversides significantly reduced growth of juvenile crappie, but after their third year predation on silversides resulted in increased growth rates and larger sizes in adult fish (14). The maximum age for black crappie seems to be about 13 years, and

the maximum size about 2.2 kg, although fish more than 6 years old and weighing more than 1 kg are unusual (15). The angling record from California, a fish from New Hogan Reservoir (Calavaras County), weighed 1.9 kg.

Black crappie mature in their second or third year at 10–20 cm TL. Spawning begins in March or April as temperatures exceed 14–17°C and may continue into July. Peak spawning typically occurs when temperatures are 18–20°C (16). Nests are shallow depressions 20–23 cm in diameter fanned out by males in mud or gravel bottoms in water less than 1 m deep near or in beds of aquatic plants. A male generally constructs a nest 1–2 m from his nearest neighbor, so spawning fish form a loose colony. Reproductive behavior is similar to that of white crappie, although it has not been described in as much detail. Each female lays up to 188,000 eggs, depending in part on size, with 3- to 4-year-old fish producing 33,000–42,000 eggs (16).

The newly hatched fry are guarded for a short period of time by males, but they soon rise off the nests and spend the next few weeks drifting in open water, feeding on zooplankton (17).

Status IID. Black crappie are abundant and popular game fish in California lakes and reservoirs. They are present in most waters that can support them, but no doubt new locations will be found for them. Black crappie are ecologically fairly similar to Sacramento perch, and once they are established in a location perch disappear (18). As early as the 1920s it was noticed that, as black crappie became abundant in the Delta region, Sacramento perch declined (1). Possibly a major interaction between the two species is competition for breeding sites, because they have roughly equivalent requirements; however, the crappie is probably much more aggressive.

References 1. Dill and Cordone 1997. 2. Scott and Crossman 1973. 3. Lee et al. 1980. 4. Lever 1996. 5. Houston 1982. 6. Baker and Heidinger 1996. 7. Guy et al. 1992. 8. Keast and Webb 1966. 9. Keast 1968a. 10. Turner 1966b. 11. Loshbaugh 1978. 12. Keast 1968b. 13. Goodson 1966. 14. Li et al. 1976. 15. Jenkins and Burkhead 1994. 16. Becker 1983. 17. Faber 1967. 18. Aceituno and Nicola 1976. 19. PG&E 1985. 20. K. Marine, University of California, Davis, unpubl. study 1983.

Largemouth Bass, *Micropterus salmoides* (Lacépède)

Identification Largemouth bass are the heaviest bodied of the California "black" basses, which in general are more elongate than sunfishes, although they become deeper bodied with age. They are distinguished by their large mouth, with

maxillae that extend to or past the hind margin of their eyes in fish longer than 10 cm TL. The two parts of the dorsal fin are nearly separated, with 9 spines in the first part and 12–14 rays in the second. The spinous dorsal fin has its longest spine in the middle, and that spine is more than twice as long as the shortest spine that follows. There is a single, more or less con-

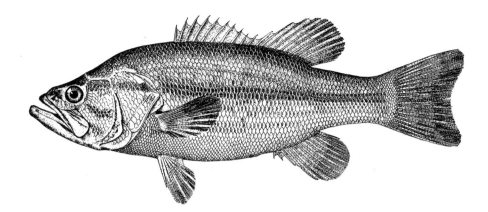

Figure 132. Largemouth bass, 23 cm SL, Potomac River, Virginia. USNM 14143. Drawing by H. L. Todd.

tinuous, heavy black lateral stripe on each side. Their anal fins have 3 spines and 11–12 rays; pectoral fins, 13–17 rays (usually 14–15); and lateral lines, 58–72 scales. The scales on their cheeks, in 9–12 rows, are about the same size as the scales on the operculum. Scales are absent from the bases of the dorsal and anal fins, and there are no teeth on the tongue. Their pyloric cecae are forked. They tend to be olive gray to shiny green on the back and sides and white on the belly, with the stripe in between and no other conspicuous markings. The lower sides may be speckled but lack rows of small spots. The eyes are brown. Juveniles lack any orange in the caudal fin, which is usually bicolored (but may be without strong banding), and have a lateral stripe that is more or less continuous (as opposed to being a series of distinct blotches).

Taxonomy In much of the older literature largemouth bass are placed in the genus *Huro,* separate from other basses, a mark of their distinctiveness. There are two subspecies, northern largemouth bass (*M. s. salmoides*), from most of their native range, and Florida largemouth bass (*M. s. floridanus*) from peninsular Florida (1). The two forms are different enough genetically that they quite likely should be listed as separate species. Both forms have been introduced into California and hybridize where they are found together (which seems to be most places), with the Florida-strain phenotype often becoming dominant. Largemouth bass

also hybridize with smallmouth bass and will form introgressed populations (40), although there are no records of this phenomenon yet in California.

Names All members of *Micropterus* are commonly referred to as black bass, hence largemouth black bass. Lacépède based his description of the genus on a single specimen of smallmouth bass with a deformed dorsal fin in which the last few rays were separated from the fin, giving the appearance of a separate fin. Thus *Micro-pterus* means short fin (2). *Salmoides* means troutlike.

Distribution Historically largemouth bass ranged from northeastern Mexico through much of the Mississippi and affiliated drainages, north into southern Ontario and Quebec, although they were apparently absent from the Atlantic seaboard north of South Carolina (3, 4). They have been introduced into all the continental United States (and Hawaii) and most provinces of Canada. They are also established in scattered locations throughout Europe, Japan, Korea, Algeria, Botswana, South Africa, Swaziland, Tanzania, Uganda, Zimbabwe, all Central American countries, Bolivia, Brazil, Colombia, Mauritius, New Caledonia, and no doubt many other countries (5). Northern largemouth bass were brought into California from Illinois in 1891 and planted in both Cuyamaca Reservoir (San Diego County) and the Feather River (Colusa County) (6). Subsequently they were spread statewide by eager anglers and agency biologists. Florida largemouth bass were first planted in San Diego County in 1959, on the assumption that they would grow larger and be harder to catch than northern largemouth bass. They were subsequently spread to northern California reservoirs, where they have hybridized with northern largemouth bass (7). The most recent expansion of their range has been into Lake Tahoe, where they are now common in warm, shallow areas, such as the Tahoe Keys (35).

Life History Warm, shallow (<6 m) waters of moderate clarity and beds of aquatic plants are the usual habitat of largemouth bass. They are abundant in farm ponds, lakes, reservoirs, sloughs, and river backwaters where other nonnative fish are abundant as well, especially species such as bluegill, redear sunfish, black and brown bullheads, golden shiners, threadfin shad, and mosquitofish. For example, in low-elevation streams above the Central Valley they occur mostly in disturbed areas where there are large, permanent pools with heavy growths of aquatic plants and 2–5 other introduced species (8, 9, 10). Stream populations are often maintained by continuous colonization from upstream sources, usually farm ponds or reservoirs (11). During periods of high flow bass may be flushed out of streams, although they do have an astonishing capacity to persist on their own, by finding shelter in flooded areas. They quickly recolonize such streams and build up populations during periods of low flow. Their persistence in isolated stream pools during droughts or in polluted waters is due to their ability to withstand adverse water quality conditions. They can persist in waters that approach 36–37°C during the day with dissolved oxygen levels as low as 1 mg/liter (12, 13). Optimal temperatures for growth of bass over 10 cm SL are 25–30°C, although growth will occur within a much wider range (10–35°C) (12). Given a choice, adult bass will hang out at around 27°C, and movements away from shallow-water feeding areas are noted when temperatures exceed this value (12). Juvenile bass, however, prefer temperatures of 30–32°C (12), perhaps as a mechanism to reduce cannibalism by adults and to ensure rapid growth in food-rich shallow-water habitats. Northern largemouth bass seem able to withstand extremes of temperature (high and low) better than Florida largemouth bass (14).

In their native habitats largemouth bass are known to live in estuarine conditions with salinities up to 16 ppt (15), but in California it is unusual to find them in water with salinities much higher than 3 ppt, and they seem to actively avoid salinities higher than 5 ppt (39). They are abundant, however, in the tidally influenced freshwater sloughs of the Delta. Bass also have a hard time persisting in alkaline waters, and this characteristic has fortuitously prevented their establishment in some waters of the state. Thus during a wet period in the early 1900s when Eagle Lake (Lassen County) had high water levels and comparatively low alkalinity, largemouth bass were introduced and a fishery established. After lake levels dropped, however, and pH increased to over 9, the bass died out, leaving the lake to its native fishes and birds.

Adult bass are solitary hunters. Each individual may either remain in a relatively restricted area centered around a submerged rock or branch (16) or wander widely. Certain places in large lakes repeatedly yield large bass to fishermen at intervals, indicating that each fish may establish a "home range" for a number of days at one spot and then move on to a new area. In reservoirs and lakes they remain close to shore and seem to be most abundant in water 1–3 m deep. Young-of-year bass and yearling bass also stay close to shore in schools that cruise near or above beds of aquatic plants. Schools of juveniles tend to stay in limited areas and are most active during the day (17). In crowded ponds older and larger bass may also school. Bass of all sizes are active most of the day and during moonlit nights. Usually, however, they become quiescent after dark, following intense foraging at dusk. Foraging is most efficient at low to moderate light levels, when prey have a harder time seeing the approaching predator (18).

Largemouth bass, with their large gape and roving body shape, are admirably suited for capturing the abundant fishes and large invertebrates that occur with them. Behaviorally they are very flexible and can capture prey by mechanisms as diverse as pursuit or ambush. They are also capable of changing foraging behavior in accordance with prey availability, type of habitat, experience, and body size (19). However, individual bass tend to specialize somewhat in their prey, at least over short periods of time (19). For the first month or two following hatching, fry feed mainly on rotifers and small crustaceans, but by the time they reach 50–60 mm SL they are feeding largely on aquatic insects and fish fry, including those of their own species (7, 20, 21, 22). In ponds, if one keeps track of an individual school of nestmates for a month or more after hatching, it soon becomes obvious that, as the schools become smaller, one or two members of each school become noticeably larger than the rest by feeding on their fellow bass fry. Once they exceed 100–125 mm SL they usually subsist primarily on fish. However, adults occasionally prefer crayfish, tadpoles, or frogs to fish, and they also prefer one fish species to another (23). The preferred prey can vary from year to year, and the apparent preference cannot always be explained in terms of the relative abundance of prey organisms. Thus in Clear Lake (Lake County) small bluegill have been abundant in shallow water since the 1920s, yet in 1948 largemouth bass measuring more than 12 cm FL fed mostly on Sacramento blackfish (7, 24). In 1956–1958, when blackfish were uncommon, they switched to feeding on bluegill. By 1973 they had switched again to feeding on the recently introduced inland silverside, although silverside importance in the diet varied tremendously with year and habitat (7). In California reservoirs they feed largely on threadfin shad, golden shiners, and bluegill (25, 26).

The flexible foraging strategies of largemouth bass and their wide environmental tolerances have made them a keystone predator in many bodies of water (19). A keystone predator is a species whose activities can cause changes throughout the ecosystem, usually by changing abundances of favored prey. In small lakes largemouth bass can reduce numbers of plankton-feeding fishes; this outcome allows large zooplankton to flourish, and they in turn graze on algae in the water column. As grazing by large zooplankton increases and algae decline, the lakes become clear (27). In California these changes have been poorly documented, but the decline and even disappearance of native minnows following bass introductions have presumably had impacts that have cascaded through local systems, such as Clear Lake.

However, largemouth bass do not appear to play a keystone role under the fluctuating conditions of reservoirs. In some situations their numbers may be regulated by the abundance of their prey. In central California reservoirs where threadfin shad were introduced to provide better forage for largemouth bass, shad actually depress survival of young bass by reducing zooplankton populations needed as food during early life history stages (28). In the absence of such competitive effects (e.g., in Colorado River reservoirs), threadfin shad introductions have improved largemouth bass fishing because adult bass grow faster and larger on a diet of shad. The variable response to shad introductions appears to be related to how abundant shad are during the early summer when bass juveniles are feeding on zooplankton. Winter die-offs of shad, during cold winters, may promote better survival of young bass (28). Too great a dependence on shad can also be bad for bass. In an isolated backwater of the Colorado River a population of mostly young-of-year bass grew rapidly, first on zooplankton and insects and then on threadfin shad. During the second year these bass measured 25–35 cm TL and were too large to consume postlarval shad and too small to capture adult shad; as a result they starved to death (29).

Growth in largemouth bass is highly variable, depending on genetic background, food availability, inter- and intraspecific competition, temperature regimes, and other limnological factors. Thus they can reach 5–20 cm in their first year, 7–32 cm in their second year, 15–37 cm in their third year, and 20–41 cm in their fourth year. The maximum size anywhere seems to be 76 cm TL or 10.5 kg, and the maximum age, 16 years (30, 31). In California reservoirs large bass (35–45 cm TL, 0.6–2.2 kg) are usually 4–5 years old (30), a growth rate that compares favorably with that of bass from Midwestern states. The state angling record (1991) is a 9.9-kg bass from Castaic Reservoir (Los Angles County). The largest bass caught in recent years have been Florida largemouth bass or hybrids, indicating that Florida bass grow larger or survive better than northern largemouth bass.

They spawn for the first time at 18–21 cm TL in males, 20–25 cm in females, usually during their second or third season. The first noticeable spawning activity is nest building by males, which starts when water temperatures reach 15–16°C, usually in March (southern California) or April (22, 30, 32). Spawning will often continue through June at temperatures up to 24°C. Nests are generally shallow de-pressions up to 1 m in diameter created by males in sand, gravel, or debris-littered bottoms at depths of 0.5–2 m. Rising waters in reservoirs may cause active nests to be located as deep as 4–5 m (32). Nests are often built next to submerged objects, such as logs or boulders. A number of bass nests may be located in one general area, but they are widely dispersed, usually at least 2 m apart (33). Nest sites are defended vigorously from other bass and potential predators, but sites may be abandoned when persistently disturbed by large carp (34). Spawning and parental behavior is similar to that of smallmouth bass. Each female lays, in multiple nests, a total of 2,000–94,000 or more eggs, the number depending on her size.

Embryos adhere to the nest substrate and hatch in 2–7 days. Sac fry then usually spend 5–8 days in the nest until they begin actively feeding. The small greenish transparent fry continue to be guarded in a swarm by the parental male for 2–4 weeks. During this period they have relatively poorly developed predator avoidance behavior (36). As fry grow the swarm expands, and it is eventually abandoned by the guarding male (34, 36). At this point small fish form schools, which cruise along the edge of the vegetation, feeding on zooplankton and small invertebrates and suffering heavy predation losses.

Status IID. Largemouth bass are a favorite game fish in California reservoirs and sloughs and as a consequence have been placed in almost all waters that will support them. Many reservoirs and farm ponds provide excellent bass fishing, with sizable populations of large, fast-growing fish. They even support large fishing tournaments in which professional and amateur bass anglers compete for big prizes, aided by high-powered boats, sophisticated electronic fish finders, and rods and lures made from the latest high-tech materials. The effectiveness of these anglers in catching fish has made it necessary for all tournament anglers to keep their fish alive for later release. The enormous popularity of bass fishing means that CDFG devotes considerable effort to finding ways to improve it, such as the introduction of Florida largemouth bass and the issuance of angling restrictions. Yet bass fishing may nonetheless decline, for three main reasons: overfishing, reservoir aging, and competition from threadfin shad and other fishes.

Overfishing. Largemouth bass, being voracious predators, are extremely vulnerable to angling, which is one of the main reasons they are such popular game fish. This means, however, that in many reservoirs at least half the population of legal-size fish is caught each year. If such fishing pressure is sustained for a number of years, the catch rate declines and the fish caught are, on average, smaller. For this reason size and bag limits on bass are increasingly restrictive, and catch-and-release fishing is encouraged.

Reservoir aging. In many reservoirs a decline in bass

populations occurs regardless of fishing pressure. Such declines are often associated with reservoir aging. For a variety of reasons, new reservoirs often develop outstanding populations of bass and other game fishes, which gradually decline as the reservoir matures. In some situations the manipulation of reservoir water levels to increase food availability or spawning success may maintain relatively large populations of bass (37). Such manipulation, however, is seldom possible because it is likely to conflict with uses for which the reservoir water was originally intended, such as irrigation and power production.

Competition. It is ironic that plankton-feeding fishes, particularly threadfin shad, which were introduced in part to provide forage for largemouth bass, have also contributed to their decline in some reservoirs, as discussed previously. The interactions between bass and their prey are sensitive to many manipulations because a competitor at early life history stages may become important prey for larger fish.

Although largemouth bass have been a major success as sport fish in California, native fishes have paid dearly for this success. Typically, when bass are abundant native fishes are absent, although there are some exceptions. In desert springs and other isolated systems pupfish and other native fishes can be rapidly driven to extinction by bass predation, as indicated in several of the species accounts. In larger systems, such as the lowlands of the Central Valley or Big Valley in the Pit River drainage, native minnows do not persist in the presence of bass, even with continual colonization from upstream areas. It is likely that largemouth bass were a factor in extinction of thicktail chub and Clear Lake splittail. Of course, bass introductions are generally made along with introductions of other species, especially other cen-

trarchids, and with major habitat change, and these factors also contribute to declines. In the lower Colorado River largemouth bass are regarded as part of the complex of predatory exotic fishes that prevent the reestablishment of native minnows and suckers. In southern California streams they prey heavily on endangered species, such as the tidewater goby (39). A developing problem is in the Delta, where largemouth bass populations (and those of other centrarchid basses) are expanding, apparently in response to increased habitat provided by the invasion of an aquatic weed, *Egeria densa*. The increased numbers of bass may be increasing predation rates on juvenile salmon and native minnows. Likewise, they have recently invaded Lake Tahoe, where they may be having negative effects on native minnows through predation on juveniles in shallow water. In general, largemouth bass create problems for native fishes wherever they are introduced (5, 38).

References 1. Bailey and Hubbs 1949. 2. Jordan and Evermann 1896. 3. Lee et al. 1980. 4. Jenkins and Burkhead 1994. 5. Lever 1996. 6. Dill and Cordone 1997. 7. Moyle and Holzhauser 1978. 8. Moyle and Nichols 1973. 9. Brown and Moyle 1993. 10. Moyle and Daniels 1982. 11. Smith 1982. 12. Coutant 1975. 13. Smale and Rabeni 1995. 14. Clellguest 1985. 15. Peterson 1988. 16. Lewis and Flickinger 1967. 17. Elliott 1976. 18. McMahon and Holanov 1995. 19. Schindler et al. 1997. 20. Keast 1966. 21. Applegate et al. 1967. 22. Weaver and Ziebell 1976. 23. Lewis et al. 1961. 24. Murphy 1949. 25. Goodson 1965. 26. Minckley 1982. 27. Mittelbach et al. 1995. 28. Von Geldern and Mitchill 1975. 29. Saiki and Tash 1978. 30. Emig 1966. 31. Scott 1967. 32. K. Miller and Kramer 1971. 33. Heidinger 1975. 34. Becker 1983. 35. S. Lehr, CDFG, pers. comm. 2001. 36. J. A. Brown 1984. 37. Heman et al. 1969. 38. Whittier et al. 1997. 39. Swift et al. 1997. 40. Morizot et al. 1991.

Smallmouth Bass, *Micropterus dolomieu* Lacépède

Identification Smallmouth bass are fairly streamlined for a bass, but have stocky bodies and mouths (maxillae) that do not reach the hind margin of their eyes. The spiny (9–10 spines) portion of their dorsal fin is only slightly rounded and broadly joined with the soft (13–15 rays) portion. Body color is greenish brown to bronze, with no conspicuous horizontal stripes on the sides but often faint, vertical, dark mottled bars. The anal fin has 3 spines and 10–12 rays; the pectoral fins, 16–18 rays; and the lateral line, 66–78 scales. Scales on the cheeks are in 14 or more rows and are smaller than those on the opercula. Small scales are present near the bases of the soft portions of the dorsal and anal fins. The belly is white and three dark, faintly iridescent bands radiate from the reddish eyes and mouth on the side of the head.

Young-of-year are darker than adults and can be distinguished from other juvenile bass by their plain coloration, tricolored tail (usually yellow or orange in the center, followed by a black band and a white fin edge), and 13–15 dorsal fin rays.

Taxonomy Two subspecies have been recognized, one from the Arkansas River (1), but their validity has been questioned (2). The Arkansas River form (Neosho smallmouth bass) was brought into California in the 1970s but did not survive the experience (3). In California smallmouth bass hybridize with spotted bass and redeye bass, as they do elsewhere (37, 38).

Names Dolomieu is after M. Dolomieu, a French mineralogist who was a colleague of Lacépède. It was formerly spelled

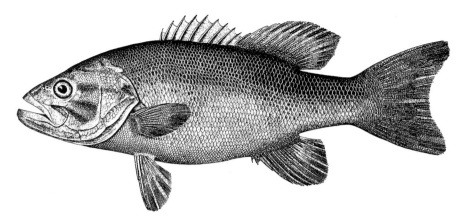

Figure 133. Smallmouth bass, 33 cm SL, Sandusky, Ohio. USNM 10323. Drawing by H. L. Todd.

rocky bottoms and overhanging trees (5). In California they are most abundant in larger tributaries to the Sacramento and San Joaquin Rivers at elevations between 100 and 1,000 m. They are consequently often associated with native minnows and suckers and, occasionally, rainbow trout (6). Their stream habitat has been considerably diminished by the flooding of many suitable areas by reservoirs. They have managed to become established in a number of reservoirs, however, where they are usually most abundant near the upstream ends. They tend to concentrate in narrow bays or in areas along shore where rocky shelves project under water, at depths of 1–10 m.

Optimum temperatures for adult growth seem to be 25–27°C, but rapid growth has been seen in the wild at temperatures as high as 29°C. Curiously, temperatures of 27–31°C are selected by adults under laboratory conditions when food is abundant, suggesting that the coolwater habitats generally selected by smallmouth bass may partially reflect prey and habitat preferences (7). Young-of-year will select temperatures as high as 29–31°C, reflecting a preference for shallow-water habitats where small prey are abundant and cannibalistic adults absent (8). For the most part, temperatures greater than 35°C are stressful to bass, and those above 38°C are lethal. Populations are rarely established where water temperatures do not exceed 19°C in summer for extended periods, and most smallmouth populations in California waters seem to be in places were summer temperatures are typically 21–22°C. Their general preference for a rather narrow range of cool temperatures is an important factor for establishing flow regimes in regulated rivers to favor them (34). Associated with their preference for cool, flowing water is the need for dissolved oxygen levels in excess of 6 mg/liter for growth and 1–3 mg/liter for survival (5, 9). They can live at a wide range (5.7–9.0) of pH values (5). In their native range smallmouth bass will enter the upper reaches of estuaries (10), but they rarely do so in California.

Social behavior of smallmouth bass is similar to that of largemouth bass, although they have less of a tendency to wander and rarely school. In lakes local populations often

with an *i* at the end, but the International Commission on Zoological Nomenclature decided it was not needed. Other names are as for largemouth bass.

Distribution Smallmouth bass are native to most of the upper Mississippi River drainage, south roughly through Arkansas, as well as to the Great Lakes watershed including the edge of southeastern Canada. They have been introduced into suitable waters throughout the United States (including Hawaii) and other parts of the world, including France, Sweden, Vietnam, Mexico, Belize, and South Africa (4). They were first introduced into central California (San Mateo County) in 1874 and have since been spread to most of the larger streams and reservoirs in the Central Valley, as well as the Pit River (Britton Reservoir), Russian River, Mad River, Freshwater Lagoon (North Coast), Trinity River, Carmel River, Colorado River, and various streams in southern California. They are also present in the lower reaches of the Truckee, Carson, and Walker Rivers, mainly in Nevada. A populations has also become established in Lake Tahoe (36).

Life History Waters preferred by smallmouth bass are large, clear lakes and clear streams and rivers with abundant cover and cool (20–27°C) summer temperatures. They are most abundant in streams with moderate gradients (0.75–4.70 m/km) and complex habitats of pools, riffles, and runs with develop that seldom exchange members with other populations in the same lake (11). Most adult smallmouth have small home ranges and return to their home areas if displaced (12). Often a single bass can be found occupying the same pool or run in a stream throughout summer. They are most active in evening and early morning, although feeding can be observed at any time of day or night, depending on conditions. Under some conditions smallmouth bass of all sizes may display aggressive behavior toward one another, suggesting that competition for food and cover is sometimes an issue (13).

Smallmouth bass fry feed largely on crustaceans and aquatic insects until they reach 3–5 cm TL, when larger prey, especially crayfish and fish, start becoming more important. Such prey rarely dominate the diet until the bass measure 10–15 cm TL. Young-of-year bass in Clare Engle Reservoir, on the Trinity River, feed mainly on chironomid midge larvae and pupae (14). In Big Chico Creek (Tehama County) they fed mainly on large mayfly larvae (15), whereas in Deer Creek (Tehama County) juveniles fed on both chironomid and mayfly larvae, as well as a wide variety of other aquatic insects (16). Smallmouth of all sizes are frequently cannibalistic. Crayfish, amphibians, and insects often become the dominant foods of local populations, at least seasonally (17, 18, 19, 20). Adult bass, especially in California, frequently feed mainly on crayfish, which in California are mostly introduced as well. Fish, amphibians, small mammals, and other items have been found in their stomachs (16). In streams where crayfish are harvested by humans, such harvesting may have a negative impact on bass populations (21).

Growth seems less variable than for largemouth bass, presumably because smallmouth bass are more restricted in their habitats. At the end of their first year, they measure 6–18 cm TL; at the end of their second year, 14–27 cm; at the end of their third year, 19–27 cm; and at the end of their fourth year, 25–41 cm. Growth in Central Valley reservoirs is excellent, so 4-year-old fish typically measure 35–39 cm FL and older fish longer than 40 cm are not uncommon (18). However, growth in Clair Engle Reservoir became progressively slower as the reservoir aged, especially for young-of-year (14). The largest smallmouth bass caught in California, from Clair Engle Reservoir in 1976, weighed 4.1 kg. In streams growth is slower than in reservoirs, especially for larger fish. Thus in Deer Creek bass measured 80 mm at the end of their first year, 138 mm at the end of their second year, 187 mm at the end of their third year, 223 m at the end of their fourth year, and 275 mm at the end of their fifth year (16).

Smallmouth bass usually become sexually mature during their third or fourth year. Their reproductive behavior is probably the best known of any member of *Micropterus* (22, 23). As water warms up in late spring, they move into shallow water (<1.5 m) in lakes or into quiet areas of streams. Some lake populations of bass migrate short distances up a stream to spawn. In northern California reservoirs most spawning takes place in May and June, but spawning in streams may occur into July, depending on flows and temperatures. Males start fanning out nest depressions 30–60 cm in diameter with their fins when water temperatures reach 13–16°C. They usually build nests on rubble, gravel, or sand bottoms at depths of around 1 m near submerged logs, boulders, or other cover. Nests, however, have been recorded on substrates ranging from organic debris to roots to large rocks at depths ranging from 0.5 to 5 m. They are solitary nesters, although in favorable areas nests may be located 1.5–2 m apart and defended from other males. Favorable nesting sites may be reused in successive years by individual males (24). In streams nesting and reproduction can be disrupted by high flows, either because embryos and fry are washed out of nests or because lower temperatures reduce spawning activity (25, 26).

Males defend their nest sites vigorously from other males and potential predators. Ripe females eventually convince males of their identities by their persistence in returning to nests, by changing color so that mottled markings on their sides become very distinct, and by maintaining a distinctive head-down posture. The male then leads the female to the nest site and the pair begins slowly circling above the nest, the male nipping the female on her sides. The female occasionally sinks into the nest and rubs her abdomen on the bottom. At the nest, the pair circle the perimeter slowly, the male always on the outside. When they are ready to spawn, the pair settles into the nest, the male parallel with the bottom, the female at variable angles. Both fish stiffen, quiver, and release the sex products. The female releases 10–50 eggs at a time at 4- to 45-second intervals, until she has released most of her eggs. When spawning is finished, the female leaves the nest or is chased away by the male. For the most part, each pair is monogamous (27). Females lay 2,000–21,000 eggs, depending on size.

The larger males spawn earlier in the season than smaller males, and they are preferred by larger females (which have higher fecundities). As a result nests of large males contain a disproportionate number of developing embryos (27). Early-spawning bass also have more opportunities to mate a second time. The fry from early-spawning males have a longer growing season and hence are likely to be larger (and have higher survival rates) at the end of the season.

The male guards the nest and developing young vigorously and continuously fans developing embryos to provide them with oxygen. This activity takes place night and day and has a high energy cost, especially in combination with limited opportunities for feeding (33). The weight loss incurred by guarding bass may be substantial, but individuals with the most vigorous defense of their young also have the highest reproductive success, a classic bioenergetic trade-off (35). The aggressiveness of the male's defense increases as the young develop from embryos to "wrigglers" to fry confined to the nest (28). This period of active defense lasts 1–2 weeks, depending on temperature. Fry remain on the bottom of the nest for 3–4 days before they start to become active and rise off the bottom. The male then herds them into a dense shoal, which he continues to guard for 1–4 weeks, although less vigorously than before (28). By the time fry reach 2–3 cm TL they are too active for the male to herd, and they soon disperse into shallow water. Mortality of young from predation is high at this stage. In streams if current velocities over the nest are in excess of 8 mm/sec, the young get swept away as they emerge, presumably not surviving in the absence of parental protection (32). Optimal current speeds for young-of-year bass appear to be 80–130 mm/sec (32).

Status IID. Smallmouth bass have been spread widely in California and probably occur in most waters that can support them. Populations in the upper reaches of reservoirs such as Pine Flat, Millerton, Folsom, Shasta, and Clair Engle provide excellent fishing for large, moderately fast-growing fish. Rivers like the Merced, Stanislaus, and Russian also have substantial populations of smallmouth bass, although the bass tend to be smaller than those found in reservoirs.

The effects of smallmouth bass on native fishes are poorly understood. In the Central Valley they have invaded many streams that support native fishes and often coexist with them, as long as smallmouth bass densities remain low. This outcome may be partly related to smallmouth bass feeding on crayfish, also introduced. Hardhead, however, also prey on crayfish, and their numbers typically decline in the presence of smallmouth bass (29). In South Yuba River hardhead and pikeminnow seem able to reproduce successfully only above a natural barrier that excludes smallmouth bass. Although large individuals of both species were found below the barrier, young-of-year were found only above it, suggesting that bass predation was limiting their survival (30). There is also evidence that smallmouth bass predation

can contribute to local extirpation of native frogs and other amphibians (31).

The streams where smallmouth bass coexist with native fish and amphibians mostly have natural flow regimes or something like them. Where flows are reduced, water temperatures may be warmer early in the season, favoring smallmouth bass spawning. During drought years, even in natural streams, smallmouth bass often show an increase in numbers for similar reasons. In "normal" or wet years, however, native fishes typically spawn a couple of months before smallmouth bass can spawn. It is possible that the large numbers of young-of-year pikeminnows that develop in shallows may reduce the success of bass spawning by preying on bass fry. In streams where there is a strong interest in protecting native fishes and amphibians, a removal fishery for large bass should be encouraged, not only because the largest fish are the most effective predators, but also because they produce the most and largest young through their early spawning behavior. In reservoirs, where conservation of native fishes is usually not a consideration, smallmouth bass populations may be enhanced by regulating the take of the largest fish.

References 1. Hubbs and Bailey 1940. 2. Jenkins and Burkhead 1994. 3. Dill and Cordone 1997. 4. Lever 1996. 5. Edwards et al. 1983. 6. Brown 2000. 7. Armour 1993. 8. Coble 1975. 9. Smale and Rabeni 1995. 10. Schmidt and Stillman 1998. 11. Forney 1961. 12. Becker 1983. 13. Sabo et al. 1995. 14. Okeyo and Hassler 1985. 15. Travanti 1990. 16. K. Baker and R. White, University of California, Davis, unpubl. rpt. 1986. 17. Webster 1954. 18. Emig 1966. 19. Applegate et al. 1967. 20. Mullan and Applegate 1968. 21. Roell and Orth 1993. 22. Breder and Rosen 1966. 23. Ridgway et al. 1989. 24. Ridgway et al. 1991a,b. 25. Graham and Orth 1986. 26. Lukas and Orth 1995. 27. Wiegmann et al. 1992. 28. Ridgway 1988. 29. Brown and Moyle 1993. 30. Gard 1994. 31. Kiesecker and Blaustein 1998. 32. Simonson and Swenson 1990. 33. Hinch and Collins 1991. 34. Lambert and Handley 1984. 35. Gillooly and Baylis 1999. 36. S. Lehr, CDFG, pers. comm. 2001. 37. Pipas and Bulow 1998. 38. Morizot et al. 1991.

Spotted Bass, *Micropterus punctulatus* (Rafinesque)

Identification Spotted bass look very much like largemouth bass, with single, irregular, black horizontal stripes on each side (made up of connecting blotches) and maxillae that extend past the middle of the eye. They can be distinguished by the following suite of characters in fish over 10 cm TL:

1. The break between the two parts of the dorsal fin is not as deeply incised. That is, the first dorsal fin is not strongly convex, so the shortest spine on the rear half of the first dorsal fin is more than half the length of the longest spine.

2. The upper jaw rarely extends beyond the rear margin of the eye.

3. The lower sides have rows of distinct black spots, as opposed to fine speckles.

4. There are teeth in a rectangular patch on the middle of the tongue.

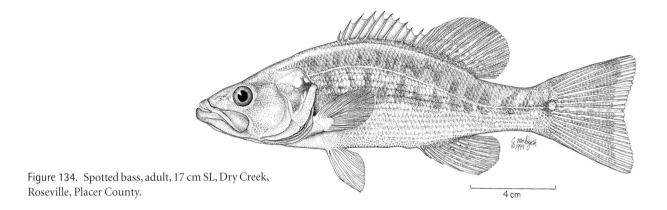

Figure 134. Spotted bass, adult, 17 cm SL, Dry Creek, Roseville, Placer County.

4 cm

5. Narrow scales are present on the bases of the soft portions of the dorsal and anal fins.

6. There is usually a distinct spot at the end of the lateral band, at the base of the tail.

7. The pyloric cecae are not forked.

Juveniles can be distinguished by the combination of a dark irregular lateral band, a tricolored tail (pale on tips, black band in middle, orange at base), and teeth on the tongue. Dorsal fins have 9–11 spines in the anterior half, which is not deeply notched where it attaches to the posterior half (9–11 rays). Their anal fins have 3 spines and 9–11 rays (usually 10); pectoral fins, 14–17 rays; and lateral lines, 55–72 (usually more than 60) scales. Scales on the cheeks are arranged in 12–17 rows (usually 13–16). Coloration is olivaceous on the back and white on the belly, with the blotched stripe in between. The caudal fin of young-of-year has a black spot at its base and is orangish with a black tip.

Taxonomy Three subspecies of spotted bass have been described (1): northern spotted bass (*M. p. punctulatus*) in most of its native range; Alabama spotted bass (*M. p. henshalli*) from Georgia, Alabama, and Mississippi; and Wichita spotted bass (*M. p. wichitae*) from the Wichita Mountains in Oklahoma.

The Alabama spotted bass may be distinct enough to merit recognition as a separate species. The northern and Alabama forms have both been introduced into California and have probably hybridized. Spotted bass hybridize with smallmouth bass in some areas, resulting in introgressed populations (2). In California apparent hybrids with both smallmouth and redeye bass occur in Oroville Reservoir, but the extent of introgression is not known (3).

Names Punctulatus means "with small spots," referring to the rows of small spots on each side. Other names are as for largemouth bass.

Distribution Spotted bass are native to the central and lower Mississippi basin (north to southern Illinois) and in Gulf Coast drainages from northwestern Florida to western Texas (4). Their range in this region has been considerably expanded by introductions. In the West they have become established in California, New Mexico, Arizona, and Nevada (5, 6). They have also been successfully introduced into South Africa and Zimbabwe (7).

Northern spotted bass were brought into California from Ohio in 1933, propagated at the Friant Hatchery (Fresno County), and, starting in 1937, widely planted in foothill rivers of the Sacramento and San Joaquin valleys (8). Alabama spotted bass, from Alabama, were first successfully planted in Perris Reservoir (Riverside County) in 1974 and were then widely introduced into southern California and the Central Valley. Alabama spotted bass were introduced because of their ability to spawn successfully in fluctuating reservoirs (9). Today they are established in most of the larger foothill and coast range reservoirs in the Central Valley (including Shasta and Oroville) and streams associated with them, in Whiskeytown Reservoir (Shasta County) on the upper Trinity River (10), and in the lower Pit River, including Britton Reservoir. It is not certain which form, Alabama or northern spotted bass, is the predominant subspecies or species in most areas, or if they have hybridized.

Life History In California spotted bass do well mainly in moderate-size, clear, low-gradient (<0.5 m/km) sections of rivers and reservoirs (11). In streams they are secretive pool dwellers, avoiding riffles and backwaters with heavy growths of aquatic plants. They like slower, more turbid water than smallmouth bass and faster water than largemouth bass. In reservoirs they are often most common along steep, rocky banks, usually toward the upstream end. They prefer water with summer temperatures of 24–31°C (12) and have relatively low tolerance for brackish water, although they have been found at salinities up to 10 ppt (13).

In reservoirs adults tend to live at moderate depths (1–4 m), often just above the thermocline, while juveniles generally remain near shore in shallow water. Young-of-year are usually found in small shoals; larger fish tend to be solitary. Each adult frequently remains in one limited area for most of the year, such as a single stream pool, but spawning migrations are common in spring (14). In reservoirs they may seek out deep water (30–40 m) once temperatures become more uniform in autumn (11). Reservoir fish also move up into inflowing rivers in summer and occupy the deep, slow pools and runs (15).

Like other basses, spotted bass are predators on larger invertebrates and fish that occur with them. Their diet changes with size, reflecting differences in both mouth size and habitat across life history stages. For fry the first foods are typically zooplankton or small insects associated with quiet waters. In streams in their native range bass smaller than 75 mm TL feed mostly on aquatic insects and crustaceans; fish measuring 75–150 mm consume, roughly in order of importance, aquatic insects, fish, crayfish, and terrestrial insects. Crayfish and (secondarily) fish are increasingly important for larger fish (16, 17). In reservoirs bass smaller than 50 mm TL feed mostly on zooplankton and then on terrestrial or aquatic insects; larger fish feed heavily on crayfish and fish, and to a lesser extent on aquatic insects (18, 19). The most common fish prey in reservoirs are various sunfish, crappie, and threadfin shad. Spotted bass also prey on their own young and those of other bass species.

Growth rates vary with habitat; fastest rates are typically achieved in fairly new warmwater reservoirs, slowest rates in cool streams. They reach 65–170 mm TL in their first year, 150–325 mm in their second year, 205–405 mm in their third year, 245–435 mm in their fourth year, 315–505 mm in their fifth year, 280–565 mm in their sixth year, and 315–610 mm in their seventh year (19). Few fish live longer than 4–5 years, so bass over 40 cm TL are unusual. Growth rates of California populations have not been recorded, but it is likely that fish in reservoir populations in northern California have growth rates toward the middle of the ranges given. The angling record for California is a 4.3-kg bass from Pine Flat Reservoir (1996), which measured about 45 cm TL.

Maturity sets in during their second or third year, and they spawn in late spring when water temperatures rise to 15–18°C (9, 20). In Perris Reservoir (Riverside County) the first sign of spawning is movement of males into shallow water in late March and early April, when temperatures are 14–15°C (9). The males then begin to construct nests in areas of large rocks, rubble, or gravel, at depths of 0.5–4.6 m (average depth, 2.5–3 m). Spawning continues through late May and early June, until temperatures reach 22–23°C. In streams nests are constructed in low-current areas on bottoms ranging from debris to gravel. Nests are 40–80 cm in diameter and generally located near cover of some sort (21). Breeding and parental behavior is similar to that of smallmouth bass (9, 20, 21). They are apparently monogamous, but some males will have more than one nest during the season. Each nest contains only 2,000–14,000 young in a similar stage of development (9, 20). Embryos and larvae are tended and vigorously defended by males for up to 4 weeks. Bluegill are common predators on embryos, which are devoured by fish that dash into a nest while the male is chasing other fish. Fry rise off the nest and form dense shoals in the vicinity, which are guarded by males until they disperse, at lengths up to 30 mm TL (21).

Status IIE. Spotted bass were originally introduced into California to occupy foothill river and reservoir habitat intermediate between that preferred by largemouth bass and that of smallmouth bass. They have been a major success in some reservoirs (e.g., Oroville Reservoir), providing much of the bass fishery. Although they are already widespread, they are likely to appear in other reservoirs as a result of official and unofficial introductions. Their impact on native fishes is not known, but it is probably low because they primarily occupy water supply reservoirs. However, their ability to colonize stream sections upstream of reservoirs on at least a seasonal basis means that they can have a considerable impact on native fishes in these reaches. They may also be detrimental to native fishes that thrive in hydroelectric reservoirs, such as the chain of reservoirs on the lower Pit River, where they feed on rough sculpins and other fishes.

An interesting problem with spotted bass is their tendency to hybridize with other bass in California reservoirs, especially smallmouth bass and redeye bass. The effects of such hybridization on fisheries and on the viability of populations of each species would be worth investigating.

References 1. Hubbs and Bailey 1940. 2. Koppleman 1994. 3. E. See, DWR, pers. comm. 1999. 4. Lee et al. 1980. 5. Sigler and Sigler 1987. 6. Sublette et al. 1990. 7. Lever 1996. 8. Dill and Cordone 1997. 9. Aasen and Henry 1981. 10. D. Lee, CDFG, pers. comm. 1999. 11. McKechnie 1966. 12. Houston 1982. 13. Peterson 1988. 14. Gerking 1953. 15. Payne and Associates 1998. 16. Smith and Page 1969. 17. Scalet 1977. 18. Mullan and Applegate 1968, 1970. 19. Vogele 1975b. 20. Howland 1931. 21. Vogele 1975a.

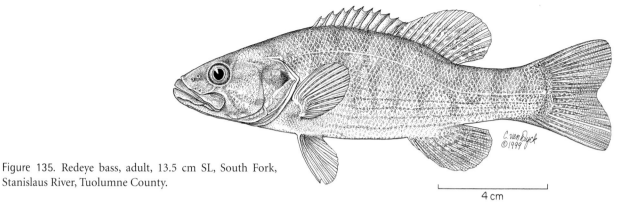

Figure 135. Redeye bass, adult, 13.5 cm SL, South Fork, Stanislaus River, Tuolumne County.

4 cm

Redeye Bass, *Micropterus coosae* Hubbs and Bailey

Identification Redeye bass are brightly colored with a distinct purplish or greenish cast to the sides and a distinct white band on the upper and lower edges of the caudal fin (1, 8). Their eyes are reddish. There may or may not be a row of diamond-shaped dark bars along the midline, but there are rows of dark spots on the lower half of each side. Live fish appear stongly patterned, including irregular blotching on the back. Opercular and basal caudal spots are usually not visible. The upper jaw (maxilla) extends to about the middle of the eye. They have 9–11 dorsal spines and 11–13 (usually 12) dorsal soft rays in shallowly notched dorsal fins. Their anal fins have 3 spines and 9–11 rays (usually 10); their pectoral fins, 14–17 rays (usually 15–16); and their lateral lines, 64–73 scales. The scales on their cheeks, usually arranged in 14 rows, are smaller than the opercular scales. Scales are usually present on the bases of the soft portions of the dorsal and anal fins; there is a patch of teeth on the tongue; and the pyloric cecae are not forked. Young-of-year can be distinguished from other basses by the distinct vertical bands on the sides that extend below the lateral line, the usual 11–12 dorsal fin rays, and the rusty red base of the caudal fin.

Taxonomy Prior to Hubbs and Bailey's revision of *Micropterus* (2), redeye bass were considered a small form of smallmouth bass. In California and elsewhere they hybridize with smallmouth bass, as they do elsewhere (13), and probably with spotted bass as well. Redeye bass are confused in the literature with the recently (1999) described shoal bass (*M. cataractae*) from Alabama, Georgia, and Florida, which is sympatric with redeye bass in the Chattahoochee River, Georgia (12).

Names *Coosae* is after the Coosa River system in Georgia, where the type specimens were collected. Other names are as for largemouth bass.

Distribution Redeye bass are native to headwaters of the Savannah, Altahama, and Mobile River basins, in Georgia, Alabama, Tennessee, North Carolina, and South Carolina. Their native range has been expanded by introductions in the region, and they have also been introduced into Puerto Rico (3). In 1962 and 1964 bass from Tennessee and Georgia were planted in Alder Creek (Sacramento County), South Fork Stanislaus River (Tuolumne County), Dry Creek (Nevada County), Santa Ana River (Riverside County), Sisquoc River (Santa Barbara County), and Santa Margarita River (San Diego County) (4). The South Fork Stanislaus River population dominates a short stretch of the river (5, but see the Status section) and has colonized New Melones Reservoir. Small numbers are apparently further spread downstream into the Delta. The Alder Creek population still existed in 1988. The Santa Margarita River population is well established in the canyon reaches (11). The other introductions apparently failed to become established, although a Sisquoc River population did exist for a few years. In 1969 redeye bass raised in the CDFG's Central Valley Hatchery were planted in Oroville Reservoir (Butte County), where they became established and have hybridized with smallmouth and spotted bass. Redeye bass also have invaded the Cosumnes River basin, where they are abundant in the foothill reaches of the river and its forks.

Life History Redeye bass are adapted for living in small, clear, upland streams (7). They were originally introduced into California because of the superficial resemblance of many foothill and coastal streams to those of their native region. These streams support mainly native fishes and so were thought to need improvement for angling purposes. The streams in which redeye bass exist in California are clear and warm (summer temperatures of 26–28°C), and the bass are typically one of the most abundant fish (5, 11). They favor pools, pockets of water near boulders, and undercut banks. They will also establish populations in reservoirs (14), but their presence is unusual, especially when they co-occur with smallmouth bass and spotted bass. In Oroville Reservoir they are widely distributed although uncommon compared with spotted bass and smallmouth bass, and they apparently hybridize with them (10).

Redeye bass are opportunistic predators that feed on the surface, in the water column, and on the bottom. Like some trout, they depend heavily on terrestrial insects, although aquatic insects, fish, crayfish, and salamanders are frequently part of their diets. They are voracious predators that cruise about looking for prey or ambush prey from cover. In either situation they are capable of making extraordinarily rapid rushes in pursuit of prey. Small juveniles feed mainly on aquatic insects, but I have observed them successfully preying on mosquitofish in shallow water. In streams they are surprisingly bold, approaching people in the water with apparent curiosity and readily taking lures cast out by anglers.

They are extremely slow-growing in streams, reaching 4.5–6.5 cm TL in their first year, averaging as few as 2–3 cm/year, and taking 9–10 years to reach 25 cm TL (6, 14). Growth in California streams is probably similar; an October sample from South Fork Stanislaus River contained young-of-year at 3–5 cm FL and older fish measuring 10–19 cm FL (5). In the Cosumnes River the largest redeye bass I have observed measured 25–30 cm FL. These were presumably 5 or 6 years old. In Oroville Reservoir length frequencies indicate that redeye bass grow fairly rapidly, reaching about 7–8 cm in their first year, 9–11 cm in their second year, and 12–14 cm in their third year. Larger fish, presumably 4–6 years old, ranged from 20 to 35 cm TL (10). This growth is considerably slower than that reported for bass in reservoirs in their native range, which average about 22 cm TL at the end of their second year (14). These fish, however, measure only about 32 cm TL at the end of their fifth year (14), suggesting genetically based size limitations. Fish up to 41 cm TL have been recorded from Oroville Reservoir (10), but these fish are larger than the maximum size achieved by redeye bass in their native range. (Angling records of 4-kg redeye bass from Florida are in actuality shoal bass [12].) It is likely that apparent redeye bass over 35 cm TL are smallmouth-redeye bass hybrids (8). They become mature at about 12–13 cm TL, at 2–4 years of age. Fecundities are high, considering the size of the females: 2,084 and 2,334 eggs in females measuring 15 cm and 21 cm TL, respectively (9).

Redeye bass move up small tributary streams or to the heads of pools in larger streams to spawn in late spring when temperatures rise to 17–21°C. Males construct nests in beds of gravel. Spawning and parental behavior is presumably similar to that of smallmouth bass.

Status IIC. The IIC rating of redeye bass is conservative because their takeover of large stretches of the Cosumnes River indicates that they are capable of invading many foothill streams in the Central Valley. In many reaches of the Cosumnes 99 percent of the fish are redeye bass (15). Native minnows and suckers are largely gone from these reaches and are present mainly in areas where redeye bass are absent (15). The disastrous success of their invasion of the Cosumnes was largely unappreciated because the redeye bass had been misidentified for years as smallmouth bass (15). It is likely that they are more widespread than presently recognized in the Stanislaus River and elsewhere in San Joaquin Valley streams. The introduction of redeye bass was unfortunate, because it was done deliberately to displace native fishes from free-flowing streams to provide a fishery (4). Although most introductions failed, the bass have demonstrated a capacity to live in both foothill streams and reservoirs. They have spread farther than was once suspected—a quiet invasion based on their confusion with smallmouth bass. Their small adult size, aggressive behavior, and generalized habitat and feeding requirements presumably allow them to dominate some foothill streams so completely. They have not provided much of a fishery because most California anglers do not know they exist and they are rather small and slow growing for a game fish. Their abundance in the Cosumnes and Stanislaus Rivers indicates that they have considerable capacity to displace native fishes, presumably through predation on juveniles. They are likely to spread to other streams and reservoirs and are highly likely to become one more major problem for conservation of native fishes and invertebrates.

References 1. Parsons 1953. 2. Hubbs and Bailey 1940. 3. Lever 1996. 4. Dill and Cordone 1997. 5. Lambert 1980. 6. Catchings 1977. 7. Parsons and Crittenden 1959. 8. Etnier and Starnes 1993. 9. Hurst et al. 1975. 10. E. See, DWR, unpubl. data 1999. 11. C. C. Swift and R. Fisher, pers. comm. 1999. 12. Williams and Burgess 1999. 13. Pipas and Bulow 1998. 14. Barwick and Moore 1983. 15. P. K. Crain, P. B. Moyle, and K. Whitener, University of California, Davis, unpubl. data.

Perches, Percidae

Perches are confined to the fresh waters of temperate North America and Eurasia. None are native west of the Rocky Mountains in North America. They are readily distinguished from other freshwater spiny-rayed fishes with ctenoid scales and thoracic pelvic fins by their 2 well-separated dorsal fins (the first composed entirely of spines) and by the presence of only 1–2 spines in the anal fin.

There are three distinct groups of perches native to North America: darters, pikeperches, and yellow perches. The darters (mainly *Etheostoma* and *Percina*) are an abundant (over 110 species) and colorful group of small, slender, bottom fishes, native only to eastern North America. Like sculpins they lack a functional air bladder. One darter, bigscale logperch, has been introduced into California. The pikeperches, walleye (*Stizostedion vitreum*) and sauger (*S. glaucum*), are piscivorous inhabitants of lakes and large rivers. Walleye are a favorite sport and commercial fish in midwestern North America. Because of this popularity they have been introduced throughout the United States. Introductions into California reservoirs have been unsuccessful. The yellow perch, a favorite prey of walleye where the two species occur together naturally, is the only member of the perch family besides bigscale logperch that has been successfully introduced into California.

Bigscale Logperch, *Percina macrolepida* Stevenson

Identification Bigscale logperch are long and slender, readily recognized by their pointed, projecting snouts; yellowish, almost tubular bodies with 14–16 complete, dark, vertical stripes and a dark spot at the base of the tail; 2 well-separated dorsal fins; and small size (usually <9 cm SL).

Scales are ctenoid and cover the breast, cheeks, operculars, and nape. A few scales are sometimes present on top of the head (supraoccipital region), but they are usually lacking in California fish. There are a row of large spiny scales on the belly and 77–90 scales along the lateral line. The first dorsal fin has 13–15 spines; the second dorsal fin, 12–15 rays; the anal fin, 2 spines and 7–10 rays; and the pelvic fins, 1 spine and 5 rays each. The pectoral fins are enlarged and fan-shaped with 12–14 rays. Breeding males may become dark on their heads and sides and have an orange bar on their dorsal fins. Males can also be identified by a ridge of 25 modified scales on the midline of the belly, behind the pectoral fins.

Taxonomy Bigscale logperch were described from Texas in 1971 (1). They were formerly considered part of the logperch (*P. caprodes*) species complex. Bigscale logperch in California were thus originally identified as *P. caprodes* (2).

Names Percina means little perch; *macro-lepida* means large scales. In clear streams other members of the genus

409

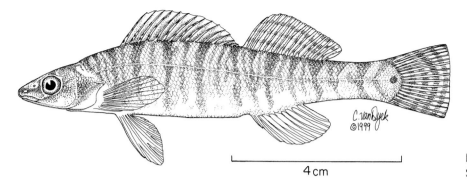

Figure 136. Bigscale logperch, 9 cm SL, Putah Creek, Yolo County.

can be observed resting on submerged logs, hence log-perch.

Distribution Bigscale logperch occur in a number of Gulf Coast river systems, from the Sabine River on the Texas-Louisiana border through Oklahoma, New Mexico, and northeastern Mexico (1, 3). They were carelessly imported into California from the Trinity River, Texas, in 1954 by USFWS (3). They were apparently mixed with a shipment of largemouth bass and bluegill planted in three small lakes on Beale Air Force Base (Yuba County) (4). During wet years these lakes overflow into the Bear River, which flows into the Feather River (5). Within 15 years bigscale logperch had become widespread in the Sacramento–San Joaquin watershed, from Oroville on the Feather River to the Delta, to sloughs in the San Joaquin Valley, including upstream lo-calities such as Putah and Cache Creeks (5, 6, 7). They are now also present in reservoirs fed by the California Aque-duct, including Del Valle Reservoir, Alameda County, and Castiac, Silverwood, Pyramid, and Irvine Reservoirs in southern California. Bigscale logperch may also be moved around by bait fishermen (9), and this may explain their presence in Berryessa Reservoir (Napa County) (9).

Life History Bigscale logperch are found in a variety of lake and stream habitats. They are most common in slower-moving stretches of warm, clear streams or in shallow wa-ters of reservoirs on bottoms of mud, gravel, rock, sticks, or large pieces of debris. In central California they are often abundant in muddy-bottomed, turbid sloughs and ditches as well as warm lowland streams. In Putah Creek they are most abundant in reaches characterized by deep pools, moderate clarity, warm summer temperatures, and bottoms with fine substrates (silt to gravel) (10). In Suisun Marsh they have been collected at salinities up to 4.2 ppt. They are often found along edges in emergent vegetation. The species that co-occur with bigscale logperch in California are mostly nonnatives: common carp, fathead minnow, various catfish, inland silverside, bluegill, largemouth bass, black crappie, and native Sacramento blackfish (10).

Bigscale logperch spend much of their time motionless on the bottom, where their barred color pattern makes them very difficult to see, even in clear water. They move only for short distances, usually propelling themselves with quick, short sweeps of the pectoral fins, although such ac-tivity may be nearly continuous when they are actively searching for food. In aquaria they bury themselves in loose gravel, with only the head or the tip of the snout showing, emerging to forage. They also dig small pits with their tails, in which they sit motionless. They are most active during the day. Although they commonly occur in small groups, neither shoaling nor territorial behavior seems to be well developed, at least outside the breeding season. However, in aquaria apparent dominance hierarchies can become estab-lished among groups of logperch, with aggressive behavior shown by the erection of fins and one fish butting another with its snout, until the subdominant fish moves.

When feeding, bigscale logperch visually inspect the bot-tom around them for food organisms, occasionally flipping over twigs, leaves, and small rocks with their projecting snouts. They will also rise quickly from the bottom to snap up small, free-swimming organisms. They are highly op-portunistic in their feeding. Usually whatever insect larvae are most abundant dominate the diet, together with am-phipods and planktonic crustaceans. Planktonic crus-taceans are most important in the diet of young logperch. Examination of the stomachs of 121 logperch from sloughs of the Delta in winter and spring of 1973 revealed many in-sect larvae (chironomid midge, mayfly beetle, stonefly, damselfly, dragonfly) as well as crustaceans (copepods, cladocerans, amphipods, opossum shrimp). Fish eggs were found in a number of fish (11). Logperch collected from a recently flooded grassy area were feeding on earthworms; those collected in small sloughs were feeding heavily (50% by volume) on copepods.

In the Delta 1-year-old fish measure 48–81 mm SL (mean, 63 mm) and 2-year-olds, 75–102 mm SL (mean, 90 mm). A single 3-year-old fish measured 104 mm SL (11). Larger fish (up to 125 mm SL) may represent older individuals.

Logperch usually mature in their second year, and each

female produces 150–400 eggs. Spawning of recently captured fish has been observed in aquaria by J. Sturgess (17) in late February:

> The largest male in the tank took station at the base of a hornwort plant. A female swam up and settled to the bottom, parallel to the male. Using her caudal and pectoral fins and always maintaining herself a few inches from the stationary male, the female swam forward and backward several times and then proceeded headfirst into a bushy portion of the hornwort. She then backed out and returned to the side of the male. After standing on her tail several times, she finally got the male to respond. The two fish approached each other head on and then rose and pressed against one another ventrally, beating their pectoral fins rapidly. This lasted about ten seconds, after which the fish returned to a horizontal position and quivered for a several seconds. . . . The eggs were deposited singly, attached to the plants.

Other observations in aquaria suggest that bigscale logperch may also spawn in small gravel pits much like other logperch. The vertical spawning behavior is different from that of other *Percina* species, which spawn in gravel riffles (12), and may explain why bigscale logperch have managed to become so abundant in sloughs and sluggish streams.

In Putah Creek logperch spawn from late February through mid-July, as indicated by the presence of larvae (10). However, in a warm downstream section, spawning began in February or March and peaked in April and May, whereas in a cooler upstream section it began in late March but did not peak until mid-June and July (10). Similar results have been found for other locations (13, 14). Larvae are pelagic and probably drift in streams for a couple of days before washing into side channels and settling down (15). They are common in larval fish samples from the Delta (15).

Status IID. Bigscale logperch demonstrate the rapidity with which an introduced fish species can spread through central and southern California via the aqueduct system and natural rivers. Within 15 years of their introduction they were widespread in the Central Valley, and within 25 years they had colonized reservoirs in southern California. They are now a common fish in lowland streams and reservoirs and are near the limits of their range in the state, although they could become established in other reservoirs. The effect of bigscale logperch on native and desirable game fishes in California is not known, but it is likely to be minimal because they almost exclusively occupy highly disturbed habitats. McKechnie (16, p. 531) stated that "[Bigscale] logperch add nothing to our fauna and do not benefit our fisheries." However, they do make interesting and attractive aquarium fish. Ironically, they are considered to be an endangered species in New Mexico, at the periphery of their natural range (8), although their populations elsewhere do not seem to be in trouble.

References 1. Stevenson 1971. 2. Sturgess 1976. 3. Stevenson and Thomson 1978. 4. Dill and Cordone 1997. 5. Boles 1976. 6. Farley 1972. 7. Moyle et al. 1974. 8. Sublette et al. 1990. 9. L. Wycoff, CDFG, Yountville, pers. comm. 1999. 10. Marchetti 1998. 11. J. Sturgess and R. Hobbs, University of California, Davis, unpubl. data. 12. Winn 1958a,b. 13. Wang 1986. 14. Rockriver 1998. 15. Simon and Kaskey 1992. 16. McKechnie 1966. 17. J. Sturgess, pers. comm. 1973.

Yellow Perch, *Perca flavescens* (Mitchill)

Identification Yellow perch are recognized by their fairly compressed yellow bodies with 6–9 dark vertical bars or saddles on each side, 2 well-separated dorsal fins, 2 anal fin spines (with 6–8 rays), and a forked tail. Their bodies are moderately deep (standard length is 3 times depth) and their heads large (25% or more of total length). There is a single small spine on each operculum. The first dorsal fin has 13–15 spines; the second dorsal fin, 2–3 spines and 12–14 rays; and the pelvic fins, 1 spine and 5 rays each. The scales are large (52–61 in the lateral line) and ctenoid. The pectoral and pelvic fins, usually yellow, may become reddish-orange in spawning males.

Names Perch (hence *Perca*) is derived from the ancient Greek word for dusky, perhaps a reference to the dusky back and bars of the similar European perch (*P. perca*). *Flavescens* means yellow.

Distribution Yellow perch are native to the northern half of North America east of the Rocky Mountains, north as far as the Mackenzie River in Canada, south through the Great

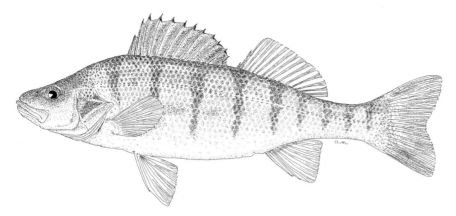

Figure 137. Yellow perch, 19 cm SL, Iron Gate Reservoir, Siskiyou County. Drawing by A. Marciochi.

Lakes region, and down the Atlantic coast to South Carolina. They have been introduced into most regions in North America from which they were historically absent, including most states in the western United States.

In California yellow perch have established themselves in a number of areas despite repeated failed introductions (1). In 1891 6,000 small perch were brought in from Illinois; about half were planted in the lower Feather River and half in Cuyamaca Reservoir (San Diego County). The Cuyamaca introduction lasted about 5 years. The Feather River introduction resulted in a population that apparently gradually spread to the Delta and sloughs of the San Joaquin Valley and was supplemented with additional fish imported in 1908. Through the early 1920s yellow perch were locally abundant in the Delta region, but they gradually became rare, and the last authentic record was two fish caught in 1951.

In 1946 yellow perch were discovered in Copco Reservoir on the Klamath River (2). These fish were presumably descendants of perch planted in the upper Klamath River in Oregon. They have since spread downstream into dredger ponds and backwaters along the river, although they are not particularly common. Yellow perch are now common in the upper Klamath Basin, especially in Copco and Iron Gate Reservoirs. They appear to be absent from or rare in the Lost River and its reservoirs (3), although they were once present there (2).

In 1984 yellow perch were found in Lafayette Reservoir (Contra Costa County), in which they are still present. Fortunately this reservoir does not have an outlet into any major waterways. In 2001 a population of yellow perch was discovered in Van Norden Reservoir (Nevada and Placer Counties) in the headwaters of the South Yuba River (11). Although these perch are a long way from any suitable habitat downstream, their eventual spread is possible. Perch have been illegally introduced into other localities in northern California on occasion, but they have not become established (1).

Life History Yellow perch usually inhabit weedy backwaters of rivers, shallow waters of lakes, and large ponds. They do best in warmwater situations, but occasionally large populations of stunted fish become established in lakes cold enough to support trout. In lakes they are almost always associated with heavy growths of aquatic plants and tend to occur in loose schools on or just above plant beds at depths of 1–10 m. Optimal summer temperatures for growth appear to be 22–27°C, but they can survive temperatures up to 32–33°C (4, 5). They can also survive dissolved oxygen levels of less than 1 mg/liter and salinities up to 10 ppt (5). They are nevertheless most abundant in areas with high water clarity because they are visual-feeding, shoaling fish that require beds of aquatic plants for spawning. The disappearance of beds of submerged aquatic plants as water quality declined may be one of the reasons perch died out in the Central Valley in the 1920s.

Shoaling, often in schools, is the typical social behavior pattern of adult perch. Even in aquaria they swim together and seldom exhibit any aggressive interactions. Compact schools are typical of immature perch (<10 cm TL) because they inhabit open waters, usually at depths of 1–4 m, rather than being associated with beds of aquatic plants.

Feeding habits of perch change with size and thus with habitat. Larval and juvenile perch are primarily zooplankton feeders; the variety of zooplankton consumed increases with the size of the fish. As schools of young perch move into shallow water, invertebrates associated with the bottom and with aquatic plants gradually become more important in their diet, especially aquatic insect larvae, snails, and various crustaceans. Adult perch browse methodically among aquatic plants and along the bottom, selecting larger invertebrates, such as crayfish, dragonfly larvae, and snails. Small fish may also be important. Their terminal, scooplike mouths, with protrusible lips and small teeth, are well suited for capturing such prey (6). In the Klamath River their main foods are small crustaceans, snails, aquatic insect larvae, and fish, mostly minnows, suckers, and sunfish (2). Yellow perch are capable of capturing small salmon and occasionally do so in the lower Klamath River (1). Most feeding takes place during the day, with peaks of activity in the morning and at dusk (7). They become quiescent at night (5).

Growth of perch in the Klamath River is similar to that observed in other waters. They average 9 cm TL by the end of the first year, 15 cm by the second year, 20 cm by the third year, 23 cm by the fourth year, and 27 cm by the fifth year (2). Elsewhere yellow perch may (rarely) reach 53 cm TL, 1.9 kg, and 13 years, although a perch exceeding 30 cm TL, 0.4 kg, and 5 years would be unusual in California. It is not unusual for large populations of stunted perch to develop in small bodies of water, where overcrowding reduces growth through severe intraspecific competition.

Yellow perch are usually ready to spawn during their second year. Spawning takes place over submerged beds of aquatic plants in quiet water at temperatures from 7 to 19°C (5). The first sign of spawning, in April and May in the Klamath River, is the presence of large schools of ripe adult perch over plant beds. Prior to spawning, females become restless and swim slowly around the spawning area. Males in small groups periodically swim up to a cruising female and follow her for a short distance, nudging her vent. When the female is ready to spawn she makes a series of rapid turns or other quick movements. Two to three males quickly approach her and start jockeying for a position immediately below her vent. The female then starts swimming rapidly, releasing a long string of eggs enclosed in a gelatinous sheath. As eggs are released, males release a cloud of sperm, enveloping the eggs (8). The strands of eggs may be as long as 2 m, but they are more typically 30–50 cm long. They are draped over the aquatic plants. Each female lays 4,000–121,000 eggs, the number being proportional to length (5). Egg masses are not eaten by potential predators, suggesting that they are unpalatable (9).

Embryos hatch in 10–20 days. Larvae (about 6 mm TL) may start to feed on zooplankton soon after hatching, but they possess some reserve food in the yolk sac until they reach about 7 mm TL (10). Larvae are attracted to light, so they swim into surface waters. Because they are weak swimmers, they are at the mercy of lake and stream currents for the first few weeks.

Status IIC. Yellow perch are clearly favored by some anglers with a longing for standard eastern panfishes, because they keep appearing in odd places. They are fairly easy to catch, and their firm, white flesh is quite tasty, although often riddled with parasites. They are not particularly desirable for California because they are smaller and slower growing than most other game fishes. They can also survive and reproduce in some trout lakes, reducing growth and survival of trout by competing with them for food.

Although they are present in reservoirs and ponds along the Klamath River, the fishery for them is small. Most yellow perch caught there are taken incidentally by trout, salmon, and catfish anglers. They are known to prey on small salmon and probably also prey on juveniles of endangered suckers in the Upper Klamath basin.

References 1. Dill and Cordone 1997. 2. Coots 1956. 3. Buettner and Scoppetone 1991. 4. Houston 1982. 5. Becker 1983. 6. Keast and Webb 1966. 7. Keast and Welsh 1968. 8. Hergenrader 1969. 9. Newsome and Tompkins 1985. 10. Houde 1969. 11. J. I. Hiscox, CDFG, unpubl. rept.

Cichlids, Cichlidae

Cichlids are an abundant and diverse group of sunfish-like fishes native to Africa, the Middle East, southern India, South America, and Central America north to the Rio Grande in Texas. They are recognized as the most species-rich of all fish families, with 1,300 species and an equal number that have not been described. They are favored as aquarium fishes because of their fascinating breeding habits (many incubate their eggs in the mouth), bright colors, and pugnacious behavior. Many cichlids of the "genus" *Tilapia* are fast growing and prolific under extreme conditions in small ponds and polluted streams. As a result they have been promoted throughout the tropical and subtropical world as a source of low-cost protein. In the United States they have been promoted as game fish and for aquatic weed control. Consequently cichlids have been released throughout the southeastern and southwestern United States. As Hubbs (1968) points out, most of these releases have been ill advised, without enough consideration given to long-term effects. The main thing that has saved the temperate world from a plague of tilapias is that most cannot live for long when water temperatures drop below 10–12°C.

Cichlids established in California are moderately deep-bodied fishes with two lateral lines, a long dorsal fin, and two or more rows of jaw teeth. They are easy to recognize from their profile: large head and anal and dorsal fins that taper off to long symmetrical points. In California there is considerable uncertainty as to how many species are present and how much they hybridize with one another. In this book I treat four species for which there is reasonable evidence of introduction and establishment: redbelly tilapia,

Mozambique tilapia, Nile tilapia, and blue tilapia. Other tilapias that may be present are wami tilapia (*O. urolepis* = *O. hornorum*) and redbreast tilapia (*T. randalli*). Wami tilapia are or have been found in the lower Colorado River, where they have extensively hybridized with Mozambique and other tilapias (Barrett 1983). There have also been some tentative identifications of redbreast tilapia from the lower Colorado River and perhaps other species as well (Dill and Cordone 1997). Apparently the lower Colorado River is a soup of tilapias with limited origins, which may extensively hybridize, so identifying individual fish is often problematic, and keys such as the one in this book are of little help for identification. As a consequence most studies of lower Colorado River and Salton Sea fish simply refer to them as "tilapia." Costa-Pierce and Doyle (1997) biochemically identified three species in southern California, redbelly tilapia, Nile tilapia, and Mozambique tilapia, but they found that the Nile and Mozambique tilapias were so different from their African ancestors that they recommended calling them regional strains labeled "California Nile tilapia" and "California Mozambique tilapia." Only one of their samples of fish was from the Colorado River, however.

The descriptions presented here are of the original forms, as described by Trewavas (1983), modified somewhat from data in Barrett (1983). The life history data are also largely from non-California sources and are presented mainly to give some idea of the potential ecological role of each form, assuming that it resembles that of ancestral fish. All information should be treated with a great deal of caution.

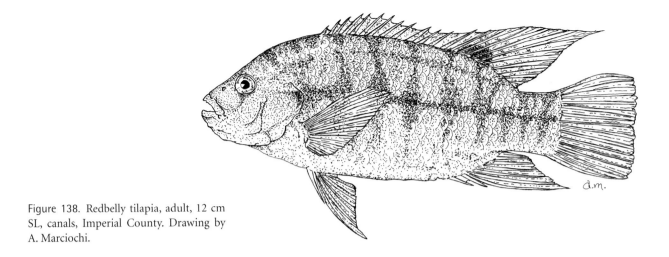

Figure 138. Redbelly tilapia, adult, 12 cm SL, canals, Imperial County. Drawing by A. Marciochi.

Redbelly Tilapia, *Tilapia zillii* (Gervais)

Identification This description may not be adequate for identifying California redbelly tilapias. This tilapia has the typical cichlid body—elongate, yet deep and laterally compressed—and long dorsal and anal fins. In adults the mouth is horizontal, or nearly so, and is situated in a head that is wider than the body. The dorsal fin has 14–16 spines and 10–13 rays, many of the latter considerably longer than the spines. The anal fin has 3–4 spines and 7–10 rays, and the pectoral fins, 14–15 rays each. There are 28–30 cycloid scales in the lateral series. The 8–12 gill rakers on the first arch are very short. The tail is rounded. Typical nonbreeding coloration is dark olive on the back and light olive or yellow-brown on the sides. The sides often have an iridescent sheen and 6–7 poorly defined vertical bars. The belly is yellow to white and the fins are brown to yellow. The dorsal fin has a dark "eye-spot" on the soft-rayed portion, often outlined in yellow, along with numerous small yellow spots on the entire fin. The operculum also has a distinct dark spot. Spawning fish become shiny dark green on the back and sides, with a bright red throat and belly and distinct vertical bands on the sides. Their heads turn dark blue-black and mottled with blue-green spots.

Taxonomy Redbelly tilapia are very closely related to redbreast tilapia (*T. randalli*) (1), and it is possible that the form in California is actually derived from *T. randalli* (17). They will hybridize with other substrate-brooding tilapias, such as spotted tilapia (*T. mariae*) (2).

Names The fish is named for M. Zill, a naturalist who provided Paul Gervais with the specimen upon which the species description was based. It is sometimes called Zill's cichlid as a result. It is one of the most abundant fish in Lake Kinneret (Sea of Galilee), Israel, and was probably one of the species caught by disciples of Jesus Christ who were fishermen, including Peter. Hence it is sometimes called St. Peter's fish. *Tilapia* is discussed in the account of Mozambique tilapia.

Distribution Redbelly tilapia are native to coastal drainages of North and West Africa north to the Jordan Valley in the Middle East, although their distribution in this region has been considerably expanded by humans. They were introduced into some of the warmer parts of the world for aquaculture and aquatic weed control, including Okinawa, Madagascar, Mexico, Guam, Hawaii, Mauritius, and New Caledonia (3). In the United States they have been extensively planted for aquatic weed control but have apparently became established only in Texas, North Carolina, California, and Arizona (3). They were first released into California waters in 1971, in canals of the Imperial Valley, for aquatic weed control. California fish originated from three pairs brought to Arizona in 1965 (15). By natural colonization and further introductions they spread into the Coachella Valley, drains of the Salton Sea, and backwaters of the lower Colorado River (4). Although tilapia in the drains of the Salton Sea region are usually not identified to species, they are most likely redbelly tilapia (19). Attempts to introduce them elsewhere in California have so far failed, although they apparently were established for a while in some southern California coastal localities (4).

Life History Redbelly tilapia normally inhabit large lakes and rivers, but they adapt well to ponds, irrigation ditches, and other artificial habitats (5). They are particularly well suited to warm, saline irrigation return waters such as exist in the Imperial Valley; populations there have been reported in waters of 29–35 ppt salinity. They are also capable of living at a wide range of temperatures. Optimal temperatures for growth are 20–32°C (6, 7), but they regularly live at temperatures up to 38°C. Temperatures above 40°C are usually lethal, and those above 42°C are always lethal (6, 8). In California they are limited by low temperatures. Although they are able to survive temperatures as low as 6.5°C for short periods, long-term (over 2 weeks) exposure to temperatures below 13°C is usually lethal (6). They also do not reproduce in waters colder than 20°C. In Napa County redbelly tilapia survived two winters in ponds that dropped to 10°C but died during the third winter when temperatures dropped even lower for an extended period of time (5). They appear to be more tolerant of low temperatures in saline waters than in fresh water (9).

Redbelly tilapia are usually associated with aquatic plants and algae that form a major part of their adult diet. Their dentition is well adapted to feeding on plants. The inner rows of jaw teeth are small and multicuspid for holding on to leaves, while teeth in the outer row are sharp and incisorlike for shearing leaves. Leaves are ground up by stout pharyngeal teeth and digested with a fairly high efficiency (10). Despite their specialized dentition and ability to live entirely on a diet of aquatic plants, redbelly tilapia will consume invertebrates as well, especially those associated with aquatic plants. Occasionally they will even take other fish, usually individuals that are already dead or dying (5). Young are more carnivorous than adults and depend on small crustaceans during the first few months of life.

Individuals in long-established populations typically do not grow very fast, even in their native Africa, where they reach 5–12 cm TL in the first year. They can double in size in the next 1–2 years, reaching 14–25 cm by the end of their third year. Males grow faster than females. However, in irrigation drains of the Imperial Valley with low fish densities and abundant food, redbelly tilapia grow exceptionally fast, with males reaching an average of 17 cm TL in their first year and 25 cm in their second. The largest 2-year-old fish from these ditches measured 315 mm TL and weighed 708 g (11). In Africa fish more than 35 cm TL and 800 g are rare and are likely to be more than 6 years old. Under crowded pond conditions, however, growth is often exceedingly slow, and the fish may take 2 years to reach 7 cm TL, a size at which they will breed. Normally they measure 13–14 cm TL

before breeding, although in Imperial Valley ditches males and females may become mature at 7–8 cm and 11–12 cm, respectively (6, 7, 11).

Courtship and pair formation begin after the water has warmed to about 20°C. When they are ready to breed, redbelly tilapia seek out shallow (<1.5 m), protected areas with soft bottoms, although they are capable of breeding on substrates from rocks to sand to mud (12). Breeding often takes place in large colonies, in which fish in the center of the colony spend less time defending nests from predators than pairs along the colony edge (18). Once a pair bond has formed, the pair build a nest (or nests), defend the territory around it, court, and spawn (13, 18). In clay substrates nests may be burrows up to 85 cm deep that adults defend from predators by plugging the tunnel with their bodies (12). In some cases these nests have multiple brood chambers for embryos and young. The eggs are often laid in rows of 50–1,200 and then fertilized. Egg laying continues until 1,000–6,000 eggs have been deposited. One or both members of the pair tend the embryos by fanning a current of water across them or by picking out debris and dead eggs. Embryos hatch in 2–3 days. During incubation the pair may construct one or more small depressions nearby. After hatching the young are transferred by mouth or by fin fanning to the depressions, where they remain for 3–4 days until the yolk sac is absorbed and they become free-swimming (13). The shoal of fry is initially herded by the parents but is abandoned in a few days. Within a month the pair may spawn again. Under optimal temperature and food conditions, redbelly tilapia will breed throughout the year, but in southern California water temperatures are probably suitable for breeding for less than 6 months. Even so, each pair would presumably be capable of producing 4–6 broods in this time.

Status IIC. Redbelly tilapia were originally established in southern California for aquatic weed control in canals. They proved only moderately successful because winter die-offs required that they be reared in large numbers to make control possible (4). Triploid grass carp were found to be more effective because of their ability to survive periods of low temperature (4). In addition, in low numbers they seem to be rather selective in the plants they will eat. Thus the main effect of wild populations may be to change the composition of the aquatic plant community without affecting plant density.

Without annual planting, redbelly tilapia seem to be in decline, and they are present today mainly in scattered drains around the Salton Sea and the Imperial and Coachella Valleys, and in a few shallow areas along the lower Colorado River (4, 14). All fish in California are descended from three breeding pairs, so there is speculation that this limited genetic heritage may affect their ability to adapt to changing environments (15). However, they are still a problem along the Salton Sea, where they apparently compete with desert pupfish, an endangered species. When redbelly tilapia decline in the drains, usually as a result of cold winters, pupfish become abundant; when the tilapia increase, pupfish decline (16).

In short redbelly tilapia is another example of a poorly considered introduction into California, fueled by optimism and ignorance, that has provided few benefits and has done some damage. It has been abandoned as a weed con-trol agent and provides little to fisheries or aquaculture, so its extirpation from the state would be welcome.

References 1. Trewavas 1983. 2. Taylor et al. 1986. 3. Lever 1996. 4. Dill and Cordone 1997. 5. Pelzman 1973. 6. Hauser 1977. 7. Platt and Hauser 1978. 8. Stauffer et al. 1989. 9. Herbold 1979. 10. Buddington 1979. 11. Hauser 1975. 12. Bruton and Gophen 1992. 13. Fryer and Iles 1972. 14. W. L. Minckley, pers. comm. 15. Costa-Pierce and Doyle 1997. 16. Schoenherr 1985, 1988. 17. B. Costa-Pierce, pers. comm. 1999. 18. Loiselle 1977. 19. Costa-Pierce and Riedel 2000.

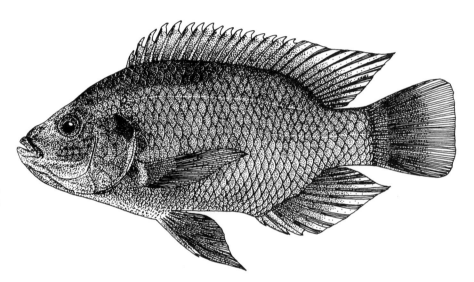

Figure 139. Mozambique tilapia, adult, 18 cm SL, Florida. From Lee et al. (1980).

Mozambique Tilapia,
***Oreochromis mossambicus* (Peters)**

Identification Mozambique tilapia in California cannot be identified with certainty without using biochemical techniques. The snout of Mozambique tilapia is bluntly pointed, and their large oblique mouth usually extends to or past the outer margin of the eye. The scales are cycloid, numbering 29–33 in the lateral series (16–23 along the first lateral line), and are arranged in 2–3 rows on the cheek below the eye. The dorsal fin has 14–18 spines and 7–13 rays; the anal fin, 2–5 spines (usually 3) and 6–13 rays (1). The pectoral fins have 14–15 rays each. The caudal fin is rounded. Gill rakers are short, numbering 14–20 on the lower limb of the first arch (18–25 total). Color is highly variable owing to interbreeding of fish of different origins. However, females and nonbreeding males are normally pale gray to washed-out yellow, with 3–4 dark spots on the sides. Spawning males have enlarged mouths and blue, thickened upper lips; their bodies turn black, often with mottling or an iridescent blue tinge; the throat and lower part of the head are pale; the dorsal fins are black with a red border; the caudal fin has a wide red band at the end; and the pectoral fins are red. Fins are free of distinct spotting.

Taxonomy The cichlids in general and the tilapiine cichlids in particular are a taxonomically complex group, because they are morphologically plastic, prone to rapid evolution in new environments, and likely to hybridize with other cichlids under the influence of humans. The Mozambique tilapia is one of the most widely introduced tilapias and consequently appears in many forms, both "pure" and hybridized. In California they differ from the original African form in part because they have a limited ancestry (perhaps five fish) and have undergone selection for aquaculture (27). California fish have hybridized to a small degree with wami tilapia (*O. urolepis*), which was once considered to be just a variety of Mozambique tilapia and deliberately introduced with it. Costa-Pierce and Doyle (27) suggest calling this biochemically distinctive form the "California Mozambique tilapia," essentially treating it as a new species. Mozambique tilapia may have also hybridized with blue tilapia (*O. aureus*) (1). The great cichlid systematist Ethelwynn Trewavas (2) removed Mozambique tilapia from the genus *Tilapia* and placed it in *Oreochromis,* along with other maternal tilapia, including the blue, wami, and Nile tilapias. Trewavas reserved the genus *Tilapia* for substrate spawners such as redbelly tilapia and placed all paternal or biparental mouth brooders in the genus *Sarotherodon.* Because of the confusion caused by the name changes, the American Fisheries Society (3) continues to use *Tilapia* for all tilapias, for the sake of simplicity.

Names Tilapia is derived from the native African (Bechuana) word *thlape,* meaning fish. The official common name is Mozambique mouthbrooder (3), but I prefer Mozambique tilapia as being more consistent with other common names. *Chromis,* derived from the Greek word for color, was the original generic name given to many cichlids. When Albert Gunther described a new species from a crater lake on Mt. Kilimanjaro in 1889, he placed it in a separate genus, naming it for the Greek word for mountain (*Oreo*) but also retaining *chromis* to indicate the relationship (2). *Oreochromis* therefore means "mountain *Chromis.*" *Mossambicus* describes the geographic area to which they are native.

Distribution Mozambique tilapia are native to coastal streams of southeast Africa, including the lower Zambezi, Mozambique, Limpopo, and Pongola Rivers, south to Algoa Bay, South Africa (2). They have been widely introduced into tropical and subtropical regions of Africa, Asia, South and Central America, Australia, New Guinea, and many Pacific islands, including Hawaii (4). In the United States Mozambique tilapia are widespread in Florida, Texas, Arizona, and California and are no doubt present at least temporarily in hot springs and ponds in other states. In California they first became established in irrigation ditches of the Imperial, Palo Verde, and Bard Valleys, Imperial and Riverside Counties (5, 6); in the Salton Sea; and in the lower Colorado River, although most of these fish may now be Mozambique-wami hybrids (27).

Mozambique tilapia now in California evolved in Africa but were established at an unknown time in Indonesia, for

aquaculture (27). In 1951 fish were brought to Hawaii from Asia, and the strain that was established may have originated from as few as five fish. From Hawaii tilapia were sent in 1953 to Steinhart Aquarium in San Francisco, which served as a distribution point for the mainland. One shipment went to Auburn University in Alabama, which started a major rearing program. In 1963 Arizona began releasing fish that originated both from Hawaii and from Auburn University (via Oklahoma). The Auburn fish, however, had hybridized with wami tilapia. From Arizona the tilapia moved via irrigation canals into California, where they were first found in 1968 (27). In addition, in 1964 some tilapia that were escapees from a tropical fish farm became established (7). In 1972 and 1973 Mozambique tilapia were deliberately introduced into the lower Santa Ana, San Gabriel, and Los Angeles Rivers and into various flood control channels in the region. They seem to be well established and to have spread naturally, through salt water, to other local estuaries (7, 8). A small population is also present in High Rock Spring (Lassen County). In 1991 two tilapia were collected in Success Reservoir, Tulare County (9). In 1994 a population of tilapia, enough to support a fishery, was established in Mormon Slough near Stockton. Presumably the latter two records are for populations that were only temporarily established in a region with winter water temperatures too cold for persistence of tilapia.

Life History Although Mozambique tilapia are native to tropical Africa, they can survive temperatures as low as 5.5–10°C for short periods, with the optimum for growth around 25–30°C (10, 11, 28). Below 15°C they become sluggish and highly susceptible to fungal and parasitic infections (28). Heavy or complete mortality is suffered at 11–14°C if exposure lasts more than a few days (2, 12). Thus when temperatures dropped to 12°C in the Salton Sea in winter 1999, a huge die-off of tilapia occurred (28). Temperatures greater than 37°C are normally avoided, and prolonged exposure to high temperature may be lethal (13). Thus the thermal tolerance zone for Mozambique tilapia has been characterized as 15–37°C (13). They normally inhabit fresh water but will live and breed in seawater, including that of southern California (2, 14). They can tolerate salinities up to 120 ppt (15, 28) and grow and reproduce at salinities up to 69 ppt (28). They are the most abundant fish in the Salton Sea, where salinities exceed 45 ppt, but are uncommon in freshwater drains that flow into it.

Their preferred habitat seems to be warm, weedy ponds, canals, and river backwaters, although they will thrive in any place with abundant food and adequate temperatures. In estuaries they become most abundant in areas sheltered from strong tidal currents, with relatively low salinities, high temperature, and beds of aquatic plants (15). In the Salton Sea they occur in all habitats but are most abundant close to shore or in the estuaries of inflowing streams (28). During the hot season (June–October) tilapia concentrate in huge numbers in inshore areas, apparently because deeper, offshore waters become anoxic (28). Stress due to overcrowding, high temperatures, and anoxic water results in major fish kills in summer, such as the 8 million tilapia that died in August 1999 (28). Within a lake or pond there may be habitat segregation by male and female tilapia. Mature males seek soft-bottomed areas, suitable for nest building, in turbid water, whereas females (and juveniles) seek hard-bottomed areas where food is more abundant (16).

Mozambique tilapia are true omnivores, feeding much of the time on planktonic algae, aquatic plants, and detritus (2) but capable of switching to invertebrates and fish when the opportunity arises (28). They can digest plant material because the extreme acidity of their stomach fluids (pH < 2) causes cells to lyse (16) and their long intestine provides plenty of surface area for digestion. Aquatic invertebrates always make up at least part of their diet, and zooplankton are especially important to young fish. Fish raised in an effluent pond near San Diego fed mostly on aquatic insect larvae, amphipods, and some adult insects (10). In the lower Colorado River they feed on a mixture of algae, detritus, chironomid midge larvae, and small molluscs (17). In the Salton Sea tilapia feed on aquatic plants, diatoms, rotifers, barnacle larvae, and copepods, although the single most important food item is pile worms (*Neanthes succinea*) (29). Their diet varies with season and place of capture (29). Given the opportunity, tilapia will capture and eat small fishes. Their omnivory may explain in part why they have not been particularly effective in controlling either aquatic plants or mosquito larvae (10, 18) except in small experimental ponds (19). Their ability to act as biological control agents is further complicated by the territorial behavior of adult males. Because males drive all other tilapia out of their territories, wild populations seldom reach densities needed for weed or insect control (20). An exception seems to be a concrete river channel in the Los Angeles basin, where warm effluent from a power plant creates dense growths of algae and chironomid larvae. The chironomids are the principal food of a large population of hybrid (Mozambique-wami) tilapia, and tilapia predation seems to keep the midges from reaching pest proportions (21). Tilapia are the principal fish in this bizarre habitat because they can survive the combination of warm effluent and periodic inundation by salt water as the result of tidal action.

Their ability to grow extremely rapidly in ponds is one of the main reasons they have been distributed throughout much of the tropical world since the late 1930s. In experimental ponds in Alabama they can grow in 18 weeks from an average of 73 mm TL and 5.5 g to 166 mm and 42.8 g (22). In California all-male Mozambique-wami hybrids grew in experimental ponds at rates of 25–61 mm *per month*

(10). Even in more natural situations they grow up to 23–28 mm/month (12). However, these rapid growth rates are achieved mainly in managed or heavily harvested populations. In natural populations growth rates are much slower, and fish measuring 16–18 cm TL are typically 3–5 years old (2). In the Salton Sea tilapia reach 25–32 TL in a year, with very little growth occurring during their second summer; very few live longer than 2 years or grow larger than 36 mm TL (28). Curiously, 80–90 percent of the fish in the Salton Sea are male (28). In some situations large populations of stunted fish (<10 cm TL) will develop. Males grow faster and become larger than females, achieving a maximum size of about 39 cm TL and a maximum age of 11 years (2). The maximum size achieved so far in California seems to be about 38 cm TL (23).

Mozambique tilapia usually become mature at lengths of 6–14 cm TL, a size that can be reached in some populations 3–6 months after hatching. They will breed continually as long as temperatures are above 20°C. In the Salton Sea mature fish are apparently mostly in their second summer but reproduce continuously, April–October (28). Breeding begins when males form aggregations. Mature individuals leave groups of nonbreeding fish and move inshore to establish territories in weedy areas. Each male clears an area about 15–140 cm in diameter (typically around 30 cm) and then digs a shallow pit or two. From that time on the males are continually active, digging, courting, spawning, feeding, and fighting with neighboring fish (24). The territory, however, is abandoned at night, and the male moves into deeper water (2). Courtship begins when a male approaches a school of females with special invitation displays. If a female is ready to spawn she follows the male back to his pit and the two fish circle around, the female occasionally biting at the bottom. During the circling the female suddenly releases a number of eggs. She then turns around and takes them into her mouth. The male ejects milt at the spot where the eggs were dropped, and the female takes the milt into her mouth as well, fertilizing the eggs with "mumbling movements" of the jaw (2, 24). The act is repeated until the female has laid 100–600 eggs, the number depending on her size. Females can produce up to 1,800 eggs, but it is uncertain if they can actually incubate so many embryos. Once spawning is completed, the male chases the female out of the territory and starts courting other females.

For the next 11–12 days the female goes into hiding while she incubates the eggs. By the end of the incubation period young are capable of swimming by themselves and are ejected by the female. They then form a shoal that follows the female around. When a threatening fish approaches, the female attacks it. If unsuccessful in driving the intruder away, the female will call the young to her mouth with special movements. They then enter it or cluster around her head. This behavior ceases after 4–8 days, and the young are on their own (24). In another 10–30 days the female may spawn again. Young then move into shallow water. In the Salton Sea young-of-year "were observed numbering in the millions in dense clouds in . . . nearshore waters during the hot season from April to November when surface water temperatures were above 25°C" (28, p. 7).

Status IIE. Mozambique tilapia (in reality a hybrid form) are now one of the most common fish in the Salton Sea, the lower Colorado River, and the lower reaches of some southern California streams. Because of their abundance, willingness to take any bait, and excellent flavor, they are popular as game fish, despite their often small size. In most areas where they are abundant their impact on native fishes is not known. It is likely they had little to do with native fish declines, because in most areas the natives were already gone. They may, however, stand in the way of native fish recovery because of their abundance and high reproductive rates. When introduced into warmwater springs they are capable of driving endemic species to extinction. Thus when they invaded High Rock Spring, Lassen County, from a legal aquaculture operation, they drove an endemic tui chub to extinction, presumably by preying on embryos and larvae (25). Unfortunately, such events could become more common because tilapia are favored for aquaculture where water temperatures are warm and because they seem to be continually moved about by anglers, aquaculturists, and others who like them. So far they do not appear to have become established in the Central Valley because winter temperatures there are too low. It is distinctly possible that a cold-hardy strain could develop, with potentially disastrous results for Central Valley fishes.

A particular problem with Mozambique tilapia exists in the Salton Sea, where they are the most abundant fish in the sea itself. They periodically die in enormous numbers, and the stench makes areas along shore uninhabitable to people. They carry heavy parasite loads and apparently incubate a form of botulism in their guts, both of which kill birds that eat them, especially pelicans (26). The die-off of waterfowl in the Salton Sea is regarded as a huge conservation problem because the sea is a major overwintering area for migratory waterfowl. Nevertheless, tilapia are likely to continue to be the dominant fish in the sea until the salinity reaches about 60 ppt (28), in about 2040 if present trends continue. Extremely high catches in gill nets (up to 11 kg/hr) suggest that the sea could support a large commercial fishery for tilapia, perhaps yielding as much as 3,600 kg/ha per year (28). An intense fishery could reduce the incidence of large die-offs, reduce the incidence of disease (which spreads to birds), and result in export of nutrients from an increasingly eutrophic system (30).

References 1. Barrett 1983. 2. Trewavas 1983. 3. American Fisheries Society 1991. 4. Lever 1996. 5. St. Amant 1966a. 6. Hoover and St. Amant 1970. 7. Dill and Cordone 1997. 8. Swift et al. 1993. 9. Heyne et al. 1991. 10. St. Amant 1966b. 11. Price et al. 1985. 12. Hoover 1971. 13. Stauffer 1986. 14. Knaggs 1977. 15. Whitfield and Blaber 1979. 16. Bowen 1976. 17. Minckley 1982. 18. Avault et al. 1966. 19. Legner and Medved 1973. 20. Legner et al. 1973. 21. Legner et al. 1980. 22. Kelly 1957. 23. W. Hauser, pers. comm. 1974. 24. Baerends and Baerends-Van Roon 1950. 25. Moyle et al. 1995. 26. Kaiser 1999. 27. Costa-Pierce and Doyle 1997. 28. Costa-Pierce and Riedel 2000. 29. R. Riedel and B. A. Costa-Pierce, unpubl. manuscript 2001. 30. Gonzalez et al. 1998.

Blue Tilapia, *Oreochromis aureus* (Steindacher)

Identification Blue tilapia in California cannot be identified with certainty without using biochemical techniques. As in the related Mozambique tilapia, the snout of blue tilapia is bluntly pointed, but their slightly oblique mouth rarely extends to or past the outer margin of the eye. Scales are cycloid, numbering 16–23 along the first lateral line and 11–16 along the second lateral line, and are arranged in 2–3 rows on the cheek below the eye. The dorsal fin has 14–17 spines and 9–14 rays; the anal fin, 1–3 spines (usually 3) and 7–12 rays (1). The caudal fin is usually truncate. The gill rakers number 18–26 on the lower limb of the first arch (22–32 total). Coloring is variable in live fish. Females and nonbreeding males are normally a pale gray to washed-out yellow, with a series of bands on the sides. Spawning males have an iridescent blue-gray head and dark blue chin but lack enlarged mouths and thickened upper lips. Their bodies are pale blue, the dorsal fins have a red border, and the caudal fins are plain with a pink upper margin.

Taxonomy Blue tilapia were placed in the genus *Oreochromis* by Trewavas (2) and are closely related to Nile tilapia (*O. nilotica*). In California and Arizona they may have hybridized extensively with Mozambique tilapia, Nile tilapia, or both. In any case, tilapia in the lower Colorado River are apparently a hybrid swarm (1). However, the fish mostly look like blue tilapia (8), although blue tilapia and Nile tilapia are very similar. Thus the fish called blue tilapia may be mostly of Nile tilapia ancestry (see the account of Nile tilapia).

Names *Aureus* means blue. This species has also been called St. Peter's fish, as for redbelly tilapia. Other names are as for Mozambique tilapia.

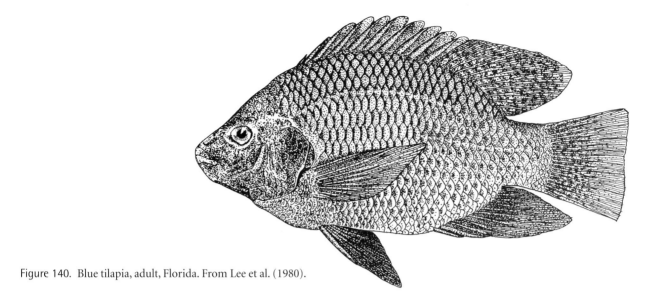

Figure 140. Blue tilapia, adult, Florida. From Lee et al. (1980).

Distribution Blue tilapia are native to much of northeastern Africa, including the Senegal, middle Niger, and Nile Rivers, north into the Jordan River drainage in the Middle East (2). They have been widely planted around the world, but most introductions have failed, except in the United States, Mexico, and a number of South American countries (3). In the United States, they were originally reared at Auburn University and spread like a disease from there to other states, including Florida, North Carolina, Oklahoma, Texas, Pennsylvania (in thermal discharges from a power plant), and Arizona (3). From Arizona blue tilapia were introduced into the Colorado River in the early 1960s (4). They (or hybrids with other tilapia) are now apparently the dominant fish in the lower Colorado River. They have not yet spread into canal systems of the Imperial or Coachella Valleys or into the Salton Sea, but they can be expected there. They are likely to be spread elsewhere by inconsiderate anglers or aquaculture operations.

Life History Blue tilapia occur in a wide variety of habitats in the lower Colorado River, thriving in the sloughs, back-

waters, canals, and reservoirs. They are mainly a fresh- or brackish-water fish but can survive in water that is 2–3 times the salinity of seawater (2). They seem to be able to tolerate slightly colder temperatures than other tilapia (down to 5°C). They are basically a subtropical to tropical species, however, and do not breed until water reaches 20–22°C (2).

Blue tilapia are largely herbivorous, and this characteristic has led to their use in ponds for aquatic weed control. In ponds they prefer certain aquatic plants and filamentous algae (5). In their native habitats adults also feed on phytoplankton and zooplankton, which they strain by pumping water across their gill rakers (6). Juvenile fish, in contrast, feed by picking individual particles, mainly zooplankton, from the water. It becomes energetically most advantageous for these fish to switch to filter feeding when they reach 6–7 cm SL (6). Large populations of filter-feeding blue tilapia are capable of altering zooplankton and phytoplankton communities (7).

Growth of blue tilapia in their native lakes is rapid: 8–16 cm TL in their first year, 16–27 cm in their second year, and 22–31 cm in their third year. Maximum size seems to be around 37 cm SL and maximum weight around 1 kg (2). Both sexes can breed in their first year, once they have reached 10–18 cm TL, and the breeding cycle can be repeated every 33–59 days. Breeding behavior and development in these mouthbrooders are similar to those of Mozambique tilapia (2).

Status IIC. Blue tilapia—or at least fish that look like blue tilapia (perhaps blue-Nile tilapia hybrids)—became the dominant tilapia in the lower Colorado River in the 1980s, displacing a population made up largely of Mozambique-wami tilapia hybrids (1, 4). This situation implies that the tilapias present in the river are more tolerant of extreme environmental conditions, perhaps especially low temperatures, than the other tilapia species. They may also be more aggressive when spawning than other tilapias and centrarchids, displacing the other species from favored spawning sites (although this has not been demonstrated). In any case they are now part of the complex of alien species in the lower Colorado River that suppresses native fishes. They are likely to become more widespread through illegal introductions and aquaculture operations, which will be unfortunate because their spread is likely to have further negative impacts on native fishes and ecosystems.

Studies on the biology of California tilapia and their role in the Colorado River and Salton Sea ecosystems are needed (see Nile tilapia).

References 1. Barrett 1983. 2. Trewavas 1983. 3. Lever 1996. 4. Dill and Cordone 1997. 5. Schwartz and Maughn 1984. 6. Yowell and Vinyard 1993. 7. Vinyard et al. 1988. 8. W. L. Minckley, pers. comm. 1999.

Nile Tilapia, *Oreochromis niloticus* (Linneaus)

Identification Not pictured; similar to blue tilapia. *The description provided here is of "classic" Nile tilapia and may not be helpful in differentiating these fish from other tilapia in California without the use of biochemical techniques.* As in the related Mozambique and blue tilapias, the snout of Nile tilapia is bluntly pointed, but their slightly oblique mouth rarely extends to or past the outer margin of the eye. The scales are cycloid, numbering 30–34 in the lateral series (1). The dorsal fin has 15–18 spines and 12–14 rays; the anal fin has 3 spines and 7–12 rays. The caudal fin is usually truncate. The gill rakers number 19–26 on the lower limb of the first arch. Coloring is variable in live fish, but the caudal fin typically has distinct vertical stripes on it. The dorsal fin normally has a gray to black margin, often suffused with red. Females and nonbreeding males are normally a pale gray to washed-out yellow, with a series of bands on the sides. Spawning males are typically flushed with red throughout the head and sides, but lack enlarged mouths and thickened upper lips.

Taxonomy Even within its native range the Nile tilapia is a highly variable species; it is termed a "superspecies" by Trewavas (1). They are closely related to blue tilapia, from which they are hard to distinguish, although the two species occur together in Africa and rarely hybridize naturally.

Their possible presence in the lower Colorado River was first noted by Barrett (2), who thought that some blue tilapia he examined were possibly hybridized with Nile tilapia. Costa-Pierce and Doyle (3) found two samples of tilapia in California that biochemically resembled Nile tilapia, one from a private fish farm and one from the Colorado River at Blythe. They suggest that these fish are different enough from African Nile tilapia that they should be labeled as a regional strain: "California Nile tilapia." It is thus possible that most of the fish in the lower Colorado River are Nile tilapia rather than blue tilapia. The stocks present probably had their ultimate origin in fish from the Niger River region of Africa (4).

Names Nile tilapia are named for the Nile River in the center of their range. Other names are as for Mozambique tilapia.

Distribution Nile tilapia are widely distributed in sub-Saharan Africa and the Nile River basin, and other watersheds north to Israel and south to Lake Tanganyika. They are increasingly the tilapia of choice for aquaculture operations and are consequently already widely distributed around the world (5). They are likely to rival Mozambique tilapia in distribution in the near future. Their distribution in California is not certain because they are widely confused with blue tilapia. However, they are a legal aquaculture species in Arizona and probably have escaped or been released into the lower Colorado River and its tributaries since the mid-1960s (3). Therefore, they are likely to be widely distributed throughout the lower Colorado River.

Life History Nothing is known about the ecology of Nile tilapia in the Colorado River, unless the dominant tilapia present are Nile, rather than blue, tilapia or hybrids between the two species. Nile tilapia are favored for aquaculture because of their rapid growth, hardiness, and herbivory, characteristics that should serve them well in the polluted waters of the lower Colorado.

They apparently are less tolerant of high salinities than blue tilapia, but hybrids have a high tolerance (1). They can live in water with temperatures up to 39–40°C, but optimal temperatures are probably 23–28°C (1, 5). Breeding seems to require minimum temperatures of 18–19°C.

Newly emerged fry are omnivores, feeding heavily on zooplankton and small insects. By the time they reach 5–6 cm TL, they are entirely herbivorous (1). The larger fish feed mainly on phytoplankton, including cyanobacteria (blue-green algae), although they will also feed on aquatic plants, small invertebrates, and bacteria. Their ability to digest superabundant cyanobacteria with acidic secretions in the gut is a major reason for their success (1). Apparently they feed during the day and then digest much of the ingested material at night.

Growth rates are slightly higher than those for blue tilapia, perhaps because Nile tilapia delay reproduction until they are larger (18–30 cm TL), although they can mature at smaller sizes under unfavorable conditions. They also can achieve larger sizes (40–65 cm TL) than blue tilapia, and weights up to 4–7 kg (1). Reproduction is similar to that of Mozambique tilapia.

Status IIC. If blue tilapia are not the dominant fish in the lower Colorado River and associated canal systems, then Nile tilapia or blue-Nile hybrids probably play that role. It may be that a new taxon has developed as a result of multiple hybridization events working on fish of limited genetic diversity. A survey of tilapias in the lower Colorado River and Salton Sea region should be made, using biochemical techniques to examine the origins of the fish. Studies should also be carried out to elucidate the ecological role of these fish and to examine salinity and temperature tolerances. These studies could estimate the likelihood of the spread of local tilapia strains to other areas and their potential for ecological damage. They could also provide the information needed to assist fisheries agencies in granting permits for various species for aquaculture. Nile tilapia in particular are in high demand because of their faster growth rates and larger potential sizes compared with most other tilapia (3). However, new imports of any tilapia species should not be allowed into California until it is clear what impact the new fish might have on local fish populations. The basic assumption must be made that no facility is completely secure, especially a production aquaculture facility, so that new strains of fish will sooner or later make their way into natural waters (or what pass for natural waters in southern California).

References 1. Trewavas 1983. 2. Barrett 1983. 3. Costa-Pierce and Doyle 1997. 4. Dill and Cordone 1997. 5. Lever 1996.

Surfperches, Embiotocidae

"A country which furnishes such novelties in our days, bids fair to enrich science with many other unexpected facts, and what is emphatically true of California, is in some measure equally true of all our waters. This ought to stimulate to renewed exertions not only our naturalists, but all the lovers of nature and science in this country." This rapturous statement was written by Louis Agassiz in 1853 as part of his description of the family Embiotocidae and two of its species. He was astonished by the discovery of a family of perchlike fish in which females became pregnant and gave birth to large fully formed young. Interest in this family led to description in 1854 of the tule perch, the only freshwater member of the family. This description by W. P. Gibbons was published in *The Daily Placer Times and Transcript,* a San Francisco newspaper. Because it was the first description of a new species of freshwater fish from California, Evermann and Clark (1931) assign 18 May 1854 as the birthday of freshwater ichthyology in California.

The surfperches continue to fascinate anglers, divers, and biologists along the Pacific coast from Baja California to Alaska, where they are among the most common inshore fishes. Delightful accounts of most of the 20 marine species found along the Pacific coast are collected in Love (1996). An additional three species are found along the Pacific coast of Asia. Although tule perch occur exclusively in fresh water, shiner perch are often found in brackish water and occasionally in the reaches of streams.

Surfperches are deep-bodied, spiny-rayed fishes, recognizable by a scaled ridge that runs along the base of the dorsal fin. However, their most distinguishing features are related to life history strategies associated with live bearing (Baltz 1984). Embryos develop in enlarged ovaries of females and initially obtain nourishment by absorbing the rich ovarian fluid that surrounds them. The dorsal and anal fins of embryos become large and vascular and lie in close contact with the convoluted, vascularized ovarian wall, indicating that they are used to obtain oxygen and nutrients from the blood of the mother fish. Males of some species become sexually mature soon after they are born and begin actively courting females. Females typically store sperm for several months before fertilizing their eggs and carry young fathered by several males.

Tule Perch, *Hysterocarpus traski* Gibbons

Identification Tule perch are deep-bodied fish with small terminal mouths. Adults (usually <15 cm TL) often have a pronounced hump between the head and the dorsal fin. The dorsal fin has a distinct ridge of scales at its base, 15–19 spines, and 9–15 rays. The rays run 24–38 percent of the length of the dorsal fin. The anal fin has 3 spines and 20–26 rays; in males it has an anterior fleshy enlargement that covers the spines. The pectoral fins have 17–19 rays, and the lateral line has 34–43 scales. The color of living fish is variable, but the back is generally dark, often a bluish or purplish, and the belly color ranges from white to yellow. There are three color phases related to the barring pattern on the sides: unbarred, broad-barred, and narrow-barred.

Taxonomy John Hopkirk (1, 2) described three subspecies of tule perch, *H. t. lagunae* from Clear Lake, *H. t. pomo* from the Russian River, and *H. t. traski* from the rest of the Sacramento–San Joaquin system. Studies of morphology (3), genetics (4), and life histories (5) confirm the distinctiveness of the three forms. One of the more curious differ-

Figure 141. Tule perch, females, each 6 cm SL, Sacramento River, showing color phases: unbarred (top), narrow-barred (bottom left), and broad-barred (bottom right).

ences among the three subspecies is the relative proportion of the three color variants: unbarred, broad-barred, and narrow-barred. Unbarred fish occur only among Sacramento tule perch (43% of fish examined), and broad-barred fish are common (27%) only among Clear Lake tule perch (3).

Names Hystero-carpus means womb-fruit, referring to their viviparity; *traski* is in honor of Dr. J. B. Trask, who sent the first specimens to W. P. Gibbons. Tule is the Aztec word, modified by the Spaniards, for bulrush (*Scirpus* spp.), with which tule perch are commonly associated.

Distribution Historically the Sacramento–San Joaquin subspecies of tule perch occurred in most lowland rivers and creeks in the Central Valley up to major canyons or waterfalls; it apparently occurred from the Kings River (Fresno County) in the south to Pit Falls on the Pit River (Shasta County) in the north. It also occurred in most of the larger tributaries to the San Francisco Estuary, including the Petaluma River in the north and Coyote Creek in the south. Their distribution in the Sacramento Valley and estuarine tributaries is similar today. In the San Joaquin drainage they are found mainly in the Stanislaus River, but they are occasionally found in the San Joaquin River near the Delta and in the lower Tuolumne River. In the San Francisco Bay region they are still present in the Delta, Suisun Marsh, the Napa River, Sonoma Creek, Alameda Creek, and Coyote Creek, and in the estuarine reaches of the Petaluma River and Green Valley Creek (6). Tule perch from the estuary have been carried via the California Aqueduct to southern California, and populations are established in Silverwood and Pyramid Reservoirs (7, 33).

The Russian River tule perch was, and apparently still is, found throughout the mainstem Russian River and the lower reaches of its major tributaries (Mendocino County). The Clear Lake tule perch is confined to Clear Lake and upper and lower Blue Lakes (Lake County). Tule perch also once occurred in the Pajaro and Salinas Rivers (Santa Cruz and Monterey Counties) but disappeared from them in the mid-20th century. A population apparently exists in Laguna Grande, near Monterey, the result of an introduction (30).

Life History Tule perch occur in a wide variety of lowland habitats, including lakes, estuarine sloughs, and clear streams and rivers. Despite their deep bodies, they can forage in fast water by taking advantage of eddies behind submerged boulders and logs, moving in from slower-moving backwaters and edges. In rivers they are typically associated with beds of emergent aquatic plants, deep pools, and banks with complex cover, such as overhanging bushes, fallen trees, and undercutting. They can also be common in riprap. In Deer Creek (Tehama County) tule perch use pools and runs 0.5–1 m deep, foraging close the bottom where water velocities are 1–14 cm/sec (8). In Putah Creek, Yolo and Solano Counties, they are found in similar habitats and are strongly associated with permanent flows and well-developed riparian habitat (9). Pregnant females are typically concealed in slower-moving areas or backwaters with beds of aquatic plants or with dense cover created by tree branches. In the lower Pit, Yuba, and Stanislaus Rivers they are abundant along the edges of large deep (2–4 m) pools in association with boulders and other cover (10, 11). Large shoals of tule perch can also be found in Britton Reservoir and other flow-through reservoirs on the Pit River, where they are most abundant close to the bottom in riverine sections (12). In Clear Lake they are most common near or in large tule beds where bottoms are sand or gravel; they are rare in areas that have been heavily modified by humans (e.g., bulkheads and channels) (13). In lakes their most common associates are bluegill and other alien centrarchids, but in streams they are associated mainly with other native fishes.

Tule perch generally require cool, well-oxygenated water for survival (14). With the exception of Clear Lake, they are rarely found in water that is warmer than 25°C for extended periods of time and generally prefer temperatures below 22°C (28). They have high salinity tolerance, however. In Suisun Marsh they live at salinities that range annually from 0 to 19 ppt. In Napa Slough they occur at salinities as high as 30 ppt and possibly higher (29).

Tule perch are gregarious, especially when feeding. In rivers small groups can be observed strung out in a line in current, moving slowly upstream while periodically picking at the bottom. In lakes and reservoirs they school in large numbers, especially off tule beds and overhanging trees. Females that are ready to give birth often aggregate in dense cover; this provides a haven for newborns and for the females themselves when their swimming ability is impaired by pregnancy.

Tule perch are adapted for feeding on small invertebrates associated with bottom or aquatic plants, although they will also feed in midwater on zooplankton. The deep body shape and maneuverable fins, combined with the large eyes and protrusible premaxillary bones of the upper jaw, allow the fish to suck or pick up their prey. Both jaw teeth and pharyngeal plates are large, for crushing prey. In the San Francisco Estuary tule perch feed mostly on small amphipods (especially *Corophium*), and secondarily on other benthic prey, such as midge larvae (Chironomidae), small clams, brachyuran crabs, and mysid shrimp (1, 15, 16, 17). In Clear Lake they seem to be primarily midwater feeders, concen-

trating on zooplankton in the cold season and chironomid midge and mayfly larvae in the warm season (18). In the Russian and American Rivers tule perch feed on a wide variety of bottom- and plant-dwelling invertebrates, but the most important are larvae of chironomid midges, baetid and ephemerellid mayflies, and other aquatic insects (19, 31).

Breeding behavior is most noticeable when males begin actively courting females in late summer. In the Russian River groups of males were observed defending small territories under overhanging branches or plants close to shore. Each male defends its territory against other males as well as against other fish. The apparent purpose of the territories is to attract females for mating, because courtship behavior was observed within them. However, each male does not appear to hold a territory for more than a day or so, and courtship and mating can also occur away from territories. This may indicate a lek-type breeding system in which a few males do most of the mating (20).

One such instance of nonterritorial courtship and mating was observed by John Norton (32) in August 1973:

> Two tule perch each about 75 mm SL were observed at a depth of 1 m behind a few boulders about 2.5 m from shore. The darker of the two fish (probably the male) pecked at the operculum of the other, who did not flee. This opercular nipping continued for nearly 5 minutes. The two fish then went into a short tail chase, head to tail, for a single revolution. They separated and opercular nipping was resumed for another minute. This was followed by both fish opening and closing their mouths, with the lips coming in contact. Next the fish attempted to align themselves side by side, facing the same direction. After several attempts, they pressed their caudal fins and peduncles together, rotating until the anal fins were in contact. Together, the bodies of the two fish formed a V, with the anal region forming the vertex. Next the anal fins were pressed close together, followed by a thrust, a jerk, and separation. Activity from the time the caudal fins were first pressed together until the fish separated took approximately 15 seconds. This mating sequence was then repeated and followed by fighting.

Similar behavior has been observed in the laboratory, where courting males approached females swimming stiffly, while rapidly moving their mouths and opercula (21).

Presumably when anal fins are pressed together the male injects sperm into the female using modified anal fin spines. Each female mates multiple times (22). Although mating occurs from July through September, fertilization of eggs does not take place immediately after intromission (23). Instead, females store sperm until about January, when fertilization occurs. The stored sperm is typically from several males, and broods show evidence of multiple paternity (22). The young develop slowly for the first few months and then grow extremely rapidly in the last 2 months before birth

(24). They are born, mostly tail first, in May or June, when food is abundant. The young begin to form aggregations soon after birth. Because males become mature within a few months after birth, some of the "territorial" males observed in the Russian River in August were probably young-of-year.

The number of young produced per female increases with size but varies among subspecies, reflecting life histories adapted to different environments. In the highly variable Russian River, where few fish live longer than 2 years, females can reproduce in their first year (70–80 mm SL) and give birth to 12–45 young (mean, about 22); a female measuring 100 mm SL produces about 42 young (5). In Clear Lake, a highly predictable environment where females commonly live 4–7 years, females delay reproduction until their second or third year and produce a relatively small number of large young. A female measuring 100 mm SL produces about 25 young, although larger fish can produce up to 60 young. Tule perch from the San Francisco Estuary are intermediate in most respects, including mortality rates, fecundity (a female measuring 100 mm SL produces about 33 young; larger fish, a maximum of about 60), and age of first reproduction (mostly age 2). Curiously, female tule perch in upper Blue Lake, closely related to Clear Lake fish, have a very low fecundity (10–20 young). Analysis of growth rates indicates that this is a stunted population (maximum size about 90 mm SL), presumably from low food availability (5).

Growth in tule perch is most rapid during the first 18 months after birth, when they are 25-45 mm SL (5, 23). However, the growth rate varies from population to population. At the end of their first summer, Russian River fish reach 5–8 cm SL; Clear Lake fish, 6–9 cm; and Delta fish, 8–10 cm (23). The growth differentials are maintained so that by the end of the third year, they will be about 10–11 cm, 11–15 cm, and 14–16 cm, respectively. Most tule perch measure less than 16 cm SL and are under 5 years of age, but a few exceed 20 cm and live 7–8 years. The largest fish on record is one measuring 238 mm FL (about 22 cm SL) from Napa Slough (29).

Status IE (as a species). Tule perch are abundant in many areas but are also absent from portions of their historical range (e.g., records in Evermann and Clark [25]). They are highly adaptable, and their viviparity seems to reduce vulnerability to competition and predation from alien fishes. Their absence from habitats dominated by nonnative species seems to be the result of poor water quality, as reflected in high temperatures, low dissolved oxygen levels, and low water clarity.

Sacramento tule perch. ID. This subspecies has responded to changes in Central Valley and Monterey Bay streams in a number of different ways. They have been extirpated from the Pajaro and Salinas Rivers, most of the San Joaquin basin, and a number of smaller streams. The causes of extirpation seem to be poor water quality and toxic chemicals. Tule perch in the Pajaro River were wiped out by a single pesticide spill in 1968 (26). In some regulated streams, such as Putah Creek and the Stanislaus River, small populations maintain themselves below dams, but they are isolated and so in continual danger of extinction. However, they seem to be able to persist in small numbers as long as suitable cover and water quality are present. The population in Coyote Creek, Santa Clara County, was long thought to be extinct but was rediscovered in 1998 as a tiny population below Anderson Dam (6). In other regulated streams, such as the American River and mainstem Sacramento River, tule perch are abundant, often one of the commonest fish in areas with heavy cover or beds of aquatic plants. In the Pit River tule perch are among the most abundant fish in Britton and other reservoirs managed for hydropower. In water supply reservoirs they rarely persist in numbers, but they have managed to colonize at least two reservoirs in southern California, via the California Aqueduct (7, 33). In the San Francisco Estuary tule perch appear to be in long-term decline, perhaps in response to increased populations of centrarchids.

Clear Lake tule perch. ID. In Clear Lake tule perch are one of the few native fishes that have managed to maintain populations in the face of continual onslaughts of new alien species, changes in water quality, and loss of tule-dominated habitats. Despite its commonness, the future of this fish in the lake is by no means secure, in the face of changes in land and shore use and the onslaught of new introductions of alien species.

Russian River tule perch. IC. Russian River tule perch seem to be much less abundant today then they were in the early 1970s when they were the subject of studies by students in University of California, Davis, field courses. A seine survey of the river during June–October 1988 found they were uncommon compared with other native and alien fishes in the river (27). The limited distribution, short life span, and low numbers of this subspecies make it susceptible to extinction from a variety of causes, but probably the most important are alterations of habitat and water quality. Coyote and Warm Springs Dams now control flows in the river, resulting in increased turbidity and decreased water quality. Increased development of agricultural land in the watershed, especially vineyards, is also contributing to a decline in water quality and increased potential for pesticide spills and other disasters. Tule perch populations in the Russian River should be monitored as a sensitive indicator of long-term water quality.

References 1. Hopkirk 1962. 2. Hopkirk 1973. 3. Baltz and Moyle 1981. 4. Baltz and Loudenslager 1984. 5. Baltz and Moyle 1982. 6. Leidy 1984 and pers. comm. 1999. 7. L. Pardy, CDFG, pers. comm. 1991. 8. Moyle and Baltz 1985. 9. Marchetti 1998.

10. Moyle and Daniels 1982. 11. Brown 2000. 12. Vondracek et al. 1989. 13. Week 1982. 14. Cech et al. 1990. 15. Turner 1966d. 16. Herbold 1987. 17. Feyrer 1999. 18. Cook 1964. 19. P. B. Moyle, unpubl. data. 20. Phelps 1989a. 21. R. Kurth, University of California, Davis, unpubl. obs. 22. Phelps et al. 1995. 23. Bundy 1970. 24. Bryant 1977. 25. Evermann and Clark 1931. 26. Smith 1982. 27. Phelps 1989b. 28. Knight 1985. 29. R. L. Leidy, USEPA, pers. comm. 1999. 30. R. N. Lea, CDFG, pers. comm. 1999. 31. J. Merz, unpubl. data. 32. J. Norton, pers. comm. 1973. 33. Swift et al. 1993.

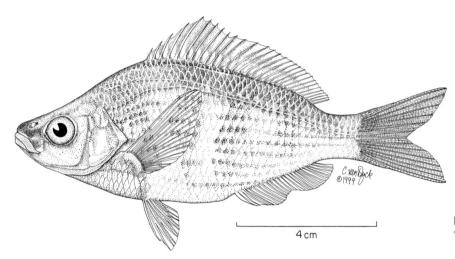

4 cm

Figure 142. Shiner perch, 10 cm SL, Tomales Bay, Marin County.

Shiner Perch, *Cymatogaster aggregata* Gibbons

Identification Shiner perch are similar to tule perch in shape and size, but are not quite as deep bodied. The dorsal fin has only 8–9 spines and 18–23 rays, but the soft–rayed portion makes up more than 45 percent of the fin. The anal fin has 3 spines and 22–25 rays, with a fleshy enlargement of the spiny portion in males. The pectoral fins have 19–21 rays; the pelvic fins, 1 spine and 5 rays each; and the lateral line 36–46 scales. The basic body color ranges from silvery to gray, and it is overlaid by a series of fine horizontal bars and by three yellow vertical bars. Breeding males may turn nearly black, obscuring the bars with dark speckles.

Names Cymato-gaster is derived from the Greek words for fetus and belly, referring to their viviparity; *aggregata* means crowded together, an apt description of their shoaling habits.

Distribution Shiner perch range along the Pacific coast from Baja California to Port Wrangell, Alaska (1). They can be expected in lagoons, estuaries, and lower reaches of streams anywhere along the California coast.

Life History Shiner perch are one of the most common fishes living in the shallow marine waters, bays, and estuaries of California. They are abundant in areas with soft bottoms and beds of eel grass, but they will aggregate around piers and other structures as well. They are tolerant of low salinities and so will often move up into tidal areas of coastal streams. In the Navarro River in August 1973 individuals were found at salinities as low as 1–3 ppt, but large shoals were noticeable only in regions where salinity seldom dropped below 9–10 ppt. The further downstream and the higher the salinity, the more numerous the shiner perch. A similar pattern exists in San Francisco Bay, where shiner perch are found at salinities of 0–34 ppt but are rare below 10 ppt (9). Most perch found at low salinities are young-of-year (4–6 cm SL), although adults (8–14 cm SL) are commonly taken in San Pablo Bay when salinities drop to 9–14 ppt (2, 9). Temperatures where they occur are 7–26°C, but they are uncommon where temperatures exceed 24–25°C. They seem to prefer temperatures less than 19°C, although (depending on acclimation history) they survive those up to 30°C (4). In San Francisco Bay the perch move into deeper water in winter, and many adults emigrate to the ocean, returning in spring to give birth. The maximum depth at which they have been collected is 146 m (9).

Peak populations occur in estuaries during summer months because young are born from May to August. Pregnant females seek out shallow water when giving birth, to give the young some measure of protection from predators, and they also place them in areas with abundant food and warm temperatures (9). Populations in San Pablo and Suisun Bays drop during years of high riverine outflow, perhaps because of long-term reductions in salinity (3). A salinity-temperature interaction may also affect their distribution. Thus they may leave estuaries and streams when temperatures become too warm or water too fresh.

In estuaries shiner perch feed mainly on zooplankton, especially copepods, but may switch seasonally (winter) or opportunistically to benthic prey (4, 5). Small (41–55 mm SL) perch in the Navarro River in August 1973 were bottom feeding mostly (95% by volume) on small euryhaline amphipods, but tipulid and chironomid midge larvae were also taken. Adults feed mainly during the day but juveniles will also feed at night.

The reproductive behavior of shiner perch is similar to that of tule perch (6). Courtship and mating take place in March–May in California, when large aggregations of perch move into shallow water, where the females give birth and then mate. Females store sperm from several males until the eggs mature in November–December, when they are fertilized (7). After 4–6 months of gestation the young are born, each measuring 3–4 cm SL. The number of young increases with size and age: fish measuring 8–10 cm SL give birth to 6–10 young, while fish measuring 11–12 cm SL give birth to 15–20 young (8).

After their first year of life, growth is relatively slow. Age 1 females average 82 mm SL; age 2, 99 mm; age 3, 105 mm; age 4, 116 mm; age 5, 119 mm, and age 6, 122 mm (8). However, few fish seem to live longer than 2 years (9). The maximum age is about 7 years; the maximum size recorded is 235 mm FL (9).

After birth, young remain in shallow water, often in large shoals, where they are major prey of fish-eating birds. Presumably their tendency to live in shallow water and the upper reaches of estuaries reflects both the abundance of food and the fact that such habitat provides a refuge from predatory fish.

Status IE. Shiner perch are abundant and widely distributed in California bays and estuaries, where they provide major entertainment for kids fishing from piers. They are an interesting, if minor, component of the tidal zone of coastal streams. In bays their abundance may be an indicator of environmental quality. In San Francisco Bay their numbers declined in the 1980s and have generally remained relatively low since then (with a brief resurgence in 1997), perhaps as a result of degradation of shallow-water nursery habitat (9).

References 1. Miller and Lea 1972. 2. Ganssle 1966. 3. Herbold et al. 1992. 4. Emmett et al. 1991. 5. Odenweller 1975. 6. Shaw and Allen 1977. 7. Darling et al. 1980. 8. Baltz 1984. 9. Baxter et al. 1999.

Gobies, Gobiidae

Gobies are adapted for bottom living in shallow and intertidal waters. Their most distinctive feature is the ventral cone-shaped suction cup formed by the complete union of the pelvic fins. This cup can be used for clinging to rocks in the face of backwash from waves or strong tidal currents and for climbing over waterfalls on island streams. Their bodies are usually elongate, their heads blunt, their mouths terminal, their two dorsal fins separated, and their eyes close to the top of the head. They are small (few grow longer than 10 cm TL) and include the smallest vertebrates known (8–10 mm as adults). These features reflect fishes that live in unlikely places, such as burrows of marine invertebrates or under stones, and ambush or grab small prey.

Although there are more than 1,900 species in the goby family (circa 7% of all fishes!), most of them are tropical and marine. Comparatively few species have invaded fresh water, but freshwater gobies exist worldwide, especially on tropical oceanic islands. Many species, however, are tolerant of low salinities and are occasionally found in the upper reaches of estuaries. In California there are 16 species, 12 of them marine. Three species can spend their entire life cycle in fresh and brackish water: tidewater goby, shimofuri goby, and yellowfin goby. In addition, longjaw mudsucker occurs often enough in low-salinity regions to justify treatment in this book. The arrow goby (*Clevelandia ios*) is a tidal mud flat dweller that occasionally enters fresh water at the heads of estuaries and coastal lagoons.

The shimofuri goby and yellowfin goby are species introduced from Asia. Additional alien species of gobies can be expected from ballast water introductions. The shimofuri goby, for example, is one of three members of the genus *Tridentiger* introduced by this route. The chameleon goby, *T. trigonocephalus,* is a marine species found in San Francisco Bay and Los Angeles harbor, while the shokihaze (bearded) goby, *T. barbulatus,* is a brackish water species known only from recent (1997–2001) specimens collected in the San Francisco Estuary. It appears to have a growing population in brackish water. Because of uncertainties in its status and ecology, it is not included here.

Tidewater Goby, *Eucyclogobius newberryi* (Girard)

Identification Tidewater gobies are small (seldom larger than 50 mm SL), with elongate, blunt tails and pelvic fins united to form a sucker. The upper quarter to third of the first dorsal fin is clear to cream-colored, while the second dorsal fin is longer and symmetrical, similar in size to the anal fin. The mouth is large and oblique, the maxillary reaching past the posterior margin of the eye. The eyes are widely spaced and close to the snout. Scales are small (66–70 in the lateral line), cycloid, and absent from the head and often from the belly as well. There are 6–7 slender spines in the first dorsal fin, 9–13 elements in the second dorsal fin, 9–12 elements in the anal fin, and 8–10 gill rakers. Their bodies are gray, brown, or olive, with dark mottling or flecking on the sides and back. Living fish are nearly translucent. Dorsal fins are mottled, pelvic fins are yellow or dusky, and the anal fin is dusky. In breeding females the sides and dorsal and anal fins turn varying degrees of black; in the most extreme cases fins become velvety black and sides blue-black (3).

Taxonomy This is the only species in the genus *Eucyclogobius.* Its closest relatives are marine species, also placed in

Figure 143. Tidewater goby, 33 mm SL, Aliso Creek lagoon, Orange County. Drawing by Camm C. Swift, from Swift et al. (1989).

separate genera (*Lepidogobius lepidus, Clevelandia ios, Ilypnus gilberti, Quietula y-cauda, Gillichthys mirabilis*) (1).

Names Eu-cyclo-gobius translates as true-cycloid-goby, referring to the cycloid scales (many gobies have ctenoid scales); *newberryi* is after J. S. Newberry, a professor at Columbia University and advisor to the U.S. Geological Survey team that first collected tidewater gobies in 1854. Goby is apparently derived from *gobio,* the Latin word for gudgeon, a freshwater gobylike cyprinid of Europe. Tidewater goby is the official common name even though it is not particularly descriptive of the fish's habitat.

Distribution Tidewater gobies are endemic to California. They are found in lagoons of coastal streams from Tillas Slough, at the mouth of the Smith River (Del Norte County), south to Agua Hedionda Lagoon (San Diego County) (2, 17). They are absent from areas where the coastline is steep and streams do not form lagoons. Their northern limit on the California-Oregon border is unique and is presumably related to their difficulty in dispersing through the ocean, especially when dominant currents flow south. Historically they were present in at least 87 coastal localities and absent from about 40 deemed suitable for them (2). They are now gone from many localities, including San Francisco Bay.

Life History Tidewater gobies are adapted for life in coastal lagoons created by inflowing streams. The lagoons are blocked seasonally from the ocean by sand bars and are typically brackish and cool (summer temperatures 16–25°C) with bottoms of sand and silt (2, 3, 4, 5, 6). The gobies prefer salinities of less than 10 ppt and as a consequence are generally found in upstream portions of larger lagoons (17). However, they can live at salinities of 0–41 ppt (2) and will breed at salinities of 2–27 ppt (3). They can live and breed at temperatures of 8–25°C (2, 3). Well-oxygenated water is also required, so they disappear from lagoon areas that stagnate or stratify. Optimal lagoon habitats are shallow sandy-bottomed areas 20–100 cm deep surrounded by beds of emergent vegetation. Open areas are critical for breeding, while vegetation is critical for overwintering survival (providing a refuge from high flows) and probably feeding as well (2, 3, 4, 7).

Tidewater gobies have been found in low-gradient sections of inflowing streams as much as 12 km upstream from a lagoon (4). Breeding does not occur in the streams, and the fish present are mainly juveniles that have moved up from lagoons. In the laboratory reproduction can occur in fresh water, but in the wild it seems to occur only in brackish water. Although they are often the most abundant in their lagoon habitats, they can be associated with freshwater fishes (arroyo chub, green sunfish, mosquitofish), anadromous fishes (rainbow trout, threespine stickleback), and euryhaline fishes (staghorn sculpin, prickly sculpin, starry flounder, California killifish).

The gobies are not particularly good swimmers, moving along the bottom in short bursts, occasionally swimming up to grab passing or hovering prey in midwater along steep banks or among emergent plants. When abundant they occur in "loose aggregations of a few to several hundred individuals, with no apparent size segregation" (2, p. 6). They escape predators by swimming in quick dashes of 1–2 m, into vegetation if it is nearby. They usually do not burrow into loose substrates to escape predators, although this behavior has been observed in one population (4). Predators (or at least native predators), however, do not appear to be important regulators of population size in lagoons in southern California. Instead, populations are controlled by environmental conditions. When streams flood and lagoon barriers are breached, creating a strongly tidal environment, populations plummet, although they quickly recover in summer. In Santa Ynez lagoon Swift et al. (4) estimated that the population went from 11 million fish in January to 11,000 by the following June. In other lagoons peak mean densities of gobies of 5–30 fish per square meter have been recorded, dropping to nearly undetectable levels at some times (3, 4). Densities are highest among emergent or submerged vegetation. More northern populations are not as dense, suggesting that their numbers are regulated in part by predation from salmonids (16).

They feed mainly on small crustaceans (ostracods, gammarid amphipods, mysid shrimp) and aquatic insects, especially chironomid midge larvae (2, 7). These benthic prey are captured by three basic techniques: plucking individuals from the bottom, sediment sifting, and midwater capture (7). The most frequent method is plucking individuals from the bottom, usually with a sideways turn of the head. Major prey (*Corophium* amphipods, chironomid larvae) that live in tubes in the bottom have to be grabbed with a quick upward motion of the head to pull them from their tubes. To capture small prey, a mouthful of sediment may be grabbed, churned in the mouth, and expelled through the gill rakers, retaining the prey. Midwater capture techniques involve hopping off the bottom and swimming up to capture passing prey. At any given time dietary breadth in a goby population is low, and individuals are highly selective in feeding (7). However, diet does change with season, and the gobies opportunistically feed on abundant prey in different habitats. Juveniles and adults have similar diets, but juveniles feed more or less continuously while adults feed mainly at night (7).

Tidewater gobies are for the most part an annual species with only an occasional individual living longer than a year (2, 3, 17). This pattern is reflected in length-frequency distributions, which typically have two modes in winter, one around 15–20 mm SL and another around 30 mm SL. As the season progresses both groups increase in average size, but the abundance of smaller fish increases, while the abundance of larger fish decreases. Maximum size seems to be around 50 mm SL, and the largest fish are associated with marsh habitats (3). Fish larger than 40 mm SL are rare.

Reproduction occurs at all times of the year, as indicated by females in various stages of ovarian development (8). Nevertheless little spawning occurs from December through March. Maturity sets in at about 24–27 mm SL, and mature fish may breed more or less continuously for several months. Females produce 150–1,100 eggs for each spawning, the number increasing in a more or less linear fashion with fish size (2, 3). Swenson (3, 17) found that $F = 37SL_{mm} - 692$ was a reasonable predictor of number of ovarian eggs. Each female can spawn every 1–3 weeks for several months, up to 12 times, spawning all eggs at one time. Thus one female measuring 43 mm SL laid six clutches averaging 839

eggs each over 4 months (3). In a lifetime a female is capable of producing 2,400–4,800 eggs if she survives through two spawning seasons, 1,200–2,400 eggs if she survives through only one (17).

Spawning behavior of tidewater goby is highly unusual, because females compete intensely for males and are therefore the more brightly colored and aggressive sex (9). Males construct vertical burrows for spawning, which they defend from other males. At the same time females defend territories around their chosen male and burrow from other females, which often leads to fights involving "fin displays, tail-beating, charging, biting, jaw locking, and wrestling" (9, p. 33). Females actively court males in burrows, approaching with stiffly arched body and dark, erect dorsal and anal fins. A female then tests the readiness of a male to mate by trying to enter the burrow or sticking her head into his mouth. One response is for the male to retreat into the burrow and plug the entrance with sand. Another is to let the female enter. Males may sometimes emerge from their burrows to court females as well. Once a pair is in a burrow together, the male usually plugs the entrance with sand and the pair remains in the burrow together for 1–3 days. At the end of the period the female attaches her eggs to burrow walls in a single layer, where they are fertilized by the male. The female then leaves, and the male replugs the burrow entrance. Each female spawns with just one male each time, and each male usually spawns with just one female (17).

Swenson (9) speculates that females compete for males because the burrows allow males to incubate only a relatively small number of embryos at a time. Burrows are constructed in anoxic sediment, are narrow, and have just one entrance, which is plugged once incubation begins. Low oxygen levels likely to occur in burrows may limit the number of embryos for which a male can care, and this, combined with the time and energy males must spend taking care of the burrow and young, makes males a scarce resource.

Males guard embryos for 9–11 days without feeding. The elliptical embryos dangle from the burrow ceiling by a single thread each, until they hatch. The male fans and rubs the embryos as they develop, but once they hatch he loses interest. The larvae, measuring 4–5 mm SL, emerge from the burrow and swim upward to join the plankton. They become benthic again at 16–18 mm SL (2).

Status IB. The tidewater goby was listed as Endangered by the USFWS in 1994 and has had fully protected status from the state of California since 1987. The listing is controversial because the fish apparently has some fairly secure populations and is at times amazingly abundant. Yet the status is merited because each population is isolated from other populations and subject to extirpation through myriad factors. Somewhere between 25 and 50 percent of its populations have been lost in the past 100 years, most of them south of Point Conception (2). In San Luis Obispo County 6 of 20 populations disappeared between 1984 and 1989 (10). Although the fish have pelagic larvae, genetic studies indicate that each population rarely has contact with other populations (11), and natural recolonization becomes increasingly unlikely as more and more populations disappear and as remaining populations become more restricted in their habitats. They are also prone to local extinction because they are annual fish and become especially vulnerable at the annual low point in their population cycle. Factors that have a negative effect on tidewater goby populations are as follows:

Poor watershed management. Each population is located at the bottom of a watershed, so anything that happens upstream can affect gobies. Many of these small coastal watersheds are highly developed for either agriculture or urban-suburban use. In the northern part of the state, the watersheds are subject to heavy logging and other abuse. The general effect of these activities is to increase sedimentation of lagoons and as well as the severity of high-flow events. To a certain extent tidewater gobies should be able to adapt to sedimentation because they breed in sandy substrates. However, increased sedimentation usually creates large amounts of shallow, warm habitats that may be unsuitable for them (and suitable for exotic predators), and it can bury the productive marshlands they need. Sediment and high flows also increase flooding of surrounding areas, so that lower reaches of streams are channelized and leveed to encourage more rapid runoff to the ocean. High flows that scour areas where gobies have little protection may be one of the major sources of seasonal mortality, even in systems in reasonably good condition (4); therefore increased peak flows may cause higher than normal depression of populations and perhaps extirpation.

Diking and draining of wetlands and riparian areas. The most dense populations of tidewater gobies are found in marshes that surround the lagoons (3). In addition, gobies penetrate farthest upstream where there are well-developed riparian habitats (4). Thus diking and draining of wetlands and channelization of streams remove prime rearing and refuge habitat. Because an estimated 75–90 percent of estuarine wetlands have been lost in California, goby habitat has become increasingly restricted, and many estuaries are no longer suitable habitat for them (12). Not surprisingly, the largest and most secure goby populations occur in lagoons where estuarine wetlands have been protected or restored (3, 4). Large expanses of wetlands are also critical, because goby populations show steady declines where wetlands are reduced to small patches (18). However, small wetlands can serve as "stepping-stones" for gene flow between isolated populations and for recolonization of restored areas (18).

Breaching of sandbars at estuary mouths. It is common practice in California estuaries to artificially open up lagoons in summer for tidal flushing, to prevent flooding of adjacent land, to reduce flooded areas that produce mosquitoes, to carry away agricultural pollutants that cause anoxia, to increase the area of mud flats for birds, to admit anadromous fish, and even to improve beaches for surfing (12). The general concept is that tidal flushing is "good" for West Coast estuaries because it occurs naturally on the East Coast (12). Unfortunately, most small California estuaries historically were blocked by sandbars for much of the summer, creating the nontidal brackish lagoons preferred by tidewater gobies. Where estuaries have been permanently breached with jetties, tidewater gobies are absent. They also are either absent from estuaries that are breached annually or confined to their upper ends (12).

Pollutants. Lagoons are where toxic and organic materials from watersheds are likely to accumulate. The deeper parts of many coastal lagoons become anoxic on a regular basis from excess organic matter (e.g., sewage and agricultural waste). Pesticides and other toxic compounds presumably enter lagoons on a regular basis, causing fish kills, although they are hard to detect (3). Pollutants are thus probably a chronic but largely undocumented cause of goby declines (17).

Nonnative species. The upper, fresher reaches of goby lagoons often contain nonnative species, such as mosquitofish, green sunfish, and largemouth bass. They can at times be significant predators on the gobies; for example, most of the diet of young-of-year largemouth bass in the upper Ynez River estuary was tidewater gobies (4). Likewise, African clawed frogs (*Xenopus leavis*), present in some lagoons in southern California, prey on tidewater gobies (13). However, there is no evidence yet that these predators limit goby populations. A bigger threat may come from two exotic gobies. Yellowfin goby are gradually spreading along the coast and can occupy habitats similar to tidewater goby, perhaps competing with them for food and space. They are also much larger than tidewater gobies and are capable of preying on them. The shimofuri goby, a new invader, has not yet reached tidewater goby habitats but is likely to in the near future, by spreading downstream from reservoirs fed by the California Aqueduct. This species has similar habitat requirements and diets and, in aquaria, behaves very aggressively toward tidewater goby, with the aggression sometimes resulting in predation (7, 14). Female tidewater gobies will sometimes actively court a large male shimofuri goby, who returns the compliment with severe aggression, harming or perhaps eating the female. These species have the potential for eliminating tidewater gobies from some of their habitats.

The tidewater goby is a classic case of a small fish that seems to have little value beyond the novelty of its behavior patterns and endemism to California estuaries (17). To protect it will require a large investment in resources for acquiring and managing wetlands, managing estuaries, fixing watersheds, reducing pollution, and generally improving the environment in which it lives. Yet the goby is also a very sensitive indicator of watershed conditions and of the health of coastal lagoons. Chances are that a watershed, estuary, or lagoon without tidewater gobies is also a poorer environment for many other native species and is in a less healthy and sustainable environment for humans. The tidewater goby is also a wonderfully resilient species; it will respond almost instantly to favorable conditions, building populations numbering in the millions. Thus, provided large wetlands are present (18), it can be restored to estuar-

ies from which it has been lost (as has been demonstrated [15]) and serve as living proof that better conditions have returned. Clearly a California with many tidewater goby populations will be a better place for all of us.

References 1. Birdsong et al. 1988. 2. Swift et al. 1989. 3. Swenson 1995. 4. Swift et al. 1997. 5. Wang 1982. 6. Wang 1986. 7. Swenson and McCray 1996. 8. Goldberg 1977. 9. Swenson 1997. 10. K. Worcester, CDFG, pers. comm. 1989. 11. Crabtree 1985. 12. Capelli 1997. 13. Lafferty and Page 1997. 14. Matern 1999. 15. Swift et al. 1993. 16. C. C. Swift, pers. comm. 1999. 17. Swenson 1999. 18. Lafferty et al. 1999a,b.

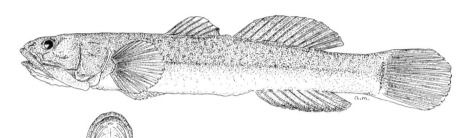

Figure 144. Longjaw mudsucker, 9.8 cm SL, California. Drawing by A. Marciochi.

Longjaw Mudsucker, *Gillichthys mirabilis* Cooper

Identification Longjaw mudsuckers are heavy-bodied gobies with exceedingly long upper jawbones that nearly reach the opercular opening in adults. Their heads are broad and flat, with small widely spaced eyes on top. The first dorsal

fin is small and low (4–8 spines); the second dorsal fin is well developed (10–17 elements). The anal fin is short, with 9–17 elements (usually 10–11 rays). The pectoral fins are rounded, with 15–23 rays. They have 10–16 gill rakers on the first arch. Scales are small, cycloid (ctenoid in juveniles), and embedded in irregular rows, with 60–100 in the lateral series. Scales are largest on the caudal peduncle, becoming smaller toward the head. The anterior portion of the belly is usually without scales. Mudsuckers are dark brown to olive on the back and sides with yellowish bellies and frequently have a row of faint vertical bars on the sides. Juveniles (<25 mm TL) have a smaller mouth (reaches to the back of the eye), eyes more on the sides of their heads, a dark blotch on the rear of the first dorsal fin, and often about 8 vertical bars on the sides.

Taxonomy Longjaw mudsuckers show some morphological and geographic variation (1, 2), but not enough to break the species into subspecies. They are part of a group of related but monogeneric gobies from the California region (3).

Names Gill-ichthys means Gill's fish, named for Theodore Gill, a 19th-century ichthyologist who worked on gobies. *Mirabilis* means wonderful, perhaps reflecting J. G. Cooper's admiration for their strange appearance and ability to live in mud flats and tolerate extreme conditions. They are well described by their common name.

Distribution Longjaw mudsuckers occur from Bahia Magdalena in Baja California north to Tomales Bay and in the northern end of the Gulf of California. The northernmost spawning populations are in Tomales Bay, Marin County (4). The population now present in the Salton Sea and associated canals was started with 500 fish planted in 1950 by CDFG (5).

Life History Longjaw mudsuckers are not true freshwater fish. The longest they usually survive in fresh water is 3–7 days. Nevertheless their presence in the Salton Sea, their occasional (if temporary) occurrence in low-salinity intertidal areas, and their use as bait fish in fresh water, particularly in the Colorado River, justifies inclusion among the inland fishes of California. They typically live in shallow, mud-bottomed tidal sloughs at the upper ends of bays and estuaries (first-order intertidal creeks) (6). When the tide is low, exposing mud flats, they retreat into burrows or tidal channels. They can survive temporary stranding on tidal flats and low oxygen levels in their burrows by gulping air into a highly vascularized chamber in their throat (buccopharyngeal chamber) (7). They can also wriggle on their bellies for short distances across exposed mud flats to reach water after being stranded (8).

In the Salton Sea mudsuckers seem to be abundant in only a few quiet, shallow areas, although they are widely distributed in the sea and have been taken as deep as 12 m (5). They can maintain populations in water with salinities as high as 82.5 ppt (2) and as low as 12 ppt (9). They can survive temperatures of at least 35°C (5), although they prefer those within the range 9–23°C (10). Thus mudsuckers can maintain populations (including breeding) in turbid salt evaporation ponds in San Francisco Bay, where temperatures range from 6 to 28°C and salinities range from 40 to 70 ppt over the course of a year (15). In southern California salt marshes their typical fish associates are California killifish, arrow goby, and topsmelt (11), whereas in northern California salt ponds they are commonly associated with topsmelt, yellowfin goby, staghorn sculpin, rainwater killifish, and threespine stickleback (15).

Longjaw mudsuckers feed on whatever invertebrates and small fish are available and their diet changes seasonally depending on prey availability (16). In the Salton Sea adults eat mostly pile worms (*Neanthus*), with lesser amounts of barnacles, aquatic insect larvae, and fish, including young mudsuckers. Large juveniles (25–90 mm SL) that concentrate in shallow areas feed mostly on brinefly larvae (*Ephydra*), waterboatmen (Corixidae), and pile worms. Small juveniles (15–25 mm SL) eat copepods, punkyfly larvae (Heleidae), and free-living nematodes (5). In coastal salt marshes they also eat a wide variety of benthic food, including algae, isopods, amphipods, and small fish (12, 15, 17). A common prey is California killifish (17). Mudsuckers forage most actively at night (13), and they will move into tidally flooded marshes to forage at this time (17).

Growth is rapid in the Salton Sea. Spring-hatched fish may reach 60–80 mm and maturity by late August. Growth slows in the winter, but by the start of their second spring most mudsuckers measure 80–120 mm SL. They live about 2 years, reaching 135–140 mm SL (5). Their basic life history is similar in southern California salt marshes (12) and in the salt marshes of San Francisco Bay (15). The maximum size seems to be about 150 mm TL (4).

Breeding takes place from December through June in San Francisco Bay (although it may occur all year round farther south) (4), but it takes place mostly in January–July (13). Each female spawns more than once, usually in response to changes in temperature, producing 8,000–27,000 eggs (1, 14). Breeding usually ceases during summer, which is a period of rapid growth and fat storage. Male mudsuckers construct burrows for breeding and defend them from other mudsuckers. The defense displays in response to an intruder are spectacular: the mouth is opened wide and long maxillary bones flare the loose buccopharyngeal skin outward, exposing a large expanse of reddish, highly vascularized interior (13).

Embryos are club shaped and attached in clusters to the side of the burrow by adhesive threads. The male guards the nest until the young hatch, usually in 10–12 days. Larvae and juveniles are quite different in appearance from the adults, having short jaws and large eyes. These differences are apparently adaptations for surviving the short period they spend living pelagically, where they presumably feed on zooplankton. Juveniles drop to the bottom at 8–12 mm TL (4).

Status IE. Longjaw mudsuckers are common throughout their range, although they may be locally depleted by collecting for use as bait. They are often most abundant in shallow, saline environments that exclude most other fishes. They have considerable value as bait fish because they can be kept for short periods of time in moist algae, are long-lived on the hook, and will not reproduce in fresh water (14).

References 1. Barlow 1961a. 2. Barlow 1963. 3. Birdsong et al. 1988. 4. Wang 1986. 5. Walker et al. 1961. 6. Desmond et al. 1999. 7. Todd and Ebeling 1966. 8. Todd 1968. 9. Courtois 1973. 10. De Vlaming 1971. 11. Williams and Zedler 1999. 12. Johnson 1999. 13. Weisel 1947. 14. Barlow and de Vlaming 1972. 15. Lonzarich and Smith 1997. 16. Hartney and Tumyan 1998. 17. West and Zedler 2000.

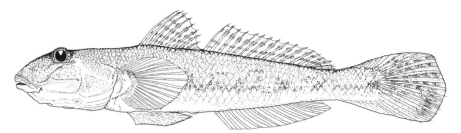

Figure 145. Yellowfin goby, 12 cm SL, San Francisco Estuary. Drawing by A. Marciochi.

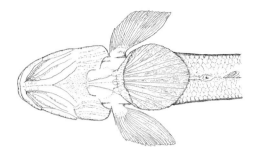

Yellowfin Goby,
Acanthogobius flavimanus (Temminck and Schlegel)

Identification Yellowfin gobies are elongated (to 30 cm TL), blunt-headed fishes, with pelvic fins united to form a sucker. There are 8 spines in the first dorsal fin, 14 rays in the second dorsal fin, 11–12 rays in the anal fin, and 55–65 ctenoid scales in the lateral series. The maxillaries do not extend beyond the center of the eyes, which are closely spaced on top of the head. The color pattern consists of "about eight diffuse dusky spots, each somewhat larger than the eye diameter, arranged in a nearly evenly spaced series down each side, the first three being concealed beneath the pectoral fin, the last forming a more prominent spot at the base of the caudal fin; the sides and back being indistinctly mottled; the upper two-thirds of the caudal fin with about 10 narrow vertical dusky, zigzag bands, the lower third plain dusky" (1, p. 302). The barring on the upper half or so of the caudal fin is a particularly good field characteristic.

Names Acanthogobius means spiny goby, referring to the large number of spines on the first dorsal fin. *Flavimanus*

means yellow hand (fin). Yellowfin goby is not a particularly descriptive common name but is a translation of the Latin name. They are also called oriental gobies and mahaze ("horse goby" in Japanese) and are sold in bait stores as "mudsuckers."

Distribution Yellowfin gobies are common in shallow coastal waters of Japan, Korea, and China. They were first collected in the San Francisco estuary in 1963, where they presumably had become established after being transported across the Pacific in the ballast water of a ship (1). They spread explosively through the estuary and are present in the Sacramento River as high as Knights Landing and in San Francisco Bay. From the Bay they have spread along the coast from Elkhorn Slough to Tomales Bay (2). They are also present on occasion in the Delta-Mendota Canal, California Aqueduct, and San Luis Reservoir (Merced County). They were first noticed in southern California in 1977, in Los Angeles Harbor, presumably the result of an independent ballast water introduction, and have subsequently been found as far south as San Diego Bay (3, 4, 14). They still seem to be spreading, so they can be expected anywhere along the California coast, especially south of San Francisco Bay. Yellowfin gobies have also become established in New South Wales, Australia (5, 18).

Life History Yellowfin gobies are found in shallow, soft-bottomed areas in fresh, brackish, and salt water. They are well adapted for estuarine living because they are capable of withstanding abrupt changes between fresh and salt water and can survive water temperatures greater than 28°C (6). In marshes in San Diego Bay they have been found at salinities of 16–40 ppt, temperatures of 15–32°C, and dissolved oxygen levels of 3–14 mg/liter (14). Although they can live in fresh water, they apparently cannot breed there, requiring salinities of at least 5 ppt (7). They can complete their entire life cycle in salt water, however. Adults living in fresh water move downstream to spawn in shallow, brackish water (mainly in San Pablo Bay in the San Francisco Estuary) in fall and winter (16). Larvae rise up in the water column to catch tidal currents to carry them inland before settling down at about 13–15 mm TL (7). Juveniles continue the upstream movement. Curiously, adults living in salt water may move upstream to spawn in brackish water and then return after spawning (12, 13).

Yellowfin gobies take most prey from ambush or by carefully searching the substrate, because they swim only short distances in a jerky manner. The smallest gobies feed on copepods and other small crustaceans; shrimp, crabs, and small fishes become more important as the size of the fish increases (8). In the lower Petaluma River, however, copepods are the principal food of all size classes, followed by mysid shrimp and amphipods (13). In Suisun Marsh yellowfin gobies of all sizes feed primarily on *Corophium* and other gammarid amphipods (9). Before the collapse of the mysid shrimp populations in the estuary, opossum shrimp were a major part of their diet (10).

Yellowfin gobies apparently only live 3 years, reaching 10–14 cm SL at the end of their first year, 16–18 cm at the end of their second year, and 17 cm or more at the end of their third year (6, 11, 13). A goby that measured 22 cm SL (27 cm TL) was over 3 years old (13). Males usually become mature after 1 year (9–10 cm SL) and females after 2 years (7, 12). They breed from December through July in the San Franscisco Estuary (7, 13), although most breeding seems to occur in March and April. Each female produces 6,000–32,000 eggs (15).

For breeding, males construct Y-shaped tunnels with two entrances in bottoms that are a mixture of mud and coarse sand (12). Pieces of pipe and other artificial materials may also be used, provided water can flow through them. Eggs are teardrop shaped and attached by adhesive filaments to the roof of the burrow in a single layer by the female. They are then fertilized by the male. Embryos are guarded by the male until they hatch (28 days at 13°C), but apparently females assist in some instances. Larvae leave the nest soon after hatching at about 5 mm TL and assume a pelagic existence for an undetermined period of time, settling on the bottom at 10–15 mm TL.

Status IID. Yellowfin goby populations in California have spread widely since the fish were first noticed in 1963. They are now one of the most abundant bottom fishes in San Francisco Bay and the Delta and are apparently still increasing their range (17). What effect this population explosion has had on native freshwater and estuarine fishes is not known. They are a common prey of piscivorous birds and fishes and so have clearly entered the food web. Given their comparatively large size and predatory habitats, they could have a negative effect on populations of other small benthic estuarine fishes, such as tidewater goby. However, no impact on staghorn sculpins in Newport Bay was detected (15). Fortunately they seem to have a hard time establishing themselves in coastal lagoons that close most of the year. During 1998 thousands of young-of-year invaded the large lagoon at the mouth of the Santa Margarita River, but only small numbers were found in nearby smaller lagoons that contained tidewater gobies (16).

On the positive side, yellowfin gobies are widely used as bait fish, at least in salt water, and catching them is a source of pocket money for some children. They are considered to be a delicacy in Japan and could no doubt be marketed here, especially as gobi sushi.

References 1. Brittan et al. 1963. 2. Miller and Lea 1972. 3. Haaker 1979. 4. Swift et al. 1993. 5. Middleton 1982. 6. Brittan et al. 1970. 7. Wang 1986. 8. Kikuchi and Yamashita 1992. 9. Feyrer 1999. 10. Herbold 1987. 11. Hoshino et al. 1993. 12. Dotsu and Mito 1955. 13. Baker 1976. 14. Williams et al. 1998. 15. Ursui 1981. 16. C. C. Swift, pers. comm. 1999. 17. Baxter et al. 1999. 18. Lever 1996.

Shimofuri Goby, *Tridentiger bifasciatus* Steindacher

Identification Shimofuri gobies are small (usually <85 mm SL) and stout with blunt, somewhat flattened heads, thick caudal peduncles, small eyes close to the top of the head, and rounded tails. In adults the maxilla extends nearly to the back of the eye. The flat region between the eyes on top of the head contains small pores that are part of the acousticolateralis system. Scales number 54–60 in the lateral series. The first dorsal fin (6–7 spines) is close to the second dorsal fin (1 spine, 11–14 rays). The anal fins have 1 spine and 9–12 rays. The pectoral fins (19–23 rays) are large, round, and transparent, with a distinct white band at the base. The uppermost ray of each pectoral fin is attached to the rest of the fin and is smooth (in contrast to that in the chameleon goby). In living fish colors are variable and changing, but most fish have tiny white speckles on the head that extend onto the ventral surface; edges of the second dorsal and anal fins that are fringed with orange and without banding; tiny spots, more or less in rows covering the tail; a distinct stripe running along the top of each side, from the top of the head to the base of the tail; and usually a distinct dark band along the midline of each side.

Taxonomy The shimofuri goby is one of seven species of goby, common in Japan and along the coast of Asia, belonging to *Tridentiger* and originally described in the 19th century. Most of the species in Japan were lumped, with little justification, into *T. trigonocephalus* (1), but the shimofuri goby was resurrected when the fish were reexamined

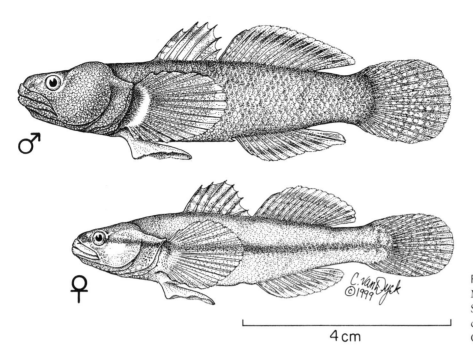

Figure 146. Shimofuri goby. Top: Male, 7 cm SL, Suisun Marsh, Solano County. Bottom: Female, 6 cm SL, Suisun Marsh, Solano County.

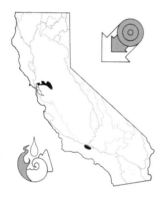

by later-Emperor Akihito and K. Sakamoto of Japan (2). Biochemical studies confirmed its distinctness (3). *T. trigonocephalus,* the chameleon goby, was found in Los Angeles harbor in 1960 and San Francisco Bay in 1962; it has become common in the Bay (4). It was therefore assumed that the first specimens of the shimofuri goby (collected in 1985) were chameleon gobies, even though they were found in fresh and brackish water, whereas the chameleon goby is a marine species. Reexamination of fish from both environments demonstrated that the fresh and brackish water forms were indeed shimofuri gobies (4).

Names The Japanese common name of this goby is "shimofuri shimahaze" (2). Shimahaze means striped goby, a name already in use in English. Shimofuri refers to beef marbled with white fat and describes the pattern of tiny white spots on the head, so it was chosen as the common name in English (4). *Tri-dentiger* refers to the teeth in the

outside row in the mouth, which have three cusps, while *bifasciatus* means two-striped.

Distribution Shimofuri goby are native to estuaries of Japan and mainland countries along the Sea of Japan. They were introduced into California, undoubtedly carried in the ballast water of a ship, sometime shortly before 1985, when the species was first collected in Suisun Marsh (4). It quickly became one of the most abundant fish in the upper San Francisco Estuary. By 1990 it was established in Pyramid Reservoir, some 513 km from the estuary via the California Aqueduct. In 1992 it was found in Piru Creek, below the reservoir (4). It can be expected in any reservoir fed by water from the California Aqueduct and in the streams and estuaries below them.

Life History Because of taxonomic confusion, shimofuri gobies have been little studied in their native habitats. Therefore most of what we know about this species comes from the study by Matern (5, 10) of the population that has invaded the San Franscico Estuary. Shimofuri gobies are widely distributed in tidal habitats in the estuary, but prefer shallow-water (<2 m) areas with complex structure—rocks, logs, tule root masses—which they use for cover and breeding. They have been observed in shallow pools at low tide with temperatures as high as 34°C that were also subjected to rapid temperature changes as the tide rose and fell. In the laboratory they can tolerate temperatures up to 37°C (5, 10). They are less tolerant of high salinities. In the estuary the gobies have been collected in water with salinities as high as 19 ppt, but in the laboratory they show extreme distress when salinities approach 17 ppt (10). They readily live and breed in fresh water, however, resulting in populations being established in reservoirs.

Shimofuri gobies apparently disperse mainly during larval and early juvenile stages, because adults appear to be fairly sedentary. Adults are quite aggressive to other members of the same species, to other benthic fishes (especially other species of gobies), and to small crabs (5). Much of this behavior is related to defense of rocks and other shelters that are suitable for breeding and refuge from predators. In aquaria they will drive smaller tidewater gobies from shelters and may occasionally devour individuals that do not retreat, such as female tidewater gobies trying to court large male shimofuri gobies (5).

The gobies' observed modes of feeding are "sideways substrate bites" and "midwater capture," typical of gobies that rely on vision for prey capture (5, 6). Biting the substrate while turning the head sideways is presumably the most common mode of feeding, based on stomach contents, because they feed entirely on benthic prey, including attached organisms. They have been found to feed heavily on attached hydroids (*Corydylophora caspia*), the cirri (exposed tentacles) of barnacles (*Balanus improvisus*), and tube-dwelling amphipods (*Corophium* spp.). Their tiny, tricuspid teeth may be an adaptation for grazing on such organisms. The fact that no other fishes in the estuary are known to feed on the abundant hydroids and barnacles, both also alien species, may be one reason why shimofuri gobies became so abundant so quickly (5). Other items in the diet are mostly crustaceans of one type or another. At any given time shimofuri goby diets are rather narrow (5).

Shimofuri gobies show wide fluctuations in abundance from year to year (7), as might be expected of a short-lived species. Most appear to live 12–18 months, although a few may live 2 years. They can grow to 35–60 mm TL in a year, at which size they begin breeding. The maximum size recorded is 105 mm TL (5). Individuals spawn repeatedly from March through August, with large males showing ragged fins and thin bodies as evidence of inadequate diet and frequent aggressive encounters, as the season progresses (8). Such males presumably die and are replaced in the limited sites available for spawning by smaller males.

Favored breeding sites are cavities under rocks or shells, where the tiny (1–1.5 mm diameter) round eggs are glued to the ceiling by the female in a single layer and fertilized by the male. They will also spawn readily in artificial cavities, such as pieces of pipe. Females spawn repeatedly and males spawn with multiple females, so thousands of embryos in various stages of development can be found in a single nest (5). Males guard and tend nests continuously, and part of their diet includes dead embryos and embryos from nests of other males. Embryos develop under a wide range of temperatures (13–34°C) and salinities (0–7 ppt). Each nest presumably produces a steady stream of larvae swimming up into the water column throughout the summer or until the male is dead of exhaustion. Any abandoned nest is quickly reoccupied, however, because suitable nesting cavities seem to be in short supply in the estuary.

Larvae are pelagic and are the smallest fish larvae in the estuary. They are often the most abundant fish larvae in the estuary, peaking in abundance in June when the water warms up, although the peak may occur earlier in dry years (11). They seem to become widely distributed through tidal currents.

Status IIE. The shimofuri goby is another one of California's explosive invaders, becoming abundant and widely distributed in less than 10 years. Its range in the state is still expanding as it moves through aqueducts to new waters. Its success can be attributed to a number of factors acting simultaneously:

1. It has a high dispersal ability, mainly through larvae, which enables it to be carried by ships across the ocean and by aqueducts to distant reservoirs.

2. It is tolerant of a wide range of environmental conditions, more so than many resident species, and this tolerance enables it to live in an ever-changing, human-altered estuary.

3. It is aggressive, enabling it to drive potential competitors out of breeding sites.

4. It has a high reproductive capacity (early maturity, high fecundity) and a high level of parental care, enabling it to quickly colonize available habitats.

5. It produces multiple, fast-hatching clutches that increase the likelihood that at least some young will enter the environment under optimal conditions of temperature, salinity, and food availability.

6. It feeds on alien invertebrates that are little used by other fishes and that are abundant in breeding areas.

7. It arrived at a time when populations of other fishes (including predatory striped bass) were in decline and other major invaders (e.g., the overbite clam, *Potamocorbula amurensis*) were also becoming established, so the system was (and still is) in flux. This presumably made it easier to invade.

Despite its abundance, there is little evidence that the shimofuri goby has had harmful interactions with native or resident species in the San Francisco estuary. This may simply be because populations of most fishes are depressed, reducing the potential for interactions. On the other hand, it may also reflect the great changes that have taken place in the estuary since the goby invasion, especially as the result of invasions of invertebrates from clams to copepods (9). An entire new food web has developed, with different energy paths and many linkages that involve only alien species. The shimofuri goby is clearly a part of this change and is very likely to be an abundant member of the ecosystem for the indefinite future. Its role as a predator or competitor, or as prey, will bear watching.

Outside the estuary its potential to cause harm to native species is high. It is highly likely to invade a number of small estuaries in southern California that contain populations of the endangered tidewater goby (4). Laboratory experiments show that, when these two species come into conflict, the tidewater goby loses (5). Shimofuri gobies can drive tidewater gobies out of cover and breeding sites, compete with them for food, and prey directly on them. Thus invasions of tidewater goby habitats should be prevented wherever possible.

References 1. Tomiyama 1936. 2. Akihito and Sakamoto 1989. 3. Mukai et al. 1996. 4. Matern and Fleming 1995. 5. Matern 1999. 6. Swenson and McCray 1996. 7. Meng et al. 1994. 8. S. A. Matern, pers. comm. 1999. 9. Cohen and Carlton 1998. 10. Matern 2001. 11. Meng and Matern 2001.

Mullets, Mugilidae

Mullets are primarily tropical and temperate marine fishes, but many species move readily into fresh and brackish water. Their distribution is worldwide. Because they school in shallow water and estuaries, feeding largely on detritus they stir up from the bottom, they are popular food fish wherever they are abundant. Only the striped mullet enters fresh water in California, although a similar species, the white mullet (*Mugil curema*), is occasionally found in marine waters (Lea et al. 1988).

Striped Mullet, *Mugil cephalus* Linnaeus

Identification Striped mullet have thick, torpedo-shaped bodies; broad, flat heads; small terminal mouths; large eyes (width greater than length of snout); deeply forked tails; widely separated spiny and soft dorsal fins; and translucent adipose eyelids that nearly cover the eyes, leaving only a narrow slit over the pupil. The maxillary is hidden when the mouth is closed, giving the mouth an inverted V shape from a head-on view. There are 4 spines in the first dorsal fin and 1 spine and 8 rays in the second. Adults have 3 spines and 8–9 rays in the anal fin, while juveniles (50 mm TL) have 2 spines and 8–9 rays. The pectoral fin has 16–17 rays, and the lateral line has 38–42 cycloid scales. In fish smaller than 50 mm TL the adipose eyelid is not evident. The intestine is long with a large gizzard, and gill rakers are long and slender. The backs of living fish are blue-green, and the sides and belly are silvery with narrow, horizontal black stripes on the upper half of the body.

Taxonomy Despite a worldwide distribution, striped mullet are definable as one species, with relatively small genetic and meristic differences among populations in different areas, compared with differences among species of *Mugil.* "However, the considerable genetic differentiation among populations, in conjunction with the extremely reduced, or nonexistent, current gene flow, suggests that at least some of them are at the stage of incipient speciation" (14, p. 217).

Names Striped mullet are known as gray mullet or sea mullet in much of the English-speaking world outside North America. *Mugil* is derived from the Latin verb meaning "to suck," referring to their feeding habits; *cephalus* is apparently derived from the Greek word for head, although the ancient Roman word for mullet was *cefalus.*

Distribution Striped mullet inhabit tropical and subtropical coastal areas around the world, including oceanic islands. Adults are found along the California coast north as far as San Francisco Bay (during El Niño years), although they are rare north of Point Conception. They are sometimes among the most abundant fish in estuaries in the San

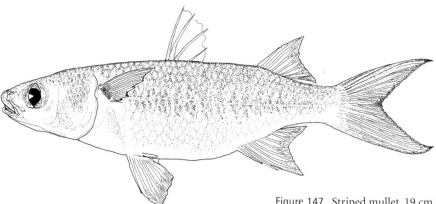

Figure 147. Striped mullet, 19 cm SL, Mexico. Drawing by A. Marciochi.

Diego region. Young-of-year striped mullet are found in lower reaches of larger streams from the Santa Clara River (Ventura County) south, although in warmer years they may be found as far north as Shuman Lagoon, Santa Barbara County (13). They have been found as far as 35 km upstream from the mouth of the Los Angeles River (1) and are present in the lower Colorado River from its mouth to Imperial Dam, about 190 km upstream. They were once the most abundant fish in the Salton Sea but are now rare there; small numbers are apparently present at the mouth of the Alamo River (10). They were present in prehistoric fillings of the sea as well (2).

Life History Striped mullet have been well studied because of their worldwide distribution and importance as food fish. However, little work on them has been done in California. Unless otherwise noted, information in this summary is based on a review of studies done elsewhere (3). Striped mullet are primarily dwellers in shallow estuaries and so have been found living at salinities of 0–75 ppt. They cannot tolerate temperatures much below 14–16°C for extended periods of time and are thus confined to waters that are warm year round. They regularly ascend sluggish rivers and may be able to complete their entire life cycle in fresh water, although they rarely do so. In the Santa Margarita River in 1998, however, they penetrated over 15 km upstream, passing riffles and rapids (13). In the Colorado River most adult mullet migrate to the Gulf of California to spawn in winter, and young gradually move back upstream during the summer of the following year. Thus most juvenile mullet in lower reaches of the river are 2–4 years old (21–37 cm SL); those just below Imperial Dam are 3–5 years old (29–46 cm SL) (4). However, postlarval mullet (28–40 mm SL), too weak to swim upstream, have been found nearly 200 km from the mouth, indicating that some freshwater spawning occurs (4).

Mullet were unable to spawn in the Salton Sea in the 1950s, even though its salinity was close to that of sea water, presumably because its ionic composition was different (5). Large populations that once existed there were recruited from young mullet moving up canals from the Colorado River or by spawning in the Alamo River or other tributaries. When the canal system was altered, making access to the sea more difficult, mullet gradually became rare, with the population maintained by occasional spawning in the Alamo River. However, some spawning in the sea may have taken place earlier in its history when salinities were lower (6).

The ability of young-of-year mullet to move upstream from estuaries may account for their abundance at times in southern California streams such as San Juan Creek or the Santa Margarita River (12), although spawning by adults cannot be completely discounted. In the Santa Margarita estuary, juvenile mullet that measured 2–3 cm TL in January grew to 20–30 cm TL by the following December (13).

Mullet are basically schooling fish, especially when young. Schools break up during feeding, and adults frequently seem to act like members more of a loosely knit aggregation than of an organized school. Spawning takes place in schools but has not been observed in great detail.

Feeding takes place on muddy bottoms in shallow water in areas where the sediment particle size is very small and that are therefore high in organic matter (7). Mullet swim at an angle, scooping up soft surface material with their stiff lower jaws. Coarse material is ejected through the mouth and gills after a mouthful of bottom mud has been ground between pharyngeal plates. Some sand is retained to help the gizzard grind up organic matter. Nutrition seems to be derived from organic detritus, diatoms, bacteria, and microinvertebrates (7). Occasionally mullet will feed on bits of algae floating close to the water surface (6). In the Colorado River their diet is mainly detritus but includes small amounts of aquatic insects (8). Most feeding takes place during the day (9).

Age and growth in striped mullet are highly variable. In the lower Colorado River fish measuring 7–13 cm SL are apparently approaching 1 year old; those 17–20 cm SL are 1–2 years old; those 21–28 cm SL, 2 years old; and those 29–37 cm SL, 4–5 years old (4). The oldest and largest mullet known were from the largely nonreproducing Salton Sea population; a female measuring 62 cm SL and a male measuring 60 cm SL were probably at least 14 years old (5). Mullet usually become mature at 2–3 years of age (23–35 cm SL). Most fish over 25–30 cm SL are females.

Status IE. Striped mullet show fluctuating abundance in California's inland waters because they are a subtropical to tropical species at the northern end of their range. In coastal southern California they appear to be permanent residents in San Diego Bay and occur in most years in the lower San Margarita River, Malibu Lagoon, Newport Bay, and similar areas (13). They are still present in the lower Colorado River despite deterioration of water flow and quality, although it is likely that numbers are lower. A small population seems to have persisted in the Salton Sea as well (10). They are of minor importance as sport fish because they can be taken only on small baited hooks on the bottom or with small flies (6). They supported a commercial fishery in the Salton Sea from 1915 to 1921 and again from 1943 to 1953, when it was forbidden in favor of the sport fishery for other species. A small commercial fishery still exists in San Diego Bay (11), as does an illegal cast net fishery (13). Their role in southern California streams and estuaries has not been investigated.

References 1. Swift et al. 1993. 2. Gobalet 1992. 3. Thomson 1963. 4. Johnson and McClendon 1970. 5. Hendricks 1961. 6. Dill 1944. 7. Odum 1968. 8. Minckley 1982. 9. Collins 1981. 10. S. Keeney, CDFG, pers. comm. 1999. 11. Lea et al. 1988. 12. R. Fisher, pers. comm. 1998. 13. C. C. Swift, pers. comm. 1999. 14. Rossi et al. 1998.

Righteye Flounders, Pleuronectidae

Righteye flounders (about 93 species) are part of a large group of marine flatfishes highly adapted for bottom living. In order to be as flat as possible, and thus blend with the bottom, flounders spend most of their lives with one side on the bottom. This has required some rather drastic morphological changes: the jaws become twisted, the dorsal and anal fins become long and nearly identical to each other, and one eye migrates during development from the bottom side of the head to the top side. Flounders swim about like normal fish fry for the first few weeks after hatching, so the changes take place when they start settling down to the bottom. Flounders are colored on their top sides to match the bottom, and most of them can change color patterns to a remarkable degree. They can match substrate patterns closely, hiding themselves from both predators and prey.

The occurrence of flounders in fresh water is rather unusual, and there are apparently no exclusively freshwater species. In North America north of Mexico only three species regularly occur in fresh water: the hogchoker, *Trinectes maculatus* (Soleidae), on the Atlantic coast; the arctic flounder, *Liopsetta glacialis*, on the Arctic Sea coast; and the starry flounder on the Pacific coast.

Starry Flounder, *Platichthys stellatus* (Pallas)

Identification Starry flounders are the only flatfish likely to be found in fresh water: they are characterized by having both eyes on the upper side of the head, a white "belly" with a single pectoral fin in the middle, pelvic fins on the dorsoventral ridge behind the operculum, and dorsal and anal fins that extend around the body on each side. They can be distinguished from other flounders that might occur in brackish water by the distinctive, alternating white to orange and black bands on the dorsal and anal fins, as well as by the roughness of their skin, caused by the star-shaped plates (modified scales). There are 52–64 rays in the dorsal fins, 38–47 in the anal fins, and 10 in each pectoral fin, as well as 6–11 gill rakers. Although they belong to the right-eyed flounder family, the eyes may be on either the right or the left side of the fish.

Names *Plat-ichthys* means flat-fish; *stellatus* means starry, after the distinctive star-shaped spiny plates on the dorsal surface. They are sometimes called rough jackets, diamond flounders, and, in Japan, swamp flounders.

Distribution Starry flounders occur along the coast of the Pacific ocean and in the lower reaches of coastal streams from the mouth of the Santa Ynez River (Santa Barbara County), north along the Alaskan coast through the Aleutian Islands and Arctic seacoast of Canada to Bathurst Inlet. In Asia they are found along the coast as far south as Tokyo Bay, Honshu Island, Japan, and Korea. In California they are

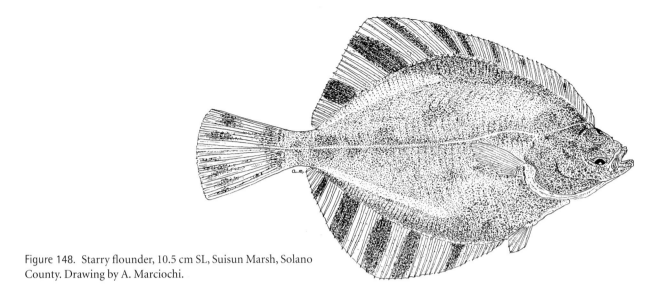

Figure 148. Starry flounder, 10.5 cm SL, Suisun Marsh, Solano County. Drawing by A. Marciochi.

common in bays and estuaries from Point Conception north. They can be expected in brackish and occasionally freshwater parts of any coastal stream with an extensive estuarine system, usually up as far as the first riffle. They are particularly common in the San Francisco Estuary and have appeared in San Luis Reservoir and O'Neill Forebay (Merced County), transported there by the California Aqueduct. They are relatively uncommon in the Delta, becoming more common in the lower parts of the estuary.

Life History Starry flounders are primarily marine or estuarine fish, but they have been found in coastal streams as far as 120 km from salt water (1). In streams they generally prefer tidal, low-gradient areas that have sandy or muddy bottoms. Most found in fresh water are young-of-year. In the San Francisco Estuary some small flounders may have resulted from spawning in the estuary, but most are apparently carried into San Francisco Bay from nearshore ocean waters by strong tidal currents along the bottom (16). These currents are strongest during years of high outflow from the rivers, and, as a consequence, juvenile starry flounder tend to be most abundant in the estuary during wet years (2, 16). Higher abundances may also be related to the greater extent of low-salinity rearing areas and the greater abundance of food organisms preferred by small flounders (3). During dry years abundances may be lower but young are more likely to be found farther upstream and to be entrained by the pumps in the south Delta. The smallest fish are generally found the farthest upstream (4), and they seek areas with higher salinity as they grow larger (16). Thus in April–June most young-of-year are living in salinities of less than 2 ppt, but by July and August they have shifted to salinities of 10–15 ppt (16). Temperatures may also influence distribution because they are usually found at 10–20°C

(16). Large flounders (<20 cm TL) encountered in fresh water seem to be mostly migrants from salt water, rather than fish that have reared there.

Although starry flounders are capable of short, swift bursts of swimming propelled by thrusts of the tail, they normally move by gliding slowly over the bottom, propelled by waves of the dorsal and anal fins (5). They rest on the edges of these same fins, with the belly slightly off the bottom. When startled, they flip sand or mud over the body with their fins, burying nearly everything except their eyes.

Adult starry flounders feed on a wide variety of bottom invertebrates, the type changing with size. Prior to metamorphosis they feed first on planktonic algae, then on planktonic crustaceans. Once on the bottom they feed by grabbing prey from the water column, rather than by picking them off the bottom (6). Thus small flounders (<20 cm TL) feed predominantly on invertebrates that emerge from bottom burrows (e.g., harpactacoid copepods, amphipods of the genus *Corophium*) or on organisms like mysid shrimp that move up and down the water column, settling periodically on or near the bottom (5, 6). Thus in San Francisco Bay small flounders fed mainly on opossum shrimp until the invasion of the overbite clam (*Potamocorbula amurensis*) caused a major reduction in shrimp abundance, forcing them to switch to a more diverse diet (4, 7, 8). In a small coastal lagoon flounders switch readily among gammarid amphipods, mysid shrimp, and other prey, depending on local abundances and conditions (9). Curiously, in fresh water they switch to feeding on insect larvae buried in soft bottoms, such as cranefly larvae (Tipulidae) (10) and annelid worms (9). Feeding in fresh water may initially put the flounder under some osmotic stress, because digestion rates are 2–3 times faster in salt water than they are in fresh water (10). Back in marine environments, large fish feed mostly on crabs, polychaete worms, and molluscs (5).

Young-of-year fish in estuaries grow at rates comparable to those living in salt water (11). At the end of their first year they measure around 12–14 cm TL; at the end of year 2, 25–28 cm; year 3, 38–40 cm; year 4, 40–46 cm; year 5, 46–48 cm, and year 6, 48–60 cm (5, 12). Larger fish, almost exclusively female, may be 7–8 years old.

In salt water males become mature at the end of the second year at about 35 cm TL, females at slightly larger sizes (over 40 cm TL). Females produce 0.9 to over 11 million eggs. Spawning occurs in near-shore marine environments, often near the mouths of estuaries, in September through March, peaking in December and January (5, 13, 14). The presence of larvae indicates that some spawning may also take place in San Francisco Bay (4, 14, 16). Embryos and larvae are pelagic and are carried inshore by currents, settling to the bottom in 39–75 days. They generally require warm, low-salinity water for rearing (16).

Status IE. Starry flounders are fairly common along the California coast and are characteristic of the more northern bays and estuaries. Estuaries such as San Francisco Bay are important rearing areas, as indicated by the decline in abundance of juveniles during extended periods of drought (2, 3). They are probably more important as sport fish than as commercial fish, although they do appear in fish markets. There has been a long-term decline in the catch of starry flounder, but it is not certain if the decline is related to changing conditions in the estuary or to changes in fishing regulations that reduce the catch (15). Populations need to be monitored both in and out of estuaries.

References 1. Gunter 1942. 2. Jassby et al. 1995. 3. Herbold et al. 1992. 4. Ganssle 1966. 5. Orcutt 1950. 6. McCall 1992. 7. Herbold 1987. 8. Feyrer 1999. 9. Martin 1995. 10. Porter 1964. 11. Radtke 1966. 12. Wydoski and Whitney 1979. 13. Love 1996. 14. Wang 1986. 15. Leet et al. 1992. 16. Baxter et al. 1999.

References

Aalto, K. R., W. D. Sharp, and P. R. Renne. 1998. ^{40}Ar/^{39}Ar dating of detrital micas from Oligocene-Pleistocene sandstones of the Olympic Peninsula, Klamath Mountains, and northern California coast ranges: province and paleodrainage patterns. Can. J. Earth Sci. 35:735–745.

Aasen, G. A., D. A. Sweetnam, and L. M. Lynch. 1998. Establishment of the wakasagi, *Hypomesus nipponensis,* in the Sacramento–San Joaquin estuary. Calif. Fish Game 84:31–35.

Aasen, K. D., and F. D. Henry, Jr. 1981. Spawning behavior and requirements of Alabama spotted bass, *Micropterus punctulatus henshalli,* in Lake Perris, Riverside County, California. Calif. Fish Game 67:118–125.

Aceituno, M. E., and S. J. Nicola. 1976. Distribution and status of the Sacramento perch, *Archoplites interruptus* (Girard), in California. Calif. Fish Game 62:246–254.

Aceituno, M. E., and C. D. Vanicek. 1976. Life history studies of the Sacramento perch, *Archoplites interruptus* (Girard), in California. Calif. Fish Game 62:5–20.

Adams, P. B., T. E. Laidig, K. R. Silberberg, M. J. Bowers, B. M. Jarvis, K. M. Sakuma, K. Baltz, and D. W. Woodbury. 1996. Historic and current presence-absence data of coho salmon (*Oncorhynchus kisutch*) in the Central Valley Evolutionary Significant Unit. NMFS SW Fish. Sci. Center Admin. Rpt. T-96-01. 24 pp.

Adams, S. B., C. A. Frissell, and B. E. Rieman. 2001. Geography of invasion in mountain streams: consequences of headwater lake introductions. Ecosystems 4:296–307.

Agassiz, L. 1853. Extraordinary fishes from California, constituting a new family. Am. J. Sci. Arts, 2nd ser. 16:380–390.

Ahmed, S. S., A. L. Linden, and J. J. Cech, Jr. 1988. A rating system and annotated bibliography for the selection of appropriate indigenous fish species for mosquito and weed control. Bull. Soc. Vector Ecol. 13:1–59.

Akihito, and K. Sakamoto. 1989. Reexamination of the status of the striped goby. Jpn. J. Ichthyol. 36:100–112.

Allen, K. O., and K. Strawn 1968. Heat tolerance of channel catfish, *Ictalurus punctatus.* Proc. 21st Annu. Conf. SE Assoc. Game Fish. Comm.: 399–412.

Allendorf, F. W., and R. S. Waples. 1996. Conservation and genetics of salmonid fishes. Pages 238–280 *in* J. C. Avise and J.

L. Hamrick, eds. Conservation genetics: case histories from nature. New York: Chapman and Hall.

Alley, D. W. 1977a. The energetic significance of microhabitat selection by fishes in a foothill Sierra stream. M.S. thesis, Univ. Calif., Davis. 267 pp.

———. 1977b. Significance of microhabitat selection for fishes in a Sierra foothill stream. Calif.-Nev. Wildl. Trans. 1977:27–33.

American Fisheries Society. 1987. Carp in North America. Bethesda, Md.: American Fisheries Society.

Anas, R. E. 1959. Three-year-old pink salmon. J. Fish. Res. Board Can. 16:91–94.

Anderson, C. A. 1936. Volcanic history of the Clear Lake area, California. Bull. Geol. Soc. Am. 97:629–664.

Anderson. N. L., D. L. Woodward, and A. E. Colwell. 1986. Pestiferous dipterans and two recently introduced aquatic species at Clear Lake. Proc. Calif. Mosq. Vector Contr. Assoc. 54:163–167.

Andreasen, J. K. 1975. Systematics and status of the family Catostomidae in southern Oregon. Ph.D. dissertation, Oreg. State Univ. 76 pp.

Anglin, D. R. 1994. Lower Klamath River instream flow study: scoping evaluation for the Yurok Indian Reservation. USFWS Lower Columbia River Fishery Resource Office, Vancouver, Wash. 46 pp.

Applegate, R. L., and J. W. Mullan. 1967. Food of the black bullhead (*Ictalurus melas*) in a new reservoir. Proc. 20th Annu. Conf. SE Assoc. Game Fish Comm.: 288–292.

Applegate, R. L., J. W. Mullan, and D. L. Morais. 1967. Food and growth of six centrarchids from shoreline areas of Bull Shoals Reservoir. Proc. 20th Annu. Conf. SE Assoc. Game Fish Comm.: 469–482.

Arendt, J. D., and D. S. Wilson. 1997. Optimistic growth: competition and an ontogenetic niche-shift select for rapid growth in pumpkinseed sunfish (*Lepomis gibbosus*). Evolution 51:1946–1954.

Arkush, K. D., M. A. Banks, D. Hedgecock, P. A. Siri, and S. Hamelberg. 1997. Winter-run chinook salmon broodstock program: progress report through April 1996. Interagency Ecol. Study Prog. Tech. Rpt. 49. 42 pp.

Armour, C., and P. L. Herrgesell. 1985. Distribution and abundance of fishes in the San Francisco Bay estuary between 1980 and 1982. Hydrobiology 129:211–227.

Armour, C. L. 1993. Evaluating temperature regimes for protection of smallmouth bass. USFWS Res. Publ. 191. 26 pp.

———. 1994. Evaluating temperature regimes for protection of brown trout. USDI Natl. Biol. Surv. Res. Publ. 201. 20 pp.

Armstrong, J. L., D. A. Armstrong, and S. B. Mathews. 1995. Food habits of estuarine staghorn sculpin, *Leptocottus armatus,* with focus on consumption of juvenile Dungeness crab, *Cancer magister.* NOAA Fish. Bull. 93:456–470.

Avault, J. W., R. O. Smitherman, and E. W. Shell. 1966. Evaluation of eight species of fish for aquatic weed control. FAO World Symp. Warmwater Pond Fish Culture. FR VII, E-3:1–14.

Avise, J. C. 1976. Genetics of plate morphology in an unusual population of threespine sticklebacks (*Gasterosteus aculeatus*). Genet. Res. Comb. 27:33–46.

Avise, J. C., and F. J. Ayala. 1976. Genetic differentiation in speciose versus depauperate phylads: evidence from the California minnows. Evolution 30:46–58.

Avise, J. C., J. J. Smith, and F. J. Ayala. 1975. Adaptive differentiation with little genic change between two native California minnows. Evolution 29:411–426.

Ayres, W. O. 1854, 1855. [Descriptions of new species of fish from San Francisco from the Daily Placer and Transcript]. Reprinted in Proc. Calif. Acad. Sci. (1857) 1:1–77.

Baerends, G. P., and J. M. Baerends-Van Roon. 1950. An introduction to the ethology of cichlid fishes. Behaviour Suppl. 11:1–292.

Bagarinao, T., and R. D. Vetter. 1993. Sulphide tolerance and adaptation in the California killifish, *Fundulus parvipinnis,* a salt marsh resident. J. Fish Biol. 42:729–748.

Baggerman, B. 1957. An experimental study of the timing of breeding and migration in the three spined stickleback. Arch. Neerl. Zool. 12:103–307.

Bagley, M. J., and G. A. E. Gall. 1998. Mitochondrial and nuclear DNA sequence variability among populations of rainbow trout (*Oncorhynchus mykiss*). Mol. Ecol. 7:945–961.

Baglin, R. E., Jr., and L. G. Hill. 1977. Fecundity of white bass, *Morone chrysops* (Rafinesque), in Lake Texoma. Am. Midl. Nat. 98:233–238.

Bailey, H. C., C. Alexander, C. Digiorgio, M. Miller, S. I. Doroshov, and D. E. Hinton. 1994. The effect of agricultural discharge on striped bass (*Morone saxatilis*) in California's Sacramento–San Joaquin drainage. Ecotoxicology 3:123–142.

Bailey, H. C., E. Hallen, T. Hampson, M. Emanuel, and B. S. Washburn. 2000. Characterization of reproductive status and spawning and rearing conditions for *Pogonichthys macrolepidotus,* a cyprinid of special concern endemic to the Sacramento–San Joaquin Estuary. Unpubl. ms., Univ. Calif., Davis.

Bailey, R. M. 1980. Comments on the classification and nomenclature of lampreys—an alternative view. Can. J. Fish. Aquat. Sci. 37:1626–1629.

———. 1982. Reply to comment on Reeve M. Bailey's view of lamprey systematics. Can. J. Fish. Aquat. Sci. 39:1217–1220.

Bailey, R. M., and C. E. Bond. 1963. Four new species of freshwater sculpins, genus *Cottus,* from western North America. Mus. Zool. Univ. Mich. Occas. Pap. 634. 25 pp.

Bailey, R. M., and C. L. Hubbs. 1949. The black basses (*Micropterus*) of Florida, with description of a new species. Mus. Zool. Univ. Mich. Occas. Pap. 516. 40 pp.

Bailey, R. M., and T. Uyeno. 1964. Nomenclature of the blue chub and the tui chub, cyprinid fishes from western United States. Copeia 1964:238–239.

Baker, F. B., T. P. Speed, and F. K. Ligon. 1995. Estimating the influence of temperature on the survival of chinook salmon smolts (*Oncorhynchus tshawytscha*) migrating through the Sacramento–San Joaquin River Delta of California. Can. J. Fish. Aquat. Sci. 52:855–863.

Baker, J. C. 1976. A contribution to the life history of the yellowfin goby (*Acanthogobius flavimanus*) in the San Francisco Bay–Delta area. M.S. thesis, Calif. State Univ., Sacramento. 37 pp.

Baker, P. H. 1967. Distribution, size composition and relative abundance of the Lahontan speckled dace, *Rhinichthys osculus robustus* (Rutter), in Lake Tahoe. Calif. Fish Game 53:165–173.

Baker, P. H., and A. J. Cordone. 1969. Distribution, size composition, and relative abundance of Piute sculpin, *Cottus beldingii* Eigenmann and Eigenmann, in Lake Tahoe. Calif. Fish Game 55:285–297.

Baker, P. H., and F. Reynolds. 1986. Life history, habitat requirements, and status of coho salmon in California. Calif. Fish Game Comm. Rpt. CDFG. 37 pp.

Baker, S. C., and R. Heidinger. 1996. Upper lethal temperature tolerance of fingerling black crappie. J. Fish Biol. 48:1123–1129.

Balling, S. S., T. Stoehr, and V. H. Resh. 1980. The effects of mosquito control recirculation ditches on the fish community of a San Francisco Bay salt marsh. Calif. Fish Game 66:25–34.

Balon, E. K. 1995. The common carp, *Cyprinus carpio:* its wild origin, domestication in aquaculture, and selection as colored nishikigoi. Guelph Ichthyol. Rev. 3:1–56.

Baltz, D. M. 1984. Life history variation among female surfperches (Perciformes: Embiotocidae). Env. Biol. Fish. 10:159–171.

Baltz, D. M., and E. J. Loudenslager. 1984. Electrophoretic variation among subspecies of tule perch (*Hysterocarpus traski*). Copeia 1984:223–227.

Baltz, D. M., and P. B. Moyle. 1981. Morphometric analysis of the tule perch (*Hysterocarpus traski*) populations in three isolated drainages. Copeia 1981:305–311.

———. 1982. Life history characteristics of tule perch (*Hysterocarpus traski*) populations in contrasting environments. Env. Biol. Fish. 7:229–242.

———. 1984. Segregation by species and size classes of rainbow trout, *Salmo gairdneri,* and Sacramento sucker, *Catostomus occidentalis,* in three California streams. Env. Biol. Fish. 10:101–110.

———. 1993. Invasion resistance to introduced species by a native assemblage of California stream fishes. Ecol. App. 3:246–255.

Baltz, D. M., P. B. Moyle, and N. J. Knight. 1982. Competitive interactions between benthic stream fishes, riffle sculpin, *Cottus gulosus,* and speckled dace, *Rhinichthys osculus.* Can. J. Fish. Aquat. Sci. 39:1502–1511.

Baltz, D. M., B. Vondracek, L. R. Brown, and P. B. Moyle. 1987. Influence of temperature on microhabitat choice by fishes in a California stream. Trans. Am. Fish. Soc. 116:12–20.

———. 1991. Seasonal changes in microhabitat selection by rainbow trout in a small stream. Trans. Am. Fish. Soc. 120:166–176.

Banks, M. A., V. K. Rashbrook, M. J. Calavetta, C. A. Dean, and D. Hedgecock. 2000. Microsatellite DNA variation in chinook salmon of California's Central Valley. Can. J. Fish. Aquat. Sci. 57:915–927.

Barlow, G. W. 1958a. High salinity mortality of desert pupfish *Cyprinodon macularius.* Copeia 1958:231–232.

———. 1958b. Daily movements of desert pupfish, *Cyprinodon macularius,* in shore pools of the Salton Sea, California. Ecology 39:580–587.

———. 1961a. Gobies of the genus *Gillichthys,* with comments on the sensory canals as a taxonomic tool. Copeia 1961:423–437.

———. 1961b. Social behavior of the desert pupfish, *Cyprinodon macularius,* in the field and in the aquarium. Am. Midl. Nat. 65:330–359.

———. 1963. Species structure of the gobiid fish *Gillichthys mirabilis* from coastal sloughs of the eastern Pacific. Pacific Sci. 17:47–72.

Barlow, G. W., and V. deVlaming. 1972. Ovarian cycling in longjaw gobies, *Gillichthys mirabilis,* from the Salton Sea. Calif. Fish Game 58:50–57.

Barnes, B. V. 1993. The landscape ecosystem approach and conservation of endangered species. End. Sp. Update 10:13–25.

Barnes, R. N. 1957. A study of the life history of the western roach, *Hesperoleucus symmetricus.* M.A. thesis, Univ. Calif., Davis. 25 pp.

Barnhart, R. A. 1986. Species profiles: life histories and environmental requirements of coastal fishes and invertebrates (Pacific Southwest)—steelhead. USFWS Biol. Rpt. 82(11.60). 21 pp.

Barnhart, R. A., M. J. Boyd, and J. E. Pequenat. 1992. The ecology of Humboldt Bay, California: an estuarine profile. USFWS Biol. Rpt. 1. 121 pp.

Barraclough, W. E. 1964. Contribution to the marine life history of the eulachon, *Thaleichthys pacificus.* J. Fish. Res. Board Can. 21:1333–1337.

Barrett, J. C., G. D. Grossman, and J. Rosenfield. 1992. Turbidity-induced changes in reactive distance of rainbow trout. Trans. Am. Fish. Soc. 121:437–443.

Barrett, P. J. 1983. Systematics of fishes of the genus *Tilapia* (Perciformes: Cichlidae) in the lower Colorado River Basin. M.S. thesis, Ariz. State Univ., Tempe. 59 pp.

Bartholomew, J. P. 1966. The effects of threadfin shad on white crappie growth in Isabella Reservoir, Kern County, California. CDFG Inland Fish. Admin. Rpt. 66-6. 11 pp.

Bartley, D. M., G. A. E. Gall, and B. Bentley. 1985. Preliminary description of the genetic structure of white sturgeon, *Acipenser transmontanus,* in the Pacific Northwest. Pages 105–109 in F. P. Binkoswki and S. I. Doroshov, eds. North American sturgeons: biology and aquaculture potential. Dordrecht: Dr. W. Junk.

———. 1990. Biochemical and genetic detection of natural and artificial hybridization of chinook and coho salmon in Northern California. Trans. Am. Fish. Soc. 119:431–437.

Bartley, D. M., B. Bentley, P. G. Olin, and G. A. E. Gall. 1992. Population genetic structure of coho salmon (*Oncorhynchus kisutch*) in California. Calif. Fish Game 78:88–104.

Barwick, D. H., and P. R. Moore. 1983. Abundance and growth of redeye bass in two South Carolina reservoirs. Trans. Am. Fish. Soc. 112:216–219.

Bateman, D. S., and H. W. Li. 2001. Nest site selection by reticulate sculpin in two streams of different geologies in the central Coast Range of Oregon. Trans. Am. Fish. Soc. 130:823–831.

Baugh, T. M. 1981. Adapting Salt Creek pupfish (*Cyprinodon salinus*) to fresh water. Great Basin Nat. 41:341–342.

———. 1982. Social behavior of the Salt Creek pupfish (*Cyprinodon salinus*) in aquaria. J. Aquaculture 2(2):25–28.

Baumgartner, J. V. 1995. Phenotypic, genetic, and environmental integration of morphology in a stream population of the threespine stickleback, *Gasterosteus aculeatus.* Can. J. Fish. Aquat. Sci. 52:1307–1317.

Baumgartner, J. V., and M. A. Bell. 1984. Lateral plate morph variation in California populations of the threespine stickleback, *Gasterosteus aculeatus.* Evolution 38:665–674.

Baxter, R. D. 1999. Status of splittail in California. Calif. Fish Game 85:28–30.

———. 2000. Splittail and longfin smelt. IEP Newslett. 13(2):19–21.

Baxter, R. D., W. Harrell, and L. Grimaldo. 1996. 1995 splittail spawning investigations. IEP Newslett. 9(4):27–31.

Baxter, R. D., K. Hieb, S. DeLeon, K. Fleming, and J. Orsi. 1999. Report on the 1980–1995 fish, shrimp, and crab sampling in the San Francisco Estuary, California. IEP Sac.–San Joaquin Estuary Tech. Rpt. 63. 503 pp.

Bay, E. C. 1966. Adaptation studies with the Argentine pearl fish, *Cynolebias belottii,* for its introduction into California. Copeia 1966:839–846.

Beacham, T. D., R. E. Withler, and A. P. Gould. 1985. Biochemical genetic stock identification of pink salmon in southern British Columbia and Puget Sound. Can. J. Fish. Aquat. Sci. 42:1474–1483.

Beamesderfer, R. C. P., D. L. Ward, and A. A. Nigro. 1996. Evaluation of the biological basis for a predator control program on northern squawfish (*Ptychocheilus oregonensis*) in the Columbia and Snake Rivers. Can. J. Fish. Aquat. Sci. 53:2898–2908.

Beamish, R. J. 1980. Adult biology of the river lamprey (*Lampetra ayresi*) and the Pacific lamprey (*Lampetra tridentata*) from the Pacific coast of Canada. Can. J. Fish. Aquat. Sci. 37:1906–1923.

———. 1987. Evidence that parasitic and nonparasitic life history types are produced by one population of lamprey. Can. J. Fish. Aquat. Sci. 44:1779–1782.

Beamish, R. J., and C. M. Neville. 1992. The importance of size as an isolating mechanism in lampreys. Copeia 1992:191–196.

———. 1995. Pacific salmon and Pacific herring mortalities in the Fraser River plume caused by river lamprey (*Lampetra ayresi*). Can. J. Fish. Aquat. Sci. 52:644–655.

Beamish, R. J., and T. G. Northcote. 1989. Extinction of a population of anadromous parasitic lamprey, *Lampetra tridentata,* upstream of an impassable dam. Can. J. Fish. Aquat. Sci. 46:420–425.

Beamish, R. J., and J. H. Youson. 1987. Life history and abundance of young adult *Lampetra ayresi* in the Fraser River and their possible impact on salmon and herring stocks in the Strait of Georgia. Can. J. Fish. Aquat. Sci. 44:525–537.

Beauchamp, D. A., B. C. Allen, R. C Richards, W. A. Wurtsbaugh, and C. R. Goldman. 1992. Lake trout spawning in Lake Tahoe: egg incubation in deepwater macrophyte beds. N. Am. J. Fish. Mgmt. 12:442–449.

Beauchamp, D. A., P. E. Budy, B. C. Allen, and J. F. Godfrey. 1994. Timing, distribution, and abundance of kokanees spawning in a Lake Tahoe tributary. Great Basin Nat. 54:130–141.

Beck, W. R., and W. H. Massmann. 1951. Migratory behavior of the rainwater killifish, *Lucania parva,* in the York River, Virginia. Copeia 1951:176.

Becker, G. C. 1983. Fishes of Wisconsin. Madison: University of Wisconsin Press.

Behnke, R. J. 1992. Native trout of western North America. Am. Fish. Soc. Mono. 6. 275 pp.

———. 1997. Evolution, systematics, and structure of *Oncorhynchus clarki clarki.* Pages 3–6 *in* J. D. Hall, P. A. Bisson, and R. E. Gresswell, eds. Sea-run cutthroat trout: biology, management and future conservation. Corvallis, Oreg.: Oregon Chapter of the American Fisheries Society.

Beinz, C. S., and J. S. Ziller. 1987. Status of three lacustrine sucker species (Catostomidae). Report to USFWS, Sacramento. 39 pp.

Beland, R. O. 1953. The effect of channelization on the fishery of the lower Colorado River. Calif. Fish Game 39:137–139.

Bell, E., W. G. Duffy, and T. D. Roelofs. 2001. Fidelity and survival of juvenile coho salmon in response to a flood. Trans. Am. Fish. Soc. 130:450–458.

Bell, M. A. 1976. Evolution of phenotypic diversity in *Gasterosteus aculeatus* superspecies on the Pacific coast of North America. Syst. Zool. 25:211–227.

———. 1978. Fishes of the Santa Clara River system, southern California. Contrib. Sci. Nat. Hist. Mus. Los Angeles Co. 295:1–20.

———. 1979. Low-plate morph of the threespine stickleback breeding in saltwater. Copeia 1979:529–533.

———. 1981. Lateral plate polymorphism and ontogeny of the complete plate morph of threespine sticklebacks (*Gasterosteus aculeatus*). Evolution 35:67–74.

———. 1982a. Differentiation of adjacent stream populations of threespine sticklebacks. Evolution 36:189–199.

———. 1982b. Melanism in a high elevation population of *Gasterosteus aculeatus.* Copeia 1982:829–835.

———. 1994. Paleobiology and evolution of threespine sticklebacks. Pages 438–471 *in* M. A. Bell and S. A. Foster, eds. The evolutionary biology of the threespine stickleback. Oxford: Oxford University Press.

Bell, M. A., and T. R. Haglund. 1978. Selective predation on threespine sticklebacks (*Gasterosteus aculeatus*) by garter snakes. Evolution 32:304–319.

Bell, M. A., and K. E. Richkind. 1981. Clinal variation of lateral plates in threespine stickleback fish. Am. Nat. 117:113–132.

Bennett, W. A., and L. Howard. 1997. El Niños and the decline of striped bass. IEP newslett. 10(4):7–10.

———. 1999. Climate change and the decline of striped bass. IEP newslett. 12(2):53–56.

Bennett, W. A., and P. B. Moyle. 1996. Where have all the fishes gone? Interactive factors producing fish declines in the Sacramento–San Joaquin estuary. Pages 519–541 *in* J. T. Hollibaugh, ed. San Francisco Bay: the Ecosystem. Pacific Division, AAAS, San Francisco.

Bennett, W. A., D. J. Ostrach, and D. E. Hinton. 1995. Larval striped bass condition in a drought-stricken estuary: evaluating pelagic food limitation. Ecol. Apps. 5:680–692.

Benson, S. B., and R. J. Behnke. 1961. *Salmo evermanni,* a synonym of *Salmo clarki henshawi.* Calif. Fish Game 37:257–259.

Berejikian, B. A. 1995. The effects of hatchery and wild ancestry and experience on the relative ability of steelhead trout fry (*Oncorhynchus mykiss*) to avoid a benthic predator. Can. J. Fish. Aquat. Sci. 52:2476–2482.

Berg, W. J. 1987. Evolutionary genetics of rainbow trout, *Parasalmo gairdneri* (Richardson). Ph.D. dissertation, Univ. Calif., Davis. 184 pp.

Bernardi, G. 1997. Molecular phylogeny of the Fundulidae (Teleostei, Cyprinodontiformes) based on the cytochrome b gene. Pages 189–197 *in* T. Kocher and C. Stepien, eds. Molecular systematics of fishes. New York: Academic Press.

Bernardi, G., and D. Talley. 2000. Genetic evidence for limited dispersal in the coastal California killifish, *Fundulus parvipinnis.* J. Exp. Mar. Biol., Ecol. 255:187–199.

Beukema, J. 1963. Experiments on the effects of the hunger state on the risk of prey of the three-spined stickleback. Arch. Nees. Zool. 15:358–361.

Bevelhimer, M. S., and S. M. Adams. 1993. A bioenergetics analysis of diel vertical migration by kokanee salmon, *Oncorhynchus nerka.* Can. J. Fish. Aquat. Sci. 50:2336–2349.

Bilby, R. E., B. R. Fransen, P. A. Bisson, and J. K. Walter. 1998. Response of juvenile coho salmon (*Oncorhynchus kisutch*) and steelhead (*Oncorhynchus mykiss*) to the addition of salmon carcases to two streams in southwestern Washington, USA. Can. J. Fish. Aquat. Sci. 55:1909–1918.

Billard, R. 1996. Reproduction of pike: gametogenesis, gamete biology, and early development. Pages 13–43 *in* J. F. Craig, ed. Pike: biology and exploitation. London: Chapman and Hall.

Bills, F. T., and C. E. Bond. 1980. A new subspecies of tui chub (Pisces: Cyprinidae) from Cowhead Lake, California. Copeia 1980:320–322.

Bird, F. H. 1975. Biology of the blue and tui chubs in Easy and Paulina lakes, Oregon. M.S. thesis, Oreg. State Univ., Corvallis. 165 pp.

Birdsong, R. S., E. O. Murdy, and F. L. Pezold. 1988. A study of the vertebral column and median fin osteology in gobioid fishes with comments on gobioid relationships. Bull. Mar. Sci. 42:174–214.

Birstein, V. J., W. E. Bemis, and J. R. Waldeman. 1997a. The threatened status of acipenseriform fishes: a summary. Env. Biol. Fish. 48:427–435.

Birstein, V. J., R. Hanner, and R. DeSalle. 1997b. Phylogeny of the Acipenseriformes: cytogenetic and molecular approaches. Env. Biol. Fish. 48:127–155.

Bisson, P. A., K. Sullivan, and J. L. Nielsen. 1988. Channel hydraulics, habitat use, and body form of juvenile coho salmon, steelhead, and cutthroat trout in streams. Trans. Am. Fish. Soc. 117:262–273.

Bjornn, T. C., and D. W. Reiser. 1991. Habitat requirements of salmonids in streams. Pages 83–138 *in* W. R. Meehan, ed. Influences of forest and rangeland management of salmonid fishes and their habitats. Bethesda, Md.: American Fisheries Society Spec. Publ. 19.

Black, G. F. 1980. Status of the desert pupfish (*Cyprinodon mac-*

ularius Baird and Girard) in California. CDFG Inland Fish. End. Sp. Prog. Spec. Publ. 80–1. 42 pp.

Black, M. 1995. Tragic remedies: a century of failed fishery policy on California's Sacramento River. Pac. Hist. Rev. 64:37–70.

Black, T., and R. V. Bulkley. 1985. Growth rate of yearling Colorado squawfish at different temperatures. SW Nat. 30:253–257.

Blumer, L. S. 1985. The significance of biparental care in the brown bullhead, *Ictalurus nebulosus.* Env. Biol. Fish. 12:231–236.

Boccone, V. M., and T. J. Mills. 1979. Spawning behavior and spawning substrate preference of the Modoc sucker, *Catostomus microps* (Rutter). CDFG Inland Fish. End. Sp. Prog. Spec. Publ. 79-2. 33 pp.

Boles, G. L. 1976. A range extension for the logperch, *Percina macrolepida,* in California. Calif. Fish Game 62:154.

———. 1990. Food habits of juvenile wild and hatchery steelhead trout, *Oncorhynchus mykiss,* in the Trinity River, California. CDFG Inland Fish. Admin. Rpt. 90-10. 8 pp.

Bolster, B. C. 1990. Five year status report, desert pupfish. Unpubl. rpt., CDFG, Sacramento. 12 pp.

Bond, C. E. 1948. Fish management problems of Lake of the Woods. M.S. thesis, Oreg. State Univ. Corvallis. 109 pp.

———. 1961. Keys to Oregon freshwater fishes. Or. St. Univ. Agr. Exp. Sta. Tech. Bull. 58:1–42.

———. 1963. Distribution and ecology of freshwater sculpins, genus *Cottus,* in Oregon. Ph.D. dissertation, Univ. Mich., Ann Arbor. 198 pp.

———. 1966. Endangered plants and animals of Oregon. Oreg. State Univ. Agric. Exp. Stn. Spec. Rpt. 205. 8 pp.

———. 1973a. Keys to Oregon freshwater fishes. Oreg. State Univ. Agric. Exp. Stn. Tech. Bull. 58. 42 pp.

———. 1973b. Occurrence of the reticulate sculpin, *Cottus perplexus,* in California, with distributional notes on *Cottus gulosus* in Oregon and Washington. Calif. Fish Game 59:93–94.

Bond, C. E., and T. T. Kan. 1973. *Lampetra (Entosphenus) minima* n. sp., a dwarfed parasitic lamprey from Oregon. Copeia 1973:568–574.

Bond, C. E., E. Rexstad, and R. M. Hughes. 1988. Habitat use of twenty-five common species of Oregon freshwater fishes. NW Sci. 62:223–232.

Borgeson, D., and G. W. McCammon. 1967. White catfish, *Ictalurus catus,* of the Sacramento–San Joaquin Delta. Calif. Fish Game 53:254–263.

Borodin, A. L. 1984. The Red Book of the USSR: species of animals and plants in danger of extinction, vol. 1. 2nd edition. Moscow: Forest Industry.

Botsford, L. W., and J. G. Brittnacher. 1998. Viability of Sacramento River winter-run chinook salmon. Cons. Biol. 12:65–79.

Botsford, L. W., B. Vondracek, T. C. Wainwright, A. L. Linden, R. G. Kope, D. E. Reed, and J. Cech, Jr. 1987. Population development of the mosquitofish, *Gambusia affinis,* in rice fields. Env. Biol. Fish. 20:143–154.

Bowen, S. H. 1976. Mechanism for digestion of detrital bacteria by the cichlid fish *Sarotherodon mossambica* (Peters). Nature 260:137–138.

Bozek, M. A., L. J. Paulson, and G. R. Wilde. 1991. Spawning season of the razorback sucker, *Xyrauchen texanus,* in Lake Mojave, Arizona and Nevada. J. Freshw. Ecol. 6:61–73.

Bradford, D. F., S. D. Cooper, T. M. Jenkins, Jr., K. Kratz, O. Sarnelle, and A. D. Brown. 1998. Influences of natural acidity and introduced fish on faunal assemblages in California alpine lakes. Can. J. Fish. Aquat. Sci. 55:2478–2491.

Bradford, M. J., and G. C. Taylor. 1997. Individual variation in dispersal behaviour of newly emerged chinook salmon (*Oncorhynchus tshawytscha*) from the upper Fraser River, British Columbia. Can. J. Fish. Aquat. Sci. 54:1585–1592.

Bradford, R. H., and S. C. Gurtin. 2000. Habitat use by hatchery-reared adult razorback suckers released into the lower Colorado River, California-Arizona. N. Am. J. Fish. Mgmt. 20:154–167.

Branner, J. C. 1907. A drainage peculiarity of the Santa Clara Valley affecting fresh-water faunas. J. Geol. 15:1–10.

Brauer, C. O. 1971. A study of the western sucker *Catostomus occidentalis* Ayres in lower Hat Creek, California. M.S. thesis, Humboldt State Univ., Arcata. 41 pp.

Breder, C. M., and D. E. Rosen. 1966. Modes of reproduction in fishes. New York: American Museum of Natural History.

Brennan, J. S., and G. M. Cailliet. 1991. Age determination and validation studies of white sturgeon, *Acipenser transmontanus,* in California. Pages 209–234 *in* P. Williot, ed. Acipenser. Bordeaux, France: CEMAGREF.

Brice, J. C. 1953. Geology of Lower Lake Quadrangle, California. Calif. Div. Mines Bull. 166. 72 pp.

Bridcut, E. E., and P. S. Giller. 1995. Diet variability and foraging strategies in brown trout (*Salmo trutta*): an analysis from subpopulations to individuals. Can. J. Fish. Aquat. Sci. 52:2543–2552.

Briggs, J. O. 1953. The behavior and reproduction of salmonid fishes in a small coastal stream. CDFG Fish Bull. 94:1–62.

Brittan, M., A. Albrecht, and J. Hopkirk. 1963. An oriental goby collected in the San Joaquin River delta near Stockton, California. Calif. Fish Game 49:302–304.

Brittan, M., J. Hopkirk, J. Connors, and M. Martin. 1970. Explosive spread of the oriental goby, *Acanthogobius flavimanus* in the San Francisco Bay–Delta region of California. Proc. Calif. Acad. Sci. 38:207–214.

Broadway, J. E., and P. B. Moyle. 1978. Aspects of the ecology of the prickly sculpin, *Cottus asper* Richardson, a persistent native species in Clear Lake, Lake County, California. Env. Biol. Fish. 3:337–343.

Brocksen, R. W., and R. E. Cole. 1972. Physiological responses of three species of fishes to various salinities. J. Fish. Res. Board Can. 29:399–405.

Brodour, R. D., and W. G. Pearcy. 1987. Diel feeding chronology, gastric evacuation, and estimated daily ration of juvenile coho salmon, *Oncorhynchus kisutch* (Walbaum), in the coastal marine environment. J. Fish Biol. 31:465–477.

Broughton, J. M. 1994. Late Holocene resource intensification in the Sacramento Valley, California: the vertebrate evidence. J. Arch. Sci. 21:501–514.

Browman, H. I., and W. J. O'Brien. 1992. The ontogeny of search behavior in the white crappie, *Pomoxis annularis.* Env. Biol. Fish. 34:181–195.

Brown, B. E., and J. S. Dendy. 1961. Observations on the food habits of flathead and blue catfish in Arkansas. Proc. 15th Annu. Conf. SE Assoc. Game Fish Comm.: 210–222.

Brown, C. J. D. 1971. Fishes of Montana. Bozeman, Mont.: Big Sky Books.

Brown, G. E., and S. Brennan. 2000. Chemical alarm signals in juvenile green sunfish (*Lepomis cyanellus,* Centrarchidae). Copeia 2000:1079–1082.

Brown, G. E., D. P. Chivers, and R. J. F. Smith. 1995. Localized defecation by pike: a response to labeling by cyprinid alarm pheromone? Behav. Ecol. Sociobiol. 36:105–110.

Brown, J. A. 1984. Parental care and the ontogeny of predator-avoidance in two species of centrarchid fish. Anim. Behav. 32:113–119.

Brown, J. H. 1971. The desert pupfish. Sci. Am. 225:104–110.

Brown, J. H., and C. R. Feldmeth. 1971. Evolution in constant and fluctuating environments: thermal tolerances of desert pupfish, *Cyprinodon.* Evolution 25:390–398.

Brown, J. R., A. T. Beckenbach, and M. J. Smith. 1992a. Influence of Pleistocene glaciations and human intervention upon mitochondrial DNA diversity in white sturgeon (*Acipenser transmontanus*) populations. Can. J. Fish. Aquat. Sci. 49:358–367.

Brown, L. B. 1988. Factors determining the distribution of three species of sculpin (*Cottus*) from the Pit River drainage, California. Ph.D. dissertation, Univ. Calif., Davis. 128 pp.

———. 1989. Temperature preferences and oxygen consumption of three species of sculpin (*Cottus*). Env. Biol. Fish. 26:223–236.

———. 1990. Age, growth, feeding, and behavior of Sacramento squawfish (*Ptychocheilus grandis*) in Bear Creek, Colusa County, California. SW Nat. 35:249–260.

———. 1991. Differences in habitat choice and behavior among three species of sculpin (*Cottus*). Copeia 1991:810–818.

———. 2000. Fish communities and their associations with environmental variables, lower San Joaquin River drainage, California. Env. Biol. Fish. 57:251–269.

Brown, L. R., and A. M. Brasher. 1995. Effect of predation by Sacramento squawfish (*Ptychocheilus grandis*) on habitat choice of California roach (*Lavinia symmetricus*) and rainbow trout (*Oncorhynchus mykiss*) in artificial streams. Can. J. Fish. Aquat. Sci. 52:1639–1646.

Brown, L. R., and P. B. Moyle. 1981. The impact of squawfish on salmonid populations: a review. N. Am. J. Fish. Mgmt. 1:104–111.

———. 1991. Changes in habitat and microhabitat partitioning within an assemblage of stream fishes in response to predation by Sacramento squawfish (*Ptychocheilus grandis*). Can. J. Fish. Aquat. Sci. 43:849–856.

———. 1992. Native fishes of the San Joaquin drainage: Status of a remnant fauna and its habitats. Pages 89–98 *in* D. L. Williams, S. Byrne, and T. A. Rado, eds. Endangered and sensitive species of the San Joaquin Valley, California. Sacramento, Calif.: California Energy Commission.

———.1993. Distribution, ecology, and status of the fishes of the San Joaquin River drainage, California. Calif. Fish Game 79:96–113.

———. 1996. Invading species in the Eel River, California: successes, failures, and relationships with resident species. Env. Biol. Fish. 49:271–291.

Brown, L. R., S. A. Matern, and P. B. Moyle. 1995. Comparative ecology of prickly sculpin, *Cottus asper,* and coastrange sculpin, *C. aleuticus,* in the Eel River, California. Env. Biol. Fish. 42:329–343.

Brown, L. R., P. B. Moyle, and R. M. Yoshiyama. 1994. Status of coho salmon (*Oncorhynchus kisutch*) in California. N. Am. J. Fish. Mgmt. 14:237–261.

Brown, L. R., P. B. Moyle, W. A. Bennett, and B. D. Quelvog. 1992b. Implications of morphological variation among populations of California roach *Lavinia symmetricus* (Cyprinidae) for conservation policy. Biol. Cons. 62:1–10.

Bruton, M. N., and M. Gophen. 1992. The effect of environmental factors on the nesting and courtship behaviour of *Tilapia zillii* in Lake Kinneret (Israel). Hydrobiology 239:171–178.

Bryan, J. E., and P. A. Larkin. 1972. Food specialization by individual trout. J. Fish. Res. Board Can. 29:1615–1624.

Bryant, G. L. 1977. Fecundity and growth of tule perch, *Hysterocarpus traski,* in the lower Sacramento–San Joaquin Delta. Calif. Fish Game 63:140–156.

Buchanan, D.V., R. M. Hooton, and J. R. Morin. 1981. Northern squawfish (*Ptychocheilus oregonensis*) predation on juvenile salmonids in sections of the Willamette River basin, Oregon. Can. J. Fish. Aquat. Sci. 38:360–364.

Buddington, R. K. 1979. Digestion of an aquatic macrophyte by *Tilapia zillii* (Gervais). J. Fish Biol. 15:449–455.

Buettner, M. E., and G. G. Scoppetone.1991. Distribution and information on the taxonomic status of the shortnose sucker (*Chasmistes brevirostris*) and Lost River sucker (*Deltistes luxatus*) in the Klamath River basin, California. USFWS (Reno) Completion Rpt. CDFG FG-8304. 101 pp.

Bundy, D. S. 1970. Reproduction and growth of the tule perch, *Hysterocarpus traskii* (Gibbons), with notes on its ecology. M.S. thesis, Univ. of Pacific, Stockton, Calif. 52 pp.

Bureau of Reclamation. 1983. Predation of anadromous fish in the Sacramento River, California. Central Valley Fish and Wildlife Management Study, Spec. Rpt. USDI-BR, Mid-Pacific Region, Sacramento. 77 pp.

Burns, D. C. 1974. Feeding by mature steelhead in a spawning stream. Calif. Fish Game 60:205–206.

Burns, J. W. 1966. Various species accounts. Pages 510–529 *in* A. Calhoun, ed. Inland fisheries management. Sacramento: California Department of Fish and Game.

———. 1972. Some effects of logging and associated road construction on northern California streams. Trans. Am. Fish. Soc. 101:1–17.

Bury, R. B. 1972. The effects of diesel fuel on a stream fauna. Calif. Fish Game 58:291–295.

Busack, C. A., and G. A. E. Gall. 1980. Ancestry of artificially propagated California rainbow trout strains. Calif. Fish Game 66:17–24.

———. 1981. Introgressive hybridization in populations of Paiute cutthroat trout (*Salmo clarki seleniris*). Can. J. Fish. Aquat. Sci. 38:939–951.

Busack, C. A., G. H. Thorgaard, M. P. Bannon, and G. A. E. Gall. 1980. An electrophoretic, karyotypic and meristic characterization of the Eagle Lake trout, *Salmo gairdneri aquilarum.* Copeia 1980:418–424.

Busby, M. S., and R. A. Barnhart. 1995. Potential food sources and feeding ecology of juvenile fall chinook salmon in California's Mattole River Lagoon. Calif. Fish Game 81:133–146.

Buth, D. G., and C. B. Crabtree. 1982. Genetic variability and population structure of *Catostomus santaanae* in the Santa Clara drainage. Copeia 1982:439–444.

Buth, D. G., R. W. Murphy, and L. Ulmer. 1987. Population differentiation and introgressive hybridization of the flannelmouth sucker and of hatchery and native stocks of the razor-back sucker. Trans. Am. Fish. Soc. 116:103–110.

Buth, D. G., C. B. Crabtree, R. D. Orton, and W. J. Rainboth. 1984. Genetic differentiation between the freshwater subspecies of *Gasterosteus aculeatus* in southern California. Biochem. Syst. Ecol. 12:423–432.

Butler, R. L., and D. P. Borgeson. 1965. California "catchable" trout fisheries. CDFG Fish Bull. 127:1–47.

Butler, V. L. 1996. Tui chub taphonomy and the importance of marsh resources in the western Great Basin of North America. Am. Antiquity 61:699–717.

Byers, S., and G. L. Vinyard. 1990. The effects on the plankton community of filter-feeding Sacramento blackfish, *Orthodon microlepidotus.* Oecologia 83:352–357.

Calhoun, A. J. 1940. Note on a hybrid minnow, *Apocope X Richardsonius.* Copeia 1940:142–143.

———. 1944. The food of the black-spotted trout (*Salmo clarki henshawi*) in two Sierra-Nevada lakes. Calif. Fish Game 30:80–85.

California Advisory Committee on Salmon and Steelhead Trout. 1988. Restoring the balance. Rpt. 120-J. 84 pp.

California Department of Fish and Game. 1987a. Factors affecting striped bass abundance in the Sacramento–San Joaquin River System. CDFG exhibit 25, State Water Res. Control Board 1987 Water Quality/Water Rights Proc. San Francisco Bay/Sacramento–San Joaquin Delta. 149 pp.

———. 1987b. Final environmental impact report: white bass management program. Sacramento: California Department of Fish and Game. 153 pp.

———. 1990. Status and management of spring-run chinook salmon. Report to Fish and Game Commission. Sacramento: California Department of Fish and Game.

———. 1991. Attachment to comments on the Northern Spotted Owl recovery plan. Appendix C. Consideration of other species and ecosystem concerns. Submitted to Secretary for Resources, California, November 26.

———. 1992a. Status report: California salmon. Report to Fish and Game Commission, Sacramento. 72 pp.

———. 1992b. Impact of water management on splittail in the Sacramento–San Joaquin estuary. State Water Resources Control Board hearing for setting interim standards for the Delta. WRINT-CDFG-Exhibit 5. 7 pp.

———. 1998. A status review of the spring-run chinook salmon (*Oncorhynchus tshawytscha*) in the Sacramento River drainage. Candidate Species Rpt. 98-01. 150 pp.

California Fish Commission 1880. 6th biennial report to the Commissioners of Fisheries of the State of California. Sacramento.

Campagna, C. G., and J. J. Cech, Jr. 1981. Gill ventilation and respiratory efficiency of Sacramento blackfish, *Orthodon microlepidotus* Ayres, in hypoxic environments. J. Fish Biol. 19:581–591.

Campbell, E. A., and P. B. Moyle. 1991. Historical and recent population sizes of spring-run chinook salmon in California. Pages 155–216 *in* T. Hassler, ed. Northeast Pacific chinook and coho salmon workshop. Arcata, Calif.: American Fisheries Society.

Campbell, J. B. 1882. Notes on the McCloud River, California, and some of its fishes. Bull. U.S. Fish. Comm. 1:44–46.

Campbell, R. D., and B. A. Branson. 1978. Ecology and popula-

tion biology of the black bullhead, *Ictalurus melas* (Rafinesque) in central Kentucky. Tulane Stud. Zool. Bot. 20:99–136.

Capelli, M. H. 1997. Tidewater goby (*Eucyclogobius newberryi*) management in California estuaries. Proc. Calif. World Ocean Conf.: 1–18.

Carl, G. C., and G. Clemens. 1953. The freshwater fishes of British Columbia. 2nd edition. British Columbia Province Museum Handbook 5. Victoria. 136 pp.

Carlander, K. D. 1969. Handbook of freshwater fishery biology, vol.1. Ames: Iowa State University Press.

Carney, D. A., and L. M. Page. 1990. Meristic variation and zoogeography of the genus *Ptychocheilus* (Teleostei: Cyprinidae). Copeia 1990:171–181.

Carpanzano, C. M. 1996. Distribution and habitat associations of different age classes and mitochondrial genotypes of *Oncorhynchus mykiss* in streams in southern California. M.A. thesis, Univ. Calif., Santa Barbara. 88 pp.

Carpelin, L. H. 1955. Tolerance of the San Francisco topsmelt, *Atherinops affinis affinis,* to conditions in salt-producing ponds bordering San Francisco Bay. Calif. Fish Game 41:279–284.

Carter, J. G., V. A. Lamarra, and R. J. Ryel. 1986. Drift of larval fishes in the upper Colorado River. J. Freshw. Biol. 3:567–577.

Casselman, J. M. 1996. Effects of environmental factors on growth, survival, activity and exploitation of northern pike. Pages 114–128 *in* J. F. Craig, ed. Pike: biology and exploitation. London: Chapman and Hall.

Casteel, R. W. 1976. Fish remains in archaeology. San Francisco: Academic Press.

Casteel, R. W., and J. H. Hutchison. 1973. *Orthodon* (Actinopterygii, Cyprinidae) from the Pliocene and Pleistocene of California. Copeia 1973:358–361.

Casteel, R. W., and M. J. Rymer. 1981. Pliocene and Pleistocene fishes from the Clear Lake area. USGS Professional Pap. 1141:232–234.

Casteel, R. W., D. P. Adam, and J. D. Sims. 1977. Late Pleistocene and Holocene remains of *Hysterocarpus traski* (tule perch) from Clear Lake, California, and inferred Holocene temperature fluctuations. Quat. Res. 7:133–143.

Castleberry, D. T., and J. J. Cech, Jr. 1986. Physiological responses of a native and an introduced desert fish to environmental stressors. Ecology 67:912–918.

———. 1993. Critical thermal maxima and oxygen minima of five fishes from the upper Klamath Basin. Calif. Fish Game 78:145–152.

Castleberry, D. T., J. E. Williams, G. M. Sato, T. E. Hopkins, A. M. Brasher, and M. S. Parker. 1990. Status and management of Shoshone pupfish, *Cyprinodon nevadensis shoshone* (Cyprinodontidae), at Shoshone Spring, Inyo County, California. Bull. S. Calif. Acad. Sci. 89:19–25.

Catching, E. D. 1977. Age and growth of redeye bass in Shoal and Little Shoal creeks, Alabama. Proc. Annu. Conf. SE Assoc. Fish Wildl. Agencies 32:380–390.

Cavender, T. M. 1978. Taxonomy and distribution of bull trout, *Salvelinus confluentus* (Suckley), from the American Northwest. Calif. Fish Game 64:139–174.

———. 1980. Systematics of *Salvelinus* from the North Pacific basin. Pages 295–322 *in* E. K. Balon, ed. Charrs: salmonid fishes of the genus *Salvelinus.* The Hague: Dr. W. Junk.

————. 1997. Morphological distinction of bull trout from the McCloud River system of northern California. Pages 271–281 in W. C. Mackay, M. K. Brewin, and M. Monita, eds. Friends of the bull trout conference proceedings. Calgary, Alberta: Trout Unlimited, Canada.

Caywood, M. L. 1974. Contributions to the life history of the splittail Pogonichthys macrolepidotus (Ayres). M.S. thesis, Calif. State Univ., Sacramento. 77 pp.

Cech, J. J. Jr., and A. L. Linden. 1987. Comparative larvivorous performances of mosquitofish, Gambusia affinis, and juvenile Sacramento blackfish, Orthodon microlepidotus, in experimental paddies. J. Am. Mosq. Contr. Assoc. 3:35–41.

Cech, J. J. Jr., M. J. Massingill, and H. Stern. 1982. Growth of juvenile Sacramento blackfish, Orthodon microlepidotus (Ayres). Hydrobiologia 97:75–80.

Cech, J. J. Jr., S. J. Mitchell, and M. J. Massingill. 1979. Respiratory adaptations of Sacramento blackfish, Orthodon microlepidotus (Ayres) for hypoxia. Comp. Biochem. Physiol. 63A:411–415.

Cech, J. J. Jr., S. J. Mitchell, D. T. Castleberry, and M. McEnroe. 1990. Distribution of California stream fishes: influence of environmental temperature and hypoxia. Env. Biol. Fish. 29:95–105.

Chadwick, H. E. 1959. California sturgeon tagging studies. Calif. Fish Game 45:297–301.

Chadwick, H. E., C. E. von Geldern, and M. L. Johnson. 1966. White bass. Pages 412–422 in A. Calhoun, ed. Inland fisheries management. Sacramento: California Department of Fish and Game.

Chapman, D. W., and R. C. Bjornn. 1969. Distribution of salmonids in streams, with special reference to food and feeding. Pages 153–176 in T. G. Northcote, ed. Symposium on salmon and trout in streams. H. R. MacMillan Lectures in Fisheries. Vancouver: University of British Columbia.

Chapman, F. A., J. P. Van Eenennaam, and S. I. Doroshov. 1996. The reproductive condition of white sturgeon, Acipenser transmontanus, in San Francisco Bay, California. Fishery Bull. 94:628–634.

Chernoff, B., J. V. Conner, and C. F. Bryan. 1981. Systematics of the Menidia beryllina complex (Pisces: Atherinidae) from the Gulf of Mexico and its tributaries. Copeia 1981:319–335.

Chigbu, P., and T. H. Sibley. 1994. Relationship between abundance, growth, fecundity, and egg size in a land-locked population of longfin smelt (Spirinchus thaleichthys). J. Fish Biol. 45:1–15.

————. 1998. Feeding ecology of longfin smelt (Spirinchus thaleichthys Ayres) in Lake Washington. Fish. Res. 38:109–119.

Chilton, E. W., II, and M. I. Moeneke. 1992. Biology and management of grass carp (Ctenopharyngodon idella, Cyprinidae) for vegetation control: a North American perspective. Rev. Fish Biol. Fish. 2:283–320.

Chivers, D. P., and R. J. F. Smith. 1995a. Chemical recognition of risky habitats is culturally transmitted among fathead minnows, Pimephales promelas (Osteicthyes, Cyprinidae). Ethology 99:286–296.

————. 1995b. Free-living fathead minnows rapidly learn to recognize pike as predators. J. Fish Biol. 46:949–954.

Christenson, D. P. 1986. Status of Little Kern golden trout. Pages 26–30 in D. P. Lee, ed. The natural (and unnatural) history of California trout. Shingle Springs, Calif.: California-Nevada Chapter, American Fisheries Society.

Christophel, D. B. 1988. Contributions to the life history of the redear sunfish (Lepomis microlophus) in Rancho Seco Reservoir, Sacramento County, California. M.S. thesis, Calif. State Univ., Sacramento. 64 pp.

Clark, A. L., and W. D. Pearson. 1978. Early piscivory in postlarvae of the white bass. Proc. Annu. Conf. SE Assoc. Fish Wildl. Agencies 32:409–414.

Clark, G. H. 1938. Weight and age determination of striped bass. Calif. Fish Game 24:176–177.

Clellguest, W. 1985. Temperature tolerance of Florida and northern largemouth bass: effects of subspecies, fish size, and season. Texas J. Sci. 37:75–84.

Clemens, H. P., and K. F. Sneed. 1957. The spawning behavior of the channel catfish, Ictalurus punctatus. USFWS Spec. Sci. Rpt. Fish. 29:1–11.

Cobern, M. M., and T. M. Cavender. 1992. Interrelationships of North American cyprinid fishes. Pages 328–373 in R. L. Mayden, ed. Systematics, historical ecology, and North American freshwater fishes. Stanford, Calif.: Stanford University Press.

Coble, D. W. 1975. Smallmouth bass. Pages 21–33 in R. H. Stroud, ed. Black bass biology and management. Washington, D.C.: Sport Fishing Institute.

Cohen, A. N., and J. T. Carlton. 1998. Accelerating invasion rate in a highly invaded estuary. Science 279:555–558.

Coleman, G. A. 1929. A biological survey of the Salton Sea. Calif. Fish Game 15:218–227.

————. 1930. A biological survey of Clear Lake, Lake County. Calif. Fish Game 16:221–227.

Collins, M. R. 1981. The feeding periodicity of striped mullet, Mugil cephalus L., in two Florida habitats. J. Fish Biol. 19:307–315.

Colwell, A. E., N. L. Anderson, and D. L. Woodward. 1997. Monitoring of dipteran pests and associated organisms in Clear Lake (California). Proc. First Annu. Clear Lake Science and Mgmt. Symp., UC Davis-CLERC: 15–32.

Contreras, G. P. 1973. Distribution of the fishes of the Lost River System, California-Oregon, with a key to the species present. M.S. thesis, Univ. Nev., Reno. 61 pp.

Contreras-MacBeath, T., and H. R. Espinoza. 1996. Some aspects of the reproductive strategy of Poeciliopsis gracilis (Osteichthyes: Poeciliidae) in the Cuautla River, Mexico. J. Freshw. Biol. 11:327–338.

Cook, S. F., Jr. 1964. The potential of two native California fish in the biological control of chironomid midges (Diptera: Chironomidae). Mosq. News 24:332–333.

Cook, S. F., Jr., and R. L. Moore. 1970. Mississippi silversides, Menidia audens (Atherinidae), established in California. Trans. Am. Fish. Soc. 99:70–73.

Cook, S. F., Jr., J. D. Connors, and R. L. Moore. 1964. The impact of the fishery on the midges of Clear Lake, Lake County, California. Ann. Entomol. Soc. Am. 57:701–707.

Cook, S. F., Jr., R. L. Moore, and J. D. Connors. 1966. The status of the native fishes of Clear Lake, Lake County, California. Wassmann J. Biol. 24:141–160.

Cooper, J. J. 1982. Observations on the reproduction and embryology of the Lahontan tui chub, Gila bicolor, in Walker Lake, Nevada. Great Basin Nat. 42:60–64.

————. 1983. Distributional ecology of native and introduced

fishes in the Pit River system, northeastern California, with notes on the Modoc sucker. Calif. Fish Game 69:39–53.

———. 1985. Age, growth, and food habits of tui chub, *Gila bicolor,* in Walker Lake, Nevada. Great Basin Nat. 45:784–788.

Coots, M. 1955. The Pacific lamprey, *Entosphenus tridentatus,* above Copco Dam, Siskiyou County, California. Calif. Fish Game 41:118–119.

———. 1956. The yellow perch, *Perca flavescens* (Mitchill), in the Klamath River. Calif. Fish Game 42:219–228.

———. 1965. Occurrences of the Lost River sucker, *Deltistes luxatus* (Cope), and shortnose sucker, *Chasmistes brevirostris* (Cope), in northern California. Calif. Fish Game 51:68–73.

Cope, E. D. 1874. On the Plagopterinae and the ichthyology of Utah. Proc. Am. Philos. Soc. 14:129–139.

———. 1879. The fishes of Klamath Lake, Oregon. Am. Nat. 13:784–785.

———. 1883. On the fishes of the recent and Pliocene lakes of the western part of the Great Basin, and of the Idaho Pliocene lake. Proc. Phila. Acad. Nat. Sci. 35:134–166.

Cope, E. D., and H. C. Yarrow. 1875. Report upon the collections of fishes made in portions of Nevada, Utah, Colorado, New Mexico, and Arizona during the years 1871, 1872, 1873, and 1874. Rpt. Geog. Geol. Expl. Surv., N 100th Mer. (Wheeler Survey) 5:635–703.

Cordone, A., and T. C. Frantz. 1966. The Lake Tahoe sport fishery. Calif. Fish Game 52:240–274.

Cordone, A., S. Nicola, P. Baker, and T. Frantz. 1971. The kokanee salmon in Lake Tahoe. Calif. Fish Game 57:28–43.

Cornelius, R. H. 1969. The systematics and zoogeography of *Rhinichthys osculus* (Girard) in Southern California. M.A. thesis, Calif. State Univ., Fullerton. 194 pp.

Costa-Pierce, B. A., and R. W. Doyle. 1997. Genetic identification and status of tilapia regional strains in southern California. Pages 1–17 *in* B. A. Costa-Pierce and J. E. Rakocy, eds. Tilapia aquaculture in the Americas, vol. 1. Baton Rouge, La.: World Aquaculture Society.

Costa-Pierce, B. A., and R. Riedel. 2000. Fisheries ecology of the tilapias in subtropical lakes of the United States. Pages 1–20 *in* B. A. Costa-Pierce and J. E. Rakocy, eds. Tilapia aquaculture in the Americas, vol. 2. Baton Rouge, La.: World Aquaculture Society.

Cote, I. M., and M. R. Gross. 1993. Reduced disease in offspring: a benefit of coloniality in sunfish. Behav. Ecol. Sociobiol. 33:269–274.

Courtois, L. A. 1973. The effects of temperature, availability of oxygen, and salinity upon the metabolism of the longjaw mudsucker, *Gillichthys mirabilis.* M.A. thesis, Calif. State Univ., Hayward. 32 pp.

Coutant, C. C. 1975. Responses of bass to natural and artificial temperature regimes. Pages 272–285 *in* H. Clepper, ed. Black bass biology and management. Washington, D.C.: Sport Fishing Institute.

Cowles, R. B. 1934. Notes on the ecology and breeding habits of the desert minnow, *Cyprinodon macularius* Baird and Girard. Copeia 1934:40–42.

Cox, T. J. 1966. A behavioral and ecological study of the desert pupfish (*Cyprinodon macularius*) in Quitobaquito Springs, Organ Pipe Cactus National Monument, Arizona. Ph.D. thesis, Univ. Ariz., Tucson. 102 pp.

———. 1972. The food habits of the desert pupfish (*Cyprinodon macularius*) in the Quitobaquito Springs, Organ Pipe National Monument, Arizona. J. Ariz. Acad. Sci. 7:25–27.

Crabtree, C. B. 1985. Allozyme variability in the tidewater goby, *Encyclogobius newberryi* (Pisces, Gobiidae). Isozyme Bull. 18:70.

Crabtree, C. B., and D. W. Buth. 1981. Gene duplication and diploidization in tetraploid catostomid fishes, *Catostomus fumeiventris* and *C. santaanae.* Copeia 1981:705–708.

Crain, P. K., and D. M. Corcoran. 2000. Age and growth of the tui chub in Eagle Lake, California. Calif. Fish Game 86:149–155.

Crawford, S. S., and E. K. Balon. 1994. Alternative life histories in the genus *Lucania.* 3. An ecomorphological explanation of altricial (*L. parva*) and precocial (*L. goodei*) species. Env. Biol. Fish. 41:369–402.

Crear, D., and I. Haydock. 1970. Laboratory rearing of the desert pupfish, *Cyprinodon macularius.* Fish. Bull. 69:151–156.

Crosby, A. W. 1986. Ecological imperialism: the ecological expansion of Europe, 900–1900. Cambridge: Cambridge University Press.

Cross, Frank B. 1967. Handbook of fishes of Kansas. Univ. Kans. Mus. Nat. Hist. Misc. Publ. 45. 357 pp.

Culver, G. B., and C. L. Hubbs. 1917. The fishes of the Santa Ana System streams in southern California. Lorquina 1:82–83.

Curry, R. A., and D. L. G. Noakes. 1995. Groundwater and the selection of spawning sites by brook trout (*Salvelinus fontinalis*). Can. J. Fish. Aquat. Sci. 52:1733–1740.

Curry, R. A., D. L. G. Noakes, and G. E. Morgan. 1995. Groundwater and the incubation and emergence of brook trout (*Salvelinus fontinalis*). Can. J. Fish. Aquat. Sci. 52:1741–1749.

Curtis, B. 1934. The golden trout of Cottonwood Lakes (*Salmo aqua-bonita* Jordan). Trans. Am. Fish. Soc. 64:259–265.

Daniels, R. A. 1987. Comparative life histories and microhabitat use in three sympatric sculpins (Cottidae: *Cottus*) in northeastern California. Env. Biol. Fish. 19:93–110.

Daniels, R. A., and L. A. Courtois. 1982. Status and proposed management of the rough sculpin, *Cottus asperrimus* Rutter, in California. CDFG Inland Fish. End. Sp. Prog. Spec. Publ. 82–1. 20 pp.

Daniels, R. A., and P. B. Moyle. 1978. Biology, distribution, and status of *Cottus asperrimus* in the Pit River drainage, northeastern California. Copeia 1978:673–679.

———. 1983. Life history of splittail (Cyprinidae: *Pogonichthys macrolepidotus*) in the Sacramento–San Joaquin estuary. NOAA Fishery Bull. 84:105–117.

———. 1984. Geographic variation and a taxonomic reappraisal of the marbled sculpin, *Cottus klamathensis.* Copeia 1984:949–959.

Danielsen, T. L. 1968. Differential predation on *Culex pipiens* and *Anopheles albimanus* mosquito larvae by two species of fish (*Gambusia affinis* and *Cyprinodon nevadensis*) and the effects of simulated reeds on predation. Ph.D. thesis, Univ. Calif., Riverside. 115 pp.

Danylchuk, A. J., and M. G. Fox. 1996. Size- and age-related variation in the seasonal timing of nesting activity, nest characteristics, and female choice of parental male pumpkinseed sunfish (*Lepomis gibbosus*). Can. J. Zool. 74:1834–1840.

Darling, J. D. S., M. L. Noble, and E. Shaw. 1980. Reproductive strategies in the surfperches. I. Multiple insemination in natural populations of the shiner perch, *Cymatogaster aggregata.* Evolution 34:271–277.

Darnell, R. M., and R. R. Meierotto. 1965. Diurnal periodicity in the black bullhead, *Ictalurus melas* (Rafinesque). Trans. Am. Fish. Soc. 94:1–8.

Davies, R. W., and G. W Thompson. 1976. Movements of mountain whitefish (*Prosopium williamsoni*) in the Sheep River watershed, Alberta. J. Fish. Res. Board Can. 33:2395–2401.

Davis, K. A. 1973. The food habits of redear sunfish in two central California farm ponds. M.S. thesis, Calif. State Univ., Sacramento. 35 pp.

Deacon, J. E. 1988. The endangered woundfin and water management in the Virgin River, Arizona. Fisheries 13:18–29.

Deacon, J. E., and W. G. Bradley. 1972. Ecological distribution of fishes of Moapa (Muddy) River in Clark County, Nevada. Trans. Am. Fish Soc. 101:408–468.

Deacon, J. E., and C. D. Williams. 1991. Ash Meadows and the legacy of the Devil's Hole pupfish. Pages 69–92 *in* W. L. Minckley and J. E. Deacon, eds. Battle against extinction: native fish management in the American West. Tucson: University of Arizona Press.

Deacon, J. E., and J. E. Williams. 1984. Annotated list of the fishes of Nevada. Proc. Biol. Soc. Wash. 97:103–118.

Decker, L. M. 1989. Coexistence of two species of sucker, *Catostomus*, in Sagehen Creek, California and notes on their status in the western Lahontan Basin. Great Basin Nat. 49: 540–551.

Decker, L. M., and D. C. Erman. 1992. Short-term seasonal changes in composition and abundance of fish in Sagehen Creek, California. Trans. Am. Fish. Soc. 121:297–306.

DeHaven, R. W. 1977. Striped bass ecology in the American and Feather rivers, California. Unpubl. ms. 36 pp.

Deinstadt, J. M. 1998. Survival, growth, and yield of brown trout stocked as fingerlings in Hot Creek, California. CDFG Inland Fish. Admin. Rpt. 98-1. 21 pp.

Deinstadt, J. M., E. J. Pratt, F. G. Hoover, and S. Sasaki. 1988. Survey of fish populations in southern California streams: 1987. CDFG Inland Fish. Admin. Rpt. 88-5. 85 pp.

Deinstadt, J. M., G. F. Sibbald, J. D. Knarr, and D. M. Wang. 1986. Survey of fish populations in streams of the Owens River drainage: 1985. CDFG Inland Fish. Admin. Rpt. 86-3. 71 pp.

Dennison, S. G., and R. V. Bulkley. 1972. Reproductive potential of the black bullhead, *Ictalurus melas*, in Clear Lake, Iowa. Trans. Am. Fish. Soc. 101:483–487.

Desmond, J. S., J. B. Zedler, G. D. Williams, and J. B. Zedler. 2000. Fish use of tidal creek habitats in two southern California salt marshes. Ecol. Engin. 14:233–252.

Dethloff, G. M. 1998. Assessing the status of rainbow trout exposed to sublethal concentrations of metals using bioindicators of exposure and immune system function. Ph.D. dissertation, Univ. Calif., Davis. 172 pp.

Dettman, D. H. 1976. Distribution, abundance, and microhabitat segregation of rainbow trout and Sacramento squawfish in Deer Creek, California. M.S. thesis, Univ. Calif., Davis. 47 pp.

deVlaming, V. L. 1971. Thermal selection behavior in the estuarine goby, *Gillichthys mirabilis* Cooper. J. Fish Biol. 1971: 277–286.

DeVries, D. R., and R. A. Stein. 1990. Manipulating shad to enhance sport fisheries in North America: an assessment. N. Am. J. Fish. Mgmt. 10:209–223.

DeVries, D. R., R. A. Stein, J. G. Miner, and G. G. Mittelbach. 1991. Stocking threadfin shad: consequences for young-of-year fishes. Trans. Am. Fish. Soc. 120:368–381.

DeWitt, J. W. 1954. A survey of the coast cutthroat trout, *Salmo clarki clarki* Richardson in California. Calif. Fish Game 40:329–335.

Dickerson, B. E., and G. L. Vinyard. 1999a. Effects of high chronic temperatures and diel temperature cycles on the survival and growth of Lahontan cutthroat trout. Trans. Am. Fish. Soc. 128:516–521.

———. 1999b. Effects of high levels of total dissolved solids in Walker Lake, Nevada, on survival and growth of Lahontan cutthroat trout. Trans. Am. Fish. Soc. 128:507–516.

Dietsch, E. L. 1959. The ecology and food habits of the sculpin (*Cottus beldingi*) in relation to the eastern brook trout (*Salvelinus fontinalis*). M.A. thesis, Univ. Calif., Berkeley. 63 pp.

Dill, W. A. 1938. The rough fish problem in California. Unpubl. report, Calif. Dept. Fish Game, Sacramento. 16 pp.

———. 1944. The fishery of the lower Colorado River. Calif. Fish Game 30:109–211.

———. 1946. A preliminary report on the fishery of Millerton Lake, California. Calif. Fish Game 32:49–70.

Dill, W. A., and A. J. Cordone. 1997. History and status of introduced fishes in California, 1871–1996. CDFG Fish Bull. 178. 414 pp.

Dobie, J., L. Meecham, S. F. Snieszko, and G. N. Washburn. 1956. Raising bait fishes. U.S. Fish Wildl. Serv. Circ. 35. 124 pp.

Docker, M. F., J. H. Youson, R. J. Beamish, and R. H. Devlin. 1999. Phylogeny of the lamprey genus *Lampetra* inferred from mitochondrial cytochrome b and ND3 gene sequences. Can. J. Fish. Aquat. Sci. 56:2340–2349.

Dominey, W. J. 1980. Female mimicry in male bluegill sunfish—a genetic polymorphism? Nature 284:546–548.

———. 1981. Anti-predator function of bluegill sunfish nesting colonies. Nature 290:586–588.

Doroshov, S. I., G. P. Moberg, and J. P. Van Eenennaam. 1997. Observations on the reproductive cycle of cultured white sturgeon, *Acipenser transmontanus*. Env. Biol. Fish. 48:265–278.

Dotsu, Y., and S. Mito. 1955. On the breeding habits, larvae, and young of a goby, *Acanthogobius flavimanus* (Temminck and Schlegal). Jpn. J. Ichthyol. 4:153–161.

Douglas, P. L. 1995. Habitat relationships of oversummering rainbow trout (*Oncorhynchus mykiss*) in the Santa Ynez River drainage. M.A. thesis, Univ. Calif., Santa Barbara. 76 pp.

Dowling, T. E., and B. D. DeMarais. 1993. Evolutionary significance of introgressive hybridization in cyprinid fishes. Nature 362:444–446.

Dowling, T. E., and C. L. Secor. 1997. The role of hybridization and introgression in the diversification of animals. Ann. Rev. Ecol. Syst. 28:583–619.

Dryfoos, R. L. 1965. The life history and ecology of the longfin smelt in Lake Washington. Ph.D. dissertation, Univ. Wash. 229 pp.

Dunham, J. B., and W. L. Minckley. 1998. Allozymic variation in desert pupfish from natural and artificial habitats: genetic conservation in fluctuating populations. Biol. Cons. 84:7–16.

Dunham, J. B., G. L. Vinyard, and B. E. Rieman. 1998. Habitat fragmentation and extinction risk of Lahontan cutthroat trout. N. Am. J. Fish. Mgmt. 17:1126–1133.

Dunham, J. B., M. M. Peacock, B. E. Rieman, R. E. Schroeter, and

G. L. Vinyard. 1999. Local and geographic variability in the distribution of stream-living Lahontan cutthroat trout. Trans. Am. Fish. Soc.128:875–889.

Dyer, B. S., and B. Chernoff. 1996. Phylogenetic relationships among atheriniform fishes (Teleostei: Atherinomorpha). Zool. J. Linn. Soc. 117:1–69.

Ebert, V. W., and R. C. Summerfelt. 1969. Contributions to the life history of the Piute sculpin, *Cottus beldingii* Eigenmann and Eigenmann, in Lake Tahoe. Calif. Fish Game 55:100–120.

Echelle, A. A., and T. E. Dowling. 1992. Mitchondrial DNA variation and evolution of the Death Valley pupfishes (*Cyprinodon,* Cyprinodontidae). Evolution 46:193–206.

Echelle, A. A., and A. F. Echelle. 1997. Patterns of abundance and distribution among members of a unisexual-bisexual complex of fishes (Atherinidae: *Menidia*). Copeia 1997:249–259.

Echelle, A. A., R. A. Van Den Busssche, T. P. Malloy, Jr., M. L. Haynie, and C. O. Minckley. 2000. Mitochondrial DNA variation in pupfishes assigned to the species *Cyprinodon macularius* (Atherinomorpha, Cyprinodontidae): taxonomic implications and conservation genetics. Copeia 2000:353–364.

ECO Northwest 1994. The potential economic consequences of critical habitat designation for the Lost River sucker and Shortnose sucker. USFWS, Portland Field Office Rpt. 100 pp.

Ecological Analysts, Inc. 1982. Fisheries studies at Millerton Lake, 1979–1982. San Ramon, Calif.: PG&E.

Edwards, E. A., G. Gebhart, and O. E. Maughan. 1983. Habitat suitability information: smallmouth bass. USFWS FWS/OBS-82/10.36. 50 pp.

Ehlinger, T. J. 1989. Foraging mode switches in the golden shiner, *Notemigonus crysoleucas.* Can. J. Fish. Aquat. Sci. 46:1250–1254.

Ehrenfeld, D. W. 1981. The arrogance of humanism. New York: Oxford University Press.

Eigenmann, C. H. 1890. The food fishes of California's fresh waters. Biennial Rpt. Calif. Fish Comm. 1888–1890:53–65.

Eigenmann, C. H., and R. S. Eigenmann. 1889. Description of a new species of *Cyprinodon.* Proc. Calif. Acad. Sci. 1:270.

———. 1893. Additions to the fauna of San Diego. Proc. Calif. Acad. Sci. 3:1–24.

Eigenmann, R. S. 1891. Description of a new species, *Catostomus rex,* from Oregon. Am. Nat. 25(part 2):667–668.

Elliott, G. V. 1976. Diel activity and feeding of schooled largemouth bass fry. Trans. Am. Fish. Soc. 105:624–627.

Elliott, G. V., and T. M. Jenkins. 1972. Winter food of trout in three high elevation Sierra Nevada lakes. Calif. Fish Game 58:231–237.

Elliott, J. M. 1994. Quantitative ecology and the brown trout. Oxford: Oxford University Press.

Ellis, S. L. N. 1922. Bits of history of Tulare Lake fishing. Calif. Fish Game 8:206–208.

Ellison, J. P. 1980. Diets of mountain whitefish, *Prosopium williamsoni* (Girard), and brook trout, *Salvelinus fontinalis* (Mitchill), in the Little Walker River, Mono County, California. Calif. Fish Game 66:96–104.

Elser, J. J., C. Luecke, M. T. Brett, and C. R. Goldman. 1995. Effects of food web compensation after manipulation of rainbow trout in an oligotrophic lake. Ecology 76:52–69.

Elston, R., and B. Bachen. 1976. Diel feeding cycles and some effects of light on the feeding intensity of the Mississippi silverside, *Menidia audens,* in Clear Lake, California. Trans. Am. Fish. Soc. 105:84–88.

Elvira, B., G. G. Nicola, and A. Almodovar. 1996. Pike and red swamp crayfish: a new case on predator-prey relationship between aliens in central Spain. J. Fish. Biol. 48:437–446.

Emig, J. W. 1966. Largemouth bass, pages 332–353; smallmouth bass, pages 354–365; bluegill, pages 375–392; redear sunfish, pages 392–398; brown bullhead, pages 463–475, *in* A. Calhoun, ed. Inland fisheries management. Sacramento: California Department of Fish and Game.

Emlen, J. M., R. R. Reisenbichler, A. M. McGie, and T. E. Nickelson. 1990. Density dependence at sea for coho salmon (*Oncorhynchus kisutch*). Can. J. Fish. Aquat. Sci. 47:1765–1772.

Emmett, R. L., D. R. Miller, and T. H. Blahn. 1986. Food of juvenile chinook, *Oncorhynchus tshawytscha,* and coho, *O. kisutch,* salmon off northern Oregon and southern Washington coasts, May–September 1980. Calif. Fish Game 72:38–46.

Emmett, R. L., S. A. Hinton, S. L. Stone, and M. E. Monaco. 1991. Distribution and abundance of fishes and invertebrates in west coast estuaries, vol. 2: Species life history summaries. Rockville, Md.: NOAA/NOS Strategic Env. Assess. Div. ELMR Rpt. 8. 329 pp.

Eng, L. L., D. Belk, and C. H. Eriksen. 1990. Californian Anostraca: distribution, habitat, and status. J. Crustacean Biol. 10:247–277.

ENTRIX, Inc. 1996. Results of fish passage monitoring at the Vern Freeman Diversion Facility, Santa Clara River, 1996. Project 324402. United Water Conservation District, Santa Paula Rpt. 43 pp.

Erkkila, L. F., J. W. Moffett, O. B. Cope, B. R. Smith, and R. S. Nelson. 1950. Sacramento–San Joaquin Delta fishery resources: effects of Tracy pumping plant and delta cross channel. USFWS Spec. Sci. Rpt. Fish. 56. 109 pp.

Erman, D. C. 1986. Long-term changes of fish populations in Sagehen Creek, California. Trans. Am. Fish. Soc. 115:682–692.

Erman, D. C., E. D. Andrews, and M. Yoder-Williams. 1988. Effects of winter floods on fishes in the Sierra Nevada. Can. J. Fish. Aquat. Sci. 45:2195–2200.

Erman, D. C., L. DeLain, and M. Myers. 1983. Limnological conditions and fish interactions in Lake Davis, California. Unpubl. rpt., California Department of Fish and Game. 100 pp.

Eschmeyer, W. N., E. S. Herald, and H. Hammann. 1983. A field guide to Pacific coast fishes of North America. Boston: Houghton Mifflin.

Etnier, D. A., and W. C. Starnes. 1993. The fishes of Tennessee. Knoxville: University of Tennessee Press.

Evans, D. H. 1969. Life history studies of the Lahontan redside, *Richardsonius egregius,* in Lake Tahoe. Calif. Fish Game 55:197–222.

Everest, F. H. 1973. Ecology and management of summer steelhead in the Rogue River. Oreg. State Game Comm. Fish. Res. Rpt. 7, Proj. AFS-312. 48 pp.

Everest, F. H., and D. W. Chapman. 1972. Habitat selection and spatial interaction by juvenile chinook salmon and steelhead trout in two Idaho streams. J. Fish. Res. Board Can. 29:91–100.

Evermann, B. W. 1893. Description of a new sucker, *Pantosteus jordani,* from the upper Missouri basin. Bull. U.S. Fish. Comm. 12:51–56.

———. 1905. The golden trout of the southern high Sierras. Bull. U.S. Bur. Fish. 25:1–51.

———. 1916. Fishes of the Salton Sea. Copeia 34:61–63.

Evermann, B. W., and H. W. Clark. 1931. A distributional list of the species of freshwater fishes known to occur in California. CDFG Fish Bull. 35. 67 pp.

Evermann, B. W., and C. Rutter. 1894. The fishes of the Colorado basin. Bull. U.S. Fish. Comm. 14:473–486.

Faber, D. J. 1967. Limnetic larval fish in northern Wisconsin lakes. J. Fish. Res. Board Can. 24:927–937.

Falter, M. A., and J. J. Cech, Jr. 1991. Maximum pH tolerance of three Klamath basin fishes. Copeia 1990:1109–1111.

Farley, D. 1972. A range extension of the logperch. Calif. Fish Game 58:248.

Farringer, R. T., A. A. Echelle, and S. F. Lehtinen. 1979. Reproductive cycle of the red shiner, Notropis lutrensis, in central Texas and south central Oklahoma. Trans. Am. Fish. Soc. 108:271–276.

Fausch, K. D. 1993. Experimental analysis of microhabitat selection by juvenile steelhead (Oncorhynchus mykiss) and coho salmon (O. kisutch) in a British Columbia stream. Can. J. Fish. Aquat. Sci. 50:1198–1207.

Feldmeth, C. R., and J. P. Waggoner. 1972. Field measurements of tolerance to extreme hypersalinity in the California killifish, Fundulus parvipinnis. Copeia 1972:592–594.

Felley, J. 1980. Analysis of morphology and asymmetry in bluegill sunfish (Lepomis macrochirus) in the southeastern United States. Copeia 1980:18–29.

Feyrer, F. V. 1999. Food habits of common Suisun Marsh fishes in the Sacramento–San Joaquin estuary, California. M.S. thesis, Calif. State Univ., Sacramento. 53 pp.

Feyrer, F. V., and R. D. Baxter. 1998. Splittail fecundity and egg size. Calif. Fish Game 84:119–126.

Fiedler, P. L., R. A. Leidy, R. D. Laven, N. Gershenz, and L. Saul. 1993. The contemporary paradigm in ecology and its implications for endangered species conservation. End. Sp. Update 10:7–12.

Fine, M. L., J. P. Friel, D. McElroy, C. B. King, K. E. Loesser, and S. Newton. 1997. Pectoral spine locking and sound production in the channel catfish, Ictalurus punctatus. Copeia 1997:777–790.

Finger, T. R. 1983. Interactive segregation among three species of sculpin (Cottus). Copeia 1982:680–694.

Fisher, F. W. 1973. Observations on the spawning of Mississippi silversides, Menidia audens Hay. Calif. Fish Game 59:315–316.

———. 1994. Past and present status of Central Valley salmon. Cons. Biol. 8:870–873.

Fite, K. R. 1973. Feeding overlap between roach and juvenile steelhead in the Eel River. M.S. thesis, Humboldt. State Univ., Arcata. 38 pp.

Fleming, I. A., and M. R. Gross. 1994. Breeding competition in a Pacific salmon (coho: Oncorhynchus kisutch): measures of natural and sexual selection. Evolution 48:637–657.

Flint, R. A., W. L. Somer, and J. Trumbo. 1998. Silver King Creek Paiute cutthroat restoration 1991 through 1993. CDFG Inland Fish Admin. Rpt. 98-7. 37 pp.

Follett, W. I. 1937. Prey of the weasel and mink. J. Mammal. 18:365.

Foin. T. C., C. M. Efferson, M. F. Coe, R. O. Spenst, L. M. Veilleux, and J. D. Flicker. 2001. Predicting invasion potential of a major predator: northern pike (Esox lucius) in California. Unpubl. ms., Univ. Calif., Davis. 51 pp.

Fong, S., and T. Takagi. 1979. Competition among four centrarchids in Clear Lake, Lake County, California. Unpubl. rpt., Univ. Calif., Davis. 41 pp.

Foote, C. J., and G. S. Brown. 1998. Ecological relationship between freshwater sculpins (genus Cottus) and beach-spawning sockeye salmon (Oncorhynchus nerka) in Iliamna Lake, Alaska. Can. J. Fish. Aquat. Sci. 55:1524–1533.

Ford, T. 1977. Status summary report on the Modoc sucker. Unpubl. rpt., USFS, Modoc Natl. Forest, Alturas, Calif. 44 pp.

Forney, J. L. 1961. Growth, movements, and survival of smallmouth bass (Micropterus dolomieui) in Oneida Lake, New York. N.Y. Fish Game J. 8:88–105.

Fortune, J. D., Jr., A. R. Gerlach, and C. J. Hanel. 1966. A study to determine the feasibility of establishing salmon and steelhead in the upper Klamath basin. Unpubl. rpt., Oregon State Game Comm. 23 pp.

Foster, N. R. 1967. Comparative studies on the biology of killifishes. Ph.D. dissertation, Cornell Univ., Ithaca, N.Y. 369 pp.

Fowler, H. W. 1913. Notes on catostomid fishes. Proc. Acad. Nat. Sci. Phila. 65:45–71.

Fox, M. G., and A. J. Crivelli. 1998. Body size and reproductive allocation in a multiple spawning centrarchid. Can. J. Fish. Aquat. Sci. 55:737–748.

Fox, R. S. 1988. Growth of young-of-the-year Sacramento blackfish Orthodon microlepidotus (Ayres). M.S. thesis, Humboldt State Univ., Arcata. 211 pp.

Franklin, J. F. 1993. Preserving biodiversity: species, ecosystems, or landscapes? Ecol. App. 3:202–205.

Fraley, J. J., and B. B. Shepard. 1989. Life history, ecology, and population status of migratory bull trout (Salvelinus confluentus) in the Flathead lake and river system, Montana. NW Sci. 63:133–143.

Frantz, T. C. 1979–1981. Job progress reports: Lake Tahoe. Nev. Dept. Wildl. F-20-R-16–17. 82 pp.

Frantz, T. C., and A. J. Cordone. 1967. Observations on deepwater plants in Lake Tahoe, California and Nevada. Ecology 48:709–714.

———. 1970. Food of lake trout in Lake Tahoe. Calif. Fish Game 56:21–35.

Fritz, E. S. 1975. The life history of the California killifish, Fundulus parvipinnis Girard, in Anaheim Bay, California. Pages 91–106 in E. D. Lane and C. W. Hill, eds. The marine resources of Anaheim Bay. CDFG Fish Bull. 165.

Frost, W. E., and M. E. Brown. 1967. The trout. London: Collins.

Fry, D. H. 1936. Life history of Hesperoleucas venustus Snyder. Calif. Fish Game 22:65–98.

———. 1961. King salmon spawning stocks of the California Central Valley, 1940–1959. Calif. Fish Game 47:55–71.

———. 1967. A 1955 record of pink salmon, Oncorhynchus gorbuscha, spawning in the Russian River. Calif. Fish Game 53:210–211.

———. 1973. Anadromous fishes of California. Sacramento: California Department of Fish and Game.

Fryer, G., and T. D. Iles. 1972. The cichlid fishes of the Great Lakes of Africa: biology and evolution. Neptune City, N.J.: TFH Publications.

Galat, D. L., and N. Vucinich. 1983a. Food of larval tui chubs, Gila bicolor, in Pyramid Lake, Nevada. Great Basin Nat. 43:175–178.

———. 1983b. Food partitioning between young of the year of

two sympatric tui chub morphs. Trans. Am. Fish. Soc. 112:486–497.

Gale, D. B., T. R. Hayden, L. S. Harris, and H. N. Voight. 1998. Assessment of anadromous fish stocks in Blue Creek, lower Klamath River, California, 1994–1996. Yurok Tribal Fisheries Program, Habitat Assess., Biol. Monit. Div. Tech. Rpt. 4. 101 pp.

Gamradt, S. C., and L. B. Kats. 1996. Effects of introduced crayfish and mosquitofish on California newts. Cons. Biol. 10:1155–1162.

Ganssle, D. 1966. Fishes and decapods of San Pablo and Suisun Bay. Pages 64–94 in D. W. Kelley, ed. Ecological studies of the Sacramento–San Joaquin Estuary. Part 1. CDFG Fish Bull. 33:64–94.

Gard, M. F. 1994. Biotic and abiotic factors affecting native stream fishes in the south Yuba River, Nevada County, California. Ph.D. dissertation, Univ. Calif., Davis. 179 pp.

Gard, R., and G. A. Flittner. 1974. Distribution and abundance of fishes in Sagehen Creek, California. J. Wildl. Mgmt. 38:347–358.

Gard, R., and D. W. Seegrist. 1965. Persistence of the native rainbow trout type following introduction of hatchery trout. Copeia 1965:182–185.

Gartz, R. 1998. Density dependent growth and diet changes in young-of-the-year striped bass (*Morone saxatilis*) in the Sacramento–San Joaquin estuary. IEP Newslett. 12(1):22–24.

———. 2000. Young-of-the-year striped bass, American shad, and threadfin shad abundance and distribution. IEP Newslett. 13(2):38–41.

Geary, R. E. 1978. Life history of the Clear Lake hitch (*Lavinia exilicauda chi*). M.S. thesis, Univ. Calif., Davis. 27 pp.

Geary, R. E., and P. B. Moyle. 1980. Aspects of the ecology of the hitch, *Lavinia exilicauda* (Cyprinidae), a persistent native cyprinid in Clear Lake, California. SW Nat. 25:385–390.

Gerald, J. W. 1971. Sound production during courtship in six species of sunfish (Centrarchidae). Evolution 25:75–87.

Gerking, S. D. 1953. Evidence for the concepts of home range and territory in stream fishes. Ecology 34:347–365.

Gerking, S. D., and R. M. Lee. 1983. Thermal limits for growth and reproduction in the desert pupfish, *Cyprinodon n. nevadensis*. Physiol. Zool. 56:1–9.

Gerking, S. D., R. M. Lee, and J. B. Shrode. 1979. Effects of generation-long temperature acclimation on reproductive performance of the desert pupfish, *Cyprinodon n. nevadensis*. Physiol. Zool. 52:113–121.

Gerstung, E. R. 1980. 1979 annual report of the Threatened Salmonids Project. Unpubl. rpt., California Department of Fish and Game, Sacramento. 8 pp.

———. 1981. Status and management of the coast cutthroat trout (*Salmo clarki clarki*) in California. Calif. Nev. Wildl. Trans. 1981:25–32.

———. 1988. Status, life history, and management of Lahontan cutthroat trout. Am. Fish. Soc. Symp. 4:93–106.

———. 1997. Status of coastal cutthroat trout in California. Pages 43–56 in J. D. Hall, P. A. Bisson, and R. E. Gresswell, eds. Sea-run cutthroat trout: biology, management and future conservation. Corvallis, Oreg.: Oregon Chapter of the American Fisheries Society.

Giger, R. D. 1972. Ecology and management of coastal cutthroat trout in Oregon. Oregon State Game Comm. Fish. Res. Rpt. 6. 61 pp.

Gilbert, C. H. 1893. Report on the fishes of the Death Valley expedition collected in southern California and Nevada in 1891, with descriptions of new species. North Am. Fauna 7:229–234.

———. 1898. The fishes of the Klamath Basin. Bull. U.S. Fish. Comm. 17:1–13.

Gilbert, C. H., and B. W. Evermann. 1894. A report upon investigations in the Columbia River basin, with descriptions of four new species of fishes. Bull. U.S. Fish Comm. 14:169–204.

Gilbert, C. H., and N. B. Scofield. 1898. Notes on a collection of fishes from the Colorado Basin in Arizona. Proc. U.S. Natl. Mus. 20:487–499.

Gill, T. N. 1861. Notes on some genera of fishes of the western coast of North America. Proc. Acad. Nat. Sci. Phila. 13:164–168.

Gillooly, J. F., and J. R. Baylis. 1999. Reproductive success and the energetic cost of parental care in male smallmouth bass. J. Fish Biol. 54:573–584.

Gingras, M. 1997. Mark/recapture experiments at Clifton Court Forebay to estimate pre-screening loss to juvenile fishes: 1976–1993. IEP Tech. Rpt. 55. 22 pp.

Girard, C. 1854a. Descriptions of new fishes, collected by Dr. A. L. Heermann, naturalist attached to the survey of the Pacific Railroad Route, under Lieut. R. S. Williamson, U.S.A. Proc. Acad. Nat. Sci. Phila. 7:129–140.

———. 1854b. Description of *Sebastes ruber, Sebastes ruber* var. *parvus, Sebastes variabilis,* and *Centrarchus maculosus.* The Pacific 3:182.

———. 1856. Researches upon the cyprinid fishes inhabiting the fresh waters of the United States of America west of the Mississippi Valley, from specimens in the Museum of the Smithsonian Institution. Proc. Acad. Nat. Sci. 1856:165–209.

———. 1859. Ichthyological notes. Proc. Acad. Nat. Sci. Phila. 11:56–68.

Gleason, E. V. 1982a. A review of the life histories of striped bass (*Morone saxatilis*) and white bass (*M. chrysops*) hybrids and an evaluation of their suitability for stocking. CDFG Inland Fish. Admin. Rpt. 82-9. 19 pp.

———. 1982b. A review of the life history of the red shiner, *Notropis lutrensis,* and a reassessment of its desirability for use as live bait in central and northern California. CDFG Inland Fish. Admin. Rpt. 82-1. 16 pp.

Gobalet, K. W. 1990. Prehistoric status of freshwater fishes of the Pajaro-Salinas River system of California. Copeia 1990: 680–685.

———. 1992. Colorado River fishes of Lake Cahuilla, Salton Basin, southern California: a cautionary tale for zooarchaeologists. Bull. S. Calif. Acad. Sci. 91:70–83.

———. 1994. Additional archaeological evidence for Colorado River fishes in the Salton Basin of Southern California. Bull. S. Calif. Acad. Sci. 93:8–41.

Gobalet, K. W., and G. L. Fenenga. 1993. Terminal Pleistocene–Early Holocene fishes from Tulare Lake, San Joaquin Valley, California, with comments on the evolution of Sacramento squawfish (*Ptychocheilus grandis:* Cyprinidae). Paleobios 15(1):1–8.

Gobalet, K. W., and T. L. Jones. 1995. Prehistoric Native American fisheries of the Central California coast. Trans. Am. Fish. Soc. 124:813–823.

Gobalet, K. W., and T. W. Wake. 2000. Archaeological and paleontological fish remains from the Salton Basin, southern California. SW Nat. 45:514–520.

Gold, J. R. 1977. Systematics of western North American trout (*Salmo*), with notes on the redband trout of Sheepheaven Creek, California. Can. J. Zool. 55:1858–1873.

Gold, J. R., and G. A. E. Gall. 1975. The taxonomic structure of six California high Sierra golden trout (*Salmo aquabonita*) populations. Proc. Calif. Acad. Sci. 40:243–263.

———. 1981. Systematics of golden trout, *Salmo aguabonita*, from the Sierra Nevada. Calif. Fish Game 67:204–230.

Goldberg, S. R. 1977. Seasonal ovarian cycle of the tidewater goby, *Eucyclogobius newberryi* (Gobiidae). SW Nat. 22:557–559.

Gomez, R., and H. L. Lindsay. 1972. Occurrence of Mississippi silversides, *Menidia audens* (Hay) in Keystone Reservoir and the Arkansas River. Proc. Okla. Acad. Sci. 52:16–18.

Gonzalez, M. R., C. M. Hart, J. R. Verfaillie, and S. H. Hurlbert. 1998. Salinity and fish effects on Salton Sea microecosystems: water chemistry and nutrient cycling. Hydrobiologia 381:105–128.

Goodell, J. A., and L. B. Kats. 1999. Effects of introduced mosquitofish on Pacific treefrogs and the role of alternative prey. Cons. Biol. 13:921–924.

Goodson, L. F. 1965. Diets of four warmwater game fishes in a fluctuating, steep-sided California reservoir. Calif. Fish Game 51:259–269.

———. 1966. Crappie, pages 312–332; landlocked striped bass, pages 402–411 *in* A. Calhoun, ed. Inland fisheries management. Sacramento: California Department of Fish and Game.

Goodyear, C. P., C. E. Boyd, and R. J. Beyers. 1972. Relationships between primary productivity and mosquitofish (*Gambusia affinis*) production in large microcosms. Limnol. Oceanogr. 17:445–450.

Gosline, W. A. 1966. The limits of the fish family Serranidae, with notes on other lower percoids. Proc. Calif. Acad. Sci. 33:91–112.

Graham, R. J., and D. J. Orth. 1986. Effects of temperature and streamflow on time and duration of spawning by smallmouth bass. Trans. Am. Fish. Soc. 115:693–702.

Grant, G. C. 1992. Selected life history aspects of Sacramento squawfish and hardhead in Pine Creek, Tehama County, California. M.S. thesis, Calif. State Univ., Chico. 86 pp.

Grant, G. C., and P. E. Maslin. 1997. Movements and reproduction of hardhead and Sacramento squawfish in a small California stream. SW Nat. 44:296–310.

Grant, J. W. A., and D. L. G. Noakes. 1988. Aggressiveness and foraging mode of young-of-year brook charr, *Salvelinus fontinalis* (Pisces, Salmonidae). Behav. Ecol. Sociobiol. 22:435–445.

Greenbank, J. 1957. Aggregational behavior in a freshwater sculpin. Copeia 1957:157.

Greenfield, D. W., and G. D. Deckert. 1973. Introgressive hybridization between *Gila orcutti* and *Hesperoleucus symmetricus* (Pisces: Cyprinidae) in the Cuyama River basin, California. II. Ecological Aspects. Copeia 1973:417–427.

Greenfield, D. W., and T. Greenfield. 1972. Introgressive hybridization between *Gila orcutti* and *Hesperoleucus symmetricus* (Pisces: Cyprinidae) in the Cuyama River Basin, California. I. Meristics, morphometrics and breeding. Copeia 1972:849–859.

Greenfield, D. W., S. T. Ross, and G. D. Deckert. 1970. Some aspects of the life history of the Santa Ana sucker, *Catostomus (Pantosteus) santaanae* (Snyder). Calif. Fish Game 56:166–179.

Gregory, R. S. 1994. The influence of ontogeny, perceived risk of predation, and visual ability on the foraging behavior of juvenile chinook salmon. Pages 271–284 *in* D. J. Stouder, K. L. Fresh, and R. J. Feller, eds. Theory and application in fish feeding ecology. Columbia: University of South Carolina Press.

Gregory, R. S., and C. D. Levings. 1998. Turbidity reduces predation on migrating juvenile chinook salmon. Trans. Am. Fish. Soc. 127:275–285.

Griffith, J. S. 1978. Effects of low temperature on the behavior and survival of threadfin shad, *Dorosoma petenense.* Trans. Am. Fish. Soc. 107:63–70.

Grimm, M. P., and M. Klinge. 1996. Pike and some aspects of its dependence on vegetation. Pages 125–156 *in* J. F. Craig, ed. Pike: biology and exploitation. London: Chapman and Hall.

Grimaldo, L., B. Ross, and D. Sweetnam. 1998. Preliminary results on the age and growth of delta smelt (*Hypomesus transpacificus*) from different areas of the estuary using otolith microstructure analysis. IEP Newslett. 11(1):25–28.

Grinstead, B. G. 1969. The vertical distribution of the white crappie in the Buncombe Creek arm of Lake Texoma. Univ. Okla. Fish. Res. Lab. Bull. 3. 37 pp.

Gross, M. R. 1979. Cuckoldry in sunfishes (*Lepomis*: Centrarchidae). Can. J. Zool. 57:1507–1509.

———. 1982. Sneakers, satellites and parentals: polymorphic mating strategies in North American sunfishes. Z. Tierpsychol. 60:1–26.

———. 1991. Evolution of alternative reproductive strategies: frequency-dependent sexual selection in male bluegill sunfish. Phil. Trans. R. Soc. Lond. B 332:59–66.

Gross, M. R., and A. M. MacMillan. 1991. Predation and the evolution of colonial nesting in bluegill sunfish (*Lepomis macrochirus*). Behav. Ecol. Sociobiol. 8:163–174.

Grost, R. T., W. A. Hubert, and T. A. Wesche. 1990. Redd site selection by brown trout in Douglas Creek, Wyoming. J. Freshw. Ecol. 5:365–371.

Groves, A. B., G. B. Collins, and P. S. Trefethen. 1968. Roles of olfaction and vision in choice of spawning site by homing adult chinook salmon (*Oncorhynchus tshawytscha*). J. Fish. Res. Board Can. 25:867–876.

Gunter, G. 1942. A list of the fishes of the mainland of North and South America recorded from both freshwater and seawater. Am. Midl. Nat. 28:305–356.

Guy, C. S., R. M. Neumann, and D. W. Willis. 1992. Movement patterns of adult black crappie, *Pomoxis nigromaculatus,* in Brant Lake, South Dakota. J. Freshw. Biol. 7:137–147.

Haaker, P. L. 1979. Two Asiatic gobiid fishes, *Tridentiger trigonocephalus* and *Acanthogobius flavimanus,* in southern California. Bull. S. Calif. Acad. Sci. 78:56–61.

Haas, G. R., and J. D. McPhail. 1991. Systematics and distributions of Dolly Varden (*Salvelinus malma*) and bull trout (*S. confluentus*) in North America. Can. J. Fish. Aquat. Sci. 48:2191–2211.

Hagen, D. W. 1967. Isolating mechanisms in three-spine sticklebacks (*Gasterosteus*). J. Fish. Res. Board Can. 24:1637–1692.

Hagen, D. W., and L. G. Gilbertson. 1972. Geographic variation and environmental selection in *Gasterosteus aculeatus* L. in the Pacific Northwest, America. Evolution 26:32–51.

Haglund, T. R., and D. G. Buth. 1988. Allozymes of the unarmored threespine stickleback (*Gasterosteus aculeatus williamsoni*) and identification of the Shay Creek population. Isozyme Bull. 21:196.

Hall, D. J., E. E. Werner, J. F. Gilliam, G. G. Mittelbach, D. Howard, C. G. Doner, J. Dickerman, and A. J. Stewart. 1979. Diel foraging behavior and prey selection in the golden shiner (*Notemigonus chrysoleucas*). J. Fish. Res. Board Can. 36:1029–1039.

Hall, L. W., Jr. 1991. A synthesis of water quality and contaminant data on early life history stages of striped bass, *Morone saxatilis*. Rev. Aquat. Sci. 4:261–288.

Hallock, R. J., and D. H. Fry. 1967. Five species of salmon, *Oncorhynchus*, in the Sacramento River, California. Calif. Fish Game 53:5–22.

Hallock, R. J., W. F. Van Woert, and L. Shapovalov. 1961. An evaluation of stocking of hatchery-reared steelhead rainbow trout (*Salmo gairdnerii gairdnerii*) in the Sacramento River system. CDFG Fish. Bull. 114:1–74.

Hamada, K. 1961. Taxonomic and ecological studies of the genus *Hypomesus* of Japan. Mem. Fac. Fish. Hokkaido Univ. 9:1–56.

Hansen, D. F. 1951. Biology of the white crappie in Illinois. Ill. Nat. Hist. Surv. Bull. 25:211–265.

———. 1965. Further observations on nesting of white crappie. Trans. Am. Fish. Soc. 94:182–184.

Hanson, J. A., and A. J. Cordone. 1967. Age and growth of lake trout, *Salvelinus namaycush* (Walbaum), in Lake Tahoe. Calif. Fish Game 53:68–87.

Hanson, J. A., and R. H. Wickwire. 1967. Fecundity and age at maturity of lake trout, *Salvelinus namaycush* (Walbaum), in Lake Tahoe. Calif. Fish Game 53:154–164.

Hard, J. J., R. G. Kope, W. S. Grant, F. W. Waknitz, L. T. Parker, and R. S. Waples. 1996. Status review of pink salmon from Washington, Oregon, and California. USDC NOAA Tech. Mem. NMFS-NWFSC-25. 131 pp.

Hardisty, M. W., and I. C. Potter, eds. 1971. The biology of lampreys, vol. 1. London: Academic Press.

Hare, S. R., N. J. Mantua, and R. C. Francis. 1999. Inverse production regimes: Alaska and West Coast Pacific salmon. Fisheries (Bethesda) 24(1):6–14.

Harrington, R. W., and E. S. Harrington. 1961. Food selection among fishes invading a high subtropical salt marsh from onset of flooding through the progress of a mosquito brood. Ecology 42:646–666.

Harry, R. R. 1951. The embryonic and early larval stages of the tui chub, *Siphatales bicolor* (Girard) from Eagle Lake, California. Calif. Fish Game 37:129–132.

Hart, J. 1996. Storm over Mono. Berkeley: University of California Press.

Hart, J. L. 1973. Pacific Fishes of Canada. Fish. Res. Board Can. Bull. 180. 740 pp.

Hart, J. L., and J. L. McHugh. 1944. The smelts (Osmeridae) of British Columbia. Fish. Res. Board Can. Bull. 54. 27 pp.

Hart, P., and S. F. Hamrin. 1988. Pike as a selective predator. Effects of prey size, availability, cover, and pike jaw dimensions. Oikos 51:220–226.

Hartman, G. F. 1965. The role of behavior in the ecology and interaction of underyearling coho salmon (*Oncorhynchus kisutch*) and steelhead trout (*Salmo gairdneri*). J. Fish. Res. Board Can. 22:1035–1081.

Hartney, K. B., and L. Tumyan. 1998. Temporal changes in diet and foraging habitat of California killifish (*Fundulus parvipinnis*) in Marina del Rey, California. Bull. S. Calif. Acad. Sci. 97:1–8.

Hartzell, L. L. 1992. Hunter-gatherer adaptive strategies and lacustrine environments in the Buena Vista Lake Basin, Kern County, California. Ph.D. dissertation, Univ. Calif., Davis. 365 pp.

Harvey, B. C. 1986. Effects of suction gold dredging on fish and invertebrates in two California streams. N. Am. J. Fish. Mgmt. 6:401–409.

———. 1998. Influence of large woody debris on retention, immigration, and growth of coastal cutthroat trout (*Oncorhynchus clarki clarki*) in stream pools. Can. J. Fish. Aquat. Sci. 55:1902–1908.

Harvey, B. C., and T. E. Lisle. 1999. Scour of chinook salmon redds on suction dredge tailings. N. Am. J. Fish. Mgmt. 19:613–617.

Harvey, B. C., and R. J. Nakamoto. 1996. Effects of steelhead density on growth of coho salmon in a small coastal California stream. Trans. Am. Fish. Soc. 125:237–243.

———. 1997. Habitat-dependent interactions between two size-classes of juvenile steelhead in a small stream. Can. J. Fish. Aquat. Sci. 54:27–31.

———. 1999. Diel and seasonal movements by adult Sacramento pikeminnow (*Ptychocheilus grandis*) in the Eel River, northwestern California. Ecol. Freshw. Fish 8:209–215.

Harwood, R. H. 1972. Diurnal feeding rhythm of *Notropis lutrensis* Baird and Girard. Texas J. Sci. 24:97–99.

Haslam. G. 1989. The lake that will not die. Pacific Discovery, Spring 1989:28–37.

Hasler, A. D., E. S. Gardella, H. F. Henderson, and R. M. Horrall. 1969. Open-water orientation of white bass, *Roccus chrysops*, as determined by ultrasonic tracking methods. J. Fish. Res. Board Can. 26:2173–2192.

Hassler, T. J. 1987. Coho salmon. USFWS Ser. Biol. Rpt. 82 (11.70). 19 pp.

———. 1988. Striped bass. USFWS Biol. Rpt. 82 (11.82). 29 pp.

Hatch, J. T. 1988. Ontogenetic shifts in feeding habits of mixed cohort schools of larval golden shiners. Verh. Int. Verein. Limnol. 23:1704–1709.

Hauser, W. J. 1969. Life history of the mountain sucker, *Catostomus platyrhynchus*, in Montana. Trans. Am. Fish. Soc. 98:209–224.

———. 1975. An unusually fast growth rate for *Tilapia zillii*. Calif. Fish Game 61:54–56.

———. 1977. Temperature requirements of *Tilapia zillii*. Calif. Fish Game 61:228–233.

Hayden, T. R., and D. B. Gale. 1999. Juvenile salmonid emigration monitoring from the Hunter Creek Basin, lower Klamath River, California, 1996–1997. Yurok Tribal Fish. Prog. Habitat Assess, Biol. Monit. Div. Tech. Rpt. 6. 63 pp.

Haynes, J. L., and R. C. Cashner. 1995. Life history and population dynamics of the western mosquitofish: a comparison of natural and introduced populations. J. Fish Biol. 46:1026–1041.

Hazel, C. R. 1969. Limnology of Upper Klamath Lake, Oregon, with emphasis on benthos. Ph.D. dissertation, Oreg. State Univ., Corvallis. 184 pp.

Healey, M. C. 1991. Life history of chinook salmon (*Oncorhynchus tshawytscha*). Pages 311–394 *in* C. Groot and L.

Margolis, eds. Pacific salmon life histories. Vancouver: University of British Columbia Press.

Heard, W. R. 1965. Limnetic cottid larvae and their utilization as food by juvenile sockeye salmon. Trans. Am. Fish. Soc. 99:191–193.

———. 1991. Life history of pink salmon. Pages 110–230 in C. Groot and L. Margolis, eds. Pacific salmon life histories. Vancouver: University of British Columbia Press.

Hearn, B. C., Jr., R. J. McLaughlin, and J. M. Donnely-Nolan 1988. Tectonic framework of the Clear Lake Basin, California. Pages 9–20 in J. D. Sims, ed. Late quaternary climate, tectonism, and sedimentation in Clear Lake. USGS Spec. Pap. 214. 225 pp.

Heidinger, R. C. 1975. Life history and biology of the largemouth bass. Pages 11–20 in H. Clepper, ed. Black bass biology and management. Washington, D.C.: Sport Fishing Institute.

Heman, M. L., R. S. Campbell, and L. C. Redmond. 1969. Manipulation of fish populations through reservoir drawdowns. Trans. Am. Fish. Soc. 98:293–304.

Hendricks, L. J. 1961. The threadfin shad, Dorosoma petenense (Gunther); the striped mullet, Mugil cephalus Linnaeus. Pages 95–103 in B. D. Walker, ed. The ecology of the Salton Sea, California, in relation to the sport fishery. CDFG Fish Bull. 113.

Hendrickson, D. A., and A. V. Romero. 1989. Conservation status of desert pupfish, Cyprinodon macularius, in Mexico and Arizona. Copeia 1989:478–483.

Hendrickson, S., and M. Hendrickson. 1993. Project kokanee—"good news for anglers." Outdoor California, July–August:6–8.

Herbold, B. 1979. Some effects of salinity and temperature on the cichlid fish Tilapia zilli. M.S. thesis, Calif. State Univ., Los Angeles. 36 pp.

———. 1987. Patterns of co-occurrence and resource use in a non-coevolved assemblage of fishes. Ph.D. dissertation, Univ. of Calif., Davis. 81 pp.

———. 1994. Habitat requirements of the delta smelt. IEP Newslett. Winter 1994:1–3.

Herbold, B., and P. B. Moyle.1986. Introduced species and vacant niches. Am. Nat. 128:751–760.

Herbold, B., A. Jassby, and P. B. Moyle. 1992. Status and trends report on aquatic resources in the San Francisco estuary. USEPA San Francisco Estuary Project, San Francisco. 257 pp.

Hergenrader, G. L. 1969. Spawning behavior of Perca flavescens in aquaria. Copeia 1969:839–841.

Heyne, T., B. Tribbey, and J. Smith. 1991. First record of Mozambique tilapia in the San Joaquin Valley, California. Calif. Fish Game 77:53–54.

Heyward, L. D., T. I. J. Smith, and W. E. Jenkins. 1995. Survival and growth of white bass, Morone chrysops, reared at different salinities. J. World Maricult. Soc. 26:475–479.

Hill, K. A., and J. D. Webber. 1999. Butte Creek spring-run chinook salmon, Oncorhynchus tshawytscha, juvenile outmigration and life history, 1995–1998. CDFG Inland Fish. Admin. Rpt. 99-5. 46 pp.

Hinch, S. G., and N. C. Collins. 1991. Importance of diurnal and nocturnal nest defense in the energy budget of male smallmouth bass: insights from direct video observations. Trans. Am. Fish. Soc. 120:657–663.

Hinds, N. E. A. 1952. Evolution of the California landscape. Calif. Div. Mines Bull. 158. 240 pp.

Hiscox, J. I. 1979. Feeding habits and growth of stocked salmonids in a California reservoir. CDFG Inland Fish. Admin. Rpt. 79-3. 22 pp.

Hiss, J. M. 1984. Diet of age-0 steelhead trout and speckled dace in Willow Creek, Humboldt County, California. M.S. thesis, Humboldt State Univ., Arcata. 51 pp.

Hocutt, C. H., S. E. Seibold, R. M. Harrell, R. V. Jesien, and W. H. Bason. 1990. Behavioral observations of striped bass (Morone saxatilis) on the spawning grounds of the Choptank and Nanticoke rivers, Maryland, USA. J. Appl. Ichthyol. 6: 211–222.

Hodges, C. A. 1966. Geomorphic history of Clear Lake, California. Ph.D. dissertation, Stanford Univ., Stanford, Calif.

Hoetker, G. M., and K. W. Gobalet. 1999. Fossil razorback sucker (Pisces: Catostomidae, Xyrauchen texanus) from southeastern California. Copeia 1999:755–759.

Hokanson, K. E. F., C. F. Kleiner, and T. W. Thorslund. 1977. Effects of constant temperatures and diel temperature fluctuations on specific growth and mortality rates and yield of juvenile rainbow trout, Salmo gairdneri. J. Fish. Res. Board Can. 34:639–648.

Holanov, S. H., and J. C. Tash. 1978. Particulate and filter feeding in threadfin shad, Dorosoma petenense at different light intensities. J. Fish Biol. 13:619–625.

Holden, P. B. 1991. Ghosts of the Green River: impacts of Green River poisoning on management of native fishes. Pages 43–54 in W. L. Minckley and J. E. Deacon, eds. Battle against extinction: native fish management in the American West. Tucson: University of Arizona Press.

Holden, P. B., and C. B. Stalnaker. 1970. Systematic studies of the cyprinid genus Gila in the upper Colorado River basin. Copeia 1970:409–429.

Holway, R. S. 1907. Physiographic changes bearing on the faunal relationships of the Russian and Sacramento Rivers, California. Science 26:382–383.

Hoogland, R., D. Morris, and N. Tinbergen. 1956. The spines of sticklebacks (Gasterosteus and Pygosteus) as means of defense against predators (Perca and Esox). Behaviour 10:205–236.

Hoopaugh, D. A. 1974. Status of redband trout (Salmo sp.) in California. CDFG Inland Fish. Admin. Rpt. 74-7. 11 pp.

Hooton, B. 1997. Status of coastal cutthroat trout in Oregon. Pages 57–67 in J. D. Hall, P. A. Bisson, and R. E. Gresswell, eds. Sea-run cutthroat trout: biology, management and future conservation. Corvallis, Oreg.: Oregon Chapter of the American Fisheries Society.

Hoover, F. G. 1971. Status report on Tilapia mossambica (Peters) in southern California. CDFG Inland Fish. Admin. Rpt. 71-16. 31 pp.

Hoover, F. G., and J. A. St. Amant. 1970. Establishment of Tilapia mossambica (Peters) in Bard Valley, Imperial County, California. Calif. Fish Game 56:70–71.

Hopelain, J. S. 1998. Age, growth, and life history of Klamath River basin steelhead trout (Oncorhynchus mykiss irideus) as determined from scale analysis. CDFG Inland Fish. Admin. Rpt. 98-3. 19 pp.

Hopkirk, J. D. 1962. Morphological variation in the freshwater embiotocid Hysterocarpus traskii Gibbons. M.A. thesis, Univ. Calif., Berkeley. 159 pp.

———. 1973. Endemism in Fishes of the Clear Lake Region. Univ. Calif. Publ. Zool. 96. 160 pp.

Hopkirk, J. D., and R. J. Behnke. 1966. Additions to the known native fish fauna of Nevada. Copeia 1966:134–136.

Hoshino, N., T. Kinoshita, and Y. Kanno. 1993. Age, growth, and ecological characteristics of the goby, *Acanthogobius flavimanus*, in Hakodate Bay, Hokkaido, Japan. Bull. Fac. Fisheries, Hokkaido Univ. 44:147–157.

Houde, E. D. 1969. Distribution of larval walleyes and yellow perch in a bay of Oneida Lake and its relation to water currents and zooplankton. N.Y. Fish Game J. 16:184–205.

House, F. 1999. Totem salmon. Boston: Beacon Press.

Houston, A. H. 1982. Thermal effects upon fishes. Natl. Res. Coun. Canada Publ. 18566. 200 pp.

Houston, J. J. 1988. Status of the green sturgeon, *Acipenser medirostris*, in Canada. Can. Field Nat. 102:286–290.

Howes, G. 1984. Phyletics and biogeography of the aspinine cyprinid fishes. Bull. Brit. Mus. Nat. Hist. (Zool.) 47:283–303.

Howland, J. W. 1931. Studies on the Kentucky black bass (*Micropterus pseudaplites* Hubbs). Trans. Am. Fish. Soc. 61:89–94.

Hoy, J. B., E. E. Kaufmann, and A. G. O'Berg. 1972. A large-scale field test of *Gambusia affinis* and Chlorpurifos for mosquito control. Mosq. News 32:163–171.

Hubbell, P. M. 1966. Pumpkinseed sunfish, pages 402–404; warmouth, pages 405–407 *in* A. Calhoun, ed. Inland fisheries management. Sacramento: California Department of Fish and Game.

Hubbs, C. L. 1921. The latitudinal variation in the number of vertical fin-rays in *Leptocottus armatus*. Mus. Zool. Univ. Mich. Occas. Pap. 94. 7 pp.

———. 1925. The life cycle and growth of lampreys. Papers Mich. Acad. Sci. Arts Ltrs. 4:587–603.

———. 1946. Wandering of pink salmon and other salmonid fishes into southern California. Calif. Fish Game 32:81–86.

———. 1953. *Eleotris picta* added to the fish fauna of California. Calif. Fish Game 39:69–76.

———. 1954. Establishment of a forage fish, the red shiner (*Notropis lutrensis*), in the lower Colorado River system. Calif. Fish Game 40:287–294.

———. 1961. Isolating mechanisms in the speciation of fishes. Pages 537–560 *in* F. Blair, ed. Vertebrate speciation. Austin: University of Texas Press.

———. 1967. Occurrence of the Pacific lamprey, *Entosphenus tridentatus*, off Baja California and in streams of southern California, with remarks on its nomenclature. Trans. San Diego Nat. Hist. Soc. 14:301–312.

———. 1971. *Lampetra (Entosphenus) lethophaga*, new species, the nonparasitic derivative of the Pacific lamprey. Trans. San Diego Nat. Hist. Soc. 16:125–164.

Hubbs, C. L., and R. M. Bailey. 1940. A revision of the black basses (*Micropterus* and *Huro*), with descriptions of four new forms. Univ. Mich. Mus. Zool. Misc. Publ. 48:1–51.

Hubbs, C. L., and K. Lagler. 1958. Fishes of the Great Lakes Region. Ann Arbor: University of Michigan Press.

Hubbs, C. L., and R. R. Miller. 1943. Mass hybridization between two genera of cyprinid fishes in the Mohave Desert, California. Pap. Mich. Acad. Sci. Arts Lett. 28:343–378.

———. 1948. The Great Basin, with emphasis on glacial and postglacial times. II. The zoological evidence: correlation between fish distribution and hydrographic history in the desert basins of western United States. Bull. Univ. Utah 38:18–166.

———. 1951. *Catostomus arenarius*, a Great Basin fish, synonymized with *C. tahoensis*. Copeia 1951:299–300.

———. 1952. Hybridization in nature between the fish genera *Catostomus* and *Xyrauchen*. Pap. Mich. Acad. Sci. Arts Lett. 38:207–233.

———. 1965. Studies of cyprinodont fishes. XXII. Variation in *Lucania parva*, its establishment in western United States, and description of a new species from an interior basin in Coahuila, Mexico. Univ. Mich. Mus. Zool. Misc. Publ. 127. 104 pp.

Hubbs, C. L., and I. C. Potter. 1971. Distribution, phylogeny and taxonomy. Pages 1–65 *in* N. W. Hardisty and I. C. Potter, eds. The biology of lampreys, vol. 1. London: Academic Press.

Hubbs, C. L., and O. L. Wallis. 1948. The native fish fauna of Yosemite National Park and its preservation. Yosemite Nat. Notes 27:131–144.

Hubbs, C. L., W. I. Follett, and L. J. Dempster. 1979. List of the fishes of California. Occas. Pap. Calif. Acad. Sci. 133:1–51.

Hubbs, C. L., L. C. Hubbs, and R. E Johnson. 1943. Hybridization in nature between species of catostomid fishes. Contr. Lab. Vert. Biol. Univ. Mich. 22:1–69.

Hubbs, C. L., R. R. Miller, and L. C. Hubbs. 1974. Hydrographic history and relict fishes of the northcentral Great Basin. Mem. Calif. Acad. Sci. 7. 259 pp.

Hubbs, Clark. 1947. Mixture of marine and freshwater fishes in the lower Salinas River, Calif. Copeia 1947:147–149.

———. 1968. An opinion on the effects of cichlid releases in North America. Trans. Am. Fish. Soc. 97:197–198.

———. 1982. Life history dynamics of *Menidia beryllina* from Lake Texoma. Am. Midl. Nat. 107:1–12.

Hubbs, Clark, and H. H. Bailey. 1977. Effects of temperature on the termination of breeding season of *Menidia audens*. SW Nat. 22:544–547.

Hubbs, Clark, H. B. Sharp, and J. F. Schneider. 1971. Developmental rates of *Menidia audens* with notes on salt tolerance. Trans. Am. Fish. Soc. 100:603–610.

Hung, S. O., P. B. Lutes, F. S. Conte, and T. Storebakken. 1989. Growth and feeding efficiency of white sturgeon (*Acipenser transmontanus*) subyearlings at different feeding rates. Aquaculture 80:147–153.

Hunt, W. G., J. M. Jenkins, R. E. Jackman, and E. G. Thelander. 1988. Foraging ecology of Bald Eagles on a regulated river. J. Raptor Res. 26:243–256.

Hunter, J. R. 1963. The reproductive behavior of the green sunfish, *Lepomis cyanellus*. Zoologica 48:13–24.

Hurlbert, S. H., J. Zedler, and D. Fairbanks. 1972. Ecosystem alteration by mosquitofish (*Gambusia affinis*) predation. Science 175:639–641.

Hurst, H., G. Bass, and C. Hubbs. 1975. The biology of the Guadalupe, Suwannee, and redeye basses. Pages 34–55 *in* H. Clepper, ed. Black bass biology and management. Washington, D.C.: Sport Fishing Institute.

Hutchings, J. A. 1994. Age- and size-specific costs of reproduction within populations of brook trout, *Salvelinus fontinalis*. Oikos 70:12–20.

———. 1996. Adaptive phenotypic plasticity in brook trout, *Salvelinus fontinalis*, life histories. Ecoscience 3:25–32.

Hynes, H. B. N. 1950. The food of freshwater sticklebacks (*Gasterosteus aculeatus* and *Pygosteus pungitius*) with a review of methods used in the study of the food of fishes. J. Anim. Ecol. 19:36–58.

Ihnat, J. M., and R. V. Bulkley. 1984. Influence of acclimation temperature and season on acute temperature preference of adult mountain whitefish, *Prosopium williamsoni.* Env. Biol. Fish. 11:29–40.

Ikusemiju, K. 1975. Aspects of the ecology and life history of the sculpin, *Cottus aleuticus* (Gilbert) in Lake Washington. J. Fish Biol. 7:235–245.

Imler, R. L., D. T. Weber, and O. L. Fyock. 1975. Survival, reproduction, age, growth, and food habits of Sacramento perch, *Archoplites interruptus* (Girard), in Colorado. Trans. Am. Fish. Soc. 104:232–236.

Itzkowitz, M. 1971. Preliminary study of the social behavior of male *Gambusia affinis* (Gaird and Girard) (Pisces: Poeciliidae) in aquaria. Chesapeake Sci. 12:219–224.

Iwanaga, P. M., and J. D. Hall. 1973. Effects of logging on growth of juvenile coho salmon. USEPA Ecol. Res. Ser. EPA-R3-73-006. 35 pp.

Jackman, R. E., W. G. Hunt, J. M. Jenkins, and P. J. Detrich. 1999. Prey of nesting bald eagles in northern California. J. Raptor Res. 33:87–96.

Jameson, R. J., and K. W. Kenyon. 1977. Prey of sea lions in the Rogue River, Oregon. J. Mammal. 58:672.

Jassby, A. D., W. J. Kimmerer, S. G. Monismith, C. Armor, J. E. Cloern, T. M. Powell, J. R. Schubel, and T. J. Vendlinksi. 1995. Isohaline position as a habitat indicator for estuarine populations. Ecol. App. 5:272–289.

Jenkins, R. E., and N. M. Burkhead. 1994. Freshwater fishes of Virginia. Bethesda. Md.: American Fisheries Society.

Jenkins, T. M., Jr. 1969. Social structure, position choice and microdistribution of two trout species (*Salmo trutta* and *Salmo gairdneri*) resident in mountain streams. Anim. Behav. Monogr. 2:57–123.

Jenkins, T. M., Jr., S. Deihl, K. W. Kratz, and S. D. Cooper. 1999. Effects of population density on individual growth of brown trout in streams. Ecology 80:941–956.

Jennings, M. R. 1996. Past occurrence of eulachon, *Thaleichthys pacificus,* in streams tributary to Humboldt Bay, California. Calif. Fish Game 82:147–148.

Jennings, M. R., and M. K. Saiki. 1990. Establishment of red shiner, *Notropis lutrensis,* in the San Joaquin Valley, California. Calif. Fish Game 76:46–57.

Jhingran, V. G. 1948. A contribution to the biology of the Klamath black dace, *Rhinichthys osculus klamathensis* (Evermann and Meek). Ph.D. dissertation, Stanford Univ., Stanford, Calif. 94 pp.

Johannes, M. R. S., D. J. McQueen, T. J. Stewart, and J. R. Post. 1989. Golden shiner (*Notemigonus crysoleucas*) population abundance: correlation with food and predators. Can. J. Fish. Aquat. Sci. 46:810–817.

John, K. R. 1963. The effect of torrential rains on the reproductive cycle of *Rhinichthys osculus* in the Chiricahua Mountains, Arizona. Copeia 1963:286–291.

———. 1964. Survival of fish in intermittent streams of the Chiricahua Mountains, Arizona. Ecology 45:112–119.

Johnson, D. W., and E. L. McClendon. 1970. Differential distribution of the striped mullet, *Mugil cephalus* Linnaeus. Calif. Fish Game 56:2138–2139.

Johnson, J. E. 1970. Age, growth, and population dynamics of threadfin shad, *Dorosoma petenense* (Günther), in Central Arizona reservoirs. Trans. Am. Fish. Soc. 99:739–753.

———. 1971. Maturity and fecundity of threadfin shad, *Dorosoma petenense* (Günther), in Central Arizona reservoirs. Trans. Am. Fish. Soc. 100:74–85.

Johnson, J. H. 1985. Comparative diets of Paiute sculpin, speckled dace, subyearling steelhead trout, in tributaries to the Clearwater River, Idaho. NW Sci. 59:1–9.

Johnson, J. H., and D. S. Dropkin. 1995. Effects of prey density and short term food deprivation on the growth and survival of American shad larvae. J. Fish Biol. 46:872–879.

Johnson, J. M. 1999. Fish use of a southern California salt marsh. M.S. thesis, Calif. State Univ., San Diego. 71 pp.

Johnson, O. W., M. H. Ruckelhaus, W. S. Grant, F. W. Waknitz, A.M. Garrett, G. J. Bryant, K. Neely, and J. J. Hard. 1999. Status review of coastal cutthroat trout from Washington, Oregon, and California. NOAA Tech. Mem. NMFS-NWFSC-37. 292 pp.

Johnson, P. C., and G. L. Vinyard. 1987. Filter-feeding behavior and particle retention efficiency in the Sacramento blackfish. Trans. Am. Fish. Soc. 116:634–640.

Johnson, S. L. 1988. The effect of the 1983 El Nino on Oregon's coho (*Oncorhynchus kisutch*) and chinook (*Oncorhynchus tshawytscha*) salmon. Fish. Res. 6:105–123.

Johnson, S. R. 1976. Age and growth of hitch (*Lavinia exilicauda*) and blackfish (*Orthodon microlepidotus*) from San Luis Reservoir, Merced County, California. Unpubl. rpt., Calif. Poly. Univ., San Luis Obispo. 18 pp.

Jones, A. C. 1962. The biology of the euryhaline fish *Leptocottus armatus* Girard (Cottidae). Univ. Calif. Publ. Zool. 67:321–367.

———. 1972. Contributions to the life history of the Piute sculpin in Sagehen Creek, California. Calif. Fish Game 58:285–290.

Jonez, A., and R. C. Sumner. 1954. Lakes Mead and Mohave investigations. Unpubl. rpt., Nev. Fish Game Comm. Wildl. Restor. Div. 174 pp.

Jordan, D. S. 1879. Notes on a collection of fishes from Clackamus River, Oregon. Proc. U.S. Natl. Mus. 1:69–85.

———. 1892. Salmon and trout of the Pacific Coast. California State Fish Commission 12th biennial report for the years 1891–1892. Sacramento: California State Fish Commission: 44–58.

———. 1893. A description of the golden trout of Kern River, California, *Salmo mykiss agua-bonita.* U.S. Natl. Mus. Proc. 15:481–483.

———. 1894. Notes on the fresh-water species of San Luis Obispo County, California. Bull. U.S. Fish. Comm. 19: 191–192.

Jordan, D. S., and B. W. Evermann. 1896. Fishes of North and Middle America. Bull. U.S. Natl. Mus. 47(1–4):1–3705.

———. 1923. American Food and Game Fishes. New York: Doubleday and Company.

Jordan, D. S., and C. H. Gilbert. 1882. Synopsis of the fishes of North America. Bull. U.S. Natl. Mus. 16:1–1018.

———. 1894. List of fishes inhabiting Clear Lake, California. Bull. U.S. Fish Comm. 14:139–140.

Jordan, D. S., and J. O. Snyder. 1906. A synopsis of the sturgeons (Acipenseridae) of Japan. Proc. U.S. Natl. Mus. 30:397–398.

Kaiser, J. 1999. Battle over a dying sea. Science 284:28–30.

Karas, N. 1997. Brook trout. New York: Lyons and Burford.

Keast, A. 1966. Trophic interrelationships in the fish fauna of a small stream. Univ. Mich. Great Lakes Res. Div. Publ. 15:51–79.

———. 1968a. Feeding biology of the black crappie, *Pomoxis nigromaculatus.* J. Fish. Res. Board Can. 24:285–297.

———. 1968b. Feeding of some Great Lakes fishes at low temperatures. J. Fish. Res. Board Can. 24:1199–1218.

Keast, A., and D. Webb. 1966. Mouth and body form relative to feeding ecology in the fish fauna of a small lake, Lake Opinicon, Ontario. J. Fish. Res. Board Can. 23:1845–1874.

Keast, A., and L. Welsh. 1968. Daily feeding periodicities, food uptake rates, and dietary changes with hour of day in some lake fishes. J. Fish. Res. Board Can. 25:1133–1149.

Kegley, S., L. Neumeister, and T. Martin. 1999. Disrupting the balance: ecological impact of pesticides in California. San Francisco: Californians for Pesticide Reform.

Keen, W. H. 1982. Behavioral interactions and body size differences in competition for food among juvenile brown bullhead (*Ictalurus nebulosus*). Can. J. Fish. Aquat. Sci. 39: 316–320.

Kelly, H. D. 1957. Preliminary studies on *Tilapia mossambica* Peters relative to experimental pond culture. Proc. 10th Annu. Conf. SE Assoc. Fish Comm.: 139–149.

Kendall, A. W., and F. J. Schwartz. 1968. Lethal temperature and salinity tolerances of the white catfish, *Ictalurus catus,* from the Patuxent River, Maryland. Chesapeake Sci. 9:103–108.

Kennedy, C. H. 1916. A possible enemy of the mosquito. Calif. Fish Game 2:179–182.

Kennedy, J. L., and P. A. Kucera. 1978. The reproductive ecology of the Tahoe sucker, *Catostomus tahoensis,* in Pyramid Lake, Nevada. Great Basin Nat. 38:181–186.

Kesner, W. D., and R. A. Barnhart. 1972. Characteristics of fall-run steelhead trout (*Salmo gairdneri gairdneri*) of the Klamath River system, with emphasis on the half-pounder. Calif. Fish Game 58:204–220.

Keys, A. B. 1931. A study of the selective action of decreased salinity and of asphyxiation on the Pacific killifish, *Fundulus parvipinnis.* Bull. Scripps Inst. Oceanogr. Univ. Calif. 2: 417–490.

Kiesecker, J. M., and A. R. Blaustein. 1998. Effects of introduced bullfrogs and smallmouth bass on habitat use, growth, and survival of native red-legged frogs (*Rana aurora*). Cons. Biol. 12:776–787.

Kikuchi, T., and Y. Yamashita. 1992. Seasonal occurrence of gobiid fish and their food habits in a small mud flat in Amakusa. Publ. Amakusa Mar. Lab. 11:73–93.

Kimmerer, W. J., J. H. Cowan, Jr., L. W. Miller, and K. A. Rose. 2000. Analysis of an estuarine striped bass (*Morone saxatilis*) population: influence of density-dependent mortality between metamorphosis and recruitment. Can. J. Fish. Aquat. Sci. 57:478–486.

Kimsey, J. B. 1950. Some Lahontan fishes in the Sacramento River Drainage, California. Calif. Fish Game 36:438–439.

———. 1954. The life history of the tui chub, *Siphateles bicolor* (Girard), from Eagle Lake, California. Calif. Fish Game 40: 395–410.

———. 1960. Observations on the spawning of Sacramento hitch in a lacustrine environment. Calif. Fish Game 46: 211–215.

Kimsey, J. B., and R. R. Bell. 1955. Observations on the ecology of the largemouth black bass and the tui chub in Big Sage Reservoir, Modoc County. CDFG Inland Fish. Admin. Rpt. 55-75. 17 pp.

———. 1956. Notes on the status of the pumpkinseed sunfish, *Lepomis gibbosus,* in the Susan River, Lassen County, California. CDFG Inland Fish. Admin. Rpt. 56-1. 20 pp.

Kimsey, J. B., and L. O. Fisk. 1960. Keys to the freshwater and anadromous fishes of California. Calif. Fish Game 46: 453–479.

———. 1964. Freshwater nongame fishes of California. Sacramento: California Department of Fish and Game.

Kimsey, J. B., D. W. Kelley, R. Hagy, and G. McCammon. 1956. A survey of the fish populations of Pardee Reservoir, Amador/ Calaveras counties. CDFG Inland Fish. Admin. Rpt. 56-18. 13 pp.

Kinne, O. 1960. Growth, food intake, and food conversion in a euryplastic fish exposed to different temperatures and salinities. Physiol. Zool. 33:288–317.

Kisanuki, T. T. 1980. Age and growth of the Humboldt sucker (*Catostomus occidentalis humboldtianus*) and the golden shiner (*Notemigonus chrysoleucas*) from Ruth Reservoir, California. M.S. thesis, Humboldt State Univ., Arcata. 67 pp.

Kitchell, J. F., and J. T. Windell. 1970. Nutritional value of algae to bluegill sunfish, *Lepomis macrochirus.* Copeia 1970: 186–190.

Kjelson, M. A. 1971. Selective predation by a freshwater planktivore, the threadfin shad, *Dorosoma petenense.* Ph.D. dissertation, Univ. Calif., Davis. 123 pp.

Kjelson, M. A., P. F. Raquel, and F. W. Fisher. 1982. Life history of fall-run juvenile chinook salmon, *Oncorhynchus tshawytscha,* in the Sacramento–San Joaquin estuary, California. Pages 393–410 *in* V. S. Kennedy, ed. Estuarine comparisons. New York: Academic Press.

Kline, K. 1978. Aspects of digestion in stomachless fishes. Ph.D. dissertation, Univ. Calif., Davis. 78 pp.

Klingbeil, R. A., R. D. Sandell, and A. W. Wells. 1975. An annotated checklist of the elasmobranchs and teleosts of Anaheim Bay. Pages 79–90 *in* E. D. Lane and C. W. Hill, eds. The marine resources of Anaheim Bay. CDFG Fish Bull. 165.

Kljukanov, V. A. 1970. Morphological basis of the classification of smelts of the genus *Hypomesus.* Zoolog. Zh. 49:1534–1542.

Knaggs, E. H. 1977. Status of the genus *Tilapia* in California's estuarine and marine waters. Calif. Nev. Wildl. Trans. 1977: 60–67.

Knapp, R. A., and T. L. Dudley. 1990. Growth and longevity of golden trout, *Oncorhynchus aguabonita,* in their native streams. Calif. Fish Game 76:161–173.

Knapp, R. A., and K. R. Matthews. 1996. Livestock grazing, golden trout, and streams in the Golden Trout Wilderness, California: impacts and management implications. N. Am. J. Fish. Mgmt. 16:805–820.

———. 2000. Non-native fish introductions and the decline of the mountain yellow-legged frog from within protected areas. Cons. Biol. 14:428–438.

Knapp, R. A., and V. T. Vredenburg. 1996. Spawning by California golden trout: characteristics of spawning fish, season and daily timing, redd characteristics, and microhabitat preferences. Trans. Am. Fish. Soc. 125:519–531.

Knapp, R. A., P. S. Corn, and D. E. Schindler. 2001. The introduction of non-native fish into wilderness lakes: good intentions, conflicting mandates, and unintended consequences. Ecosystems 4:275–278.

Knight, N. J. 1985. Microhabitats and temperature requirements of hardhead (*Mylopharodon conocephalus*) and Sacramento squawfish (*Ptychocheilus grandis*), with notes for some other native California stream fishes. Ph.D. dissertation, Univ. Calif., Davis. 161 pp.

Knudsen, D., and T. Mills. 1980. Copco Lake, Siskiyou County, fish sampling, Spring 1980. File rpt., CDFG End. Sp. Proj., Sacramento. 12 pp.

Kodric-Brown, A. 1981. Variable breeding systems in pupfishes (Genus *Cyprinodon*): adaptations to changing environments. Pages 205–236 in R. J. Naiman and D. L. Soltz, eds. Fishes in North American deserts. New York: John Wiley.

Koehn, R. 1965. Development and ecological significance of nuptial tubercles of the red shiner, *Notropis lutrensis*. Copeia 1965:462–467.

Kohlhorst, D. W. 1976. Sturgeon spawning in the Sacramento River in 1973, as determined by distribution of larvae. Calif. Fish Game 62:32–40.

———. 1999. Status of striped bass in the Sacramento–San Joaquin estuary. Calif. Fish Game 85:31–36.

Kohlhorst, D. W., L. W. Miller, and J. J. Orsi. 1980. Age and growth of white sturgeon collected in the Sacramento–San Joaquin estuary, California: 1965–1970 and 1973–1976. Calif. Fish Game 66:83–95.

Kohlhorst, D. W., L. W. Botsford, J. S. Brennan, and G. M. Cailliet. 1991. Aspects of the structure and dynamics of an exploited central California population of white sturgeon (*Acipenser transmontanus*). Pages 277–293 in P. Williot, ed. Acipenser. Bordeaux, France: CEMAGREF.

Kondolf, G. M. 1994. Livestock grazing and habitat for a threatened species: land-use decisions under scientific uncertainty in the White Mountains, U.S.A. Env. Mgmt. 18:501–509.

Konecki, J. T., C. A. Woody, and T. P. Quinn. 1995. Influence of temperature on incubation rates of coho salmon (*Oncorhynchus kisutch*) from ten Washington populations. NW Sci. 69:126–132.

Kope, R. G., and L. W. Botsford. 1990. Determination of factors affecting recruitment of chinook salmon *Oncorhynchus tshawytscha* in central California. USDA Fish. Bull. 88: 257–269.

Koppleman, J. B. 1994. Hybridization between smallmouth bass, *Micropterus dolomieu*, and spotted bass, *M. punctulatus*, in the Missouri River system, Missouri. Copeia 1994:204–209.

Kottcamp, G. 1973. Variation, behavior and ecology of the prickly sculpin (*Cottus asper* Richardson) from the San Joaquin River, California. M.A. thesis, Calif. State Univ., Fresno. 56 pp.

Kottcamp, G., and P. B. Moyle. 1972. Use of disposable beverage cans by fish in the San Joaquin Valley. Trans. Am. Fish. Soc. 101:566.

Kramer, R. H., and L. L. Smith. 1960. Utilization of the nests of largemouth bass, *Micropterus salmoides*, by golden shiner, *Notemigonus crysoleucas*. Copeia 1960:73–74.

Kramer, V. L., R. Garcia, and A. E. Colwell. 1987. An evaluation of the mosquitofish, *Gambusia affinis*, and the inland silverside, *Menidia beryllina*, as mosquito control agents in California wild rice fields. J. Am. Mosq. Contr. Assoc. 3:626–632.

Kratz, K., and G. L. Vinyard. 1981. Mechanisms of prey selectivity in the Piute sculpin, *Cottus beldingi*. Calif.-Nev. Wildl. Trans. 1981:11–18.

Kresja, R. J. 1965. The systematics of the prickly sculpin, *Cottus asper*: an investigation of genetic and non-genetic variation within a polytypic species. Ph.D. dissertation, Univ. British Columbia, Vancouver. 109 pp.

———. 1970. The systematics of the prickly sculpin, *Cottus asper*: an investigation of genetic and non-genetic variation within a polytypic species. Part I. Synonymy, nomenclatural history and distribution. Pac. Sci. 21:241–251.

Kroeber, A. L., and S. A. Barrett. 1960. Fishing among the Indians of Northwestern California. Anthropol. Rec. Univ. Calif. 21. 270 pp.

Krumholz, L. A. 1948. Reproduction in the western mosquitofish *Gambusia affinis affinis* (Baird and Girard) and its use in mosquito control. Ecol. Monogr. 18:1–43.

Kubicek, P. F., and D. G. Price. 1976. An evaluation of water temperature and its effect on juvenile steelhead trout in geothermally active areas of Big Sulphur Creek. Calif.-Nev. Wildl. Trans. 1976:1–24.

Kukowski, G. E. 1972. A checklist of the fishes of the Monterey Bay area including Elkhorn Slough, the San Lorenzo, Pajaro, and Salinas rivers. Moss Landing Mar. Lab. Tech. Publ. 72-2. 69 pp.

LaBue, C. P., and M. A. Bell. 1993. Phenotypic manipulation by the cestode parasite *Schistocephalus solidus* of its intermediate host, *Gasterosteus aculeatus*, the threespine stickleback. Am. Nat. 142:725–735.

LaBounty, J. F., and J. E. Deacon. 1972. *Cyprinodon milleri*, a new species of pupfish (family Cyprinodontidae) from Death Valley, California. Copeia 1972:769–780.

Lafferty, K. D., and C. J. Page. 1997. Predation on the endangered tidewater goby by the introduced African clawed frog, with notes on the frog's parasites. Copeia 1997:589–592.

Lafferty, K, D., C. C. Swift, and R. F. Ambrose. 1999a. Extirpation and recolonization in a metapopulation of an endangered fish, the tidewater goby. Cons. Biol. 13:1447–1453.

———. 1999b. Postflood persistence and recolonization of endangered tidewater goby populations. N. Am. J. Fish. Mgmt. 19:618–622.

Lambert, T. R. 1980. Status of redeye bass, *Micropterus coosae*, in the South Fork Stanislaus River, California. Calif. Fish Game 66:240–242.

Lambert, T. R., and J. M. Handley 1984. Setting an instream flow requirement for smallmouth bass (*Micropterus dolomieui*). Pages 387–395 in A. Lillehammer and S. J. Saltveit, eds. Regulated rivers. Oslo: Universitetsforlaget As.

Lambou, V. W. 1965. Observations on the size distribution and spawning behavior of threadfin shad. Trans. Am. Fish. Soc. 94:385–386.

Lanse, R. 1965. The occurrence of warmouth, *Chaenobryttus gulosus* (Cuvier), in the lower Colorado River. Calif. Fish Game 51:123.

Lappalainen, A., A. Shurukhin, G. Alekseev, and J. Rinne. 2000. Coastal fish communities along the northern coast of the Gulf of Finland, Baltic Sea: responses to salinity and eutrophication. Int. Rev. Hydrobiol. 85:687–696.

Larimore, R. W. 1957. Ecological life history of the warmouth (Centrarchidae). Bull. Ill. Nat. Hist. Surv. 27. 83 pp.

La Rivers, I. 1962. Fish and fisheries of Nevada. Reno: Nevada State Fish and Game Commission.

Larson, J., J. McKeon, T. Salamunovich, and T. D. Hofstra. 1983. Water quality and productivity of the Redwood Creek estuary. Pages 190–199 in C. Van Riper, L. D. Whittig, and M. L. Murphy, eds. Proc. 1st Biennial Conf. Res. Calif. Natl. Parks. Davis, Calif.: National Park Service.

Lau, S., and C. Boehm. 1991. A distribution survey of desert pupfish (*Cyprinodon macularius*) around the Salton Sea, California. Unpubl. rpt., Proj. EF90XIII-1. CDFG, Sacramento. 21 pp.

Lauder, G. V. 1983. Functional and morphological bases of trophic specialization in sunfishes (Teleostei, Centrarchidae). J. Morph. 178:1–21.

Lea, R. N., C. C. Swift, and R. J. Lavenberg. 1988. Records of *Mugil curema*, the white mullet, from southern California. Bull. S. Calif. Acad. Sci 87:31–34.

Lee, D. S., C. R. Gilbert, C. H. Hocutt, R. E. Jenkins, D. E. McAllister, and J. R. Stauffer, Jr. 1980. Atlas of North American freshwater fishes. Raleigh: North Carolina Museum of Natural History.

Lee, R. M., and S. D. Gerking. 1980. Survival and reproductive performance of the desert pupfish, *Cyprinodon n. nevadensis* (Eigenmann and Eigenmann), in acid waters. J. Fish Biol. 17:507–515.

Leet, W. S., C. M. Dewees, and C. W. Haugen. 1992. California's living marine resources and their utilization. Calif. Sea Grant Publ. UCSGEP-92-12. 257 pp.

Leggett, W. C. 1973. The migrations of the shad. Sci. Am. 228:92–100.

Leggett, W. C., and J. E. Carscadden. 1978. Latitudinal variation in reproductive characteristics of American shad (*Alosa sapidissima*): evidence for population specific life history strategies in fish. J. Fish. Res. Board Can. 35:1469–1478.

Legner, E. F., and R. A. Medved. 1973. Predation of mosquitoes and chironomid midges in ponds by *Tilapia zillii* and *T. mossambica* (Teleostei: Cichlidae). Proc. Annu. Conf. Calif. Mosq. Contr. Assoc. 41:119–121.

Legner, E. F., T. W. Fisher, and R. A. Medved. 1973. Biological control of aquatic weeds in the lower Colorado River basin. Proc. Annu. Conf. Calif. Mosq. Contr. Assoc. 41:115–117.

Legner, E. F., R. A. Medved, and F. Pelsue. 1980. Changes in chironomid breeding patterns in a paved river channel following adaptation of cichlids of the *Tilapia mossambica-hornorum* complex. Ann. Ent. Soc. Am. 73:293–299.

Leidy, R. A. 1983. Distribution of fishes in streams of the Walnut Creek basin, California. Calif. Fish Game 69:23–32.

———. 1984. Distribution and ecology of stream fishes in the San Francisco Bay drainage. Hilgardia 52:1–175.

Leidy, R. A., and P. L. Fiedler. 1985. Human disturbance and patterns of fish species diversity in the San Francisco Bay drainage, California. Biol. Cons. 33:247–267.

Leipzig, P., and J. M. Deinstadt. 1997. Food habits of brown trout in the East Walker River, California. CDFG Inland Fish. Admin. Rpt. 97-11. 27 pp.

Lemke, M. J., and S. H. Bowen. 1998. The nutritional value of organic detrital aggregate in the diet of fathead minnows. Freshw. Biol. 39:447–453.

Leslie, A. J., J. R. Cassani, and R. J. Wattendorf. 1996. An introduction to grass carp biology and management in the United States. Pages 1–13 in J. Cassani, ed. Managing aquatic vegetation with grass carp: a guide for resource managers. Bethesda, Md.: American Fisheries Society.

Lever, C. 1996. Naturalized fishes of the world. San Diego: Academic Press.

Levesque, R. C., and R. J. Reed. 1972. Food availability and consumption by young Connecticut River shad, *Alosa sapidissima*. J. Fish. Res. Board Can. 29:1495–1499.

Lewis, W. M., and S. Flickinger. 1967. Home range tendency of the largemouth bass (*Micropterus salmoides*). Ecology 48:1020–1023.

Lewis, W. M., G. E. Gunning, E. Lyles, and W. L. Bridges. 1961. Food choice of largemouth bass as a function of availability and vulnerability of food items. Trans. Am. Fish. Soc. 90:277–280.

Li, H. W., and P. B. Moyle. 1976. Feeding ecology of the Pit sculpin, *Cottus pitensis*, in Ash Creek, California. Bull. S. Calif. Acad. Sci. 75:111–118.

Li, H. W., P. B. Moyle, and R. L. Garrett. 1976. Effects of the introduction of the Mississippi silverside (*Menidia audens*) on the growth of black crappie (*Pomoxis nigromaculatus*) and white crappie (*P. annularis*) in Clear Lake, California. Trans. Am. Fish. Soc. 105:404–408.

Limburg, K. E., and R. M. Ross. 1995. Growth and mortality rates of larval American shad, *Alosa sapidissima*, at different salinities. Estuaries 18:335–340.

Lindberg, J., B. Baskerville-Bridges, J. Van Eenennaam, and S. Doroshov. 1999. Development of delta smelt culture techniques: year-end report 1999. Unpubl. rpt., Univ. Calif., Davis. 27 pp.

Lindquist, A. W., C. D. Deonier, and J. E. Hanley. 1943. The relationship to fish of the Clear Lake gnat, Clear Lake, California. Calif. Fish Game 29:196–202.

Lindstrom, S. 1996. Great Basin fisherfolk: optimal diet breadth modeling the Truckee River aboriginal subsistence fishery. Pages 114–179 in M. G. Plew, ed. Prehistoric hunter-gatherer fishing strategies. Boise, Id.: Boise State University.

Liu, R. K. 1969. The comparative behavior of allopatric species (Teleostei-Cyprinodontidae: *Cyprinodon*). Ph.D. dissertation, Univ. Calif., Los Angeles. 185 pp.

Lobon-Cervia, J., C. G. Utrilla, P. A. Rincon, and F. Amezcua. 1997. Environmentally induced spatio-temporal variations in the fecundity of brown trout *Salmo trutta* L.: tradeoffs between egg size and number. Freshw. Biol. 38:277–288.

Lockington, W. N. 1878. Report upon the food fishes of San Francisco. Calif. Fish. Comm. Rpt. 1878-78:17–58.

Loeb, H. A. 1964. Submergence of brown bullheads in bottom sediment. N.Y. Fish Game J. 11:119–124.

Loggins, R. E. 1997. Comparison of four populations of California roach, *Lavinia symmetricus*, using RAPD markers. M.S. thesis, Calif. State Univ., Chico. 58 pp.

Lohr, S. C., and K. D. Fausch. 1996. Effects of green sunfish (*Lepomis cyanellus*) predation on survival and habitat use of plains killifish (*Fundulus zebrinus*). SW Nat. 41:155–160.

Loiselle, P. V. 1977. Colonial breeding by an African substratum-spawning cichlid fish, *Tilapia zilli* (Gervais). Biol. Behav. 2:129–142.

———. 1982. Male spawning-partner preference in an arena-breeding teleost *Cyprinodon macularius californiensis* Girard (Atherinomorpha: Cyprinodontidae). Am. Nat. 120:721–732.

Lollock, D. L. 1968. An evaluation of the fishery resources of the Pajaro River Basin. Unpubl. rpt., California Department of Fish and Game, Monterey. 60 pp.

Long, C. W. 1968. Diel movement and vertical distribution of juvenile anadromous fish in turbine intakes. Fishery Bull. USFWS 66:599–609.

Lonzarich, D. G., and J. J. Smith. 1997. Water chemistry and community structure of saline and hypersaline salt evaporation ponds in San Francisco Bay, California. Calif. Fish Game 83:89–104.

Lorion, C. M., D. F. Markle, S. B. Reid, and M. F. Docker. 2000. Redescription of the presumed-extinct Miller Lake lamprey, *Lampretra minima.* Copeia 2000:1019–1028.

Lorz, H. W., and T. G. Northcote. 1965. Factors affecting stream location, and timing and intensity of entry by spawning kokanee (*Oncorhynchus nerka*) into an inlet of Nicola Lake, British Columbia. J. Fish. Res. Board Can. 22:665–687.

Loshbaugh. D. F. 1978. Prey selection by the black crappie (Centrarchidae: *Pomoxis nigromaculatus*). M.S. thesis, Univ. Calif., Santa Barbara. 70 pp.

Lott, J. 1998. Feeding habits of juvenile and adult delta smelt from the Sacramento–San Joaquin estuary. IEP Newslett. 11(1):14–19.

Loud, L. L. 1929. Notes on the northern Piute. Univ. Calif. Publ. Arch. Ethnol. 25:152–164.

Lougheed, V. L., B. Crosbie, and P. Chow-Fraser. 1997. Predictions on the effect of common carp (*Cyprinus carpio*) exclusion on water quality, zooplankton, and submergent macrophytes in a Great Lakes wetland. Can. J. Fish. Aquat. Sci. 55:1189–1197.

Love, M. 1996. Probably more than you want to know about the fishes of the Pacific Coast. 2nd edition. Santa Barbara, Calif.: Really Big Books.

Lowe, C. H., and W. G. Heath. 1969. Behavioral and physiological responses to temperature in the desert pupfish, *Cyprinodon macularius.* Physiol. Zool. 42:53–59.

Lowe, C. H., D. S. Hinds, and E. A. Halpern. 1967. Experimental catastrophic selection and tolerances to low oxygen concentrations in native Arizona freshwater fishes. Ecology 48:1013–1017.

Ludwig, H. R., Jr., and J. A. Leitch. 1996. Interbasin transfer of aquatic biota via anglers' bait buckets. Fisheries 21(7):14–18.

Lukas, J. A., and D. J. Orth. 1995. Factors affecting nesting success of smallmouth bass in a regulated Virginia streams. Trans. Am. Fish. Soc. 124:726–735.

Lundberg, J. G. 1982. The comparative anatomy of the toothless blindcat, *Trogloglanis pattersoni* Eigenmann, with a phylogenetic analysis of ictalurid catfishes. Misc. Publ. Mus. Zool. Univ. Mich. 163:1–85.

Mabee, P. M. 1993. Phylogenetic interpretation of ontogenetic change: sorting out the actual and artefactual in an empirical case study of centrarchid fishes. Zool. J. Linn. Soc. 107:175–291.

MacCrimmon, H. R., and J. S. Campbell. 1969. World distribution of brook trout, *Salvelinus fontinalis.* J. Fish. Res. Board Can. 26:1699–1725.

MacCrimmon, H. R., T. C. Marshall, and B. L. Gots. 1970. World distribution of brown trout, *Salmo trutta:* further observations. J. Fish. Res. Board Can. 27:811–818.

Macedo, R. 1994. Swimming upstream without a hitch. Outdoor California 55(1):1–5.

MacKenzie, C., L. S. Weiss-Glanz, and J. R. Moring. 1985. Species profiles: life histories and environmental requirements of coastal fishes and invertebrates (mid-Atlantic). American shad. USFWS Biol. Rpt. 82(11.37). 18 pp.

Magee, A., C. A. Myrick, and J. J. Cech, Jr. 1999. Thermal preference of female threespine sticklebacks under fed and food-deprived conditions. Calif. Fish Game 85:102–112.

Mager, R. C. 1996. Gametogenesis, reproduction, and artificial propagation of delta smelt, *Hypomesus transpacificus.* Ph.D. dissertation, Univ. Calif., Davis. 125 pp.

Malcolm, J. R. 1992. Supporting information for a petition to list as endangered or threatened: Shay Creek stickleback, *Gasterosteus* sp. Pages 213–222 *in* P. B. Moyle and R. M. Yoshiyama, eds. Fishes, aquatic diversity management areas, and endangered species: a plan to protect California's native aquatic biota. Berkeley: California Policy Seminar, University of California.

Marchetti, M. P. 1998. Ecological effects of non-native fish species in low elevation streams of the Central Valley, California. Ph.D. dissertation, Univ. Calif., Davis. 84 pp.

———. 1999. An experimental study of competition between native Sacramento perch (*Archoplites interruptus*) and introduced bluegill (*Lepomis macrochirus*). Invasion Biol. 1:55–65.

Marchetti, M. P., and P. B. Moyle. 2000. Spatial and temporal ecology of native and introduced fish larvae in lower Putah Creek, California. Env. Biol. Fish. 58:75–87.

———. 2001. Effects of flow regime and habitat structure on fish assemblages in a regulated California stream. Ecol. App. 11:530–539.

Marine, K. R. 1997. Effects of elevated water temperature on some aspects of the physiological and ecological performance of juvenile chinook salmon (*Oncorhynchus tshawytscha*). M.S. thesis, Univ. Calif., Davis. 63 pp.

Markham, J. L., D. L. Johnson, and R. W. Petering. 1991. White crappie summer movements and habitat use in Delaware Reservoir, Ohio. N. Am. J. Fish. Mgmt. 11:504–512.

Markle, D. F. 1982. Evidence of bull trout x brook trout hybrids in Oregon. Pages 58–67 *in* P. J. Howell and D.V. Buchanan, eds. Proceedings of the Gearhart Mountain bull trout workshop. Corvallis: Oregon Chapter of the American Fisheries Society.

Markley, M. H. 1940. Notes on the food habits and parasites of the stickleback *Gasterosteus aculeatus* (Linnaeus) in the Sacramento River, California. Copeia 1940:223–225.

Marrin, D. L. 1980. Food selectivity and habitat use by introduced trouts and native nongame fishes in sub-alpine lakes. M.S. thesis, Univ. Calif., Berkeley. 87 pp.

———. 1983. Ontogenetic changes and intraspecific resource partitioning in the Tahoe sucker, *Catostomus tahoensis.* Env. Biol. Fish. 8:39–47.

Marrin, D. L., and D. C. Erman. 1982. Evidence against competition between trout and nongame fishes in Stampede Reservoir, California. N. Am. J. Fish. Mgmt. 2:262–269.

Marrin, D. L., D. C. Erman, and B. Vondracek. 1984. Food availability, food habits, and growth of Tahoe sucker, *Catostomus tahoensis,* from a reservoir and a natural lake. Calif. Fish Game 70:4–10.

Marsh, P. C. 1985. Effect of incubation temperature on survival of embryos of native Colorado River fishes. SW Nat. 30: 129–140.

———. 1987. Digestive tract contents of adult razorback suckers in Lake Mohave, Arizona-Nevada. Trans. Am. Fish. Soc. 116:117–118.

Marsh, P. C., and J. E. Brooks. 1989. Predation by ictalurid catfishes as a deterrent to re-establishment of hatchery-reared razorback suckers. SW Nat. 34:188–195.

Marsh, P. C., and D. R. Langhorst. 1988. Feeding and fate of wild larval razorback sucker. Env. Biol. Fish. 21:59–67.

Marsh, P. C., and W. L. Minckley. 1989. Observations on recruitment and ecology of razorback sucker, lower Colorado River, Arizona-California-Nevada. Great Basin Nat. 49:71–78.

Martel, G. 1996. Growth rate and influence of predation risk on territoriality in juvenile coho salmon (*Oncorhynchus kisutch*). Can. J. Fish. Aquat. Sci. 53:660–669.

Martin, B. A., and M. K. Saiki. 1999. Effects of ambient water quality on the endangered Lost River sucker in Upper Klamath Lake, Oregon. Trans. Am. Fish. Soc. 128:953–961.

———. 2001. Gut contents of juvenile Chinook salmon from the upper Sacramento River, California, during spring 1998. Calif. Fish Game 87:38–43.

Martin, J. A. 1995. Food habits of some estuarine fishes in a small seasonal central California lagoon. M.S. thesis, Calif. State Univ., San Jose. 48 pp.

Martin, M. 1967. The distribution and morphology of the North American catostomid fishes of the Pit River system, California. M.A. thesis, Sacramento State Coll. 60 pp.

———. 1972. Morphology and variation of the Modoc sucker, *Catostomus microps* Rutter, with notes on feeding adaptations. Calif. Fish Game 58:277–284.

Maslin, P., M. Lennox, J. Kindopp, and W. McKinney. 1997. Intermittent streams as rearing habitat for Sacramento River chinook salmon (*Oncorhynchus tshawytscha*). Unpubl. rpt., Calif. State Univ., Chico. 89 pp.

Matern, S. A. 1999. The invasion of the shimofuri goby (*Tridentiger bifasciatus*) into California: establishment, potential for spread, and likely effects. Ph.D. dissertation, Univ. Calif., Davis. 167 pp.

———. 2001. Using temperature and salinity tolerances to predict the success of the shimofuri goby, a recent invader into California. Trans. Am. Fish. Soc. 130:592–599.

Matern, S. A., and K. J. Fleming. 1995. Invasion of a third Asian goby, *Tridentiger bifasciatus,* into California. Calif. Fish Game 81:71–76.

Mathews, S. B. 1962. The ecology of the Sacramento perch, *Archoplites interruptus,* from selected areas of California and Nevada. M.A. thesis, Univ. Calif., Berkeley. 93 pp.

———. 1965. Reproductive behavior of the Sacramento perch, *Archoplites interruptus.* Copeia 1965:224–228.

Mathias, A., and R. J. F. Smith. 1992. Avoidance of areas marked with a chemical alarm substance by fathead minnows (*Pimephales promelas*) in a natural habitat. Can J. Zool. 70:1473–1476.

Mathur, D. 1972. Seasonal food habits of adult white crappie, *Pomoxis annularis* Rafinesque in Conowingo Reservoir. Am. Midl. Nat. 87:236–241.

Mathur, D., and T. W. Robbins. 1971. Food habits and feeding chronology of young white crappie, *Pomoxis annularis* Rafi-nesque in Conowingo Reservoir. Trans. Am. Fish. Soc. 100:307–311.

Matthews, K. R. 1996a. Diel movement and habitat use of California golden trout in the Golden Trout Wilderness, California. Trans. Am. Fish. Soc. 125:78–86.

———. 1996b. Habitat selection and movement patterns of California golden trout in degraded and recovering stream sections in the Golden Trout Wilderness, California. N. Am. J. Fish. Mgmt. 16:579–590.

Matthews, K. R., and N. H. Berg. 1997. Rainbow trout responses to water temperature and dissolved oxygen in two southern California stream pools. J. Fish Biol. 50:50–67.

Matthews, W. J., and L. G. Hill. 1977. Tolerance of the red shiner, *Notropis lutrensis* (Cyprinidae) to environmental parameters. SW Nat. 22:89–98.

———. 1979. Age-specific differences in the distribution of red shiners, *Notropis lutrensis,* over physiochemical ranges. Am. Midl. Nat. 102:366–372.

Mayden, R. L. 1989. Phylogenetic studies of North American minnows, with emphasis on the genus *Cyprinella* (Teleostei, Cypriniformes). Univ. Kansas Mus. Nat. Hist. Misc. Publ. 80. 189 pp.

Mayden, R. L., W. J. Rainboth, and D. G. Buth. 1991. Phylogenetic systematics of the cyprinid genera *Mylopharodon* and *Ptychocheilus* comparative morphology. Copeia 1991:819–834.

McAfee, W. B. 1966. Rainbow trout, pages 192–215; Golden trout, pages 216–224; Eagle Lake rainbow trout, pages 221–225; Lahontan cutthroat trout, pages 225–230; Piute cutthroat trout, pages 231–233; eastern brook trout, pages 242–260; lake trout, pages 260–271; Dolly Varden trout, pages 271–274; landlocked king salmon, pages 294–295; mountain whitefish, pages 299–304 *in* A. Calhoun, ed. Inland fisheries management. Sacramento: California Department of Fish and Game.

McAllister, D. E. 1963. A revision of the smelt family, Osmeridae. Bull. Natl. Mus. Can. 191. 53 pp.

McCabe, G. T., Jr., and C. A. Tracy. 1994. Spawning and early life history of white sturgeon, *Acipenser transmontanus,* in the lower Columbia River. NOAA Fish. Bull. 92:760–772.

McCabe, G. T., Jr., R. L. Emmett, and S. A. Hinton. 1993. Feeding ecology of juvenile white sturgeon (*Acipenser transmontanus*) in the lower Columbia River. NW Sci. 67:170–180.

McCall, J. N. 1992. Source of harpactacoid copepods in the diet of juvenile starry flounder. Mar. Ecol. Prog. Ser. 86:41–50.

McCammon, G. W., D. L. Faunce, and C. M. Seeley. 1964. Observations on the food of fingerling largemouth bass in Clear Lake, Lake County, California. Calif. Fish Game 50:158–169.

McCammon, G. W., and C. M. Seeley. 1961. Survival, mortality, and movements of white catfish and brown bullheads in Clear Lake, California. Calif. Fish Game 47:237–255.

McCarraher, D. B. 1972. Survival of some freshwater fishes in the alkaline eutrophic waters of Nebraska. J. Fish. Res. Board Can. 28:1811–1814.

McCarraher, D. B., and R. W. Gregory. 1970. Adaptability and status of introductions of Sacramento perch, *Archoplites interruptus,* in North America. Trans. Am. Fish. Soc. 99:700–707.

McCarthy, M. S., and W. L. Minckley. 1987. Age estimation from razorback sucker (Pisces: Catostomidae) from Lake Mohave, Arizona and Nevada. J. Ariz. Nev. Acad. Sci. 21:87–97.

McClanahan, L. L., C. R. Feldmeth, J. Jones, and D. L. Soltz. 1986.

Energetics, salinity, and temperature tolerance in the Mohave tui chub, *Gila bicolor mohavensis.* Copeia 1986:45–52.

McCoid, M. J., and J. A. St. Amant. 1980. Notes on the establishment of the rainwater killifish, *Lucania parva,* in California. Calif. Fish Game 66:124–125.

McCosker, J. E. 1989. Freshwater eels (Anguillidae) in California: current conditions and future scenarios. Calif. Fish Game 75:4–10.

McCrimmon, H. R. 1968. Carp in Canada. Fish. Res. Board Can. Bull. 165. 93 pp.

McCullough, D. A. 1999. A review and synthesis of effects of alterations to the water temperature regime on freshwater life stages of salmonids, with special reference to chinook salmon. USEPA Rpt. 910–R-00–010, Seattle, Wash. 279 pp.

McDonaugh, T. A., and J. P. Buchanan. 1991. Factors affecting abundance of white crappies in Chickamauga Reservoir, Tennessee, 1970–1989. N. Am. J. Fish. Mgmt. 11:513–524.

McDowall, R. M., B. M. Clark, G. J. Wright, and T. G. Northcote. 1993. *Trans-2-cis-6-*nonadienal: the cause of cucumber odor in osmerid and retropinnid smelts. Trans. Am. Fish. Soc. 122:144–147.

McEnroe, M., and J. J. Cech, Jr. 1987. Osmoregulation in white sturgeon: life history aspects. Am. Fish. Soc. Symp. 1:191–196.

McEwan, D. 1988. Microhabitat selection and some aspects of life history of the Owens tui chub (*Gila bicolor snyderi*) in the Hot Creek headsprings, Mono County, California. CDFG Contract Rpt. C-1467. 58 pp.

———. 2001. Central Valley steelhead. Pages 1–44 *in* R. L. Brown, ed. Contributions to the biology of Central Valley salmonids. CDFG Fish Bull. 179.

McIntyre, J. D. 1969. Spawning behavior of the brook lamprey, *Lampetra planeri.* J. Fish. Res. Board Can. 26:3252–3254.

McKechnie, R. J. 1966. Spotted bass, pages 366–370; golden shiner, pages 488–492; logperch, pages 530–531 *in* A. Calhoun, ed. Inland fisheries management. Sacramento: California Department of Fish and Game.

McKechnie, R. J., and R. B. Fenner, 1971. Food habits of white sturgeon, *Acipenser transmontanus,* in San Pablo and Suisun bays, California. Calif. Fish Game 57:209–212.

McKechnie, R. J., and R. C. Tharratt. 1966. Green sunfish, 399–402 *in* A. Calhoun, ed. Inland fisheries management. Sacramento: California Department of Fish and Game.

McLarney, W. O. 1968. Spawning habits and morphological variation in the coast range sculpin, *Cottus aleuticus,* and the prickly sculpin, *Cottus asper.* Trans. Am. Fish. Soc. 97:46–48.

McLaughlin, R. L., J. W. A. Grant, and D. L. Kramer. 1994. Foraging movements in relation to morphology, water-column use, and diet for recently emerged brook trout (*Salvelinus fontinalis*) in still-water pools. Can. J. Fish. Aquat. Sci 51:268–279.

McMahon, T. E., and S. H. Holanov. 1995. Foraging success of largemouth bass at different light intensities: implications for time and depth of feeding. J. Fish Biol. 46:759–767.

McMahon, T. E., and R. R. Miller. 1985. Status of the fishes of the Rio Sonoyta basin, Arizona and Sonora, Mexico. Proc. Desert Fishes Counc. 4(1982):53–59.

McMahon, T. E., and J. C. Tash. 1988. Experimental analysis of the roles of emigration in population regulation of desert pupfish. Ecology 69:1871–1883.

McPeek, M. A. 1992. Mechanisms of sexual selection operating on body size in the mosquitofish (*Gambusia holbrooki*). Behav. Ecol. 3:1–12.

McPhail, J. O., and C. C. Lindsey. 1970. Freshwater fishes of northwestern Canada and Alaska. Fish. Res. Board Can. Bull. 173. 381 pp.

McPhee, J. 1993. Water war. New Yorker, 26 April:120.

Mearns, A. J. 1975. *Poeciliopsis gracilis* (Heckel), newly introduced poeciliid fish in California. Calif. Fish Game 61:251–253.

Meffe, G. K., and M. L. Crump. 1987. Possible growth and reproductive benefits of cannibalism in the mosquitofish. Am. Nat. 129:203–212.

Meffe, G. K., and F. F. Snelson, Jr., eds. 1989. Ecology and evolution of livebearing fishes (Poeciliidae). Englewood Cliffs, N.J.: Prentice-Hall.

Meng, L. 1993a. Estimating food requirements of striped bass larvae: an energetics approach. Trans. Am. Fish. Soc. 122:244–251.

———. 1993b. Sustainable swimming speeds of striped bass larvae. Trans. Am. Fish. Soc. 122:702–708.

Meng, L., and S. A. Matern. 2001. Native and introduced larval fishes of Suisun Marsh, California: the effects of freshwater flow. Trans. Am. Fish. Soc. 130:750–765.

Meng, L., and P. B. Moyle. 1995. Status of splittail in the Sacramento–San Joaquin Estuary. Trans. Am. Fish. Soc. 124:538–549.

Meng, L., and J. J. Orsi. 1991. Selective predation by larval striped bass on native and introduced copepods. Trans. Am. Fish. Soc. 120:187–192.

Meng, L., P. B. Moyle, and B. Herbold. 1994. Changes in abundance and distribution of native and introduced fishes of Suisun Marsh. Trans. Am. Fish. Soc. 123:498–507.

Mense, J. B. 1967. Ecology of the Mississippi silverside, *Menidia audens* Hay, in Lake Texoma. Bull. Okla. Fish. Res. Lab. 6:1–32.

Merz, J. E. 1998. Juvenile chinook salmon feeding habits in the lower Mokelumne River, California. Unpubl. rpt., E. Bay Muni. Utility Dist., Lodi, Calif. 13 pp.

Merz, J. E., and C. D. Vanicek. 1996. Comparative feeding habits of juvenile chinook salmon, steelhead, and Sacramento squawfish in the lower American River, California. Calif. Fish Game 82:149–159.

Mesick, C. F. 1995. Response of brown trout to streamflow, temperature, and habitat restoration in a degraded stream. Rivers 5:75–95.

Messersmith, J. D. 1965. Southern range extension for chum and silver salmon. Calif. Fish Game 51:220.

Michael, J. H., Jr. 1983. Contribution of cutthroat trout in headwater streams to the sea-run populations. Calif. Fish Game 69:68–76.

———. 1984. Additional notes on the repeat spawning by Pacific lamprey. Calif. Fish Game 70:186–188.

Middaugh, D. P., and J. M. Shenker. 1988. Salinity tolerance of young topsmelt, *Atherinops affinis,* cultured in the laboratory. Calif. Fish Game 74:232–235.

Middaugh, D. P., M. J. Hemmer, and Y. Lamadrid-Rose. 1986. Laboratory spawning cues in *Menidia beryllina* and *M. peninsulae* (Pisces: Atherinidae) with notes on survival and growth of larvae at different salinities. Env. Biol. Fish. 15:107–117.

Middleton, M. J. 1982. The oriental goby, *Acanthogobius flavimanus* (Temmick and Schlegel), an introduced fish in the coastal waters of New South Wales, Australia. J. Fish Biol. 21:513–523.

Miller, D. J., and R. N. Lea. 1972. Guide to the coastal marine fishes of California. CDFG Fish Bull. 157:1–235.

Miller, D. L., P. M. Leonard, R. M. Hughes, J. R. Karr, P. B. Moyle, L. H. Schrader, B. A. Thompson, R. A. Daniels, K. D. Fausch, G. A. Fitzhugh, J. R. Gammon, D. B. Halliwell, P. L. Angermeier, and D. J. Orth. 1988. Regional applications of an index of biotic integrity for use in water resource management. Fisheries (Bethesda) 13:12–20.

Miller, E. E. 1966. White catfish, pages 430–440; channel catfish, pages 440–463; black bullhead, pages 476–479; yellow bullhead, pages 479–480 *in* A. Calhoun, ed. Inland fisheries management. Sacramento: California Department of Fish and Game.

———. 1970. The age and growth of centrarchid fishes in Millerton and Pine Flat reservoirs, California. CDFG Inland Fish. Admin. Rpt. 71-4. 17 pp.

Miller, K. D., and R. H. Kramer. 1971. Spawning and early life history of largemouth bass (*Micropterus salmoides*) in Lake Powell. Pages 73–83 *in* G. E. Hall, ed. Reservoir fisheries and limnology. Am. Fish. Soc. Spec. Publ. 8.

Miller, L. M., L. Kallemeyn, and W. Senanan. 2001. Spawning-site and natal-site fidelity by northern pike in a large lake: mark-recapture and genetic evidence. Trans. Am. Fish. Soc. 130:307–316.

Miller, L. W. 1972a. White sturgeon population characteristics in the Sacramento–San Joaquin estuary as measured by tagging. Calif. Fish Game 58:94–101.

———. 1972b. Migrations of sturgeon tagged in the Sacramento–San Joaquin estuary. Calif. Fish Game 58:102–106.

Miller, L. W., and R. J. McKechnie. 1968. Observation of striped bass spawning in the Sacramento River. Calif. Fish Game 54:306–307.

Miller, R. B. 1957. Permanence and size of home territory in stream-dwelling cutthroat trout. J. Fish. Res. Board Can. 14:687–691.

Miller, R. G. 1951. The natural history of Lake Tahoe fishes. Ph.D. dissertation, Stanford Univ., Stanford, Calif. 160 pp.

Miller, R. R. 1939. Occurrence of the cyprinodont fish *Fundulus parvipinnis* in freshwater in San Juan Creek, southern California. Copeia 1939:168.

———. 1943a. Further data on freshwater populations of the Pacific killifish, *Fundulus parvipinnis.* Copeia 1943:51–52.

———. 1943b. *Cyprinodon salinus,* a new species of fish from Death Valley, California. Copeia 1943:69–78.

———. 1943c. The status of *Cyprinodon macularius* and *Cyprinodon nevadensis,* two desert fishes of western North America. Univ. Mich. Mus. Zool. Occas. Pap. 473. 25 pp.

———. 1945a. A new cyprinid fish from Southern Arizona, and Sonora, Mexico, with description of a new subgenus of *Gila* and a review of related species. Copeia 1945:104–110.

———. 1945b. The status of *Lavinia ardesiaca,* a cyprinid fish from the Pajaro-Salinas Basin, California. Copeia 1945: 197–204.

———. 1946. *Gila cypha,* a remarkable new species of fish from the Colorado River in Grand Canyon, Arizona. Wash. Acad. Sci. J. 36:403–415.

———. 1948. The cyprinodont fishes of the Death Valley System of eastern California and southwestern Nevada. Univ. Mich. Mus. Zool. Misc. Publ. 68. 155 pp.

———. 1952. Bait fishes of the lower Colorado River from Lake Mead, Nevada, to Yuma, Arizona with a key for their identification. Calif. Fish Game 38:7–42.

———. 1959. Origin and affinities of the freshwater fish fauna of western North America. Pages 187–222 *in* C. L. Hubbs, ed. Zoogeography. Washington, D.C.: AAAS.

———. 1961. Speciation rates in some freshwater fishes of western North America. Pages 537–560 *in* F. Blair, ed. Vertebrate Speciation. Austin: University of Texas Press.

———. 1963. Synonymy, characters, and variation of *Gila crassicauda,* a rare Californian minnow, with an account of its hybridization with *Lavinia exilicauda.* Calif. Fish Game 49: 20–29.

———. 1965. Quarternary freshwater fishes in western North America. Pages 569–581 *in* H. E. Wright and D. G. Frey, eds. The Quaternary of the United States. Princeton, N.J.: Princeton University Press.

———. 1968. Records of some native freshwater fishes transplanted into various waters of California, Baja California, and Nevada. Calif. Fish Game 54:170–179.

———. 1973. Two new fishes, *Gila bicolor snyderi* and *Catostomus fumeiventris,* from the Owens River Basin, California. Mus. Zool. Univ. Mich. Occas. Pap. 667. 19 pp.

———. 1981. Coevolution of deserts and pupfishes (Genus *Cyprinodon*) in the American southwest. Pages 39–94 *in* R. J. Naiman and D. L. Soltz, eds. Fishes in North American deserts. New York: John Wiley.

Miller, R. R., and L. A. Fuiman. 1987. Description and status of *Cyprinodon macularius eremus,* a new subspecies of pupfish from Organ Pipe National Monument, Arizona. Copeia 1987:593–609.

Miller, R. R., and C. L. Hubbs. 1969. Systematics of *Gasterosteus aculeatus* with particular reference to intergradation and introgression along the Pacific Coast of North America: a commentary on a recent contribution. Copeia 1969:52–69.

Miller, R. R., and E. P. Pister. 1971. Management of the Owens pupfish, *Cyprinodon radiosus,* in Mono County, California. Trans. Am. Fish. Soc. 100:502–509.

Miller, R. R., and G. R. Smith. 1967. New fossil fishes from Plio-Pleistocene Lake Idaho. Mus. Zool. Univ. Mich. Occas. Pap. 654. 24 pp.

———. 1981. Distribution and evolution of *Chasmistes* (Pisces: Catostomidae) in western North America. Mus. Zool. Univ. Mich. Occas. Pap. 696. 46 pp.

Miller, R. R., C. Hubbs, and F. H. Miller. 1991. Ichthyological exploration of the American West: The Hubbs-Miller era, 1915–1950. Pages 19–42 *in* W. L. Minckley and J. E. Deacon, eds. Battle against extinction: native fish management in the American West. Tucson: University of Arizona Press.

Millikan, A. E. 1968. The life history and ecology of *Cottus asper* Richardson and *Cottus gulosus* (Girard) in Conner Creek, Washington. M.S. thesis, Univ. Wash., Seattle. 81 pp.

Mills, T. J. 1979. High Rock Spring, Lassen County, with notes on the species comprising *Gila bicolor,* in California. Unpubl. rpt., California Department of Fish and Game.

———. 1980. Life history, status, and management of the Modoc sucker, *Catostomus microps* (Rutter), in California, with a recommendation for endangered classification. CDFG Inland Fish. End. Sp. Prog. Spec. Publ. 80-6. 35 pp.

———. 1983. Utilization of the Eel River tributary streams by anadromous salmonids. Appendix H *in* F. L. Reynolds, ed. 1983 Status Rpt. Calif. Wild and Scenic Rivers: Salmon and Steelhead Fisheries. California Department of Fish and Game. 57 pp.

Mills, T. J., and K. A. Mamika. 1980. The thicktail chub, *Gila crassicauda,* an extinct California fish. CDFG Inland Fisheries End. Sp. Prog. Spec. Publ. 80-2. 20 pp.

Mills, T. J., D. McEwan, and M. R. Jennings. 1997. California salmon and steelhead: beyond the crossroads. Pages 91–111 *in* D. J. Strouder, P. A. Bisson, R. J. Naiman, eds. Pacific salmon and their ecosystems. New York: Chapman and Hall.

Minckley, W. L. 1959. Fishes of the Blue River Basin, Kansas. Univ. Kansas Mus. Nat. Hist. Publ. 11:401–402.

———. 1973. Fishes of Arizona. Tucson: Ariz. Dept. Fish Game.

———. 1982. Trophic interrelations among introduced fishes in the lower Colorado River, southwestern United States. Calif. Fish Game 68:78–89.

———. 1991a. Native fishes of the Grand Canyon: an obituary? Pages 124–177 *in* Colorado River ecology and dam management. Washington, D.C.: National Academy Press.

———. 1991b. Native fishes of arid lands: a dwindling resource of the American Southwest. USDA Forest Serv. Gen. Tech. Rpt. RM-206. 45 pp.

———. 1999. Ecological review and management recommendations for recovery of the endangered Gila topminnow. Great Basin Nat. 59:230–244.

Minckley, W. L., and J. E. Deacon. 1959. Biology of flathead catfish in Kansas. Trans. Am. Fish. Soc. 88:344–355.

———. 1968. Southwestern fishes and the enigma of "endangered species." Science 159:1424–1431.

Minckley, W. L., and B. D. DeMarais. 2000. Taxonomy of chubs (Teleostei, Cyprinidae, genus *Gila*) in the American Southwest, with comments on conservation. Copeia 2000:251–256.

Minckley, W. L., D. G. Buth, and R. L. Mayden. 1989. Origin of brood stock and allozyme variation in hatchery-reared bonytail, an endangered North American cyprinid fish. Trans. Am. Fish. Soc. 118:131–137.

Minckley, W. L., D. A. Hendrickson, and C. E. Bond. 1986. Geography of western North American freshwater fishes: description and relationships to intercontinental tectonism. Pages 519–614 *in* C. H. Hocutt and E. O. Wiley, eds. The zoogeography of North American freshwater fishes. New York: John Wiley.

Minckley, W. L., P. C. Marsh, J. E. Brooks, J. E. Johnson, and B. L. Jensen. 1991a. Management towards recovery of the razorback sucker. Pages 303–358 *in* W. L. Minckley and J. E. Deacon, eds. Battle against extinction: native fish management in the American West. Tucson: University of Arizona Press.

Minckley, W. L., G. K. Meffe, and D. L. Soltz 1991b. Conservation and management of short-lived fishes: the cyprinodontids. Pages 247–282 *in* W. L. Minckley and J. E. Deacon, eds. Battle against extinction: native fish management in the American West. Tucson: University of Arizona Press.

Minshall, G. W., S. E. Jensen, and W. S. Platts. 1989. The ecology of stream and riparian habitats of the Great Basin Region: a community profile. USFWS Biol. Rpt. 85(7.24). 142 pp.

Mire, J. B. 1993. Behavioral ecology and conservation biology of the Owens pupfish, *Cyprinodon radiosus.* Ph.D. dissertation, Univ. Calif., Berkeley. 221 pp.

Mire, J. B., and L. Millett. 1994. Size of mother does not determine size of eggs or fry in the Owens pupfish, *Cyprinodon radiosus.* Copeia 1994:100–107.

Mittelbach, G. G. 1983. Optimal foraging and growth in bluegill. Oecologia 59:157–162.

Mittelbach, G. G., C. W. Osenberg, and P. C. Wainwright. 1992. Variation in resource abundance affects diet and feeding morphology in the pumpkinseed sunfish (*Lepomis gibbosus*). Oecologia 90:8–13.

Mittelbach, G. G., A. M. Turner, D. J. Hall, J. E. Rettig, and C. W. Osenberg. 1995. Perturbation and resilience: a long-term whole-lake study of predator extinction and reintroduction. Ecology 76:2347–2360.

Modde, T., and G. B. Haines. 1996. Brook stickleback (*Culea inconstans* [Kirtland 1841]), a new addition to the upper Colorado River basin fish fauna. Great Basin Nat. 56:281–282.

Modde, T., K. P. Burham, and E. J. Wick. 1996. Population status of the razorback sucker in the Middle Green River (U.S.A.). Cons. Biol. 10:110–119.

Modde, T., A. T. Scholz, J. H. Williamson, G. B. Haines, B. D. Burdick, and F. K. Pfeifer. 1995. An augmentation plan for razorback sucker in the upper Colorado River basin. Am. Fish. Soc. Symp. 15:102–111.

Moffett, J. W., and S. H. Smith. 1950. Biological investigations of the fishery resources of Trinity River, California. USFWS Spec. Sci. Rpt. Fish. 12. 71 pp.

Monaco, G. A., R. L. Brown, and G. A. E. Gall. 1981. Exploring the aquaculture potential of sub-surface agriculture drainage water. Unpubl. rpt., Univ. Calif., Davis Aquaculture Prog. 141 pp.

Monaco, M. E., D. M. Nelson, R. L. Emmett, and S. A. Hinton. 1991. Distribution and abundance of fishes and invertebrates in west coast estuaries, vol. 1: data summaries. NOAA/NOS Strategic Env. Assess. Div., Rockville, Md., ELMR Rpt. 4. 240 pp.

Moore, K. M. S., and S. V. Gregory. 1988. Summer habitat utilization and ecology of cutthroat trout fry (*Salmo clarki*) in Cascade Mountain streams. Can. J. Fish. Aquat. Sci. 45: 1921–1930.

Morizot, D. C., S. W. Calhoun, L. L. Clepper, M. E. Schmidt, J. H. Williamson, and G. J. Carmichael. 1991. Multispecies hybridization among native and introduced centrarchid basses in central Texas. Trans. Am. Fish. Soc. 120:283–289.

Morris, R. W. 1960. Temperature, salinity, and southern limits of three species of Pacific cottid fishes. Limnol. Oceanogr. 5:175–179.

———. 1977. An analysis of some meristic characters of the staghorn sculpin *Leptocottus armatus* Girard. Pac. Sci. 31:259–277.

Morrow, J. E. 1980. The freshwater fishes of Alaska. Anchorage: Alaska Northwest Publishing.

Moser, M. L., A. F. Olson, and T. P. Quinn. 1991. Riverine and estuarine migratory behavior of coho salmon (*Oncorhynchus kisutch*) smolts. Can. J. Fish. Aquat. Sci. 48:1670–1678.

Mount, J. F. 1995. California rivers and streams: the conflict between fluvial process and land use. Berkeley: University of California Press.

Moyle, J. B., and J. Kuehn. 1964. Carp, a sometimes villain. Pages

635–642 *in* J. P. Linduska, ed. Waterfowl tomorrow. Washington, D.C.: U.S. Department of the Interior.

Moyle, P. B. 1969. Comparative behavior of young brook trout of domestic and wild origin. Prog. Fish Cult. 31:51–56.

———. 1970. Occurrence of king (chinook) salmon in the Kings River, Fresno County. Calif. Fish Game 56:314–315.

———. 1976. Some effects of channelization on the fishes and invertebrates of Rush Creek, Modoc County, California. Calif. Fish Game 62:179–186.

———. 1977. In defense of sculpins. Fisheries 2:20–23.

———. 1980. Delta smelt, page 123; hitch, page 199; California roach, page 200; Sacramento blackfish, page 329; Clear Lake splittail, page 345; Sacramento splittail, page 346; Sacramento squawfish, page 347; Modoc sucker, page 384; Sacramento sucker, page 385; Tahoe sucker, page 391; Sacramento perch, page 502; tule perch, page 777; rough sculpin, page 603; marbled sculpin, page 815; Pit sculpin, page 819 *in* D. S. Lee et al., eds. Atlas of North American freshwater fishes. Raleigh, N.C.: North Carolina Museum of Natural History.

———. 1984. America's carp. Natural History 93(9):42–51.

———. 1992. True smelts. Pages 75–78 *in* W. S. Leet, C. M. Dewees, and C. W. Haugen, eds. California's marine resources and their utilization. Calif. Sea Grant Ext. Publ. UCSGEP-92-12.

———. 1995. Conservation of native freshwater fishes in the Mediterranean type climate of California, U.S.A.: a review. Biol. Cons. 72:271–280.

———. 2000. Restoring aquatic ecosystems is a matter of values. Calif. Agriculture 54(2):16–25.

Moyle, P. B., and D. M. Baltz. 1985. Microhabitat use by an assemblage of California stream fishes: Developing criteria for instream flow determinations. Trans. Am. Fish. Soc. 114:695–704.

Moyle, P. B., and J. J. Cech, Jr. 2000. Fishes: an introduction to ichthyology. 4th edition. Saddle River, N.J.: Prentice-Hall.

Moyle, P. B., and R. A. Daniels. 1982. Fishes of the Pit River system, and Surprise Valley region. Univ. Calif. Publ. Zool. 115:1–82.

Moyle, P. B., and J. Ellison. 1991. A conservation-oriented classification system for California's inland waters. Calif. Fish Game 77:161–180.

Moyle, P. B., and B. Herbold.1987 Life-history patterns and community structure in stream fishes of western North America: comparisons with eastern North America and Europe. Pages 25–32 *in* W. J. Matthews and D. C. Heins, eds. Community and evolutionary ecology of North American stream fishes. Norman: University of Oklahoma Press.

Moyle, P. B., and N. J. Holzhauser. 1978. Effects of the introduction of Mississippi silverside (*Menidia audens*) and Florida largemouth bass (*Micropterus salmoides floridanus*) on the feeding habits of young-of-year largemouth bass in Clear Lake, California. Trans. Am. Fish. Soc.107:575–582.

Moyle, P. B., and D. Koch, eds. 1975. Trout-nongamefish relationships in streams. Univ. Nev. Cent. Water Res. Misc. Publ. 17. 45 pp.

Moyle, P. B., and H. W. Li. 1979. Community ecology and predator-prey relationships in warmwater streams. Pages 171–180 *in* H. W. Clepper, ed. Predator-prey systems in fisheries management. Washington, D.C.: Sport Fishing Institute.

Moyle, P. B., and T. Light. 1996. Biological invasions of fresh water: empirical rules and assembly theory. Biol. Cons. 78:149–162.

Moyle, P. B., and M. P. Marchetti. 1998. Applications of indices of biotic integrity to California streams and watersheds. Pages 367–380 *in* T. P. Simon, ed. Assessing the sustainability and biological integrity of water resources using fish assemblages. Boca Raton, Fla.: CRC Press.

Moyle, P. B., and A. Marciochi. 1975. Biology of the Modoc sucker, *Catostomus microps* (Pisces: Catostomidae) in northeastern California. Copeia 1975:556–560.

Moyle, P. B., and M. Massingill. 1981. Hybridization between hitch, *Lavinia exilicauda,* and Sacramento blackfish, *Orthodon microlepidotus,* in San Luis Reservoir, California. Calif. Fish Game 67:196–198.

Moyle, P. B., and P. R. Moyle. 1995. Endangered fishes and economics: intergenerational obligations. Env. Biol. Fish. 43:29–37.

Moyle, P. B., and R. D. Nichols. 1973. Ecology of some native and introduced fishes of the Sierra Nevada foothills in central California. Copeia 1973:478–490.

———. 1974. Decline of the native fish fauna of the Sierra-Nevada foothills, central California. Am. Midl. Nat. 92:72–83.

Moyle, P. B., and P. J. Randall. 1998. Evaluating the biotic integrity of watersheds in the Sierra Nevada, California. Cons. Biol. 12:1318–1326.

Moyle, P. B., and G. M. Sato. 1991. On the design of preserves to protect native fishes. Pages 155–169 *in* W. L. Minckley and J. E. Deacon, eds. Battle against extinction: native fish management in the American West. Tucson: University of Arizona Press.

Moyle, P. B., and J. J. Smith. 1995. Freshwater fishes of the Central California coast. Pages 17–22 *in* N. Chiariello and R. F. Dasmann, eds. Symposium of biodiversity of the Central California Coast. San Francisco: Association for the Golden Gate Biosphere Reserve.

Moyle, P. B., and B. Vondracek. 1985. Structure and persistence of the fish assemblage in a small Sierra stream. Ecology 66:1–13.

Moyle, P. B., and J. E. Williams. 1990. Loss of biodiversity in the temperate zone: decline of the native fish fauna of California. Cons. Biol. 4:275–284.

Moyle, P. B., and R. M. Yoshiyama. 1992. Fishes, aquatic diversity management areas, and endangered species: a plan to protect California's native aquatic biota. Berkeley: California Policy Seminar.

———. 1994. Protection of aquatic biodiversity in California: A five-tiered approach. Fisheries (Bethesda) 19(2):6–18.

Moyle, P. B., F. W. Fisher, and H. W. Li. 1974. Mississippi silversides and logperch in the Sacramento–San Joaquin River system. Calif. Fish Game 60:145–147.

Moyle, P. B., P. J. Foley, and R. M. Yoshiyama. 1992. Status of green sturgeon in California. Unpubl. rpt., NOAA-NMFS, Terminal Island, Calif. 11 pp.

Moyle, P. B., H. W. Li, and B. A. Barton. 1986. The Frankenstein effect: impact of introduced fishes on native fishes in North America. Pages 415–426 *in* R. H. Stroud, ed. Fish culture in fisheries management. Bethesda, Md.: American Fisheries Society.

Moyle, P. B., S. B. Mathews, and N. Bonderson. 1974. Feeding habits of the Sacramento perch, *Archoplites interruptus.* Trans. Am. Fish. Soc. 103:399–402.

Moyle, P. B., B. Vondracek, and G. D. Grossman. 1983. Response of fish populations in the North Fork of the Feather River, California to treatment with fish toxicants. N. Am. J. Fish. Mgmt. 3:48–60.

Moyle, P. B., R. A. Daniels, B. Herbold, and D. M. Baltz. 1985. Patterns in distribution and abundance of a noncoevolved assemblage of estuarine fishes in California. NOAA Fish. Bull. 84:105–117.

Moyle, P. B., B. Herbold, D. E. Stevens, and L. W. Miller. 1992. Life history and status of delta smelt in the Sacramento–San Joaquin Estuary, California. Trans. Am. Fish. Soc. 77:67–77.

Moyle, P. B., T. Kennedy, D. Kuda, L. Martin, and G. Grant. 1991. Fishes of Bly Tunnel, Lassen County, California. Great Basin Nat. 51:267–270

Moyle, P. B., M. P. Marchetti, J. Baldrige, and T. L. Taylor. 1998. Fish health and diversity: justifying flows for a California stream. Fisheries (Bethesda) 23(7):6–15.

Moyle, P. B., J. J. Smith, R. A. Daniels, and D. M. Baltz. 1982. Distribution and ecology of stream fishes of the Sacramento–San Joaquin Drainage System, California: a review. Univ. Calif. Publ. Zool. 115:225–256.

Moyle, P. B., R. M. Yoshiyama, J. E. Williams, and E. D. Wikramanayake. 1995. Fish species of special concern of California. 2nd edition. Sacramento: California Department of Fish and Game.

Mueller, G., P. C. Marsh, G. Knowles, and T. Wolters. 2000. Distribution, movements, and habitat use of razorback suckers (*Xyrauchen texanus*) in a lower Colorado River reservoir, Arizona-Nevada. West. N. Am. Nat. 60:180–187.

Muir, W. D., R. L. Emmett, and R. J. McConnell. 1988. Diet of juvenile and subadult white sturgeon in the lower Columbia River and its estuary. Calif. Fish Game 74:49–54.

Mukai, T., T. Sato, K. Naruse, K. Inaba, A. Shima, M. Morisawa. 1996. Genetic relationships of the genus *Tridentiger* (Pisces: Gobiidae) based on allozyme polymorphism. Zool. Sci. 13:175–183.

Mullan, J. W., and R. L. Applegate. 1968. Centrarchid food habits in a new and old reservoir during the following bass spawning. Proc. 21st Annu. Conf. SE Assoc. Game Fish Comm.: 332–342.

Mullan, J. W., and R. L. Applegate. 1970. Food habits of five centrarchids during the filling of Beaver Reservoir 1965–1966. U.S. Bur. Sport Fish. Wildl. Tech. Pap. 50. 16 pp.

Mulligan, M. J. 1975. The ecology of fish populations in Mill Flat Creek: tributary to the Kings River. M.S. thesis, Calif. State Univ., Fresno. 135 pp.

Murphy, G. I. 1943. Sexual dimorphism in the minnows *Hesperoleucus* and *Rhinichthys*. Copeia 1943:187–188.

———. 1948a. A contribution to the life history of the Sacramento perch (*Archoplites interruptus*) in Clear Lake, Lake County, California. Calif. Fish Game 34:93–100.

———. 1948b. Notes on the biology of the Sacramento hitch (*Lavinia e. exilicauda*) of Clear Lake, California. Calif. Fish Game 34:101–110.

———. 1948c. Distribution and variation of the roach (*Hesperoleucus*) in the coastal region of California. M.A. thesis, Univ. Calif., Berkeley. 55 pp.

———. 1949. The food of young largemouth bass (*Micropterus salmoides*) in Clear Lake, California. Calif. Fish Game 35:159–163.

———. 1950. The life history of the greaser blackfish (*Orthodon microlepidotus*) of Clear Lake, Lake County, California. Calif. Fish Game 36:119–133.

———. 1951. The fishery of Clear Lake, Lake County, California. Calif. Fish Game 37:439–484.

Murphy, G. I., and J. W. DeWitt. 1951. Notes on the fishes and fisheries of the lower Eel River, Humboldt County, California. CDFG Admin. Rpt. 51-9. 28 pp.

Muth, R. T., and D. E. Snyder. 1995. Diets of young Colorado squawfish and other small fish in backwaters of the Green River, Colorado and Utah. SW Nat. 55:95–104.

Myers, G. S. 1965. *Gambusia*, the fish destroyer. Aust. Zool. 13:102.

Myers, J. M., R. G. Kope, G. J. Bryant, D. Teel, L. J. Lierheimer, T. C. Wainwright, W. S. Grant, F. W. Waknitz, K. Neely, S. T. Lindley, and R. S. Waples. 1998. Status review of chinook salmon from Washington, Idaho, Oregon, and California. NOAA Tech. Mem. NMFS-NWFSC-35. 443 pp.

Myrick, C. A. 1996. The application of bioenergetics to the control of fish populations below reservoirs: California stream fish swimming performances. M.S. thesis, Univ. Calif., Davis. 35 pp.

Naiman, R. J. 1975. Food habits of the Amargosa pupfish in a thermal stream. Trans. Am. Fish. Soc. 104:536–538.

———. 1976. Productivity of a herbivorous pupfish population (*Cyprinodon nevadensis*) in a warm desert stream. J. Fish Biol. 9:125–137.

Naiman, R. J., H. Decamps, and M. Pollock. 1993. The role of riparian corridors in maintaining regional biodiversity. Ecol. Appl. 3:209–212.

Naiman, R. J., S. D. Gerking, and T. D. Ratcliffe. 1973. Thermal environment of a Death Valley pupfish. Copeia 1973:366–369.

Nakamoto, R. J., T. T. Kisanuki, and G. H. Goldsmith. 1995. Age and growth of Klamath River green sturgeon (*Acipenser medirostris*). USFWS, Arcata, unpubl. rpt. Proj. 93–FP-13. 19 pp.

Nakamura, K. 1994. Air breathing abilities of the common carp. Fish. Sci. 60:271–274.

National Marine Fisheries Service. 1996. Proposed endangered status for five ESUs of steelhead and proposed threatened status for five ESUs of steelhead in Washington, Oregon, Idaho, and California. Fed. Reg. 61(155):41541–41561.

———. 1996. NMFS proposed recovery plan for the Sacramento River winter-run chinook salmon. NMFS SW Region, Long Beach, Calif. ca. 250 pp.

Needham, P. R., and R. Gard. 1959. Rainbow trout in Mexico and California, with notes on the cutthroat series. Univ. Calif. Publ. Zool. 67. 123 pp.

Needham, P. R., and A. A. Hanson. 1935. U.S. Bur. Fish., unpubl. rpt. 55 pp.

Needham, P. R., and A. C. Jones. 1959. Flow, temperature, solar radiation and ice in relation to activities of fishes in Sagehen Creek, California. Ecology 40:465–474.

Needham, P. R., and T. M. Vaughan. 1952. Spawning of the Dolly Varden, *Salvelinus malma*, in Twin Creek, Idaho. Copeia 1952:197–199.

Needham, P. R., and E. H. Vestal. 1938. Notes on growth of golden trout (*Salmo agua-bonita*) in two high sierra lakes. Calif. Fish Game 24:273–279.

Nehlson, W., J. E. Williams, and J. A. Lichatowich. 1991. Pacific salmon at the crossroads; stocks at risk from California, Oregon, Idaho, and Washington. Fisheries (Bethesda) 16(2): 4–21.

Nelson, J. S. 1994. Fishes of the world. 3rd edition. New York: John Wiley.

Nelson, S. M., and S. A. Flickinger. 1992. Salinity tolerance of Colorado squawfish, *Ptychocheilus lucius* (Pisces: Cyprinidae). Hydrobiologia 246:165–168.

Netboy, A. 1974. The salmon: their fight for survival. Boston: Houghton Mifflin.

Newsome, G. E., and J. Tompkins. 1985. Yellow perch egg masses deter predators. Can J. Zool. 64:282–284.

NHI (Natural Heritage Institute). 1992. Petition for listing under the Endangered Species Act: longfin smelt and Sacramento splittail. Submitted to USFWS, Sacramento Field Office, 5 November.

Nichols, F. H., J. K. Thompson, and L. R. Scheme 1990. Remarkable invasion of San Francisco Bay by the Asian Clam, *Potamocorbula amurensis*. II. Displacement of a former community. Mar. Ecol. Prog. Series 66:95–101.

Nichols, J. T. 1943. The fresh-water fishes of China. New York: American Museum of Natural History.

Nickelson, T. E., J. D. Rodgers, S. L. Johnson, and M. F. Solazzi. 1992. Seasonal changes in habitat use by juvenile coho salmon (*Oncorhynchus kisutch*) in Oregon coastal streams. Can J. Fish. Aquat. Sci. 49:783–789.

Nickelson, T. E., M. F. Solazzi, J. D. Rodgers, and S. L. Johnson. 1992. Effectiveness of selected stream improvement techniques to create suitable summer and winter rearing habitat for juvenile coho salmon (*Oncorhynchus kisutch*) in Oregon coastal streams. Can J. Fish. Aquat. Sci. 49:790–794.

Nicola, S. J. 1974. The life history of the hitch, *Lavinia exilicauda* Baird and Girard, in Beardsley Reservoir, California. CDFG Inland Fish. Admin. Rpt. 74-6. 16 pp.

Nielsen, J. L. 1992a. Microhabitat specific foraging behavior, diet, and growth of juvenile coho salmon. Trans. Am. Fish. Soc. 121:617–634.

———. 1992b. The role of cold-pool refuge in the freshwater fish assemblage in northern California rivers. Pages 79–88 *in* H. M. Kerner, ed. Proceedings of the symposium on biodiversity of northwestern California. Davis: Univ. Calif. Wildland Resources Center Rpt. 29.

———. 1994. Invasive cohorts: impact of hatchery-reared coho salmon on the trophic, developmental, and genetic ecology of wild stocks. Pages 361–386 *in* D. L. Stouder, K. L. Fresh, and R. J. Feller, eds. Theory and application in fish feeding ecology. Columbia: University of South Carolina Press.

———. 1999. The evolutionary history of steelhead (*Oncorhynchus mykiss*) along the U.S. Pacific Coast: developing a conservation strategy using genetic diversity. ICES J. Marine Sci. 56:449–458.

Nielsen, J. L., and M. C. Fountain. 1999. Microsatellite diversity in sympatric reproductive ecotypes of Pacific steelhead (*Oncorhynchus mykiss*) from the Middle Fork Eel River, California. Ecol. Freshw. Fish 8:159–168.

Nielsen, J. L., K. D. Crow, and M. C. Fountain. 1999. Microsatellite diversity and conservation of a relic trout population: McCloud River redband trout. Mol. Ecol. 8(suppl. 1): S129–142.

Nielsen, J. L., C. Carpanzano, M. C. Fountain, and J. M. Wright. 1997. Mitochondrial DNA and nuclear microsatellite diversity in hatchery and wild *Oncorhynchus mykiss* from freshwater habitats in southern California. Trans. Am. Fish. Soc. 126:397–427.

Nielsen, J. L., M. C. Fountain, J. C. Favela, K. Cobble, and B. J. Jensen. 1998. *Oncorhynchus* at the southern extent of their range: a study of mtDNA control-region sequence with special reference to an undescribed subspecies of *O. mykiss* from Mexico. Env. Biol. Fish. 51:7–23.

Nobriga, M. 1998a. Evidence of food limitation in larval delta smelt. IEP Newslett. 11(1):20–24.

———. 1998b. Trends in the food habits of larval delta smelt, *Hypomesus transpacificus*, in the Sacramento–San Joaquin estuary, California, 1992–1994. M.S. thesis, Calif. State Univ., Sacramento. 49 pp.

Northcote, T. G., and G. L. Ennis. 1994. Mountain whitefish biology and habitat use in relation to compensation and improvement possibilities. Rev. Fish. Sci. 2:347–371.

Northcote, T. G., and G. F. Hartman. 1959. A case of "schooling" behavior in the prickly sculpin, *Cottus asper* Richardson. Copeia 1959:158–159.

Norton, B. G. 1987. Why preserve natural variety? Princeton, N.J.: Princeton University Press.

Noss, R. F. 1992. Issues of scale in conservation biology. Pages 239–250 *in* P. L. Fiedler and S. K. Jain, eds. Conservation biology: the theory and practice of nature conservation, preservation, and management. New York: Chapman and Hall.

O'Brien, W. J., B. I. Evans, and G. L. Howick. 1986. A new view of the predation cycle of a planktivorous fish, white crappie, *Pomoxis annularis*. Can. J. Fish. Aquat. Sci. 43:1894–1899.

Odemar, M. W. 1964. Southern range extension of the eulachon, *Thaleichthys pacificus*. Calif. Fish Game 50:304–307.

Odenweller, D. B. 1975. The life history of the shiner surfperch, *Cymatogaster aggregata* Gibbons, in Anaheim Bay. CDFG Fish Bull. 165:107–115.

Odion, D. C., T. L. Dudley, and C. M. D'Antonio. 1988. Cattle grazing in southeastern Sierran meadows: ecosystem change and prospects for recovery. Pages 277–292 *in* C. A. Hall and V. Doyle-Jones, eds. Plant biology of eastern California. White Mountain Research Station, University of California, Los Angeles.

Odum, W. E. 1968. The ecological significance of fine particle selection by the striped mullet, *Mugil cephalus*. Limnol. Oceanogr. 13:92–98.

Ogura, M., and Y. Ishida. 1995. Homing behavior and vertical movements of four species of Pacific salmon (*Oncorhynchus* spp.) in the central Bering Sea. Can. J. Fish. Aquat. Sci. 52:532–540.

Ohmart, R. D., B. W. Anderson, and W. C. Hunter. 1988. The ecology of the lower Colorado River from Davis Dam to the Mexico–United States international boundary: a community profile. USFWS Biol. Rpt. 85(7.19). 296 pp.

Okeyo, D. O., and T. J. Hassler. 1985. Growth, food and habitat of age 0 smallmouth bass in Clair Engle Reservoir, California. Calif. Fish Game 71:76–87.

Olson, M. D. 1988. Upstream changes in native fish abundance after reservoir impoundment in California streams of the Lahontan Basin. M.S. thesis, Univ. Calif., Berkeley. 34 pp.

Olson, M. D., and D. C. Erman. 1987. Distribution and abundance of mountain sucker, *Catostomus platyrhynchus*, in five California streams of the Western Lahontan Basin. Unpubl. rpt., CDFG Contract C-2057. 65 pp.

Omel'chenko, V. T., and G. P. Vyalova. 1990. Population structure of pink salmon. Soviet J. Marine Biol. 16:1–10.

Onuf, C. P. 1987. The ecology of Mugu Lagoon, California: an estuarine profile. USFWS Biol. Rpt. 85(7.15). 122 pp.

Orcutt, H. G. 1950. The life history of the starry flounder *Platichthys stellatus* (Pallas). CDFG Fish Bull. 78:1–64.

Oregon Department of Fish and Wildlife. 1991. Status report: Columbia River fish runs and fisheries 1960–1990. Unpubl. rpt. 154 pp.

Orians, G. H. 1993. Endangered at what level? Ecol. Appl. 3:206–208.

Osenberg, C. W., E. E. Werner, G. G. Mittelbach, and D. J. Hall. 1988. Growth patterns in bluegill (*Lepomis macrochirus*) and pumpkinseed (*L. gibbosus*) sunfish: environmental variation and the importance of ontogenetic niche shifts. Can. J. Fish. Aquat. Sci. 45:17–26.

Osmundson, D. B., and K. P. Burnham. 1998. Status and trends of the endangered Colorado squawfish in the upper Colorado River. Trans. Am. Fish. Soc. 127:957–970.

Osmundson, D. B., R. J. Ryel, M. E. Tucker, B. D. Burdick, W. R. Emblad, and T. E. Chart. 1998. Dispersal patterns of subadult and adult Colorado squawfish in the upper Colorado River. Trans. Am. Fish. Soc. 127:943–956.

Overton, C. K., G. L. Chandler, and J. A. Pisano. 1994. Northern/intermountain regions' fish habitat inventory: grazed, rested, and ungrazed reference stream reaches, Silver King Creek, California. USFS Intermountain Res. Stn. Gen. Tech. Rpt. INT-GTR-311. 27 pp.

PG&E. 1985. Bald Eagle and Fish Study. Pit 3, 4 and 5 Project. Biosystems Analysis, Inc. and University of California, Davis. Final rpt. ca. 250 pp.

Page, L. M., and B. M. Burr. 1991. A field guide to freshwater fishes, North America north of Mexico. Boston: Houghton Mifflin.

Painter, R., L. Wixom, and M. Meinz. 1979. American shad management plan for the Sacramento River drainage. Final rpt., CDFG, Sacramento, Anadromous Fish Conservation Act Proj. AFS-17, Job 5. 17 pp.

Panek, F. M. 1987. Biology and ecology of carp. Pages 1–16 *in* Carp in North America. Bethesda, Md.: American Fisheries Society.

Papoulias, D., and W. L. Minckley. 1992. Effects of food availability on survival and growth of larval razorback suckers in ponds. Trans. Am. Fish. Soc. 121:340–355.

Parsons, J. W. 1953. Growth and habits of the redeye bass. Trans. Am. Fish. Soc. 83:202–211.

Parsons, J. W., and E. Crittenden. 1959. Growth of the redeye bass in Chipola River, Florida. Trans. Am. Fish. Soc. 88:191–192.

Patten, B. G. 1971. Spawning and fecundity of seven species of northwest American *Cottus*. Am. Midl. Nat. 85:493–506.

Patten, B. G., and D. T. Rodman. 1969. Reproductive behavior of the northern squawfish, *Ptychocheilus oregonensis*. Trans. Am. Fish. Soc. 98:108–111.

Paukert, C. P., J. A. Klammer, R. B. Bruce, and T. D. Simonson. 2001. An overview of northern pike regulations in North America. Fisheries 26(8):6–13.

Pauley, G. B., K. L. Bowers, and G. L. Thomas. 1988. Chum salmon. USFWS Biol. Rpt. 82(11.81). 17 pp.

Pauley, G. B., K. Oshima, K. L. Bowers, and G. L. Thomas. 1989. Sea-run cutthroat trout. USFWS Biol. Rpt. 82(11.86). 21 pp.

Payne, T. R., and Associates. 1998. Recovery of fish populations in the upper Sacramento River following the Cantara Spill of July 1991. 1997 Annu. Rpt. CDFG. 88 pp.

Pearcy, W. G. 1992. Ocean ecology of north Pacific salmonids. Seattle: University of Washington Press.

Pearsons, T. N., H. W. Li, and G. A. Lamberti. 1992. Influence of habitat complexity on resistance to flooding and resilience of stream fish assemblages. Trans. Am. Fish. Soc. 121:427–436.

Pease, R. W. 1965. Modoc country: a geographic time continuum of the California volcanic tableland. Univ. Calif. Publ. Geog. 17. 304 pp.

Peden, A. 1972. The function of gonopodial parts and behavioral pattern during copulation by *Gambusia* (Poeciliidae). Can. J. Zool. 50:955–968.

Pelzman, R. J. 1971. The blue catfish. CDFG Inland Fish. Admin. Rpt. 71-11. 7 pp.

———. 1973. A review of the life history of *Tilapia zillii* with a reassessment of its desirability in California. CDFG Inland Fish. Admin. Rpt. 74-1. 9 pp.

Perez-Espana, H., F. Galvan-Magana, and L. A. Abitia-Cardenas. 1998. Growth, consumption, and productivity of the California killifish in Ojo de Liebre Lagoon, Mexico. J. Fish Biol. 52:1068–1077.

Perkins, D. L., and G. G. Scoppetone. 1996. Spawning and migration of Lost River suckers (*Deltistes luxatus*) and shortnose suckers (*Chasmistes brevirostris*) in the Clear Lake drainage, Modoc County, California. Natl. Biol. Serv., Calif. Field Office, Reno, Nev. Rpt. CDFG Contract FG1494. 52 pp.

Perrow, M. R., A. J. D. Jowitt, and S. R. Johnson. 1996. Factors affecting habitat selection of tench in a shallow eutrophic lake. J. Fish Biol. 48:859–870.

Perry, W. G. 1968. Distribution and relative abundance of blue catfish, *Ictalurus furcatus*, and channel catfish with relation to salinity. Proc. 21st Annu. Conf. SE Assoc. Game Fish Comm.: 436–444.

Perry, W. G., and J. W. Avault. 1969. Culture of blue, channel, and white catfish in brackish water ponds. Proc. 23rd Annu. Conf. SE Assoc. Game Fish Comm.: 1–15.

Pert, H. A. 1993. Winter food habits of coastal juvenile steelhead and coho salmon in Pudding Creek, Northern California. M.S. thesis, Univ. Calif., Berkeley. 65 pp.

Peters, E. J., R. S. Holland, M. A. Callam, and D. L. Bunnell. 1989. Habitat utilization, preference and suitability index criteria for fish and aquatic invertebrates in the lower Platte River. Nebraska Game, Parks Tech. Ser. 17. 135 pp.

Peterson, M. S. 1988. Comparative physiological ecology of centrarchids in hyposaline environments. Can. J. Fish. Aquat. Sci. 45:827–833.

Peterson, M. S., D. E. Gustafson, and F. R. Moore. 1987. Orientation behaviour of *Lepomis macrochirus* Rafinesque to salinity fluctuations. J. Fish. Biol. 30:451–458.

Petrusso, P. A., and D. B. Hayes. 2001a. Condition of juvenile chinook salmon in the upper Sacramento River, California. Calif. Fish Game 87:19–37.

———. 2001b. Invertebrate drift and feeding habits of juvenile chinook salmon in the Upper Sacramento River, California. Calif. Fish Game 87:1–18.

Phelps, A. 1989a. Behavioral ecology of tule perch: life history and reproductive behavior. Ph.D. dissertation, Univ. Calif., Davis. 89 pp.

———. 1989b. Distribution and abundance of tule perch, *Hysterocarpus traski*, in the Russian River, California. Unpubl. rpt., Univ. Calif., Davis. 10 pp.

Phelps, A., D. Bartley, and D. Hedgecock 1995. Electrophoretic evidence for multiple mating in tule perch. Calif. Fish Game 81:147–154.

Phillips, R. W., and E. W. Claire. 1966. Intragravel movement of the reticulate sculpin, *Cottus perplexus*, and its potential as a predator on salmonid embryos. Trans. Am. Fish. Soc. 95: 210–212.

Pintler, H. E., and W. C. Johnson. 1958. Chemical control of rough fish in the Russian River drainage, California. Calif. Fish Game 44:91–124.

Pipas, J. C., and F. J. Bulow. 1998. Hybridization between redeye bass and smallmouth bass in Tennessee streams. Trans. Am. Fish. Soc. 127:141–146.

Pisano, M. S., M. J. Inansci, and W. L. Minckley. 1983. Age and growth and length-weight relationship for flathead catfish, *Pylodictis olivaris*, from Coachella Canal, southeastern California. Calif. Fish Game 69:124–128.

Pister, E. P. 1974. Desert fishes and their habitats. Trans. Am. Fish. Soc. 103:531–540.

———. 1990. Pure Colorado trout saved by California. Outdoor California 51(1):12–15.

———. 1991. The Desert Fishes Council: catalyst for change. Pages 55–68 *in* W. L. Minckley and J. E. Deacon, eds. Battle against extinction: native fish management in the American West. Tucson: University of Arizona Press.

———. 1993. Species in a bucket. Nat. Hist. 102(1):14–19.

Platt, S., and J. Hauser. 1978. Optimum temperature for feeding and growth of *Tilapia zillii*. Prog. Fish. Cult. 40:105–107.

Pontius, R. W., and M. Parker. 1973. Food habits of the mountain whitefish, *Prosopium williamsoni* (Girard). Trans. Am. Fish. Soc. 102:764–773.

Porter, R. G. 1964. Food and feeding of staghorn sculpin (*Leptocottus armatus* Girard) and starry flounders (*Platichthys stellatus* Pallas) in euryhaline environments. M.S. thesis, Humboldt State Coll., Arcata. 84 pp.

Powell, J. A. 1984. Observations of cleaning behavior in bluegill (*Lepomis macrochirus*), a centrarchid. Copeia 1984:996–998.

Pratt, K. L. 1982. A review of bull trout life history. Pages 5–9 *in* P. J. Howell and D.V. Buchanan, eds. Proceedings of the Gearhart Mountain bull trout workshop. Corvallis, Oreg.: Oregon Chapter of the American Fisheries Society.

Price, E. E., J. R. Stauffer, Jr., and M. C. Swift. 1985. Effect of temperature on growth of juvenile *Oreochromis mossambicus* and *Sarotherodon melanotheron*. Env. Biol. Fish. 13:149–152.

Priegel, G. R. 1970. Food of the white bass, *Roccus chrysops*, in Lake Winnebago, Wisconsin. Trans. Am. Fish. Soc. 99: 440–443.

———. 1971. Age and rate of growth of the white bass in Lake Winnebago, Wisconsin. Trans. Am. Fish. Soc. 100:567–590.

Ptacek, M. B., and F. Breden. 1998. Phylogenetic relationships among the mollies (Poeciliidae: *Poecilia: Mollinesia* group) based on mitochondrial DNA sequences. J. Fish Biol. 53: 64–81.

Ptacek, M. B., and J. Travis. 1997. Mate choice in the sailfin molly, *Poecilia latipinna*. Evolution 51:1217–1231.

Quinn, T. P., and D. J. Adams. 1996. Environmental changes affecting the migratory timing of American shad and sockeye salmon. Ecology 77:1151–1162.

Quinn, T. P., J. L. Nielsen, C. Gan, M. J. Unwin, R. Wilmot, C. Guthrie, and F. M. Utter. 1996. Origin and genetic structure of chinook salmon, *Oncorhynchus tshawytscha*, transplanted from California to New Zealand: allozyme and mtDNA evidence. NOAA Fish. Bull. 94:506–521.

Radtke, L. D. 1966. Distribution of smelt, juvenile sturgeon and starry flounder in the Sacramento–San Joaquin Delta. Pages 115–119 *in* S. L. Turner and D. W. Kelley, eds. Ecological studies of the Sacramento–San Joaquin Delta, Part II. CDFG Fish Bull. 136.

Railsback, S. F., and K. A. Rose. 1999. Bioenergetics modeling of stream trout growth: temperature and food consumption effects. Trans. Am. Fish. Soc. 128:241–256.

Raney, E. 1952. The life history of the striped bass, *Roccus saxatilis* (Walbaum). Bull. Bingham Oceanogr. Collect. Yale Univ. 14:5–97.

Raquel, P. F. 1986. Juvenile blue catfish in the Sacramento–San Joaquin Delta of California. Calif. Fish Game 72:186–187.

Rauschenberger, M. 1989. Annotated list of species of the subfamily Poeciliinae. Pages 359–368 *in* G. K. Meffe and F. F. Snelson, Jr., eds. Ecology and evolution of livebearing fishes (Poeciliidae). Englewood Cliffs, N.J.: Prentice-Hall.

Rawstron, R. R. 1964. Spawning of the threadfin shad, *Dorosoma petenense*, at low water temperatures. Calif. Fish Game 50:58.

———. 1967. Harvest, mortality, and movement of selected warmwater fishes in Folsom Lake, California. Calif. Fish Game 53:40–48.

Rees, B. 1958. Attributes of the mosquitofish in relation to mosquito control. Proc. 26th Annu. Conf. Calif. Mosq. Contr. Assoc.: 71–75.

Reeves, G. H., and J. R. Sedell. 1992. An ecosystem approach to the conservation and management of freshwater habitat for anadromous salmonids in the Pacific Northwest. Trans. 57th N.A. Wildl. and Nat. Res. Conf.: 408–415.

Reeves, G. H., F. H. Everest, and J. D. Hall. 1987. Interactions between redside shiner (*Richardsonius balteatus*) and steelhead (*Salmo gairdneri*) in western Oregon: the influence of water temperature. Can. J. Fish. Aquat. Sci. 44:1603–1613.

Reeves, J. E. 1964. Age and growth of hardhead minnow, *Mylopharodon conocephalus* (Baird and Girard), in the American River basin of California, with notes on its ecology. M.S. thesis, Univ. Calif., Berkeley. 90 pp.

Reiman, B. E., and R. C. Beamesderfer. 1990. Dynamics of a northern squawfish population and the potential to reduce predation on juvenile salmonids in a Columbia River reservoir. N. Am. J. Fish. Mgmt. 10:228–241.

Reimers, N. 1958. Conditions of existence, growth and longevity of brook trout in a small high-altitude lake of the eastern Sierra Nevada. Calif. Fish Game 44:319–333.

———. 1979. A history of a stunted brook trout population in an alpine lake: a life span of 24 years. Calif. Fish Game 65:196–215.

Reisner, M. 1986. Cadillac desert: the American West and its disappearing water. New York: Viking Penguin.

Reist, J. D. 1983. Behavioral variation in pelvic phenotypes of brook stickleback, *Culea inconstans*, in response to predation by northern pike, *Esox lucius*. Env. Biol. Fish. 8:255–267.

Renfro, W. C. 1960. Salinity relations of some fishes in the Arkansas River, Texas. Tulane Stud. Zool. 8:83–91.

Retallack, A. 1998. Striped bass: revitalizing a fishery. Outdoor California 59:9–12.

Reznick, D. N., and B. Braun. 1987. Fat cycling in the mosquitofish (*Gambusia affinis*): fat storage as a reproductive adaptation. Oecologia 73:401–413.

Richards, C., and D. L. Soltz. 1986. Feeding of rainbow trout (*Salmo gairdneri*) and arroyo chubs (*Gila orcutti*) in a California mountain stream. SW Nat. 31:250–253.

Richards, J. E., and F. W. H. Beamish. 1981. Initiation of feeding and salinity tolerance in the Pacific lamprey, *Lampetra tridentata*. Marine Biol. 63:73–77.

Richards, J. E., R. J. Beamish, and F. W. H. Beamish. 1982. Descriptions and keys for ammocoetes of lampreys from British Columbia, Canada. Can. J. Fish. Aquat. Sci. 39:1484–1495.

Richards, R. C., C. R. Goldman, E. Byron, and C. Lavation. 1991. The mysid and lake trout of Lake Tahoe: a 25-year history of changes in fertility, plankton, and fishery of an alpine lake. Am. Fish. Soc. Symp. 9:30–38.

Richardson, M. J., F. G. Whoriskey, and L. H. Roy. 1995. Turbidity generation and biological impacts of an exotic fish, *Carassius auratus*, introduced into shallow seasonally anoxic ponds. J. Fish Biol. 47:576–585.

Ricker, S. J. 1997. Evaluation of salmon and steelhead spawning habitat quality in the Shasta River Basin, 1997. CDFG Inland Fish. Admin. Rpt. 97-9. 13 pp.

Ricker, W. E. 1960. A population of dwarf coastrange sculpins (*Cottus aleuticus*). J. Fish. Res. Board Can. 17:929–932.

Ridgway, M. S. 1988. Developmental stage of offspring and brood defense in smallmouth bass (*Micropterus dolomieui*). Can. J. Zool. 66:1722–1728.

Ridgway, M. S., G. P. Goff, and M. H. A. Keenleyside. 1989. Courtship and spawning behavior in smallmouth bass (*Micropterus dolomieui*). Am. Midl. Nat. 122:209–231.

Ridgway, M. S., J. A. MacClean, and J. C. MacLeod. 1991a. Nest-side fidelity in a centrarchid fish, the smallmouth bass (*Micropterus dolomieui*). Can. J. Zool. 69:3103–3105.

Ridgway, M. S., B. J. Shuter, and E. E. Post 1991b. The relative influence of body size and territorial behaviour on nesting asynchrony in male smallmouth bass, *Micropterus dolomieui* (Pisces: Centrarchidae). J. Anim. Ecol. 60:665–681.

Riggs, C. C. 1955. Reproduction of the white bass, *Morone chrysops*. Invest. Indiana Lakes Streams 4:87–110.

Rincon, P. A., J. C. Velasco, N. Gonzalez-Sanchez, and C. Pollo. 1990. Fish assemblages in small streams in western Spain: the influence of an introduced predator. Arch. Hydrobiol. 118:81–91.

Ringler, N. H. 1979. Selective predation by drift-feeding brown trout (*Salmo trutta*). J. Fish. Res. Board Can. 46:392–403.

Robertson, C. H. 1957. Survival of precociously maturing male salmon parr (*Oncorhynchus tshawyscha*) after spawning. Calif. Fish Game 43:119–129.

Robins, C. R., and R. R. Miller. 1957. Classification, variation, and distribution of the sculpins, genus *Cottus*, inhabiting pacific slope waters in California and southern Oregon, with a key to the species. Calif. Fish Game 43:213–233.

Robins, C. R., R. M. Bailey, C. E. Bond, J. R. Brooker, E. A. Lachner, R. N. Lea, and W. B. Scott. 1991. A list of common and scientific names of fishes from the United States and Canada. 5th edition. Am. Fish. Soc. Spec. Publ. 20. 183 pp.

Rochard, E., G. Castelnaud, and M. Lepage. 1990. Sturgeons (Pisces: Acipenseridae): threats and prospects. J. Fish Biol. 37(suppl. A):123–132.

Rockriver, A. K. C. 1998. Spatial and temporal distribution of larval fish in littoral habitats of the Central Valley, California. M.S. thesis, Calif. State Univ., Sacramento. 83 pp.

———. 2001. Delta smelt. IEP Newsletter 14(2):19–21.

Rode, M. 1990. Bull trout, *Salvelinus confluentus* Suckley, in the McCloud River: status and recovery recommendations. CDFG Inland Fish. Admin. Rpt. 90–15. 43 pp.

Roedel, P. M. 1953. Common fishes of the California coast. CDFG Fish Bull. 91:1–184.

Roell, M. J., and D. J. Orth. 1993. Trophic basis of production of stream-dwelling smallmouth bass, rock bass, and flathead catfish in relation to invertebrate bait harvest. Trans. Am. Fish. Soc. 122:46–62.

Roelofs, T. D. 1983. Current status of California summer steelhead (*Salmo gairdneri*) stocks and habitat, and recommendations for their management. USDA For. Serv. Reg. Rpt. 5. 77 pp.

Rogers, D. A. 1984. Seasonal variation in freshwater smelt abundance and size and their use as forage by other fishes in Lake Shastina, Siskiyou County, 1972–1973. CDFG Inland Fish. Admin. Rpt. 84-2. 8 pp.

Rogers, D. W. 1974. Chum salmon observations in four north coast streams. Calif. Fish Game 60:148.

Rogers, K. B., L. E. Bergsted, and E. P. Bergersen. 1996. Standard weight equation for mountain whitefish. N. Am. J. Fish. Mgmt. 16:207–209.

Roos, J. F., P. Gilhousen, S. R. Killick, and E. R. Zyblut. 1973. Parasitism on juvenile Pacific salmon (*Oncorhynchus*) and Pacific herring (*Clupea harengeus pallasi*) in the Strait of Georgia by the river lamprey (*Lampetra ayersi*). J. Fish. Res. Board Can. 30:565–568.

Rosato, P., and D. Ferguson. 1968. The toxicity of endrin-resistant mosquitofish to eleven species of vertebrates. Bioscience 18:783–784.

Roscoe, T. J. 1993. Life history aspects of California roach (*Hesperoleucas symmetricus*) affected by changes in flow regime in the North Fork Stanislaus River, Calaveras Co., California. M.S. thesis, Calif. State Univ., Sacramento. 57 pp.

Rosen, D. E., and R. M. Bailey. 1963. The poeciliid fishes (Cyprinodontiformes), their structure, zoogeography, and systematics. Bull. Am. Mus. Nat. Hist. 126:1–126.

Rosenfield, J. A., 1998. Detection of natural hybridization between pink salmon and chum salmon in the Laurentian Great Lakes using meristic, morphological, and color evidence. Copeia 1998:706–714.

Ross, S. T. 1973. The systematics of *Gasterosteus aculeatus* (Pisces: Gasterosteidae) in central and southern California. Contrib. Sci. Nat. Hist. Mus. Los Angeles Co. 243:1–20.

Rossi, A. R., M. Capula, D. Crosetti, D. E. Campton, and L. Sola. 1998. Genetic divergence and phylogenetic inferences in five species of Mugilidae (Pisces: Perciformes). Marine Biol. 131:13–18.

Ruiz-Campos, G. 1995. First occurrence of the yellow bullhead (Amierus natalis) in the lower Colorado River, Baja California, Mexico. Calif. Fish Game 81:80–81.

Ruiz-Campos, G., and S. Gonzalez-Guzman. 1996. First freshwater record of Pacific lamprey, *Lampetra tridentata,* from Baja California, Mexico. Calif. Fish Game 82:144–146.

Ruppert, J. R., R. T. Muth, and T. P. Nester. 1993. Predation on fish larvae by adult red shiner, Yampa and Green rivers, Colorado. SW Nat. 38:397–399.

Rutter, C. 1903. Notes on fishes from streams and lakes of northeastern California not tributary to the Sacramento Basin. Bull. U.S. Fish Comm. 22:143–148.

———. 1908. The fishes of the Sacramento–San Joaquin basin, with a study of their distribution and variation. Bull. U.S. Bur. Fish. 27:103–152.

Ryan, J. H., and S. J. Nicola. 1976. Status of the Paiute cutthroat trout, *Salmo clarki seleneris* Snyder, in California. CDFG Inland Fish. Admin. Rpt. 76-3. 56 pp.

Rybock, J. T., H. F. Horton, and J. L. Kessler. 1975. Use of otoliths to separate juvenile steelhead trout from juvenile rainbow trout. NOAA Fish. Bull. 73:654–659.

Sabo, M. J., E. J. Pert, and K. O. Winemiller. 1995. Agonistic behavior of juvenile largemouth and smallmouth bass. J. Freshw. Biol. 11:115–118.

Sada, D. W. 1989. Status, distribution, and morphological variation of speckled dace in the Owens River system. CDFG Inland Fish. Rpt. Contract FG7343. 77 pp.

———. 1990. Factors affecting structure of a Great Basin stream fish assemblage. Ph.D. dissertation, Univ. of Nevada, Reno. 129 pp.

Sada, D. W., and J. E. Deacon. 1995. Spatial and temporal variability of pupfish (genus *Cyprinodon*) habitat and populations at Salt Creek and Cottonball Marsh, Death Valley National Park, California. Unpubl. rpt., U.S. Natl. Park Serv. Agreement 8000-2-9003. 76 pp.

Sada, D. W., H. B. Britten, and P. F. Brussard. 1993. Morphometric and genetic differentiation among Death Valley System *Rhinichthys osculus.* CDFG Inland Fish. Rpt. Contract FG0524. 51 pp.

———. 1995. Desert aquatic ecosystems and the genetic and morphological diversity of Death Valley System speckled dace. Pages 350–359 *in* J. L. Nielsen, ed. Evolution and the aquatic ecosystem. AFS Symposium 17. Bethesda, Md.: American Fisheries Society.

Sada, D. W., K. Pindal, D. L. Threloff, and J. E. Deacon. 1997. Spatial and temporal variability of pupfish (genus *Cyprinodon*) habitat and populations at Saratoga Springs and the lower Amargosa River, Death Valley National Park, California. Unpubl. rpt., U.S. Natl. Park Serv. Agreement 8000-2-9003. 99 pp.

Sagar, P. M., and G. J. Glova. 1988. Diel feeding periodicity, daily ration and prey selection of a riverine population of juvenile chinook salmon, *Oncorhynchus tshawytscha* Walbaum. J. Fish Biol. 33:643–653.

Saiki, M. K. 1984. Environmental conditions and fish faunas in low elevation rivers on the irrigated San Joaquin Valley floor, California. Calif. Fish Game 70:145–157.

———. 1997. Survey of small fishes and environmental condi-

tions in Mugu Lagoon, California, and tidally influenced reaches of its tributaries. Calif. Fish Game 83:153–167.

Saiki, M. K., and R. S. Ogle. 1995. Evidence of impaired reproduction by western mosquitofish inhabiting seleniferous agricultural drainwater. Trans. Am. Fish. Soc. 124:578–587.

Saiki, M. K., and J. C. Tash. 1978. Unusual population dynamics in largemouth bass, *Micropterus salmoides* (Lacepede), caused by a seasonally fluctuating food supply. Am. Midl. Nat. 100:116–125.

Saiki, M. K., M. R. Jennings, and R. H. Wiedmeyer. 1992. Toxicity of agricultural subsurface drainwater from the San Joaquin Valley, California, to juvenile chinook salmon and striped bass. Trans. Am. Fish. Soc. 121:78–93.

Saiki, M. K., D. P. Monda, and B. L. Bellerud. 1999. Lethal levels of selected water quality variables to larval and juvenile Lost River and shortnose suckers. Env. Pollution 100:1–8.

St. Amant, J. A. 1966a. Addition of *Tilapia mossambica* Peters to the California fauna. Calif. Fish Game 52:54–55.

———. 1966b. Progress report of the culture of *Tilapia mossambica* Peters hybrids in southern California. CDFG Admin. Rpt. 66-9. 25 pp.

———. 1970. Addition of Hart's rivulus, *Rivulus harti* (Boulenger) to the California fauna. Calif. Fish Game 56:138.

St. Amant, J. A., and F. G. Hoover. 1969. Addition of *Misgurnus anguillicaudatus* (Cantor) to the fish fauna of California. Calif. Fish Game 55:330–331.

Saksena, K. 1962. The post-hatching stages of the red shiner, *Notropis lutrensis.* Copeia 1962:539–544.

Salo, E. O. 1991. Life history of chum salmon (*Oncorhynchus keta*). Pages 231–310 *in* C. Groot and L. Margolis, eds. Pacific salmon life histories. Vancouver: University of British Columbia Press.

Sandercock, F. K. 1991. Life history of coho salmon. Pages 395–446 *in* C. Groot and L. Margolis, eds. Pacific salmon life histories. Vancouver: University of British Columbia Press.

Sanderson, S. L., and J. J. Cech, Jr. 1992. Energetic cost of suspension versus particulate feeding by juvenile Sacramento blackfish. Trans. Am. Fish. Soc. 121:149–157.

———. 1995. Particle retention during respiration and particulate feeding in the suspension-feeding Sacramento blackfish, *Orthodon microlepidotus.* Can. J. Fish. Aquat. Sci. 52:2534–2542.

Sanderson, S. L., J. J. Cech, Jr., and M. L. Patterson. 1991. Fluid dynamics in suspension-feeding blackfish. Science 251:1346–1348.

Sanford, J. R. 1975. The age and growth of the American shad, *Alosa sapidissima,* in the Yuba River, 1969. M.S. thesis, Calif. State Univ., Sacramento. 45 pp.

Saruwatari, T., J. A. Lopez, and T. W. Pietsch. 1997. A revision of the osmerid genus *Hypomesus* Gill (Teleostei: Salmoniformes), with description of a new species from the southern Kuril Islands. Species Diversity 2:59–82.

Sasaki, S. 1966. Distribution and food habits of king salmon, *Oncorhynchus tshawytscha,* and steelhead rainbow trout, *Salmo gairdnerii,* in the Sacramento–San Joaquin Delta. CDFG Fish Bull. 133:108–114.

Savitz, J., and J. Janssen. 1982. Utilization of green sunfish and bluegills by largemouth bass: influence of ingestion time. Trans. Am. Fish. Soc. 111:462–464.

Scalet, C. G. 1977. Summer food habits of sympatric stream populations of spotted bass, *Micropterus punctulatus,* and large-mouth bass, *M. salmoides* (Osteichthyes: Centrarchidae). SW Nat. 21:493–501.

Schaffter, R. G. 1997a. Growth of white catfish in California's Sacramento–San Joaquin Delta. Calif. Fish Game 83:57–67.

———. 1997b. White sturgeon spawning migrations and location of spawning habitat in the Sacramento River, California. Calif. Fish Game 83:1–20.

Schaffter R. G., and D. W. Kohlhorst. 1997. Mortality rates of white catfish in California's Sacramento–San Joaquin Delta. Calif. Fish Game 83:45–57.

———. 1999. Status of white sturgeon in the Sacramento–San Joaquin estuary. Calif. Fish Game 85:37–41.

Schindler, D. E., J. R. Hodgson, and J. F. Kitchell. 1997. Density-dependent changes in individual foraging specialization of largemouth bass. Oecologia 110:592–600.

Schindler, D. E., R. A. Knapp, and P. R. Leavitt. 2001. Alteration of nutrient cycles and algal production resulting from fish introductions into mountain lakes. Ecosystems 4:308–320.

Schluter, D., and L. M. Nagel. 1995. Parallel speciation by natural selection. Am. Nat. 146:292–301.

Schmidt, R. E., and T. Stillman. 1998. Evidence of potamodromy in an estuarine population of smallmouth bass (*Micropterus dolomieu*). J. Freshw. Biol. 13:155–163.

Schoenherr, A. A. 1979. Niche separation within a population of freshwater fishes in an irrigation drain near the Salton Sea, California. Bull. S. Calif. Acad. Sci. 78:46–55.

———. 1985. Replacement of *Cyprinodon macularius* by *Tilapia zillii* in an irrigation drain near the Salton Sea. Proc. Desert Fishes Counc. 13–15:65–66.

———. 1988. A review of the life history and status of the desert pupfish, *Cyprinodon macularius.* Bull. S. Calif. Acad. Sci. 87:104–134.

———. 1992. The effect of a flash flood on the Salt Creek, Riverside County, population of the endangered desert pupfish, *Cyprinodon macularius.* Proc. Desert Fishes Counc. 12–13: 53–60.

Schreck, C. B., and R. J. Behnke. 1971. Trouts of the upper Kern River Basin, California, with reference to systematics and evolution of western North American *Salmo.* J. Fish. Res. Board Can. 28:987–998.

Schultz, L. P. 1930. The life history of *Lampetra planeri* Bloch, with a statistical analysis of the rate of growth of the larvae from western Washington. Mus. Zool. Univ. Mich. Occas. Pap. 221. 39 pp.

———. 1933. The age and growth of *Atherinops affinis oregonia* Jordan and Snyder, and of subspecies of baysmelt along the Pacific coast of the United States. Univ. Wash. Publ. Biol. 2:45–102.

Schultz, L. P., and A. C. DeLacey. 1935. Fishes of the American Nothwest: a catalogue of the fishes of Washington and Oregon, with distributional records and a bibliography. J. Pan-Pacific Res. Inst. 10:365–380.

Schulz, P. D. 1980. Fish remains. Pages 1–18 *in* M. Praetzellis, A. Praetzellis, and M. R. Brown, eds. Historical archaeology at the Golden Eagle site. Sonoma State University, Rohnert Park, Anthropological Studies Center.

———. 1995. Prehistoric fish remains, including thicktail chub, from the Pajaro River system. Calif. Fish Game 81:82–84.

Schulz, P. D., and D. D. Simons. 1973. Fish species diversity in a prehistoric central California Indian midden. Calif. Fish Game 59:107–113.

Schwartz, D. P., and O. E. Maughn. 1984. The feeding preferences of *Tilapia aurea* (Steindacher) for five aquatic plants. Proc. Okla. Acad. Sci. 64:14–16.

Scofield, E. C. 1931. The striped bass of California (*Roccus lineatus*). CDFG Fish Bull. 29:1–84.

Scofield, N. B. 1916. The humpback and dog salmon taken in San Lorenzo River. Calif. Fish Game 2:41.

Scoppetone, G. G. 1988. Growth and longevity of the cui-ui and other catostomids and cyprinids in western North America. Trans. Am. Fish. Soc. 117:301–307.

Scoppetone, G. G., and G. Vinyard. 1991. Life history and management of four endangered suckers. Pages 303–358 *in* W. L. Minckley and J. E. Deacon, eds. Battle against extinction: native fish management in the American West. Tucson: University of Arizona Press.

Scoppetone, G. G., J. E. Harvey, S. P. Shea, B. Nielsen, and P. H. Rissler. 1992. Ichthyofaunal survey and habitat monitoring of streams inhabited by the Modoc sucker (*Catostomus microps*). Unpubl. rpt., Natl. Biol. Surv., Reno, Nev. 51 pp.

Scoppetone, G. G., S. Shea, and M. E. Buettner. 1995. Information on population dynamics and life history of shortnose sucker (*Chasmistes brevirostris*) and Lost River sucker (*Deltistes luxatus*) in Tule and Clear lakes. Unpubl. rpt., Natl. Biol. Surv., NW Biol. Cent., Reno, Nev. 79 pp.

Scott, D., J. Hewitson, and J. C. Fraser. 1978. The origins of rainbow trout, *Salmo gairdneri* Richardson, in New Zealand. Calif. Fish Game 64:307–308.

Scott, W. B. 1967. Freshwater fishes of eastern Canada. Toronto: University of Toronto Press.

Scott, W. B., and E. J. Crossman. 1973. Freshwater fishes of Canada. Fish. Res. Board Can. Bull. 184. 966 pp.

Seaburg, K. G., and J. B. Moyle. 1964. Feeding habits, digestion rates, and growth of some Minnesota warmwater fishes. Trans. Am. Fish. Soc. 93:269–285.

Seale, A. 1896. Note on *Deltistes,* a new genus of catostomid fish. Proc. Natl. Acad. Sci. 6:269.

Seeley, C. M., and G. W. McCammon. 1966. Kokanee. Pages 274–294 *in* A. Calhoun, ed. Inland fisheries management. Sacramento: California Department of Fish and Game.

Seki, H., A. Hamada, T. Iwami, and R. J. LeBrasseur. 1981. Hobikiami sail trawling in Japan. Fisheries 6(6):2–15.

Setzler-Hamilton, E. M., J. A. Whipple, and B. Macfarlane. 1988. Striped bass populations in Chesapeake and San Francisco bays: two environmentally impacted estuaries. Marine Poll. Bull. 19:466–477.

Shao, B. 1997a. Effects of golden shiner (*Notemigonus crysoleucas*) nest association on host pumpkinseeds (*Lepomis gibbosus*): evidence for a non-parasitic relationship. Behav. Ecol. Sociobiol. 41:399–406.

———. 1997b. Nest association of pumpkinseed, *Lepomis gibbosus,* and golden shiner, *Notemigonus crysoleucas.* Env. Biol. Fish. 50:41–48.

Shapovalov, L. 1941. The freshwater fish fauna of California. Proc. Sixth Pac. Sci. Congr. 3:441–446.

———. 1944. The tench in California. Calif. Fish Game 30: 54–57.

Shapovalov, L., and W. A. Dill. 1950. A check list of the freshwater and anadromous fishes of California. Calif. Fish Game 36:382–391.

Shapovalov, L., and A. C. Taft. 1954. The life histories of the steelhead rainbow trout (*Salmo gairdneri gairdneri*) and silver salmon (*Oncorhynchus kisutch*). CDFG Fish Bull. 98:1–275.

Shapovalov, L., W. A. Dill, and A. J. Cordone. 1959. A revised checklist of the freshwater and anadromous fishes of California. Calif. Fish Game 45:155–180.

Shaw, E., and J. Allen. 1977. The productive behavior of the female shiner perch, *Cymatogaster aggregata,* family Embiotocidae. Mar. Biol. 40:81–86.

Shaw, T., C. Jackson, D. Nehler, and M. Marshall. 1997. Klamath River (Iron Gate Dam to Seiad Creek) life stage periodicities for chinook, coho, and steelhead. Draft rpt., USFWS Coastal Calif. Fish Wildl. Off., Arcata. 54 pp.

Sheldon, A. L. 1968. Drift, growth, and mortality of juvenile sculpins in Sagehen Creek, California. Trans. Am. Fish. Soc. 97:495–496.

Shiraishi, Y. 1957. Feeding habits of pond smelt, *Hypomesus olidus,* and plankton succession in Lake Suwa. Tokyo Bull. Freshw. Fish. Res. Lab. 7:33–55.

Sholes, W. H., and R. J. Hallock. 1979. An evaluation of rearing fall-run chinook salmon, *Oncorhynchus tshawytscha,* to yearlings at Feather River Hatchery, with a comparison of returns from hatchery and downstream releases. Calif. Fish Game 65:239–255.

Shoup, D. E., and L. G. Hill. 1997. Ecomorphological diet predictions: an assessment using inland silverside (*Menidia beryllina*) and longear sunfish (*Lepomis megalotis*) from Lake Texoma. Hydrobiologia 350:87–98.

Shrode, J. B. 1975. Developmental temperature tolerance of a Death Valley pupfish (*Cyprinodon nevadensis*). Physiol. Zool. 48:378–389.

Shrode, J. B., and S. D. Gerking. 1977. Effects of constant and fluctuating temperatures on reproductive performance of a desert pupfish, *Cyprinodon n. nevadensis.* Physiol. Zool. 50:1–10.

Siefert, R. E. 1968. Reproductive behavior, incubation and mortality of eggs, and postlarval food selection of white crappie. Trans. Am. Fish. Soc. 97:252–259.

Sigler, W. F. 1951. The life history and management of the mountain whitefish, *Prosopium williamsoni* (Girard) in Logan River, Utah. Utah State Ag. Coll. Bull. 347. 21 pp.

———. 1958. The ecology and use of carp in Utah. Utah State Univ. Agric. Exp. Stn. Bull. 405. 63 pp.

Sigler, W. F., and R. R. Miller. 1963. Fishes of Utah. Salt Lake City: Utah Dept. Fish Game.

Sigler, W. F., and J. W. Sigler. 1987. Fishes of the Great Basin: natural history. Reno: University of Nevada Press.

Sigler, W. F., W. T. Helm, P. A. Kucera, S. Vigg, and G. W. Workman. 1983. Life history of the Lahontan cutthroat trout, *Salmo clarki henshawi,* in Pyramid Lake, Nevada. Great Basin Nat. 43:1–29.

Simon, D. C., and D. F. Markle. 1997. Interannual abundance of nonnative fathead minnows (*Pimephales promelas*) in upper Klamath Lake, Oregon. Great Basin Nat. 57:142–148.

Simon, T. P., and J. B. Kaskey. 1992. Description of the eggs, larvae, and early juveniles of the bigscale logperch, *Percina macrolepida* Stevenson, from the west fork of the Trinity River Basin, Texas. SW Nat. 37:28–34.

Simons, A. M., and R. L. Mayden. 1998. Phylogenetic relationships of the western North American phoxinins (Actinopterygii, Cyprinidae) as inferred from Mitochondrial 12S and 16S ribosomal RNA sequences. Mol. Phylo. Evol. 9:308–329.

Simonson, T. D., and W. A. Swenson. 1990. Critical stream velocity for young-of-year smallmouth bass in relation to habitat use. Trans. Am. Fish. Soc. 119:902–910.

Skillman, R. A. 1969. The population, organization, and dispersal of redear sunfish in Lake Berryessa. Ph.D. thesis, Univ. Calif., Davis. 101 pp.

Skinner, J. E. 1962. An historical view of the fish and wildlife resources of the San Francisco Bay area. CDFG Water Projects Branch Rpt. 1. 225 pp.

———. 1972. Ecological studies of the Sacramento–San Joaquin Estuary. CDFG Delta Fish Wildl. Prot. Stud. Rpt. 8. 94 pp.

Slater, D. W. 1963. Winter-run chinook salmon in the Sacramento River, California, with notes on water temperature requirements at spawning. USFWS Spec. Sci. Rpt. Fish. No. 461. 9 pp.

Smale, M. A., and C. F. Rabeni. 1995. Hypoxia and hyperthermia tolerances in headwater stream fishes. Trans. Am. Fish. Soc. 124:698–710.

Smith, C. L., and C. R. Powell. 1971. The summer fish communities of Brier Creek, Marshall County, Oklahoma. Am. Mus. Nov. 2548:1–30.

Smith, G. E., and M. E. Aceituno. 1987. Habitat preference criteria for brown, brook, and rainbow trout in eastern Sierra Nevada streams. CDFG Stream Eval. Rpt. 87-2. 104 pp.

Smith, G. E., J. M. Deinstadt, and A. C. Knutson, Jr. 1999. Changes in a wild trout fishery following chemical treatment and fish barrier construction on lower Hat Creek, California: 1968–73. CDFG Inland Fish. Admin. Rpt. 99-6. 27 pp.

Smith, G. R. 1966. Distribution and evolution of the North American catostomid fishes of the subgenus *Pantosteus,* genus *Catostomus.* Univ. Mich. Mus. Zool. Misc. Publ. 129:1–33.

———. 1981. Late Cenozoic freshwater fishes of North America. Annu. Rev. Ecol. Syst.12:163–193.

———. 1992. Phylogeny and biogeography of the Catostomidae, freshwater fishes from North America and Asia. Pages 778–826 *in* R. L. Mayden, ed. Systematics, historical ecology, and North American freshwater fishes. Stanford, Calif.: Stanford University Press.

Smith, G. R., and R. F. Stearley. 1989. The classification and scientific names of rainbow and cutthroat trouts. Fisheries (Bethesda) 14(1):4–10.

Smith, J. J. 1977. Distribution, movements, and ecology of the fishes of the Pajaro River system, California. Ph.D. dissertation, Univ. Calif., Davis. 230 pp.

———. 1982. Fishes of the Pajaro River system. Univ. Calif. Publ. Zool. 115:83–170.

Smith, J. J., and H. W. Li. 1983. Energetic factors influencing foraging tactics of juvenile steelhead trout, *Salmo gairdneri.* Pages 173–180 *in* D. L. G. Noakes et al., eds. Predators and prey in fishes. The Hague: Dr. W. Junk.

Smith, P. W., and L. M. Page. 1969. The food of spotted bass in streams of the Wabash River Drainage. Trans. Am. Fish. Soc. 98:647–651.

Snelson, F. F., Jr. 1982. Indeterminate growth in males of the sailfin molly, *Poecilia latipinna*. Copeia 1982:296–304.

———. 1984. Seasonal maturation and growth of males in a natural population of *Poecilia latipinna*. Copeia 1984: 252–255.

———. 1985. Size and morphological variation in males of the sailfin molly, *Poecilia latipinna*. Env. Biol. Fish. 13:35–47.

Snelson, F. F., Jr., and J. D. Wetherington. 1980. Sex ratio in the sailfin molly, *Poecilia latipinna*. Evolution 34:308–319.

Snider, W. M., and E. Gerstung. 1986. Instream flow requirements of the fish and wildlife resources of the Lower American River, Sacramento, California. CDFG Stream Eval. Rpt. 86-1. 32 pp.

Snider, W. M., and A. Linden. 1981. Trout growth in California streams. CDFG Inland Fish. Admin. Rpt. 81-1. 11 pp.

Snyder, D. E. 1983. Identification of catostomid larvae in Pyramid Lake and the Truckee River, Nevada. Trans. Am. Fish. Soc. 112:333–348.

Snyder, G. 1990. The practice of the wild. San Francisco: North Point Press.

Snyder, J. O. 1905. Notes on the fishes of the streams flowing into San Francisco Bay. Rpt. U.S. Bur. Fish. 5:327–338.

———. 1908a. Relationships of the fish fauna of the lakes of southeastern Oregon. Bull. U.S. Bur. Fish. 27:69–102.

———. 1908b. Description of *Pantosteus santa-anae*, a new species of fish from the Santa Ana River, California. Proc. U.S. Natl. Mus. 34:33–34.

———. 1908c. The fishes of the coastal streams of Oregon and northern California. Bull. U.S. Bur. Fish. 27:153–189.

———. 1908d. The fauna of the Russian River, California, and its relation to that of the Sacramento. Science 27:269–271.

———. 1913. The fishes of the streams tributary to Monterey Bay, California. Bull. U.S. Bur. Fish. 32:49–72.

———. 1916. The fishes of the streams tributary to Tomales Bay, California. Bull. U.S. Bur. Fish. 34:375–381.

———. 1918. The fishes of the Lahontan system of Nevada and northeastern California. Bull. U.S. Bur. Fish. 35:31–86.

———. 1919a. An account of some fishes from the Owens River, California. Proc. U.S. Natl. Mus. 54:201–205.

———. 1919b. The fishes of Mohave River, California. Proc. U.S. Natl. Mus. 54:297–299.

———. 1931. Salmon of the Klamath River, California. CDFG Fish Bull. 34. 129 pp.

———. 1933a. Description of *Salmo seleneris*, a new California trout. Proc. Calif. Acad. Sci. 20(11):471–472.

———. 1933b. California trout. Calif. Fish Game 19:81–112.

Snyder, R. J. 1984. Seasonal variation in the diet of the threespine stickleback, *Gasterosteus aculeatus*, in Contra Costa County, California. Calif. Fish Game 70:167–172.

———. 1990. Clutch size of anadromous and freshwater threespine sticklebacks: a reassessment. Can. J. Zool. 68:2027–2030.

———. 1991a. Migration and life histories of the threespine stickleback: evidence for adaptive variation in growth rate between populations. Env. Biol. Fish. 31:381–388.

———. 1991b. Quantitative genetic analysis of life histories in two freshwater populations of the threespine stickleback. Copeia 1991:526–529.

Snyder, R. J., and H. Dingle. 1989. Adaptive, genetically based differences in life history between estuary and freshwater three-spine sticklebacks (*Gasterosteus aculeatus* L.). Can. J. Zool. 67:2448–2454.

———. 1990. Effects of freshwater and marine overwintering environments on life histories of threespine sticklebacks: evidence for adaptive variation between anadromous and resident freshwater fish populations. Oecologia 84:386–390.

Soltz, D. L., and R. J. Naiman. 1978. The natural history of native fishes in the Death Valley system. Nat. Hist. Mus. Los Angeles Co. Sci. Ser. 30:1–76.

Sommer, T., R. Baxter, and B. Herbold. 1997. Resilience of splittail in the Sacramento–San Joaquin estuary. Trans. Am. Fish. Soc. 126:961–976.

Sommer, T. R., B. Harrell, M. Nobriga, R. Brown, P. Moyle, W. Kimmerer, and L. Schemel. 2001a. California's Yolo Bypass: evidence that flood control can be compatible with fisheries, wetlands, wildlife, and agriculture. Fisheries 26(8):6–16.

Sommer, T. R., M. L. Nobriga, W. C. Harrell, W. Batham, and W. J. Kimmerer. 2001b. Floodplain rearing of juvenile chinook salmon: evidence of enhanced growth and survival. Can. J. Fish. Aquat. Sci. 58:325–333.

Sorenson, P. W., J. R. Cardwell, T. Essington, and D. E. Weigel. 1995. Reproductive interactions between sympatric brook and brown trout in a small Minnesota stream. Can. J. Fish. Aquat. Sci. 52:1958–1965.

Soule, M. 1951. Power development of the Kings River drainage, Fresno County, California. Rpt. No. 3., Junction development of the Kings River and its Middle and South Forks. CDFG Inland Fish. Admin. Rpt. 51. 17 pp.

Spina, A. P. 2001. Incubation discharge and aspects of brown trout populations dynamics. Trans. Am. Fish. Soc. 130: 322–327.

Staley, C. S. 1980. Life history aspects of the Sacramento blackfish, *Orthodon microlepidotus* (Ayres) in the Beach/Stone lakes basin, California. M.S. thesis, Calif. State Univ., Sacramento. 55 pp.

Staley, J. 1966. Brown trout. Pages 233–242 *in* A. Calhoun, ed. Inland fisheries management. Sacramento: California Department of Fish and Game.

Stanford, J. A., and J. V. Ward. 1986. Fish of the Colorado system. Pages 385–402 *in* B. R. Davies and K. F. Walker, eds. The ecology of river systems. Dordrecht: Dr. W. Junk.

Stanley, S. E., P. B. Moyle, and H. B. Shaffer. 1995. Allozyme analysis of delta *smelt, Hypomesus transpacificus*, and longfin smelt, *Spirinchus thaleichthys*, in the Sacramento–San Joaquin estuary. Copeia 1995:390–396.

Stauffer, J. R., Jr. 1986. Effects of salinity on preferred and lethal temperatures of the Mozambique tilapia, *Oreochromis mossambicus* (Peters). Water Res. Bull. 22:205–208.

Stauffer, J. R., Jr., J. M. Boltz, and S. E. Boltz. 1989. Temperature preference of the redbelly tilapia, *Oreochromis zilli* (Gervais). Arch. Hydrobiol. 114:453–456.

Stauffer, J. R., Jr., C. H. Hocutt, and W. F. Goodfellow. 1985. Effects of sex and maturity on preferred temperatures: a proximate factor for increased survival of young *Poecilia latipinna?* Arch. Hydro. 103:129–132.

Stearley, R. F. 1992. Historical ecology of Salmoninae, with special reference to *Oncorhynchus*. Pages 622–658 *in* R. L. Mayden, ed. Systematics, historical ecology, and North American freshwater fishes. Stanford, Calif.: Stanford University Press.

Stearley, R. F., and G. R. Smith. 1993. Phylogeny of Pacific trouts and salmons (*Oncorhynchus*) and genera of the family Salmonidae. Trans. Am. Fish. Soc. 122:1–33.

Stearns, S. C. 1983. The genetic basis of differences in life-history traits among six populations of mosquitofish (*Gambusia affinis*) that shared ancestors in 1905. Evolution 37:618–627.

Stefferud, J. A. 1993. Spawning season and microhabitat use by California golden trout (*Oncorhynchus mykiss aguabonita*) in the southern Sierra Nevada. Calif. Fish Game 79:133–144.

Stein, R. A., P. E. Reimers, and J. D. Hall. 1972. Social interaction between juvenile coho (*Oncorhynchus kisutch*) and fall chinook salmon (*O. tshawytscha*) in Sixes River, Oregon. J. Fish. Res. Board Can. 29:1737–1748.

Steiner, W. P. 1996. A history of the salmonid decline in the Russian River. Potter Valley, Calif.: Steiner Environmental Consulting.

Stephens, J. S., Jr. 1990. Maximum salinity tolerance of the egg, larvae, and adult life stages of Salton Sea sport fish. CDFG Proj. Rpt. F-51-R-1. 8 pp.

Stevens, D. E. 1966a. Distribution and food habits of the American shad, *Alosa sapidissima*, in the Sacramento–San Joaquin Estuary. Pages 97–107 *in* J. L. Turner and D. W. Kelley, eds. Ecological studies of the Sacramento–San Joaquin Estuary, part II: fishes of the Delta. CDFG Fish Bull. 136.

———. 1966b. Food habits of striped bass (*Roccus saxatilis*) in the Sacramento–San Joaquin Delta. Pages 68–96 *in* J. L. Turner and D. W. Kelley, eds. Ecological studies of the Sacramento–San Joaquin Estuary, part II: fishes of the Delta. CDFG Fish Bull. 136.

———. 1977. Striped bass (*Morone saxatilis*) year class strength is relation to river flow in the Sacramento–San Joaquin Estuary, California. Trans. Am. Fish. Soc. 106:34–42.

Stevens, D. E., and L. W. Miller. 1970. Distribution and abundance of sturgeon larvae in the Sacramento–San Joaquin River system, Calif. Fish Game 56:80–86.

———. 1983. Effects of river flow on abundance of young chinook salmon, American shad, longfin smelt, and Delta smelt in the Sacramento–San Joaquin river system. N. Am. J. Fish. Mgmt. 3:425–437.

Stevens, D. E., H. K. Chadwick, and R. E. Painter. 1987. American shad and striped bass in California's Sacramento–San Joaquin river system. Am. Fish. Soc. Symp. 1:66–78.

Stevens, D. E., D. W. Kohlhorst, L. W. Miller, and D. W. Kelley. 1985. The decline of striped bass in the Sacramento–San Joaquin estuary, California. Trans. Am. Fish. Soc. 114:12–30.

Stevenson, M. M. 1971. *Percina macrolepida* (Pisces, Percidae, Etheostomatinea), a new percid fish of the subgenus *Percina* from Texas. SW Nat. 16:65–83.

Stevenson, M. M., and B. A. Thomson. 1978. Further distribution records for the bigscale logperch, *Percina macrolepida* (Osteichthyes: Percidae) from Oklahoma, Texas and Louisiana, with notes on its occurrence in California. SW Nat. 23:303–313.

Stier, D. J., and J. H. Crance. 1985. Habitat suitability index models and instream suitability curves: American shad. USFWS Biol. Rpt. 82 (10.88). 34 pp.

Stone, L. 1876. Report of operations in California in 1873: Clear Lake. U.S. Fish. Comm. Rpt. 1873–1874, 1874–1875. Appendix B: 377–381.

Strange, E. M. 1995. Pattern and process in stream fish community organization: field study and simulation modeling. Ph.D. dissertation, Univ. Calif., Davis. 147 pp.

Strange, E. M., P. B. Moyle, and T. C. Foin. 1992. Interactions between stochastic and deterministic processes in stream fish community assembly. Env. Biol. Fish. 36:1–15.

Stubbs, K., and R. White. 1993. Lost River (*Deltistes luxatus*) and shortnose (*Chasmistes brevirostris*) sucker recovery plan. Portland: USFWS.

Sturgess, J. A. 1976. Taxonomic status of *Percina* in California. Calif. Fish Game 62:79–81.

Sturgess, J. A., and P. B. Moyle. 1978. Biology of rainbow trout (*Salmo gairdneri*), brown trout (*Salmo trutta*) and interior Dolly Varden (*Salvelinus confluentus*) in the McCloud River, California, in relation to management. Calif. Nev. Wildl. 1978:239–350.

Sublette, J. E., M. D. Hatch, and M. Sublette. 1990. The fishes of New Mexico. Albuquerque: University of New Mexico Press.

Svalastog, D. 1991. A note on maximum age of brown trout, *Salmo trutta* L. J. Fish Biol. 38:967–968.

Swanson, C., and J. J. Cech, Jr. 1995. Environmental tolerances and requirements of the delta smelt, *Hypomesus transpacificus*. Univ. Calif., Davis, final rpt. to Calif. Dept. Water Resources. 77 pp.

Swanson, C., J. J. Cech, Jr., and R. H. Piedrahita. 1996. Mosquitofish: biology, culture, and use in mosquito control. Elk Grove, Calif.: California Mosquito Vector Control Association.

Swanson, C., T. Reid, P. S. Young, and J. J. Cech, Jr. 2000. Comparative environmental tolerances of threatened delta smelt (*Hypomesus transpacificus*) and introduced wakasagi (*H. nipponensis*) in an altered California estuary. Oecologia 123: 384–390.

Swanson, C., P. S. Young, and J. J. Cech, Jr. 1998. Swimming performance and behavior of delta smelt: maximum performance and behavioral and kinematic limitations of swimming at submaximal velocities. J. Exp. Biol. 201:333–345.

Swe, W., and W. R. Dickinson. 1970. Sedimentation and thrusting of late Mesozoic rocks in the coast ranges near Clear Lake, California. Geol. Soc. Am. Bull. 81:165–187.

Sweetnam, D. A. 1995. Field identification of delta smelt and wakasagi. IEP Newslett. Spring 1995:1–3.

———. 1999. Status of delta smelt in the Sacramento–San Joaquin estuary. Calif. Fish. Game 85:22–27.

Sweetnam, D. A., and D. E. Stevens. 1993. A status review of the delta smelt (*Hypomesus transpacificus*) in California. Report to Fish and Game Commission, Candidate Species Rpt. 93-DS. 98 pp.

Swenson, R. O. 1995. The reproductive behavior and ecology of the tidewater goby *Eucyclogobius newberryi* (Pisces: Gobiidae). Ph.D. dissertation, Univ. Calif., Berkeley. 230 pp.

———. 1997. Sex-role reversal in the tidewater goby, *Eucyclogobius newberryi*. Env. Biol. Fish. 50:27–40.

———. 1999. The ecology, behavior, and conservation of the tidewater goby, *Eucyclogobius newberryi*. Env. Biol. Fish. 55:99–114.

Swenson, R. O., and A. T. McCray. 1996. Feeding ecology of the tidewater goby. Trans. Am. Fish. Soc. 125:956–970.

Swift, C. 1965. Early development of the hitch, *Lavinia exilicauda*, of Clear Lake, California. Calif. Fish Game 51:74–80.

Swift, C. C., P. Duangsitti, C. Clemente, K. Hasserd, and L. Valle. 1997. Biology and distribution of the tidewater goby, *Eucyclogobius newberryi*, on Vandenberg Air Force Base, Santa Barbara County, California. Final rpt., USNBS Cooperative Agreement 1445-007-94-8129. 121 pp.

Swift, C. C., J. L. Nelson, C. Maslow, and T. Stein. 1989. Biology and distribution of the tidewater goby, *Eucyclogobius newberryi* (Pisces: Gobiidae), of California. Los Angeles Co. Mus. Nat. Hist. Contribs. Sci. 404:1–19.

Swift, C. C., T. R. Haglund, M. Ruiz, and R. N. Fisher. 1993. The status and distribution of the freshwater fishes of southern California. Bull. S. Calif. Acad. Sci. 92:101–167.

Taft, A. C. 1938. Pink salmon in California. Calif. Fish Game 24:197–198.

Taft, A. C., and G. I. Murphy. 1950. Life history of the Sacramento squawfish (*Ptychocheilus grandis*). Calif. Fish Game 36: 147–164.

Talbot, G. B., and J. E. Sykes. 1958. Atlantic coast migrations of American shad. Fishery Bull. 58:473–490.

Tasto, R. N. 1975. Aspects of the biology of the Pacific staghorn sculpin, *Leptocottus armatus* Girard, in Anaheim Bay. Pages 123–135 in E. D. Lane and C. W. Hill, eds. The marine resources of Anaheim Bay. CDFG Fish Bull. 165.

Tate, A. E., and L. A. Helfrich. 1998. Off-season spawning of sunshine bass (*Morone chrysops x M. saxatilis*) exposed to 6- or 9-month phase-shifted photothermal cycles. Aquaculture 167:67–83.

Taucher, C. 1987. How to apply for a state fishing record. Outdoor California 48(6):18–19.

Taylor, E. B. 1988. Water temperature and velocity as determinants of microhabitats of juvenile chinook and coho salmon in a laboratory stream channel. Trans. Am. Fish. Soc. 117:22–28.

———. 1990. Phenotypic correlates of life-history variation in juvenile chinook salmon, *Oncorhynchus tshawytscha*. J. Anim. Ecol. 59:455–468.

———. 1991. Behavioural interactions and habitat use in juvenile chinook, *Oncorhynchus tshawytscha,* and coho, *O. kisutch,* salmon. Anim. Behav. 42:729–744.

Taylor, F. R., R. R. Miller, J. W. Pedretti, and J. E. Deacon. 1988. Rediscovery of the Shoshone pupfish, *Cyprinodon nevadensis shoshone* (Cyprinodontidae), at Shoshone Springs, Inyo County, California. Bull. S. Calif. Acad. Sci. 87:67–73.

Taylor, J. N., D. B. Snyder, and W. R. Courtenay, Jr. 1986. Hybridization between two introduced, substrate-spawning tilapias (Pisces: Cichlidae) in Florida. Copeia 1986:903–909.

Taylor, T. L. 1993. A survey of 1992 brown trout spawning in Rush and Lee Vining creeks, Mono County, California. Unpubl. rpt., Trihey and Associates. 45 pp.

Taylor, T. L., P. B. Moyle, and D. G. Price. 1982. Fishes of the Clear Lake Basin. Univ. Calif. Publ. Zool. 115:171–224.

Thomas, J. L. 1967. The diet of juvenile and adult striped bass, *Roccus saxatilis,* in the Sacramento–San Joaquin river system. Calif. Fish Game 53:49–62.

Thomas, J. W., and the Scientific Analysis Team. 1993. Viability assessments and management considerations for species associated with late-successional and old-growth forests of the Pacific Northwest. Portland: USFS.

Thomson, J. M. 1963. Synopsis of biological data on the grey mullet, *Mugil cephalus* Linneaus, 1758. CSIRO Fish. Synopsis 1. 65 pp.

Thompson, G. E., and R. W. Davies. 1976. Observations on the age, growth, reproduction, and feeding of mountain white-

fish (*Prosopium williamsoni*) in the Sheep River, Alberta. Trans. Am. Fish. Soc. 105:208–219.

Tinbergen, N. 1953. Social behavior in animals. London: Methuen.

Tippets, W. E., and P. B. Moyle. 1978. Epibenthic feeding by rainbow trout (*Salmo gairdneri*) in the McCloud River, California. J. Anim. Ecol. 47:549–559.

Titus, R. G. 1990. Seasonal condition of Lahontan cutthroat trout at Heenan Lake, California, with fecundity and spawning-female size estimates. CDFG Inland Fish. Admin. Rpt. 90-9. 12 pp.

Titus, R. G., D. C. Erman, and W. M. Snider. 1994. History and status of steelhead in California coastal drainages south of San Francisco Bay. Unpubl. ms. 219+ pp.

Todd, E. S. 1968. Terrestrial sojourns of the longjaw mudsucker, *Gillichthys mirabilis.* Copeia 1968:192–199.

Todd, E. S., and A. W. Ebeling. 1966. Aerial respiration in the longjaw mudsucker *Gillichthys mirabilis* (Teleostei: Gobiidae). Biol. Lab. Woods Hole 130:265–288.

Tomiyama, I. 1936. Gobiidae of Japan. Jpn. J. Zool. 7:37–112.

Tranah, G. J. 2001. The development and application of unique molecular genetic markers for discerning population structure and detecting hybridization of endangered fish species. Ph.D. dissertation, Univ. Calif., Davis. 117 pp.

Trautman, M. B. 1957. Fishes of Ohio. Columbus: Ohio State University Press.

Travanti, L. 1990. The effects of piscicidal treatment on the fish community of a northern California stream. M.S. thesis, Calif. State Univ., Chico. 67 pp.

Trenham, P. C., H. B. Shaffer, and P. B. Moyle. 1998. Biochemical identification and assessment of population subdivision in morphologically similar native and invading smelt species (*Hypomesus*) in the Sacramento–San Joaquin Estuary, California. Trans. Am. Fish. Soc. 127:417–424.

Tres, J. A. 1992. Breeding biology of the arroyo chub, *Gila orcutti* (Pisces: Cyprinidae). M.S. thesis, Calif. State Poly. Univ., Pomona. 73 pp.

Trewavas, E. 1983. Tilapiine fishes of the genera *Sarotherodon, Orechromis,* and *Danakilia.* London: British Museum (Natural History).

Trexler, J. C. 1985. Variation in degree of viviparity in the sailfin molly, *Poecilia latipinna.* Copeia 1985:999–1004.

———. 1989. Phenotypic plasticity in poeciliid life histories. Pages 201–216 in G. K. Meffe and F. F. Snelson, Jr., eds. Ecology and evolution of livebearing fishes (Poeciliidae). Englewood Cliffs, N.J.: Prentice-Hall.

Trexler, J. C., R. C. Tempe, and J. Travis. 1994. Size-selective predation of sailfin mollies by two species of heron. Oikos 69:250–258.

Trihey and Associates. 1996a. Instream flow requirements for tribal trust species in the Klamath River. Concord, Calif. 43 pp.

———. 1996b. Lagunitas Creek anadromous fish monitoring report, Fall, 1996. Concord, Calif. 25 pp.

Trotter, P. C. 1997. Sea-run cutthroat trout life history profile. Pages 7–15 in J. D. Hall, P. A. Bisson, and R. E. Gresswell, eds. Sea-run cutthroat trout: biology, management and future conservation. Corvallis, Oreg.: Oregon Chapter of the American Fisheries Society.

Tucker, M. E., C. M. Williams, and R. R. Johnson. 1998. Abundance, food habits, and life history aspects of Sacramento squawfish and striped bass at the Red Bluff Diversion Complex, California, 1994–1996. Red Bluff, Calif.: USFWS Red Bluff Research Pumping Plant Rpt. 4. 54 pp.

Turner, B. J. 1983. Genic variation and differentiation of remnant natural populations of the desert pupfish, *Cyprinodon macularius.* Evolution 37:690–700.

Turner, J. L. 1966a. Distribution and food habits of ictalurid fishes in the Sacramento–San Joaquin Delta. Pages 130–143 *in* J. L. Turner and D. W. Kelley, eds. Ecological Studies of the Sacramento–San Joaquin Delta, part II. CDFG Fish Bull. 136.

———. 1966b. Distribution and food habits of centrarchid fishes in the Sacramento–San Joaquin Delta. Pages 144–151 *in* J. L. Turner and D. W. Kelley, eds. Ecological Studies of the Sacramento–San Joaquin Delta, part II. CDFG Fish Bull. 136.

———. 1966c. Distribution of cyprinid fishes in the Sacramento–San Joaquin Delta. CDFG Fish Bull. 136.

———. 1966d. Distribution of threadfin shad, *Dorosoma petenense,* tule perch, *Hysterocarpus traskii,* and crayfish spp. in the Sacramento–San Joaquin Delta. CDFG Fish Bull. 136.

———. 1972. Striped bass. Pages 36–43 *in* J. E. Skinner, ed. Ecological studies of the Sacramento–San Joaquin Estuary. CDFG Delta Fish Wildl. Prot. Stud. Rpt. 8.

Turner, J. L., and D. W. Kelley. 1966. Ecological studies of the Sacramento–San Joaquin Delta. CDFG Fish Bull. 136:1–168.

Turner, J. M. 1989. Age and growth of bluegill (*Lepomis macrochirus*) in the Sacramento–San Joaquin Delta, California. M.S. thesis, Calif. State Univ., Sacramento. 35 pp.

Turner, P. R., and R. C. Summerfelt. 1971. Reproductive biology of the flathead catfish, *Pylodictis olivaris* (Rafinesque), in a turbid Oklahoma reservoir. Pages 107–119 *in* G. E. Hall, ed. Reservoir fisheries and limnology. Bethesda, Md.: American Fisheries Society Spec. Publ. 8.

Tyus, H. M. 1987. Distribution, reproduction, and habitat use of razorback sucker, *Xyrauchen texanus,* in the Green River, Utah, 1979–1986. Trans. Am. Fish. Soc. 116:111–116.

Tyus, H. M. 1991a. Ecology and management of Colorado squawfish. Pages 379–402 *in* W. L. Minckley and J. E. Deacon, eds. Battle against extinction: native fish management in the American West. Tucson: University of Arizona Press.

———. 1991b. Movements and habitat use of young Colorado squawfish in the Green River, Utah. J. Freshw. Ecol. 6:43–51.

Tyus, H. M., and J. M. Beard. 1990. *Esox lucius* (Esocidae) and *Stizostedion vitreum* (Percidae) in the Green River Basin, Colorado and Utah. Gr. Basin Nat. 50:33–39.

Tyus, H. M., and C. A. Karp. 1989. Habitat use and streamflow needs of rare and endangered fishes, Yampa River, Colorado. USFWS Biol. Rpt. 89(14). 26 pp.

Underwood, T. J. 1983. Age and growth of the western sucker (*Catostomus occidentalis*) in Whale Rock Reservoir, California. Senior thesis, Biol. Sci. Dept., Calif. Poly. Univ., San Luis Obispo. 14 pp.

University of California–Mexus Border Water Project. 1999. Alternative futures for the Salton Sea. Issue Paper No. 1. 23 pp.

Ursui, C. A. 1981. Behavioral, metabolic, and seasonal size comparisons of an introduced gobiid fish, *Acanthogobius flavimanus,* and a native cottid, *Leptocottus armatus,* from Upper Newport Bay, California. M.S. thesis, Calif. State Univ., Fullerton. 52 pp.

U.S. Commission for Fish and Fisheries. 1892. Report to the Commissioner for 1888, part XVI. Washington, D.C.

———. 1894. Report of the Commissioner for the year ending June 30, 1892. Part 18: LVII–LVIII. Washington, D.C.

U.S. Fish and Wildlife Service. 1979. Klamath River fisheries investigations: progress, problems, and prospects. Annual Report. Arcata, Calif. 49 pp.

———. 1984a. Recovery plan for the Mohave tui chub, *Gila bicolor mohavensis.* Portland. 56 pp.

———. 1984b. Recovery plan for Owens pupfish, *Cyprinodon radiosus.* Portland. 47 pp.

———. 1994. Lahontan cutthroat trout, *Oncorhynchus clarki henshawi,* recovery plan. Portland. 147 pp.

———. 1996. Recovery plan for the Sacramento/San Joaquin Delta native fishes. Portland. 195 pp.

———. 1999. Determination of threatened status for the Sacramento splittail. Fed. Reg. 64(25):5963–5981.

Utoh, H. 1988. Life history and fishery of the smelt *Hypomesus transpacificus nipponensis* McAllister. Jpn. J. Limnol. 49: 296–299.

Uyeno, T. 1966. Osteology and phylogeny of the American cyprinid fishes, allied to the genus *Gila.* Ph.D. dissertation, Univ. Mich., Ann Arbor. 173 pp.

Vane-Wright, R. I., C. J. Humphries, and P. H. Williams. 1991. What to protect?—systematics and the agony of choice. Biol. Cons. 55:235–254.

Vanicek, C. D. 1970. Distribution of Green River fishes in Utah and Colorado following closure of Flaming George Dam. SW Nat. 19:297–315.

———. 1980. Decline of the Lake Greenhaven Sacramento perch population. Calif. Fish Game 66:178–183.

Vanicek, C. D., and H. Kramer. 1969. Life history of the Colorado squawfish, *Ptychocheilus lucius,* and the Colorado chub, *Gila robusta,* in the Green River in Dinosaur National Monument, 1964–1966. Trans. Am. Fish. Soc. 98:192–208.

Varley, M. E. 1967. British freshwater fishes. London: Fishing News.

Vicker, C. E. 1973. Aspects of the life history of the Mojave chub, *Gila bicolor mohavensis* (Snyder) from Soda Lake, California. M.A. thesis, Calif. State Univ., Fullerton. 27 pp.

Vigg, S. 1978. Vertical distribution of adult fish in Pyramid Lake, Nevada. Great Basin Nat. 38:417–428.

———. 1980. Seasonal benthic distribution of adult fish in Pyramid Lake, Nevada. Calif. Fish Game 66:49–58.

———. 1981. Species composition and relative abundance of adult fish in Pyramid Lake, Nevada. Great Basin Nat. 41: 395–408.

Vigg, S., and T. J. Hassler. 1982. Distribution and relative abundance of fish in Ruth Reservoir, California, in relation to environmental variables. Great Basin Nat. 42:529–540.

Villa, N. A. 1985. Life history of the Sacramento sucker, *Catostomus occidentalis,* in Thomes Creek, Tehama County, California. Calif. Fish Game 71:88–106.

Vincent, D. T. 1968. The influence of some environmental factors on the distribution of fishes in upper Klamath Lake. M.S. thesis, Oreg. State Univ., Corvallis. 75 pp.

Vincent, R. E., and W. H. Miller. 1969. Altitudinal distribution of brown trout and other fishes in a headwater tributary of the south Platte River, Colorado. Ecology 50:464–466.

Vinyard, G. L. 1982. Variable kinematics of Sacramento perch

(*Archoplites interruptus*) capturing evasive and nonevasive prey. Can. J. Fish. Aquat. Sci. 39:208–211.

Vinyard, G. L., R. W. Drenner, M. Gophen, U. Pollingher, D. L. Winkelman, K. D. Hambright. 1988. An experimental study of the plankton community impacts of two omnivorous filter-feeding cichlids, *Tilapia galilaea* and *Tilapia aurea.* Can. J. Fish. Aquat. Sci. 45:685–690.

Vladykov, V. D. 1973. *Lampetra pacifica,* a new nonparasitic species of lamprey (Petromyzontidae) from Oregon and California. J. Fish. Res. Board Can. 30:205–213.

Vladykov, V. D., and W. I. Follett. 1958. Redescription of *Lampetra ayresii* (Gunther) of western North America, a species of lamprey (Petromyzontidae) distinct from *Lampetra fluviatilis* (Linnaeus) of Europe. J. Fish. Res. Board Can. 15:47–77.

———. 1962. The teeth of lampreys (Petromyzonidae): their terminology and use in a key to the holarctic genera. J. Fish. Res. Board Can. 24:1067–1075.

Vladykov, V. D., and E. Kott. 1976a. A new nonparasitic species of lamprey of the genus *Entosphenus* Gill, 1862 (Petromyzonidae) from south central California. Bull. S. Calif. Acad. Sci. 75:60–67.

———. 1976b. A second nonparasitic species of *Entosphenus* Gill, 1862 (Petromyzonidae) from Klamath River System, California. Can. J. Zool. 54:974–989.

———. 1979. A new parasitic species of the holarctic lamprey genus *Entosphenus* Gill, 1862 (Petromyzonidae) from Klamath River, in California and Oregon. Can. J. Zool. 57:808–823.

———. 1984. A second record for California and additional morphological information on *Entosphenus hubbsi* Vladykov and Kott 1976 (Petromyzonidae). Calif. Fish Game 70:121–127.

Vogel, D. A., and K. R. Marine. 1991. Guide to upper Sacramento River chinook salmon life history. Redding, Calif.: USBR Central Valley Project and CH2M Hill. ca. 55 pp.

Vogele, L. E. 1975a. Reproduction of spotted bass, *Micropterus punctulatus,* in Bull Shoals Reservoir, Arkansas. USFWS Tech. Pap. 84. 21 pp.

———. 1975b. The spotted bass. Pages 21–33 *in* H. Clepper, ed. Black bass biology and management. Washington, D.C.: Sport Fishing Institute.

Voight, H. N., and T. R. Hayden. 1997. Direct observation of coastal cutthroat trout abundance and habitat utilization in the South Fork Smith River Basin, California. Pages 175–176 *in* J. D. Hall, P. A. Bisson, and R. E. Gresswell, eds. Sea-run cutthroat trout: biology, management and future conservation. Corvallis, Oreg.: Oregon Chapter of the American Fisheries Society.

Vondracek, B. 1987. Digestion rates and gastric evacuation times in relation to temperature of the Sacramento squawfish, *Ptychocheilus grandis.* NOAA Fish. Bull. 85:159–163.

Vondracek, B., and D. R. Longanecker. 1993. Habitat selection by rainbow trout *Oncorhynchus mykiss* in a California stream: implications for the instream flow incremental methodology. Ecol. Freshw. Fish 2:173–186.

Vondracek, B., J. J. Cech, Jr., and D. Longanecker. 1982. Effect of cycling and constant temperatures on the respiratory metabolism of the Tahoe sucker, *Catostomus tahoensis* (Pisces: Catostomidae). Comp. Biochem. Physiol. 73A:11–14.

Vondracek, B., J. J. Cech, Jr., and R. K. Buddington. 1989. Growth, growth efficiency, and assimilation efficiency of the Tahoe sucker in cyclic and constant temperature. Env. Biol. Fish. 24:151–156.

Vondracek, B., W. A. Wurtsbaugh, and J. J. Cech, Jr. 1988a. Growth and reproduction of the mosquitofish, *Gambusia affinis,* in relation to temperature and ration level: consequences for life history. Env. Biol. Fish. 21:45–57.

Vondracek, B., D. M. Baltz, L. R. Brown, and P. B. Moyle. 1988b. Spatial, seasonal, and diel distribution of fishes in a California reservoir dominated by native fishes. Fish. Res. 7:31–53.

Von Geldern, C. E. 1964. Distribution of white catfish, *Ictalurus catus,* and *Salmo gairdneri,* in Folsom Lake, California, as determined by gill netting from February through November, 1961. CDFG Inland Fish. Admin. Rpt. 64-15. 9 pp.

———. 1965. Evidence of American shad reproduction in a landlocked environment. Calif. Fish Game 51:212–213.

Von Geldern, C. E., and D. F. Mitchill. 1975. Largemouth bass and threadfin shad in California. Pages 436–449 *in* H. Clepper, ed. Black bass biology and management. Washington, D.C.: Sport Fishing Institute.

Wahrshaftig, C., and J. H. Birman. 1965. The Quaternary of the Pacific mountain system in California. Pages 299–338 *in* H. E. Wright and D. G. Frey, eds. The Quaternary of the United States. Princeton, N.J.: Princeton University Press.

Wainwright, P. C., C. W. Osenberg, and G. G. Mittelbach. 1991. Trophic polymorphism in the pumpkinseed (*Lepomis gibbosus* Linnaeus): effects of environment on ontogeny. Funct. Ecol. 5:40–55.

Waldman. J., ed. 1998. Stripers: an angler's anthology. Camden, Me.: Ragged Mountain Press.

Waldvogel, J. 1988. Fall chinook salmon spawning escapement estimate for a tributary of the Smith River, California. 2nd Interim Report (1980–1987). Univ. Calif. Sea Grant Extension Prog. Rpt. 88-5. 21 pp.

Wales, J. H. 1939. General report of investigations on the McCloud River drainage in 1938. CDFG 25:272–309.

———. 1946. The hardhead problem in the Sacramento River above Shasta Lake. CDFG Inland Fish. Admin. Rpt. 46-1. 4 pp.

———. 1951. The decline of the Shasta River king salmon run. Redding, Calif.: California Department of Fish and Game Unpubl. rpt. 82 pp.

———. 1962. Introduction of pond smelt from Japan into California. Calif. Fish Game 48:141–142.

Walker, B. W., ed. 1961. The Ecology of the Salton Sea, California, in relation to the sport fishery. CDFG Fish Bull. 113:1–204.

Walker, B. W., R. R. Whitney, and G. W. Barlow. 1961. The fishes of the Salton Sea. Pages 77–92 *in* B. W. Walker, ed. The ecology of the Salton Sea, California, in relation to the sport fishery. CDFG Fish Bull. 113.

Wallace, C. R. 1967. Observations on the reproductive behavior of black bullhead (*Ictalurus melas*). Copeia 1967:852–853.

Wallace, M., and B. W. Collins. 1997. Variation is use of the Klamath River Estuary by juvenile chinook salmon. Calif. Fish Game 83:132–143.

Walters, L. L., and E. F. Legner. 1980. Impact of desert pupfish, *Cyprinodon macularius,* and *Gambusia affinis affinis* on fauna in pond ecosystems. Hilgardia 48(3):1–18.

Walton, I. 1653 [reprint 1965]. The compleat angler. New York: Collier Books.

Wang, J. C. S. 1982. Early life history and protection of the tide-water goby *Eucyclogobius newberryi:* (Girard) in the Rodeo Lagoon of the Golden Gate National Recreation Area. Natl. Park Serv. Contrib. SCSU/UCD 022/3 Tech. Rpt. 7. 24 pp.

———. 1986. Fishes of the Sacramento–San Joaquin estuary and adjacent waters, California: a guide to the early life histories. IEP Tech. Rpt. 9. ca. 800 pp.

———. 1991. Early life stages and early life history of the delta smelt, *Hypomesus transpacificus,* in the Sacramento–San Joaquin estuary, with comparison with the early life history of longfin smelt, *Spirinchus thaleichthys.* IEP Tech. Rpt. 28. 52 pp.

———. 1995. Observations of early life stages of splittail (*Pogonichthys macrolepidotus*) in the Sacramento–San Joaquin estuary, 1988 to 1994. IEP Tech. Rpt. 43. 22 pp.

———. 1996. Observations of larval smelt and splittail in a dry year and in a wet year. IEP Newslett. 9(3):31–35.

Waples, R. S. 1991a. Definition of "species" under the Act: application to Pacific salmon. NOAA Tech. Mem. NMFS F/NWC-194. 29 pp.

———. 1991b. Genetic interactions between hatchery and wild salmonids: lessons from the Pacific Northwest. Can. J. Fish. Aquat. Sci. 48(suppl. 1):124–133.

Ward, D. L., and R. A. Fritzsche. 1987. Comparison of meristic and morphometric characters among and within subspecies of the Sacramento sucker (*Catostomus occidentalis* Ayres). Calif. Fish Game 73:175–187.

Warner, G. 1991. Remember the San Joaquin. Pages 61–69 *in* A. Lufkin, ed. California's salmon and steelhead: the struggle to restore an imperiled resource. Berkeley: University of California Press.

Weatherley, A. H. 1959. Some features of the biology of the tench, *Tinca tinca* (Linnaeus) in Tasmania. J. Anim. Ecol. 28:73–87.

Weaver, R. O., and C. D. Ziebell. 1976. Ecology and early life history of largemouth bass and bluegill in Imperial Reservoir, Arizona. SW Nat. 21:145–149.

Webber, L. B., and T. H. Suchanek, eds. 1998. Second Annual Clear Lake Science and Management Symposium. Lakeport, Calif.: CLERC. 138 pp.

Webster, D. W. 1954. Smallmouth bass, *Micropterus dolomieui,* in Cayuga Lake, part I: life history and environment. Cornell Univ. Agric. Exp. Stat. Mem. 327. 38 pp.

Week, L. E. 1982. Habitat selectivity of littoral zone fishes at Clear Lake, California. CDFG Inland Fish. Admin. Rpt. 82-7. 31 pp.

Weisel, G. F. 1947. Breeding behavior and early development of the mudsucker, a gobiid fish of California. Copeia 1947:77–85.

Weitcamp, L. A., T. C. Wainwright, G. J. Bryant, G. B. Milner, D. J. Teel, R. G. Kope, and R. S. Waples. 1995. Status review of coho salmon from Washington, Oregon, and California. NOAA Tech. Mem. NMFS-NWFSC-24. 258 pp.

Welch, D. W. 1997. Anatomical specialization in the gut of Pacific salmon (*Oncorhynchus*): evidence for oceanic limits to salmon production? Can. J. Zool. 75:936–942.

Wells, A. W., and J. S. Diana. 1975. Survey of the freshwater fishes and their habitats in the coastal drainages of southern California. Rpt. to CDFG, Inland Fish. Branch from L.A. County Mus. Nat. Hist. 360 pp.

Welsh, H. H., Jr., G. R. Hodgson, B. C. Harvey, and M. F. Roche. 2001. Distribution of juvenile coho salmon in relation to water temperatures in tributaries of the Mattole River, California. N. Am. J. Fish. Mgmt. 21:464–470.

Werner, E. E., and D. J. Hall. 1977. Competition and habitat shift in two sunfishes (Centrarchidae). Ecology 58:869–876.

Werner, R. G. 1967. Intralacustrine movements of bluegill fry in Crane Lake, Indiana. Trans. Am. Fish. Soc. 96:416–420.

———. 1969. Ecology of limnetic bluegill (*Lepomis macrochirus*) in Crane Lake, Indiana. Am. Midl. Nat. 81:164–181.

West, J. M., and J. B. Zedler. 2000. Marsh-creek connectivity: fish use of a tidal salt marsh in southern California. Estuaries 23:699–710.

Wheeler, A. 1969. The Fishes of the British Isles and northwest Europe. East Lansing: Michigan State University Press.

White, H. C., J. C. Medcof, and L. R. Day. 1965. Are killifish poisonous? J. Fish. Res. Board Can. 22:635–638.

White, J. L., and B. C. Harvey. 1999. Habitat separation of prickly sculpin, *Cottus asper,* and coastrange sculpin, *Cottus aleuticus,* in the mainstem Smith River, Northwestern California. Copeia 1999:371–375.

———. 2001. Effects of an introduced piscivorous fish on native benthic fishes in a coastal river. Freshwater Biol. 46:987–995.

White, M. D. 1993. Morphological characteristics of threespine stickleback (*Gasterosteus aculeatus*) from the Sweetwater River, San Diego County, California. Western Proc. 73rd Annu. Conf. Fish Wildl. Agencies: 219–224.

Whitfield, A. K., and S. J. M. Blaber. 1979. The distribution of the freshwater cichlid *Sarotherodon mossambicus* in estuarine systems. Env. Biol. Fish. 4:77–81.

Whittier, T. R., D. B. Halliwell, and S. G. Paulsen. 1997. Cyprinid distributions in Northeast U.S.A. lakes: evidence of regional-scale minnow diversity loss. Can. J. Fish. Aquat. Sci. 54:1593–1607.

Wiegmann, D. D., J. R. Baylis, and M. H. Hoff. 1992. Sexual selection and fitness variation in a population of smallmouth bass, *Micropterus dolomieui* (Pisces: Centrarchidae). Evolution 46:1740–1753.

Wigglesworth, K. A. 1975. Some life history aspects of the smelt (*Hypomesus transpacificus nipponensis*) in Lake Shastina, California. M.S. thesis, Calif. State Univ., Sacramento. 22 pp.

Wilbur, R. L. 1969. The redear sunfish in Florida. Fla. Game Freshw. Fish Comm. Fish Bull. 5. 64 pp.

Wilcox, T. P., and D. J. Hornbach. 1991. Macrobenthic community response to carp (*Cyprinus carpio*) foraging. J. Freshw. Ecol. 6:171–183.

Williams, C. D., T. P. Hardy, and J. E. Deacon. 1982. Distribution and status of fishes of the Amargosa River Canyon, California. Unpubl. rpt., USFWS Endangered Species Office, Sacramento, Calif. 115 pp.

Williams, G. D., and J. B. Zedler. 1999. Fish assemblage composition in constructed and natural tidal marshes of San Diego Bay: relative influence of channel morphology and restoration history. Estuaries 22:702–716.

Williams, G. D., J. S. Desmond, and J. B. Zedler. 1998. Extension of 2 nonindigenous fishes, *Acanthogobius flavimanus* and *Poecilia latipinna,* into San Diego Bay marsh habitats. Calif. Fish Game 84:1–17.

Williams, J. D., and G. H. Burgess. 1999. A new species of bass, *Micropterus cataractae* (Teleostei: Centrarchidae), from the Apalachicola River basin in Alabama, Florida, and Georgia. Bull. Fla. Mus. Nat. Hist. 42:80–114.

Williams, J. E. 1991. Preserves and refuges for native western fishes: history and management. Pages 171–190 *in* W.L. Minckley and J. E. Deacon, eds. Battle against extinction: native fish management in the American West. Tucson: University of Arizona Press.

Williams, J. E., and B. C. Bolster. 1989. Observation on Salt Creek pupfish mortality during a flash flood. Calif. Fish Game 75:57–59.

Williams, J. E., and D. W. Sada. 1985. Status of two endangered fishes, *Cyprinodon nevadensis mionectes* and *Rhinichthys osculus nevadensis,* from two springs in Ash Meadows, Nevada. SW Nat. 30:475–484.

Willsrud, T. 1971. A study of the Tahoe sucker, *Catostomus tahoensis* Gill and Jordan. M.S. thesis, San Jose State Coll., San Jose. 96 pp.

Winn, H. E. 1958a. Comparative reproductive behavior and ecology of fourteen species of darters (Pisces-Percidae). Ecol. Monogr. 28:155–191.

———. 1958b. Observations on the reproductive habits of darters (Pisces-Percidae). Am. Midl. Nat. 59:190–212.

———. 1960. Biology of the brook stickleback, *Eucalia inconstans.* Am. Midl. Nat. 63:424–440.

Winn, H. E., and R. R. Miller. 1954. Native post-larval fishes of the lower Colorado River Basin with a key to their identification. Calif. Fish Game 40:273–285.

Winter, B. D. 1987. Racial identification of juvenile summer and winter steelhead and resident rainbow trout (*Salmo gairdneri* Richardson). CDFG Inland Fish. Admin. Rpt. 87-1. 32 pp.

Wood, C. C., and C. J. Foote. 1996. Evidence for sympatric genetic divergence of anadromous and nonanadromous morphs of sockeye salmon (*Oncorhynchus nerka*). Evolution 50:1265–1279.

Wooten, M. C., K. T. Scribner, and M. H. Smith. 1988. Genetic variability and systematics of *Gambusia,* in the southeastern United States. Copeia 1988:283–289.

Wooten, R. J. 1976. The biology of sticklebacks. New York: Academic Press.

———. 1984. A functional biology of sticklebacks. London: Croom Helm.

Wurtsbaugh, W. A., and J. J. Cech, Jr. 1983. Growth and activity of juvenile mosquitofish: temperature and ration effects. Trans. Am. Fish. Soc. 112:653–660.

Wurtsbaugh, W., and H. Li. 1985. Diel migrations of a zooplanktivorous fish (*Menidia beryllina*) in relation to the distribution of its prey in a large eutrophic lake. Limnol. Ocean. 30:565–576.

Wurtsbaugh, W. A., R. W. Brocksen, and C. R. Goldman. 1975. Food and distribution of underyearling brook and rainbow trout in Castle Lake, California. Trans. Am. Fish. Soc. 104: 88–95.

Wydoski, R. S. 2001. Life history and fecundity of mountain whitefish from Utah streams. Trans. Am. Fish. Soc. 130: 692–698.

Wydoski, R. S., and R. R. Whitney. 1979. Inland fishes of Washington. Seattle: University of Washington Press.

Wynne-Edwards, V. C. 1932. The breeding habits of the black-headed minnow (*Pimephales promelas* Raf.). Trans. Am. Fish. Soc. 62:382–383.

Yamagiwa, M. 1996. 1996 range monitoring report, Modoc National Forest. Unpubl. rpt., USFS. 58 pp.

Yoshiyama, R. M. 1999. A history of salmon and people in the Central Valley region of California. Rev. Fish. Sci. 7:197–239.

Yoshiyama, R. M., F. W. Fisher, and P. B. Moyle. 1998. Historical abundance and decline of chinook salmon in the Central Valley region of California. N. Am. J. Fish. Mgmt. 18:487–521.

———. 2000. Chinook salmon in the California Central Valley: an assessment. Fisheries (Bethesda) 25(2):6–20.

Yoshiyama, R. M., E. R. Gerstung, F. W. Fisher, and P. B. Moyle. 2001. Historical and present distribution of chinook salmon in the Central Valley drainage of California. Pages 71–76 *in* R. L. Brown, ed. Contributions to the biology of Central Valley salmonids. CDFG Fish Bull. 179.

Young, C. 1991. Record size Colorado squawfish found in Utah. USFWS Newsletter for Recovery Program for the endangered fishes of the upper Colorado, Summer 1991:6.

Young, G., C. L. Brown, R. S. Nishioka, L. C. Folmar, M. Andrews, J. R. Kashman, and H. A. Bern. 1994. Histology, blood chemistry, and physiological status of normal and moribund striped bass involved in summer mortality ("die-off") in the Sacramento–San Joaquin Delta, California. J. Fish Biol. 44:491–512.

Young, K. L., and P. C. Marsh. 1990. Age and growth of flathead catfish in four southwestern rivers. Calif. Fish Game 76: 224–233.

Young, P. S., and J. J. Cech, Jr. 1996. Environmental tolerances and requirements of splittail. Trans. Am. Fish. Soc. 125:664–678.

Yowell, D. W., and G. L. Vinyard. 1993. An energy-based analysis of particulate-feeding and filter-feeding by blue tilapia, *Tilapia aurea.* Env. Biol. Fish. 36:65–72.

Zastrow, C. E., E. D. Houde, and E. H. Saunders. 1989. Quality of striped bass (*Morone saxatilis*) eggs in relation to river source and female weight. Pages 34–42 *in* J. H. S. Baxter, J. C. Gamble, and H. von Westernhagen, eds. The early life history of fish. Copenhagen: Conseil International pour l'Exploration de la Mer.

Zengel, S. A., and E. P. Glenn. 1996. Presence of the endangered desert pupfish (*Cyprinodon macularius,* Cyprinodontidae) in Cienega de Santa Clara, Mexico, following an extensive marsh dry-down. SW Nat. 41:73–78.

Ziebell, C. D., J. C. Tash, and R. L. Barefield. 1986. Impact of threadfin shad on macrocrustacean zooplankton in two Arizona lakes. J. Freshw. Ecol. 3:399–406.

Ziller, J. S. 1992. Distribution and relative abundance of bull trout in the Sprague River subbasin, Oregon. Pages 18–29 *in* P. J. Howell and D. V. Buchanan, eds. Proceedings of the Gearhart Mountain bull trout workshop. Corvallis, Oreg.: Oregon Chapter of the American Fisheries Society.

Zimmerman, F. E. 1995. A survey of the fishes of Auburn Ravine Creek, with emphasis on the hitch (*Lavinia exilicauda*). M.S. thesis, Calif. State Univ., Sacramento. 49 pp.

Zwick, P. 1992. Stream habitat fragmentation—threat to biodiversity. Biodiver. Cons. 1:80–97.

Zydlewski, J., and S. D. McCormick. 1997. The ontogeny of salinity tolerance in the American shad, *Alosa sapidissima.* Can. J. Fish. Aquat. Sci. 54:182–189.

Index

Page numbers followed by *f* and *t* indicate figures and tables, respectively. Additional information on place names of interest (e.g., rivers, lakes, dams) may be found by checking the species accounts in the text, particularly on the pages indicated by the index subentry "distribution" under the species entry.

in Colorado River province, 46

oriental. *See* goby, yellowfin

in Sacramento-San Joaquin province, 35, 39

shimofuri, 35, 39, 437–440, 438*f*
    distribution, 10*t*, 19*t*, 438, 438*f*
    identification, 93, 437
    interaction with tidewater goby, 433, 440
    introduction, 59*t*
    life history, 438–439
    names, 438
    status, 59*t*, 439–440
    taxonomy, 437–438

shokihaze, 59*t*, 93, 430

in South Coast province, 20

tidewater, 20, 430–434, 431*f*
    distribution, 3*t*, 7*t*, 10*t*, 19*t*, 20, 431, 431*f*
    identification, 93, 430
    interaction, 433, 437, 440
    interaction with other species, 437
    life history, 431–432
    names, 431
    status, 3*t*, 432–434
    taxonomy, 430–431

yellowfin, 436–437, 436*f*
    distribution, 7*t*, 10*t*, 19*t*, 436, 436*f*
    identification, 93, 436
    interactions, 433, 437
    introduction, 59*t*
    life history, 436–437
    names, 436
    status, 59*t*, 437

goldfish, 170–172, 171*f*. *See also* minnow
    family (Cyprinidae)
    distribution, 6*t*, 9*t*, 16*t*, 18*t*, 170–171, 170*f*
    identification, 86, 170
    introduction, 59*t*, 62
    life history, 171–172
    names, 170
    status, 59*t*, 172
    taxonomy, 170

Goose Lake subprovince, 4*f*, 13, 73
    California roach in, 141
    chub in, 124, 125–126
    conservation strategies in, 73
    fish species in, 9*t*–10*t*, 11
    lampreys in, 5, 96, 97, 99
    rainbow trout in, 11, 274–276, 282
    Sacramento suckers in, 186, 188
    speckled dace in, 11, 162
    tui chub in, 124–126

grayling, Arctic, 59*t*, 88, 89*f*, 243

Great Basin fishless areas, 20

Great Basin subprovince, 1, 2*t*–3*t*, 3, 4*f*, 5
    Amargosa River, 20, 161, 162, 164, 332–335
    Amargosa subprovince, 4*f*
        fish species in, 16*t*–17*t*, 20
        pupfish in, 20, 332–335
    Eagle Lake subprovince
        ecology of, 44*t*, 55
        fish species in, 15, 16*t*–17*t*
    ecology of
        Eagle Lake, 44*t*, 55
        Lahontan streams, 41*f*
        Lake Tahoe, 42–44, 43*f*
    fish species in, 10, 14–20, 16*t*–17*t*
    Lahontan subprovince, 4*f*

ecology of, 40–42, 41*f*
    fish species in, 10, 15
    geography and geological history of, 15
    Mojave subprovince, 4*f*, 16*t*–17*t*, 20
    Owens subprovince, 4*f*, 15
    suckers in, 16*t*, 41, 42, 44, 44*t*
    Surprise Valley subprovince, 4*f*, 15
    trout in, 15, 16*t*–17*t*, 20, 40–42, 44, 44*t*

Green River, 159

grilse, 242

Gualala River, 6*t*–7*t*, 14, 38, 143, 273

gunnel
    penpoint, 91, 92*f*
    saddleback, 91, 92*f*

gunnel family (Pholididae), 91, 92*f*
    identification, 82, 91

guppy, 62, 91, 317

habitat. *See* ecology

half-pounder, 276

hardhead, 8, 151–154, 152*f*, Plate 8
    anatomy (frenum), 87*f*
    distribution, 2*t*, 6*t*, 8, 9*t*, 11, 14, 50*t*, 152, 152*f*
    ecology of, 2*t*, 25–29, 26*f*–30*f*, 152
    identification, 86, 151
    interaction with smallmouth bass, 404
    life history, 152–153
    names, 152
    status, 2*t*, 153–154
    taxonomy, 152

hatcheries, 63–66. *See also* trout, rainbow

Hat Creek, 12, 31, 53, 74, 124, 360, 361

Havasu Reservoir, 207

head length, 80

herring family (Clupeidae), 114–120. *See also*
    herring, Pacific; shad, American;
    shad, threadfin
    adaptations, 114
    identification, 82, 85, 114

herring, Pacific, 85, 114

Hidden Valley Dam, 52

High Rock Spring, 58, 126, 420

hitch, 136–139, 136*f*
    distribution, 2*t*, 6*t*, 8, 9*t*, 12–14, 16*t*, 18*t*, 32, 50*t*, 136*f*, 137
    identification, 86, 136
    life history, 137–138
    life style, 2*t*
    names, 137
    status, 2*t*, 138–139
    taxonomy, 136–137

Honey Lake, 15, 124, 388

Horton Creek, 74

Humboldt Bay, 14, 235, 237, 252, 291

Humboldt River, Nevada, 15

hybridization, 63

Hypomesus nipponensis. *See* wakasagi

Hypomesus pretiosus. *See* smelt, surf

Hypomesus transpacificus. *See* smelt, delta

Hysterocarpus traski. *See* perch, tule

Ictaluridae. *See* catfish family (Ictaluridae)

Ictalurus furcatus. *See* catfish, blue

Ictalurus punctatus. *See* catfish, channel

Ictiobus cyprinellus. *See* sucker, bigmouth
    buffalo

identification of fishes, 79–94. *See also* species
    accounts

Imperial Valley, 45, 49, 82, 322, 415, 416

Indian Creek, 74

introduced species
    in Colorado River province, 19*t*
    in Great Basin province, 17*t*
    halting invasion by, 70–71
    and hatcheries, impact of, 63–66
    human intervention and, 23
    impact of, 62–63
    increase in, 48
    in Klamath province, 7*t*
    methods and motivations of introduction, 58–60
    in North Coast province, 7*t*
    origin and year of introduction, 59*t*
    in Sacramento-San Joaquin province, 10*t*
    in South Coast province, 19*t*
    status, 59*t*

Iron Gate Dam, 3, 40

Iron Gate Reservoir, 40, 124, 129, 196, 199

Iron Mountain Mine, 58

Isabella Reservoir, 13, 395

jacks, 242

jacksmelt, 312

June Lake, 194

Kaweah Reservoir, 58, 61, 374, 375

Kaweah River, 8, 31, 103

Kelsey Creek, 13, 105

kelt, 242

Kerckoff Dam, 103

Kerckoff Reservoir, 37

Kern Brook, lampreys in, 97, 103–104

Kern Lake, 29, 54

Kern River, 4*f*, 8
    fish species in, 9*t*–10*t*, 13–14, 25, 29–31, 61
    golden trout in, 13–14, 283–286
    hardhead in, 152
    suckers in, 186

killifish
    Baja, 314
    California, 313–315, 314*f*
        distribution, 2*t*, 18*t*, 20, 314, 314*f*
        identification, 90, 313
        life history, 314–315
        life style, 2*t*
        names, 314
        status, 2*t*
        taxonomy, 314
    rainwater, 315–316, 316*f*
        distribution, 9*t*, 18*t*, 315–316, 315*f*
        identification, 90, 315
        introduction, 59*t*
        life history, 316
        names, 315
        status, 59*t*, 316

killifish family (Fundulidae), 83, 90, 313–316
    in South Coast province, 20

Oncorhynchus nerka. See salmon, sockeye (kokanee)
Oncorhynchus tshawytscha. See salmon, chinook
Oreochromis aureus. See tilapia, blue
Oreochromis mossambicus. See tilapia, Mozambique
Oreochromis niloticus. See tilapia, Nile
Oreochromis urolepis. See tilapia, wami
Oroville Reservoir
    bass in, 79, 405–408
    kokanee in, 265
    northern pike in, 223
    trout in, 304
    wakasagi in, 233
Orthodon microlepidotus. See blackfish, Sacramento
Osmeridae. *See* smelt family (Osmeridae)
Owens Lake, 15, 330
Owens River, 4f. *See also* Owens subprovince
    catfish in, 211
    ecology of, 53
    fish species in, 15, 16t–17t
    geological history of, 15
    Owens pupfish in, 330, 331
    Owens sucker in, 194, 195
    Sacramento perch in, 377, 380
    speckled dace in, 161, 162, 163
    threespine stickleback in, 340
    trout in, 295, 296
    tui chub in, 124, 126
Owens subprovince, 4f, 15. *See also* Owens River
Owens Valley, 48, 60
Owens Valley Native Fishes Sanctuary, 330, 331

Pajaro River, 13, 23
    California roach in, 141
    carp in, 173
    ecology of, 57
    hitch in, 137
    Sacramento perch in, 427
    Sacramento pikeminnows in, 155
    Sacramento suckers in, 186
    speckled dace in, 162
    thicktail chub in, 127
parr, 242
Pascifasticus fortis. See crayfish, Shasta
pearlfish, Argentine, 313
Perca flavescens. See perch, yellow
perch
    Sacramento, 8, 376–381, 377f, Plate 36
        distribution, 3t, 7t, 10t, 14, 17t, 26f, 29–31, 377, 377f, 378t, 380, 427
        identification, 91, 376
        interaction with other species, 372, 380, 384, 397
        life history, 377–380
        life style, 3t
        names, 377
        in Sacramento-San Joaquin province, 26f, 29–31
        status, 3t, 380–381
        taxonomy, 376–377
    shiner, 428–429, 428f
        distribution, 3t, 7t, 10t, 19t, 428, 428f

        identification, 93, 428
        life history, 428–429
        life style, 3t
        names, 428
        status, 3t, 429
    tule, 424–428, 425f, Plate 37
        distribution, 3t, 7t, 8, 10t, 11–14, 17t, 19t, 26f, 27–31, 27f, 49t, 50t, 425, 425f
        identification, 93, 424
        life history, 425–427
        life style, 3t
        names, 425
        in Sacramento-San Joaquin province, 26f, 27–31, 27f
        status, 3t, 427
        taxonomy, 424–425
    yellow, 411–413, 412f
        distribution, 7t, 10t, 411–412, 411f
        identification, 93, 411
        introduction, 59t
        life history, 412–413
        names, 411
        status, 59t, 413
perch and darter family (Percidae), 93, 409–413
    in Sacramento-San Joaquin province, 10t, 11–14, 32, 35, 36, 38, 39, 49t, 50t
Percichthyidae. *See* temperate bass family (Moronidae)
Percidae. *See* perch and darter family (Percidae)
Perciformes, 364
Percina macrolepida. See logperch, bigscale
Perris Reservoir, 60f, 381, 405, 406
Pescadero Creek, 13
pesticides, 71
Petaluma River, 14, 147, 437
Petromyzontidae. *See* lampreys (Petromyzontidae)
pets, and introduced species, 62
pharyngeal tooth counts, 81
Pholididae. *See* gunnel family (Pholididae)
Pholis ornata. See gunnel, saddleback
pike, northern, 83f, 222–225
    control of, 58–60
    distribution, 83, 222–223
    identification, 222–225
    introduction, 59t, 61
    life history, 223–225
    names, 222
    status, 59t
    taxonomy, 222
pike family (Esocidae), 83, 222–225
pikeminnow
    Colorado, 158–160, 158f, Plate 6
        distribution, 2t, 18t, 21, 45, 45t, 158–159, 158f
        identification, 86, 158
        interaction with pupfish, 328
        life history, 159
        life style, 2t
        names, 158
        status, 2t, 160
        taxonomy, 158
    in Colorado River province, 21, 45, 46t
    in North Coast province, 14
    Sacramento, 154–158, 155f, Plate 7

        distribution, 2t, 6t, 8–11, 9t, 13, 14, 18t, 49t, 50t, 155, 155f
        ecology of, 2t, 24–31, 26f–30f, 36–38
        identification, 86, 154
        life history, 155–157
        names, 155
        status, 2t, 157–158
        taxonomy, 154–155
        in Sacramento-San Joaquin province, 9t, 11, 13, 14, 25–29, 26f–29f, 31, 36–38, 49t, 50t
pikeperches, 409
Pillsbury Reservoir, 105, 155
Pimephales promelas. See minnow, fathead
Pine Creek, 44, 55
Pine Flat Dam, 103
Pine Flat Reservoir, 61, 157, 374, 383, 404, 406
pipefish, bay, 39, 82
pipefishes (Syngnathidae), identification, 82
Piru Creek, 438
Pismo Creek, 162
Pit River, 3, 4f, 5, 8
    bass in, 402, 405, 406
    brook lampreys in, 12, 99
    California roach in, 141, 143
    chinook salmon in, 253
    ecology of, 25, 31–32, 53
    fish species in, 9t–10t, 11–12, 23, 30f, 31
    geological history of, 11
    hardhead in, 11, 152
    rainbow trout in, 25, 275, 276
    redear sunfish in, 385
    Sacramento pikeminnow in, 155
    Sacramento suckers in, 189–190
    sculpins in, 10, 11, 351, 353, 356, 357, 360, 361
    speckled dace in, 162
    tui chub in, 124, 126
    tule perch in, 11, 425, 427
    white sturgeon in, 107, 109
Platichthys stellatus. See flounder, starry
platyfish, 317
    variable, 91
Pleuronectidae. *See* flounder (Pleuronectidae)
Poecilia latipinna. See molly, sailfin
Poecilia mexicana. See molly, shortfin
Poecilia reticulata. See guppy
Poeciliidae. *See* livebearer family (Poeciliidae)
Poeciliopsis gracilis. See livebearer, porthole
Poecilistes pleurospilus. See livebearer, porthole
Pogonichthys ciscoides. See splittail, Clear Lake
Pogonichthys macrolepidotus. See splittail, Sacramento
pollution, 57–58, 71
Pomoxis annularis. See crappie, white
Pomoxis nigromaculatus. See crappie, black
poolfish, Pahrump, 334
Porter-Cologne Act, 370
Potamocorbula amurensis. See clam, overbite
predation, effects of, 63, 310, 328, 376, 401
preserves, 74–75
Prosopium williamsoni. See whitefish, mountain
protection measures. *See* conservation strategies

Designer:	Barbara Jellow
Compositor:	Princeton Editorial Associates, Inc., Scottsdale, Arizona
Text:	Minion
Display:	Rotis Semi Sans
Printer and binder:	Friesens, Altona, Manitoba, Canada